Lecture Notes in Artificial Intelligence 11053

Subseries of Lecture Notes in Computer Science

LNAI Series Editors

Randy Goebel
University of Alberta, Edmonton, Canada
Yuzuru Tanaka
Hokkaido University, Sapporo, Japan
Wolfgang Wahlster
DFKI and Saarland University, Saarbrücken, Germany

LNAI Founding Series Editor

Joerg Siekmann
DFKI and Saarland University, Saarbrücken, Germany

More information about this series at http://www.springer.com/series/1244

Ulf Brefeld · Edward Curry
Elizabeth Daly · Brian MacNamee
Alice Marascu · Fabio Pinelli
Michele Berlingerio · Neil Hurley (Eds.)

Machine Learning and Knowledge Discovery in Databases

European Conference, ECML PKDD 2018
Dublin, Ireland, September 10–14, 2018
Proceedings, Part III

Springer

Editors
Ulf Brefeld
Leuphana University
Lüneburg, Germany

Alice Marascu
Nokia (Ireland)
Dublin, Ireland

Edward Curry
National University of Ireland
Galway, Ireland

Fabio Pinelli
Vodafone
Milan, Italy

Elizabeth Daly
IBM Research - Ireland
Dublin, Ireland

Michele Berlingerio
IBM Research - Ireland
Dublin, Ireland

Brian MacNamee ⓘ
University College Dublin
Dublin, Ireland

Neil Hurley ⓘ
University College Dublin
Dublin, Ireland

ISSN 0302-9743   ISSN 1611-3349   (electronic)
Lecture Notes in Artificial Intelligence
ISBN 978-3-030-10996-7   ISBN 978-3-030-10997-4   (eBook)
https://doi.org/10.1007/978-3-030-10997-4

Library of Congress Control Number: 2018965756

LNCS Sublibrary: SL7 – Artificial Intelligence

This Springer imprint is published by the registered company Springer Nature Switzerland AG
The registered company address is: Gewerbestrasse 11, 6330 Cham, Switzerland

Foreword to Applied Data Science, Demo, and Nectar Tracks

We are pleased to present the proceedings of the Applied Data Science (ADS), Nectar, and Demo Tracks of ECML PKDD 2018.

The ADS track aims to bring together participants from academia, industry, governments, and non-governmental organizations and highlights the practical, real-world applications of machine learning, knowledge discovery, and data mining. Novel and practical ideas, open problems in applied data science, description of application-specific challenges, and unique solutions adopted in bridging the gap between research and practice are some of the relevant topics for which papers were submitted and accepted in this track.

This year's ADS Track included 37 accepted paper presentations from 143 submissions (an acceptance rate of just 26%). This represents a 40% increase in interest in this track over the previous edition of the conference, indicating a strong growth in interest in the application of data science in industry.

The unexpected high volume of submissions meant that our ADS Program Committee had to work hard to ensure that every paper received at least three reviews. We are very thankful to the committee for their commitment to ensuring an interesting ADS program. A wide range of applications are covered in our papers, such as the use of data analytics to monitor mental health in the health sector; to the assessment of driver behavior in the transport sector; to learning novel flight itineraries in the travel sector, to name just a few.

The Demo Track at ECML-PKDD is an important forum for presenting state-of-the-art data mining and machine learning systems and research prototypes. The evaluation criteria for papers submitted to the demo track encompassed innovation and technical advances, meeting novel challenges, and potential impact and interest for researchers and practitioners. In 2018, 22 submissions were received and 12 were accepted for presentation at the conference and inclusion in the proceedings. These papers covered the application of a wide range of machine learning and data mining techniques in a diverse set of real-world application domains. We thank all authors for submitting their work, and our Demo Track Program Committee for volunteering their time and expertise.

The Nectar Track offers researchers from other disciplines a place to present their work to the ECML-PKDD community and to raise the community's awareness of data analysis results and open problems in their field. We received 23 submissions that were reviewed by at least three reviewers. According to the reviews, five papers were selected for presentation in a dedicated Nectar Track session at the conference.

According to the spirit of the Nectar Track, those five papers covered a range of topics including music recommendation, mood predication, and question-answering sessions.

September 2018

Ulf Brefeld
Edward Curry
Elizabeth Daly
Brian MacNamee
Alice Marascu
Fabio Pinelli

Preface

We are delighted to introduce the proceedings of the 2018 edition of the European Conference on Machine Learning and Principles and Practice of Knowledge Discovery in Databases (ECML-PKDD 2018). The conference was held in Dublin, Ireland, during September 10–14, 2018. ECML-PKDD is an annual conference that provides an international forum for the discussion of the latest high-quality research results in all areas related to machine learning and knowledge discovery in databases, including innovative applications. This event is the premier European machine learning and data mining conference and builds upon a very successful series of ECML-PKDD conferences.

We would like to thank all participants, authors, reviewers, and organizers of the conference for their contribution to making ECML-PKDD 2018 a great scientific event.

We would also like to thank the Croke Park Conference Centre and the student volunteers. Thanks to Springer for their continuous support and Microsoft for allowing us to use their CMT software for conference management and providing support throughout. Special thanks to our many sponsors and to the ECML-PKDD Steering Committee for their support and advice.

Finally, we would like to thank the organizing institutions: the Insight Centre for Data Analytics, University College Dublin (Ireland), and IBM Research, Ireland.

September 2018
Michele Berlingerio
Neil Hurley

Organization

ECML PKDD 2018 Organization

General Chairs

Michele Berlingerio   IBM Research, Ireland
Neil Hurley   University College Dublin, Ireland

Program Chairs

Michele Berlingerio   IBM Research, Ireland
Francesco Bonchi   ISI Foundation, Italy
Thomas Gärtner   University of Nottingham, UK
Georgiana Ifrim   University College Dublin, Ireland

Journal Track Chairs

Björn Bringmann   McKinsey & Company, Germany
Jesse Davis   Katholieke Universiteit Leuven, Belgium
Elisa Fromont   IRISA, Rennes 1 University, France
Derek Greene   University College Dublin, Ireland

Applied Data Science Track Chairs

Edward Curry   National University of Ireland Galway, Ireland
Alice Marascu   Nokia Bell Labs, Ireland

Workshop and Tutorial Chairs

Carlos Alzate   IBM Research, Ireland
Anna Monreale   University of Pisa, Italy

Nectar Track Chairs

Ulf Brefeld   Leuphana University of Lüneburg, Germany
Fabio Pinelli   Vodafone, Italy

Demo Track Chairs

Elizabeth Daly   IBM Research, Ireland
Brian Mac Namee   University College Dublin, Ireland

PhD Forum Chairs

Bart Goethals   University of Antwerp, Belgium
Dafna Shahaf   Hebrew University of Jerusalem, Israel

Discovery Challenge Chairs

Martin Atzmüller   Tilburg University, The Netherlands
Francesco Calabrese   Vodafone, Italy

Awards Committee

Tijl De Bie   Ghent University, Belgium
Arno Siebes   Uthrecht University, The Netherlands
Bart Goethals   University of Antwerp, Belgium
Walter Daelemans   University of Antwerp, Belgium
Katharina Morik   TU Dortmund, Germany

Professional Conference Organizer

Keynote PCO   Dublin, Ireland

www.keynotepco.ie

ECML PKDD Steering Committee

Michele Sebag   Université Paris Sud, France
Francesco Bonchi   ISI Foundation, Italy
Albert Bifet   Télécom ParisTech, France
Hendrik Blockeel   KU Leuven, Belgium and Leiden University, The Netherlands
Katharina Morik   University of Dortmund, Germany
Arno Siebes   Utrecht University, The Netherlands
Siegfried Nijssen   LIACS, Leiden University, The Netherlands
Chedy Raïssi   Inria Nancy Grand-Est, France
João Gama   FCUP, University of Porto/LIAAD, INESC Porto L.A.,
                          Portugal
Annalisa Appice   University of Bari Aldo Moro, Italy
Indré Žliobaité   University of Helsinki, Finland
Andrea Passerini   University of Trento, Italy
Paolo Frasconi   University of Florence, Italy
Céline Robardet   National Institute of Applied Science Lyon, France

Jilles Vreeken   Saarland University, Max Planck Institute for Informatics,
                        Germany
Sašo Džeroski   Jožef Stefan Institute, Slovenia
Michelangelo Ceci   University of Bari Aldo Moro, Italy
Myra Spiliopoulu   Magdeburg University, Germany
Jaakko Hollmén   Aalto University, Finland

Area Chairs

Michael Berthold   Universität Konstanz, Germany
Hendrik Blockeel   KU Leuven, Belgium and Leiden University, The Netherlands
Ulf Brefeld   Leuphana University of Lüneburg, Germany
Toon Calders   University of Antwerp, Belgium
Michelangelo Ceci   University of Bari Aldo Moro, Italy
Bruno Cremilleux   Université de Caen Normandie, France
Tapio Elomaa   Tampere University of Technology, Finland
Johannes Fürnkranz   TU Darmstadt, Germany
Peter Flach   University of Bristol, UK
Paolo Frasconi   University of Florence, Italy
João Gama   FCUP, University of Porto/LIAAD, INESC Porto L.A.,
                        Portugal
Jaakko Hollmén   Aalto University, Finland
Alipio Jorge   FCUP, University of Porto/LIAAD, INESC Porto L.A.,
                        Portugal
Stefan Kramer   Johannes Gutenberg University Mainz, Germany
Giuseppe Manco   ICAR-CNR, Italy
Siegfried Nijssen   LIACS, Leiden University, The Netherlands
Andrea Passerini   University of Trento, Italy
Arno Siebes   Utrecht University, The Netherlands
Myra Spiliopoulu   Magdeburg University, Germany
Luis Torgo   Dalhousie University, Canada
Celine Vens   KU Leuven, Belgium
Jilles Vreeken   Saarland University, Max Planck Institute for Informatics,
                        Germany

Conference Track Program Committee

Carlos Alzate	Roberto Bayardo	Indrajit Bhattacharya
Aijun An	Martin Becker	Marenglen Biba
Fabrizio Angiulli	Srikanta Bedathur	Silvio Bicciato
Annalisa Appice	Jessa Bekker	Mario Boley
Ira Assent	Vaishak Belle	Gianluca Bontempi
Martin Atzmueller	Andras Benczur	Henrik Bostrom
Antonio Bahamonde	Daniel Bengs	Tassadit Bouadi
Jose Balcazar	Petr Berka	Pavel Brazdil

Dariusz Brzezinski
Rui Camacho
Longbing Cao
Francisco Casacuberta
Peggy Cellier
Loic Cerf
Tania Cerquitelli
Edward Chang
Keke Chen
Weiwei Cheng
Silvia Chiusano
Arthur Choi
Frans Coenen
Mário Cordeiro
Roberto Corizzo
Vitor Santos Costa
Bertrand Cuissart
Boris Cule
Tomaž Curk
James Cussens
Alfredo Cuzzocrea
Claudia d'Amato
Maria Damiani
Tijl De Bie
Martine De Cock
Juan Jose del Coz
Anne Denton
Christian Desrosiers
Nicola Di Mauro
Claudia Diamantini
Uwe Dick
Tom Diethe
Ivica Dimitrovski
Wei Ding
Ying Ding
Stephan Doerfel
Carlotta Domeniconi
Frank Dondelinger
Madalina Drugan
Wouter Duivesteijn
Inŝ Dutra
Dora Erdos
Fabio Fassetti
Ad Feelders
Stefano Ferilli
Carlos Ferreira

Cesar Ferri
Răzvan Florian
Eibe Frank
Elisa Fromont
Fabio Fumarola
Esther Galbrun
Patrick Gallinari
Dragan Gamberger
Byron Gao
Paolo Garza
Konstantinos Georgatzis
Pierre Geurts
Dorota Glowacka
Nico Goernitz
Elsa Gomes
Mehmet Gönen
James Goulding
Michael Granitzer
Caglar Gulcehre
Francesco Gullo
Stephan Günnemann
Tias Guns
Sara Hajian
Maria Halkidi
Jiawei Han
Mohammad Hasan
Xiao He
Denis Helic
Daniel Hernandez-Lobato
Jose Hernandez-Orallo
Thanh Lam Hoang
Frank Hoeppner
Arjen Hommersom
Tamas Horvath
Andreas Hotho
Yuanhua Huang
Eyke Hüllermeier
Dino Ienco
Szymon Jaroszewicz
Giuseppe Jurman
Toshihiro Kamishima
Michael Kamp
Bo Kang
Andreas Karwath
George Karypis
Mehdi Kaytoue

Latifur Khan
Frank Klawonn
Jiri Klema
Tomas Kliegr
Marius Kloft
Dragi Kocev
Levente Kocsis
Yun Sing Koh
Alek Kolcz
Irena Koprinska
Frederic Koriche
Walter Kosters
Lars Kotthoff
Danai Koutra
Georg Krempl
Tomas Krilavicius
Yamuna Krishnamurthy
Matjaz Kukar
Meelis Kull
Prashanth L. A.
Jorma Laaksonen
Nicolas Lachiche
Leo Lahti
Helge Langseth
Thomas Lansdall-Welfare
Christine Largeron
Pedro Larranaga
Silvio Lattanzi
Niklas Lavesson
Binh Le
Freddy Lecue
Florian Lemmerich
Jiuyong Li
Limin Li
Jefrey Lijffijt
Tony Lindgren
Corrado Loglisci
Peter Lucas
Brian Mac Namee
Gjorgji Madjarov
Sebastian Mair
Donato Malerba
Luca Martino
Elio Masciari
Andres Masegosa
Florent Masseglia

Ernestina Menasalvas
Corrado Mencar
Rosa Meo
Pauli Miettinen
Dunja Mladenic
Karthika Mohan
Anna Monreale
Joao Moreira
Mohamed Nadif
Ndapa Nakashole
Jinseok Nam
Mirco Nanni
Amedeo Napoli
Nicolo Navarin
Benjamin Negrevergne
Benjamin Nguyen
Xia Ning
Kjetil Norvag
Eirini Ntoutsi
Andreas Nurnberger
Barry O'Sullivan
Dino Oglic
Francesco Orsini
Nikunj Oza
Pance Panov
Apostolos Papadopoulos
Panagiotis Papapetrou
Ioannis Partalas
Gabriella Pasi
Dino Pedreschi
Jaakko Peltonen
Ruggero Pensa
Iker Perez
Nico Piatkowski
Andrea Pietracaprina
Gianvito Pio
Susanna Pirttikangas
Marc Plantevit
Pascal Poncelet
Miguel Prada
Philippe Preux
Buyue Qian
Chedy Raissi
Jan Ramon
Huzefa Rangwala
Zbigniew Ras

Chotirat Ratanamahatana
Jan Rauch
Chiara Renso
Achim Rettinger
Fabrizio Riguzzi
Matteo Riondato
Celine Robardet
Juan Rodriguez
Fabrice Rossi
Celine Rouveirol
Stefan Rueping
Salvatore Ruggieri
Yvan Saeys
Alan Said
Lorenza Saitta
Tomoya Sakai
Alessandra Sala
Ansaf Salleb-Aouissi
Claudio Sartori
Pierre Schaus
Lars Schmidt-Thieme
Christoph Schommer
Matthias Schubert
Konstantinos Sechidis
Sohan Seth
Vinay Setty
Junming Shao
Nikola Simidjievski
Sameer Singh
Andrzej Skowron
Dominik Slezak
Kevin Small
Gavin Smith
Tomislav Smuc
Yangqiu Song
Arnaud Soulet
Wesllen Sousa
Alessandro Sperduti
Jerzy Stefanowski
Giovanni Stilo
Gerd Stumme
Mahito Sugiyama
Mika Sulkava
Einoshin Suzuki
Stephen Swift
Andrea Tagarelli

Domenico Talia
Letizia Tanca
Jovan Tanevski
Nikolaj Tatti
Maguelonne Teisseire
Georgios Theocharous
Ljupco Todorovski
Roberto Trasarti
Volker Tresp
Isaac Triguero
Panayiotis Tsaparas
Vincent S. Tseng
Karl Tuyls
Niall Twomey
Nikolaos Tziortziotis
Theodoros Tzouramanis
Antti Ukkonen
Toon Van Craenendonck
Martijn Van Otterlo
Iraklis Varlamis
Julien Velcin
Shankar Vembu
Deepak Venugopal
Vassilios S. Verykios
Ricardo Vigario
Herna Viktor
Christel Vrain
Willem Waegeman
Jianyong Wang
Joerg Wicker
Marco Wiering
Martin Wistuba
Philip Yu
Bianca Zadrozny
Gerson Zaverucha
Bernard Zenko
Junping Zhang
Min-Ling Zhang
Shichao Zhang
Ying Zhao
Mingjun Zhong
Albrecht Zimmermann
Marinka Zitnik
Indré Žliobaité

Applied Data Science Track Program Committee

Oznur Alkan
Carlos Alzate
Nicola Barberi
Gianni Barlacchi
Enda Barrett
Roberto Bayardo
Srikanta Bedathur
Daniel Bengs
Cuissart Bertrand
Urvesh Bhowan
Tassadit Bouadi
Thomas Brovelli
Teodora Sandra
Berkant Barla
Michelangelo Ceci
Edward Chang
Soumyadeep Chatterjee
Abon Chaudhuri
Javier Cuenca
Mahashweta Das
Viktoriya Degeler
Wei Ding
Yuxiao Dong
Carlos Ferreira
Andre Freitas
Feng Gao
Dinesh Garg
Guillermo Garrido
Rumi Ghosh
Martin Gleize
Slawek Goryczka
Riccardo Guidotti
Francesco Gullo
Thomas Guyet
Allan Hanbury

Sidath Handurukande
Souleiman Hasan
Georges Hebrail
Thanh Lam
Neil Hurley
Hongxia Jin
Anup Kalia
Pinar Karagoz
Mehdi Kaytoue
Alek Kolcz
Deguang Kong
Lars Kotthoff
Nick Koudas
Hardy Kremer
Helge Langseth
Freddy Lecue
Zhenhui Li
Lin Liu
Jiebo Luo
Arun Maiya
Silviu Maniu
Elio Masciari
Luis Matias
Dimitrios Mavroeidis
Charalampos
  Mavroforakis
James McDermott
Daniil Mirylenka
Elena Mocanu
Raul Moreno
Bogdan Nicolae
Maria-Irina Nicolae
Xia Ning
Sean O'Riain
Adegboyega Ojo

Nikunj Oza
Ioannis Partalas
Milan Petkovic
Fabio Pinelli
Yongrui Qin
Rene Quiniou
Ambrish Rawat
Fergal Reid
Achim Rettinger
Stefan Rueping
Elizeu Santos-Neto
Manali Sharma
Alkis Simitsis
Kevin Small
Alessandro Sperduti
Siqi Sun
Pal Sundsoy
Ingo Thon
Marko Tkalcic
Luis Torgo
Radu Tudoran
Umair Ul
Jan Van
Ranga Vatsavai
Fei Wang
Xiang Wang
Wang Wei
Martin Wistuba
Erik Wittern
Milena Yankova
Daniela Zaharie
Chongsheng Zhang
Yanchang Zhao
Albrecht Zimmermann

Nectar Track Program Committee

Annalisa Appice
Martin Atzmüller
Hendrik Blockeel
Ulf Brefeld
Tijl De
Kurt Driessens

Peter Flach
Johannes Fürnkranz
Joao Gama
Andreas Hotho
Kristian Kersting
Sebastian Mair

Ernestina Menasalvas
Mirco Musolesi
Franco Maria
Maryam Tavakol
Gabriele Tolomei
Salvatore Trani

Demo Track Program Committee

Gustavo Carneiro	Brian Mac Namee	Konstantinos Skiannis
Derek Greene	Susan McKeever	Jerzy Stefanowski
Mark Last	Joao Papa	Luis Teixeira
Vincent Lemaire	Niladri Sett	Grigorios Tsoumakas

Contents – Part III

ADS Engineering and Design

ADS Financial/Security

ADS Health

ADS Sensing/Positioning

Nectar Track

Demo Track

Contents – Part I

Deep Learning

Ensemble Methods

Evaluation

Contents – Part II

Kernel Methods

Learning Paradigms

Matrix and Tensor Analysis

Probabilistic Models and Statistical Methods

Recommender Systems

Transfer Learning

ADS Data Science Applications

Neural Article Pair Modeling
for Wikipedia Sub-article Matching

Muhao Chen[1]([⊠]), Changping Meng[2], Gang Huang[3], and Carlo Zaniolo[1]

[1] University of California, Los Angeles, CA, USA
{muhaochen,zaniolo}@cs.ucla.edu
[2] Purdue University, West Lafayette, IN, USA
meng40@purdue.edu
[3] Google, Mountain View, CA, USA
ganghuang@google.com

Abstract. Nowadays, editors tend to separate different subtopics of a long Wiki-pedia article into multiple *sub-articles*. This separation seeks to improve human readability. However, it also has a deleterious effect on many Wikipedia-based tasks that rely on the *article-as-concept* assumption, which requires each entity (or concept) to be described solely by one article. This underlying assumption significantly simplifies knowledge representation and extraction, and it is vital to many existing technologies such as automated knowledge base construction, cross-lingual knowledge alignment, semantic search and data lineage of Wikipedia entities. In this paper we provide an approach to match the scattered sub-articles back to their corresponding main-articles, with the intent of facilitating automated Wikipedia curation and processing. The proposed model adopts a hierarchical learning structure that combines multiple variants of neural document pair encoders with a comprehensive set of explicit features. A large crowdsourced dataset is created to support the evaluation and feature extraction for the task. Based on the large dataset, the proposed model achieves promising results of cross-validation and significantly outperforms previous approaches. Large-scale serving on the entire English Wikipedia also proves the practicability and scalability of the proposed model by effectively extracting a vast collection of newly paired main and sub-articles. Code related to this paper is available at: https://github.com/muhaochen/subarticle.

Keywords: Article pair modeling · Sub-article matching
Text representations · Sequence encoders · Explicit features
Wikipedia

1 Introduction

Wikipedia has been the essential source of knowledge for people as well as computing research and practice. This vast storage of encyclopedia articles for real-world entities (or concepts) has brought along the automatic construction of

This work is accomplished at Google, Mountain View.

U. Brefeld et al. (Eds.): ECML PKDD 2018, LNAI 11053, pp. 3–19, 2019.
https://doi.org/10.1007/978-3-030-10997-4_1

Fig. 1. The main-article *Bundeswehr* (armed forces of Germany) and its sub-articles.

knowledge bases [24,28] that support knowledge-driven computer applications with vast structured knowledge. Meanwhile, Wikipedia also triggers the emerging of countless AI-related technologies for semantic web [29,44,50], natural language understanding [4,33,45], content retrieval [1,22], and other aspects.

Most existing automated Wikipedia techniques assume the one-to-one mapping between entities and Wikipedia articles [13,26,28]. This so-called *article-as-concept* assumption [26] regulates each entity to be described by at most one article in a language-specific version of Wikipedia. However, recent development of Wikipedia itself is now breaking this assumption, as rich contents of an entity are likely to be separated in different articles and managed independently. For example, many details about the entity "Harry Potter" are contained in other articles such as "Harry Potter Universe", "Harry Potter influences and analogues", and "Harry Potter in translation". Such separation of entity contents categorizes Wikipedia articles into two groups: the *main-article* that summarizes an entity, and the *sub-article* that comprehensively describes an aspect or a subtopic of the main-article. Consider another example: for the main-article *Bundeswehr* (i.e. unified armed forces of Germany) in English Wikipedia, we can find its split-off sub-articles such as *German Army, German Navy, German Airforce, Joint Support Service of Germany*, and *Joint Medical Service of Germany* (as shown in Fig. 1). This type of sub-article splitting is quite common on Wikipedia. Around 71% of the most-viewed Wikipedia entities are split-off to an average of 7.5 sub-articles [26].

While sub-articles may enhance human readabilities, the violation of the article-as-concept assumption is problematic to a wide range of Wikipedia-based technologies and applications that critically rely on this assumption. For instance, Wikipedia-based knowledge base construction [24,28,32] assumes the article title as an entity name, which is then associated with the majority of relation facts for the entity from the infobox of the corresponding article. Split-off of articles sow confusion in a knowledge base extracted with these techniques. Clearly, explicit [15,17] and implicit semantic representation techniques [8,37,46] based on Wikipedia are impaired due to that a part of links and text features utilized by these approaches are now likely to be isolated from the entities, which further affects NLP tasks based on these semantic representations such as semantic relatedness analysis [27,33,40], relation extraction [7,32], and named entity disambiguation [47]. Multilingual tasks such as knowledge alignment [5,6,41,45] and cross-lingual Wikification [42] become challenging for entities with multiple articles, since these tasks assume that we have a one-to-one match between articles in both languages. Semantic search [3,29,50] is also affected due to the diffused nature of the associations between entities and articles.

To support the above techniques and applications, it is vital to address the *sub-article matching* problem, which aims to restore the complete view of each entity by matching the sub-articles back to the main-article. However, it is nontrivial to develop a model which recognizes the implicit relations that exist between main and sub-articles. A recent work [26] has attempted to tackle this problem by characterizing the match of main and sub articles using some explicit features. These features focus on measuring the symbolic similarities of article and section titles, the structural similarity of entity templates and page links, as well as cross-lingual co-occurrence. Although these features are helpful to identify the sub-article relations among a small scale of article pairs, they are still far from fully characterizing the sub-article relations in the large body of Wikipedia. And more importantly, the semantic information contained in the titles and text contents, which is critical to the characterization of semantic relations of articles, has not been used for this task. However, effective utilization of the semantic information would also require a large collection of labeled main and sub-article pairs to generalize the modeling of such implicit features.

In this paper, we introduce a new approach for addressing the sub-article matching problem. Our approach adopts neural document encoders to capture the semantic features from the titles and text contents of a candidate article pair, for which several encoding techniques are explored with. Besides the semantic features, the model also utilizes a set of explicit features that measure the symbolic and structural aspects of an article pair. Using a combination of these features, the model decides whether an article is the sub-article of another. To generalize the problem, massive crowdsourcing and strategic rules are applied to create a large dataset that contains around 196k Wikipedia article pairs, where around 10% are positive matches of main and sub-articles, and the rest comprehensively cover different patterns of negative matches. Held-out estimation proves effectiveness of our approach by significantly outperforming previous baselines, and reaching near-perfect precision and recall for detecting positive main and sub-article matches from all candidates. To show the practicability of our approach, we also employ our model to extract main and sub-article matches in the entire English Wikipedia using a 3,000-machine MapReduce [11]. This process has produced a large collection of new high-quality main and sub-article matches, and are being migrated into a production knowledge base.

2 Related Work

A recent work [26] has launched the first attempt to address the Wikipedia sub-article matching problem, in which the authors have defined the problem into the binary classification of candidate article pairs. Each article pair is characterized based on a group of explicit features that lies in three categories: (1) symbolic similarity: this includes token overlapping of the titles, maximum token overlapping among the section titles of the two articles, and term frequencies of the main-article titles in the candidate sub-article contents; (2) structural similarity: this includes structure similarity of article templates, link-based centrality measures and the Milne-Witten Index [31]; (3) cross-lingual co-occurrence: these features consider the proportion of languages where the given article pairs have been identified as main and sub-articles, and the relative multilingual "globalness" measure of the candidate main-article. Although some statistical classification models learnt on these explicit features have offered satisfactory accuracy of binary classification on a small dataset of 3k article pairs that cover a subset of the most-viewed Wikipedia articles, such simple characterization is no-doubt far from generalizing the problem. When the dataset scales up to the range of the entire Wikipedia, it is very easy to find numerous counterfactual cases for these features. Moreover, the cross-lingual co-occurrence-based features are not generally usable due to the incompleteness of inter-lingual links that match the cross-lingual counterparts of Wikipedia articles. Some recent works have even pointed out that such cross-lingual alignment information only covers less than 15% of the articles [5,24,43]. More importantly, we argue that the latent semantic information of the articles should be captured, so as to provide more generalized and comprehensive characterization of the article relation.

Sentence or article matching tasks such as textual entailment recognition [18,39,49] and paraphrase identification [48,49] require the model to identify content-based discourse relations of sentences or paragraphs [23], which reflect logical orders and semantic consistency. Many recent efforts adopt different forms of deep neural document encoders to tackle these tasks, where several encoding techniques have been widely employed, including convolutional neural networks [18,48], recurrent neural networks [39], and attentive techniques [20,36,49]. Detection of the sub-article relations requires the model to capture a high-level understanding of both contents and text structuring of articles. Unlike the previously mentioned discourse relations, the sub-article relations can be reflected from different components of Wikipedia articles including titles, text contents and link structures. To tackle new and challenging task of sub-article matching, we incorporate neural document encoders with explicit features in our model, so as to capture the sub-article relation based on both symbolic and semantic aspects of the Wikipedia article pairs. Meanwhile, we also take efforts to prepare a large collection of article pairs that seek to well generalize the problem.

3   Modeling

In this section, we introduce the proposed model for the Wikipedia sub-article matching task. We begin with the denotations and problem definition.

3.1   Preliminaries

Denotations. We use W to denote the set of Wikipedia articles, in which we model an article $A_i \in W$ as a triple $A_i = (t_i, c_i, s_i)$. t_i is the title, c_i the text contents, and s_i the miscellaneous structural information such as templates, sections and links. $t_i = \{w_{t1}, w_{t2}, ..., w_{tl}\}$ and $c_i = \{w_{c1}, w_{c2}, ..., w_{cm}\}$ thereof are both sequences of words. In practice, we use the first paragraph of A_i to represent c_i since it is the summary of the article contents. For each word w_i, we use bold-faced $\mathbf{w}_i$ to denote its embedding representation. We use $F(A_i, A_j)$ to denote a sequence of explicit features that provide some symbolic and structural measures for titles, text contents and link structures, which we are going to specify in Sect. 3.3. We assume that all articles are written in the same language, as it is normally the case of main-articles and their sub-articles. In this paper, we only consider English articles $w.l.o.g.$

Problem Definition. *Sub-article matching* is defined as a binary classification problem on a set of candidate article pairs $P \subseteq W \times W$. Given a pair of articles $p = (A_i, A_j) \in P$, a model should decide whether A_j is the sub-article of A_i. The problem definition complies with the previous work that first introduces the problem [26], and is related to other sentence matching problems for discourse relations such as text entailment and paraphrase identification [35,39,48].

The sub-article relation is qualified based on two criteria, i.e. A_j is a sub-article of A_i if (1) A_j describes an aspect or a subtopic of A_i, and (2) c_j can be inserted as a section of A_i without breaking the topic summarized by t_i. It is noteworthy that the sub-article relation is anti-symmetric, i.e. if A_j is a sub-article of A_i then A_i is not a sub-article of A_j. We follow these two criteria in the crowdsourcing process for dataset creation, as we are going to explain in Sect. 4. To address the sub-article matching problem, our model learns on a combination of two aspects of the Wikipedia articles. Neural document encoders extract the implicit semantic features from text, while a series of explicit features are incorporated to characterize the symbolic or structural aspects. In the following, we introduce each component of our model in detail.

3.2   Document Encoders

A neural document encoder $E(X)$ encodes a sequence of words X into a latent vector representation of the sequence. We investigate three widely-used techniques for document encoding [18,36,39,48,49], which lead to three types of encoders for both titles and text contents of Wikipedia articles, i.e. convolutional encoders (CNN), gated recurrent unit encoders (GRU), and attentive encoders.

Convolutional Encoders. A convolutional encoder employs the 1-dimensional weight-sharing convolution layer to encode an input sequence. Given the input sequence $X = \{x_1, x_2, ..., x_l\}$, a convolution layer applies a kernel $\mathbf{M}_c \in \mathbb{R}^{h \times k}$ to generate a latent representation $\mathbf{h}_i^{(1)}$ from a window of the input vector sequence $\mathbf{x}_{i:i+h-1}$ by

$$\mathbf{h}_i^{(1)} = \tanh(\mathbf{M}_c \mathbf{x}_{i:i+h-1} + \mathbf{b}_c)$$

for which h is the kernel size and $\mathbf{b}_c$ is a bias vector. The convolution layer applies the kernel to all consecutive windows to produce a sequence of latent vectors $\mathbf{H}^{(1)} = \{\mathbf{h}_1^{(1)}, \mathbf{h}_2^{(1)}, ..., \mathbf{h}_{l-h+1}^{(1)}\}$, where each latent vector leverages the significant local semantic features from each h-gram of the input sequence. Like many other works [18,21,38], we apply n-max-pooling to extract robust features from each n-gram of the convolution outputs by $\mathbf{h}_i^{(2)} = \max(\mathbf{h}_{i:n+i-1}^{(1)})$.

Gated Recurrent Unit Encoder. The gated recurrent unit (GRU) is an alternative of the long-short-term memory network (LSTM) that has been popular for sentence (sequence) encoders in recent works [12,19]. Each unit consists of two types of gates to track the state of sequences without using separated memory cells, i.e. the reset gate $\mathbf{r}_i$ and the update gate $\mathbf{z}_i$. Given the vector representation $\mathbf{x}_i$ of an incoming item x_i, GRU updates the current state $\mathbf{h}_i^{(3)}$ of the sequence as a linear interpolation between previous state $\mathbf{h}_{i-1}^{(3)}$ and the candidate state $\tilde{\mathbf{h}}_i^{(3)}$ of new item x_i, which is calculated as below.

$$\mathbf{h}_i^{(3)} = \mathbf{z}_i \odot \tilde{\mathbf{h}}_i^{(3)} + (1 - \mathbf{z}_i) \odot \mathbf{h}_{i-1}^{(3)}$$

The update gate $\mathbf{z}_i$ that balances between the information of the previous sequence and the new item is updated as below, where $\mathbf{M}_z$ and $\mathbf{N}_z$ are two weight matrices, $\mathbf{b}_z$ is a bias vector, and σ is the sigmoid function.

$$\mathbf{z}_i = \sigma\left(\mathbf{M}_z \mathbf{x}_i + \mathbf{N}_z \mathbf{h}_{i-1}^{(3)} + \mathbf{b}_z\right)$$

The candidate state $\tilde{\mathbf{h}}_i^{(3)}$ is calculated similarly to those in a traditional recurrent unit as below, where $\mathbf{M}_s$ and $\mathbf{N}_s$ are two weight matrices, and $\mathbf{b}_s$ is a bias vector.

$$\tilde{\mathbf{h}}_i^{(3)} = \tanh\left(\mathbf{M}_s \mathbf{x}_i + \mathbf{r}_i \odot (\mathbf{N}_s \mathbf{h}_{i-1}^{(3)}) + \mathbf{b}_s\right)$$

The reset gate $\mathbf{r}_i$ thereof controls how much information of the past sequence contributes to the candidate state

$$\mathbf{r}_i = \sigma\left(\mathbf{M}_r \mathbf{x}_i + \mathbf{N}_r \mathbf{h}_{i-1}^{(3)} + \mathbf{b}_r\right)$$

The above defines a GRU layer which outputs a sequence of hidden state vectors given the input sequence X. Unlike CNN that focuses on leveraging local semantic features, GRU focuses on capturing the sequence information of the language. Note that LSTM generally performs comparably to GRU in sequence encoding tasks, but is more complex and require more computational resources for training [9].

Attentive Encoder. An attentive encoder imports the self-attention mechanism to the recurrent neural sequence encoder, which seeks to capture the overall meaning of the input sequence unevenly from each encoded sequence item. The self-attention is imported to GRU as below.

$$\mathbf{u}_i = \tanh\left(\mathbf{M}_a\mathbf{h}_i^{(3)} + \mathbf{b}_a\right); \qquad a_i = \frac{\exp\left(\mathbf{u}_i^\top\mathbf{u}_X\right)}{\sum_{x_i \in X} \exp\left(\mathbf{u}_i^\top\mathbf{u}_X\right)}; \qquad \mathbf{h}_i^{(4)} = |X|a_i\mathbf{u}_i$$

$\mathbf{u}_i$ is the intermediary latent representation of GRU output $\mathbf{h}_i^{(3)}$, $\mathbf{u}_X = \tanh(\mathbf{M}_a\mathbf{h}_X^{(3)} + \mathbf{b}_a)$ is the intermediary latent representation of the averaged GRU output $\mathbf{h}_X^{(3)}$ that can be seen as a high-level representation of the entire input sequence. By measuring the similarity of each $\mathbf{u}_i$ with $\mathbf{u}_X$, the normalized attention weight a_i for $\mathbf{h}_i^{(3)}$ is produced through a softmax function. Note that a coefficient $|X|$ (the length of the input sequence) is applied along with a_i to $\mathbf{u}_i$ to obtain the weighted representation $\mathbf{h}_i^{(4)}$, so as to keep $\mathbf{h}_i^{(4)}$ from losing the original scale of $\mathbf{h}_i^{(3)}$.

Document Embeddings. We adopt each one of the three encoding techniques, i.e. the convolution layer with pooling, the GRU, and the attentive GRU, to form three types of document encoders respectively. Each encoder consists of one or a stack of the corresponding layers depending on the type of the input document, and encodes the document into a embedding vector. The document embedding is obtained by applying an affine layer to the average of the output $\mathbf{H}$ from the last pooling, GRU, or attentive GRU layer, i.e. $E(X) = \mathbf{M}_d\left(\frac{1}{|\mathbf{H}|}\sum_{i=1}^{|\mathbf{H}|}\mathbf{h_i}\right) + \mathbf{b}_d$.

3.3 Explicit Features

In addition to the implicit semantic features provided by document encoders, we define explicit features $F(A_i, A_j) = \{r_{tto}, r_{st}, r_{indeg}, r_{mt}, f_{TF}, I_{MW}, r_{outdeg}, d_{te}, r_{dt}\}$. A portion of the explicit features are carried forward from [26] to provide some token-level and structural measures of an article pair (A_i, A_j):

- r_{tto}: token overlap ratio of titles, i.e. the number of overlapped words between t_i and t_j divided by $|t_i|$.
- r_{st}: the maximum of the token overlap ratios among the section titles of A_i and those of A_j.
- r_{indeg}: the in-degree ratio, which is the number of incoming links in A_i divided by that of A_j. r_{indeg} measures the relative centrality of A_i with regard to A_j.
- r_{mt}: the maximum of the token overlap ratios between any anchor title of the main-article template[1] of A_i and t_j, or zero if the main-article template does not apply to A_i.

---

[1] https://en.wikipedia.org/wiki/Template:Main.

- f_{TF}: normalized term frequency of t_i in c_j.
- d_{MW}: Milne-Witten Index [31] of A_i and A_j, which measures the similarity of incoming links of two articles via the Normalized Google Distance [10].

In addition to the above features, we also include the following features.

- r_{outdeg}: the out-degree ratio, which measures the relative centrality of A_i and A_j similar to r_{indeg}.
- d_{te}: the average embedding distance of tokens in titles t_i and t_j.
- r_{dt}: token overlap ratio of c_i and c_j, which is used in [25] to measure the document relatedness. The calculation of r_{dt} is based on the first paragraphs.

We normalize the distance and frequency-based features (i.e. f_{TF} and d_{te}. d_{MW} is already normalized by its definition.) using min-max rescaling. Note that, we do not preserve the two cross-lingual features in [26]. This is because, in general, these two cross-lingual features are not applicable when the candidate space scales up to the range of the entire Wikipedia, since the inter-lingual links that match articles across different languages are far from complete [5,24].

3.4 Training

Learning Objective. The overall architecture of our model is shown in Fig. 2. The model characterizes each given article pair $p = (A_i, A_j) \in P$ in two stages.

1. Four document encoders (of the same type) are used to encode the titles and text contents of A_i and A_j respectively, which are denoted as $E_t^{(1)}(t_i)$, $E_t^{(2)}(t_j)$, $E_c^{(1)}(c_i)$ and $E_c^{(2)}(c_j)$. Two logistic regressors realized by sigmoid multi-layer perceptrons (MLP) [2] are applied on $E_t^{(1)}(t_i) \oplus E_t^{(2)}(t_j)$ and $E_c^{(1)}(c_i) \oplus E_c^{(2)}(c_j)$ to produce two confidence scores s_t and s_c for supporting A_j to be the sub-article of A_i.
2. The two semantic-based confidence scores are then concatenated with the explicit features $(\{s_t, s_c\} \oplus F(A_i, A_j))$, to which another linear MLP is applied to obtain the two confidence scores $\hat{s}_p^+$ and $\hat{s}_p^-$ for the boolean labels of positive prediction l^+ and negative prediction l^- respectively. Finally, $\hat{s}_p^+$ and $\hat{s}_p^-$ are normalized by binary softmax functions $s_p^+ = \frac{\exp(\hat{s}_p^+)}{\exp(\hat{s}_p^+)+\exp(\hat{s}_p^-)}$ and $s_p^- = \frac{\exp(\hat{s}_p^-)}{\exp(\hat{s}_p^+)+\exp(\hat{s}_p^-)}$.

The learning objective is to minimize the following binary cross-entropy loss.

$$L = -\frac{1}{|P|} \sum_{p \in P} \left(l^+ \log s_p^+ + l^- \log s_p^- \right)$$

Annotated Word Embeddings. We pre-train the Skipgram [30] word embeddings on the English Wikipedia dump to support the input of the article titles

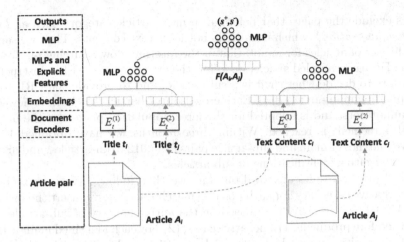

Fig. 2. Learning architecture of the model.

and text contents to the model, as well as the calculation of the feature d_{te}. We parse all the inline hyperlinks of Wikipedia dump to the corresponding article titles, and tokenize the article titles in the plain text corpora via Trie-based maximum token matching. This tokenization process aims at including Wikipedia titles in the vocabulary of word embeddings. Although this does not ensure all the titles to be involved, as some of them occur too rarely in the corpora to meet the minimum frequency requirement of the word embedding model. This tokenization process is also adopted during the calculation of d_{te}. After pre-training, we fix the word embeddings to convert each document to a sequence of vectors to be fed into the document encoder.

4 Experiments

In this section, we present the experimental evaluation of the proposed model. We first create a dataset that contains a large collection of candidate article pairs for the sub-article matching problem. Then we compare variants of the proposed model and previous approaches based on held-out estimation on the dataset. Lastly, to show the practicability of our approach, we train the model on the full dataset, and perform predictions using MapReduce on over 108 million candidate article pairs extracted from the entire English Wikipedia.

Dataset Preparation. We have prepared a new dataset, denoted WAP196k, in two stages. We start with producing positive cases via crowdsourcing. In detail, we first select a set of articles, where each article title concatenates two Wikipedia entity names directly or with a proposition, e.g. *German Army* and *Fictional Universe of Harry Potter*. We hypothesize that such articles are more likely to be a sub-article of another Wikipedia article. Note that this set of articles exclude the pages that belong to a meta-article category such as *Lists*[2] and *Disambiguation*[3], which usually do not have text contents. Then we sample from this set of articles for annotation in the internal crowdsourcing platform of Google. For each sampled article, we follow the criteria in Sect. 3.1 to instruct the annotators to decide whether it is a sub-article, and to provide the URL to the corresponding main-article if so. Each crowdsourced article has been reviewed by three annotators, and is adopted for the later population process of the dataset if total agreement is reached. Within three months, we have obtained 17,349 positive matches of main and sub-article pairs for 5,012 main-articles, and around 4k other negative identifications of sub-articles.

Based on the results of crowdsourcing, we then follow several strategies to create negative cases: (1) For each positive match (A_i, A_j), we insert the inverted pair (A_j, A_i) as a negative case based on the anti-symmetry of sub-article relations, therefore producing 17k negative cases; (2) For each identified main-article, if multiple positively matched sub-articles coexist, such sub-articles are paired into negative cases as they are considered as "same-level articles". This step contributes around 27k negative cases; (3) We substitute A_i with other articles that are pointed by an inline hyperlink in c_j, or substitute A_j with samples from the 4k negative identifications of sub-articles in stage 1. We select a portion from this large set of negative cases to ensure that each identified main-article has been paired with at least 15 negative matches of sub-articles. This step contributes the majority of negative cases. In the dataset, we also discover that around 20k negative cases are measured highly (> 0.6) by at least one of the symbolic similarity measures r_{tto}, r_{st} or f_{TF}.

The three strategies for negative case creation seek to populate the WAP196k dataset with a large amount of negative matches of articles that represent different counterfactual cases. The statistics of WAP196k are summarized in Table 1, which indicate it to be much more large-scale than the dataset used by previous approaches [26] that contains around 3k article pairs. The creation of more negative cases than positive cases is in accord with the general circumstances of the Wikipedia where the sub-article relations hold for a small portion of the article pairs. Hence, the effectiveness of the model should be accordingly evaluated by how precisely and completely it can recognize the positive matches from all candidate pairs from the dataset. As we have stated in Sect. 3.1, we encode the first paragraph of each article to represent its text contents.

4.1   Evaluation

We use a held-out estimation method to evaluate our approach on WAP196k. Besides three proposed model variants that combine a specific type of neural document encoders with the explicit features, we compare several statistical classification algorithms that [26] have trained on the explicit features. We also compare with three neural document pair encoders without explicit features that represent the other line of related work [18, 36, 39, 48].

---

[2] https://en.wikipedia.org/wiki/Category:Lists.
[3] https://en.wikipedia.org/wiki/Wikipedia:Disambiguation.

Table 1. Statistics of the dataset.

# Article pairs	# Positive cases	# Negative cases	# Main-articles	# Distinct articles
195,960	17,349	178,611	5,012	32,487

Table 2. Cross-validation results on WAP196k. We report precision, recall and F1-scores on three groups of models: (1) statistical classification algorithms based on explicit features, including logistic regression, Naive Bayes Classifier (NBC), Linear SVM, Adaboost (SAMME.R algorithm), Decision Tree (DT), Random Forest (RF) and k-nearest-neighbor classifier (kNN); (2) three types of document pair encoders without explicit features; (3) the proposed model in this paper that combines explicit features with convolutional document pair encoders (CNN+F), GRU encoders (GRU+F) or attentive encoders (AGRU+F).

Model	Explicit features						
	Logistic	NBC	Adaboost	LinearSVM	DT	RF	kNN
Precision (%)	82.64	61.78	87.14	82.79	87.17	89.22	65.80
Recall (%)	88.41	87.75	85.40	89.56	84.53	84.49	78.66
F1-score	0.854	0.680	0.863	0.860	0.858	0.868	0.717

Model	Semantic features			Model	Explicit+Semantic		
	CNN	GRU	AGRU		CNN+F	GRU+F	AGRU+F
Precision (%)	95.83	95.76	93.98	Precision (%)	**99.13**	98.60	97.58
Recall (%)	90.46	87.24	86.47	Recall (%)	**98.06**	88.47	86.80
F1-score	0.931	0.913	0.901	F1-score	**0.986**	0.926	0.919

Model Configurations. We use AdaGrad to optimize the learning objective function and set the learning rate as 0.01, batchsize as 128. For document encoders, we use two convolution/GRU/attentive GRU layers for titles, and two layers for the text contents. When inputting articles to the document encoders, we remove stop words in the text contents, zero-pad short ones and truncate overlength ones to the sequence length of 100. We also zero-pad short titles to the sequence length of 14, which is the maximum length of the original titles. The dimensionality of document embeddings is selected among {100, 150, 200, 300}, for which we fix 100 for titles and 200 for text contents. For convolutional encoders, we select the kernel size and the pool size from 2 to 4, with the kernel size of 3 and 2-max-pooling adopted. For pre-trained word embeddings, we use context size of 20, minimum word frequency of 7 and negative sampling size of 5 to obtain 120-dimensional embeddings from the tokenized Wikipedia corpora mentioned in Sect. 3.4. Following the convention, we use one hidden layer in MLPs, where the hidden size averages those of the input and output layers.

Evaluation Protocol. Following [26], we adopt 10-fold cross-validation in the evaluation process. At each fold, all models are trained till converge. We aggregate *precision*, *recall* and *F1-score* on the positive cases at each fold of testing, since the objective of the task is to effectively identify the relatively rare article

Table 3. Ablation on feature categories for CNN+F.

Features	Precision	Recall	F1-score
All features	99.13	98.06	0.986
No titles	98.03	85.96	0.916
No text contents	98.55	95.78	0.972
No explicit	95.83	90.46	0.931
Explicit only	82.64	88.41	0.854

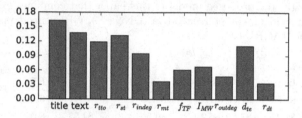

Fig. 3. Relative importance (RI) of features analyzed by Garson's algorithm. RI of each feature is aggregated from all folds of cross-validation.

relation among a large number of article pairs. All three metrics are preferred to be higher to indicate better performance.

Results. Results are reported in Table 2. The explicit features alone are helpful to the task, on which the result by the best baseline (Random Forest) is satisfactory. However, the neural encoders for document pairs, even without the explicit features, outperform Random Forest by 4.76% on precision, 1.98% on recall and 0.033 on F1-score. This indicates that the implicit semantic features are critical for characterizing the matching of main and sub-articles. Among the three types of document encoders, the convolutional encoder is more competent than the rest two sequence encoders, which outperforms Random Forest by 6.54% of precision, 8.97% of recall and 0.063 of F1-score. This indicates that the convolutional and pooling layers that effectively capture the local semantic features are key to the identification of sub-article relations, while such relations appear to be relatively less determined by the sequence information and overall document meanings that are leveraged by the GRU and attention encoders. The results by the proposed model which combines document pair encoders and explicit features are very promising. Among these, the model variant with convolutional encoders (CNN+F) obtained close to perfect precision and recall.

Meanwhile, we perform ablation on different categories of features and each specific feature, so as to understand their significance to the task. Table 3 presents the ablation of feature categories for the CNN-based model. We have already shown that completely removing the implicit semantic features would noticeably impair the precision. Removing the explicit features moderately hinders both precision and recall. As for the two categories of semantic features, we find that

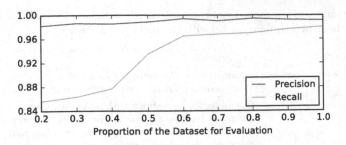

Fig. 4. Sensitivity of CNN+F on the proportion of dataset for evaluation.

removing either of them would noticeably impair the model performance in terms of recall, though the removal of title embeddings has much more impact than that of text content embeddings. Next, we perform Garson's algorithm [14, 34] on the weights of the last linear MLP of CNN+F to analyze the relative importance (RI) of each specific feature, which are reported as Fig. 3. It is noteworthy that, besides the text features, the explicit features r_{tto}, r_{st} and d_{te} that are related to article or section titles also show high RI. This is also close to the practice of human cognition, as we humans are more likely to be able to determine the semantic relation of a given pair of articles based on the semantic relation of the titles and section titles than based on other aspects of the explicit features.

Furthermore, to examine how much our approach may rely on the large dataset to obtain a generic solution, we conduct a sensitivity analysis of CNN+F on the proportion of the dataset used for cross-validation, which is reported in Fig. 4. We discover that, training the model on smaller portions of the dataset would decrease the recall of predictions by the model, though the impact on the precision is very limited. However, even using 20% of the data, CNN+F still obtains better precision and recall than the best baseline Random Forest that is trained solely on explicit features in the setting of full dataset.

To summarize, the held-out estimation on the WAP196k dataset shows that the proposed model is very promising in addressing the sub-article matching task. Considering the large size and heterogeneity of the dataset, we believe the best model variant CNN+F is close to a well-generalized solution.

4.2 Mining Sub-articles from the Entire English Wikipedia

For the next step, we move on to putting the proposed model into production by serving it to identify the main and sub-article matching on the entire body of the English Wikipedia. The English Wikipedia contains over 5 million articles, which lead to over 24 trillion ordered article pairs. Hence, instead of serving our model on that astronomical candidate space, we simplify the task by predicting only for each article pair that forms an inline hyperlink across Wikipedia pages, except for those that appear already in the *main-article* templates. This reduces our candidate space to about 108 million article pairs.

Table 4. Examples of recognized main and sub-article matches. The italicize sub-article titles are without overlapping tokens with the main article titles.

Main article	Sub-articles
Outline of government	*Bicameralism, Capitalism, Dictatorship, Confederation, Oligarchy, Sovereign state*
Computer	Computer for operations with functions, Glossary of computer hardware terms, Computer user, *Timeline of numerical analysis after 1945*, Stored-program computer, Ternary computer
Hebrew alphabet	Romanization of Hebrew
Recycling by material	Drug recycling, *Copper, Aluminium*
Chinese Americans	History of Chinese Americans in Dallas-Fort Worth, History of Chinese Americans in San Francisco, Anti-Chinese Violence in Washington
Genetics	Modification (Genetics), Theoretical and Applied Genetics, Encyclopedia of Genetics
Service Rifle	United States Marine Corps Squad Advanced Marksman Rifle, United States Army Squad Designated Marksman Rifle
Transgender rights	LGBT rights in Panama, LGBT rights in the United Arab Emirates, Transgender rights in Argentina, History of transgender people in the United States
Spectrin	Spectrin Repeat
Geography	Political Geography, Urban geography, Visual geography, *Colorado Model Content Standards*
Nuclear Explosion	Outline of Nuclear Technology, International Day Against Nuclear Tests
Gay	*LGBT Rights by Country or Territory*, Philadelphia Gay News, Troll (gay slang), Gay literature
FIBA Hall of Fame	*Šarūnas Marčiulionis*
Arve Isdal	*March of the Norse, Between Two Worlds*
Independent politician	*Balasore (Odisha Vidhan Sabha Constituency)*
Mathematics	Hierarchy (mathematics), *Principle part*, Mathematics and Mechanics of Complex Systems, *Nemytskii operator, Spinors in three dimensions, Continuous functional calculus*, Quadrature, Table of mathematical symbols by introduction date, *Hasse invariant of an algebra*, Concrete Mathematics
Homosexuality	*LGBT rights in Luxembourg*, List of Christian denominational positions on homosexuality
Bishop	*Roman Catholic Diocese of Purnea, Roman Catholic Diocese of Luoyang*
Lie algebra	Radical of a Lie algebra, Restricted Lie algebra, *Adjoint representation*, Lie Group

We train the best model variant CNN+F from the previous experiment for serving. We carry forward the model configurations from the previous experiment. The model is trained on the entire WAP196k dataset till converge.

The extraction of the candidate article pairs as well as the serving of the model is conducted via MapReduce on 3,000 machines, which lasts around 9 h in total. We select the 200,000 positive predictions with highest confidence scores s_p^+, based on which human evaluation on three turns of 1,000 sampled results estimates a 85.7% of $P@200k$ (precision at top 200,000 predictions). Examples of identified main and sub-article matches are listed in Table 4. Based on the selected positive predictions, the number of sub-articles per main-article is estimated as 4.9, which is lower than 7.5 that is estimated on the 1,000 most viewed articles by [26]. There are also around 8% of sub-articles that are paired with more than one main-articles. Based on the promising results from the large-scale model serving, our team is currently working on populating the identified sub-article relations into the backend knowledge base for our search engine.

5   Conclusion and Future Work

In this paper, we have proposed a neural article pair model to address the sub-article matching problem in Wikipedia. The proposed model utilizes neural document encoders for titles and text contents to capture the latent semantic features from Wikipedia articles, for which three types of document encoders have been considered, including the convolutional, GRU and attentive encoders. A set of explicit features are incorporated into the learning framework that comprehensively measured the symbolic and structural similarity of article pairs. We have created a large article pair dataset WAP196k from English Wikipedia which seeks to generalize the problem with various patterns of training cases. The experimental evaluation on WAP196k based on cross-validation shows that the document encoders alone are able to outperform the previous models using only explicit features, while the combined model based on both implicit and explicit features is able to achieve near-perfect precision and recall. Large-scale serving conducted on the entire English Wikipedia is able to produce a large amount of new main and sub-article matches with promising quality. For future work, it is natural to apply the proposed model to other language-versions of Wikipedia for production. It is also meaningful to develop an approach to differentiate the sub-articles that describe refined entities and those that describe abstract sub-concepts. Meanwhile, we are interested in extending our approach to populate the incomplete cross-lingual alignment of Wikipedia articles using bilingual word embeddings such as Bilbowa [16], and a different set of explicit features.

References

1. Ackerman, M.S., Dachtera, J., et al.: Sharing knowledge and expertise: the CSCW view of knowledge management. CSCW **22**, 531–573 (2013)
2. Bengio, Y.: Learning deep architectures for AI. Found. Trends Mach. Learn. **2**(1), 1–127 (2009)
3. Cai, Z., Zhao, K., et al.: Wikification via link co-occurrence. In: CIKM (2013)

4. Chen, D., Fisch, A., et al.: Reading Wikipedia to answer open-domain questions. In: ACL (2017)
5. Chen, M., Tian, Y., et al.: Multilingual knowledge graph embeddings for cross-lingual knowledge alignment. In: IJCAI (2017)
6. Chen, M., Tian, Y., et al.: Co-training embeddings of knowledge graphs and entity descriptions for cross-lingual entity alignment. In: IJCAI (2018)
7. Chen, M., Tian, Y., et al.: On2Vec: embedding-based relation prediction for ontology population. In: SDM (2018)
8. Chen, M., Zaniolo, C.: Learning multi-faceted knowledge graph embeddings for natural language processing. In: IJCAI (2017)
9. Chung, J., Gulcehre, C., et al.: Empirical evaluation of gated recurrent neural networks on sequence modeling. arXiv (2014)
10. Cilibrasi, R.L., Vitanyi, P.M.: The Google similarity distance. TKDE **19**(3), 370–383 (2007)
11. Dean, J., Ghemawat, S.: MapReduce: simplified data processing on large clusters. Comm. ACM **51**(1), 107–113 (2008)
12. Dhingra, B., Liu, H., et al.: Gated-attention readers for text comprehension. In: ACL (2017)
13. Dojchinovski, M., Kliegr, T.: Entityclassifier.eu: real-time classification of entities in text with Wikipedia. In: Blockeel, H., Kersting, K., Nijssen, S., Železný, F. (eds.) ECML PKDD 2013. LNCS (LNAI), vol. 8190, pp. 654–658. Springer, Heidelberg (2013). https://doi.org/10.1007/978-3-642-40994-3_48
14. Féraud, R., Clérot, F.: A methodology to explain neural network classification. Neural Netw. **15**(2), 237–246 (2002)
15. Gabrilovich, E., Markovitch, S., et al.: Computing semantic relatedness using Wikipedia-based explicit semantic analysis. In: IJCAI (2007)
16. Gouws, S., Bengio, Y., Corrado, G.: Bilbowa: fast bilingual distributed representations without word alignments. In: ICML (2015)
17. Hecht, B., Carton, S.H., et al.: Explanatory semantic relatedness and explicit spatialization for exploratory search. In: SIGIR (2012)
18. Hu, B., Lu, Z., et al.: Convolutional neural network architectures for matching natural language sentences. In: NIPS, pp. 2042–2050 (2014)
19. Jozefowicz, R., Zaremba, W., et al.: An empirical exploration of recurrent network architectures. In: ICML (2015)
20. Kadlec, R., Schmid, M., et al.: Text understanding with the attention sum reader network. In: ACL, vol. 1 (2016)
21. Kim, Y.: Convolutional neural networks for sentence classification. In: EMNLP (2014)
22. Kittur, A., Kraut, R.E.: Beyond Wikipedia: coordination and conflict in online production groups. In: CSCW (2010)
23. Lascarides, A., Asher, N.: Temporal interpretation, discourse relations and commonsense entailment. Linguist. Philos. **16**(5), 437–493 (1993)
24. Lehmann, J., Isele, R., Jakob, M., Jentzsch, A., et al.: DBpedia–a large-scale, multilingual knowledge base extracted from Wikipedia. Seman. Web **6**(2), 167–195 (2015)
25. Lin, C.Y., Hovy, E.: From single to multi-document summarization: a prototype system and its evaluation. In: ACL (2002)
26. Lin, Y., Yu, B., et al.: Problematizing and addressing the article-as-concept assumption in Wikipedia. In: CSCW (2017)
27. Liu, X., Xia, T., et al.: Cross social media recommendation. In: ICWSM (2016)

28. Mahdisoltani, F., Biega, J., Suchanek, F., et al.: Yago3: a knowledge base from multilingual Wikipedias. In: CIDR (2015)
29. Meij, E., Balog, K., Odijk, D.: Entity linking and retrieval for semantic search. In: WSDM (2014)
30. Mikolov, T., Sutskever, I., et al.: Distributed representations of words and phrases and their compositionality. In: NIPS (2013)
31. Milne, D., Witten, I.H.: Learning to link with Wikipedia. In: CIKM (2008)
32. Mousavi, H., Atzori, M., et al.: Text-mining, structured queries, and knowledge management on web document Corpora. SIGMOD Rec. **43**(3), 48–54 (2014)
33. Ni, Y., Xu, Q.K., et al.: Semantic documents relatedness using concept graph representation. In: WSDM (2016)
34. Olden, J.D., Jackson, D.A.: Illuminating the "black box": a randomization approach for understanding variable contributions in artificial neural networks. Ecol. Model. **154**(1–2), 135–150 (2002)
35. Poria, S., Cambria, E., et al.: Deep convolutional neural network textual features and multiple Kernel learning for utterance-level multimodal sentiment analysis. In: EMNLP (2015)
36. Rocktäschel, T., Grefenstette, E., et al.: Reasoning about entailment with neural attention (2016)
37. Schuhmacher, M., Ponzetto, S.P.: Knowledge-based graph document modeling. In: WSDM (2014)
38. Severyn, A., Moschitti, A.: Twitter sentiment analysis with deep convolutional neural networks. In: SIGIR (2015)
39. Sha, L., Chang, B., et al.: Reading and thinking: re-read LSTM unit for textual entailment recognition. In: COLING (2016)
40. Strube, M., Ponzetto, S.P.: Wikirelate! Computing semantic relatedness using Wikipedia. In: AAAI (2006)
41. Suchanek, F.M., Abiteboul, S., et al.: Paris: probabilistic alignment of relations, instances, and schema. In: PVLDB (2011)
42. Tsai, C.T., Roth, D.: Cross-lingual Wikification using multilingual embeddings. In: NAACL (2016)
43. Vrandečić, D.: Wikidata: a new platform for collaborative data collection. In: WWW (2012)
44. Vrandečić, D., Krötzsch, M.: Wikidata: a free collaborative knowledge base. Comm. ACM **57**(10), 78–85 (2014)
45. Wang, Z., Li, J., et al.: Cross-lingual knowledge linking across Wiki knowledge bases. In: WWW (2012)
46. Xie, R., Liu, Z., et al.: Representation learning of knowledge graphs with entity descriptions. In: AAAI (2016)
47. Yamada, I., Shindo, H., et al.: Joint learning of the embedding of words and entities for named entity disambiguation. In: CoNLL (2016)
48. Yin, W., Schütze, H.: Convolutional neural network for paraphrase identification. In: NAACL (2015)
49. Yin, W., Schütze, H., et al.: Abcnn: Attention-based convolutional neural network for modeling sentence pairs. TACL **4**(1), 259–272 (2016)
50. Zou, L., Huang, R., et al.: Natural language question answering over RDF: a graph data driven approach. In: SIGMOD (2014)

LinNet: Probabilistic Lineup Evaluation Through Network Embedding

Konstantinos Pelechrinis[(✉)]

University of Pittsburgh, Pittsburgh, USA
kpele@pitt.edu

Abstract. Which of your team's possible lineups has the best chances
against each of your opponent's possible lineups? To answer this question,
we develop LinNet (which stands for LINeup NETwork). LinNet exploits
the dynamics of a directed network that captures the performance of
lineups during their matchups. The nodes of this network represent the
different lineups, while an edge from node B to node A exists if lineup
λ_A has outperformed lineup λ_B. We further annotate each edge with the
corresponding performance margin (point margin per minute). We then
utilize this structure to learn a set of latent features for each node (i.e.,
lineup) using the **node2vec** framework. Consequently, using the latent,
learned features, LinNet builds a logistic regression model for the probability of lineup λ_A outperforming lineup λ_B. We evaluate the proposed
method by using NBA lineup data from the five seasons between 2007–08
and 2011–12. Our results indicate that our method has an out-of-sample
accuracy of 68%. In comparison, utilizing simple network centrality metrics (i.e., PageRank) achieves an accuracy of just 53%, while using the
adjusted plus-minus of the players in the lineup for the same prediction
problem provides an accuracy of only 55%. We have also explored the
adjusted lineups' plus-minus as our predictors and obtained an accuracy
of 59%. Furthermore, the probability output of LinNet is well-calibrated
as indicated by the Brier score and the reliability curve. One of the main
benefits of LinNet is its generic nature that allows it to be applied in
different sports since the only input required is the lineups' matchup
network, i.e., not any sport-specific features are needed.

Keywords: Network science · Network embedding · Sports analytics
Probabilistic models

1 Introduction

During the past decade or so, the availability of detailed sports data in conjunction with the success enjoyed by early adopters, has led to the explosion
of the field of sports analytics. Part of this can be attributed to the advancements in computing technologies that have facilitated the collection of detailed
(spatio-temporal) data that can shed light to aspects of the sport(s) in question
that were not possible before. For example, since 2013 a computer vision system

© Springer Nature Switzerland AG 2019
U. Brefeld et al. (Eds.): ECML PKDD 2018, LNAI 11053, pp. 20–36, 2019.
https://doi.org/10.1007/978-3-030-10997-4_2

installed in all NBA stadiums collects the location of all players on the court and the ball 25 times every second. Using this information the Toronto Raptors were able to (manually) identify the optimal position for defenders given the offensive scheme. These *optimal* defenders are called *ghosts* and can be used to evaluate defensive skills, an aspect of the game severely underrepresented in traditional boxscore statistics [8]. Since then automated ways for ghosting, and in general for analyzing and understanding fine-grained in-game behavior, have been developed - in various sport - relying on the advancements in representation (deep) learning (e.g., [7,10,16,19]).

However, representation learning can also help answer more *traditional* questions in a new way. For example, one of the decisions that a basketball coach has to constantly make during a game (or even for game preparation) is what lineup to play in order to maximize the probability of outperforming the opponent's lineup currently on the court. This lineup evaluation problem has been traditionally addressed through ranking lineups. More specifically, player and lineup ratings based on (adjusted) plus/minus-like approaches [18], or efficiency ratings (i.e., points scored/allowed/net per 100 possessions [1]) have been used to rank lineups. This ranking can then be used to evaluate which lineup is *better*. Nevertheless, these ratings do not explicitly account for the game situation/context. For instance, a lineup that outperformed its opponent by 10 points in *garbage* time does not provide us with the same information as compared to the same lineup outperforming its opponent during the start of the game. Hence, to account for this the use of in-game win probability models has been proposed [12]. In this case, instead of computing the net points scored in every stint[1] we calculate the win probability added by each of the two lineups.

To the best of our knowledge apart from these basic metrics that are used to rank lineups, there exist no studies in the public sphere that evaluate the predictive power of these metrics and/or introduce other ways of evaluating and predicting lineup matchups[2]. In the current study, we propose a completely different approach that is based on representation learning on networks. In particular, we first define the matchup network $\mathcal{G}$:

Definition 1.1: Matchup Network

The matchup network $\mathcal{G} = (\mathcal{V}, \mathcal{E}, \mathcal{W})$, is a weighted directed network where nodes represent lineups. An edge $e_{i,j} \in \mathcal{E}$ points from node $i \in \mathcal{V}$ to node $j \in \mathcal{V}$ iff lineup j has outperformed lineup i. The edge weight $w_{e_{i,j}}$ is equal to the performance margin of the corresponding matchup.

Using the structure of this network we can learn a vector representation of the nodes. For this purpose we utilize a network embedding, which projects the

---

[1] Stint refers to a time-period during the game when no substitutions happen by either team.

[2] Of course, we expect professional teams to perform their own analysis - potentially beyond simply ranking - but their proprietary nature makes it impossible to study and evaluate.

network nodes on a latent space $\mathcal{X}$. In our study we adopt the **node2vec** [6] framework for learning the latent space. Simply put, the embedding learns a set of features $\mathbf{x}_u$ for node u. These features are then utilized to build a logistic regression model that models the probability of lineup λ_i outperforming lineup λ_j, $\Pr[\lambda_i \succ \lambda_j | \mathbf{x}_{\lambda_i}, \mathbf{x}_{\lambda_j}]$. Figure 1 visually represents the **LinNet** framework.

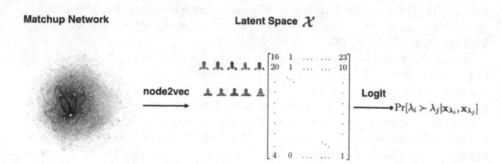

Fig. 1. The **LinNet** lineup evaluation framework

Our evaluations indicate that **LinNet** can predict the outcome of a lineup matchup correctly with approximately 68% accuracy, while the probabilities are well-calibrated with a Brier score of 0.19. Furthermore, the probability validation curve of **LinNet** is statistically indistinguishable from the $y = x$ line, i.e., the predicted matchup probability is equal to the *actual* probability (see Fig. 3). Hence, the logistic regression model on the latent space $\mathcal{X}$ captures accurately the lineups' matchup probabilities. In comparison, we evaluate the following three baseline methods inspired both from current approaches in ranking lineups as well as network ranking; (i) a PageRank-based ranking using the same matchup lineup network $\mathcal{G}$, (ii) a model based on the adjusted plus/minus of the players that are part of each lineup, and (iii) a model based on the adjusted plus/minus of the lineups. The lineup adjusted plus/minus has the best performance among the baselines, but still worse than **LinNet**, with an accuracy of 59%.

The main contribution of our work is twofold:

– We introduce and evaluate a novel approach for evaluating basketball lineups in a probabilistic way using representation learning on networks.
– The proposed method is generic, i.e., it can be adopted in other sports without the need to incorporate sport-specific information.

We also hope that this study will trigger more interest and research in the applications of network science in sports. While network science methods have been used in the literature to study and answer sports analytics questions, these studies are primarily focused on analyzing the most straightforward network structure in sports, namely, passing networks, i.e., *who-passes-to-whom struc-tures* (e.g., [4,5,14]). However, networks can also be used to represent complex structures that might not be *visible* directly – such as the win-loss relationships of teams or lineups that we use in our current study – and can provide new and novel insights.

The rest of the paper is organized as following. In Sect. 2 we present in details the operations of LinNet as well as the datasets we used. Section 3 presents our results, while Sect. 4 concludes our study and discusses the implications and limitations of our work.

2 Materials and Methods

In this section we will present in detail (a) the design of LinNet, (b) the baseline methods for comparison, and (c) the datasets we used for our evaluations.

2.1 LinNet

The first step of LinNet is defining the matchup network $\mathcal{G}$. There is flexibility in choosing the performance margin that one can use for the edge weights. In the current implementation of LinNet, the weights of $\mathcal{G}$ correspond to the point margin per minute for the two lineups.

Once the network is obtained the next step is to learn the network embedding. As our network embedding mechanism we will utilize the approach proposed by Grover and Leskovec [6], namely, node2vec. node2vec utilizes (2nd order) random walks on the network in order to learn the latent features of the nodes, i.e., a function $f : \mathcal{V} \to \Re^d$, where d is the dimensionality of the latent space. Starting from node u in the network and following the random walk strategy R the network neighborhood $N_R(u)$ of u is defined. Then node2vec learns the network embedding f by solving the following optimization problem:

$$\max_f \sum_{u \in \mathcal{V}} \log(\Pr[N_R(u)|f(u)]) \tag{1}$$

where $\Pr[N_R(u)|f(u)]$ is the (conditional) probability of observing $N_R(u)$ as the network neighborhood for node u. Simply put, the network embedding maximizes the log-likelihood of observing a network neighborhood $N_R(u)$ for node u conditioned on the network embedding f. To keep the optimization tractable, node2vec makes use of two standard assumptions; (i) conditional independence of the nodes in $N_R(u)$, and (ii) the probability of each source-neighborhood node pair is modeled through a softmax of the dot product of their features f (to be learned). When two nodes are *similar*, they are expected to appear within the same random walk often, and the optimization problem (1) ensures that they will be close in the embedding space.

The random walk strategy - which implicitly defines the *similarity* of two nodes - is defined by two parameters, p and q, that offer a balance between a purely breadth-first search walk and a purely depth-first search walk. In particular, the random walk strategy of node2vec includes a bias term α controlled by parameters p and q. Assuming that a random walk is on node u (coming from node v), the unnormalized transition probability $\pi_{ux} = \alpha_{pq}(v, x) \cdot w_{ux}$. With d_{ux} being the shortest path distance between u and x we have:

$$\pi_{ux} = \begin{cases} 1/p, & if\ d_{ux} = 0 \\ 1, & if\ d_{ux} = 1 \\ 1/q, & if\ d_{ux} = 2 \end{cases}$$

As alluded to above, parameters p and q control the type of network neighborhood $N_R(u)$ we obtain. Different sampling strategies will provide different embeddings. For example, if we are interested in having a set of nodes that are tightly connected in the original network, to be close to each other in the latent space, p and q need to be picked in such a way that allows for "local" sampling. In our application we are interested more in identifying structurally equivalent nodes, i.e., nodes that are similar because their connections in the network are similar (not necessarily close to each other with respect to network distance). This requires a sampling strategy that allows for the network neighborhood of a node to include nodes that are further away as well. Given this objective and the recommendations by Grover and Leskovec [6] we choose $q = 3$ and $p = 0.5$ for our evaluations. Furthermore, we generate 3,000 walks for each network, of 3,500 hops each, while, we choose as our latent space dimensionality, $d = 128$. Increasing the dimensionality of the space improves the quality of the embedding as one might have expected, however, our experiments indicate that increasing further the dimensionality beyond $d = 128$ we operate with diminishing returns (with regards to computational cost and improvement in performance).

Once the latent space $\mathcal{X}$ is obtained, we can build a logistic regression model for the probability of lineup λ_i outperforming λ_j. In particular, we use the Bradley-Terry model [2]. The Bradley-Terry model is a method for (probabilistically) ordering a given set of items based on their characteristics and understanding the impact of these characteristics on the ranking. In our case the set of items are the lineups and the output of the model for items i and j provides us essentially with the probability of lineup λ_i outperforming λ_j. In particular, the Bradley-Terry model is described by [2]:

$$\Pr(\lambda_i \succ \lambda_j | \pi_i,\ \pi_j) = \frac{e^{\pi_i - \pi_j}}{1 + e^{\pi_i - \pi_j}} \tag{2}$$

where π_i is λ_i's *ability*. Given a set of lineup-specific explanatory variables $\mathbf{z}_i$, the difference in the ability of lineups λ_i and λ_j can be expressed as:

$$\pi_i - \pi_j = \sum_{r=1}^{d} \alpha_r (z_{ir} - z_{jr}) + U \tag{3}$$

where $U \sim N(0, \sigma^2)$. The Bradley-Terry model is then a generalized linear model that can be used to predict the probability of λ_i outperforming λ_j. In our case, the explanatory variables are the latent features learned for each lineup, $\mathbf{x}_{\lambda_i}$.

Previously Unseen Lineups: One of the challenges (both in out-of-sample evaluations as well as in a real-world setting), is how to treat lineups that we have not seen before, and hence, we do not have their latent space representation. In the current design of LinNet we take the following simple approach. In particular, for each lineup λ_i of team $\mathcal{T}$ we define the similarity in the players' space $\sigma_{\lambda_i,\lambda_j}$ of λ_i with $\lambda_j \in \mathcal{L}_{\mathcal{T}}$, as the number of common players between the two lineups (i.e., $\sigma_{\lambda_i,\lambda_j} \in \{0,\ldots,4\}$). One might expect that lineups with high overlap in the players' space, should also reside closely in the embedding space. In order to get a feeling of whether this is true or not, we calculated for every team and season the correlation between the similarity between two lineups in the players' space (i.e., $\sigma_{\lambda_i,\lambda_j}$) and the distance for the corresponding latent features (i.e., $\mathtt{dist}(\mathbf{x}_i, \mathbf{x}_j)$). As we can see from Fig. 2 all teams exhibit negative correlations (all correlations are significant at the 0.001 level), which means the more common players two lineups have, the more closely they will be projected in the embedding space. Of course, the levels of correlation are moderate at best since, the embedding space is obtained by considering the performance of each lineup, and two lineups that differ by only one player might still perform completely differently on the court. With this in mind, once we obtain the lineup similarity values, we can assign the latent feature vector for the previously unseen lineup λ_i as a weighted average of the lineups in $\mathcal{L}_{\mathcal{T}}$ (with σ being the weighting factor):

$$
\mathbf{x}_{\lambda_i} = \frac{\displaystyle\sum_{\lambda_j \in \mathcal{L}_{\mathcal{T}}} \sigma_{\lambda_i,\lambda_j} \cdot \mathbf{x}_j}{\displaystyle\sum_{\lambda_j \in \mathcal{L}_{\mathcal{T}}} \sigma_{\lambda_i,\lambda_j}}
\tag{4}
$$

It should be evident that this is simply a heuristic that is currently implemented in LinNet. One could think of other ways to approximate the latent space features of a lineup not seen before.

2.2 Baselines

For comparison purposes we have also evaluated three baseline approaches for predicting lineup matchup performance. The first one is based on network ranking that operates directly on the matchup network (i.e., without involving any embedding of the network), while the rest two are based on the adjusted plus/minus rating of the players that belong to the lineup, as well as, the lineup itself.

Network Ranking. In our prior work we have shown that ranking teams through centrality metrics - and in particular PageRank - of a win-loss network, achieves better matchup prediction accuracy as compared to their win-loss record [13]. The intuition behind the team network ranking is that nodes (teams) with high PageRank have outperformed many more teams or have outperformed *good* teams that themselves have performed many other or *good* teams etc. Therefore,

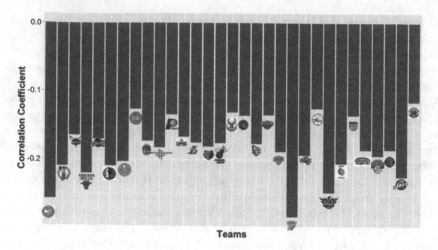

Fig. 2. Lineups with higher overlap in terms of players exhibit smaller distance in the latent embedding space $\mathcal{X}$

we follow a similar approach using the lineup matchup network and rank lineups based on their PageRank score. The PageRank of $\mathcal{G}$ is given by:

$$r = D(D - \alpha A)^{-1}\mathbf{1} \tag{5}$$

where A is the adjacency matrix of $\mathcal{G}$, α is a parameter (a typical value of which is 0.85) and D is a diagonal matrix where $d_{ii} = \max(1, k_{i,out})$, with $k_{i,out}$ being the out-degree of node i. Using the PageRank score differential $\Delta r_{ij} = r_{\lambda_i} - r_{\lambda_j}$ as our independent variable we build a logistic regression model for the probability: $\Pr(\lambda_i \succ \lambda_j | \Delta r_{ij})$.

Player Adjusted Plus/Minus (PAPM). The APM statistic of a player is a modern NBA statistic - and for many people the best single statistic we currently have - for rating players [15]. It captures the additional points that the player is expected to add with his presence in a lineup consisting of league average players matching up with a lineup with league average players. APM captures the impact of a player beyond pure scoring. For instance, a player might impact the game by performing good screens that lead to open shots, something not captured by current box score statistics. The other benefit of APM is that it controls for the rest of the players in the lineups. More specifically the APM for a player is calculated through a regression model. Let us consider that lineup λ_i has played against λ_j, and has outscored the latter by y points per 48 min. y is the dependent variable of the model, while the independent variable is a binary vector $\mathbf{p}$, each element of which represents a player. All elements of $\mathbf{p}$

are 0 except for the players in the lineups. Assuming λ_i is the home lineup[3], $p_n = 1$, $\forall p_n \in \lambda_i$, while for the visiting lineup, $p_n = -1$, $\forall p_n \in \lambda_j$. Then these data are used to train a regression model:

$$y = \mathbf{a}^T \cdot \mathbf{p} \qquad (6)$$

where $\mathbf{a}$ is the vector of regression coefficients. Once obtaining this vector, the APM for player p_n is simply a_{p_n}. The rating of lineup λ_i, ρ_{λ_i} is then the average APM of its players:

$$\rho_{\lambda_i} = \frac{a_{p_n}}{5}, \ \forall p_n \in \lambda_i \qquad (7)$$

Using the lineup rating differential $\Delta \rho_{ij} = \rho_{\lambda_i} - \rho_{\lambda_j}$ as our independent variable we again build a logistic regression model for the probability: $\Pr(\lambda_i \succ \lambda_j | \Delta \rho_{ij})$.

Lineup Adjusted Plus/Minus (LAPM). The above baseline method assumes that the lineup is simply the sum of its individual parts in a vacuum. However, this is certainly not true in many cases (if not in most/all of the cases). Players can help each other boost their performance, or they might not be in sync and hence, not perform as expected. For example, one should expect that a lineup that includes your best player (e.g., the highest APM) should perform better than one where he is substituted. However, this is not necessarily true. For instance, Skinner [17] used network theory to show that a lineup that does not include the best player of the team, might perform better as compared to lineups including this payer. Thus, simply summing up the individual players' contribution can overestimate or underestimate a lineup's performance. For this reason, we examine another baseline that considers the adjusted plus/minus of the lineups (as opposed to individual players). More specifically, we follow the same approach as with PAPM but our independent variable binary vector now represents lineups rather than individual players. The corresponding regression coefficient represents the adjusted plus/minus of the lineup l_{λ_i}. Using the LAPM differential $\Delta l_{ij} = l_{\lambda_i} - l_{\lambda_j}$ we further build a logistic regression model for the probability: $\Pr(\lambda_i \succ \lambda_j | \Delta l_{ij})$.

2.3 Datasets

In order to evaluate LinNet we used lineup data during the 5 NBA seasons between 2007–08 and 2011–12 obtained through basketballvalue.com. This dataset includes information for all the lineup matchups for each of the 5 seasons. In particular, for each pair of lineups (e.g., λ_i, λ_j) that matched up on the court we obtain the following information:

---

[3] If this information is not available - e.g., because the input data include the aggregate time the lineups matched up over multiple games - without loss of generality we can consider the home lineup to be the one with lower ID number for reference purposes. This is in fact the setting we have in our dataset.

1. Total time of matchup
2. Total point differential
3. Players of λ_i
4. Players of λ_j.

We would like to note here that these data include aggregate information for matchups between lineups. For example, if lineup λ_A played against λ_B over multiple stints - either during the same game or across different games - the performance over these matchups will be aggregated. The benefit of this approach is that we now have a longer, and potentially more robust (see Sect. 4), observation period for the matchup between λ_A and λ_B. On the other hand, aggregating information does not allow us to account for home-field advantage. Nevertheless, the latter is typically considered to be 3 points per game (i.e., per 48 min) in the NBA [18], which means that during a 5 min stint there will be an approximately 0.3 points adjustment missed. Hence, we should not expect a big impact on our final results. Furthermore, in our dataset only approximately 10% of the lineup pairs have matched-up over separate stints.

We used these data in order to obtain both the matchup network as well as to calculate the APM for every player in each season. Using these data we build the lineup matchup networks. Table 1 depicts some basic statistics for these networks. Note here that the dataset for the 2011–12 season includes only approximately 75% of that season's games and this is why the network is smaller. All networks have similar size and density and exhibit similar diameter. Furthermore, they all have right-skewed degree distributions. Table 1 also presents the power-law exponent obtained for every network after fitting a power-law distribution, i.e., $P(k) \propto k^{-\gamma}$, where k is the node degree.

Table 1. Basic network statistics for the lineup matchup networks used.

Season	Nodes	Edges	Diameter	Power-law exponent γ
2007–08	10,380	50,038	15	3.5
2008–09	10,004	48,414	16	2.8
2009–10	9,979	49,258	15	2.6
2010–11	10,605	49,694	18	2.5
2011–12	8,498	35,134	17	2.8

3   Analysis and Results

We now turn our attention to evaluating LinNet. Our focus is on evaluating the accuracy of LinNet in predicting future lineup matchups, as well as the calibration of the inferred probabilities. For every season, we build LinNet (both the network embedding as well as the Bradley-Terry model) using 80% of the matchups and we evaluate them on the remaining 20% of the matchups. Our evaluation metrics include: (i) prediction accuracy, (ii) Brier score and (iii) the probability calibration curve.

3.1   Prediction Accuracy

Table 2 presents the accuracy of each method predicting the outcome of lineup matchups over all seasons. As we can see LinNet outperforms all the baselines during all five seasons. Out of the three baselines we evaluated, LAPM performs the best. This further indicates that the performance of a lineup cannot be simply described by the sum of its individual parts; metrics that evaluate each player individually cannot capture well the performance of a lineup.

Table 2. LinNet outperforms all three baselines with respect to accuracy. LAPM performs the best among the baselines.

Season	Page rank	PAPM	LAPM	LinNet
2007–08	52%	55%	59%	67%
2008–09	53%	56%	57%	69%
2009–10	52%	54%	58%	68%
2010–11	54%	55%	59%	68%
2011–12	53%	56%	58%	67%

3.2   Probability Calibration

Accuracy figures cannot fully evaluate a probabilistic model as it does not provide any insight on how well-calibrated and accurate the output probabilities are. To evaluate the probability calibration of LinNet we rely on the Brier score and the reliability curve.

Brier Score. In a probabilistic model, its classification accuracy paints only part of the picture. For example, two models M_1 and M_2 that both predict lineup λ_A will outperform λ_B will exhibit the same accuracy. However, if $\text{Pr}_{M_1}(\lambda_A \succ \lambda_B) = 0.9$ and $\text{Pr}_{M_2}(\lambda_A \succ \lambda_B) = 0.55$, the two models have different probability calibration. The latter can be evaluated by calculating the Brier score [3] of each model, which can essentially be thought of as a cost function. In particular, for the case of binary probabilistic prediction, the Brier score is calculated as:

$$\beta = \frac{1}{N} \sum_{i=1}^{N} (\pi_i - y_i)^2 \tag{8}$$

where N is the number of observations, π_i is the probability assigned to instance i being equal to 1 and y_i is the actual (binary) value of instance i. The Brier score takes values between 0 and 1 and as alluded to above evaluates the calibration of these probabilities, that is, the level of confidence they provide. The lower the value of β the better calibrated the output probabilities are – recall that

Brier score is essentially a cost function. Continuing on the example above a 0.9 probability is better calibrated compared to a 0.55 probability (when the ground truth is label 1) and hence, even though M_1 and M_2 have the same accuracy, M_1 is better calibrated (lower Brier score – 0.01 compared to 0.2025). The Brier scores for LinNet and the baselines examined are presented in Table 3.

Table 3. LinNet exhibits better probability calibration compared to the baselines examined.

Season	Page rank	PAPM	LAPM	LinNet
2007–08	0.23	0.22	0.22	0.19
2008–09	0.23	0.23	0.22	0.19
2009–10	0.23	0.23	0.21	0.19
2010–11	0.23	0.23	0.22	0.19
2011–12	0.23	0.22	0.21	0.18

As we can see LinNet exhibits a lower Brier score as compared to the baselines. Furthermore, typically the Brier score of a model is compared to a *climatology* model [9]. A climatology model assigns the same probability to every observation, which is equal to the fraction of positive labels in the whole dataset, i.e., a base rate. Therefore, in our case the climatology model assigns a probability of 0.5 to each observation. As alluded to above we do not have information about home and visiting lineup so our model estimates the probability of the lineup with the smaller ID outperforming the one with the larger ID. Given that the lineup ID has no impact on this probability the climatology model probability is 0.5. The Brier score for this reference model is $\beta_{climatology} = 0.25$, which is of lower quality as compared to LinNet and also slightly worse than our baselines.

Reliability Curve. Finally, we evaluate the accuracy of the probability output of LinNet by deriving the probability validation curves. In order to compute the accuracy of the predicted probabilities we would ideally want to have every matchup played several times. If the favorite lineup were given a 75% probability of outperforming the opposing lineup, then if the matchup was played 100 times we would expect the favorite to *win* approximately 75 of them. However, this is clearly not realistic and hence, in order to evaluate the accuracy of the probabilities we will use all the matchups in our dataset. In particular, if the predicted probabilities were accurate, when considering all the matchups where the favorite was predicted to win with a probability of x%, then the favorite should have outperformed the opponent in (approximately) x% of these matchups. Given the continuous nature of the probabilities we quantize them into groups that cover a 5% probability range. Figure 3 presents the predicted win probability for the reference lineup (i.e., the lineup with the smaller ID) on the x-axis, while the y-axis

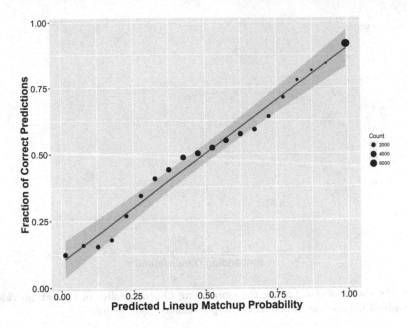

Fig. 3. The LinNet probability validation curve is very close to the $y = x$ line, translating to fairly accurate probability estimations (for $d = 128$).

presents how many of these matchups this reference lineup won. Furthermore, the size of the points represents the number of instances in each situation. As we can see the validation curve is very close to the $y = x$ line, which practically means that the predicted probabilities capture fairly well the actual matchup probabilities. In particular, the linear fit has an intercept of 0.1 and a slope of 0.85.

3.3 Dimensionality of LinNet

One of the parameters that we must choose for the network embedding, which forms the core of LinNet is its dimensionality d, i.e., how long is the vector representation of each lineup/node in $\mathcal{G}$. In all of our experiments above we have used a dimensionality of $d = 128$. However, we have experimented with different embedding dimensionality values and the accuracy results[4] are presented in Fig. 4. As we can see, low dimensionality does not provide any significant benefits with respect to the accuracy of the model over the baselines. Increasing the dimensionality further, improves the model performance. However, for values higher than $d = 128$ we see a plateau in the performance. In fact, we even see a slight decrease for dimensionality greater than 128. Higher dimensions lead to solutions that might not be as robust, since there are many more variables to

---

[4] The Brier score exhibits similar qualitatively behavior but the differences are much smaller compared to the model accuracy and hence, we omit their presentation.

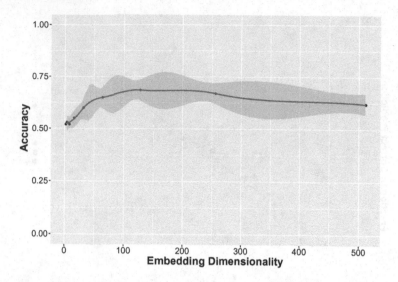

Fig. 4. The choice of $d = 128$ for the embedding dimensionality of `LinNet` provides a good tradeoff between accuracy and (computational) complexity.

optimize for the same amount of data. This can also lead to overfitting, which consequently degrades the out-of-sample performance.

3.4   Season Win-Loss Record and Lineup Performance

How well can lineup *ratings* obtained from `LinNet` explain the win-loss record of a team? One should expect that there is a correlation between `LinNet` lineup ratings and the record of a team - which as we will see indeed is the case. However, this correlation should not be expected to be perfect, since it relies also on coaching decisions as well as availability of the lineups (e.g., a lineup can be unavailable due to injuries). In order to examine this we focus on lineups that played for a total of more than a game (i.e., 48 min) during the season. Let p_{λ_i} be the average probability of lineup λ_i (of team τ) to outperform each of the opponent's lineups. I.e.,

$$p_{\lambda_i} = \frac{\sum_{\lambda_j \in \mathcal{L} \setminus \mathcal{L}_\tau} \Pr(\lambda_i \succ \lambda_j)}{|\mathcal{L} \setminus \mathcal{L}_\tau|} \tag{9}$$

where $\mathcal{L}_\tau$ is the set of all lineups of team τ and $\mathcal{L}$ is the set of all league lineups. Then the `LinNet` team rating of team τ is:

$$r(\tau) = \frac{\sum_{\lambda_i \in \mathcal{L}_\tau} \gamma_i \cdot p_{\lambda_i}}{\sum_{\lambda_i \in \mathcal{L}_\tau} \gamma_i} \tag{10}$$

where γ_i is the total time lineup λ_i has been on the court over the whole season. Our results are presented in Fig. 5. The linear regression fit has a statistically significant slope (p-value < 0.001), which translates to a statistically important relationship. However, as we can see there are outliers in this relationship, such as the 2008–09 Cavaliers and the 2011–12 Nets. The linear relationship explains 27% of the variability at the win-loss records of the teams. This might be either because teams do not choose (due to various reasons) their best lineup to matchup with the opponent, or because the time that a lineup is on court is important for its performance (we discuss this in the following section), something that LinNet currently does not account for (see Sect. 4). Overall, the correlation coefficient between the LinNet team rating and the win-loss record is moderate adn equal to 0.53 (p-value < 0.0001).

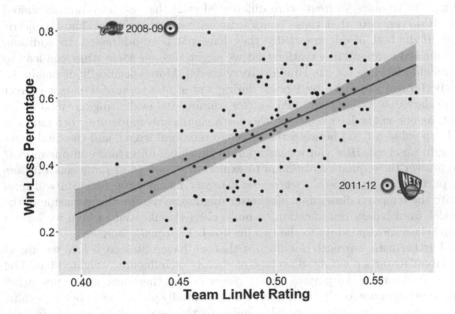

Fig. 5. The team ratings we obtain from LinNet explain 27% of the win-loss variability of teams.

4   Discussion and Conclusions

In this work, we presented LinNet, a network embedding approach for evaluating lineups. Our evaluations indicate that the probability output from LinNet is well calibrated and more accurate than traditional lineup evaluation methods. More importantly, while we have evaluated LinNet using basketball lineup data, the proposed method is sport-agnostic and not specific to basketball, i.e., there are no basketball-related features used. In particular, LinNet can be used to evaluate lineups in other sports as long as they involve frequent substitutions (e.g., hockey, volleyball etc.) and an appropriate performance metric is defined. Furthermore, it can also be used for evaluating and rating teams, as well as, predicting future games. In this case network G will capture the win-loss relationships between teams rather than lineups.

However, there are still open issues with the design of LinNet. More specifically, a matchup between lineups might last only for a few minutes (or even just a couple of possessions). This creates a reliability issue with any prediction that one tries to perform with similar information. Even though we adjust the performance margin on a per minute basis, it is not clear that a lineup can keep up its performance over a larger time span. It is a very plausible hypothesis that a lineup has its own skill curve, similar to the players' *skill* curves introduced by Oliver [11]. Other contextual factors can also impact the performance of a lineup (e.g., foul troubles, current score differential etc.) that we have not accounted for. However, note that these issues exist with every lineup evaluation metric and to the best of our knowledge they have not been addressed. In addition, the limited temporal observation that we might have for some stints can lead to unreliable labeling for the Bradley-Terry model. More specifically, if lineup λ_A outperformed lineup λ_B by 1 point during a stint of a total of 90 s, is it correct to declare that λ_A outperformed λ_B (for training and evaluating our model)? In fact, as one might have expected there is a significant, moderate, correlation of 0.44 (p-value < 0.01) between the matchup temporal length and the final point margin observed. How can we incorporate this label uncertainty in our model? To answer these questions we plan to explore the concept of *fuzzy classification* as part of our future work, where the category membership function will integrate the temporal dimension. The latter might also require the extension of the model from binary classification to multi-class classification, where we have a third class corresponding to the two lineups being *equally matched*.

Furthermore, currently for lineups that we have not seen before we use as its latent features a weighted average of already seen lineups, weighted based on their similarity in the players' space. Nevertheless, there are other approaches that one might use for this task that could potentially provide even better results. For example, a regression model (similar to the one used for calculating the adjusted plus/minus) can be used to infer the latent features based on the players in the lineup.

Finally, currently LinNet utilizes a generic network embedding framework from the network science literature (i.e., node2vec), with a number of parameters that need to be tuned and optimized[5]. However, optimizing the neighborhood objective that node2vec does might not be the most appropriate objective for evaluating lineups. Thus, a task-specific embedding might perform better than a generic framework. For example, one of the problems in ranking sports teams (and lineups) is the several intransitivity relationships (e.g., lineup λ_A outperforms lineup λ_B, lineup λ_B outperforms lineup λ_C, but lineup λ_C outperformed

---

[5] In the current version parameters p, and q, as well as, the size and number of random walks have not been necessarily optimally chosen.

lineup λ_A). These relationships manifest themselves as triangles in the matchup network. An objective function that incorporates these cycles might be more appropriate. Moreover, modeling the point performance margin between two lineups is also of interest, since in many cases a lineup needs to outscore its opponent more than just one point in order for the team to win or obtain the lead. All these are promising directions for future research on the usage of network science and representation learning for basketball analytics in general, and on evaluating lineups in particular. Despite these open issues, we firmly believe that our current study makes a solid contribution in the problem of evaluating lineups and a strong case for the use of network science methods and tools in the field of sports analytics in general.

References

1. NBA advanced stats: Lineup efficiency. https://stats.nba.com/lineups/advanced/. Accessed 04 Apr 2018
2. Agresti, A.: An Introduction to Categorical Data Analysis. Wiley Series in Probability and Statistics. Wiley-Interscience, Hoboken (2007)
3. Brier, G.W.: Verification of forecasts expressed in terms of probability. Mon. Weather Rev. **78**(1), 1–3 (1950)
4. Fewell, J.H., Armbruster, D., Ingraham, J., Petersen, A., Waters, J.S.: Basketball teams as strategic networks. PloS One **7**(11), e47445 (2012)
5. Gonçalves, B., Coutinho, D., Santos, S., Lago-Penas, C., Jiménez, S., Sampaio, J.: Exploring team passing networks and player movement dynamics in youth association football. PloS One **12**(1), e0171156 (2017)
6. Grover, A., Leskovec, J.: Node2vec: scalable feature learning for networks. In: Proceedings of the 22nd ACM SIGKDD International Conference on Knowledge Discovery and Data Mining, pp. 855–864. ACM (2016)
7. Le, H.M., Yue, Y., Carr, P., Lucey, P.: Coordinated multi-agent imitation learning. In: International Conference on Machine Learning, pp. 1995–2003 (2017)
8. Lowe, Z.: Lights, cameras, revolution (2013). http://grantland.com/features/the-toronto-raptors-sportvu-cameras-nba-analytical-revolution/
9. Mason, S.J.: On using "climatology" as a reference strategy in the Brier and ranked probability skill scores. Mon. Weather Rev. **132**(7), 1891–1895 (2004)
10. Mehrasa, N., Zhong, Y., Tung, F., Bornn, L., Mori, G.: Deep learning of player trajectory representations for team activity analysis (2018)
11. Oliver, D.: Basketball on Paper: Rules and Tools for Performance Analysis. Potomac Books, Lincoln (2004)
12. Pelechrinis, K.: Lineup evaluations through in-game win probability models and bayesian adjustment. Technical report (2018). https://www.pitt.edu/~kpele/TR-SCI-PITT-032018.pdf
13. Pelechrinis, K., Papalexakis, E., Faloutsos, C.: SportsNetRank: network-based sports team ranking. In: ACM SIGKDD Workshop on Large Scale Sports Analytics (2016)
14. Peña, J., Touchette, H.: A network theory analysis of football strategies. In: Clanet, C. (ed.), Sports Physics: Proceedings of 2012 Euromech Physics of Sports Conference, pp. 517–528, Editions de l'Ecole Polytechnique, Palaiseau (2013). ISBN: 978-2-7302-1615-9

15. Rosenbaum, D.: Measuring how NBA players help their teams win (2004). http://www.82games.com/comm30.htm. Accessed 5 June 2018
16. Seidl, T., Cherukumudi, A., Hartnett, A., Carr, P., Lucey, P.: Bhostgusters: real-time interactive play sketching with synthesized NBA defenses (2018)
17. Skinner, B.: The price of anarchy in basketball. J. Quant. Anal. Sports **6**(1), 3 (2010)
18. Winston, W.L.: Mathletics: How Gamblers, Managers, and Sports Enthusiasts Use Mathematics in Baseball, Basketball, and Football. Princeton University Press, Princeton (2012)
19. Zhan, E., Zheng, S., Yue, Y., Lucey, P.: Generative multi-agent behavioral cloning. arXiv preprint arXiv:1803.07612 (2018)

Improving Emotion Detection with Sub-clip Boosting

Ermal Toto[⊠], Brendan J. Foley, and Elke A. Rundensteiner

Computer Science Department, Worcester Polytechnic Institute,
100 Institute Road, Worcester, MA 01609, USA
{toto,bjfoley,rundenst}@wpi.edu

Abstract. With the emergence of systems such as Amazon Echo, Google Home, and Siri, voice has become a prevalent mode for humans to interact with machines. Emotion detection from voice promises to transform a wide range of applications, from adding emotional-awareness to voice assistants, to creating more sensitive robotic helpers for the elderly. Unfortunately, due to individual differences, emotion expression varies dramatically, making it a challenging problem. To tackle this challenge, we introduce the Sub-Clip Classification Boosting (SCB) Framework, a multi-step methodology for emotion detection from non-textual features of audio clips. SCB features a highly-effective sub-clip boosting methodology for classification that, unlike traditional boosting using feature subsets, instead works at the sub-instance level. Multiple sub-instance classifications increase the likelihood that an emotion cue will be found within a voice clip, even if its location varies between speakers. First, each parent voice clip is decomposed into overlapping sub-clips. Each sub-clip is then independently classified. Further, the Emotion Strength of the sub-classifications is scored to form a sub-classification and strength pair. Finally we design a FilterBoost-inspired "Oracle", that utilizes sub-classification and Emotion Strength pairs to determine the parent clip classification. To tune the classification performance, we explore the relationships between sub-clip properties, such as length and overlap. Evaluation on 3 prominent benchmark datasets demonstrates that our SCB method consistently outperforms all state-of-the art-methods across diverse languages and speakers. Code related to this paper is available at: https://arcgit.wpi.edu/toto/EMOTIVOClean.

Keywords: Classification · Emotion · Boosting · Sub-clip Sub-classification

1 Introduction

Motivation. In recent years voice activated assistants such as Amazon Echo[1], Google Home[2], and Siri[3], have become a prevalent mode of human-machine

---

[1] https://wwww.amazon.com/.
[2] https://madeby.google.com/home/.
[3] http://www.apple.com/ios/siri/.

U. Brefeld et al. (Eds.): ECML PKDD 2018, LNAI 11053, pp. 37–52, 2019.
https://doi.org/10.1007/978-3-030-10997-4_3

interaction. While prior findings [20] indicate that human-machine interactions are inherently social, current devices lack any empathic ability. If only we could automatically extract and then appropriately respond to the emotion of the speaker, this would empower a large array of advanced human-centric applications, from customer service [30] and education [19], to health care [24]. As it is best expressed by the advertising industry, emotion detection is key to "Capturing the heart" [23] of users and customers. Thus, as machines grow more complex, emotions can no longer be ignored.

Challenges. Emotion detection from voice continues to be a challenging problem and thus an active area of study [3,12,13,18,34,35]. To emphasize the challenges of this task, it has been observed that even humans experience difficulty in distinguishing reliably between certain classes of emotion - especially for speakers from other nationalities [8].

In addition, the location of emotion cues can vary both between and within a voice clip. That is, in voice clips of different content, cues will naturally be in different locations. However, even for clips with the same textual content, cues will vary due to stark differences in expressing emotion between individuals [2].

Limitations of the State-of-the-Art. Emotion detection is an active area of research with studies ranging from text analysis [36] to facial pattern recognition [11]. Sentiment analysis from text suffers from problems specific to each language, even though humans share a set of universal emotions [1]. In addition, even when controlling for differences in language, emotion expression is highly non-verbal [15]. Thus, methods that consider facial expressions in addition to voice are at times utilized [11]. However, such multi-modal data types are in many cases either too obtrusive, too costly or simply not available.

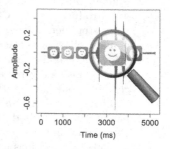

Fig. 1. Sub-clip classification boosting can be compared to an emotion search over the length of the clip, which overcomes the fuzzy transition boundaries between emotions and neutral states

To avoid the above pitfalls, our work concentrates in emotion detection from non-verbal voice features. In this context, despite the existence of universal emotions [6], state-of-the-art methods for emotion recognition still do not generalize well between benchmark datasets [7] (Table 3). Some approaches [16,17] attempt to address the problem of generalization by eliminating clip outliers, and thus do not succeed to classify all data instances.

Simple multi-class regression methods have been widely used, but fail to achieve practical accuracy [25,31]. Core work in emotion detection from voice uses a mixture of GMM (Gaussian Mixture Models) [14], and HMM (Hidden Markov Models) [5]. Most recently, with the increased popularity of deep learning (DL), we see the rise of attempts to apply DL to this problem, in particular, Random Deep Belief Networks [35], Recurrent Neural Networks (RNN) [13], and Convolutional Neural Networks (CNN) [3,12,34]. Convolutional Neural Networks (CNN) [3,12,34] achieve better results than other state-of-the-art methods, but still fail to generalize to all benchmark datasets and leave room for improvement. Due to these shortcomings, more work is needed in voice-based emotion recognition.

Our Proposed Approach. To solve this problem, we set out to design a novel Sub-clip Classification Boosting methodology (SCB). SCB is guided by our hypothesis that emotion tends not to be sustained over the entire length of the clip – not even for clips that have been carefully created for a benchmark. Therefore, the prediction of the emotion state of a clip needs to carefully consider its sub-clips. Sub-clipping (Sect. 5.1) effectively works as a moving lens (Fig. 1) over the sound signal - providing the first step to solving the undetermined boundary problem observed in a prior survey [2].

Next, a high dimensional non-verbal feature extraction from these voice sub-clips is tamed by prioritized feature selection [21] to reduce data dimensionality from more than 1500 features to a few hundred. As we will demonstrate, a dataset independent feature selection approach guided by the emotion classification problem itself, achieves better generalization across diverse benchmarks and emotion classes [7], which is one of the challenges observed with current state-of-the-art. In addition, utilizing non-verbal features overcomes classification challenges due to different languages.

To account for the unequal distribution of emotion cues between sub-clips, an Emotion Strength (ES) score is assigned to each sub-clip classification (SC), forming a SC, ES pair. Finally, inspired in part by FilterBoost [4], SCB implements an "Oracle" that utilizes all the Sub-Classification (SC), ES pairs (of a voice clip) to determine the parent clip classification.

To evaluate SCB, we have conducted an extensive study using three popular and widely used benchmark datasets described in Sect. 3. The results demonstrate that compared to the state-of-art algorithms, SCB consistently boosts the accuracy of predictions in all these benchmark datasets. For repeatability[4], we have posted our SCB code base, associated data, and meta-data. To summarize, our main contributions include:

- We design the SCB framework for the robust detection of emotion classes from voice across speakers, and languages.
- We design an Oracle-based algorithm inspired by FilterBoost that utilizes sub-classification (SC), Emotion Strength (ES) pairs to classify the parent clip.
- We tune our models, and explore the relationships between sub-clip properties, such as length and overlap, and the classification accuracy.
- Lastly, we evaluate the resulting SCB framework using 3 benchmark datasets and demonstrate that SCB consistently achieves superior classification accuracy improvements over all state-of-the-art algorithms.

---

[4] https://arcgit.wpi.edu/toto/EMOTIVOClean.

The rest of the paper is organized as follows. Section 2 situates our work in the context of Affective Computing, current methodologies for Emotion Detection, and Boosting Strategies. Section 3 introduces preliminary notation, and our problem definition. Section 4 introduces the Sub-clip Classification Boosting (SCB) Framework. Section 5 explains the SCB method and its sub-components. Section 6 describes our evaluation methodology and experimental results on benchmarks, while Sect. 7 concludes our work.

2 Related Work

Affective Computing. Affective computing [22] refers to computing that relates to, arises from, or deliberately influences emotion. Given that emotion is fundamental to human experience, it influences everyday tasks from learning, communication, to rational decision-making. Affective computing research thus develops technologies that consider the role of emotion to address human needs. This promises to enable a large array of services, such as better customer experience [30], intelligent tutoring agents [19] and better health care services [24]. Recent related research in psychology [27] points to the existence of universal basic emotions shared throughout the world. In particular it has been shown that at least six emotions are universal [6], namely: Disgust, Sadness, Happiness, Fear, Anger, and Surprise.

Emotion Detection from Voice. Detection of universal emotions from voice continues to be a challenging problem. Current state-of-the-art methods for emotion recognition do not generalize well from one benchmark database [7] to the next, as summarized in Table 3.

Some approaches [16,17] attempt to address the problem of generalization by eliminating clip outliers belonging to particular emotion classes, and thus providing an incomplete solution.

SVM (Support Vector Machines) have been identified as an effective method, especially when working with smaller datasets [31]. However their accuracy remains low - not sufficient for practical purposes (Table 3). Softmax regression [25] is similar to SVM, in that it is a multi-class regression that determines the probability of each emotion class. The resulting, Softmax accuracy is higher than SVM, but still lower than the more recent state-of-the-art methods described below.

GMM (Gaussian Mixture Models) [14] are better at identifying individual differences through the discovery of sub-populations using normally distributed features.

HMM (Hidden Markov Models) [5] are a popular method utilized for voice processing, that relies on building up a probabilistic prediction from very small sub-clips of a voice clip. In contrast to SCB, the sub-clips considered by HMM are too small to contain information that can be accurately classified on its own.

More recently, with the increase in popularity of Deep Learning, Random Deep Belief Networks [35], Recurrent Neural Networks (RNN) [13], and Convolutional Neural Networks (CNN) [3,12,34] have been used for the emotion classification task. RNN [13] and Random Deep Belief Networks [35] have been shown to have low accuracy for this task (see also our Table 3). CNN [3,12,34] on the other hand has achieved some of the higher accuracy predictions for specific benchmark datasets. However, this method fails to generalize to other sets [3,12,34]. This draw back may come in part from the lack of large available datasets with labeled human emotions, which are hard to obtain given the unique nature of this task. As our in-depth analysis reveals, the accuracy of CNN is comparable to previously discussed and less complex methods, while still leaving room for improvement.

Boosting Strategies. SCB is a meta-model approach, such as Boosting, Bagging, and Stacking [26] more specifically a type of Boosting. Boosting methods have been shown to be effective in increasing the accuracy of machine learning models. Unlike other boosting methods that work with subsets of features or data, SCB utilizes sub-instances (sub-clips) to boost the classification accuracy. Results of a direct comparison with current meta-models are given in Table 2.

3   Preliminaries

We work with **three benchmark datasets** for emotion detection on voice clips, namely: RML Emotion Database[5] [33], an emotion dataset, containing 710 audio-visual clips by 8 actors in 6 languages, displaying 6 universal [6] emotions, namely: Disgust, Sadness, Happiness, Fear, Anger, and Surprise. Berlin Emotion Database[6] [32], a German emotion dataset, containing 800 audio clips by 10

Table 1. Preliminary notation

Notation	Definition
$\mathcal{V}$	An ordered set of voice clips
v_i	The ith voice clip. To distinguish between entities with known labels (current data) and new data to be classified, we use the prime notation (E.g. v_i' is a new voice clip)
$|v_i|$	Length of v_i
N	Number of sub-clips
Ω	ratio of overlap between adjacent sub-clips
$v_{i,j}^{N,\Omega}$	The jth sub-clip of v_i, given N and Ω
e_i	Emotion label of v_i and all sub-clips of v_i
$\hat{e}_i'$	The predicted emotion class of sub-clip $v_{i,j}'^{N,\Omega}$
$f_{i,j}'^{N,\Omega}$	Feature vector extracted from $v_{i,j}'^{N,\Omega}$
$\hat{\mathcal{E}}_i'^{N,\Omega}$	Ordered set of sub-classification labels for all sub-clips of v_i'
$\hat{\mathcal{W}}_i'^{N,\Omega}$	Ordered set of Emotion Strength scores for all sub-clips of v_i'

---

[5] http://www.rml.ryerson.ca/rml-emotion-database.html.
[6] http://www.emodb.bilderbar.info/.

actors (5 male, 5 female), displaying 6 universal emotions and Neutral. SAVEE[7] [10], an English emotion dataset, containing 480 audio-visual clips by 4 subjects, displaying 6 universal emotions and Neutral. Table 1 explicitly highlights key notation used in the paper.

Problem Definition. Given a set of voice clips $v_i' \in \mathcal{V}$ with unknown emotion states, the problem is to predict the class e_i' for each instance v_i' with high accuracy.

4   The SCB Framework

To tackle the above emotion detection problem, we introduce the SCB framework (Sub-clip Classification Boosting) as shown in Fig. 2. Below we describe the main processes of the SCB framework.

Sub-clip Generation module takes as input a voice clip, the desired number of sub-clips N, and a percent overlap parameter Ω, to generate N sub-clips of equal length. During model training sub-clips are labeled after their parent clip. This process is further described in Sect. 5.1.

Feature Extraction is performed using openSMILE[8] [9] which generates a set of over 1500 features. Due to the high dimensionality of the data generated by openS-MILE, at model training a feature selection mechanism is applied to reduce the complexity of further steps – allowing us to instead focus on the core classification technology.

Feature Selection is applied during model training. For this task we explored several strategies for dimension selection, all equally applica-

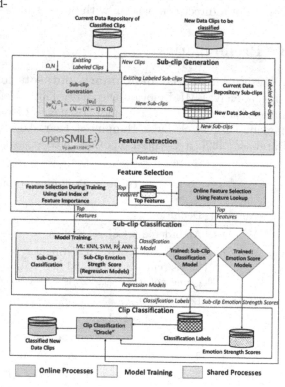

Fig. 2. SCB framework

ble. Finally, in our experiments, we work with one found to be particularly effective in our context, namely, the maximization of the mean decrease in node

---

[7] http://kahlan.eps.surrey.ac.uk/savee/.
[8] http://audeering.com/technology/opensmile/.

impurity based on the Gini Index for Random Forests [21]. Based on this metric, a vector with the importance of each attribute is produced. Guided by this importance criteria, the system selects a set of top features. For the purposes of this research, we select the top 350 features, a more than 4 fold reduction from the original set of features. However, our approach is general, and alternate feature selection strategies could equally be plugged into this framework.

Cached Feature Selection (On-line). During model deployment the cached list of features selected during training is looked-up to reduce the complexity of Feature Extraction from new clips.

Sub-clip Classification module trains sub-classification models to assign (universal and neutral) emotion labels to future sub-clips. Several ML algorithms are evaluated for this task (Sect. 5.2).

Sub-clip Emotion Strength Scores are computed for each sub-classification. To reduce bias, SCB utilizes a non-linear ensemble of multiple regression models that generate scores between 0 and 1, where a score close to 0 indicates low Emotion Strength or a weak sub-classification, while a score closer to 1 indicates high Emotion Strength or a strong sub-classification (Sect. 5.2).

Clip Classification "Oracle" evaluates SC, and ES pairs to classify the parent clip. Several "Oracle" implementations are evaluated, namely, Cumulative Emotion Strength, Maximum Emotion Strength, and Sub-clip Voting. These are discussed in more detail in Sect. 5.3.

5   Sub-clip Classification Boosting

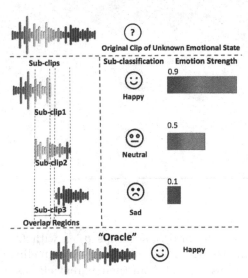

Fig. 3. Overview of SCB. Each voice clip is assigned a sub-classification, emotion strength pair. An "Oracle" uses these pairs to classify the parent clip

Our proposed Sub-clip Classification Boosting (SCB) method (Algorithm 1) is based on the intuition that strong emotion cues are concentrated unevenly within a voice clip. Thus, our algorithm works as a moving lens to find sub-clips of a data instance that can be used to boost the classification accuracy of the whole. Each sub-clip is independently classified, and for each sub-classification (SC) an Emotion Strength score (ES) is also predicted. To combine SC, ES pairs into the parent clip classification, SCB implements a FilterBoost [4] inspired "Oracle" (Sect. 5.3).

A high level description of the SCB algorithm is given in Fig. 3, while the actual method is summarized in Algorithm 1, which takes as inputs: a voice clip v'_i, the number of sub-clips N, sub-clip overlap Ω, and a pre-selected set of features $\mathcal{F}$. The length of the sub-clips is than determined in Line 1. Next, the parent voice clip $\boldsymbol{v}_i'$ is cut into sub-clips (Line 6) which are processed one at a time. Then, openSMILE (Line 9) extracts a preselected set of features $\mathcal{F}$, as described in Sect. 4, from each individual sub-clip. These features are stored as a vector $\boldsymbol{f}_{i,j}'^{N,\Omega}$. In Line 10 a previously trained sub-classification model C (Sect. 5.2) labels the sub-clip based on the feature vector $\boldsymbol{f}_{i,j}'^{N,\Omega}$. In Line 12 the ES score is computed by a regression model R (Sect. 5.2). Thus, each sub-clip receives a SC, ES pair of a label and a score. In the example in Fig. 3, the pairs are (Happy, 0.9), (Neutral, 0.5), and (Sad, 0.1). SC labels, and ES scores are respectively saved in the $\hat{\mathcal{E}}_i'^{N,\Omega}$ and $\hat{\mathcal{W}}_i'^{N,\Omega}$ ordered sets.

Finally, the Classification "Oracle" evaluates the $\hat{\mathcal{E}}_i'^{N,\Omega}$ and $\hat{\mathcal{W}}_i'^{N,\Omega}$ ordered sets, and outputs the parent clip classification (Line 16 as described in Sect. 5.3.

Algorithm 1. The SCB Method

Data: $\boldsymbol{v}_i'$, N, Ω, $\mathcal{F}$
Result: $\hat{e}_i'$

1 $SubClipLength \leftarrow \frac{|\boldsymbol{v}_i'|}{(N-(N-1)\times\Omega)}$;

2 $j \leftarrow 0$;

3 $Start \leftarrow 0$;

4 $End \leftarrow SubClipLength$;

5 **while** $j < N$ **do**

6 $\quad$ $\boldsymbol{v}_{i,j}'^{N,\Omega} \leftarrow Subclip(\boldsymbol{v}_i', Start, End)$;

7 $\quad$ $Start \leftarrow End - \Omega \times SubClipLength$;

8 $\quad$ $End \leftarrow Start + SubClipLength$;

9 $\quad$ $\boldsymbol{f}_{i,j}'^{N,\Omega} \leftarrow openSMILE(\boldsymbol{v}_{i,j}'^{N,\Omega}, \mathcal{F})$;

10 $\quad$ $\hat{e}_{i,j}'^{N,\Omega} \leftarrow C(\boldsymbol{f}_{i,j}'^{N,\Omega})$;

11 $\quad$ $\hat{\mathcal{E}}_i'^{N,\Omega}.push(\hat{e}_{i,j}'^{N,\Omega})$;

12 $\quad$ $\hat{w}_{i,j}'^{N,\Omega} \leftarrow R(\boldsymbol{f}_{i,j}'^{N,\Omega}|\hat{e}_{i,j}'^{N,\Omega})$;

13 $\quad$ $\hat{\mathcal{W}}_i'^{N,\Omega}.push(\hat{w}_{i,j}'^{N,\Omega})$;

14 $\quad$ $j \leftarrow j + 1$;

15 **end**

16 $\hat{e}_i' = Oracle(\hat{\mathcal{E}}_i'^{N,\Omega}, \hat{\mathcal{W}}_i'^{N,\Omega})$;

17 **return** $\hat{e}_i'$;

5.1 Sub-clip Generation

To generate sub-clips, each voice clip $\boldsymbol{v}_i$ is split into N sub-clips $\boldsymbol{v}_{i,j}^{N,\Omega}$ with Ω overlap, where $0 \leq \Omega < 1$ is relative to the length of the sub-clips. The sub-clip length $|\boldsymbol{v}_{i,j}^{N,\Omega}|$ is computed by Eq. 1. Sub-clips are cut from their parent clip as described in Algorithm 1 (Lines 1–8) and illustrated in Fig. 3.

$$|\boldsymbol{v}_{i,j}^{N,\Omega}| = \frac{|\boldsymbol{v}_i|}{(N-(N-1)\times\Omega)}. \tag{1}$$

Tuning the number of clips and their overlap has interesting implications for the accuracy of the parent clip classification (Sects. 6.5.2 and 6.5.1). Increasing the number of sub-clips, increases the number of independent classifications per clip, however it also reduces the amount of information per sub-clip. To mitigate the information reduction due to sub-clipping, sub-clip overlap adds additional, but redundant information to each sub-clip. This redundancy needs to be controlled during model training and testing in order to avoid over-fitting. Our results show that a certain amount of overlap increases parent-clip classification accuracy. By tuning these parameters we discovered optimal zones for both N and Ω, making these hyper parameters important to model tuning.

5.2   Sub-classification and Emotion Strength Pairs

Sub-classification (SC). A sub-clip classification Model C is trained using labeled feature vectors of audio sub-clips, so that it can predict the label of a new instance represented by a feature vector $f_{i,j}'^{N,\Omega}$. Therefore, this process can be expressed as $\hat{e}_{i,j}'^{N,\Omega} = C(f_{i,j}'^{N,\Omega})$, where $\hat{e}_{i,j}'^{N,\Omega}$ is the SC label of the sub-clip $v_{i,j}'^{N,\Omega}$. To implement these models, we evaluate several machine learning algorithms suitable for classification, such as SVM, RandomForest, kNN, and naiveBayes.

Emotion Strength (ES). Given a predicted SC label $\hat{e}_{i,j}'^{N,\Omega}$, the ES score $0 \leq \hat{w}_{i,j}'^{N,\Omega} \leq 1$ of the sub-clip $v_{i,j}'^{N,\Omega}$ is predicted by a previously trained regression model such that $\hat{w}_{i,j}'^{N,\Omega} = R(f_{i,j}'^{N,\Omega}|\hat{e}_{i,j}'^{N,\Omega})$. Since SC labels are unknown prior to sub-clip classification, separate ES regression models need to be trained for each of the 7 emotions (6 universal emotions, and Neutral) that are present in the datasets. During training, categorical labels are converted to binary values, with 1 indicating the presence of the emotion for which the regression model is being trained to generate an ES score, and 0 for all other emotions. Several machine learning algorithms are tested for this task, such as SVM, KNN, ANN, and Linear Models. A non-linear ensemble of models such as $\hat{w}_{i,j}'^{N,\Omega} = R_{SVM}(f_{i,j}'^{N,\Omega}|\hat{e}_{i,j}'^{N,\Omega}) \times R_{kNN}(f_{i,j}'^{N,\Omega}|\hat{e}_{i,j}'^{N,\Omega})$, was found optimal for reducing bias. Intuitively, ES scores closer to 1 are viewed as strong support for the SC label of the (SC,ES) pair, while low ES scores indicate low support for the pair's SC label.

5.3   Clip Classification "Oracle"

We evaluate three "Oracle" implementations for the classification of the parent clips from SC, ES pairs, namely, Cumulative Emotion Strength Sub-clip Classification, Maximum Sub-clip Emotion Strength Classification and Sub-clip Voting Classification. In each of the implementations, the "Oracle" receives $\hat{\mathcal{E}}_i'^{N,\Omega}$ and $\hat{\mathcal{W}}_i'^{N,\Omega}$ ordered sets containing SC labels and ES scores, and makes a decision on the classification of the parent clip as described below.

5.3.1 Cumulative Emotion Strength

(Equation 2) is an "Oracle" implementation based on the intuition that all sub-clips have a weight in the final clip classification. For example, if a parent clip is divided into 3 sub-clips, with the following SC, ES pairs: (Fear, 0.7), (Happiness, 0.4), (Happiness, 0.6), the emotion with the highest cumulative ES (0.4+0.6 = 1) is Happiness.

$$\hat{e}'_i = \underset{e \in \mathcal{E}}{\operatorname{argmax}} \sum_{\forall j | \hat{e}'^{N,\Omega}_{i,j} = e} \hat{w}'^{N,\Omega}_{i,j}. \tag{2}$$

ES scores, play no role in sub-classification, and are only used during the parent clip classification. In case of a tie, one of multiple possible labels is chosen at random.

5.3.2 Maximum Emotion Strength

(Equation 3) is an "Oracle" implementation based on the intuition that certain sub-clips best represent the overall emotion of the parent clip. Therefore, this method promotes the sub-classification with the highest ES score. To illustrate, if a target clip is divided into 3 sub-clips, with the following SC, ES pairs: (Fear, 0.7), (Happiness, 0.4), (Happiness, 0.6), the parent clip would be classified as Fear based on the maximum emotion strength. In case of a tie, one of multiple possible labels is chosen at random.

$$\hat{e}'_i = \underset{e \in \mathcal{E}}{\operatorname{argmax}}(\max(\hat{w}'^{N,\Omega}_{i,j} \mid \hat{e}'^{N,\Omega}_{i,j} = e)). \tag{3}$$

5.3.3 Sub-clip Voting Classification

(Equation 4) is an "Oracle" implementation based on the intuition that all sub-classifications contribute equally to the parent classification, in essence a voting ensemble. To illustrate, if a parent clip is divided into 5 sub-clips, of which three are classified as *Fear* one as *Happiness* and one as *Anger*, the clip will be classified as *Fear*. In case of a tie, one of multiple possible labels is chosen at random.

$$\hat{e}'_i = \underset{e \in \mathcal{E}}{\operatorname{argmax}}(\operatorname{count}(\hat{e}'^{N,\Omega}_{i,j} = e)). \tag{4}$$

6 Experimental Evaluation

To evaluate the performance of our SCB framework we conducted comprehensive experiments using three benchmark datasets that are commonly used to compare state-of-the-art methods. The experimental results show that SCB can achieve up to 4% improvement in classification accuracy over current state-of-the-art methods, while consistently achieving better generalization between datasets.

6.1   Benchmark Data Sets

SCB was evaluated with three popular benchmark datasets for emotion detection on voice clips described in Sect. 3. These dataset are chosen for their complete coverage of the six universal emotions [6], namely: Disgust, Sadness, Happiness, Fear, Anger, and Surprise. In addition, SAVEE and the Berlin Emotion Database, contain voice clips labeled as Neutral. These datasets also provide clips in different languages and from diverse actors. Further, they are widely used by state-of-the-art methods, thus provide solid baseline evaluations.

6.2   Comparison of Alternate Methods

To validate SCB we utilized a two fold comparison approach. First, we compare against common baseline methods deployed within the SCB framework as replacements for the SCB method (Algorithm 1). For this task, we compare against common classification algorithms such as kNN, SVM and randomForest, as well as boosting and ensemble methods (summarized in Table 2).

Second, we compare against state-of-the-art methods that classify emotion from the same benchmark datasets, namely, Hidden Markov Model [31], Gaussian Mixture Model [37], Orthogonal and Mahalanobis Distance Based [17], Random Deep Belief Networks [35], Simple SVM [29], Adaptive Boosting [28], Softmax Regression [25], CNN [3,12,34], and RNN [13]. These methods are briefly described in Sect. 2. Results of comparisons with state-of-the-art methods are summarized in Table 3.

6.3   Evaluation Settings

To directly compare to state-of-the-art methods, we utilize parent clip classification accuracy defined as the ratio of True Clip Classifications (TC) and the Total Number of classifications, which includes False Classifications (FC), $\frac{TC}{TC+FC}$.

For consistency, all models are trained and validated using a n-fold cross-validation approach, where $n = 5$. It is important to note that when $\Omega > 0$ adjacent sub-clips overlap, thus share common information. If not controlled, overlap will cause partial over-fitting and artificially higher model accuracy. Therefore, we operate under the constrain that it is critical that sub-clips of the same clip are kept in one fold during model training and testing procedures, thus avoiding any over-fitting.

6.4   Evaluating "Oracle" Implementations

Different "Oracle" implementations discussed in Sect. 5.3 are evaluated for their relative classification accuracy. The results are illustrated in Fig. 4. Overall, classification based on the sub-clip Maximum Emotion Strength yields the best and most consistent results. This finding supports our hypothesis that emotion cues are localized. It addition, it was also observed that SVM was the best performing sub-classification algorithm. Results are further improved after hyper parameter tuning in Sect. 6.5.

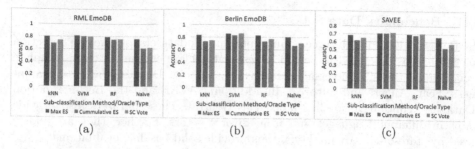

Fig. 4. Comparing "Oracle" implementations, using different sub-classification algorithms. Maximum emotion strength "Oracle" consistently beats other implementations. SVM is the over all winner as the choice of sub-classification algorithm

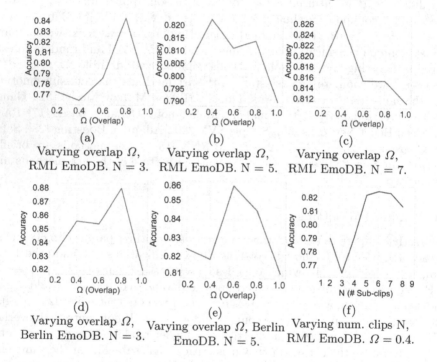

Fig. 5. Classification accuracy when varying overlap Ω and number of clips N. The charts indicate the existence of optimal values for Ω and N across datasets

6.5   Hyper Parameter Tuning

6.5.1   Sub-clip Overlap

In the examples in Figs. 5a through e, overlap Ω displays optimal ranges that present tuning opportunities. Further, the optimal range for Ω, interacts with N, as seen for $N = 3$ (Fig. 5a), $N = 5$ (Fig. 5b), and $N = 7$ (Fig. 5c, EmoDb). These finding are consistent across datasets as also seen for the Berlin EmoDB for $N = 3$ (Fig. 5d), and $N = 5$ (Fig. 5e).

6.5.2 Number of Sub-clips

One example of tuning for optimal number of sub-clips, is given in Fig. 5f. In this example, parent clip classification accuracy peaks between $5 \leq N \leq 7$, while $\Omega = 0.4$. This is consistent with the intuition that more sub-clips increase classification opportunities for the parent clip, but decrease the amount of information in each sub-clip, therefore a trade-off is expected. In addition, the amount of information in each clip also depends on the amount of overlap Ω, the optimization of which is explored in Sect. 6.5.1.

Table 2. Comparison of SCB to baseline methods. When used as baselines, classification algorithms (kNN, RF, SVM and Naive Bayes) directly classify the parent clip using it's feature vector. In the context of AdaBoost, multiple models using the same algorithm are created using feature subsets to boost the model accuracy. In Stacking, multiple models from different algorithms are combined. In the context of SCB, classification algorithms are used to generate sub-classification (SC) labels

Meta model	Dataset	kNN	RF	SVM	Naive Bayes
Baseline	Berlin EmoDb	0.781	0.789	0.873	0.776
	SAVEE	0.675	0.688	0.735	0.619
	RML	0.666	0.714	0.761	0.645
SCB	Berlin EmoDb	0.859	0.857	**0.880**	0.833
	SAVEE	0.739	0.760	**0.789**	0.714
	RML	0.839	0.825	**0.843**	0.797
AdaBoost	Berlin EmoDb	0.735	0.815	0.867	0.791
	SAVEE	0.638	0.700	0.727	0.663
	RML	0.638	0.710	0.726	0.645
Stacking	Berlin EmoDb	0.787	0.821	0.813	0.787
Baseline = MetaModel	SAVEE	0.669	0.700	0.723	0.688
MetaModel(Stack: kNN, SVM)	RML	0.696	0.712	0.731	0.690

6.6 Overall Results

In the context of emotion detection SCB is superior to Baseline (Table 2) and state-of-the-art algorithms (Table 3) for all three of the evaluated datasets. We also note that some of the state-of-the-art methods are not directly comparable. In the case of [17], the method eliminates 5% of the data that is labeled as outliers. Similarly, the accuracy for [25] was adjusted based on the number of instances per class. Coincidentally, in the RML EmoDB dataset *Happiness* had the highest classification error rate, but a lower number of instances. In that case, applying this [25] approach would artificially increase accuracy, however when applied to the SAVEE dataset which has a balanced number of instances, the same method failed to produce similar accuracy improvements. Therefore, in comparable settings SBC consistently outperforms all state-of-the-art algorithms and baselines.

Table 3. Accuracy evaluation of the SCB framework. *Method eliminates 5% of the data. Accuracy not directly comparable. **Accuracy was adjusted based on number of instances per class, and controlling for speaker, therefore not comparable. ∗ ∗ ∗Across speakers accuracy varied from 0.84 to 0.845 for the Berlin EMODb. One study [12] used the SAVEE dataset in addition to Berlin EMODb

Method	Berlin	RML	SAVEE
SCB	**0.880**	**0.843**	**0.789**
Hidden Markov Model [31]	0.830		
Gaussian Mixture Model [37]	0.81		
Orthogonal and Mahalanobis Distance Based [17]	0.820*	0.810*	0.830*
Random Deep Belief Networks [35]	0.823		0.734
Simple SVM [29]	0.563		
Adaptive Boosting [28]	0.708		
Softmax Regression [25]	0.790 (0.879**)		0.734
CNN [3,12,34]	0.845***		0.69 [12]
RNN [13]	0.695		

7   Conclusion and Future Work

In this study we introduced Sub-clip Classification Boosting (SCB), a novel boosting method based on sub-clip classification. Further we discovered interesting trade-off patterns related to the number of sub-clips and sub-clip overlap. We also introduced and evaluated, several "Oracle" implementations to determine parent clip classifications from its sub-clips. Lastly, we evaluated our SCB framework using 3 benchmark datasets and achieved significant parent clip classification accuracy improvements over state-of-the-art algorithms, as well as consistent generalization across datasets.

References

1. Abbasi, A., Chen, H., Salem, A.: Sentiment analysis in multiple languages: feature selection for opinion classification in web forums. ACM Trans. Inf. Syst. (TOIS) **26**(3), 12 (2008)
2. Anagnostopoulos, C.N., Iliou, T., Giannoukos, I.: Features and classifiers for emotion recognition from speech: a survey from 2000 to 2011. Artif. Intell. Rev. **43**(2), 155–177 (2015)
3. Badshah, A.M., Ahmad, J., Rahim, N., Baik, S.W.: Speech emotion recognition from spectrograms with deep convolutional neural network. In: 2017 International Conference on Platform Technology and Service, PlatCon, pp. 1–5. IEEE (2017)
4. Bradley, J.K., Schapire, R.E.: FilterBoost: regression and classification on large datasets. In: NIPS, pp. 185–192 (2007)
5. Chenchah, F., Lachiri, Z.: Speech emotion recognition in acted and spontaneous context. Proc. Comput. Sci. **39**, 139–145 (2014)

6. Ekman, P.: Strong evidence for universals in facial expressions: a reply to Russell's mistaken critique (1994)
7. El Ayadi, M., Kamel, M.S., Karray, F.: Survey on speech emotion recognition: features, classification schemes, and databases. Pattern Recogn. **44**(3), 572–587 (2011)
8. Eyben, F., Unfried, M., Hagerer, G., Schuller, B.: Automatic multi-lingual arousal detection from voice applied to real product testing applications. In: 2017 IEEE International Conference on Acoustics, Speech and Signal Processing, ICASSP, pp. 5155–5159. IEEE (2017)
9. Eyben, F., Weninger, F., Gross, F., Schuller, B.: Recent developments in openS-MILE, the Munich open-source multimedia feature extractor. In: Proceedings of the 21st ACM International Conference on Multimedia, pp. 835–838. ACM (2013)
10. Haq, S., Jackson, P., Edge, J.: Audio-visual feature selection and reduction for emotion classification. In: Proceedings of International Conference on Auditory-Visual Speech Processing, AVSP 2008, Tangalooma, Australia, September 2008
11. Hossain, M.S., Muhammad, G., Alhamid, M.F., Song, B., Al-Mutib, K.: Audio-visual emotion recognition using big data towards 5G. Mobile Netw. Appl. **21**(5), 753–763 (2016)
12. Huang, Z., Dong, M., Mao, Q., Zhan, Y.: Speech emotion recognition using CNN. In: Proceedings of the 22nd ACM International Conference on Multimedia, pp. 801–804. ACM (2014)
13. Kerkeni, L., Serrestou, Y., Mbarki, M., Raoof, K., Mahjoub, M.A.: A review on speech emotion recognition: case of pedagogical interaction in classroom. In: 2017 International Conference on Advanced Technologies for Signal and Image Processing, ATSIP, pp. 1–7. IEEE (2017)
14. Kishore, K.K., Satish, P.K.: Emotion recognition in speech using MFCC and wavelet features. In: 2013 IEEE 3rd International Advance Computing Conference, IACC, pp. 842–847. IEEE (2013)
15. Knapp, M.L., Hall, J.A., Horgan, T.G.: Nonverbal Communication in Human Interaction. Cengage Learning, Boston (2013)
16. Kobayashi, V., Calag, V.: Detection of affective states from speech signals using ensembles of classifiers. In: FIET Intelligent Signal Processing Conference (2013)
17. Kobayashi, V.: A hybrid distance-based method and support vector machines for emotional speech detection. In: Appice, A., Ceci, M., Loglisci, C., Manco, G., Masciari, E., Ras, Z.W. (eds.) NFMCP 2013. LNCS, vol. 8399, pp. 85–99. Springer, Cham (2014). https://doi.org/10.1007/978-3-319-08407-7_6
18. Kraus, M.W.: Voice-only communication enhances empathic accuracy. Am. Psychol. **72**(7), 644 (2017)
19. Litman, D.J., Silliman, S.: ITSPOKE: an intelligent tutoring spoken dialogue system. In: Demonstration Papers at HLT-NAACL 2004, pp. 5–8. Association for Computational Linguistics (2004)
20. Nass, C., Moon, Y.: Machines and mindlessness: social responses to computers. J. Soc. Issues **56**(1), 81–103 (2000)
21. Pal, M.: Random forest classifier for remote sensing classification. Int. J. Remote Sens. **26**(1), 217–222 (2005)
22. Picard, R.W.: Affective computing (1995)
23. Poels, K., Dewitte, S.: How to capture the heart? Reviewing 20 years of emotion measurement in advertising. J. Advert. Res. **46**(1), 18–37 (2006)
24. Riva, G.: Ambient intelligence in health care. CyberPsychol. Behav. **6**(3), 295–300 (2003)

25. Sun, Y., Wen, G.: Ensemble softmax regression model for speech emotion recognition. Multimed. Tools Appl. **76**(6), 8305–8328 (2017)
26. Todorovski, L., Džeroski, S.: Combining classifiers with meta decision trees. Mach. Learn. **50**(3), 223–249 (2003)
27. Valstar, M., et al.: AVEC 2013: the continuous audio/visual emotion and depression recognition challenge. In: Proceedings of the 3rd ACM International Workshop on Audio/Visual Emotion Challenge, pp. 3–10. ACM (2013)
28. Vasuki, P.: Speech emotion recognition using adaptive ensemble of class specific classifiers. Res. J. Appl. Sci. Eng. Technol. **9**(12), 1105–1114 (2015)
29. Vasuki, P., Vaideesh, A., Abubacker, M.S.: Emotion recognition using ensemble of cepstral, perceptual and temporal features. In: International Conference on Inventive Computation Technologies, ICICT, vol. 2, pp. 1–6. IEEE (2016)
30. Verhoef, P.C., Lemon, K.N., Parasuraman, A., Roggeveen, A., Tsiros, M., Schlesinger, L.A.: Customer experience creation: determinants, dynamics and management strategies. J. Retail. **85**(1), 31–41 (2009)
31. Vlasenko, B., Wendemuth, A.: Tuning hidden Markov model for speech emotion recognition. Fortschritte der Akustik **33**(1), 317 (2007)
32. Vogt, T., André, E., Bee, N.: EmoVoice—a framework for online recognition of emotions from voice. In: André, E., Dybkjær, L., Minker, W., Neumann, H., Pieraccini, R., Weber, M. (eds.) PIT 2008. LNCS, vol. 5078, pp. 188–199. Springer, Heidelberg (2008). https://doi.org/10.1007/978-3-540-69369-7_21
33. Wang, Y., Guan, L.: Recognizing human emotional state from audiovisual signals. IEEE Trans. Multimed. **10**(5), 936–946 (2008)
34. Weißkirchen, N., Bock, R., Wendemuth, A.: Recognition of emotional speech with convolutional neural networks by means of spectral estimates. In: 2017 Seventh International Conference on Affective Computing and Intelligent Interaction Workshops and Demos (ACIIW), pp. 50–55. IEEE (2017)
35. Wen, G., Li, H., Huang, J., Li, D., Xun, E.: Random deep belief networks for recognizing emotions from speech signals. Comput. Intell. Neurosci. **2017** (2017)
36. Yu, D., Deng, L.: Automatic Speech Recognition. SCT. Springer, London (2015). https://doi.org/10.1007/978-1-4471-5779-3
37. Zao, L., Cavalcante, D., Coelho, R.: Time-frequency feature and AMS-GMM mask for acoustic emotion classification. IEEE Signal Process. Lett. **21**(5), 620–624 (2014)

Machine Learning for Targeted Assimilation of Satellite Data

Yu-Ju Lee[1,2]($\boxtimes$) (iD), David Hall[1], Jebb Stewart[3], and Mark Govett[4]

[1] Department of Computer Science, University of Colorado Boulder,
Boulder, CO 80309-0430, USA
`Yuju.Lee@Colorado.EDU`
[2] Cooperative Institute for Research in Environmental Sciences,
NOAA/OAR/ESRL/Global Systems Division, University of Colorado Boulder,
Boulder, CO, USA
[3] Cooperative Institute for Research in the Atmosphere,
NOAA/OAR/ESRL/Global Systems Division, Colorado State University,
Boulder, CO, USA
[4] NOAA/OAR/ESRL/Global Systems Division,
325 Broadway, Boulder, CO, USA

Abstract. Optimizing the utilization of huge data sets is a challenging problem for weather prediction. To a significant degree, prediction accuracy is determined by the data used in model initialization, assimilated from a variety of observational platforms. At present, the volume of weather data collected in a given day greatly exceeds the ability of assimilation systems to make use of it. Typically, data is ingested uniformly at the highest fixed resolution that enables the numerical weather prediction (NWP) model to deliver its prediction in a timely fashion. In order to make more efficient use of newly available high-resolution data sources, we seek to identify regions of interest (ROI) where increased data quality or volume is likely to significantly enhance weather prediction accuracy. In particular, we wish to improve the utilization of data from the recently launched Geostationary Operation Environmental Satellite (GOES)-16, which provides orders of magnitude more data than its predecessors. To achieve this, we demonstrate a method for locating tropical cyclones using only observations of precipitable water, which is evaluated using the Global Forecast System (GFS) weather prediction model. Most state of the art hurricane detection techniques rely on multiple feature sets, including wind speed, wind direction, temperature, and IR emissions, potentially from multiple data sources. In contrast, we demonstrate that this model is able to achieve comparable performance on historical tropical cyclone data sets, using only observations of precipitable water.

Keywords: Numeric weather prediction · Satellite
Machine learning · Data assimilation · Tropical cyclone
Precipitable water · Water vapor · Global Forecast System (GFS)

© Springer Nature Switzerland AG 2019
U. Brefeld et al. (Eds.): ECML PKDD 2018, LNAI 11053, pp. 53–68, 2019.
https://doi.org/10.1007/978-3-030-10997-4_4

1 Introduction

Extreme weather has the potential to cause significant economic damage and loss of life. In order to minimize these losses, high precision weather predictions are needed. It is also important to make such predictions as far in advance as possible, to provide adequate advanced warning of impending storms. Improving the accuracy and time horizon of weather predictions are among the primary research objectives pursued by the National Oceanic and Atmospheric Administration (NOAA). To a significant degree, the accuracy of a numerical weather prediction is determined by the model initialization procedure, wherein data is assimilated from a variety of observational platforms. Over wide swaths of ocean or in remote areas of land, where in-situ observations are lacking, satellite data is used to augment and complete the construction of the initial conditions. As assimilation of satellite data is computationally expensive, the data resolution is typically reduced in order to accelerate its incorporation into the forecast. In the vicinity of an extreme weather event, such as a tropical cyclone, the situation can change rapidly. It is important to update the initial conditions more frequently using higher resolution data, in order to produce the most accurate forecasts. To this end, we are interested in automatically identifying specific regions of interests (ROI) where supplemental satellite observations could help increase the forecast's quality and overall impact.

At present, detection of extreme weather is primarily a manual process relying on difficult-to-quantify human expertise and experience, with no clear-cut definition for most weather phenomena. For example, it is well known that a tropical cyclone is characterized by a region of low surface-pressure surrounded by high speed winds and enhanced water vapor. However, there is no universally agreed upon combination of wind speed, pressure, and vapor that definitively identifies a tropical cyclone. If we attempt to hand-craft a definition identifying all tropical cyclones, we would have to deal with many edge cases that don't meet that definition. In addition to the challenge of constructing adequate definitions, there are also limits on the quality and quantity of observational data available. For example, the data needed for forecasting tropical cyclones is often provided by satellites in polar orbits. Those observations may be poorly timed, leading to images where the target cyclone is located in the periphery of the observed region, or absent from the image entirely (Fig. 1).

In this article, we propose a tropical cyclone detection algorithm that requires only observations of water vapor which is the primary data source to be provided by the new geostationary GOES-16 satellite. As it doesn't require measurements of wind speed or direction, we can avoid intermittent data from non-geostationary satellites. By using only high-frequency geostationary satellite data, we ensure continuous tracking. The proposed algorithm also employs a sliding window data augmentation strategy to overcome data sparsity, as discussed in Sect. 5.

(a) Water Vapor image of GOES-15

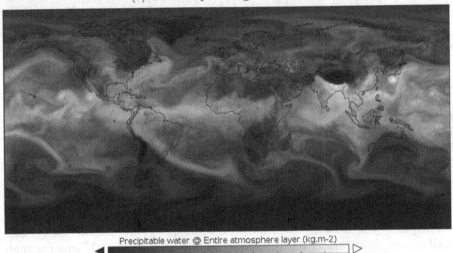

Precipitable water @ Entire atmosphere layer (kg.m-2)

0.1   17.4   34.7   52.1   69.4   86.7

(b) Precipitable water of GFS data

Fig. 1. Precipitable water is a good approximate of satellite water vapor channel for preliminary study

2   Related Work

Many researchers have investigated extreme weather detection using both remote sensing data and Numerical Weather Prediction (NWP) models. A technique introduced by Dvorak in the 1970s is a widely accepted approach for classifying the intensity of tropical cyclones [3]. This technique uses visual identification of images of tropical cyclones in the visible and infrared bands for classification.

However, cyclone images vary a great deal, and expert opinion is required in order to properly apply this method. Although Dvorak's technique has been modified, with an increased reliance on NWP models, meteorologists still rely primarily on judgment and experience to identify and locate tropical cyclones using meteorological data. Since that time, significant research has been conducted in an effort to develop improved tropical cyclone indicators [2,9,10,12,14]. Research into the estimation of tropical cyclone intensity using satellite data has also been an active area of investigation. For example, Velden et al. [13] made use of sequences of GOES-8 infrared images to infer tropical cyclone wind speed and track. Jaiswal et al. [6] suggested matching infrared images with a database of known intensities as a means to estimate tropical cyclone strength. Disadvantages of these techniques include the need for manual adjustment of the matching-index threshold, and the requirement that the cyclone to be measured is well centered in the image.

More recently, there has been an increased effort to apply machine learning techniques to automate the identification of severe weather phenomena. Liu et al. used data generated by the CAM-5 atmosphere model to automate extreme weather detection in climate simulations [8]. However, this technique also required the target object be well centered in the image, which is not well suited for use with GOES satellite images. Ho et al. identified tropical cyclones from QuickScat satellite data using support vector machines (SVMs) applied to wind speed and direction [5]. They also built a system to combine the data from QuickScat with TRMM precipitation measurements using a Kalman filter for cyclone tracking [4]. The technique of Panangadan et al. uses multiple satellite images with a graph-based algorithm to detect the eye of the cyclone and a Kalman filter or particle filter for cyclone tracking [11]. Zou et al. employed wind circulation as an additional feature for cyclone detection using QuickSCAT data [15]. Their technique used a wind speed and direction histogram to perform coarse identification, and then use the wind circulation path to refine the classification. However, common drawbacks in these techniques include their reliance on multiple data sources and their focus on wind-speed and direction for cyclone identification. In contrast, the technique described in this work achieves high accuracy identification of tropical cyclones using only water vapor images from a single geostationary satellite source.

3   Data

In this section, we describe the Global Forecast System (GFS) data and International Best Track Archive for Climate Stewardship (IBTrACS) data. From this data, we extract information covering tropical cyclones in the west pacific basin. The IBTrACS and GFS datasets are combined to form our labeled dataset used in training, validation and prediction.

3.1   Global Forecast System Data

The Global Forecast System (GFS) is a weather forecast model produced by the National Centers for Environmental Prediction (NCEP). The main goal of GFS is the production of operational forecasts for weather prediction over both short-range and long-range forecasts. GFS proves a large set of variables which have the potential to impact global weather. Examples variables include temperature, winds, precipitation, atmospheric ozone, and soil moisture. GFS data is provided in a gridded format on the regular latitude-longitude grid. GFS provides data with a horizontal resolution of 18 miles (28 km) for weather forecasts out to 16 days, and 44 mile resolution (70 km) for forecasts of up to two weeks. GFS data is produced 4 times per day at 00, 06, 12, 18 UTC time [1].

3.2   International Best Track Archive for Climate Stewardship (IBTrACS)

International Best Track Archive for Climate Stewardship (IBTrACS) is a project aimed at providing best-track data for tropical cyclones from all available Regional Specialized Meteorological Centers (RSMCs) and other agencies. It combines multiple datasets into a single product, describing the track of each tropical cyclone in latitude-longitude coordinates. This dataset makes it simple for researchers to locate a given tropical cyclone in both space and time [7] (Fig. 2).

Fig. 2. Combining visualizations of the GFS precipitable water field (left) with IBTrACS cyclone track data (right) produces a labeled dataset of tropical cyclones in the west Pacific basin.

4   Issues and Challenges

Although the atmosphere is continuous, computational constraints force use to approximate it using discrete, gridded data. In order to identify severe weather phenomenon, we naturally need to separate a large region data into small areas to analyze and prioritize data. Those areas we identify as having high potential impact become our regions of interest (ROI). We define these areas containing a cyclone center as positive samples. However, such identification can cause some challenges.

4.1   Area Size Determination Problem

There is no clear rule to decide the appropriate grid size which could clearly describe different severe weather phenomenon. If we use the smallest unit of data set, a single point of GFS data, it contains only one value and could not be used to describe or identify a weather phenomenon. However, if we use an overly large selection, it may contain too much information which is not related to a weather phenomenon. Therefore, to discuss tropical cyclones in this paper, we adapt a 32×32 image as our area size because this scale is large enough for us to capture each tropical cyclone in our GFS data.

4.2   Occasional Weather Events Problem

Weather phenomenon like tropical cyclone does not happen occur in spatial and time domain. Therefore, events are not distributed evenly throughout the entire region and time. In Fig. 3, we could find that we only have one positive sample in total 36 small areas of west pacific basin at the same time. Also, in entire IBTrACS data set, it is quite common no positive samples on some days because there are no tropical cyclones for that time period. Positive samples are extremely rare compared to negative samples. As a result, we don't have sufficient number of positive samples which are needed for analysis and causes data sparsity problem.

4.3   Weather Events Location Problem

In data preprocessing, we divide a large region into many small areas equally. Throughout tropical cyclone life cycle, location and size vary through time, therefore, center of a tropical cyclones is not permanently located in the center of area without external adjustment. In Fig. 4, we show that there are many situations, a tropical cyclone located near the edge and area doesn't cover entire tropical cyclone.

4.4   No Labeled Data for Negative Samples Problem

Best tracking data set are designed to provide information of tropical cyclones such as location, wind speed, wind direction, timestamps. However, scientists are only interested in regions with tropical cyclones, regions without tropical cyclones are not recorded in any best tracking data set. For example, we can easily find amount of tropical cyclones images from the Internet, but there are

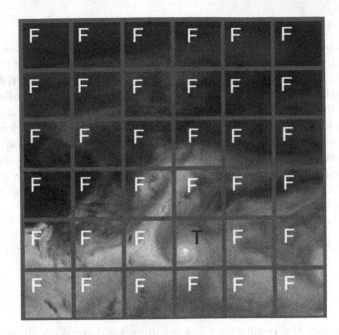

Fig. 3. The west pacific basin is divided into areas and mark a area with a tropical cyclone as true, others as false through cyclone tracking data set

no images described as no tropical cyclones from the Internet. Therefore, there is no clear definition of samples without tropical cyclones from scientists. It leads no definition of negative samples for our training set. In this paper, we randomly select areas which doesn't include cyclones at best tracking data as negative samples. In order to improve the confidence level of prediction, we increase the scale of time and space to make our random selection samples are well distributed.

5 System

Our system is designed to output probability prediction as reference for data assimilation. The input data of system only comes from precipitable water. In order to identify cyclone from a region, our system needs to apply sliding window technique to solve problems we list in Sect. 4. We use center radiation feature extraction to increase accuracy rate. After that, an ensemble algorithm helps the system to reduce the penalty for area prediction error. Figure 5 shows system flow and processing at each step.

5.1   Data Augment with Sliding Window

We want to identify tropical cyclones within an area regardless the cyclone size is large or small, cyclone image is part or all, and cyclone position is in the middle or edge of the image. In order to train our model to identify all these cases, we need to provide training set to cover as more variability as possible. When we form our labeled training set, we naturally don't have too many positive samples with high variability and generate lots or samples which cyclones locate on edge of images if we separate entire region equally. Our method is to use sliding window technique to solve these two issues. We treat the target as a small sight and then sweep the entire picture in the order of upper left, upper right, lower left and lower right. Each area we slide would be treat as a positive or negative sample. This process would produce $W \times H$ samples

$$W = H = (R - A)/S \qquad (1)$$

where, R = length of region

A = length of area

S = Interval points between area

Figure 6 shows the results after using sliding window. With sliding window process, image with a cyclone in west pacific basin produces continuous positive samples with the same cyclone in different location of areas. Continuously

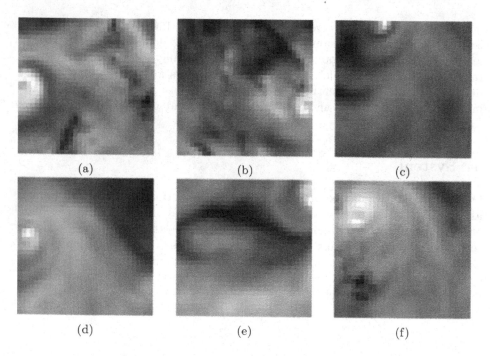

(a)   (b)   (c)

(d)   (e)   (f)

Fig. 4. Center of tropical cyclone located near or on area edge.

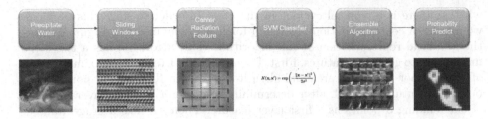

Fig. 5. System overview

slide entire region not only solving the boundary issues because at least one area including entire tropical cyclone located in center, but also generating many positive samples with variability for training. Data augment methods like rotation, flips and scaling are useful to deal few samples issues. However, for our GFS data set, it is more suitable to use sliding window technique for data augment. The reason is that GFS data set cover entire tropical cyclone life cycle and has time relationship between each data. As time changes, the location, shape and size of a tropical cyclone will change. Another advantage of sliding window technique is tropical cyclones would rotate and varies with time, it would generate different variance samples naturally. If we rotate and flips manually, it may generate non-existent tropical cyclone phenomenon.

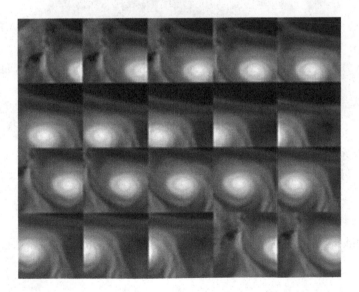

Fig. 6. Output images of tropical cyclone at west pacific basin through sliding windows.

5.2 Feature Extraction Through Center Radiation

Although separating a region into small areas could reduce difficulty to identify a tropical cyclone, it still is not enough to provide high accuracy to determine whether there is a tropical cyclone in the area. In order to solve this problem, we provide a center radiation algorithm for feature engineering. Figure 7 shows the graphical representation of this algorithm. This algorithm has a two steps mechanism to generate features. First, Use Algorithm 1 to locate maximum value of water vapor within an area and align it as a center. The central point would be a first feature. Second, after determining the center, we take average value of surrounding 8 points as a first layer feature. Then we take average value of surrounding 16 points as a second layer feature. We increase features until reach 8 layers from center. We think this feature engineering could represent the characteristic of tropical cyclones in precipitable water data set. Although a tropical cyclone changing its size and rotating over time, we assume that its shape is still roughly symmetrical. This means that its center will have the maximum value of precipitable water and decrease as the radius widens. Although a tropical cyclone can have more than one maximum value of precipitable water, these values are not randomly distributed but are gathered together. We can arbitrarily choose one of the maximum values without losing representativeness in Algorithm 1.

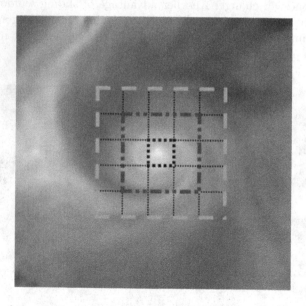

Fig. 7. Think of the center point as a feature and the surrounding point layer as a feature.

5.3   Ensemble Algorithm

With sliding window technique, we separate a region into many small areas, and we need to ensemble classification results of those areas into one probability predict result of entire region. We want to design an algorithm to reduce the cost of predicting errors and increase the weight of correct predictions. The idea is the

Data: Precipitable water value of an area of west pacific basin region

for *Each point in areas* **do**
> From Top left to Bottom right, find the maximum points;
> **if** *There is only one maximum value* **then**
> > Select that point as the center;
>
> **else**
> > Find the first point with maximum value
> > Select that point as the center;
>
> **end**

end

Algorithm 1. Determine center point in an area.

more prediction results for one point, the less predictive error will pay. Therefore, we leverage the sliding window result to give a point multiple predictions and use an algorithm to ensemble those predictions. Algorithm 2 shows how we ensemble points in the same area. At present we take $\Delta = 1$ in our algorithm.

Data: Precipitable water value of west pacific basin region

Calculating how many time each point be slided by a sliding window; Total sliding times = the number of times passed by sliding window of a point Value = the value of a point **for** *Each area within the region* **do**
> Identify each area through classifier;
> **if** *The area is identified to have a tropical cyclone* **then**
> > All points within area Value + Δ;
>
> **else**
> > All points within area Value - Δ;
>
> **end**

end
if *The Value of any point ¡ 0* **then**
> set Value of that point to 0

end
Calculation of probability for each points;
Prob of a point = Value / Total sliding times ;

Algorithm 2. Ensemble classified areas into Probability Prediction.

6   Experiment

6.1   Experiment Setup

We combined GFS precipitable water data and IBTrACS best tricking data and generated five years labeled data from 2010 to 2014. With sliding window for data preprocessing, we solved data sparsity problem and could have enough training samples. Also, we randomly selected negative samples to balance positive and negative samples. Because number of tropical cyclones per year are not equal, number of samples from 2010 to 2015 are also not equal, too. In order to prove our center radiation feature engineering and models work well for tropical cyclone characteristics and can use without worrying about concept drift over time, we treated 2015 year samples, processing with the same sliding window and feature engineering as another test set (Table 1).

Table 1. Data source, training and test data set information.

Data	Event	Resolution	Feature
GFS data	Tropical cyclones	32x32	Precipitable Water

Data Set	Year	Positive Samples	Negative Samples
Train	2010-2014	105000	105000
Test1	2010-2014	35000	35000
Test2	2015	10000	10000

6.2   Experiment Result and Discussion

In Sect. 5.2, we explained how to extract features through origin data, starting from the center point and layer by layer to extract our features. However, it is hard to estimate how many features we needed. We didn't know if more features can help us better to identify tropical cyclones. An experiment is designed to verify whether this kind of feature extraction works. This experiment began with only one feature, the maximum value we found from Algorithm 1. Then, we increased more features in each step. In order to reduce the influence of very similar points, we obtained the feature by adding two points to the radius at a time. For example, the first added feature is points with radius = 2 from their center point, the second added feature is points with radius = 4 from their center point. We use the same approach for both training and test data.

Table 2 shows the predict results of 2010–2014 test set. In general, the more feature is trained, the more accurate the prediction is for that model. More interesting, from Table 2 we can find the best number of feature when adding more layers will not cause better results. Because number of features may have a close relationship with the size of tropical cyclones, and size of each tropical cyclone varies, too many features will reduce the tolerance of the model for different tropical cyclones. In addition, the balanced training data has a large impact. In our experience, if we don't make the training samples equal, it would be very biased to negative samples as natural of meteorology data.

Meteorological data has a strong relationship with time, therefore, we designed another experiment to discuss our models with concept drift over time. In this experiment we used 2010–2014 training model to predict 2015 tropical cyclone data. As Table 3 shows, it is still consistent with our previous experimental results. The accuracy is proportional to the number of features.

Table 2. Results of prediction of 5 different training models of 2010–2014 test set.

Configuration	True/false	Predict false	Predict true	Accuracy
Center point	False	31223	3777	0.892
	True	934	9690	0.948
Center point + 1 radiation feature	False	31454	3546	0.899
	True	1732	33268	0.951
Center point + 2 radiation features	False	31572	3428	0.952
	True	1682	33318	0.952
Center point + 3 radiation features	False	31634	3366	0.902
	True	1679	33330	0.952
Center point + 4 radiation features	False	31693	3307	0.923
	True	1639	33361	0.953

Table 3. Results of prediction of 5 different training models of 2015 test set.

Configuration	True/false	Predict false	Predict true	Accuracy
Center point	False	9127	873	0.913
	True	328	9627	0.967
Center point + 1 radiation feature	False	9107	893	0.910
	True	229	9771	0.977
Center point + 2 radiation features	False	9117	883	0.912
	True	190	9810	0.981
Center point + 3 radiation features	False	9120	880	0.912
	True	193	9807	0.981
Center point + 4 radiation features	False	9134	886	0.912
	True	193	9807	0.981

We designed a system to find region of interest of serve weather phenomenon and gave these regions high probability. In our system, we used Algorithm 2 to ensemble areas prediction into region probability prediction. Figure 8 are the outcome of our system. Left side of Fig. 8 are origin precipitable water data from GFS data and right side are output probability predict results which data assimilation process would refer to. Figure 8a and c are one day at 2015 and 2016 without any tropical cyclone. On the other hand, Fig. 8e has two tropical cyclones (one is located at edge) and Fig. 8g has one tropical cyclone. Ideally, Fig. 8b and d should be all 0; however, these two figures still are affected by false predictions and have some high probability points which should not exist. In Fig. 8f and h, we covered most tropical cyclones center and regions around tropical cyclones although Fig. 8h is affected by two major false prediction in corners. Fortunately, on the premise that computing power can be handled, a small amount of false prediction will only consume more computing power

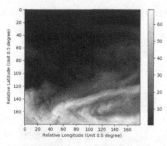

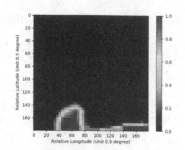

(a) Precipitable water data
without tropical cyclone.

(b) Probability predict result
without tropical cyclone.

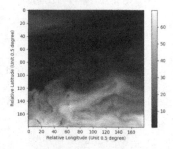

(c) Precipitable water data
without tropical cyclone.

(d) Probability predict result
without tropical cyclone.

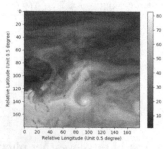

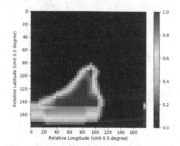

(e) Precipitable water data
with two tropical cyclones.

(f) Probability predict result
with two tropical cyclones.

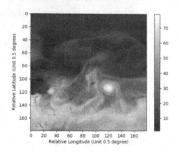

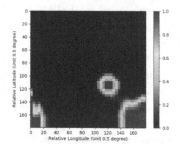

(g) Precipitable water data
with a tropical cyclone.

(h) Probability predict result
with a tropical cyclone.

Fig. 8. Probability prediction

without affecting the predict result of NWP. Our system is designed to feed high resolution satellite data into region of interesting, but still keep update original resolution data for others.

7   Conclusion

In this paper, we apply machine learning in GFS precipitable water data to identify tropical cyclones, facilitate sliding window technique to solve data sparsity and overcome the traditional limitation that tropical cyclones were needed to be put in the center of figures for identification. The center-radiation methods for feature engineering of precipitable water data was proved to achieve fairly high classification accuracy rate for 2010–2015 GFS data set around 94%. Our system produces the probability predict result as a reference of data assimilation process. The probability predict results indicate the region of interest where we may need to put high resolution satellite data and increase initial value update frequency to help NWP with better weather prediction. This successful result could be a good pilot study for further GOES-15 or GOES-16 satellite image research.

Acknowledgement. This work was carried out at NOAA Earth System Research Laboratory (ESRL), University of Colorado Boulder with funding from NOAA. The first author is supported by funding from NOAA Award Number NA14OAR4320125 and the third author is supported by funding from NOAA Award Number NA17OAR4320101. The authors thank Christina Bonfanti for her help to suggest alternative best tracking data produced by NOAA.

References

1. Global forecast system (GFS). https://www.ncdc.noaa.gov/data-access/model-data/model-datasets/global-forcast-system-gfs
2. Cecil, D.J., Zipser, E.J.: Relationships between tropical cyclone intensity and satellite-based indicators of inner core convection: 85-GHZ ice-scattering signature and lightning. Mon. Weather Rev. **127**(1), 103–123 (1999)
3. Dvorak, V.F.: Tropical cyclone intensity analysis and forecasting from satellite imagery. Mon. Weather Rev. **103**(5), 420–430 (1975)
4. Ho, S.-S., Talukder, A.: Automated cyclone discovery and tracking using knowledge sharing in multiple heterogeneous satellite data. In: Proceedings of the 14th ACM SIGKDD International Conference on Knowledge Discovery and Data Mining, pp. 928–936. ACM (2008)
5. Ho, S.-S., Talukder, A.: Automated cyclone identification from remote QuikSCAT satellite data. In: 2008 IEEE Aerospace Conference, pp. 1–9. IEEE (2008)
6. Jaiswal, N., Kishtawal, C.M., Pal, P.K.: Cyclone intensity estimation using similarity of satellite IR images based on histogram matching approach. Atmos. Res. **118**, 215–221 (2012)
7. Knapp, K.R., Kruk, M.C., Levinson, D.H., Diamond, H.J., Neumann, C.J.: The international best track archive for climate stewardship (IBTrACS) unifying tropical cyclone data. Bull. Am. Meteorol. Soc. **91**(3), 363–376 (2010)

8. Liu, Y., et al.: Application of deep convolutional neural networks for detecting extreme weather in climate datasets. arXiv preprint arXiv:1605.01156 (2016)
9. McBride, J.L., Zehr, R.: Observational analysis of tropical cyclone formation. Part II: comparison of non-developing versus developing systems. J. Atmos. Sci. **38**(6), 1132–1151 (1981)
10. Mueller, K.J., DeMaria, M., Knaff, J., Kossin, J.P., Vonder Haar, T.H.: Objective estimation of tropical cyclone wind structure from infrared satellite data. Weather Forecast. **21**(6), 990–1005 (2006)
11. Panangadan, A., Ho, S.-S., Talukder, A.: Cyclone tracking using multiple satellite image sources. In: Proceedings of the 17th ACM SIGSPATIAL International Conference on Advances in Geographic Information Systems, pp. 428–431. ACM (2009)
12. Ryan, B.F., Watterson, I.G., Evans, J.L.: Tropical cyclone frequencies inferred from gray's yearly genesis parameter: validation of GCM tropical climates. Geophys. Res. Lett. **19**(18), 1831–1834 (1992)
13. Velden, C.S., Olander, T.L., Wanzong, S.: The impact of multispectral goes-8 wind information on Atlantic tropical cyclone track forecasts in 1995. Part I: dataset methodology, description, and case analysis. Mon. Weather Rev. **126**(5), 1202–1218 (1998)
14. Zhuge, X.-Y., Guan, J., Yu, F., Wang, Y.: A new satellite-based indicator for estimation of the western North Pacific tropical cyclone current intensity. IEEE Trans. Geosci. Remote Sens. **53**(10), 5661–5676 (2015)
15. Zou, J., Lin, M., Xie, X., Lang, S., Cui, S.: Automated typhoon identification from QuikSCAT wind data. In: 2010 IEEE International Geoscience and Remote Sensing Symposium (IGARSS), pp. 4158–4161. IEEE (2010)

From Empirical Analysis to Public Policy: Evaluating Housing Systems for Homeless Youth

Hau Chan[1]($\boxtimes$), Eric Rice[2], Phebe Vayanos[2], Milind Tambe[2],
and Matthew Morton[3]

[1] University of Nebraska-Lincoln, Lincoln, NE 68588, USA
`hchan3@unl.edu`
[2] University of Southern California, Los Angeles, CA 90007, USA
[3] Chapin Hall at the University of Chicago, Chicago, IL 60637, USA

Abstract. There are nearly 2 million homeless youth in the United States each year. Coordinated entry systems are being used to provide homeless youth with housing assistance across the nation. Despite these efforts, the number of youth still homeless or unstably housed remains very high. Motivated by this fact, we initiate a first study to understand and analyze the current governmental housing systems for homeless youth. In this paper, we aim to provide answers to the following questions: (1) What is the current governmental housing system for assigning homeless youth to different housing assistance? (2) Can we infer the current assignment guidelines of the local housing communities? (3) What is the result and outcome of the current assignment process? (4) Can we predict whether the youth will be homeless after receiving the housing assistance? To answer these questions, we first provide an overview of the current housing systems. Next, we use simple and interpretable machine learning tools to infer the decision rules of the local communities and evaluate the outcomes of such assignment. We then determine whether the vulnerability features/rubrics can be used to predict youth's homelessness status after receiving housing assistance. Finally, we discuss the policy recommendations from our study for the local communities and the U.S. Housing and Urban Development (HUD).

Keywords: Housing system · Homeless youth · Classification

1 Introduction

There are nearly 2 million homeless youth in the United States each year. These are young people between the age of 13 and 24 who are homeless, unaccompanied by family, living outdoors, in places not fit for human habitation, and in emergency shelters [9]. The consequences of youth homelessness are many, including many preventable problems such as exposure to violence, trauma, substance use, and sexually transmitted disease [9]. A critical solution to improve long

© Springer Nature Switzerland AG 2019
U. Brefeld et al. (Eds.): ECML PKDD 2018, LNAI 11053, pp. 69–85, 2019.
https://doi.org/10.1007/978-3-030-10997-4_5

term outcomes for homeless youth is to quickly and efficiently help the homeless youth find safe and stable housing situations. Indeed, there are many non-profit organizations and public sector programs designed to do this. In almost all communities in the United States, the number of youth experiencing homelessness exceeds the capacity of the housing resources available to youth [3]. This situation leaves communities with the terrible predicament of trying to decide who to prioritize for the precious few spots in housing programs which are available at any given time. Most communities have moved to what is referred to as a Coordinated Entry System. In such systems, most agencies within a community pool their housing resources in a centralized system. Persons who are seeking housing are first assessed for eligibility for housing, which usually includes HUD-defined chronic homelessness, other criteria such as veteran status, and "vulnerability". Based on these assessments, individual youth are prioritized for housing and placed on waiting lists until appropriate housing becomes available in the community [3]. Despite these efforts, most of the prioritization decisions are made by humans manually working in the housing communities using some (possibly unknown) rubric. Could we understand how humans make these prioritization decisions? Could we provide important insights to the local communities using the housing assistance assignment data from the past? In this paper, we provide simple machine learning analyses and tools that could be of use to communities. Our results are the most comprehensive, non-experimental evaluation of youth coordinated entry systems of which we are aware. We view this paper as a gateway for providing policy recommendations to improve the housing systems.

1.1   Our Goal

HUD wants community housing systems to be systematic, evidence-based and grounded in research [3,4]. Despite of this, save for a few exceptions (e.g. [2]), the current housing allocation system for youth has not been evaluated for its success. As a result, the goal of this paper is to see if we can evaluate the success of the current system using the data from the HUD's Homelessness Management Information System (HMIS), the primary repository for data on homeless services delivery in the U.S. If we can uncover (which we have) new insights in the current system, there is a potential to make a major impact in policy and societal outcomes. We present the first study on evaluating such system.

Our Contribution. In additional to providing an overview of the current housing system, we provide insights that would help the communities to understand and evaluate the current housing assignment process. In particular, using the past housing assistance assignment data of homeless youth, we:

(a) Infer the decision rules of the local communities for providing youth with housing assistance;
(b) Evaluate the outcome of the current assignment process;

(c) Build and learn an interpretable classifier to predict homelessness outcome of each youth to a homelessness exit[1]; by leveraging vulnerability assessment tools;
(d) Provide public policy recommendations to improve the housing system.

Since our tools need to be understood by the housing communities, it is important for the tools to be explainable and easy to use. As such, we focus on learning interpretable classifiers such as logistic regressions and decision trees. While our analyses and the tools are simple, they are extremely impactful as evident by the fact that we are requested by the HUD to make policy recommendations based on our findings. The remainder of this paper is organized as follows. In Sect. 2, we provide an overview of the current housing systems for homeless youth. In Sect. 3, we discuss the dataset obtained from Ian De Jong (Orgcode), as part of a working group called "Youth Homelessness Data, Policy, Research" led by Megan Gibbard (A Way Home America) and Megan Blondin (MANY), which includes members of HUD, USICH, and ACF, as well as researchers from USC (Rice) and Chapin Hall at the University of Chicago (Morton). In Sect. 4, we discuss our methodology for learning the classifiers. We then infer the communities' decision rules for assigning youth to various housing programs in Sect. 5 and show the outcome of such assignment in Sect. 6. In Sect. 7, we show how we can use the vulnerability assessment tools/features to predict homelessness outcome of homeless youth. In Sect. 8, we present our policy recommendations to the HUD and housing communities based on our analysis and the summary of our results. In Sect. 9, we conclude this paper by highlighting the values of our study to the communities and HUD.

2   Current Approach for Housing Prioritization

HUD offers many mandates, guidelines, and best practice recommendations to communities on housing youth [3,4]. In most Coordinated Entry Systems for homeless youth, housing agencies within a community pool their housing resources in a centralized system. First, a homeless youth enters a centralized intake location (e.g. emergency shelters, street outreach workers, or drop-in centers) to sign up for housing support. There, they are assessed for housing eligibility and vulnerability/risk. All this information is then entered into the HMIS. Then, based on these assessments, a case manager or a team of housing navigators decide how that youth is to be prioritized for housing. The youth is then placed on a waiting list until appropriate housing becomes available.

Although communities may decide for themselves what risk/vulnerability assessment tool to use, the most frequently used tool for assessing the risk levels of youth is the Next Step Tool (NST) for Homeless Youth developed by OrgCode Consulting Inc. and Community Solutions and thus we focus our analyses on this

---

[1] There are different ways homeless youth can exit homelessness; which include: being assigned to housing programs, going back to live with family members, and finding a stable living on their own.

tool.[2] Roughly speaking, the NST is a set of multiple-choice, dichotomous, and frequency-type questions to measure a youth's vulnerability based on his/her history of housing and homelessness, risks, socialization and daily functions, and wellness. Based on the results of NST, a youth is scored from 0 to 17.

Based on the recommendations provided in the NST documentation, youth who score 8 to 17 are designated "high risk" youth and should be prioritized for Permanent Supportive Housing (PSH), a resource-intensive housing program which includes "wrap-around" social services for youth to assist them in remaining stably housed. Youth who score lower (the 4–7 range) are typically referred to Rapid Rehousing (RRH) which is a short-term rental subsidy program that infrequently has many social services attached. Some youth who score low (less that 4) may not ever receive housing resources. For many providers and communities, this step is often painful as the desire to help all homeless youth is foremost in the minds of every provider. The NST scoring recommendations are not a hard and fast set of rules, but as we show in our analyses, most communities follow these cut points when assigning housing to youth.

However, the NST is a general vulnerability measure, not tied to a particular outcome, and no research has been conducted to date which links this tool to particular outcomes, particularly long-term housing stability. As noted by many communities, the housing stability of a youth as they exit a program is often the most robust measure of success [7]. That is, they want to assign youth to the appropriate housing programs in order to maximize the youth's chances of being stably housed in the future. For instance, if a youth is placed in PSH, a successful outcome would be continuation of stay unless they transition to stable unsubsidized housing. For those receiving RRH, remaining a stable renter without further government assistance is a positive outcome. Such outcomes, however, might not have any positive correlation with the youth's risk levels.

3   Description of the Data

The dataset consists of 10,922 homeless youth registered for housing services from the HMIS database from different communities in the U.S.[3] In the context of social science and our domain, this dataset is considered to be large and valuable by many researchers and industrial partners. These records were anonymized and provided by Iain De Jong of Orgcode. Some youth have already been assigned to some housing programs while others are still waiting for housing

---

[2] The full name is Transition Age Youth - Vulnerability Index - Service Prioritization Decision Assistance Tool. The tool can be assessed at http://orgcode.nationbuilder. com/tools_you_can_use. This tool incorporated work from the TAY Triage Tool developed by Rice, which can be accessed at http://www.csh.org/wp-content/uploads/ 2014/02/TAY_TriageTool_2014.pdf.

[3] While these data come from urban, rural, and suburban communities, there are still communities in the country who are providing PSH and RRH but not in a coordinated way and we do can not speak to the effectiveness of those interventions in decentralized systems.

Table 1. Different subsets of features.

Subset	Description
DOM	Basic demographic information: Age, gender, race, LGBT status
COM	Type of communities: 16 urban, suburban, and rural communities
NSTQ	Responses to the NST questionnaires: 40 questions (1 multiple choice, 9 numerical)
NSTT	17 binary features tallying sub-responses
NSTS	1 NST score
NSTA	NSTQ + NSTT + NSTS

assignments. Each record has the youth's age, gender, LGBT status, ethnicity, type of communities, and a list of responses to the NST questions (including NST score) assessing a youth's vulnerability.

3.1   Features from the Youth

The features of the youth are divided into the following subsets as in Table 1. The DOM features are basic demographic characteristics of the youth. The COM features are the type of community in which a youth lives. The NST evaluates the vulnerability of a youth based on his/her responses to the forty questions (NSTQ) about youth's history of housing and homelessness, risks, socialization, daily functioning, and wellness components. Each component scores a youth based on the responses to the questions within the component (NSTT). The NST score (NSTS) is the sum of the scores from the components.

Table 2. Basic statistics of the data. #Y = Number of Youth, TofE = Type of Exits, #SH = Number of Youth Still Housed, and AvgNSTS = Avg. NST Scores.

#Y	TofE	#SH	AvgNSTS
1145	Self resolve	873	4.21
1259	Family	1006	4.65
1103	Unkown	N/A	6.38
2885	RRH	2209	6.52
3610	Pending	N/A	6.84
54	SSVF	28	7.11
579	PSH	474	10.24
211	Incarcerated	N/A	10.25

3.2   Types of Exits from Homelessness

For each homeless youth in the data, there are fields specifying his/her type of exit from homelessness and whether s/he is still living in the same type of exit after a fixed time period (a.k.a. Still Housed). The still-housed responses indicate whether a housing program was successful for the youth; "Yes" answers indicate a youth is still stably housed, a positive outcome; and "No" indicates a youth has exited housing assistance and returned to homelessness, a negative outcome. Table 2 lists the number of youth in each type of exit in the data.

A large number of youth is still waiting for housing assignments and/or have been lost to the housing system. In many cases, some homeless youth went to live with their family members (Family) or were able to find housing themselves (Self Resolve). There are three main types of housing programs in the dataset: supportive services for veteran families (SSVF), permanent supportive housing (PSH), and rapid re-housing (RRH).

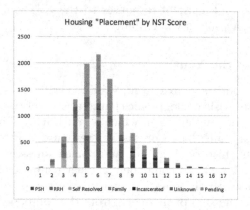

Fig. 1. Histograms of housing assignment/placement by NST scores.

Figure 1 shows the housing assignment by the NST scores. Most communities are not assigning youth to PSH with NST scores lower than 8 while most communities assign youth with NST scores between 3 and 9 to RRH.

Given the data, our prediction tasks are to (1) infer the decision rules of the local communities and (2) understand how the features of a youth affect the probability that a youth will have positive outcomes (i.e. still-housed) given different types of exits from homelessness. Due to the small sample sizes of SSVF, we focus on Family, PSH, RRH, and Self Resolve exits.

4   Methodology and Evaluation Metrics

To infer the decision rules of the local communities for assigning youth to PSH and RRH, we consider multiclass classification problem of classifying youth into

Family, PSH, RRH, or Self Resolve exits. This allows us to infer the most likely exit of the youth and the decision rules. We use the one vs all strategies where we train a classifier per class and view the same-class examples as positive examples and the other classes' examples as negative examples. To infer a youth's probabilities of success for different types of exits, we can naturally cast the problem as binary classification and learn a binary classifier for each type of exit. In this binary classification, for each type of exit, youth assigned to the exit with "Yes" and "No" still-housed responses are positive and negative examples, respectively. In both cases, we consider the following classifiers and performance measure.

4.1 Classifiers

Due to the explainability and ease of interpretation for end users, we focused on learning logistic regression and decision tree classifiers for each type of exit [5,6,8]. Moreover, we require the classifier to output class posterior probabilities for each of our classifiers.[4] To learn our classifiers, we use 80% and 20% of the data, pertaining to the type of exit, for training and testing, respectively. We use 10-fold cross validation in the training set to find the best hyperparameters to regularize the classifiers (L1-norm for logistic regression and depth for the decision tree). For constructing the decision (classification) trees, we consider the standard CART model to build a binary tree and select nodes/features and values to split based on Gini's diversity index [1]. For each split, we consider all the possible pairs of features and values. To control the depth of the tree, we use cross validation to select the best minimum number of nodes at the leaves. Finally, we prune the learned tree to obtain a simpler decision tree classifier.

Using different feature combinations, we learn logistic regression and decision tree classifiers and measure the performance using AUROC (defined below).

4.2 Performance Measure

We measure the predictive performance using the area under the receiver operating characteristic curve (AUROC). The ROC is constructed for the learned classifiers based on the true positive rate (true positive divided by true positive plus false negative) and the false positive rate (false positive divided by false positive plus true negative) points for each possible cutoff posterior probabilities in the test data. We then compute the area under the ROC. Roughly speaking, the AUROC is equal to the probability that a randomly chosen youth with a positive class/still-housed outcome ranks above (i.e, has a higher probability of success) than a randomly chosen youth with a negative class/still-housed outcome. Thus, higher AUROC indicates that the classifier is able to distinguish the classes effectively. AUROC is particularly useful in our setting because

---

[4] Logistic regression classifier returns class posterior probabilities by default, decision tree classifier can return the percentage of the majority label at the leaves. This is known as calibration, or platt scaling, in the machine learning literature.

the unbalanced nature of our data ($\approx$76–82% positive outcomes) as the standard 50% cutoffs for computing accuracy could provide us with a representative model rather than a discriminative model. We report the average AUROC over 100 different 80% and 20% splits of our data into training and testing for the learned classifiers. We omit reporting the small standard deviations for brevity.

4.3 Building the Final Model

To build the final model, we trained the classifiers using all of the available data for each exit. We highlight the important coefficients and decision nodes of the learned logistic regressions and decision trees. We can interpret the exponentiated coefficient of a predictor as the odds ratio when holding other predictors constant (i.e, a one-unit increase in the predictor value corresponds to some percentage of (multiplicity) increase in the odds of being successful).

5   Decision Rules and Youths' Most Likely Exits

In this section, we are interested in inferring the decision rules used by the communities to assign and prioritize youth for housing assistance (e.g., PSH and RRH) as well as the youths' likelihood of returning to family (e.g., Family) and finding their own housing or rental subsides (e.g., Self Resolve) without any housing assistance. Table 3 shows the AUROC of the logistic regression and decision tree classifiers for the youths' exits from homelessness using different combination of features. The learned logistic regressions and decision trees for PSH and RRH can be viewed as decision rules for assigning youth.

Table 3. AUROC of logistic regression and decision tree for each type of exits. F = Family Exit, P = PSH Exit, R = RRH Exit, S = Self-resolve Exit.

Type of exits	F		P		R		S	
Classifiers	LG	DT	LG	DT	LG	DT	LG	DT
Baseline	0.50	0.50	0.50	0.50	0.50	0.50	0.50	0.50
DOM	0.62	0.65	0.63	0.69	0.56	0.64	0.57	0.61
COM	0.64	0.62	0.53	0.53	0.57	0.57	0.52	0.51
NSTS	0.75	0.70	0.97	0.92	0.71	0.81	0.83	0.79
NSTQ	0.76	0.69	0.96	0.89	0.74	0.71	0.80	0.70
NSTT	0.77	0.72	0.97	0.90	0.75	0.78	0.84	0.76
NSTQ + NSTS	0.78	0.77	0.97	0.95	0.78	0.89	0.84	0.83
NSTT + NSTS	0.77	0.75	0.97	0.93	0.75	0.89	0.84	0.81
NSTA	0.78	0.77	0.97	0.95	0.78	0.89	0.84	0.83
NSTA + COM	0.81	0.79	0.97	0.95	0.80	0.90	0.84	0.83
NSTA + DOM + COM	0.81	0.80	0.97	0.96	0.80	0.91	0.84	0.83

5.1   Decision Rules for Assigning Youth to PSH and RRH

From Table 3, we observe that a youth's NST score (NSTS) alone is a very good predictor for predicting whether the youth will exit homelessness to PSH. The learned logistic regression with NSTS feature yields the highest AUROC than the learned decision tree with any combination of features. Moreover, its decision boundary is similar to the NST recommendation (see left of Fig. 2).

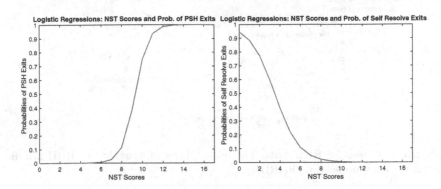

Fig. 2. Logistic regressions of the probabilities of (left) PSH exits and self resolve (right) exits. The coefficient and constant of the left (right) logistic regression are 1.58 (−0.83) and −14.74 (2.88), respectively.

On the other hand, it seems that the decision rule of RRH is more complicated – the NSTS alone is no longer the best predictor for predicting RRH exit. Indeed, the RRH decision rule can be better captured using a non-linear decision tree classifier (see Table 3). In general, the learned decision trees with high AUROC have a very similar structure – those with NST scores less than 5 or greater than 9 have almost no chance of getting RRH while those with NST scores between 5 and 9 have high chance of getting RRH subject to various additional conditions such as ages, lengths since last lived in stable housing, and violence at home between family members (see Fig. 3 as an example learned decision tree with feature set NSTA + DOM + COM). The learned decision rule for RRH is similar to the NST recommendation. However, youth seem to be selected based on additional criteria other than the NST score alone (Table 4).

5.2   Youths' Exits to Family and Self Resolve

Similarly, Table 3 shows the set of features that can help us to predict youths' chances of existing homelessness to Family and Self Resolve. To predict Self Resolve exit, the learned logistic regression that uses NSTS as the only feature is a good predictive model relative to different combination of features and decision tree classifiers. Its decision boundary is shown in (the right of) Fig. 2. Interestingly, the youth with low NST scores have high chances of self resolve.

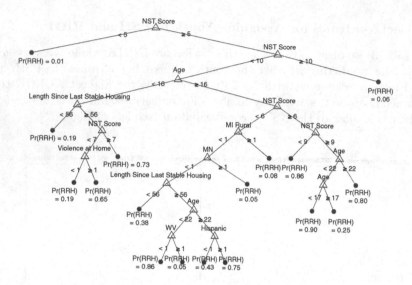

Fig. 3. Decision trees for RRH exits. The probabilities of assigning to RRH are displayed at the leave nodes (i.e., Pr(RHH)).

Table 4. Family exit: top-6 (ordered) important features of the learned logistic regression.

Weight	Description
+1.33	MN community
+1.29	WV community
+1.29	17 or younger
+1.25	MI rural community
−0.68	Northeast Florida City community
−0.51	NST (risk) score

However, youth with NST scores between 6 and 8 seem to have low chances of going to PSH and being able to Self Resolve.

While the logistic regression that uses NSTS as the only feature is a good predictive model for predicting Family exit, other combinations of features improve the AUROC. In particular, the logistic regressions learned using NSTA + COM and NSTA + DOM + COM yield the highest AUROC. We select the model with the lowest number of features and report their coefficients. Figure 4 displays the top-6 important features of the learned logistic regression with NSTA + COM. The important features are related to the communities, ages, and NST scores of the youth. Surprisingly, the youths' communities have some impact on the youths' chances of going back to Family. As a result, we list and separate the communities with positive and negative weights in Table 5. Youth in the positive communities seem to be more likely go to back to home while those in the negative communities are less likely.

Table 5. Family exit: positive and negative weighted communities from the learned logistic regression.

Positive	MN, WV, MI rural, MI suburb, Virginia suburb, Southern California metropolis SUBURB
Negative	Northeast Florida City, South Michigan City, Large East Coast City, South Florida City, FL

6   Outcome of the Current Housing Assignment Process

Now that we have have a better understanding of how the assignments are performed in practice, we discuss the outcome of such practice by looking at the number of youth still-housed for each type of exits based on the NST scores.

Figure 4 shows the percent of youth remaining housed by the housing placement based on their NST score. While Table 2 indicates that 75%–81% of youth are still-housed under the assignment, the above analysis provides us some comparative statistics when the percentages are breakdown by the NST scores.

In particular, as the NST score increases, the number of youth who is still-housed decreases. PSH seems effective for youth with high NST scores: 70% of youth with NST score of 14 and 100% of the 16 youth with NST score 7 or lower are still-housed. For RRH, 80%–70% of youth with score 4–9 remain still-housed while 57% of he 19 youth with score 10 is still-housed. For the youth that went back to their family, 90% of those with NST score of 4 or less and 80% of those with NST score of 5–6 are still-housed. However, the percentage drops significantly for higher NST score youth. For the self-resolve youth, roughly 90% of low NST score (1–4) youth is still-housed while the percentage drops below 50% for those with high NST scores. As we will show in the next section, there are additional factors beyond NST score that could be used to predict the still-housed outcome of the youth.

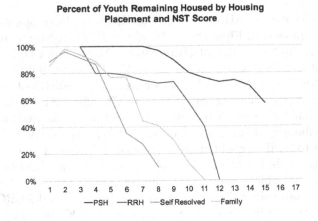

Fig. 4. Percentages of youth still-housed vs NST scores.

7   Predicting the Outcomes of Housing Assignments

In this section, we want to learn an interpretable classifier that would tell us a youth's probability of success (i.e., positive still-housed outcome) for a particular type of exit. Our plan is that communities can use our classifier as a decision aid to assist in determining the probability of success for a particular youth in a specific housing program. Such human-machine interaction will provide improved decision-making in the allocation of housing programs to youth.

Table 6. AUROC of logistic regression and decision tree for each type of exits. F = Family Exit, P = PSH Exit, R = RRH Exit, S = Self-resolve Exit.

Type of exits	F		P		R		S	
Classifiers	LG	DT	LG	DT	LG	DT	LG	DT
Baseline	0.50	0.50	0.50	0.50	0.50	0.50	0.50	0.50
DOM	0.58	0.60	0.50	0.51	0.60	0.58	0.65	0.74
COM	0.49	0.51	0.50	0.47	0.50	0.49	0.51	0.55
NSTS	0.77	0.72	0.66	0.62	0.55	0.54	0.75	0.71
NSTQ	0.72	0.61	0.60	0.57	0.71	0.69	0.68	0.62
NSTT	0.77	0.67	0.63	0.59	0.65	0.62	0.86	0.77
NSTQ + NSTS	0.77	0.73	0.65	0.64	0.73	0.69	0.79	0.72
NSTT + NSTS	0.77	0.75	0.66	0.62	0.65	0.63	0.86	0.84
NSTA	0.77	0.74	0.65	0.63	0.73	0.69	0.86	0.85
NSTA + COM	0.76	0.74	0.65	0.63	0.73	0.69	0.86	0.84
NSTA + DOM + COM	0.76	0.75	0.65	0.62	0.74	0.69	0.86	0.85

7.1   Logistic Regressions and Decision Trees

Table 6 shows the average AUROC of the classifiers for different subsets of features for each type of exit. First, the performance of the logistic regressions and decision trees is similar for most feature combinations. The COM features are not very useful and the learned models have AUROC only slightly better than random. The NST score (NSTS) alone is a reasonable predictor of the outcome. We tried different combinations of NSTQ, NSTT, and NSTS and found that we are able to build a better model with a higher/similar AUROC for all types of exits by combining all three subsets (i.e., NSTA). We also added COM and DOM to NSTA to see if we could improve the model. Unfortunately, the AUROC did not improve much. For consistency, we report the learned logistic regression and decision tree classifiers using the feature set NSTA. For all of the exit types except PSH exit, our learned models are reasonable (with good AUROC scores). For RRH (R) exit and Self Resolve (S) exit, the AUROC increased by 15% to 30% over the AUROC of the learned logistic regressions using only the NST score. This provides some important indication that other parts of NST are useful for building better predictive models. On the other hand, our models for PSH seem to be much weaker than the other models for different types of exits, and the performance does not improve with more features.

7.2   Significant Features of the Learned Models

Now that we have discussed the performance of the classifiers, let us look at important features in the models for RRH and Self-resolve exits as they benefit from more features. We select the model with the lowest number of features and train the classifiers using all of the available data for the exits. We highlight the important coefficients of the logistic regressions.

Learned Logistic Regressions. Tables 7 and 8 show the top-6 most important features (in terms of weight) of the learned logistic regressions for RRH and Self Resolve exits. In many of these classifiers, the locations where youth most frequently sleep and the NST scores are important features.

Table 7. RRH exit: top-6 (ordered) important features of the learned logistic regression.

Weight	Description
−1.59	Sleep on couch most frequently
−1.53	Sleep in trans. housing most frequently
+0.82	Physical, mental, substance abuse issues
+0.49	Sleep in shelters most frequently
−0.28	17 or younger
−0.16	NST (risk) score

For RRH exit (Table 7), having physical, mental, and substance abuse issues and sleeping in shelters most frequently are positive factors for being successful in RRH. Youth with age of 17 or younger have lowered chances for success. For Self Resolve exit (Table 8), youth that are 17 or younger or have some physical, mental, substance abuse, and medication issues have decreased chances of being able to successfully exit homelessness on their own. If a youth has used marijuana at 12 or younger, then the youth has an increased probability of being successful.

8   Policy Recommendations from Our Empirical Analysis

To obtain feedback on the findings, we shared and discussed our initial results with some members of the local communities and HUD on August and November of 2017. Our audiences (even for those that do not have any technical background) found the logistic regression and decision tree classifiers interpretable

Table 8. Self resolve exit: top-6 (ordered) important features of the learned logistic regression.

Weight	Description
−2.43	17 or younger
−0.92	NST (risk) score
+0.38	Have used marijuana at 12 or younger
−0.37	Physical, mental, substance abuse issues
−0.35	Medications
−0.34	Left program due to physical health

and understandable after some brief explanation. They found the results helpful for justifying the existing method for prioritizing youth for housing program based on the NST score. Moreover, they are intrigued and excited to see the important features for determining youths' chances of successes in different types of exits. Due to the significant of our results and findings, the housing communities and HUD asked us to make policy recommendations based on our analysis. As a result, below is our recommendations from the analysis:

1. Use the TAY-VI-SPDAT: Next Step Tool for Homeless Youth (NST) to assess vulnerability.
2. PSH is effective for almost any youth, regardless of NST score.
3. Rapid rehousing (RRH) can be an effective solution for youth who score up to 10.
4. 66% of youth who score less than 4 successfully self-resolve or return home. Family reunification and other case management services appear sufficient for many.
5. More experimentation with rapid rehousing for higher scoring youth is needed.

Given that most of the local communities is assigning homeless youth to PSH and RRH using various components of the NST tool according to our analysis in Sect. 5, the policy on using the NST tool is a natural step and is easy to incorporate into any existing local housing systems. Policy recommendations of (2) and (3) are also sensible since some of the existing assignments are being done based on these cutoffs. Our observation of (4) will allow the communities to better allocate housing resources to those of higher NST scores since lower NST score homeless youth can find successful outcome by going back to family or self-resolve. The last item in the policy recommendations aims to solicit more data from the communities so that we can provide a better evaluation of RRH.

In addition to these recommendations, Table 9 provides a summary of our results. We observe that in many situations, the assignment decisions and outcomes are not strictly depending on just the NSTS score for other types of exits except PSH. For instances, those that were placed into RRH and their assignment outcome are based on their responses to the NST questionaries, demographic information, and communities.

Table 9. Observation summary. F = Family Exit, P = PSH Exit, R = RRH Exit, S = Self-resolve Exit, NSTS = NST Score, Beyond NSTS = Features Beyond NSTS

	F	P	R	S
Assignment decision	Beyond NSTS	NSTS	Beyond NSTS	NSTS
Assignment outcome	NSTS	NSTS	Beyond NSTS	Beyond NSTS

9 Conclusion, Discussion, and Future Work

This initial study has generated a great deal of interest from the housing communities and government agencies and a set of implementable public policy recommendations. Moving forward, there is much potential for creating decision aids to improve housing systems. There is room for effective human-machine collaboration in the context of housing allocation. Our analyses show that the local communities follow the current housing system guidelines closely for assigning youth to PSH. On the other hand, local communities have used additional criteria to assign youth to RRH. Our analyses also show that NST scores have a small negative correlation when youth are given housing interventions but a profound negative trajectory as NST score increases without intervention. This suggests that assigning youth based on NST score is an effective intervention for assisting high risk youth. As such, the current housing assignment systems are providing a much needed housing resource to youth who would otherwise not achieve stable living on their own.

Moving Forward. Assignment decisions based on NST scores, however, can be greatly augmented by additional predictive analytics. As shown in Sect. 7, we can potentially improve the current housing systems by providing better interpretable and explainable tools/classifiers to estimate a youth's probability of success for each possible type of exit (i.e. PSH, RRH, Family, Self Resolve). Given these probabilities, social workers in each community can decide more precisely which housing intervention (PSH or RRH) is best or whether a given youth is likely to successfully achieve stability without help from the system (family or self-resolution). Such information could do much to aid housing providers in making more informed decisions as to where to place a youth such that he/she is most likely to succeed. Moreover, providers may feel less anxiety about providing limited resources to some youth if they have information that suggests that a youth has a high probability of self-resolution or return to family. Thus, social service efforts can be focused more comfortably on those youth who are unlikely to succeed unless given more intensive resources such as PSH or RRH.

Our decision aids can further complement a human user. The ordered important features identified by the logistic regressions can serve as "red flags" for providers. For example, youth who have been abused or traumatized are less likely to be successful in PSH. This does not mean that providers should not place youth with such histories into PSH. Rather, additional supports, perhaps mental health treatment for trauma, are needed for youth with such a history.

Likewise, youth who are 17 or younger are much less likely to be able to succeed in RRH, suggesting that youth under 17 if given rental subsidies may need added attention beyond just the basic rental subsidy in order to improve their chances of remaining stably housed.

Based on these preliminary findings, youth housing systems are an ideal setting in which to further explore the potential for decision aid systems for social service providers. These two basic additions which we have outlined here, do much to enhance the vulnerability screening currently in place and could greatly aid humans in making difficult decisions about which youth to place in which housing programs, and which youth within those programs may need additional attention in order to thrive. In the future, by continuing to work with HUD and local communities we hope to build a decision aid system that will provide humans with enhanced predictive criteria for outcomes of housing placements for particular youth. Finally, we plan to provide useful interactive assistants, such as graphical user interfaces, to facilitate and encourage the collaboration between machine (i.e., our system) and human users in the community.

Lesson Learned. Many housing providers are resistant to using tools whether based on an index or machine learning that will decide on housing placements in an automated fashion. Housing is a critical resource that profoundly impacts the well-being of youth. People working in the communities that provide housing assistance to youth feel that humans must remain a part of the decision-making process [7,9]. Many current systems are often perceived as too rigid, and future systems must make room for human-machine interaction in decision-making.

Acknowledgments. This research was supported by MURI Grant W911NF-11-1-0332.

References

1. Breiman, L., Friedman, J., Stone, C., Olshen, R.: Classification and Regression Trees. The Wadsworth and Brooks-Cole Statistics-Probability Series. Taylor & Francis, London (1984). https://books.google.com/books?id=JwQx-WOmSyQC
2. Focus Strategies: Children's Hospital Los Angeles Youth Coordinated Entry System (CES) Final Evaluation Report (2017). http://focusstrategies.net/wp-content/uploads/2016/09/Updated-Final-Youth-CES-Report-042017.pdf
3. Housing and Urban Development (HUD): Coordinated Entry Policy Brief (2015). https://www.hudexchange.info/resources/documents/Coordinated-Entry-Policy-Brief.pdf
4. Housing and Urban Development (HUD): Coordinated Entry and Youth FAQs (2016). https://www.hudexchange.info/resources/documents/Coordinated-Entry-and-Youth-FAQs.pdf
5. Lipton, Z.C.: The mythos of model interpretability. CoRR (2016)
6. Ribeiro, M.T., Singh, S., Guestrin, C.: "Why should i trust you?": explaining the predictions of any classifier. In: Proceedings of the 22nd ACM SIGKDD International Conference on Knowledge Discovery and Data Mining, KDD 2016, pp. 1135–1144. ACM, New York (2016)

7. Rice, E.: Assessment Tools for Prioritizing Housing Resources for Homeless Youth (2017). https://static1.squarespace.com/static/56fb3022d210b891156b3948/t/5887e0bc8419c20e9a7dfa81/1485299903906/Rice-Assessment-Tools-for-Youth-2017.pdf
8. Tibshirani, R.: Regression shrinkage and selection via the lasso: a retrospective. J. Roy. Stat. Soc.: Ser. B (Stat. Methodol.) **73**(3), 273–282 (2011)
9. Toro, P.A., Lesperance, T.M., Braciszewski, J.M.: The heterogeneity of homeless youth in America: examining typologies. National Alliance to EndHomelessness, Washington, DC (2011)

Discovering Groups of Signals in In-Vehicle Network Traces for Redundancy Detection and Functional Grouping

Artur Mrowca[1,2]([✉]), Barbara Moser[1], and Stephan Günnemann[2]

[1] Bayerische Motoren Werke AG, Knorrstr. 147, 80788 Munich, Germany
{artur.mrowca,barbara.moser}@bmw.de
[2] Technical University of Munich, Boltzmannstr. 3, 85748 Garching, Germany
guennemann@in.tum.de

Abstract. Modern vehicles exchange signals across multiple ECUs in order to run various functionalities. With increasing functional complexity the amount of distinct signals grew too large to be analyzed manually. During development of a car only subsets of such signals are relevant per analysis and functional group. Moreover, historical growth led to redundancies in signal specifications which need to be discovered. Both tasks can be solved through the discovery of groups. While the analysis of in-vehicle signals is increasingly studied, the grouping of relevant signals as a basis for those tasks was examined less. We therefore present and extensively evaluate a processing and clustering approach for semi-automated grouping of in-vehicle signals based on traces recorded from fleets of cars.

Keywords: In-vehicle · Clustering · Signal · Redundancy detection

1 Introduction and Related Work

Modern vehicles communicate across multiple Electronic Control Units (ECUs) in order to run various functionalities. Those are implemented by multiple domains and are incrementally optimized throughout the development process of a car. With the growing demand for security, safety and entertainment more functionality is added and with this the complexity in in-vehicle networks increased. E.g. modern premium vehicles contain over 100 million lines of source-code onboard and, per function, up to 15 ECUs communicate with more than 2 million messages transmitted per minute. This communication between ECUs is defined via signals that are sent in defined messages. Signals resemble a dimension of information transmitted over time (e.g. GPS position).

In order to optimize and diagnose behavior in such in-vehicle systems during development, traces are recorded from in-vehicle networks of test vehicles and

This work was supported by the BMW Group.

U. Brefeld et al. (Eds.): ECML PKDD 2018, LNAI 11053, pp. 86–102, 2019.
https://doi.org/10.1007/978-3-030-10997-4_6

analyzed off-board (Fig. 1). One dimension of optimization is the refinement of specifications of the communication behavior of signals, as historical system growth led to redundancy in signal specifications. I.e. potentially identical and thus, redundant information is transmitted multiple times via distinct signals. This leads to more message collisions on in-vehicle networks, which result in loss of information or jam signals that can cause ECUs to fail. Thus, it is imperative to reduce the number of signals to improve safety and stability of in-vehicle systems. Further, the number and complexity of signals aggravates subsequent data analyses per domain, as it becomes increasingly difficult to identify signals that belong together and need to be analyzed jointly. E.g. analyzing the correct functioning of the wiper may only include signals, such as the rain sensor or the wiper position, while other signals are needless to consider.

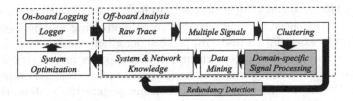

Fig. 1. Data is logged on-board and analyzed off-board. There functional groups and redundancy are detected. The scope of our contribution is marked grey here.

Finding such groupings is cumbersome and in general not possible to be done manually. Consequently, detection of redundant and interrelated signals needs to be performed automatically, e.g. by using unsupervised learning algorithms.

We therefore present and evaluate a Data Mining approach that allows for systematic clustering of interrelated signals using recorded in-vehicle network traces. Clustering in-vehicle signals is challenging due to several reasons. First, the signals to cluster are heterogeneous, i.e. data can be categorical, numerical, ordinal or binary. Massive amounts of traces are processed, with more than 10 million signals transmitted per minute. Also, different to classical scenarios the ratio of "number of target clusters" to "number of input samples" is high. Lastly, data is recorded as raw bytes which requires prior interpretation to achieve a data format that allows for subsequent clustering.

Related Work: A comparable preprocessing approach was introduced both in [2] and in [1]. However, in [2] features are extracted from multiple signals in order to classify them as normal or abnormal. Also, in [1] the focus is on finding causal relations between individual features of signals and fault types. Both, approaches group segments of signals, whereas we aim to group whole signals. Also, no heterogeneous, but rather numerical signals only are considered there. Grouping of signals was performed in [12], where supervised learning approaches were used to classify signals as internal (state of vehicle) and external (state of environment). However, for massive numbers of signals a supervised approach requires high labeling costs. To overcome this we investigate the possibility of an unsupervised scenario, where no labels are given. Also, we examine more than two target classes. Many Data Mining approaches where applied to in-vehicle signals, most of which are focused on diagnostics. In [8] diagnostic neural networks are trained for fault classification and in [4] induction motor drive faults are detected using recurrent dynamic neural networks. More recently diagnosis in in-vehicle signals was done for anomaly detection, e.g. by using condition indicators [14]. CAN signals were used for predictive maintenance [11]. In [13] vehicle signals are used to predict compressor faults, and in [16] to model the remaining useful life time of batteries in trucks.

Moreover, in-vehicle signals were used in applications, such as detection of scenarios [18] or driver workload monitoring [15].

Most existing Data Mining tasks are based on a subset of relevant signals. While Data Mining on in-vehicle signals is well studied, less attention was paid to the detection of those relevant signals. But, the growing numbers of existing signals makes it essential to find groups of signals before such techniques can be applied or to optimize their performance. This is, as investigation of irrelevant signals causes misclassification in diagnosis and increases computational complexity. We are the first ones to in-depth investigate the systematic grouping of in-vehicle signals for the purpose of functional grouping and redundancy detection.

Contributions: First, we present a concept that detects groupings of automotive signals. This is done by reducing data to relevant features which allows for local inspection of signals towards redundancy detection and domain-specific grouping. Second, we evaluated the approach using 10 real world data sets of different characteristics. With this the dependence of window size and selected features on the clustering performance is evaluated. Third, the performance of 9 clustering approaches for the task of in-vehicle signal segmentation was inspected formally and experimentally. We show that using our approach systematic analyses of in-vehicle signals is enabled. Lastly, we discuss the influence of cluster parameterization on the granularity of the grouping. Fine granularity results in better performance towards redundancy detection. Coarser parameterization is better for detection of interrelated signals, that affect common functions.

Concept Overview: Our approach for grouping interrelated signals is shown in Fig. 2. During *Preprocessing* raw traces are interpreted and prepared with the strategy proposed in [7]. Next, *Feature Engineering* is used to extract and reduce feature vectors per signal. Lastly, *Clustering* is used to find groupings.

2   Preparing Traces

In this section first, automotive traces are introduced. Next, grouping of signals from such traces requires data extraction, preprocessing and feature engineering steps which are presented in the second part of this section.

2.1   Automotive Traces

Cars implement several functionalities. Those require ECUs, sensors and actuators to exchange information across its internal vehicle network (e.g. CAN bus). This information can be decoded to yield multivariate time-series of signals. I.e. each dimension represents a defined information in time. This is called a signal and shown in Fig. 2 *ii*. Signals can be categorical (e.g. car mode: driving/parking), binary (e.g. engine: on/off), ordinal (e.g. heating: level 3) or numerical (e.g. speed). The incremental growth of vehicular systems results in an increasing number of signals. E.g. one vehicle may contain up to ten thousand CAN signals. Manual grouping of interrelated signals has become intractable, which requires automated unsupervised algorithms as presented in this work.

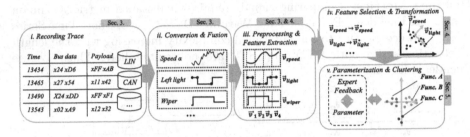

Fig. 2. Overview of the Data Mining approach with the respective sections detailing the process stated.

2.2   Data Extraction and Cleaning

Data Extraction: Data is ingested as time stamped byte sequence and is converted to signals as shown in Fig. 2 *ii*. It contain billions of entries per vehicle. Those are stored in large-scale distributed database systems, such as Apache Hive and processed with distributed engines such as Apache Spark [3].

Data Cleaning: First, raw signals contain invalid entries which may result from invalid network states, such as jammed signals or defect ECUs. Those invalid entries are identified using network specifications and are dropped. Second, depending on its data types signals require different features to be extracted. We thus categorize signals as numerical, if more then a threshold number of values in that signal is numeric and as categorical else. Third, missing data is replaced via interpolation if data is numeric and via repetition of the last valid signal value else. Fourth, numerical data is smoothed to reduce noise resulting from disturbances in the network, e.g. using Exponential Moving Average filtering. Lastly, for better comparability numerical signals need to be within a common value range. Several approaches can be used for this. However, we found Interquartile Range normalization to achieve best comparability as the resulting signals allow to compare the shape of the signals rather than their absolute values.

3   Feature Engineering

Here we describe feature extraction per signal, identification and transformation of features and formal evaluation of clustering algorithms for our scenario.

3.1   Feature Extraction

In the given scenario we assume that interrelated signals occur within common time ranges and change their value in similar time intervals. To capture such temporal-causal dependence in a feature vector, we propose the following approach.

Distance Metrics: Comparing signals requires a distance metric. Common metrics are Euclidean, Dynamic-Time-Warping (DTW) [10] or Short Time Series (STS) distance [6]. Due to computational complexity increasing with the length of the time-series both DTW and STS are not suited here. Thus, we use Euclidean distance.

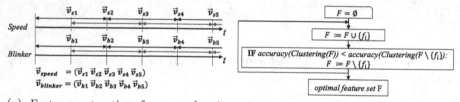

(a) Feature extraction from overlapping windows.

(b) Feature selection using forward backward search.

Fig. 3. Feature extraction and forward backward feature selection approach.

Extraction Approach: As shown in Fig. 3a, all signals are sliced according to overlapping windows. Per window w_i and per signal s_i a sub-feature vector $v_i = f(w_i, s_i)$ is extracted using $f : (w, s) \mapsto v$ and the signals data type $type(s_i)$. Signals s_i can be numerical $type(s_i) = num$ or categorical $type(s_i) = cat$. Depending on this data type, different characteristics (i.e. features) are extracted to represent the value behavior of the corresponding signal s_i. Thus, for each s_i the following features are extracted according to its $type(s_i)$. If $type(s_i) = num$, it is required to capture the shape of the signal per windows. This is done per window with features *mean, variance, skew, arch,* the *magnitude of the average slope, variance of the slope, maximal slope,* the *mean of the SAX symbol occurrences* and the *wavelet coefficients*. Those numerical features were chosen as they were successfully applied in similar tasks in [9] and [17].

If $type(s_i) = cat$ only information about occurrences and their value is available and thus, need to be extracted as appropriate features. In order to make numerical signals $(type(s_i) = num)$comparable to categorical once, numerical signals are symbolized in value using Symbolic Aggregate approXimation (SAX)

[5]. Next, per signal identical repetitions are removed, which allows the numerical signals to be additionally treated as categorical. On those categorical and discretized numerical signals extracted features are the *number of times a value changed*, the *number of times a value occurred* and the *change ratio per window*, which is the number of value changes divided by the number of samples per window. Thus, the change ratio is the weight of a window, i.e. the amount of change that occurred in it.

Choosing those features allows to compare nominal and numeric signals, while comparison among numerical signals is done on a more fine grained level using its numerical features. To now represent a signal s_i with identifier m, sliced in n windows, as a feature vector v_m, all subvectors v_{mi} are stacked as $v_m = (v_{m1}v_{m2}...v_{mi}...v_{mn})$. This representation captures temporal interrelation, as same dimensions represent same windows and value behavior is represented by each value in a dimension.

3.2   Wrapper-Based Feature Selection

A classical approach to determine most important features is to use forward-backward search in a wrapper-based evaluation. I.e. the quality of a subset of features is evaluated on a validation data set using the clustering target (e.g. redundancy or function grouping) it is optimized for. As clustering is unsupervised, in this step a ground truth was defined for the training set, which is done manually by experts. As a optimization target the precision is used, i.e. ratio of signals that were correctly clustered (according to the expert).

With this, per data set a feature subset with maximal precision is found. To avoid overfitting the search is run on several data sets with various characteristics (e.g. ratio of numeric to categorical signals), yielding an optimal feature subset per data set. Next, all features are ranked by counting subsets that contain this feature, which ranks more general features that are valid for more data sets higher. The top ranked features are used for further processing (e.g. top 50%) and are extracted per window. This process is depicted in Fig. 3b.

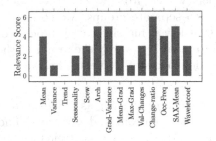 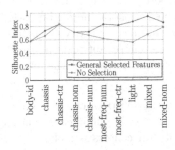

(a) Relevance score determined as number of optimal feature subsets in which a feature occurred.

(b) Clustering performance in terms of Silhouette index before and after the generalized feature selection is applied.

Fig. 4. Results of the experiments for the feature selection used.

3.3 Feature Transformation

The resulting feature vector is of high dimension, as for n windows and f features, the vector has $n \cdot f$ dimensions. However, the curse of dimensionality states the problem that with higher dimensions of feature vectors, feature vectors appear further away in terms of distance measures. Also, higher dimensions are computationally more expensive during clustering. That is why the number of dimensions needs to be reduced. For this we apply a two step approach.

First, per dimension the variance is computed to determine its amount of information. Low variance indicates less information per feature and window e.g. if a signal did not occur in a window. Therefore dimensions with variance smaller than a threshold are dropped. Second, dimension is reduced to a information maximizing space with a Principal Component Analysis (PCA). As PCA is a linear transformation, inherent properties of each signal vector are conserved (e.g. Euclidean Distance). Also, only the most informative dimensions are used. The transformed feature vector is used for clustering.

4 Clustering

The nature of in-vehicle network signals renders not all clustering approaches equally applicable for the given use-case of redundancy and correlation detection. Therefore, we investigate the suitability of multiple approaches next.

4.1 Properties and Approaches

Desirable Properties: Signals are heterogeneous, of huge size and yield very less signals per target clusters. Furthermore, even after reduction, data is of high dimension due to highly complex characteristics per signal. Consequently, a desirable property for redundancy detection is the possibility to parameterize the approach towards clusters of certain levels of granularity. E.g. at a higher level of granularity per wheel the four sensors of the rotational frequency should be grouped, while at a lower level both left wheel (front & back) and both right wheel sensors should be assigned to two separate groups. Moreover, the computational complexity needs to be still kept low as massive data is processed. Lastly, it is important to visualize the data to be able to verify the results of the clustering and to inspect the level of correlation between elements of clusters.

Clustering Approaches: Centroid-based clustering, such as k-Means and k-Medoids, represent classes relatively to a class centroid and iteratively find such optimal centers per class. In Expectation Maximization centroids can be Gaussian probability distributions, which are optimally fit over given data. In hierarchical clustering approaches groups are sequentially decomposed, which can be done bottom-up or top-down. Density based approaches (e.g. Density-Based Spatial Clustering of Applications with Noise (DBSCAN)) group data by assigning a radius around a data point and grouping overlapping neighborhoods. Raster-based approaches include WaveCluster, where a raster is defined and points in the same raster are grouped. In Affinity propagation data points are passing messages which indicate the affinity to its neighbors to determine clusters. In Self-organizing maps (SOM) the space is divided in hexagons which approach each other iteratively to form groupings.

4.2 Suitability Analysis

Clustering approaches are not equally well suited to fulfill the properties stated in Subsect. 4.1 and are a trade-off between those. We compared the most important approaches formally in terms of their applicability to in-vehicle signal clustering with results shown in Table 1 and discussed here. An experimental evaluation of those approaches is given in Sects. 5 and 6.

Centroid-Based: Granularity is settable as target clusters k. k-Means is in general suited for high dimensional data as prototypes are found as mean of all clusters and a separation is forced through k. But, only spherical clusters are possible which is contrary to signal feature vectors which can be grouped in any shape. k-Medoids and EM are less suited. In k-Medoids samples are part of the data set which shifts the centroid on a data point and thus, imbalances the center.

Hierarchical: Such approaches are independent of shape, as successive splitting or joining is performed based on neighborhoods. But, top-down clustering tends to split the biggest cluster more often. This results in many clusters of similar size which is not the target grouping in our scenario where cluster sizes may vary. Granularity can be parameterized on according splitting and joining rules.

Density-Based: Those approaches allow for multiple granularity by setting the radius per data point, while they are independent of shape as neighboring elements are found using the radius. This radius can exist in any dimension leaving this approach to be well suited for clustering of signals.

Grid-Based: Those approaches allow for multiple granularity by setting the raster size and are independent of shape as the raster can be of any shape. Above that, high dimensional clustering is possible with the limitation that dimensions need to be restricted as e.g. in WaveCluster similar Wavelet coefficients will be too far away to be assigned in one clusters (due to curse of dimensionality). With the reduction to a sufficient number of dimensions and its low computational complexity those approaches are well suited.

Affinity Propagation: Here prototypes are data points themselves, leading to similar imbalance as in k-Medoids. However, common grouping is not dependent on cluster shape as the totality of points is considered for clustering.

Table 1. Comparison of algorithms in clustering of in-vehicle signals. I.e. handling high-dim. data, detect clusters of any shape, allow multiple granularities of clusters, visual representation and computational complexity, with t iterations, maximal depth d, n examples and k classes.

Approach	High-dim.	Complex-shapes	Mult. Gran.	Visualization	Complexity
k-Means	Yes	No	Yes	No	$\mathcal{O}(nkt)$
k-Medoids	No	No	Yes	No	$\mathcal{O}(k(n-k)^2 * t)$
EM	No	No	Yes	No	$\mathcal{O}(nk * t)$
DBSCAN	Yes	Yes	Yes	No	$\mathcal{O}(n \log n)$
Agglomerative	Yes	Yes	Yes	Dendrogram	$\mathcal{O}(n^3)$
Top-down	No	Yes	Yes	Dendrogram	$\mathcal{O}(2^d * nkt)$
WaveCluster	Indirect	Yes	Yes	No	$\mathcal{O}(n)$
Affinity propagation	No	Yes	No	No	$\mathcal{O}(n^2)$
SOM	Indirect	Yes	Yes	Map	$\mathcal{O}(n * t)$

Self-organizing Maps: Due to small numbers of signals each hexagon is sparsely populated by data points making cluster detection difficult.

According to this formal evaluation we expect WaveCluster and DBSCAN to be most suited for the identification of groupings among in-vehicle signals.

5   Evaluation

Preprocessing and Clustering of in-vehicle signals requires appropriate parameterization in terms of windowing, feature engineering and selection of clustering algorithms. This is studied in this section.

5.1   Experimental Setup

Environment: Due to large raw traces, preprocessing and feature extraction were implemented on a cluster with 70 servers in Apache Spark. With the reduced data set the remaining steps (selection, transformation, clustering) were performed locally on a 64-Bit Windows 7 PC with an Intel® Core™i5-4300U processor and 8 GB of RAM using RapidMiner Studio, Python's Data Mining stack and R.

Datasets: The statistics of our data sets are shown in Table 2. To cover most characteristics of automotive in-vehicle network traces we evaluated the approach on 10 test data sets that are different in terms of signal types (e.g. chassis-nom vs. chassis-num), data points per type, signal number, association to one (e.g. chassis) or multiple (e.g. mixed) functions and resemble different excerpts of a journey. The target of our evaluation is the grouping of signals in terms of their assignment to similar functions. All approaches were parameterized per data set such that the true number of clusters is achieved and the best possible grouping (according to the expert) within this clustering is reached.

Table 2. Statistics of the datasets: total number and proportions of numerical and nominal signals, data points per set, recorded part of journey. Here, small subsets are used for evaluation, while in practice thousands of signals are considered.

Set	Signals (tot[num/nom])	Datapoints (tot[num/nom])	Part of journey
body-id	38 [1/37]	2251 [89/2162]	Complete
chassis	53 [18/35]	9999 [9896/103]	Start
chassis-nom	35 [0/35]	9896 [0/9896]	Start
chassis-num	18 [18/0]	103 [103/0]	Start
chassis-ctr	12 [11/1]	10000 [9999/1]	Mid
most-freq-num	24 [24/0]	12508 [12508/0]	Start
most-freq-ctr	22 [19/3]	11773 [11765/8]	Mid
light	39 [6/33]	10055 [2941/7114]	Start
mixed	25 [12/13]	69402 [69339/63]	Start
mixed-nom	13 [0/13]	9509 [0/9509]	Start

Evaluation Criteria: The approach is evaluated in terms of clustering quality. *Accuracy:* Accuracy is the number of samples $n_{correct}$ correctly clustered in relation to the total number of samples in the data set $n_{dataset}$ given as $acc = \frac{n_{correct}}{n_{dataset}}$. Here, the assignment of reference cluster labels to each signal as a ground truth is done manually by experts. *Silhouette index $s(i)$:* Measures a clustering assignment per data point i in terms of degree of affinity to its assigned cluster relatively to all other clusters. I.e. $a(i)$ as distance of i to all element within its cluster, $b(i)$ as average distance to all data points in all other clusters. It is optimal for $s(i) = 1$ and defined as $s(i) = \frac{b(i)-a(i)}{max\{a(i),b(i)\}}$.

5.2   Window Size

Setup: After preprocessing we split each signal in windows with 50% overlap, extract all features, transform them and perform clustering. Per data set the window size is increased successively from 0.1 s to 5000 s and the performance is measured in terms of accuracy. From this we identify the window size with highest accuracy as optimal. We evaluate k-means for clustering here, while other approaches yielded similar results. The results are shown in Table 3.

Results: If the window is too small patterns relevant for features are overseen, while for big windows feature details are simplified away. Also, as can be seen in Table 3 less frequently changing signals, e.g. with a higher number of nominal signals, require bigger windows , e.g. in body-id, light and mixed-nom, as those signals do change less often. If more frequently changing numerical signals need to be clustered smaller windows appear to be optimal which is the case in chassis, chassis-ctr, most-freq-ctr and mixed.

Table 3. Experimentally determined optimal windows per data set in seconds.

body-id	chassis	chassis-ctr	most-freq-ctr	light	mixed	mixed-nom
128.8	3.5	79.8	1.8	533.6	1.5	2147.7

5.3 Feature Selection

Setup: Feature selection is performed as described in Sect. 3.2 where feature subsets are successively searched and evaluated once per data set using clustering performance as optimization target. To find features that generalize over all data sets, we measured the number of times a feature was included in the optimal feature subset. k-Means was again used for clustering. The results are shown in Fig. 4a. The performance gain of our generalized feature selection was measured before and after the ranking selection, with results shown in Fig. 4b.

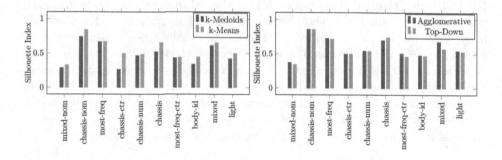

(a) Comparison of centroid-based algorithms in terms of Silhouette index.

(b) Comparison of hierarchical algorithms in terms of Silhouette index.

Fig. 5. Results of experiments for centroid and hierarchical approaches.

Results: It can be seen that for the numerical characteristics best features are the mean, skew, arch, as well as the variance and magnitude of the gradient. This shows that the fine granularity of numerical signal characteristics requires to capture noise, value and shape characteristics. For nominal characteristics all nominal features were suited. This shows that the frequency and type of a nominal/discretized numerical signal can be captured. Further, this resembles our assumption that in-vehicle signals are correlated, when they occur and change their value together. As Fig. 4b depicts a performance gain of up to 20% (e.g. at light data set) is achieved with this approach. Notably, all data sets show an improvement after the generalized selection.

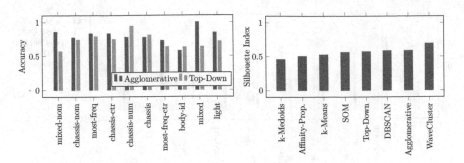

(a) Comparison of hierarchical algorithms in terms of Accuracy.

(b) Average Silhouette index over all data sets.

Fig. 6. Results of experiments for hierarchical approaches and all data sets.

5.4   Comparison of Clustering Algorithms

As stated in Sect. 4 the characteristics of in-vehicle signals require clustering algorithms that can handle high-dimensionality, different granularities and have low computational complexity.

Setup: To examine the suitability of different algorithms for grouping of in-vehicle signals, we evaluated k-Means, k-Medoids, DBSCAN, Agglomerative, WaveCluster and SOM clustering approaches on all data sets in terms of clustering quality. This is done by using the optimal feature subset as selected by our feature selection approach, parameterization with expert feedback and by consequent application of the clustering approaches.

We first compared Agglomerative against Top-Down clustering and k-Means against k-Medoids, to evaluated the characteristics of those sub types in terms of applicability to in-vehicle signals. This is followed by a general experimental comparison of all approaches.

Results - Sub Types: As illustrated in Fig. 5a, among centroid-based approaches k-means performs better than k-Medoids. This is, as taking the mean among signals for clustering avoids a shifting bias as stated in Sect. 4.

Results - Hierarchical: Among hierarchical approaches Agglomerative clustering results in better accuracy in 90% and in better Silhouette index in 70% of all cases which is shown in Figs. 5b and 6a. The best centroid-based and hierarchical approaches are evaluated with further clustering approaches giving results shown in Figs. 6b and 7.

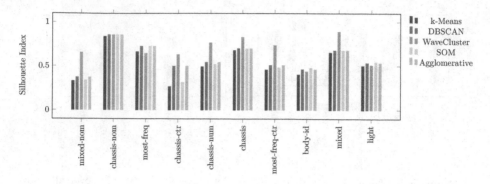

Fig. 7. Silhouette index per data set and clustering algorithm.

Results - Overall: As depicted in Fig. 6b, DBSCAN, Agglomerative clustering and WaveCluster works best if a data set contains mixed characteristics (i.e. different signal types, proportions of nominal to numerical, etc.) combined. Also, in those cases centroid-based approaches perform worse. This confirms our expectations and formal analysis of the approaches in Sect. 4. Further, as depicted in Fig. 7, WaveCluster performs best on 80% of all data sets and shows solid results in the remaining 20%. Thus, this approach seems best suited for our scenario. This is because extraction of Wavelet coefficients enables to well capture both fine and coarse grained properties of signals equally. Also, as described in Sect. 4 WaveCluster can well represent the shape and the data's high dimension.

Similarly, DBSCAN and Agglomerative Clustering are well suited to capture those properties. However, the latter approach is biased in that it tends to find clusters of nearly similar sizes which is not given in all test sets.

As deduced in Sect. 4 SOM and k-means perform slightly worse, as dimensions are reduced in SOMs and k-means cannot capture varying cluster shapes.

Conclusion: All clustering approaches have solid results in terms of cluster quality. This shows that the proposed processing and clustering approach is well suited for groupings of in-vehicle network signals. WaveCluster and DBSCAN perform best due to their ability to capture most of the heterogeneous characteristics included in such signals. As described an optimal window size depends on the structure of the processed data and thus, needs to be determined. Further, features subsets as discussed in Sect. 5.3 allow for good generalization when clustering in-vehicle signals.

6   Case Study

In this section we exemplary show how our approach can be used to detect signal redundancies and group signals of common functions.

Setup: For this case study a realistic data set was used. After the preprocessing of Sect. 2 this data set contains 419 signals and (after reduction) 20 026 065 data points recorded from one vehicle over eight days. This processing is again implemented on the Hadoop system described in Sect. 5.1, while the resulting reduced data is processed locally. *Preparation:* An optimal window size of 17.7 s was found with 7 477 windows of 50% overlap. Per window the features found in Sect. 5.3 were used resulting in more than 10 000 dimensions per signal. Reduction to less dimensions is done by filtering for dimensions with a variance bigger 0.3 and a successive PCA, resulting in 80 dimensions per signal which can be used for local clustering. For clustering we used DBSCAN.

Analogy Detection by Cluster Inspection: Depending on the parameterization of the clustering, granularity of the target can be set. I.e. if redundancies need to be detected a more fine grained target parameterization is required, while the opposite holds for grouping according to functions. This is illustrated in Fig. 8b where a coarse grouping separates signals with different data types and finer clustering extracts signals of similar functions. Finding an appropriate granularity is done through expert feedback. The extracted clusters can be inspected and successively parameterized towards a good target clustering. Experts can then asses the grouping results, e.g. decide whether a grouping signifies a redundancy.

Results: With our approach redundancies and related signals were found in the analyzed data set. E.g. we found redundancy among speed signals and signals representing the time. The further were the speed signal for the speedometer, the state of the speed in horizontal direction and the speed of the car's

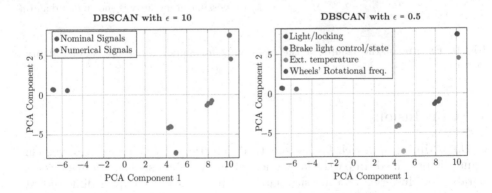

(a) Result of signal clustering with DB-SCAN and $\epsilon = 10$

(b) Result of clustering with DBSCAN and $\epsilon = 0.5$.

Fig. 8. Clustering with DBSCAN at different granularities by varying ϵ. The legend shows the signals which were grouped. E.g. the locking signal turns on the light when the car is closed.

mass center. Those are all identical as they measure the vehicle speed and thus, can be reduced to one signal in future architectures of the vehicle. An example for detected groups of similar functions are signals related to the braking function which were grouped (see Fig. 9b, red cluster). It shows that the brake light state, state of the driver braking, braking momentum on the wheels and the target braking momentum resulting from the driver pressing the pedal are grouped. In particular as Fig. 9a shows, with our approach nominal signals were grouped together with related numerical signals. Further examples of discovered functional groups are signals for automated parking (e.g. parking space, driver intervention), battery state (e.g. battery capacity, state of charge) or constant signals (e.g. air pressure, state of the belt buckle). Thus, the proposed approach is well suited to find signals of common functionality, which in turn enables successive domain-specific analyses of relevant signals and Data Mining applications on related signals.

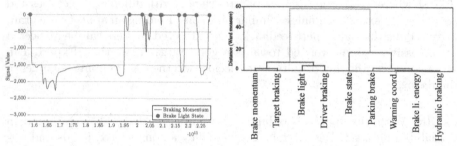

(a) The numerical signal "Braking Momentum" and the nominal signal "Brake Light State" are grouped with our approach.

(b) Dendrogram illustrating hierarchical clustering at various granularities. I.e. branches resemble possible groupings. E.g. one possible granularity is shown in red and brown.

Fig. 9. Excerpt of the results of the case-study for signal clustering. (Color figure online)

7   Conclusion

A Data Mining approach for systematical detection of groupings of in-vehicle signals is presented. In particular numerical and nominal signals are made comparable for clustering and massive data is reduced early to a lower dimensional space. We evaluated the optimal window size, a general feature subset and the suitability of different clustering algorithms for clustering of in-vehicle signals. A case-study showed that redundancies and signal groupings can successfully be found with this approach.

Lessons Learnt: We found the optimal window size depends on the character of the data set and on the features. We showed that grouping heterogeneous signals is possible by assigning nominal features to numerical signals. To handle massive traces, extraction, selection and transformation steps are performed on a cluster, while locally a tractable lower dimensional vector is analyzed with expert feedback. We formally and experimentally demonstrated, that automated clustering of in-vehicle networks is possible. There both fine and coarse grained structure needs to be captured, which is best possible with WaveCluster. The introduced approach allows for future automation of tasks like anomaly detection or situation detection. System optimization is enabled by detection of redundancy and understanding which signals are sent jointly. Future work includes reduced Feature Engineering or further automation of the process. Also, a knowledge base could be designed to capture groupings, that are verified by experts and clustering optimized by exploiting this base. Further parameter evaluation needs to be performed to get a deeper insight on the mapping between parameters and redundancy or functional grouping. With the presented clustering approach we set the basis for future in-vehicle signal analysis of modern vehicles.

References

1. Crossman, J., Guo, H., Murphey, Y., Cardillo, J.: Automotive signal fault diagnostics. I. Signal fault analysis, signal segmentation, feature extraction and quasi-optimal feature selection. IEEE Trans. Veh. Technol. **52**(4), 1063–1075 (2003). https://doi.org/10.1109/tvt.2002.807635
2. Guo, H., Crossman, J., Murphey, Y., Coleman, M.: Automotive signal diagnostics using wavelets and machine learning. IEEE Trans. Veh. Technol. **49**(5), 1650–1662 (2000). https://doi.org/10.1109/25.892549
3. Hadoop, A.: Hadoop (2009)
4. Kim, K., Parlos, A.: Induction motor fault diagnosis based on neuropredictors and wavelet signal processing. IEEE/ASME Trans. Mechatron. **7**(2), 201–219 (2002). https://doi.org/10.1109/tmech.2002.1011258
5. Lin, J., Keogh, E., Lonardi, S., Chiu, B.: A symbolic representation of time series, with implications for streaming algorithms. In: Proceedings of the 8th ACM SIGMOD Workshop on Research Issues in Data Mining and Knowledge Discovery - DMKD 2003. ACM Press (2003). https://doi.org/10.1145/882082.882086
6. Möller-Levet, C.S., Klawonn, F., Cho, K.-H., Wolkenhauer, O.: Fuzzy clustering of short time-series and unevenly distributed sampling points. In: R. Berthold, M., Lenz, H.-J., Bradley, E., Kruse, R., Borgelt, C. (eds.) IDA 2003. LNCS, vol. 2810, pp. 330–340. Springer, Heidelberg (2003). https://doi.org/10.1007/978-3-540-45231-7_31
7. Mrowca, A., Pramsohler, T., Steinhorst, S., Baumgarten, U.: Automated interpretation and reduction of in-vehicle network traces at a large scale. In: Proceedings of the 55th Annual Design Automation Conference - DAC 2018. ACM Press (2018). https://doi.org/10.1145/3195970.3196000
8. Murphey, Y.L., Masrur, M., Chen, Z., Zhang, B.: Model-based fault diagnosis in electric drives using machine learning. IEEE/ASME Trans. Mechatron. **11**(3), 290–303 (2006). https://doi.org/10.1109/tmech.2006.875568

9. Nanopoulos, A., Alcock, R., Manolopoulos, Y.: Feature-based classification of time-series data. Int. J. Comput. Res. **10**(3), 49–61 (2001)
10. Niennattrakul, V., Ratanamahatana, C.A.: On clustering multimedia time series data using k-means and dynamic time warping. In: 2007 International Conference on Multimedia and Ubiquitous Engineering (MUE 07). IEEE (2007). https://doi.org/10.1109/mue.2007.165
11. Nowaczyk, S., Prytz, R., Byttner, S.: Ideas for fault detection using relation discovery. In: The 27th Annual Workshop of the Swedish Artificial Intelligence Society (SAIS), 14–15 May 2012, Örebro, Sweden, no. 071, pp. 1–6, Linköping University Electronic Press (2012)
12. Prytz, R., Nowaczyk, S., Byttner, S.: Towards relation discovery for diagnostics. In: Proceedings of the First International Workshop on Data Mining for Service and Maintenance - KDD4Service 2011. ACM Press (2011). https://doi.org/10.1145/2018673.2018678
13. Prytz, R., Nowaczyk, S., Rgnvaldsson, T., Byttner, S.: Predicting the need for vehicle compressor repairs using maintenance records and logged vehicle data. Eng. Appl. Artif. Intell. **41**, 139–150 (2015). https://doi.org/10.1016/j.engappai.2015.02.009
14. Raptis, I.A., et al.: A particle filtering-based framework for real-time fault diagnosis of autonomous vehicles. In: Annual Conference of the Prognostics and Health Management Society (2013)
15. Taylor, P., Griffths, N., Bhalerao, A., Popham, T., Zhou, X., Dunoyer, A.: Redundant feature selection for telemetry data. In: Cao, L., Zeng, Y., Symeonidis, A.L., Gorodetsky, V., Müller, J.P., Yu, P.S. (eds.) ADMI 2013. LNCS (LNAI), vol. 8316, pp. 53–65. Springer, Heidelberg (2014). https://doi.org/10.1007/978-3-642-55192-5_5
16. Voronov, S., Jung, D., Frisk, E.: Heavy-duty truck battery failure prognostics using random survival forests. IFAC-PapersOnLine **49**(11), 562–569 (2016). https://doi.org/10.1016/j.ifacol.2016.08.082
17. Wang, X., Smith, K.A., Hyndman, R.J.: Dimension reduction for clustering time series using global characteristics. In: Sunderam, V.S., van Albada, G.D., Sloot, P.M.A., Dongarra, J. (eds.) ICCS 2005. LNCS, vol. 3516, pp. 792–795. Springer, Heidelberg (2005). https://doi.org/10.1007/11428862_108
18. Zheng, H., Zhang, H., Meng, H., Wang, X.: Qualitative modeling of vehicle behavior for scenario parsing. In: 2006 IEEE Intelligent Transportation Systems Conference. IEEE (2006). https://doi.org/10.1109/itsc.2006.1706813

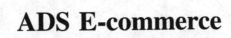

ADS E-commerce

SPEEDING Up the Metabolism in E-commerce by Reinforcement Mechanism DESIGN

Hua-Lin He(✉)[iD], Chun-Xiang Pan[iD], Qing Da[iD], and An-Xiang Zeng[iD]

Alibaba Inc., Zone Xixi, No. 969, West WenYi Road, Hangzhou 310000, China
{hualin.hhl,daqing.dq}@alibaba-inc.com, {xuanran,renzhong}@taobao.com
https://www.taobao.com

Abstract. In a large E-commerce platform, all the participants compete for impressions under the allocation mechanism of the platform. Existing methods mainly focus on the short-term return based on the current observations instead of the long-term return. In this paper, we formally establish the lifecycle model for products, by defining the *introduction*, *growth*, *maturity* and *decline* stages and their transitions throughout the whole life period. Based on such model, we further propose a reinforcement learning based mechanism design framework for impression allocation, which incorporates the first principal component based permutation and the novel experiences generation method, to maximize short-term as well as long-term return of the platform. With the power of trial-and-error, it is possible to recognize in advance the potentially hot products in the introduction stage as well as the potentially slow-selling products in the decline stage, so the metabolism can be speeded up by an optimal impression allocation strategy. We evaluate our algorithm on a simulated environment built based on one of the largest E-commerce platforms, and a significant improvement has been achieved in comparison with the baseline solutions. Code related to this paper is available at: https://github.com/WXFMAV/lifecycle_open.

Keywords: Reinforcement learning · Mechanism design · E-commerce

1 Introduction

Nowadays, E-commerce platform like Amazon or Taobao has developed into a large business ecosystem consisting of millions of customers, enterprises and start-ups, and hundreds of thousands of service providers, making it a new type of economic entity rather than enterprise platform. In such a economic entity, a major responsibility of the platform is to design economic institutions to achieve various business goals, which is the exact field of *Mechanism Design* [23]. Among all the affairs of the E-commerce platform, impression allocation is one of the key

Supported by organization Alibaba-Inc.

U. Brefeld et al. (Eds.): ECML PKDD 2018, LNAI 11053, pp. 105–119, 2019.
https://doi.org/10.1007/978-3-030-10997-4_7

strategies to achieve its business goal, while products are players competing for the resources under the allocation mechanism of the platform, and the platform is the game designer aiming to design game whose outcome will be as the platform desires.

Existing work of impression allocation in literature are mainly motivated and modeled from a perspective view of supervised learning, roughly falling into the fields of information retrieval [2,6] and recommendation [10,14]. For these methods, a Click-Through-Rate (CTR) model is usually built based on either a ranking function or a collaborative filtering system, then impressions are allocated according to the CTR scores [8]. However, these methods usually optimize the short-term clicks, by assuming that the properties of products is independent of the decisions of the platform, which may hardly hold in the real E-commerce environment. There are also a few work trying to apply the mechanism design to the allocation problem from an economic theory point of view such as [16,17,19]. Nevertheless, these methods only work in very limited cases, such as the participants play only once, and their properties is statistically known or doesn't change over time, etc., making them far from practical use in our scenario. A recent pioneer work named *Reinforcement Mechanism Design* [22] attempts to get rid of nonrealistic modeling assumptions of the classic economic theory and to make automated optimization possible, by incorporating the Reinforcement Learning (RL) techniques. It is a general framework which models the resource allocation problem over a sequence of rounds as a Markov decision process (MDP) [18], and solves the MDP with the state-of-the-art RL methods. However, by defining the impression allocation over products as the action, it can hardly scale with the number of products/sellers as shown in [3,4]. Besides, it depends on an accurate behavioral model for the products/sellers, which is also unfeasible due to the uncertainty of the real world.

Although the properties of products can not be fully observed or accurately predicted, they do share a similar pattern in terms of development trend, as summarized in the *product lifecycle theory* [5,11]. The life story of most products is a history of their passing through certain recognizable stages including *introduction, growth, maturity* and *decline* stages.

- *Introduction*: Also known as *market development* - this is when a new product is first brought to market. Sales are low and creep along slowly.
- *Growth*: Demand begins to accelerate and the size of the total market expands rapidly.
- *Maturaty*: Demand levels off and grows.
- *Decline*: The product begins to lose consumer appeal and sales drift downward.

During the lifecycle, new products arrive continuously and outdated products wither away every day, leading to a natural metabolism in the E-commerce platform. Due to the insufficient statistics, new products usually attract few attention from conventional supervised learning methods, making the metabolism a very long period.

Inspired by the product lifecycle theory as well the reinforcement mechanism design framework, we consider to develop reinforcement mechanism design while taking advantage of the product lifecycle theory. The key insight is, with the power of trial-and-error, it is possible to recognize in advance the potentially hot products in the introduction stage as well as the potentially slow-selling products in the decline stage, so the metabolism can be speeded up and the long-term efficiency can be increased with an optimal impression allocation strategy.

We formally establish the lifecycle model and formulate the impression allocation problem by regarding the global status of products as the state and the parameter of a scoring function as the action. Besides, we develop a novel framework which incorporates a first principal component based algorithm and a repeated sampling based experiences generation method, as well as a shared convolutional neural network to further enhance the expressiveness and robustness. Moreover, we compare the feasibility and efficiency between baselines and the improved algorithms in a simulated environment built based on one of the largest E-commerce platforms.

The rest of the paper is organized as follows. The product lifecycle model and reinforcement learning algorithms are introduced in Sect. 2. Then a reinforcement learning mechanism design framework is proposed in Sect. 3. Further more, experimental results are analyzed in Sect. 4. Finally, conclusions and future work are discussed in Sect. 5.

2   Preliminaries

2.1   Product Lifecycle Model

In this subsection, we establish a mathematical model of product lifecycle with noises. At step t, each product has an observable attribute vector $x_t \in \mathbb{R}^d$ and an unobservable latent lifecycle state $z_t \in \mathcal{L}$, where d is the dimension of the attribute space, and $\mathcal{L} = \{0, 1, 2, 3\}$ is the set of lifecycle stages indicating the the *introduction, growth, maturity* and *decline* stages respectively. Let $p_t \in \mathbb{R}$ be the CTR and $q_t \in \mathbb{R}$ be the accumulated user impressions of the product. Without loss of generality, we assume p_t and q_t are observable, p_t, q_t are two observable components of x_t, the platform allocates the impressions $u_t \in \mathbb{R}$ to the product. The dynamics of the system can be written as

$$\begin{cases} q_{t+1} = q_t + u_t \\ p_{t+1} = p_t + f(z_t, q_t) \\ z_{t+1} = g(x_t, z_t, t) \end{cases} \tag{1}$$

where f can be seen as the derivative of the p, and g is the state transition function over $\mathcal{L}$.

According to the product lifecycle theory and online statistics, the derivative of the CTR can be formulated as

$$f(z_t, q_t) = \begin{cases} \dfrac{(c_h - c_l)e^{-\delta(q_t)}}{(2-z)(1+e^{-\delta(q_t)})^2} + \xi, & z \in \{1,3\} \\ \xi, & z \in \{0,2\} \end{cases} \tag{2}$$

where $\xi \sim \mathcal{N}(0, \sigma^2)$ is a gaussian noise with zero mean and variance σ^2, $\delta(q_t) = (q_t - \tilde{q}_{tz} - \delta_\mu)/\delta_\sigma$ is the normalized impressions accumulated from stage z , $\tilde{q}_{tz}$ is the initial impressions when the product is firstly evolved to the life stage z, $\delta_\mu, \delta_\sigma$ are two unobservable parameters for normalization, and $c_h, c_l \in \mathbb{R}$ are the highest CTR and the lowest CTR during whole product lifecycle, determined through two neural networks, respectively:

$$c_l = h(x_t|\theta_l), \quad c_h = h(x_t|\theta_h), \tag{3}$$

where $h(\cdot|\theta)$ is a neural network with the fixed parameter θ, indicating that c_l, c_h are unobservable but relevant to attribute vector x_t. Intuitively, when the product stays in introduction or maturity stage, the CTR can be only influenced by the noise. When the product in the growth stage, f will be a positive increment, making the CTR increased up to the upper bound c_h. Similar analysis can be obtained for the product in the decline stage.

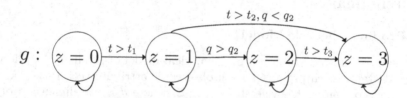

Fig. 1. State transition during product lifecycle

Then we define the state transition function of product lifecycle as a finite state machine as illustrated in Fig. 1. The product starts with the initial stage $z = 0$, and enters the growth stage when the time exceeds t_1. During the growth stage, a product can either step in to the maturity stage if its accumulated impressions q reaches q_2, or the decline stage if the time exceeds t_2 while q is less than q_2. A product in the maturity stage will finally enter the last decline stage if the time exceeds t_3. Otherwise, the product will stay in current stage. Here, t_1, t_2, t_3, q_2 are the latent thresholds of products.

We simulate several product during the whole lifecycle with different latent parameters (the details can be found in the experimental settings), the CTR curves follow the exact trend described as is shown in Fig. 2.

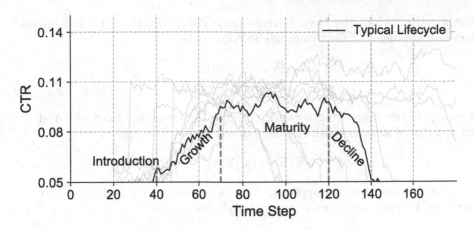

Fig. 2. CTR evolution with the proposed lifecycle model.

2.2   Reinforcement Learning and DDPG Methods

Reinforcement learning maximizes accumulated rewards by trial-and-error app-roach in a sequential decision problem. The sequential decision problem is for-mulated by MDP as a tuple of state space $\mathcal{S}$, action space $\mathcal{A}$, a conditional prob-ability distribution $p(\cdot)$ and a scalar reward function $r = R(s, a), R : \mathcal{S} \times \mathcal{A} \rightarrow \mathbb{R}$. For states $s_t, s_{t+1} \in \mathcal{S}$ and action $a_t \in \mathcal{A}$, distribution $p(s_{t+1}|s_t, a_t)$ denotes the transition probability from state s_t to s_{t+1} when action a_t is adopted in time step t, and the Markov property $p(s_{t+1}|s_t, a_t) = p(s_{t+1}|s_1, a_1, \cdots, s_t, a_t)$ holds for any historical trajectories $s_1, a_1, \cdots, s_t$ to arrive at status s_t. A future dis-counted return at time step t is defined as $R_t^\gamma = \sum_{k=t}^\infty \gamma^{k-t} R(s_k, a_k)$, where γ is a scalar factor representing the discounted. A policy is denoted as $\pi_\theta(a_t|s_t)$ which is a probability distribution mapping from $\mathcal{S}$ to $\mathcal{A}$, where different policies are distinguished by parameter θ.

The target of agent in reinforcement learning is to maximize the expected discounted return, and the performance objective can be denoted as

$$\max_\pi J = \mathbb{E}\left[R_1^\gamma | \pi\right]$$
$$= \mathbb{E}_{s \sim d^\pi, a \sim \pi_\theta}\left[R(s, a)\right] \tag{4}$$

where $d^\pi(s)$ is a discounted state distribution indicating the possibility to encounter a state s under the policy of π. An action-value function is then obtained iteratively as

$$Q(s_t, a_t) = \mathbb{E}\left[R(s_t, a_t) + \gamma \mathbb{E}_{a \sim \pi_\theta}\left[Q(s_{t+1}, a_{t+1})\right]\right] \tag{5}$$

In order to avoid calculating the gradients of the changing state distribu-tion in continuous action space, the Deterministic Policy Gradient (DPG) method [20,21] and the Deep Deterministic Policy Gradient [12] are brought forward. Gradients of the deterministic policy π is

$$\nabla_{\theta^\mu} J = \mathbb{E}_{s \sim d^\mu} \left[\nabla_{\theta^\mu} Q^w(s, a) \right]$$
$$= \mathbb{E}_{s \sim d^\mu} \left[\nabla_{\theta^\mu} \mu(s) \nabla_a Q^w(s, a)|_{a=\mu(s)} \right] \quad (6)$$

where μ is the deep actor network to approximate policy function. And the parameters of actor network can be updated as

$$\theta^\mu = \theta^\mu + \alpha \mathbb{E} \left[\nabla_{\theta^\mu} \mu(s_t) \nabla_a Q^w(s_t, a_t)|_{a=\mu(s)} \right] \quad (7)$$

where Q^w is an obtained approximation of action-value function called critic network. Its parameter vector w is updated according to

$$\min_w L = \mathbb{E}_{s \sim d^\mu} \left[y_t - Q^w(s_t, a_t))^2 \right] \quad (8)$$

where $y_t = R(s_t, a_t) + \gamma Q^{w'}(s_{t+1}, \mu'(s_{t+1}))$, μ' is the target actor network to approximate policy π, $Q^{w'}$ is the target critic network to approximate action-value function. The parameters $w', \theta^{\mu'}$ are updated softly as

$$w' = \tau w' + (1 - \tau)w$$
$$\theta^{\mu'} = \tau \theta^{\mu'} + (1 - \tau)\theta^\mu \quad (9)$$

3   A Scalable Reinforcement Mechanism Design Framework

In our scenario, at each step, the platform observes the global information of all the products, and then allocates impressions according to the observation and some certain strategy, after which the products get their impressions and update itself with the attributes as well as the lifecycle stages. Then the platform is able to get a feedback to judge how good its action is, and adjust its strategy based on all the feedbacks. The above procedures leads to a standard sequential decision making problem.

However, application of reinforcement learning to this problem encounters sever computational issues, due to high dimensionality of both action space and state space, especially with a large n. Thus, we model the impression allocation problem as a standard reinforcement learning problem formally, by regarding the global information of the platform as the state

$$s = [x_1, x_2, ..., x_n]^{\mathrm{T}} \in \mathbb{R}^{n \times d} \quad (10)$$

where n is the number of the product in the platform, and regarding the parameter of a score function as the action,

$$a = \pi(s|\theta^\mu) \in \mathbb{R}^d \quad (11)$$

where π is the policy to learn parameterize by θ^μ, and the action a can be further used to calculate scores of all products

$$o_i = \frac{1}{1 + e^{-a^T x_i}}, \quad \forall i \in \{1, 2, ..., n\} \tag{12}$$

After which the result of impression allocation over all n products can be obtained by a softmax layer as

$$u_i = \frac{e^{o_i}}{\sum_i^n e^{o_i}}, \quad \forall i \in \{1, 2, ..., n\} \tag{13}$$

Without loss of generosity, we assume at each step the summation of impressions allocated is 1, i.e., $\sum_i^n u_i = 1$. By such definition, the dimension of the action space is reduced to d, significantly alleviating the computational issue in previous work [3], where the the dimension of the action space is n.

The goal of policy is to speeded up the metabolism by scoring and ranking products under the consideration of product lifecycle, making the new products grow into maturity stage as quickly as possible and keeping the the global efficiency from dropping down during a long term period. Thus, we define the reward related to s and a as

$$R(s, a) = \frac{1}{n} \sum_i^n \left[\frac{1}{t_i} \int_{t=0}^{t_i} p(t) \frac{dq(t)}{dt} dt \right] \tag{14}$$

where t_i is the time step of the i-th product after being brought to the platform. The physical meaning of this formulation is the mathematical expect over all products in platform for the average click amount of an product during its lifecycle, indicating the efficiency of products in the platform and it can be calculated accumulatively in the online environment, which can be approximately obtained by

$$R(s, a) \approx \frac{1}{n} \sum_i^n \frac{1}{t_i} \sum_{\tau=0}^{t_i} p_\tau^i u_\tau^i \tag{15}$$

A major issue in the above model is that, in practices there will be millions or even billions of products, making combinations of all attribute vectors to form a complete system state with size $n \times d$ computationally unaffordable as referred in essays [4]. A straightforward solution is to applying feature engineering technique to generate a low dimension representation of the state as $s_l = \mathcal{G}(s)$, where $\mathcal{G}$ is a pre-designed aggregator function to generate a low dimensional representation the statistics. However, the pre-designed aggregator function is a completely subjective and highly depends on the the hand-craft features. Alternatively, we attempt to tackle this problem using a simple sampling based method. Specifically, the state is approximated by n_s products uniformly sampled from all products

$$\hat{s} = [x_1, x_2, \cdots, x_{n_s}]^T \in \mathbb{R}^{n_s \times d} \tag{16}$$

where $\hat{s}$ is the approximated state. Then, two issues arise with such sampling method:

- In which order should the sampled n_s products permutated in $\hat{s}$, to implement the *permutation invariance*?
- How to reduce the bias brought by the sampling procedure, especially when n_s is much smaller than n?

To solve these two problem, we further propose the first principal component based permutation and the repeated sampling based experiences generation, which are described in the following subsections in details.

3.1   First Principal Component Based Permutation

The order of each sampled product in the state vector has to be proper arranged, since the unsorted state matrix vibrates severely during training process, making the parameters in network hard to converge. To avoid it, a simple way for permutation is to make order according to a single dimension, such as the brought time t_i, or the accumulated impressions q_i. However, such ad-hoc method may lose information due to the lack of general principles. For example, if we sort according to a feature that is almost the same among all products, state matrix will keep vibrating severely between observations. A suitable solution is to sort the products in an order that keep most information of all features, where the first principal components are introduced [1]. We design a first principal component based permutation algorithm, to project each x_i into a scalar v_i and sort all the products according to v_i

$$e_t = \arg\max_{\|e\|=1} \left(e^{\mathrm{T}} s_t{}^{\mathrm{T}} s_t e \right) \tag{17}$$

$$\hat{e} = \frac{\beta\hat{e} + (1-\beta)(e_t - \hat{e})}{\|\beta\hat{e} + (1-\beta)(e_t - \hat{e})\|} \tag{18}$$

$$v_i = \hat{e}^{\mathrm{T}} x_i, i = 1, 2, \cdots, n_s \tag{19}$$

where e_t is the first principal component of system states in current step t obtained by the classic PCA method as in Eq. 17. $\hat{e}$ is the projection vector softly updated by e_t in Eq. 18, with which we calculate the projected score of each products in Eq. 19. Here $0 < \beta < 1$ is a scalar indicating the decay rate of $\hat{e}$. Finally, the state vector is denoted as

$$\hat{s} = [x_{k_1}, x_{k_2}, \cdots, x_{k_{n_s}}]^{\mathrm{T}} \tag{20}$$

where $k_1, k_2, \cdots, k_{n_s}$ is the order of products, sorted by v_i.

3.2   Repeated Sampling Based Experiences Generation

We adopt the classic experience replay technique [13,15] to enrich experiences during the training phase just as other reinforcement learning applications.

In the traditional experience replay technique, the experience is formulated as (s_t, a_t, r_t, s_{t+1}). However, as what we describe above, there are $C_n^{n_s}$ observations each step theoretically, since we need to sample n_s products from all the n products to approximate the global statistics. If n_s is much smaller than n, such approximation will be inaccurate.

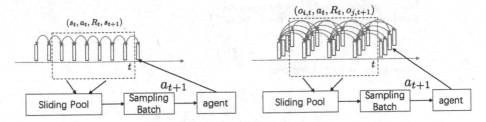

Fig. 3. Classical experiences generation(left): One experience is obtained each step by pair(s_t, a_t, r_t, s_{t+1}); Repeated sampling based experiences generation(right): m^2 experiences are obtained each step by pair$(\hat{s}_t^i, a_t, r_t, \hat{s}_{t+1}^j)$

To reduce the above bias, we propose the repeated sampling based experiences generation. For each original experience, we do repeated sampling s_t and s_{t+1} for m times, to obtain m^2 experiences of

$$(\hat{s}_t^i, a_t, r_t, \hat{s}_{t+1}^j), \quad i, j \in 1, 2, \cdots, m \tag{21}$$

as illustrated in Fig. 3.

This approach improves the stability of observation in noise environment. It is also helpful to generate plenty of experiences in the situation that millions of times repetition is unavailable.

It is worth noting that, the repeated sampling is conducted in the training phase. When to play in the environment, the action a_t is obtained through a randomly selected approximated state $\hat{s}_t$, i.e., $a_t = \pi(\hat{s}_t^1)$. Actually, since a_t does not necessarily equal to $\pi(\hat{s}_t^i), \forall i \in 1, 2, \cdots, m$, it can further help learning a invariant presentation of the approximated state observations.

The overall procedure of the algorithm is described in Algorithm 1. Firstly, a random sampling is utilized to get a sample of system states. And then the sample is permutated by the projection of the first principal components. After that, a one step action and multiple observations are introduced to enrich experiences in experience pool. Moreover, a shared convolutional neural network is applied within the actor-critic networks and target actor-critic networks to extract features from the ordered state observation [7, 24], as is shown in Fig. 4. Finally, the agent observes system repeatedly and train the actor-critic network to learn an optimized policy gradually.

Algorithm 1. The Scalable Reinforcement Mechanism Design Framework

Initialize the parameters of the actor-critic network $\theta^\mu, w, \theta^{\mu'}, w'$, Initialize the replay buffer M, Initialize m observations $\hat{s}_0^j$, Initialize the first principal component $\hat{p}$ by $\hat{s}_0$

foreach *training step t* **do**

    Select action $a_t = \mu(\hat{s}_t^1|\theta^\mu)$

    Execute action a_t and observe reward r_t

    foreach $j \in 1, 2, \cdots, m$ **do**

        Sample a random subset of n_s products

        Combine an observation in the order of $x_k^\mathrm{T}\hat{e}$

$$\hat{s}_t^j \leftarrow \left(x_{k_1}, x_{k_2}, \cdots, x_{k_{n_s}} \right)^\mathrm{T}$$

    Update first principal component

$$e_t \leftarrow \arg \max_{\|e\|=1} \left(e^\mathrm{T} \hat{s}_t^{j\mathrm{T}} \hat{s}_t^j e \right)$$

$$\hat{e} \leftarrow \mathrm{norm}\left(\beta\hat{e} + (1-\beta)(e_t - \hat{e})\right)$$

    end

    foreach $i, j \in 1, 2, \cdots, m$ **do**

$$M \leftarrow M \cup \{(\hat{s}_t^i, a_t, r_t, \hat{s}_{t+1}^j)\}$$

    end

    Sample n_k transitions from M: $(\hat{s}_k, a_k, r_k, \hat{s}_{k+1})$

    Update critic and actor networks

$$w \leftarrow w + \frac{\alpha_w}{n_k}\sum_k (y_k - Q^w(\hat{s}_k, a_k))\nabla_w Q^w(\hat{s}_k, a_k)$$

$$\theta^\mu \leftarrow \theta^\mu + \frac{\alpha_\mu}{n_k}\sum_k \nabla_{\theta^\mu}\mu(\hat{s}_k)\nabla_{a_k}Q^w(\hat{s}_k, a_k)$$

    Update the target networks

$$w' \leftarrow \tau w' + (1-\tau)w$$

$$\theta^{\mu'} \leftarrow \tau\theta^{\mu'} + (1-\tau)\theta^\mu$$

end

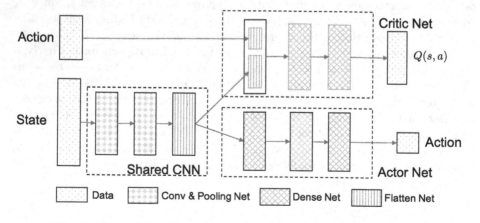

Fig. 4. Network details of the parameter shared actor-critic network

4   Experimental Results

To demonstrate how the proposed approach can help improve the long-term efficiency by speeding up the metabolism, we apply the proposed reinforcement learning based mechanism design, as well as other comparison methods, to a simulated E-commerce platform built based on the proposed product lifecycle model.

4.1   The Configuration

The simulation is built up based on product lifecycle model proposed in Sect. 2.1. Among all of the parameters, q_2 is uniformly sampled from $[10^4, 10^6]$, $t_1, t_2, t_3, \delta_\mu, \delta_\sigma$ are uniformly sampled from $[5, 30], [35, 120], [60, 180], [10^4, 10^6]$, $[2.5 \times 10^3, 2.5 \times 10^5]$ respectively, and parameter σ is set as 0.016 . The parameters c_l, c_h are generated by a fixed neural network whose parameter is uniformly sampled from $[-0.5, 0.5]$ to model online environments, with the outputs scaled into the intervals of $[0.01, 0.05]$ and $[0.1, 0.15]$ respectively. Apart from the normalized dynamic CTR p and the accumulated impressions q, the attribute vector x is uniformly sampled from $[0, 1]$ element-wisely with the dimension $d = 15$. All the latent parameters in the lifecycle model are assumed unobservable during the learning phase.

The DDPG algorithm is adopted as the learning algorithm. The learning rates for the actor network and the critic network are 10^{-4} and 10^{-3} respectively, with the optimizer ADAM [9]. The replay buffer is limit by 2.5×10^4. The most relevant parameters evolved in the learning procedure are set as Table 1.

Table 1. Parameters in learning phase.

Param	Value	Reference
n_s	10^3	Number of products in each sample
β	0.999	First principal component decay rate
γ	0.99	Rewards discount factor
τ	0.99	Target network decay rate
m	5	Repeated observation times

Comparisons are made within the proposed reinforcement learning based methods.

- **CTR-A:** The impressions are allocated in proportion to the CTR score.
- **T-Perm:** The basic DDPG algorithm, with brought time based permutation and a fully connected network to process the state
- **FPC:** The basic DDPG algorithm, with first principal component based permutation and a fully connected network to process the state.

- **FPC-CNN:** FPC with a shared two-layers convolutional neural network in actor-critic networks.
- **FPC-CNN-EXP:** FPC-CNN with the improved experiences generation method.

where CTR-A is the classic supervised learning method and the others are the proposed methods in this paper. For all the experiments, CTR-A is firstly applied for the first 360 steps to initialize system into a stable status, i.e., the distribution over different lifecycle stages are stable, then other methods are engaged to run for another 2k steps and the actor-critic networks are trained for 12.8k times.

4.2   The Results

We firstly show the discounted accumulated rewards of different methods at every step in Fig. 5. After the initialization with the CTR-A, we find that the discounted accumulated reward of CTR-A itself almost converges to almost 100 after 360 steps (actually that why 360 steps is selected for the initialization), while that of other methods can further increase with more learning steps. It is showed that all FPC based algorithms beat the T-Perm algorithm, indicating that the FPC based algorithm can find a more proper permutation to arrange items while the brought time based permutation leads to a loss of information, making a drop of the final accumulated rewards. Moreover, CNN and EXP algorithms perform better in extracting feature from observations automatically, causing a slightly improvement in speeding up the converging process. Both the three FCP based algorithms converge to same final accumulated rewards for their state inputs have the same observation representation.

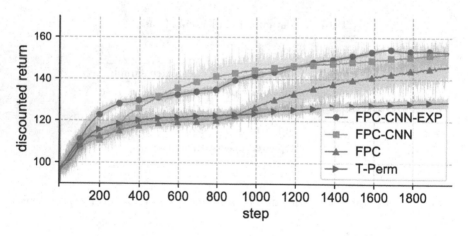

Fig. 5. Performance comparison between algorithms

Then we investigate the distribution shift of the impression allocation over the 4 lifecycle stages after the training procedure of the FPC-CNN-EXP method, as shown in Fig. 6. It can be seen that the percentage of decline stage is decreased and percentage of introduction and maturity stages are increased. By giving up the products in the decline stage, it helps the platform to avoid the waste of the impressions since these products are always with a low CTR. By encouraging the products in the introduction stage, it gives the changes of exploring more potential hot products. By supporting the products in the maturity stage, it maximizes the short-term efficiency since the they are with the almost highest CTRs during their lifecycle.

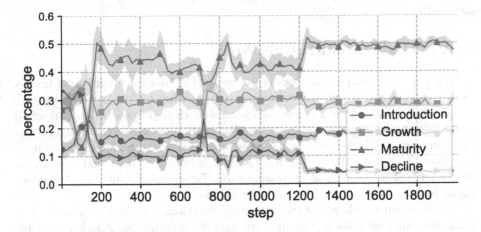

Fig. 6. Percentage of impressions allocated to different stages.

We finally demonstrate the change of the global clicks, rewards as well as the averaged time durations for a product to grow up into maturity stage from its brought time at each step, in terms of relative change rate compared with the CTR-A method, as is shown in Fig. 7. The global average click increases by

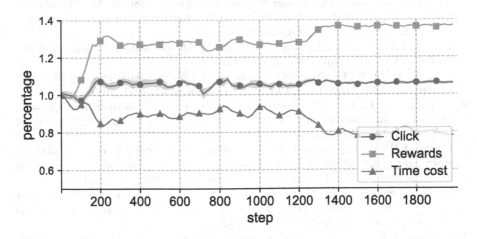

Fig. 7. Metabolism relative metrics

6% when the rewards is improved by 30%. The gap here is probably caused by the inconsistency of the reward definition and the global average click metric. In fact, the designed reward contains some other implicit objectives related to the metabolism. To further verify the guess, we show that the average time for items to growth into maturity stage has dropped by 26%, indicating that the metabolism is significantly speeded up. Thus, we empirically prove that, through the proposed reinforcement learning based mechanism design which utilizes the lifecycle theory, the long-term efficiency can be increased by speeding up the metabolism.

5 Conclusions and Future Work

In this paper, we propose an end-to-end general reinforcement learning framework to improve the long-term efficiency by speeding up the metabolism. We reduce action space into a reasonable level and then propose a first principal component based permutation for better observation of environment state. After that, an improved experiences generation technique is engaged to enrich experience pool. Moreover, the actor-critic network is improved by a shared convolutional network for better state representation. Experiment results show that our algorithms outperform the baseline algorithms.

For the future work, one of the promising directions is to develop a theoretical guarantee for first principal component based permutation. Another possible improvement is to introduce the nonlinearity to the scoring function for products.

References

1. Abdi, H., Williams, L.J.: Principal component analysis. Wiley interdisc. Rev.: Comput. Stat. **2**(4), 433–459 (2010)
2. Burges, C., et al.: Learning to rank using gradient descent. In: Proceedings of the 22nd International Conference on Machine Learning, pp. 89–96. ACM (2005)
3. Cai, Q., Filos-Ratsikas, A., Tang, P., Zhang, Y.: Reinforcement mechanism design for e-commerce. CoRR abs/1708.07607 (2017)
4. Cai, Q., Filos-Ratsikas, A., Tang, P., Zhang, Y.: Reinforcement mechanism design for fraudulent behaviour in e-commerce (2018)
5. Cao, H., Folan, P.: Product life cycle: the evolution of a paradigm and literature review from 1950–2009. Prod. Plann. Control **23**(8), 641–662 (2012)
6. Cao, Z., Qin, T., Liu, T.Y., Tsai, M.F., Li, H.: Learning to rank: from pairwise approach to listwise approach. In: Proceedings of the 24th International Conference on Machine Learning, pp. 129–136. ACM (2007)
7. Cheng, Y.H., Yi, J.Q., Zhao, D.B.: Application of actor-critic learning to adaptive state space construction. In: 2004 Proceedings of 2004 International Conference on Machine Learning and Cybernetics, vol. 5, pp. 2985–2990. IEEE (2004)
8. Deng, Y., Shen, Y., Jin, H.: Disguise adversarial networks for click-through rate prediction. In: Proceedings of the 26th International Joint Conference on Artificial Intelligence, pp. 1589–1595. AAAI Press (2017)
9. Kingma, D.P., Ba, J.: Adam: a method for stochastic optimization. CoRR abs/1412.6980 (2014). http://arxiv.org/abs/1412.6980

10. Koren, Y., Bell, R.: Advances in collaborative filtering. In: Ricci, F., Rokach, L., Shapira, B. (eds.) Recommender Systems Handbook, pp. 77–118. Springer, Boston, MA (2015). https://doi.org/10.1007/978-1-4899-7637-6_3
11. Levitt, T.: Exploit the product life cycle. Harvard Bus. Rev. **43**, 81–94 (1965)
12. Lillicrap, T.P., et al.: Continuous control with deep reinforcement learning. arXiv preprint arXiv:1509.02971 (2015)
13. Lin, L.J.: Self-improving reactive agents based on reinforcement learning, planning and teaching. Mach. Learn. **8**(3–4), 293–321 (1992)
14. Linden, G., Smith, B., York, J.: Amazon. com recommendations: item-to-item collaborative filtering. IEEE Internet comput. **7**(1), 76–80 (2003)
15. Mnih, V., et al.: Playing atari with deep reinforcement learning. arXiv preprint arXiv:1312.5602 (2013)
16. Myerson, R.B.: Optimal auction design. Math. Oper. Res. **6**(1), 58–73 (1981)
17. Nisan, N., Ronen, A.: Algorithmic mechanism design. Games Econ. Behav. **35**(1–2), 166–196 (2001)
18. Papadimitriou, C.H., Tsitsiklis, J.N.: The complexity of markov decision processes. Math. Oper. Res. **12**(3), 441–450 (1987)
19. Shoham, Y., Leyton-Brown, K.: Multiagent Systems: Algorithmic, Game-Theoretic, and Logical Foundations. Cambridge University Press, Cambridge (2008)
20. Silver, D., Lever, G., Heess, N., Degris, T., Wierstra, D., Riedmiller, M.: Deterministic policy gradient algorithms. In: Proceedings of the 31st International Conference on Machine Learning (ICML-14), pp. 387–395 (2014)
21. Sutton, R.S., McAllester, D.A., Singh, S.P., Mansour, Y.: Policy gradient methods for reinforcement learning with function approximation. In: Advances in Neural Information Processing Systems, pp. 1057–1063 (2000)
22. Tang, P.: Reinforcement mechanism design. In: Early Carrer Highlights at Proceedings of the 26th International Joint Conference on Artificial Intelligence (IJCAI, pp. 5146–5150 (2017)
23. Vickrey, W.: Counterspeculation, auctions, and competitive sealed tenders. J. Financ. **16**(1), 8–37 (1961)
24. Wu, Y., Tian, Y.: Training agent for first-person shooter game with actor-critic curriculum learning (2016)

Discovering Bayesian Market Views for Intelligent Asset Allocation

Frank Z. Xing[1]([⊠]), Erik Cambria[1], Lorenzo Malandri[2], and Carlo Vercellis[2]

[1] School of Computer Science and Engineering, Nanyang Technological University,
Singapore, Singapore
{zxing001,cambria}@ntu.edu.sg
[2] Data Mining and Optimization Research Group, Politecnico di Milano, Milan, Italy
{lorenzo.malandri,carlo.vercellis}@polimi.it

Abstract. Along with the advance of opinion mining techniques, public mood has been found to be a key element for stock market prediction. However, how market participants' behavior is affected by public mood has been rarely discussed. Consequently, there has been little progress in leveraging public mood for the asset allocation problem, which is preferred in a trusted and interpretable way. In order to address the issue of incorporating public mood analyzed from social media, we propose to formalize public mood into market views, because market views can be integrated into the modern portfolio theory. In our framework, the optimal market views will maximize returns in each period with a Bayesian asset allocation model. We train two neural models to generate the market views, and benchmark the model performance on other popular asset allocation strategies. Our experimental results suggest that the formalization of market views significantly increases the profitability (5% to 10% annually) of the simulated portfolio at a given risk level.

Keywords: Market views · Public mood · Asset allocation

1 Introduction

Sales and macroeconomic factors are some of the driving forces behind stock movements but there are many others. For example, the subjective views of market participants also have important effects. Along with the growing popularity of social media in the past decades, people tend to rapidly express and exchange their thoughts and opinions [21]. As a result, the importance of their views has dramatically risen [6]. Currently, stock movements are considered to be essentially affected by new information and the beliefs of investors [17].

Meanwhile, sentiment analysis has emerged as a new tool for analyzing the opinions shared on social media [7]. It is a branch of affective computing research that aims to classify natural language utterances as either positive or negative,

A public available dataset is released at https://github.com/fxing79/ibaa.

© Springer Nature Switzerland AG 2019
U. Brefeld et al. (Eds.): ECML PKDD 2018, LNAI 11053, pp. 120–135, 2019.
https://doi.org/10.1007/978-3-030-10997-4_8

but sometimes also neutral [9]. In the financial domain, sentiment analysis is frequently used to obtain a data stream of public mood toward a company, stock, or the economy. Public mood is the aggregation of individual sentiments which can be obtained and estimated from various sources, such as stock message boards [2,19], blogs, newspapers, and really simple syndication (RSS) feeds [34].

Recently, Twitter has become a dominant microblogging platform on which many works rely for their investigations, such as [20,23,27]. Many previous studies support the claim that public mood helps to predict the stock market. For instance, the fuzzy neural network model considering public mood achieves high directional accuracy in predicting the market index. The mood time series is also proved a Granger cause of the market index [4]. Si et al. build a topic-based sentiment time series and predict the market index better with a vector autoregression model to interactively link the two series [26]. The Hurst exponents also suggest a long-term dependency for time series of mood extracted form financial news, similar to many market indices [8].

Despite the important role in stock market prediction, we assume that public mood does not directly effect the market: it does *indirectly* through market participants' views. The actions taken by market participants as agents, are dependent on their own views, and their knowledge about other agents' views. The changes of asset prices are the consequences of such actions. These assumptions are very different from econometric research using productivity, equilibrium, and business cycle models [1], but closer to agent-based models [14]. However, the mechanism of how market views are formed from public mood is heavily overlooked even in the latter case. An intuitive hypothesis could be: the happier the public mood, the higher the stock price. In the real-world market, however, this relationship is far more complicated. Therefore, existing superficial financial applications of AI do not appear convincing to professionals.

In this paper, we attempt to fill this gap by proposing a method for incorporating public mood to form market views computationally. To validate the quality of our views, we simulate the trading performance with a constructed portfolio. The key *contributions* of this paper can be summarized as follows:

1. We introduce a stricter and easier-to-compute definition of the market views based on a Bayesian asset allocation model. We prove that our definition is compatible, and has the equivalent expressiveness as the original form.
2. We propose a novel online optimization method to estimate the expected returns by solving temporal maximization problem of portfolio returns.
3. Our experiments show that the portfolio performance with market views blending public mood data stream is *better* than directly training a neural trading model without views. This superiority is robust for different models selected with the right parameters to generate market views.

The remainder of the paper is organized as follows: Sect. 2 explains the concept of Bayesian asset allocation; following, we describe the methodologies developed for modeling market views in Sect. 3; we evaluate such methodologies by running trading simulations with various experimental settings in Sect. 4 and show the interpretability of our model with an example in Sect. 5; finally, Sect. 6 concludes the paper and describes future work.

2   Bayesian Asset Allocation

The portfolio construction framework [18] has been a prevalent model for investment for more than half a century. Given the an amount of initial capital, the investor will need to allocate it to different assets. Based on the idea of trading-off between asset returns and the risk taken by the investor, the mean-variance method proposes the condition of an efficient portfolio as follows [18,29]:

$$
\text{maximize} \quad \overbrace{\sum_{i=1}^{N} \mu_i w_i}^{\text{return item}} \; \overbrace{-\frac{\delta}{2} \sum_{i=1}^{N} \sum_{j=1}^{N} w_i \sigma_{ij} w_j}^{\text{risk item}} \tag{1}
$$

$$
\text{subject to} \quad \sum_{i=1}^{N} w_i = 1, \; i = 1, 2, \ldots, N. \quad w_i \geq 0.
$$

where δ is an indicator of risk aversion, w_i denotes the weight of the corresponding asset in the portfolio, μ_i denotes the expected return of asset i, σ_{ij} is the covariance between returns of asset i and j. The optimized weights of an efficient portfolio is therefore given by the first order condition of Eq. 1:

$$
w^* = (\delta \Sigma)^{-1} \mu \tag{2}
$$

where Σ is the covariance matrix of asset returns and μ is a vector of expected returns μ_i. At the risk level of holding w^*, the efficient portfolio achieves the maximum combinational expected return.

However, when applying this mean-variance approach in real-world cases, many problems are faced. For example, the two moments of asset returns are difficult to estimate accurately [25], as they are non-stationary time series. The situation is worsened by the fact that, the Markowitz model is very sensitive to the estimated returns and volatility as inputs. The optimized weights can be very different because of a small error in μ or Σ. To address the limitation of the Markowitz model, a Bayesian approach that integrates the additional information of investor's judgment and the market fundamentals was proposed by Black and Litterman [3]. In the Black-Litterman model, the expected returns μ_{BL} of a portfolio is inferred by two antecedents: the equilibrium risk premiums Π of the market as calculated by the capital asset pricing model (CAPM), and a set of views on the expected returns of the investor.

The Black-Litterman model assumes that the equilibrium returns are normally distributed as $r_{eq} \sim \mathcal{N}(\Pi, \tau \Sigma)$, where Σ is the covariance matrix of asset returns, τ is an indicator of the confidence level of the CAPM estimation of Π. The market views on the expected returns held by an investor agent are also normally distributed as $r_{views} \sim \mathcal{N}(Q, \Omega)$.

Subsequently, the posterior distribution of the portfolio returns providing the views is also Gaussian. If we denote this distribution by $r_{BL} \sim \mathcal{N}(\bar{\mu}, \bar{\Sigma})$, then $\bar{\mu}$ and $\bar{\Sigma}$ will be a function of the aforementioned variables (see Fig. 1).

$$[\bar{\mu}, \bar{\Sigma}] = f(\tau, \Sigma, \Omega, \Pi, Q) \tag{3}$$

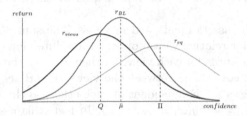

Fig. 1. The posterior distribution of the expected returns as in the Black-Litterman model, which has a mean between two prior distributions and a variance less than both of them.

The function can be induced from applying Bayes' theorem on the probability density function of the posterior expected returns:

$$pdf(\bar{\mu}) = \frac{pdf(\bar{\mu}|\Pi)\ pdf(\Pi)}{pdf(\Pi|\bar{\mu})} \tag{4}$$

Then, the optimized Bayesian portfolio weights have a similar form to Eq. 2, only substituting Σ and μ by $\bar{\Sigma}$ and $\bar{\mu}$:

$$w_{BL}^* = (\delta \bar{\Sigma})^{-1} \bar{\mu}. \tag{5}$$

The most common criticism of the Black-Litterman model is the subjectivity of investor's views. In other words, the model resorts to the good quality of the market views, while it leaves the question of how to actually form these views unanswered. In Sect. 3, we will investigate the possibility of automatically formalizing the market views from public mood distilled from the Web and the maximization of portfolio returns for each time period.

3 Methodologies

3.1 Modeling Market Views

The Black-Litterman model defines a view as a statement that the expected return of a portfolio has a normal distribution with mean equal to q and a standard deviation given by ω. This hypothetical portfolio is called a *view portfolio* [13]. In practice, there are two intuitive types of views on the market, termed *relative views* and *absolute views*, that we are especially interested in. Next, we introduce the formalization of these two types of views.

Because the standard deviation ω can be interpreted as the confidence of expected return of the view portfolio, a *relative view* takes the form of "I have ω_1 confidence that asset x will outperform asset y by $a\%$ (in terms of expected return)"; an *absolute view* takes the form of "I have ω_2 confidence that asset z will outperform the (whole) market by $b\%$". Consequently, for a portfolio consisting of n assets, a set of k views can be represented by three matrices $P_{k,n}$, $Q_{k,1}$, and $\Omega_{k,k}$.

$P_{k,n}$ indicates the assets mentioned in views. The sum of each row of $P_{k,n}$ should either be 0 (for relative views) or 1 (for absolute views); $Q_{k,1}$ is a vector comprises expected returns for each view. Mathematically, the confidence matrix $\Omega_{k,k}$ is a measure of covariance between the views. The Black-Litterman model assumes that the views are independent of each other, so the confidence matrix can be written as $\Omega = diag(\omega_1, \omega_2, \ldots, \omega_n)$. In fact, this assumption will not affect the expressiveness of the views as long as the k views are compatible (not self-contradictory). Because when $\Omega_{k,k}$ is not diagonal, we can always do spectral decomposition: $\Omega = V\Omega^\Lambda V^{-1}$. Then we write the new mentioning and new expected return matrices as $P^\Lambda = V^{-1}P$, $Q^\Lambda = V^{-1}Q$, where Ω^Λ is diagonal. Under these constructions, we introduce two important properties of the view matrices in Theorems 1 and 2.

Theorem 1 (Compatibility of Independent Views). *Any set of independent views are compatible.*

Proof. Compatible views refer to views that can hold at the same time. For example, {asset x will outperform asset y by 3%, asset y will outperform asset z by 5%, asset x will outperform asset z by 8%} is compatible. However, if we change the third piece of view to "asset z will outperform asset x by 8%", the view set becomes self-contradictory. Because the third piece of view is actually a deduction from the former two, the view set is called "not independent".

Assume there is a pair of incompatible views $\{p, q\}$ and $\{p, q'\}$, $q \neq q'$. Both views are either explicitly stated or can be derived from a set of k views. Hence, there exist two different linear combinations, such that:

$$\sum_{i=1}^{k} a_i p_i = p \qquad \sum_{i=1}^{k} a_i q_i = q$$

$$\sum_{i=1}^{k} b_i p_i = p \qquad \sum_{i=1}^{k} b_i q_i = q'$$

where $(a_i - b_i)$ are not all zeros.

Thus, we have $\sum_{i=1}^{k} (a_i - b_i) p_i = \mathbf{0}$, which means that matrix P is rank deficient and the k views are not independent. According to the law of contrapositive, the statement "all independent view sets are compatible" is true. $\square$

Theorem 2 (Universality of Absolute View Matrix). *Any set of independent relative and absolute views can be expressed with a non-singular absolute view matrix.*

Proof. Assume a matrix P with r relative views and $(k - r)$ absolute views.

$$P_{k,n} = \begin{bmatrix} p_{1,1} & p_{1,2} & \cdots & p_{1,n} \\ \vdots & \vdots & \ddots & \vdots \\ p_{r,1} & p_{r,2} & \cdots & p_{r,n} \\ \vdots & \vdots & \ddots & \vdots \\ p_{k,1} & p_{k,2} & \cdots & p_{k,n} \end{bmatrix}$$

The corresponding return vector is $Q = (q_1, q_2, \ldots, q_k)$, the capital weight vector for assets is $w = (w_1, w_2, \ldots, w_k)$. Hence, we can write $(r + 1)$ equations with regard to r new variables $\{q_1', q_2', \ldots, q_r'\}$, where $j = 1, 2, \ldots, r$:

$$1 + q_j' = \sum_{i \neq j}^{r} (1 + q_i') \frac{w_i}{\sum_{s \neq j} w_s} (1 + q_j)$$

$$\sum_{i=1}^{r} q_i' w_i + \sum_{i=r+1}^{k} q_i w_i = Qw^\mathsf{T}$$

If we consider $\{asset_{r+1}, \ldots, asset_k\}$ to be one asset, return of this asset is decided by $P_{r,n}$. Hence, r out of the $(r + 1)$ equations above are independent.

According to Cramer's rule, there exists a unique solution $Q' = (q_1', q_2', \ldots, q_r', q_{r+1}, \ldots, q_k)$ to the aforementioned $(r + 1)$ equations, such that view matrices $\{P', Q'\}$ is equivalent to view matrices $\{P, Q\}$ for all the assets considered, where

$$P_{k,n}' = \begin{bmatrix} 1 & 0 & \cdots & 0 \\ \vdots & \vdots & \ddots & \vdots \\ 0 & p_{r,r} = 1 & \cdots & 0 \\ \vdots & \vdots & \ddots & \vdots \\ p_{k,1} & p_{k,2} & \cdots & p_{k,n} \end{bmatrix}.$$

Now, $P_{k,n}'$ only consists of absolute views. By deleting those dependent views, we can have a non-singular matrix that only consists of absolute views and is compatible. $\square$

Given Theorems 1 and 2, without loss of generality, we can use the following equivalent yet stricter definition of market views to reduce computational complexity.

Definition 1. *Market views on n assets can be represented by three matrices $P_{n,n}$, $Q_{n,1}$, and $\Omega_{n,n}$, where $P_{n,n}$ is an identity matrix; $Q_{n,1} \in \mathbb{R}^n$; $\Omega_{n,n}$ is a nonnegative diagonal matrix.*

3.2   The Confidence Matrix

In the most original form of the Black-Litterman model, the confidence matrix Ω is set manually according to investors' experience. Whereas in the numerical example given by [13], the confidence matrix is derived from the equilibrium covariance matrix:

$$\hat{\Omega}_0 = diag(P(\tau\Sigma)P') \tag{8}$$

This is because $P(\tau\Sigma)P'$ can be understood as a covariance matrix of the expected returns in the views as well. Using our definition, it is easier to understand this estimation, because P is an identity matrix, $P(\tau\Sigma)P'$ is already diagonal. The underlying assumption is that the variance of an absolute view on asset i is proportional to the volatility of asset i. In this case, the estimation of Ω utilizes past information of asset price volatilities.

3.3   Optimal Market Views

We obtain the optimal market views $\{P, Q, \Omega\}$ in a hybrid way, first we adopt the confidence matrix $\hat{\Omega}_0$, then Q can be derived from the inverse optimization problem using the Black-Litterman model.

We start from the optimal portfolio weights that maximize the portfolio returns for each period t. Obviously, without short selling and transaction fees, one should re-invest his whole capital daily to the fastest-growing asset in the next time period.

The optimal holding weights for each time period t thus take the form of a one-hot vector, where $\oslash$ and $\odot$ denote element-wise division and product:

$$w_t^* = \text{argmax } w_t \oslash price_t \odot price_{t+1} \tag{9}$$

Let this w_t^* be the solution to Eq. 1, we will have:

$$w_t^* = (\delta\bar{\Sigma}_t)^{-1}\bar{\mu}_t \tag{10}$$

where the Black-Litterman model gives[1]:

$$\bar{\Sigma}_t = \Sigma_t + [(\tau\Sigma_t)^{-1} + P'\hat{\Omega}_t^{-1}P]^{-1} \tag{11}$$

$$\bar{\mu}_t = [(\tau\Sigma_t)^{-1} + P'\hat{\Omega}_t^{-1}P]^{-1}[(\tau\Sigma_t)^{-1}\Pi_t + P'\hat{\Omega}_t^{-1}Q_t] \tag{12}$$

According to Eqs. 10, 11, and 12, the optimal expected returns for our market views for each period t is:

$$
\begin{aligned}
Q_t^* &= \hat{\Omega}_{0,t}\{[(\tau\Sigma_t)^{-1} + P'\hat{\Omega}_{0,t}^{-1}P]\bar{\mu}_t - (\tau\Sigma_t)^{-1}\Pi_t\} \\
&= \delta[\hat{\Omega}_{0,t}(\tau\Sigma_t)^{-1} + \mathbb{I}]\bar{\Sigma}_t w_t^* - \hat{\Omega}_{0,t}(\tau\Sigma_t)^{-1}\Pi_t \\
&= \delta[\hat{\Omega}_{0,t}(\tau\Sigma_t)^{-1} + \mathbb{I}][\Sigma_t + [(\tau\Sigma_t)^{-1} + \hat{\Omega}_t^{-1}]^{-1}]w_t^* \\
&\quad - \hat{\Omega}_{0,t}(\tau\Sigma_t)^{-1}\Pi_t
\end{aligned}
\tag{13}
$$

---

[1] The proof of Eqs. 11 and 12 can be found from the appendix of [24].

3.4   Generating Market Views with Neural Models

Equation 13 provides a theoretical perspective on determining the expected return of optimal market views. However, computing w_t^* requires future asset prices, which is not accessible. Therefore, the feasible approach is to learn approximating Q_t^* with historical data and other priors as input. We use the time series of asset prices, trading volumes, and public mood data stream to train neural models (nn) for this approximation problem of optimal market views:

$$\hat{Q}_t = nn(prices, volumes, sentiments; Q_t^*) \qquad (14)$$

We denote the time series of asset prices $price_{t-k}, price_{t-k+1}, \ldots, price_t$ by a lag operator $\mathcal{L}^{0 \sim k} price_t$. The notation of trading volumes follows a similar form. Then the model input at each time point: $[\mathcal{L}^{0 \sim k} price_t, \mathcal{L}^{0 \sim k} volume_t, sentiment_t, capital_t]$ can be denoted by $[p, v, s, c]_t$ in short.

Two types of neural models, including a neural-fuzzy approach and a deep learning approach are trained for comparison. Figure 2 provides an illustration of the online training process using a long short-term memory (LSTM) network, where $\hat{Q}$ is the output.

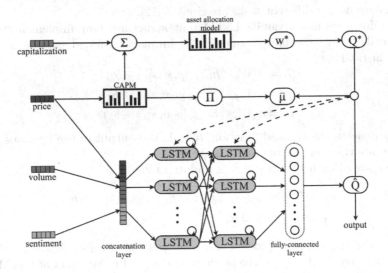

Fig. 2. Model training process (LSTM) with/without sentiment information.

Dynamic Evolving Neural-Fuzzy Inference System (DENFIS) is a neural network model with fuzzy rule nodes [16]. The partitioning of which rule nodes to be activated is dynamically updated with the new distribution of incoming data. This evolving clustering method (ECM) features the model with stability and fast adaptability. Comparing to many other fuzzy neural networks, DENFIS performs better in modeling nonlinear complex systems [32].

Considering the financial market as a real-world complex system, we learn the first-order Takagi-Sugeno-Kang type rules online. Each rule node has the form of:

$$IF \ \mathcal{L}^{0 \sim k} attribute_{t,i} = pattern_i, \ i = 1, 2, \ldots, N$$
$$THEN \ \hat{Q}_t = f_{1,2,\ldots,N}([p, v, s]_t)$$

where we have 3 attributes and $(2^N - 1)$ candidate functions to activate. In our implementation of the DENFIS model, all the membership functions are symmetrical and triangular, which can be defined by two parameters $b \pm d/2$. b is where the membership degree equals to 1; d is the activation range of the fuzzy rule. In our implementation, b is iteratively updated by linear least-square estimator of existing consequent function coefficients.

LSTM is a type of recurrent neural network with gated units. This unit architecture is claimed to be well-suited for learning to predict time series with an unknown size of lags and long-term event dependencies. Early attempts, though not very successful [11], have been made to apply LSTM to time series prediction. It is now recognized that though LSTM cells can have many variants, their performance across different tasks are similar [12].

Therefore, we use a vanilla LSTM unit structure. Our implementation of LSTM cells follows the update rules of the input gate, forget gate, and output gate as in Eq. 15:

$$i_t = \sigma(W_i \cdot [h_{t-1}, [p, v, s]_t] + b_i)$$
$$f_t = \sigma(W_f \cdot [h_{t-1}, [p, v, s]_t] + b_f) \tag{15}$$
$$o_t = \sigma(W_o \cdot [h_{t-1}, [p, v, s]_t] + b_o)$$

where σ denotes the sigmoid function, h_{t-1} is the output of the previous state, W is a state transfer matrix, and b is the bias.

The state of each LSTM cell c_t is updated by:

$$c_t = f_t \odot c_{t-1} + i_t \odot (W_c \cdot [h_{t-1}, [p, v, s]_t] + b_c)$$
$$h_{t-1} = o_t \odot \tanh(c_{t-1}) \tag{16}$$

We make the training process online as well, in a sense that each time a new input is received, we use the previous states and parameters of LSTM cells $[c_{t-1}, \mathbf{W}, \mathbf{b}]$ to initialize the LSTM cells for period t.

4   Experiments

To evaluate the quality and effectiveness of our formalization of market views, we run trading simulations with various experimental settings.

4.1 Data

The data used in this study are publicly available on the Web. We obtain the historical closing price of stocks and daily trading volumes from the Quandl API[2]; the market capitalization data from Yahoo! Finance; the daily count and intensity of company-level sentiment time series from PsychSignal[3]. The sentiment intensity scores are computed from multiple social media platforms using NLP techniques. Figure 3 depicts a segment example of the public mood data stream. The market is closed on weekends, so a corresponding weekly cycle of message volume can be observed.

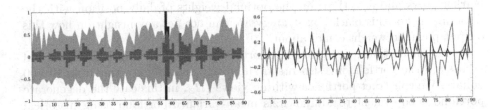

Fig. 3. The volume of daily tweets filtered by cashtag `AAPL` (blue, left); average sentiment intensity (red, left); net sentiment polarity (red, right); daily returns (black, right) in a time period of 90 days (2017-03-04 to 2017-06-04). All the series are normalized. (Color figure online)

We investigate a window of around 8 years (2800 days). All the time series are trimmed from 2009-10-05 to 2017-06-04. For missing values such as the closing prices on weekends and public holidays, we fill them with the nearest historical data to train the neural models. The lagged values we use for both price and trading volume consist of 4 previous days and a moving average of the past 30 days, that is, the input of our neural models takes the form of Eqs. 17 and 18:

$$\mathcal{L}^{0\sim k} price_t = (p_t, p_{t-1}, p_{t-2}, p_{t-3}, \frac{\sum_{i=1}^{30} p_i}{30}) \tag{17}$$

$$\mathcal{L}^{0\sim k} volume_t = (v_t, v_{t-1}, v_{t-2}, v_{t-3}, \frac{\sum_{i=1}^{30} v_i}{30}) \tag{18}$$

4.2 Trading Simulation

We construct a virtual portfolio consisting of 5 big-cap stocks: Apple Inc (`AAPL`), Goldman Sachs Group Inc (`GS`), Pfizer Inc (`PFE`), Newmont Mining Corp (`NEM`), and Starbucks Corp (`SBUX`). This random selection covers both the NYSE and NASDAQ markets and diversified industries, such as technology, financial services, health care, consumer discretionary etc. During the period investigated,

---

[2] http://www.quandl.com/tools/api.

[3] http://psychsignal.com.

there were two splits: a 7-for-1 split for AAPL on June 9th 2014, and a 2-for-1 split for SBUX on April 9th 2015. The prices per share are adjusted according to the current share size for computing all related variables, however, dividends are not taken into account. We benchmark our results with two portfolio construction strategies:

(1) The value-weighted portfolio (VW): we re-invest daily according to the percentage share of each stock's market capitalization. In this case, the portfolio performance will be the weighted average of each stock's performance. This strategy is fundamental, yet empirical study [10] shows that beating the market even before netting out fees is difficult.

(2) The neural trading portfolio (NT): we remove the construction of market views and directly train the optimal weights of daily position with the same input. For this black-box strategy, we can not get any insight on how this output portfolio weight comes about.

In the simulations, we assume no short selling, taxes, or transaction fees, and we assume the portfolio investments are infinitely divisible, starting from 10, 000 dollars. We construct portfolios with no views ($\Omega_\varnothing$, in this case the degenerate portfolio is equivalent to Markowitz's mean-variance portfolio using historical return series to estimate covariance matrix as a measure of risk), random views (Ω_r), the standard views using the construction of Black-Litterman model (Ω_0), with and without our sentiment-induced expected returns (s). The trading performances are demonstrated in Fig. 4.

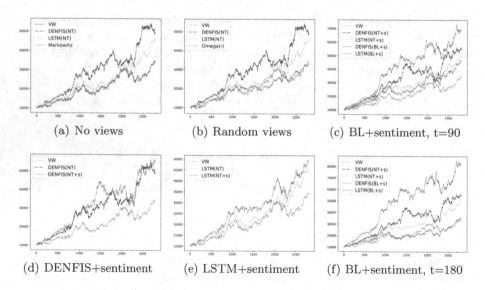

(a) No views   (b) Random views   (c) BL+sentiment, t=90

(d) DENFIS+sentiment   (e) LSTM+sentiment   (f) BL+sentiment, t=180

Fig. 4. Trading simulation performance with different experimental settings: (x-axis: number of trading days; y-axis: cumulative returns). In particular, we use a timespan of 90 and 180 days for our approach. The performance of neural trading is independent from timespan, accordingly the two neural models are compared in (d) and (e) respectively for better presentation.

Following the previous research [13], we set the risk aversion coefficient $\delta = 0.25$ and confidence level of CAPM, $\tau = 0.05$. Let the activation range of fuzzy membership function $d = 0.21$, we obtain 21 fuzzy rule nodes from the whole online training process of DENFIS. This parameter minimizes the global portfolio weight error. For the second neural model using deep learning, we stack two layers of LSTMs followed by a densely connected layer. Each LSTM layer has 3 units; the densely connected layer has 50 neurons, which is set times larger than the number of LSTM units. We use the mean squared error of vector Q as the loss function and the rmsprop optimizer [30] to train this architecture. We observe fast training error convergence in our experiments.

4.3 Performance Metrics

Diversified metrics have been proposed to evaluate the performance of a given portfolio [5, 15, 31]. We report four metrics in our experiments.

Root mean square error (RMSE) is a universal metric for approximation problems. It is widely used for engineering and data with normal distribution and few outliers. We calculate the RMSE of our realized portfolio weights to the optimal weights:

$$\text{RMSE} = \sqrt{\frac{1}{n} \sum_{i=1}^{n} \|w_i - \hat{w}_i\|^2} \tag{19}$$

Annualized return (AR) measures the profitability of a given portfolio. We calculate the geometric mean growth rate per year, which is also referred to as compound annual growth rate (CAGR) for these 2800 days.

Sharpe ratio (SR) is a risk-adjusted return measure. We choose the value-weighted portfolio as a base, consequently the Sharpe ratio of VW will be 1:

$$\text{SR} = \frac{\mathbb{E}(R_{portfolio}/R_{VW})}{\sigma(R_{portfolio})/\sigma(R_{VW})} \tag{20}$$

SR uses the standard deviation of daily returns as the measure of risk. Note that to distinguish between good and bad risk, we can also use the standard deviation of downside returns only [28]. Our results suggest that the Sortino ratios, which are not reported due to page limit, are very close to SRs and lead to the same conclusion.

The maximum drawdown (MDD) measures the maximum possible percentage loss of an investor:

$$\text{MDD} = \max_{0 < t < \tau} \left\{ \frac{Value_t - Value_\tau}{Value_t} \right\} \tag{21}$$

Asset allocation strategies with large MDD are exposed to the risk of withdrawal. Table 1 presents the metrics.

Table 1. Performance metrics for various portfolio construction strategies, timespan = 90 and 180 days. Top three metrics are in bold.

	RMSE	SR	MDD(%)	AR(%)
VW	0.8908	1.00	25.81	17.49
Markowitz90($\Omega_\varnothing$)	0.9062	1.00	25.81	17.51
Markowitz180($\Omega_\varnothing$)	0.8957	1.00	25.82	17.45
BL90(Ω_r)	0.9932	0.90	**23.47**	17.17
BL180(Ω_r)	0.9717	1.06	**20.59**	22.31
DENFIS(NT)	0.9140	**2.94**	29.84	23.09
DENFIS(NT+s)	0.9237	**4.35**	**23.07**	**25.16**
DENFIS(BL90+s)	0.9424	1.52	24.44	**28.69**
DENFIS(BL180+s)	0.9490	**1.58**	24.19	**29.49**
LSTM(NT)	**0.8726**	1.38	25.68	22.10
LSTM(NT+s)	0.8818	1.42	25.96	23.21
LSTM(BL90+s)	**0.8710**	1.34	25.90	22.33
LSTM(BL180+s)	**0.8719**	1.07	24.88	17.68

4.4   Findings

We have some interesting observations from Fig. 4 and Table 1. SR and AR are usually considered as the most important, and besides, RMSE and MDD are all very close in our experiments. The correlation between RMSE and the other three metrics is weak, though it is intuitive that if the realized weights are close to the optimal weights, the portfolio performance should be better. On the contrary, the LSTM models seem to overfit as they are trained on the mean squared error of weights or expected return of views [22]. However, as mentioned in Sect. 1, the relationship between weights and daily returns is nonlinear. Therefore, *holding portfolio weights that are close to the optimal weights does not necessarily means that the AR must be higher*. In fact, it is dangerous to use any seemingly reasonable metrics outside the study of asset allocation, such as directional accuracy of price change prediction [4,33], to evaluate the expected portfolio performance.

The Markowitz portfolio ($\Omega_\varnothing$) displays a very similar behavior to the market-following strategy. This is consistent with the inefficacy of the mean-variance approach in practice mentioned by previous studies: holding the Markowitz portfolio is holding the market portfolio. In fact, if the CAPM holds, the market portfolio already reflects the adjustments to risk premiums, that is, fewer market participants will invest on highly risky assets, for this reason their market capitalization will be smaller as well.

However, the Black-Litterman model does not always guarantee better performance over the Markowitz portfolio. "Garbage in, garbage out" still holds for this circumstance. Given random views (Ω_r), it can be worse than market-following in terms of both SR and AR. The lesson learned is that *if the investor knows nothing, it is better to hold no views and follow the market than pretending to know something.*

In our experiments, DENFIS generally performs better than LSTM models, achieving higher SRs and ARs. The reason may be LSTM models adapt faster to the incoming data, whereas financial time series are usually very noisy. The ECM mechanism provides DENFIS models with converging learning rates, which may be beneficial to the stability of memorized rules. However, it is important to note that *the ARs for both neural models improve with the blending of sentiments.* The timespan used to estimate correlation and volatility of assets seems not that critical. DENFIS models perform better with longer timespan, while LSTM models perform better with shorter timespan. The Markowitz portfolio is less affected by timespan.

5   A Story

One of the main advantages of our formalization and computing of market views is that some *transparency* is brought to the daily asset reallocation decisions. In most cases, a stock price prediction system based on machine learning algorithms cannot justify "why he thinks that price will reach that predicted point". Unlike these systems, our method can tell a story of the portfolio to professional investors and advice seekers. Take June 1st 2017 as an example:

"On June 1st 2017, we observe 164 positive opinions of polarity +1.90, 58 negative opinions of polarity −1.77 on AAPL stock; 54 positive opinions of polarity +1.77, 37 negative opinions of polarity −1.53 on GS stock; 5 positive opinions of polarity +2.46, 1 negative opinion of polarity −1.33 on PFE stock; no opinion on NEM stock; and 9 positive opinions of polarity +1.76, 5 negative opinions of polarity −2.00 on SBUX stock. Given the historical prices and trading volumes of the stocks, we have 6.29% confidence that AAPL will outperform the market by −70.11%; 23.50% confidence that GS will outperform the market by 263.28%; 0.11% confidence that PFE will outperform the market by −0.50%; 1.21% confidence that SBUX will outperform the market by 4.57%. Since our current portfolio invests 21.56% on AAPL, 25.97% on GS, 29.43% on PFE, and 23.04% on SBUX, by June 2nd 2017, we should withdraw all the investment on AAPL, 2.76% of the investment on GS, 81.58% of the investment on PFE, and 30.77% of the investment on SBUX, and re-invest them onto NEM."

6   Conclusion and Future Work

In previous studies which have considered sentiment information for financial forecasting, the role of the investor as a market participant is often absent. In this paper, we present a novel approach to incorporate market sentiment by fusing public mood data stream into the Bayesian asset allocation framework.

This work is pioneering in formalizing sentiment-induced market views. Our experiments show that the market views provide a powerful method to asset management. We also confirm the efficacy of public mood data stream based on social media for developing asset allocation strategies.

A limitation of this work is that we fixed a portfolio with five assets, though in practice the portfolio selection problem is of equal importance. How to assess the quality of sentiment data is not discussed in this paper as well. We are not at the stage to distinguish or detect opinion manipulation though concern like the open networks are rife with bots does exist. Another limitation is that survivor bias is not taken into account: the risk that assets selected in the portfolio may quit the market or suffer from a lack of liquidity. This problem can be alleviated by only including high quality assets. In the future, we will study examining the quality of sentiment data obtained using different content analysis approaches. We also plan to develop a Bayesian asset allocation model that can deal with market frictions.

References

1. Angeletos, G., La'O, J.: Sentiments. Econometrica **81**(2), 739–779 (2013)
2. Antweiler, W., Frank, M.Z.: Is all that talk just noise? The information content of internet stock message boards. J. Finance **59**(3), 1259–1294 (2004)
3. Black, F., Litterman, R.: Asset allocation: combining investor view with market equilibrium. J. Fixed Income **1**, 7–18 (1991)
4. Bollen, J., Mao, H., Zeng, X.: Twitter mood predicts the stock market. J. Comput. Sci. **2**(1), 1–8 (2011)
5. Brandt, M.W.: Portfolio choice problems. In: Handbook of Financial Econometrics, vol. 1, chap. 5, pp. 269–336. Elsevier B.V., Oxford (2009)
6. Cambria, E.: Affective computing and sentiment analysis. IEEE Intell. Syst. **31**(2), 102–107 (2016)
7. Cambria, E., Das, D., Bandyopadhyay, S., Feraco, A. (eds.): A Practical Guide to Sentiment Analysis. Springer International Publishing, Switzerland (2017). https://doi.org/10.1007/978-3-319-55394-8
8. Chan, S.W., Chong, M.W.: Sentiment analysis in financial texts. Decis. Support Syst. **94**, 53–64 (2017)
9. Chaturvedi, I., Ragusa, E., Gastaldo, P., Zunino, R., Cambria, E.: Bayesian network based extreme learning machine for subjectivity detection. J. Frankl. Inst. **355**(4), 1780–1797 (2018)
10. Fama, E.F., French, K.R.: Luck versus skill in the cross-section of mutual fund returns. J. Financ. **65**(5), 1915–1947 (2010)
11. Gers, F.A., Eck, D., Schmidhuber, J.: Applying LSTM to time series predictable through time-window approaches. In: Dorffner, G., Bischof, H., Hornik, K. (eds.) ICANN 2001. LNCS, vol. 2130, pp. 669–676. Springer, Heidelberg (2001). https://doi.org/10.1007/3-540-44668-0_93
12. Greff, K., Srivastava, R.K., Koutnik, J., Steunebrink, B.R., Schmidhuber, J.: LSTM: a search space odyssey. IEEE Trans. Neural Netw. Learn. Syst. **28**(10), 2222–2232 (2017)
13. He, G., Litterman, R.: The intuition behind black-litterman model portfolios. Goldman Sachs Working Paper (1999). https://doi.org/10.2139/ssrn.334304
14. Hommes, C.: The New Palgrave Dictionary of Economics. Interacting Agents in Finance, 2nd edn. Palgrave Macmillan, Basingstoke (2008)

15. Hyndman, R.J., Koehler, A.B.: Another look at measures of forecast accuracy. Int. J. Forecast. **22**(4), 679–688 (2006)
16. Kasabov, N.K., Song, Q.: Denfis: dynamic evolving neural-fuzzy inference system and its application for time-series prediction. IEEE Trans. Fuzzy Syst. **10**, 144–154 (2002)
17. Li, Q., Jiang, L., Li, P., Chen, H.: Tensor-based learning for predicting stock movements. In: Proceedings of the Twenty-Ninth AAAI Conference on Artificial Intelligence, pp. 1784–1790 (2015)
18. Markowitz, H.: Portfolio selection. J. Finance **7**, 77–91 (1952)
19. Nguyen, T.H., Shirai, K.: Topic modeling based sentiment analysis on social media for stock market prediction. In: Proceedings of the Annual Meeting of the Association for Computational Linguistics, pp. 1354–1364 (2015)
20. Nofer, M., Hinz, O.: Using twitter to predict the stock market: where is the mood effect? Bus. Inf. Syst. Eng. **57**(4), 229–242 (2015)
21. O'Connor, B., Balasubramanyan, R., Routledge, B.R., Smith, N.A.: From tweets to polls: linking text sentiment to public opinion time series. In: Proceedings of the Fourth International AAAI Conference on Weblogs and Social Media, pp. 122–129 (2010)
22. Pant, P.N., Starbuck, W.H.: Innocents in the forest: forecasting and research methods. J. Manag. **16**(2), 433–460 (1990)
23. Ranco, G., Aleksovski, D., Caldarelli, G., Grčar, M., Mozetič, I.: The effects of twitter sentiment on stock price returns. PLoS One **10**(9), 1–21 (2015)
24. Satchell, S., Scowcroft, A.: A demystification of the black-litterman model: managing quantitative and traditional portfolio construction. J. Asset Manag. **1**(2), 138–150 (2000)
25. Shen, W., Wang, J.: Portfolio selection via subset resampling. In: Proceedings of the Thirty-First AAAI Conference on Artificial Intelligence, pp. 1517–1523 (2017)
26. Si, J., Mukherjee, A., Liu, B., Li, Q., Li, H., Deng, X.: Exploiting topic based twitter sentiment for stock prediction. In: Proceedings of the Annual Meeting of the Association for Computational Linguistics, pp. 24–29 (2013)
27. Smailović, J., Grčar, M., Lavrač, N., Žnidaršič, M.: Predictive sentiment analysis of tweets: a stock market application. In: Holzinger, A., Pasi, G. (eds.) HCI-KDD 2013. LNCS, vol. 7947, pp. 77–88. Springer, Heidelberg (2013). https://doi.org/10.1007/978-3-642-39146-0_8
28. Sortino, F.A., Price, L.N.: Performance measurement in a downside risk framework. J. Invest. **3**, 59–64 (1994)
29. Steinbach, M.C.: Markowitz revisited: mean-varian-ce models in financial portfolio analysis. SIAM Rev. **43**(1), 31–85 (2001)
30. Tieleman, T., Hinton, G.E.: Lecture 6.5-RMSProp: divide the gradient by a running average of its recent magnitude. COURSERA: Neural Netw. Mach. Learn. **4**, 26–31 (2012)
31. Xing, F.Z., Cambria, E., Welsch, R.E.: Natural language based financial forecasting: a survey. Artif. Intell. Rev. **50**(1), 49–73 (2018)
32. Xing, F.Z., Cambria, E., Zou, X.: Predicting evolving chaotic time series with fuzzy neural networks. In: International Joint Conference on Neural Networks, pp. 3176–3183 (2017)
33. Yoshihara, A., Seki, K., Uehara, K.: Leveraging temporal properties of news events for stock market prediction. Artif. Intell. Res. **5**(1), 103–110 (2016)
34. Zhang, W., Skiena, S.: Trading strategies to exploit blog and news sentiment. In: Proceedings of the Fourth International AAAI Conference on Weblogs and Social Media, pp. 375–378 (2010)

Intent-Aware Audience Targeting
for Ride-Hailing Service

Yuan Xia[2], Jingbo Zhou[1,3]([✉]), Jingjia Cao[2,5], Yanyan Li[1,3], Fei Gao[2],
Kun Liu[2], Haishan Wu[4], and Hui Xiong[1,3]

[1] Business Intelligence Lab, Baidu Research, Beijing, China
{zhoujingbo,liyanyanliyanyan,xionghui01}@baidu.com
[2] Baidu Inc, Beijing, China
{xiayuan,caojingjia,gaofei09,liukun12}@baidu.com
[3] National Engineering Laboratory of Deep Learning Technology and Application,
Beijing, China
[4] SenSight.ai Ltd., Beijing, China
hswu85@gmail.com
[5] Beijing Jiaotong University, Beijing, China

Abstract. As the market for ride-hailing service is increasing dramatically, an efficient audience targeting system (which aims to identify a group of recipients for a particular message) for ride-hailing services is demanding for marketing campaigns. In this paper, we describe the details of our deployed system for intent-aware audience targeting on Baidu Maps for ride-hailing services. The objective of the system is to predict user intent for requesting a ride and then send corresponding coupons to the user. For this purpose, we develop a hybrid model to combine the LSTM model and GBDT model together to handle sequential map query data and heterogeneous non-sequential data, which leads to a significant improvement in the performance of the intent prediction. We verify the effectiveness of our method over a large real-world dataset and conduct a large-scale online marketing campaign over Baidu Maps app. We present an in-depth analysis of the model's performance and trade-offs. Both offline experiment and online marketing campaign evaluation show that our method has a consistently good performance in predicting user intent for a ride request and can significantly increase the click-through rate (CTR) of vehicle coupon targeting compared with baseline methods.

Keywords: Audience targeting · Location based service
Marketing campaign

1 Introduction

With the rapid development of mobile internet, an increasing number of people tend to use mobile phone applications for ride-hailing services (including booking

---

Y. Xia and J. Zhou—Co-first authors.

U. Brefeld et al. (Eds.): ECML PKDD 2018, LNAI 11053, pp. 136–151, 2019.
https://doi.org/10.1007/978-3-030-10997-4_9

rides and paying for car driver) provided by a transportation network company
(TNC) such as Uber, Lyft, Didi and GrabTaxi. Now the ride-hailing service is
also an important function in Baidu Maps and Google Maps.

Given the tremendous amount of users on Baidu Maps, we aim to attract as
many users as possible to use the ride-hailing service provided by Baidu Maps.
To achieve this purpose, a useful method is to resort to audience targeting, which
identifies a group of recipients for a particular message. Audience targeting is
well known to be very useful in message propagation and marketing promotion.
Therefore, in order to increase the overall flow of the ride-hailing platform and
attract more customers, an accurate audience targeting system for ride-hailing
services is demanding. Here, the specific problem is vehicle coupon targeting
which aims to deliver coupons to potential users.

Typically, there are mainly two approaches to audience targeting. The first
one is audience segments targeting which aims to identify a specific group of
people, e.g., the man between the ages of 18 and 25. The second approach is
audience behaviors targeting [20] which tries to push customized message according to user's online or offline behaviors. There are also mature architectures and
methods for traditional audience targeting problems, like ads targeting in Google
[19], Facebook [13] and Bing [8].

Our work mainly focuses on the area of audience targeting for ride-hailing
services. Different from the traditional general framework for ads targeting, there
are several challenges. First, the types of the coupon are diverse. For a specific
service, we usually have different coupons with a short lifetime. It is computationally expensive to train different models for each kind of coupons. Second, the
timeliness and context are also important. Users may not be interested in any
vehicle coupon if they do not have intention for requesting a ride. The vehicle
coupon targeting is also a kind of instant service, we need to push the customized
message (e.g., coupon) to the targeted user before his/her movement.

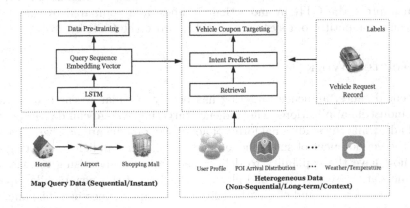

Fig. 1. Illustration of the audience targeting system for ride-hailing service

In this paper, we propose an intent-aware audience targeting system for ride-hailing services. The key idea to solve this problem is to precisely predict the user's intent for requesting a ride and then send the corresponding coupons. The overall architecture of the system is shown in Fig. 1. First, our insight is that map query data on Baidu Maps reflects the instant travel intent of a user. In order to capture user's instant intent and better handle sequential map query data, we do data pre-training and generate query sequence embedding with the user's map query data based on the Long-Short Term Memory (LSTM) [15] model. Second, for the comprehensive understanding of user's intention, our method combines the LSTM and gradient boosting decision tree (GBDT) [6] to deal with multi-source heterogeneous data (sequential, non-sequential, instant, long-term and context data). Third, to achieve real-time performance, before the intent prediction model, we develop a coarse retrieval module based on the frequent pattern mining method. The main contributions of this paper are summarized as follows:

- We propose an intent-aware audience targeting framework for ride-hailing services which first captures the user's instant intent and then delivers coupons to particular users.
- We develop a method to capture the user's instant intent by extracting implicit intention from map query sequential data. By combining the LSTM model and GBDT model, we can deal with both map query sequential data and other multi-source heterogeneous non-sequential data. We demonstrate that the combined model can significantly improve the overall prediction performance.
- We conduct extensive experiments to evaluate the effectiveness of our framework. Both offline experiments and online marketing campaign evaluation show that our model has a consistently good performance for predicting the user's intent for requesting a ride. Our method can significantly improve the CTR of vehicle coupon targeting: compared to simple baseline our system can improve the CTR by more than 7 times; and compared to the almost optimal competitor our system can still improve the CTR by 26%.

2   Related Work

In recent decades, audience targeting has been a popular research topic with many industrial applications. The audience targeting system can deliver personalized ads, while the traditional forms of advertising media are passive in nature. With the development of smartphone handsets, modern audience targeting systems now have the capability to deliver personalized advertising messages to consumers at the proper time and place. In the light of this new mobile capability, how to effectively and efficiently deliver a customized message (like an ad or a coupon) to meet consumers' needs has become a critical issue [4,17].

There are two types of targeting methods which are prevalent in industrial applications: user segment and behavioral targeting (BT). BT provides an approach to learning the targeting function from past user behavior, especially using implicit feedback (i.e., ad clicks) to match the best ads to users [5]. Therefore, BT is generally used for improving the influence of the online advertising by targeting the most relevant user for the ads being displayed and vice versa [28]. In [20], the authors investigated the effect of historical user activity for targeting. As an effective method, BT has attracted much research attention from different perspectives, such as the scalability [1,18], privacy [7], user identification [12] and ad display [25]. The human mobility prediction methods [33,34] can also be used to improve the performance of the BT. Though our system can be considered as a special case of behavior targeting, we focus on the instant intent prediction to identify the audience with potential demand for picking a ride.

3   Preliminaries

3.1   Problem Definition

We define the problem as follows. For each user $u \in \mathcal{U}$, we want to predict the user's intent for requesting a ride in the next T hours, and then we push the vehicle coupon $c \in \mathcal{C}$ through Baidu Maps app. The main purpose of our targeting system is to improve the CTR of coupon targeting and attract more customers to use Baidu Maps app platform for ride-hailing service.

In this problem, we mainly have two kinds of data: sequential data D_{sq} and non-sequential data D_{non-sq}. The D_{sq} is mainly user's map query sequence data $\{q_1, q_2, \cdots, q_t, \cdots\}$, and the non-sequential data includes user's profile data, vehicle historical request data, user's Points of Interests (POI) arrival distribution data, temperature and weather data, etc. Based on all these data, we want to build models to deal with the problem of vehicle coupon targeting. Specifically, the vehicle request record $y \in \{0,1\}$ represents the fact that whether a user has actually requested a ride in T hours. Given the sequential data x_{sq}, non-sequential data x_{non-sq} and the vehicle request record $y \in \{0,1\}$, we can model a multivariable non-linear function $\Psi(\cdot)$ to calculate the probability $p_u \in [0,1]$ which represents the user's intent for requesting a ride. After obtaining the probability, we can push the corresponding vehicle coupon according to a predefined threshold τ. The coupon type is determined by the marketing team.

3.2   Data Specification

To deal with the audience targeting for the ride-hailing service on Baidu Maps, our work mainly focuses on the following data sources:

- **User Map Query Data.** The data records the user's query search behavior on Baidu Maps app. When a user is searching for a place at Baidu Maps, it will record query word, destination POI information, user id, user's current coordinates, query destination coordinates and the current timestamp.
- **User Profile Data.** The data makes speculations on user's profile, including the user's age, gender, consumption level, job, education level, life stage, interest, etc.

- **User Vehicle Request Data.** The data records the user's online vehicle request history. It includes request id, start time, end time, duration, source, destination, source coordinates, destination coordinates, distance, price and discount.
- **POI Arrival Data.** The data records the user's POI arrival information in the recent period of time. Different from the map query data which represents the user's current intent, the POI arrival information records the user's POI arrival distribution. For example, one record can be as follows: Tom visited "Shopping Mall" 3 times, visited "Restaurant" 4 times, and visited "Hotel" 2 times on Oct, 2016.
- **Weather and Temperature Data.** The data keeps track of the weather and temperature information in the whole nation. For each item in the database, it records the city name, area name, area code, weather and temperature information. Weather data includes nearly 40 weather types, and temperature data is measured by Celsius format.

4   The Intent Prediction Model

To fully exploit the sequential and non-sequential data to achieve a good performance on audience targeting, we propose an intent-aware ride-hailing service targeting system which takes the advantages of Sequential Pattern Mining, Long Short-Term Memory (LSTM) and Gradient Boosting Decision Tree (GBDT). As for our problem, we do have heterogeneous data from different sources. For travel intent prediction, the user's sequential map query data is essential. The following subsections demonstrate the methodology of our system.

4.1   Sequential Pattern Mining

User's historical map query sequential behaviors on Baidu Maps can implicitly reveal user's intention. There are several well-known sequential mining methods available, such as *PrefixSpan* [11], *FreeSpan* [10] and *GSP Algorithm* [22]. These methods can find out sequential frequent patterns in a user map query database. However, these methods have their own limitations. First, they can only list the frequent patterns for different scenes (the scene here is vehicle coupon targeting), and they cannot give probabilistic output. Second, these methods cannot deal with uncertain data, i.e, if a pattern is unseen before, then these methods will meet problems.

While the sequential pattern mining method is not a good choice for making final predictions, we can use frequent patterns as coarse filters to our system. Given the map query sequence database S and the minimum support threshold min_sup, we first adopt the *PrefixSpan Algorithm* for sequential pattern mining to find frequent patterns $p \in \mathcal{P}$ in S. Then, we extract the ride request oriented patterns from the user's vehicle request record. Finally, by eliminating some of the query patterns which have little correlation with requesting a ride, we can boost the speed of the targeting system and get the real-time performance.

4.2   Map Query Based LSTM Method

Recurrent Neural Network (RNN) [21,26] has been broadly used to deal with sequential data. The standard recurrent neural network can map a sequence of map queries of variable length to a fixed-length vector. By recursively transforming current map query vector q_t with the previous step vector h_{t-1}, we can get the current step vector h_t. The transition function is typically a linear layer followed by a non-linearity layer such as $tanh$. Given the user's map query sequence $\{q_1, q_2, \cdots, q_t, \cdots\}$, the standard RNN computes the output sequence of q as follows:

$$h_t = tanh(W_{hq} \cdot q_t + W_{hh} \cdot h_{t-1} + b_h) \tag{1}$$

where $W_{hq} \in R^{m \times n}$, $W_{hh} \in R^{m \times m}$ and $b_h \in R^m$. m and n are dimensions of the hidden vector and the query embedding vector, respectively. Unfortunately, standard RNN suffers the problem of gradient vanishing or exploding [3,15], where gradients may grow or decay exponentially over long sequences. This makes it difficult to model long-distance correlations in a sequence.

The advent of LSTM is to deal with the weakness of standard RNN. There is one memory cell g surrounded by three gates controlling whether to input new data (the input gate i), whether to forget history (the forget gate f), and whether to produce current value (the output gate o) at each time step t. The memory cell in LSTM summarizes the information at each time step of what information has been observed up to now. Such structures are more capable to learn a complex composition of query vectors than standard RNN. The definition of the gates and cell update and output are as follows:

$$i_t = \sigma(W_{ix} \cdot q_t + W_{ih} \cdot h_{t-1} + b_i) \tag{2}$$
$$f_t = \sigma(W_{fx} \cdot q_t + W_{fh} \cdot h_{t-1} + b_f) \tag{3}$$
$$o_t = \sigma(W_{ox} \cdot q_t + W_{oh} \cdot h_{t-1} + b_o) \tag{4}$$
$$\widetilde{g}_t = \phi(W_{gx} \cdot q_t + W_{gh} \cdot h_{t-1} + b_g) \tag{5}$$
$$g_t = f_t \odot g_{t-1} + i_t \odot \widetilde{g}_t \tag{6}$$
$$h_t = o_t \odot g_t \tag{7}$$

where $\odot$ represents the product of with a gate value. σ and ϕ are non-linear activation functions. Here, $\sigma(\cdot)$ is sigmoid and $\phi(\cdot)$ is hyperbolic tangent. The W and b are the parameter matrices and bias vectors learned during the training.

For building the LSTM model with the map query, we first conduct a semantic representation transformation for the data. There are usually millions of POIs in a city. The number of the POI in a city is too large to be handled by the LSTM model. Therefore, we introduce the semantic representation transformation to reduce the vocabulary of the input space of the LSTM. For each user $u \in \mathcal{U}$, we collect his/her recently map query records in the past period of time and get the raw query sequence $\{q_r^1, q_r^2, \cdots, q_r^t, \cdots\}$, which is denoted by q_r. Then, according to the POI information, we map the raw query word into a semantic space with limited dimensions.

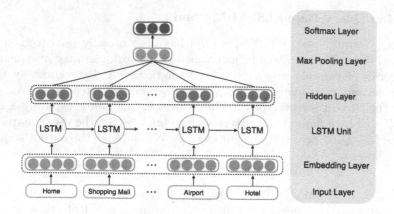

Fig. 2. Architecture of map query based LSTM model

$$q_c^t = P(q_r^t), \quad t \in \{1, 2, \cdots, l\} \tag{8}$$

where $P(\cdot)$ projects the raw query q_r^t into the semantic space by categorization methods. For the semantic space mapping, we use the Baidu Maps' API to map the POI into its category space. For example, when someone is searching for "JW Marriott Hotel", it will be projected to the semantic space with the tag "Hotel". After projecting the user's raw query sequence into the semantic space, we get $\{q_c^1, q_c^2, \cdots, q_c^t, \cdots\}$, which is denoted by q_c. For instance, a user first queried for "Beijing Capital International Airport" and then "JW Marriott Hotel", and later queried for "Bird's Nest" and "Quanjude Roast Duck". Then, after mapping the POI name into the semantic space, we get a user semantic query sequence $\{Airport \rightarrow Hotel \rightarrow Landmark \rightarrow Restaurant\}$.

Then, we put user's map query sequence data into LSTM model. The architecture of our LSTM model is illustrated in Fig. 2. We use an embedding layer to map the query category id to a fixed length vector.

$$c_t = T(q_c^t), \quad t \in \{1, 2, \cdots, l\} \tag{9}$$

where $T(\cdot)$ transforms the one-hot semantic representation of query q_c^t into the latent vector c_t using a learned embedding matrix W_c.

Then through LSTM, each input of the sequence will generate a hidden output, and at the end of the sequence, we get a final hidden output. Generally, people use the last hidden output as the representation of the whole sequence [2,23].

$$h_t = LSTM(h_{t-1}, c_t) \tag{10}$$

We believe that for a map query sequence, each query word needs to be considered in order to get a better permanence. In [24], the author added an average pooling layer before the softmax layer to their network structure for merging the sequential hidden state. By adding an extra pooling layer on top of each hidden output, we can get a final output utilizing each hidden query vector. We test two methods for best utilizing the query sequence, including average pooling and max pooling. In the experiment section, we prove that max pooling outperforms the average pooling in this problem.

$$h_o = Merge(h_1, \cdots, h_l) \tag{11}$$

Finally, through the softmax layer, the LSTM outputs the p_u, which indicates the probability of the user's intent on requesting a ride in the next T hours.

$$p_u = softmax(W_d \cdot h_o + b_d) \tag{12}$$

where W_d and b_d are parameters of the dense output layer.

4.3   Combination of LSTM and GBDT

We further propose a hybrid model to combine the LSTM and Gradient Boosting Decision Tree (GBDT) for intent prediction. It is not good enough to predict the intent of a user to order a car for travelling only by map query sequential data. We need more data to help us make predictions. In Baidu, we have all sorts of data from different sources. Long Short-Term Memory (LSTM) is proved to be a good solution for long-term dependencies sequential data problem [9]. However, recurrent neural network based algorithms cannot handle the multi-source heterogeneous data well, and they require significantly more tuning compared with decision tree based methods. However, decision tree based method, for instance, Gradient Boosting Decision Tree (GBDT), is popular when it comes to solving the problem of learning non-linear functions from data, for that it has an important advantage of dealing with heterogeneous data when different features come from different sources. In this section, we explain our hybrid model combing the LSTM and GBDT for intent prediction.

First, we extract the user's map query sequential features, user's profile features, user's query context features, user's POI arrival statistical features, weather and temperature features from different kinds of sources, and then we encode the features into an id-value format. (Please recall the example illustrated in Fig. 1). For a user's map query sequence, we get sequential features F_{sq}, and for the rest of the features, we get non-sequential features F_{non-sq}. Then, we train the LSTM model with map query sequential data. The detail has been illustrated in Sect. 4.2.

Next, the problem is how to combine the LSTM learned sequential features into the GBDT model. Here, we evaluate two kinds of methods. One is stacking [27] which involves training a learning algorithm to combine the predictions of several other learning algorithms. The other is feature embedding which is a kind of popular method in recent years. In computer vision, people use Convolutional Neural Network (CNN) [16] for image feature embedding. In the recent work [29,31,32], the author uses CNN for spatial-temporal data embedding. There are also some works [30] using knowledge base embedding for recommendation system.

Our work applies LSTM for map query sequence embedding. In the experiment section, we show that the feature embedding method outperforms the stacking method. Similar to feature extraction in image classification problem, we keep the output vector before the final softmax layer, and use this feature vector as our LSTM embedding vector E_{sq}.

$$E_{sq} = LSTM^e(F_{sq}) \tag{13}$$

Then we put the embedding output of LSTM into the GBDT model. The reason we use GDBT rather than other deep learning based method (like CNN, DNN) is that GBDT is better when dealing with heterogeneous data. Finally, the GBDT model makes predictions based on two kinds of features, i.e., sequential data and non-sequential data. F_{sq} represents the user's instant features, and F_{non-sq} involves the context features, long-term features and part of instant features.

$$p'_u = GBDT(E_{sq}, F_{non-sq}) \tag{14}$$

Here, p'_u is different from the p_u in the previous section, since p'_u not only considers instant and sequential characteristics of map query but also utilizes the heterogeneous data from different sources.

5   Experiment

In this section, we evaluate the performance of our model in a real-world dataset from Baidu Maps and compare it with other models. For offline evaluation, we mainly consider the area under the curve (AUC) as our metrics. In next section, we consider the coupon click-through rate (CTR) as a measurement of our online vehicle coupon targeting system.

5.1   Data Description

Using the user id as the key value, we combine the mentioned different sources of data and get our offline training dataset for user vehicle coupon targeting. It consists of 5,986,928 map queries, and covering 1,089,571 users. Each user's map query sequence is appended with the user's non-sequential data. Then we split the dataset as the training set (4,791,755 queries), the validation set (597,417 queries) and the test set (597,756 queries) with ratio 8:1:1. Note that, the dataset is sampled from the whole map query database. We mainly focus on queries which are connected to the user's ride historical request record, because the primary thing we want to do is vehicle coupon targeting. After connecting with the user's ride request history, we get 628,215 map queries which are correlated with ride request history. In other words, the user requests a ride in the next T hours after these query behaviors in Baidu Maps. Here, T is a parameter. We evaluate the T with grid search and find that the model performance is relative stable if T is large enough (like more than one day). In the following experiment, the T is set to 36 h. These queries are labeled as positive samples. Simply, we can throw all other queries into the training dataset and labeled as negative. However, similar to user purchase behavior in the e-commerce scene, people's behavior on requesting a ride is also quite sparse. In our data set, the negative samples are randomly sampled from the map queries which are not connected to the ride request behavior. After the sampling process, we finally get 5,358,713 negative samples, which means the pos/neg rate is near 1/10.

In the training process, we try different parameters to evaluate our model performance with validation data, and then we use the model which has the best result to evaluate the predictive performance on test data.

5.2 Model Comparison

For the problem of intent prediction, there are several candidate models available. In the next subsection, we demonstrate the reason why we choose to integrate the LSTM and GBDT model for such intent prediction problem and then prove that our model is the best choice for such problem. The following candidate models will be evaluated in the next subsection:

- **Pattern Match Prediction.** Pattern match model mainly utilizes the frequent patterns we discovered to make predictions. If a user's search queries match the ride request oriented patterns we extracted before, then we predict the user will requests a ride. How to mine the patterns is demonstrated in Sect. 4.1.
- **GBDT Model.** Gradient boosting decision tree model is a prediction model in the form of an ensemble of decision trees generated by gradient boosting.
- **DNN Model.** Deep Neural Network (DNN) [14] uses a cascade of many layers of nonlinear processing units for feature extraction and transformation.
- **LSTM Model.** LSTM is one of the variant of RNN, and its gated memory structure can deal with the weakness of standard RNN.

5.3 Experiment Results

In this subsection, first, we compare the performance of LSTM, GBDT and DNN when only dealing with sequential data, and then we compare the performance of GBDT and DNN when only dealing with non-sequential data. Note that, through feature preprocess, we can fit non-sequential data into LSTM model, but it makes no sense in the real situation. Second, we demonstrate that our proposed LSTM embedding and LSTM stacking model beat the traditional GBDT model in the real-world dataset. Third, we illustrate that utilizing the heterogeneous data from different sources, including map query data, other non-sequential long-term and context data can significantly improve the performance.

Comparison for Different Data. First, we compare the performance of LSTM, GBDT and DNN when dealing with sequential data and non-sequential data, respectively. Our LSTM model architecture is shown in Fig. 2. To better clarify the performance of LSTM model, we test the LSTM model with different kinds of architectures and different parameters. We prove that adding an extra pooling layer on top of the hidden layer can lead to better performance in our map query sequential data. We also test the parameters such as batch size, learning rate and embedding dimension. The best result is achieved when using LSTM with a max pooling extra layer, the batch size is 64, the embedding dimension is 256, and the LSTM hidden state dimension is 128. The model is trained with Adam optimizer at the learning rate of 1e-4. The best AUC for LSTM on sequential data is 0.763. We evaluate the sequential and non-sequential data with GBDT and DNN model, respectively. The best result for traditional GBDT is achieved when the depth is 6, the number of trees is 200, and the learning rate is 0.3. For DNN model, we build a three-layer neural network with the batch size being 128 and the hidden unit dimension being 128. We use dropout for the last connected layer, and the other optimizer parameters are the same with LSTM.

The best results of each model are shown in Table 1. It is obvious that LSTM outperforms GBDT and DNN model when handling with sequential data. When dealing with non-sequential heterogeneous data, GBDT and DNN model get similar performance, while GBDT model is slightly better than DNN. However, GBDT is much faster than DNN and requires less feature engineering and parameter tuning. Therefore, in our model, we chooses the combination of LSTM and GBDT.

Table 1. Single model comparison for different data

Method	Sequential (AUC)	Non-sequential (AUC)
LSTM	**0.763**	–
GBDT	0.743	**0.754**
DNN	0.739	0.748

Different Model Comparison. Second, we compare our proposed model with several other models. We evaluate our model with different trees number N and tree depth d by grid search. The number of boosting trees N is ranging from 100 to 1000, and the depth of a boosting tree d is chosen from {3, 6, 9,12}. We also try different learning rates η, and η is chosen from {0.01, 0.03, 0.1, 0.3}. Finally we set the number of trees as 200, the depth as 6 and η as 0.1.

There are two methods for integrating the LSTM model with GBDT model. The one is to use stacking, which considers the final score of LSTM as a feature; the other one is to use feature embedding method. Our method extracts the output of max pooling layer as our embedding vector. The LSTM training parameter is consistent with the previous best configuration. The evaluation results of traditional GBDT, LSTM stacking and embedding method with GBDT are shown in Fig. 3.

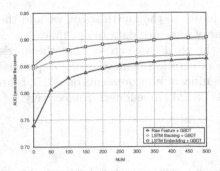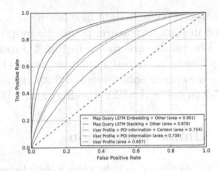

Fig. 3. Left: Receiver Operating Characteristic (ROC) curves for feature integration. The top-left two curves (blue and green curves) show that the integration of map query sequential data and other non-sequential data leads to significant improvements on AUC. Right: model comparison (depth = 6). Red triangle, green circle and blue square represent the performance of traditional GBDT, LSTM stacking and embedding with GBDT, respectively. (Color figure online)

The comparison results of different models are shown in Table 2. It shows that our model outperforms the Pattern Match prediction, traditional GBDT and DNN model. The LSTM embedding model is superior to the LSTM stacking model, probably because the former one preserves more information about learned map query sequence data.

Table 2. Different model comparison

Method	AUC score
Pattern match	0.652
DNN	0.850
GBDT	0.863
GBDT + LSTM stacking	0.878
GBDT + LSTM embedding	**0.901**

Significance of Feature Integration. Third, we prove that the model performance is significantly improved by integrating heterogeneous data from different sources. Furthermore, the map query sequential data is crucial for the overall performance. To verify above points, we do a four-stage experiment for model evaluation. All the data used in this section is discussed in Sect. 3.2. First, we only use profile data to make predictions. In the second stage, we add the POI arrival information to our model. The data records the user's POI arrival distribution information in recent one month. As the context information can affect user's decision on requesting a ride, therefore, in the third stage, we add the mentioned context data into our model. Finally, in the fourth stage, we add the most important feature, i.e, user's map query sequence data to our model.

The results are shown in Fig. 3. From the result, we can see that the user's map query is indeed important, and the integration of map query sequential data and other non-sequential data gives significant improvements on AUC, from 0.754 to 0.878 (LSTM stacking) and 0.901(LSTM embedding), respectively.

6   Online Evaluation

In this section, we evaluate our vehicle coupon targeting system for the real-time online vehicle coupon targeting. The evaluation is based on the Baidu Maps online marketing campaign for vehicle coupon pushing. The campaign aims at attracting more customers to use Baidu Maps app to request a ride (such as taxi, Didi and Baidu cars).

The marketing campaign is mainly launching at the four cities in China, including Guangzhou, Wuhan, Nanjing and Chengdu, and lasts for a month. The coupon has two kinds of types. One is for the Baidu Ordinary Car discount, and the other is for the Baidu Luxury Car discount. In general, the coupon discount on Baidu Luxury Car is higher than Baidu Ordinary Car, for that Baidu Luxury Car is expensive compared to Baidu Ordinary Car. The coupon type is selected by the marketing team. Our proposed model is primarily to tell the system which user should receive vehicle coupons.

We test following two methods as baselines, and refer Click-Through Rate (CTR) as our evaluation metric:

- **Random Targeting.** The random targeting method is the baseline method, which randomly pushes a vehicle coupon to a user.
- **Airport Enclosure.** The airport enclosure method pushes vehicle coupons based on airport environment. The mechanism is that, if the system detects that there is a user whose current location is within 5 km from the airport, then he/she will receive the vehicle coupon. Note that the airport environment has a very close correlation with vehicle coupon targeting, since people appear within airport has strong demand for requesting a ride. Therefore, it can be treated as an almost optimal competitor for our online test.

For our method, we need to determine the predefined threshold τ for pushing coupons to users. Usually, it can be determined by the marketing team according to the actual campaign requirement. However, in this evaluation, in order to make the performance of our method be comparable with the baselines, we set the threshold τ to push coupons to the same percent of users out of the studied group users as the same as the percent of the detected users in airports out of total monthly active users on Baidu Maps app.

The reasons that we use CTR rather than AUC as online evaluation metric are as follows. First, the value of AUC is calculated by its corresponding Receiver Operating Characteristic (ROC) curves, and the ROC curve is created by plotting the true positive rate against the false positive rate at various threshold settings. While our baseline methods (like Airport Enclosure) cannot output probabilistic result, the output is just binary (pushing or not pushing), which makes its ROC curve not be reasonable. Second, to some extent, the coupon pushing activity can affect the online AUC results, thus using CTR evaluation metric is better. Third, for the business of Baidu Maps, the CTR is the most meaningful metric they care about.

The coupon CTR of our system and two other marketing campaign methods is shown in Table 3. The CTR of our model is 2.67%, which is higher than all the baselines. Note that as for the scene of vehicle coupon targeting, the CTR is hard to improve, for the reason that we may have tremendous of offline data, but the online click data is rare. Our best model can get a 26.5% boost comparing to the airport enclosure baseline, which is a significant improvement.

By retrospecting the results of the campaign, we also analyze coupon CTR on different coupon types. The result is shown in Fig. 4. We can see the overall coupon CTR on Baidu Luxury Car is higher than Baidu Ordinary Car (2.79% vs 2.60%). We also find the iOS customer group has a higher CTR than Android customer group (3.05% vs 2.43%, which is not shown in Fig. 4).

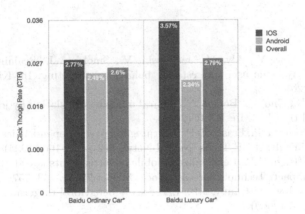

Fig. 4. Click-Trough Rate (CTR) on different coupon types. The left side is Baidu ordinary car coupon, and the right side is baidu luxury car coupon. Purple, blue and green bar stands for IOS, Android and overall average CTR, respectively. (Color figure online)

7   Conclusion

In this work, we described our deployed intent-aware audience targeting system on Baidu Maps. We demonstrated that the user's sequential map query data implicitly represents the user's intent and is significantly important for such intent prediction problem. In order to better exploit the map query sequential

Table 3. Online evaluation comparison

Methods	Online CTR	Δ CTR
Random push	0.32%	-
Airport enclosure	2.11%	0% -
Our model	**2.67%**	**26.5%** ↑

data and other multi-source heterogeneous data, we develop a method to combine the advantages of LSTM model and GBDT model to handle sequential and non-sequential data, and we showed that the combination can lead to significant improvements in the overall prediction performance. We evaluated the performance of our method over a large real world dataset and conducted a large-scale online marketing campaign over Baidu Maps app client to verify the effectiveness of our audience targeting system. The experiment results of the offline evaluation and online marketing campaign demonstrates the effectiveness of our system for predicting the user's intent on requesting a ride, and also exhibited significant improvements in the click-through rate of vehicle coupon targeting.

References

1. Ahmed, A., Low, Y., Aly, M., Josifovski, V., Smola, A.J.: Scalable distributed inference of dynamic user interests for behavioral targeting. In: KDD, pp. 114–122. ACM (2011)
2. Bahdanau, D., Cho, K., Bengio, Y.: Neural machine translation by jointly learning to align and translate. In: ICLR (2015)
3. Bengio, Y., Simard, P., Frasconi, P.: Learning long-term dependencies with gradient descent is difficult. IEEE Trans. Neural Netw. **5**(2), 157–166 (1994)
4. Chen, P.T., Hsieh, H.P.: Personalized mobile advertising: its key attributes, trends, and social impact. Technol. Forecast. Soc. Change **79**(3), 543–557 (2012)
5. Chen, Y., Pavlov, D., Canny, J.F.: Large-scale behavioral targeting. In: KDD, pp. 209–218. ACM (2009)
6. Friedman, J.H.: Greedy function approximation: a gradient boosting machine. Ann. Stat. **29**(5), 1189–1232 (2001)
7. Goldfarb, A., Tucker, C.E.: Online advertising, behavioral targeting, and privacy. Commun. ACM **54**(5), 25–27 (2011)
8. Graepel, T., Candela, J.Q., Borchert, T., Herbrich, R.: Web-scale Bayesian click-through rate prediction for sponsored search advertising in microsoft's bing search engine. In: ICML, pp. 13–20 (2010)
9. Ha, J.W., Pyo, H., Kim, J.: Large-scale item categorization in e-commerce using multiple recurrent neural networks. In: KDD, pp. 107–115. ACM (2016)
10. Han, J., Pei, J., Mortazavi-Asl, B., Chen, Q., Dayal, U., Hsu, M.C.: FreeSpan: frequent pattern-projected sequential pattern mining. In: KDD, pp. 355–359. ACM (2000)
11. Han, J., et al.: PrefixSpan: mining sequential patterns efficiently by prefix-projected pattern growth. In: ICDE, pp. 215–224 (2001)

12. Hao, T., Zhou, J., Cheng, Y., Huang, L., Wu, H.: User identification in cyber-physical space: a case study on mobile query logs and trajectories. In: SIGSPA-TIAL, p. 71. ACM (2016)
13. He, X., et al.: Practical lessons from predicting clicks on ads at Facebook. In: ADKDD, pp. 1–9. ACM (2014)
14. Hinton, G.E., Osindero, S., Teh, Y.W.: A fast learning algorithm for deep belief nets. Neural Comput. **18**(7), 1527–1554 (2006)
15. Hochreiter, S., Schmidhuber, J.: Long short-term memory. Neural Comput. **9**(8), 1735–1780 (1997)
16. Krizhevsky, A., Sutskever, I., Hinton, G.E.: ImageNet classification with deep convolutional neural networks. In: NIPS, pp. 1097–1105 (2012)
17. Li, K., Du, T.C.: Building a targeted mobile advertising system for location-based services. Decis. Supp. Syst. **54**(1), 1–8 (2012)
18. Liu, K., Tang, L.: Large-scale behavioral targeting with a social twist. In: CIKM, pp. 1815–1824. ACM (2011)
19. McMahan, H.B., et al.: Ad click prediction: a view from the trenches. In: KDD. pp. 1222–1230. ACM (2013)
20. Pandey, S., et al.: Learning to target: what works for behavioral targeting. In: CIKM, pp. 1805–1814. ACM (2011)
21. Rumelhart, D.E., Hinton, G.E., Williams, R.J.: Learning representations by back-propagating errors. Cogn. Model. **5**(3), 1 (1988)
22. Srikant, R., Agrawal, R.: Mining sequential patterns: generalizations and performance improvements. In: Apers, P., Bouzeghoub, M., Gardarin, G. (eds.) EDBT 1996. LNCS, vol. 1057, pp. 1–17. Springer, Heidelberg (1996). https://doi.org/10.1007/BFb0014140
23. Sutskever, I., Vinyals, O., Le, Q.V.: Sequence to sequence learning with neural networks. In: NIPS, pp. 3104–3112 (2014)
24. Tang, D., Qin, B., Liu, T.: Document modeling with gated recurrent neural network for sentiment classification. In: EMNLP, pp. 1422–1432 (2015)
25. Tang, J., et al.: Learning to rank audience for behavioral targeting in display ads. In: CIKM, pp. 605–610. ACM (2011)
26. Werbos, P.J.: Backpropagation through time: what it does and how to do it. Proc. IEEE **78**(10), 1550–1560 (1990)
27. Wolpert, D.H.: Stacked generalization. Neural Netw. **5**(2), 241–259 (1992)
28. Yan, J., Liu, N., Wang, G., Zhang, W., Jiang, Y., Chen, Z.: How much can behavioral targeting help online advertising? In: WWW, pp. 261–270. ACM (2009)
29. Yao, H., et al.: Deep multi-view spatial-temporal network for taxi demand prediction. In: AAAI (2018)
30. Zhang, F., Yuan, N.J., Lian, D., Xie, X., Ma, W.Y.: Collaborative knowledge base embedding for recommender systems. In: KDD, pp. 353–362. ACM (2016)
31. Zhang, J., Zheng, Y., Qi, D.: Deep spatio-temporal residual networks for citywide crowd flows prediction. In: AAAI (2017)
32. Zhang, J., Zheng, Y., Qi, D., Li, R., Yi, X.: DNN-based prediction model for spatio-temporal data. In: SIGSPATIAL, pp. 92:1–92:4. ACM (2016)
33. Zhou, J., Pei, H., Wu, H.: Early warning of human crowds based on query data from Baidu maps: analysis based on shanghai stampede. In: Shen, Z., Li, M. (eds.) Big Data Support of Urban Planning and Management. AGIS, pp. 19–41. Springer, Cham (2018). https://doi.org/10.1007/978-3-319-51929-6_2
34. Zhou, J., Tung, A.K., Wu, W., Ng, W.S.: A "semi-lazy" approach to probabilistic path prediction in dynamic environments. In: KDD, pp. 748–756. ACM (2013)

A Recurrent Neural Network Survival Model: Predicting Web User Return Time

Georg L. Grob[1(✉)], Ângelo Cardoso[2], C. H. Bryan Liu[2(✉)], Duncan A. Little[2], and Benjamin Paul Chamberlain[1,2]

[1] Imperial College London, London, UK
grobgl@gmail.com
[2] ASOS.com, London, UK
bryan.liu@asos.com

Abstract. The size of a website's active user base directly affects its value. Thus, it is important to monitor and influence a user's likelihood to return to a site. Essential to this is predicting *when* a user will return. Current state of the art approaches to solve this problem come in two flavors: (1) Recurrent Neural Network (RNN) based solutions and (2) survival analysis methods. We observe that both techniques are severely limited when applied to this problem. Survival models can only incorporate aggregate representations of users instead of automatically learning a representation directly from a raw time series of user actions. RNNs can automatically learn features, but can not be directly trained with examples of non-returning users who have no target value for their return time. We develop a novel RNN survival model that removes the limitations of the state of the art methods. We demonstrate that this model can successfully be applied to return time prediction on a large e-commerce dataset with a superior ability to discriminate between returning and non-returning users than either method applied in isolation. Code related to this paper is available at: https://github.com/grobgl/rnnsm.

Keywords: User return time · Web browse sessions
Recurrent neural network · Marked temporal point process
Survival analysis

1 Introduction

Successful websites must understand the needs, preferences and characteristics of their users. A key characteristic is *if* and *when* a user will return. Predicting user return time allows a business to put in place measures to minimize absences and maximize per user return probabilities. Techniques to do this include timely incentives, personalized experiences [16] and rapid identification that a user is losing interest in a service [2]. A related problem is user churn prediction for non-contractual services. In this case a return time threshold can be set, over

---

Â. Cardoso—Now with Vodafone Research and ISR, IST, Universidade de Lisboa.

© Springer Nature Switzerland AG 2019
U. Brefeld et al. (Eds.): ECML PKDD 2018, LNAI 11053, pp. 152–168, 2019.
https://doi.org/10.1007/978-3-030-10997-4_10

which a user is deemed to have churned. Similar prevention measures are often put in place for users with high churn risks [5].

This paper focuses on predicting user return time for a website based on a time series of user sessions. The sessions have additional features and so form a *marked temporal point processes* for each user. As some users do not return within the measurement period, their return times are regarded as *censored*. The presence of missing labels makes the application of standard supervised machine learning techniques difficult. However, in the field of survival analysis, an elegant solution to the missing label problem exists which has been transferred to a variety of settings [24,31].

Recurrent Neural Networks (RNNs) have achieved significant advances in sequence modelling, achieving state-of-art performance in a number of tasks [19,33,34]. Much of the power of RNNs lie in their ability to automatically extract high order temporal features from sequences of information. Many complex temporal patterns of web user behaviour can exist. This can include noisy oscillations of activity with periods of weeks, months, years, pay days, festivals and many more beside. Exhaustively handcrafting exotic temporal features is very challenging and so it is highly desirable to employ a method that can automatically learn temporal features.

In this paper, we predict user return time by constructing a recurrent neural network-based survival model. This model combines useful aspects of both RNNs and survival analysis models. RNNs automatically learn high order temporal features from user 'sessions' and their associated features. They can not however, be trained with examples of users who do not return or the time since a user's latest session. Survival models can include information on users who do not return (right-censored users) and the time since their last session, but cannot learn from a sequence of events.

Our main contribution is to develop a RNN-based survival model which incorporates the advantages of using a RNN and of using survival analysis. The model is trained on sequences of sessions and can also be trained with examples of non-returning users. We show that this combined model outperforms both RNNs and survival analysis employed in isolation. We also provide the code implementation for use by the wider research community.[1]

2   Background

We are interested in predicting the return times of users to a website. We select a period of time during which users must be observed visiting a site and call this the *Activity window*. We declare a separate disjoint period of time called the *Prediction window* from which we generate return time labels and make predictions and both windows are illustrated in Fig. 1. There are necessarily two types of user: returning and non-returning. We consider a user as non-returning if they do not have any sessions within the Prediction window, and a returning

---

[1]   https://github.com/grobgl/rnnsm.

user as those who do. As suggested by Wangperawong et al. [35] we record data for some time preceding the Activity window (called the *Observation window*) to avoid producing a model that would predominantly predict non-returning users. This set up allows us to define the set of users active in the activity window; $\mathcal{C}$, the set of returning users, $\mathcal{C}_{\text{ret}}$ and the set of non-returning users $\mathcal{C}_{\text{non-ret}}$.

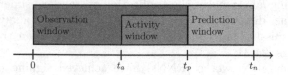

Fig. 1. Illustration of the observation window, the activity window, and the prediction window. The x-axis denotes time. We use observations occurring within the observation window $[0, t_p]$ on users active in the activity window $[t_a, t_p]$. We then predict their return times with respect to the prediction window $(t_p, t_n]$.

We follow the definition of return time suggested by Kapoor et al. [26], that is, $d_{j+1}^{(i)} = t_{j+1}^{(i)} - t_j^{(i)}$, denotes the period between the end of the ith user's jth session and the beginning of the succeeding session. A session occurs when a user browses a website or mobile app. The ith user's jth session, $s_j^{(i)} = \left(t_j^{(i)}, \boldsymbol{y}_j^{(i)} \right)$, has an associated start time $t_j^{(i)} \in [0, \infty)$ and a vector of n features $\boldsymbol{y}_j^{(i)}$. A user's browsing history can therefore be represented as a sequence of sessions, $\mathcal{S}^{(i)}$, where $\mathcal{S}^{(i)} = \left(s_1^{(i)}, s_2^{(i)}, \ldots \right)$.

2.1 Survival Analysis

Survival analysis models the time until an event of interest occurs [27]. In the context of users' return time prediction, return time is equivalent to survival time.

Harzard and Survival Functions. Here we clarify the notation and standard results used throughout the paper, as defined in [32]. T is a random variable denoting the lifetime of an individual. The probability density function for T, corresponding to the probability of the event of interest occurring at time t is written as:

$$f(t) = \lim_{\delta t \to 0} \frac{P(t < T \leq t + \delta t)}{\delta t}, \tag{1}$$

with $F(t)$ as the corresponding cumulative density function. The survival function $S(t)$, denoting the probability of the event of interest not having occurred by time t, is defined as:

$$S(t) = P(T \geq t) = 1 - F(t) = \int_t^\infty f(z) \, \mathrm{d}z. \tag{2}$$

The hazard function, which models the instantaneous rate of occurrence given that the event of interest did not occur until time t, is defined as:

$$\lambda(t) = \lim_{\delta t \to 0} \frac{P(t \leq T < t + \delta t | T \geq t)}{\delta t} = \frac{f(t)}{1 - F(t)} = \frac{f(t)}{S(t)}. \tag{3}$$

The hazard function is related to the survival function by

$$-\frac{d \log S(t)}{dt} = \lambda(t). \tag{4}$$

Censoring. Censoring occurs when labels are partially observed. Klein and Moeschberger [27] provide two definitions of censoring (1) an *uncensored observation* when the label value is observed and (2) *right-censored observation* when the label value is only known to be above an observed value.

In the context of return time prediction, some users do not return to the website during the observation period. We label these users as non-returning, but some will return to the website after the observation period. Figure 2 shows a selection of returning and non-returning users. To estimate the average return time, it is not sufficient to only include returning users as this underestimates the value for all users. Including non-returning users' time since their latest session still underestimates the true average return time.

To address this problem, we must incorporate censoring into our survival model. This is achieved using a likelihood function that has separate terms to account for censoring:

$$L(\theta) = \prod_{i \in \text{unc.}} P(T = T_i | \theta) \prod_{j \in \text{r.c.}} P(T > T_j | \theta) = \prod_{i \in \text{unc.}} f(T_i | \theta) \prod_{j \in \text{r.c.}} S(T_j | \theta),$$
$$\tag{5}$$

where θ is a vector of model parameters. *unc.* and *r.c.* denote the uncensored and right-censored observations respectively. T_i and T_j denote the exact value of the uncensored observation and the minimum value of the right-censored observation respectively. For simplicity we assume there are no observations which are subject to other types of censoring.

2.2   Cox Proportional Hazards Model

The Cox proportional hazards model [11] is a popular survival regression model. It is applied by Kapoor et al. [26] to predict the return time of users.

The model assumes that one or more given covariates have (different) multiplicative effects on a base hazard. The model can be defined in terms of its hazard function:

$$\lambda(t | x) = \lambda_0(t) \exp\left(x^T \beta\right), \tag{6}$$

where $\lambda_0(t)$ is the baseline hazard, x a vector of covariates, and β_i the multiplier for covariate x_i. The model implicitly assumes the parameters for each covariate can be estimated without considering the baseline hazard function.

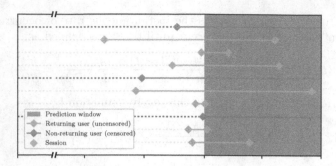

Fig. 2. Visualisation of censored return times. The horizontal axis represents time and the vertical axis different users. The shaded area represents the prediction window. The solid lines represent the return time after users' last sessions in the observation time frame – the value we are aiming to predict. The return times of users that do not return in the prediction time frame are censored. We do not know their actual return time. However, we do know that their return time spans at least across the entire prediction window.

Various methods to estimate the multipliers β exist [3,12,25], and we use that featured in the *lifelines* library [14], which maximises the Efron's partial likelihood [17] for β to obtain the estimate $\hat{\beta}$.

The estimated baseline hazard is then computed as described by Cox and Oakes [13]:

$$\hat{\lambda}_0\left(t_{(i)}\right) = \frac{d_{(i)}}{\sum_{j \in \mathcal{R}\left(t_{(i)}\right)} \exp\left(\hat{\beta}^T x_j\right)}, \tag{7}$$

where $t_{(i)}$ denotes the i unique ordered time of an event of interest, $d_{(i)}$ is the number of events of interest occurring at time $t_{(i)}$, and $\mathcal{R}\left(t_{(i)}\right)$ is the set of individuals for whom the event of interest has not occurred by time $t_{(i)}$. The users' return times can then be estimated by calculating their expected survival time.

The Cox proportional hazards model is particularly suitable for this problem as it allow us to include right-censored observations (users who appeared not returning) as training examples. Their value is only known to be above a certain value, which corresponds to the time between the last session in the observation time frame and the end of the prediction time frame.

2.3 Recurrent Neural Networks, LSTM, and Embeddings

A recurrent neural network (RNN) is a feedforward neural network where the output of a hidden unit at the current timestep is fed back into the hidden unit so that it forms part of the input for the preceding timesteps. This allows RNNs to learn from sequences of events ordered chronologically or otherwise. The power of an RNN lies in its ability to learn from the current state of the sequence within the context of what has gone before. This context is stored as an internal memory within the hidden units of the RNN. For modelling time series data, sequences of events are discretised in time [4, 6, 20].

Long Short-Term Memory (LSTM) units [23] and Gated Recurrent Units (GRUs) [8] were developed to overcome the problems associated with learning long-term dependencies in traditional RNNs [1]. LSTMs and GRUs solve this issue by learning what information they should keep from the previous time step and what information they should forget.

It is also common to add embedding layer(s) to neural networks to transform large and sparse information info dense representations before the actual training of the network [10, 18]. The embedding layer will automatically learn features (in the form of the dense representation's individual components) for the neural network's consumption.

First popularised by Mikolov et al. [30] to encode words in documents, embedding layers have been shown to encode various categorical features with a large number of possible values well [7, 28]. In this paper we use the embedding layer implementation in Keras [9] with a TensorFlow backend.

2.4   Recurrent Temporal Point Processes

Temporal point processes model times of reoccurring events, which may have markers (features) associated with them, such as click rates and duration for web sessions. Manzoor et al. [29] modeled both the timing and the category of a user's next purchase given a history of such events using a Hawkes process, which assumes the occurrence of past events increases the likelihood of future occurrences [22].

Du et al. [15] propose the recurrent marked temporal point process (RMTPP) to predict both timings and markers (non-aggregated features) of future events given a history of such events. They assume events have exactly one discrete marker and employ RNNs to find a representation for the event history, which then serves as input to the hazard function. The paper demonstrates that such process can be applied in a wide variety of settings, including return times of users of a music website.

The RMTPP model is formulated as follows. Let $\mathcal{H}_t$ be the history of events up to time t, containing event pairs $(t_j, y_j)_{j \in \mathbb{Z}^+}$ denoting the event timing and marker respectively. The conditional density function corresponds to the likelihood of an event of type y happening at time t: $f^*(t, y) = f(t, y | \mathcal{H}_t).$[2]

A compact representation $\boldsymbol{h}_j$ of the history up to the jth event is found through processing a sequence of events $\mathcal{S} = (t_j, y_j)_{j=1}^{n}$ with an RNN. This allows the representation of the conditional density of the next event time as:

$$f^*(t_{j+1}) = f(t_{j+1} | \boldsymbol{h}_j).$$ (8)

---

[2] The $*$-notation is used to denote the conditioning on the history.

Given $\boldsymbol{h}_j$, the hazard function of the RMTPP is defined as follow:

$$\lambda^*(t) = \exp\Big(\underbrace{\boldsymbol{v}^{(t)\top}\boldsymbol{h}_j}_{\substack{\text{past}\\\text{influence}}} + \underbrace{w(t-t_j)}_{\substack{\text{current}\\\text{influence}}} + \underbrace{b^{(t)}}_{\substack{\text{base}\\\text{intensity}}}\Big), \tag{9}$$

where $\boldsymbol{v}^{(t)}$ is the hidden representation in the recurrent layer in the RNN (which takes only $\boldsymbol{h}_j$, the representation of past history, into account), $t - t_j$ is the absence time at the time of prediction (the current information), w is a specified weight balancing the influence from the past history from that of the current information,[3] and $b^{(t)}$ is the base intensity (or bias) term of the recurrent layer.

The conditional density is given by swapping the terms in Eq. (3) and integrating Eq. (4):

$$f^*(t) = \lambda^*(t)\exp\left(-\int_{t_j}^{t}\lambda^*(\tau)\,d\tau\right)$$

$$= \exp\Big(\boldsymbol{v}^{(t)\top}\boldsymbol{h}_j + w(t-t_j) + b^{(t)} + \frac{1}{w}\exp\Big(\boldsymbol{v}^{(t)\top}\boldsymbol{h}_j + b^{(t)}\Big)$$

$$- \frac{1}{w}\exp\Big(\boldsymbol{v}^{(t)\top}\boldsymbol{h}_j + w(t-t_j) + b^{(t)}\Big)\Big). \tag{10}$$

The timings of the next event can then be estimated by taking the expectation of the conditional density function:

$$\hat{t}_{j+1} = \int_{t_j}^{\infty} t f^*(t)\,dt. \tag{11}$$

The architecture of the RNN is illustrated in Fig. 3. The event markers are embedded into latent space. The embedded event vector and the event timings are then fed into a recurrent layer. The recurrent layer maintains a hidden state $\boldsymbol{h}_j$ which summarises the event history. The recurrent layer uses a rectifier as activation function and is implemented using LSTM or GRUs.

The parameters of the recurrent layer ($\boldsymbol{v}^{(t)}$ and $b^{(t)}$) are learned through training the RNN using a fully connected output layer with a single neuron and linear activation, with the negative log-likelihood[4] of observing a collection of example sequences $\mathcal{C} = \left\{\left(t_j^{(i)}, y_j^{(i)}\right)_{j=1}^{n^{(i)}}\right\}_{i\in\mathbb{Z}^+}$ defined as:

$$-\ell(\mathcal{C}) = -\sum_i\sum_j \log f\left(t_{j+1}^{(i)}\Big|\boldsymbol{h}_j\right). \tag{12}$$

---

[3] Du et al. [15] specified a different fixed value for w in their models fitted under different datasets. We use Bayesian Optimisation to find the best w in our experiments.

[4] The original log-likelihood function in [15] took into account both the likelihood of the timing of the next event and the likelihood of the marker taking certain value. The later is omitted for simplicity as we do not deal with marker prediction here.

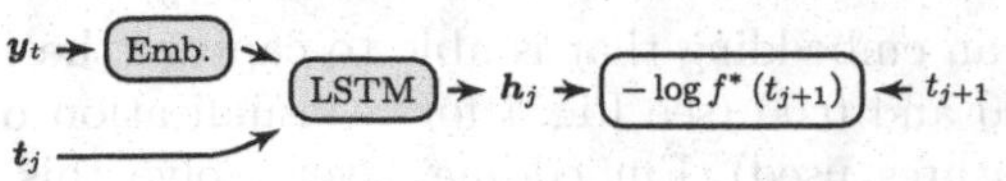

$$y_t \rightarrow \boxed{\text{Emb.}} \rightarrow \boxed{\text{LSTM}} \rightarrow h_j \rightarrow \boxed{-\log f^*(t_{j+1})} \leftarrow t_{j+1}$$
$$t_j \longrightarrow$$

Fig. 3. Architecture of the recurrent neural network used in the RMTPP model to learn a representation h_j of an event history consisting of pairs of timings and markers (t_i, y_i). t_j and y_j represent the timings and events of the history up to the jth event. h_j is learned through minimising the negative log-likelihood of the $j+1^{\text{th}}$ event occurring at time t_{j+1}. The hidden representation h_j is then used as parameter to a point process.

3   Method

Survival models can only accept a vector of features aggregated for each user as input. By using aggregated features, we discard a significant proportion of information contained in the time series of events. Unlike survival models, RNNs are capable of utilising the raw user history and automatically learning features. However, censored data can not be included. Omitting censored users causes predictions to be heavily biased towards low return times. We remove the limitations of RNNs and Cox proportional hazard models by developing a novel model that can incorporate censored data, use multiple heterogeneous markers and automatically learn features from raw time series data.

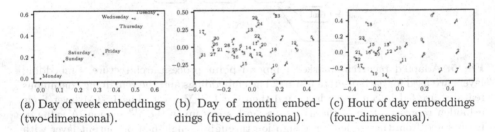

(a) Day of week embeddings (two-dimensional). (b) Day of month embeddings (five-dimensional). (c) Hour of day embeddings (four-dimensional).

Fig. 4. Embeddings for discrete features used in the simple recurrent neural network model. The embeddings translate discrete inputs into vectors of specified dimension. We can observe a clusters of late night and early morning hours in (c), and a separation of weekend days from weekdays in (a), suggesting that viable representations are found. The embeddings are found by training a model with embeddings layers as input (alongside the remaining inputs) and then evaluating the embeddings for each possible input. Dimensionality reduction through PCA is used to produce the visualisations in case the embedding vector has more than two dimensions.

3.1   Heterogeneous Markers

In many practical settings, multiple heterogeneous markers are available describing the nature of events. Markers can be be both discrete and continuous. To encode discrete markers we use an embedding layer. We also embed cyclic features, an example being the hour of an event. Instead of encoding the hour as

$\{0, 1..23\}$ we learn an embedding that is able to capture the similarity between e.g. the hours 23:00 and 0:00 (see Fig. 4 for a visualisation of the embeddings on some of the features used). Embedding layers solve this problem through mapping discrete values into a continuous vector space where similar categories are located nearby to each other. We train the network to find mappings that represent the meaning of the discrete features with respect to the task, i.e. to predict the return time. We apply a unit norm constraint to each embedding layer, enforcing the values to be of a similar magnitude to the non-categorical variables, which are also normalized.

To avoid an expensive search for a suitable number of embedding dimensions during training, we perform a preliminary simulation. We train a model with a high number of dimensions per feature and use Principal Component Analysis (PCA) to reduce the dimensionality to the minimum number required to account for more than 90% of the initial variance.

The embeddings and the non-categorical features are fed into a single dense layer, which produces the input to the LSTM. Figure 5 shows how the model processes heterogeneous input data.

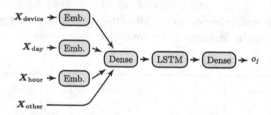

Fig. 5. Adapted recurrent marked temporal point process architecture. Embedding layers are used for each discrete feature. The embeddings and the remaining continuous features are fed into a dense layer and then the LSTM layer. The LSTM layer finds a hidden representation of the user's session history up to the jth session. A value o_j is obtained from the hidden representation through a single-neuron output layer with linear activation. The negative log-likelihood of the next session occurring at time t_{j+1} is used to train the model.

3.2   Recurrent Neural Network Survival Model (RNNSM)

We combine ideas from RNNs and survival analysis to produce a model that is able to incorporate non-returning users and automatically learn features in a survival regression setting. To achieve this we use the survival function as a factor for right-censored (non-returned user) observations. We start with the log likelihood function defined in Eq. (5)

$$\ell\left(\mathcal{C}\right) = \sum_m \sum_{i \in \text{unc.}} \log f^* \left(t_{i+1}^{(m)}\right) + \sum_n \sum_{j \in \text{r.c.}} \log S^* \left(t_{j+1}^{(n)}\right), \tag{13}$$

which is a sum over all session intervals for all users.

The first term is thus the log-likelihood for a single returning user at time t_{j+1} given an embedded representation of the user's session history up to their latest session at time t_j as h_j. This log-likelihood is defined as

$$
\begin{aligned}
\ell_{\text{ret}}\left(t_{j+1}\right) &= \log f^*\left(t_{j+1}\right) \\
&= o_j + w\left(t_{j+1} - t_j\right) + \frac{1}{w}\exp\left(o_j\right) - \frac{1}{w}\exp\left(o_j + w\left(t_{j+1} - t_j\right)\right),
\end{aligned}
\tag{14}
$$

where o_j represents the output of the fully-connected layer after jth step of the input sequence: $o_j = v^T h_j + b$. For the expression in the second term we substitute the survival function given by

$$
S^*\left(t\right) = \exp\left(-\int_{t_j}^{t}\lambda^*\left(\tau\right)d\tau\right) = \exp\left(\frac{1}{w}\exp\left(o_j\right) - \frac{1}{w}\exp\left(o_j + w\left(t - t_j\right)\right)\right)
\tag{15}
$$

from Eq. (10) to get the log-likelihood term for a single censored user:

$$
\ell_{\text{non-ret}}\left(t_{j+1}\right) = \log S^*\left(t_{j+1}\right) = \frac{1}{w}\exp\left(o_j\right) - \frac{1}{w}\exp\left(o_j + w\left(t - t_j\right)\right),
\tag{16}
$$

where t_{j+1} refers to the time between the jth user's last session in the observation window and the end of the prediction window.

We can now express a loss function that incorporates examples of non-returned users. Note that in a sequence of active days of a non-returning user, only the last return time is censored. The loss for all users is given by

$$
-\ell(\mathcal{C}) = -\sum_i \sum_j \ell\left(t_{j+1}^{(i)}\right),
\tag{17}
$$

where

$$
\ell\left(t_j^{(i)}\right) = \begin{cases} \ell_{\text{non-ret}}\left(t_j^{(i)}\right), & \text{if } i \in \mathcal{C}_{\text{non-ret}} \text{ and } j = n^{(i)} + 1 \\ \ell_{\text{ret}}\left(t_j^{(i)}\right), & \text{otherwise} \end{cases}
\tag{18}
$$

and $\mathcal{C} = \mathcal{C}_{\text{ret}} \cup \mathcal{C}_{\text{non-ret}}$ denotes the collections of all users' session histories, consisting of the histories of returning and non-returning users.

3.3   Return Time Predictions

We predict the return time, which is the time between two sessions: $d_{j+1} = t_{j+1} - t_j$ using the expectation of the return time given the session history:

$$
\hat{d}_{j+1} = \mathbb{E}\left[T|\mathcal{H}_j\right] = \int_{t_j}^{\infty} S^*\left(t\right)dt.
\tag{19}
$$

This integral does not in general have a closed form solution, but can easily be evaluated using numerical integration.

However, this expression allows the model to predict users to return before the start of the prediction window (see Fig. 2). Therefore we need to censor the predictions by finding $\mathbb{E}[T|T > t_p]$ where t_p is the start of the prediction window. We show that the conditional expected return time can be derived from the expected return time through applying the definition of the survival function.

$$\mathbb{E}[T] = P(T > t_p)\,\mathbb{E}[T|T > t_p] + P(\overline{T > t_p})\,\mathbb{E}[T|\overline{T > t_p}]$$
$$\Leftrightarrow \quad \mathbb{E}[T|T > t_p] = \frac{\mathbb{E}[T] - P(T \le t_p)\,\mathbb{E}[T|T \le t_p]}{P(T > t_p)}. \tag{20}$$

Using Eq. (2), we obtain:

$$\mathbb{E}[T|T > t_s] = \frac{\int_0^\infty S(z)\,dz - (1 - S(t_s))\int_0^{t_s} S(z)\,dz}{S(t_s)}$$
$$= \frac{\int_{t_s}^\infty S(z)\,dz}{S(t_s)} + \int_0^{t_s} S(z)\,dz. \tag{21}$$

4   Experiments

Here we compare several methods for predicting user return time, discussing the advantages, assumptions and limitations of each and providing empirical results on a real-world dataset. We experiment with six distinct models: (1) a baseline model, using the time between a user's last session in the observation time frame and the beginning of the prediction time frame ("Baseline"); (2) a simple RNN architecture ("RNN");[5] (3) a Cox proportional hazard model ("CPH") (4) a Cox proportional hazard model conditioned on absence time ("CPHA"—see Sect. 3.3); (5) a RNN survival model ("RNNSM"); (6) a RNN survival model conditioned on absence time ("RNNSMA"—see Sect. 3.3).

The dataset is a sample of user sessions from ASOS.com's website and mobile applications covering a period of one and a half years. Each session is associated with a user, temporal markers such as the time and duration, and behavioural data such as the number of images and videos viewed during a session. The dataset is split into training and test sets using split that is stratified to contain equal ratios of censored users. In total, there are 38,716 users in the training set and 9,680 users in the test set. 63.6% of users in both sets return in the prediction window. In the test set, the targeted return time of returning users is 58.04 days on average with a standard deviation of 50.3 days.

Evaluating the models based solely on the RMSE of the return time predictions is problematic because churned users can not be included. We therefore

---

[5] It consists of a single LSTM layer followed by a fully connected output layer with a single neuron. The loss function is the mean squared error using only returning users as there is no target value for non-returning users.

use multiple measures to compare the performance of return time models. These are the root mean squared error [15, 26],[6] concordance index [21],[7] non-returning AUC, and non-returning recall.[8]

Table 1. Comparison of performance of return time prediction models. The RMSE is just for returning users. Best values for each performance metric are highlighted in bold.

	Baseline	CPH	CPHA	RNN	RNNSM	RNNSMA
RMSE (days)	43.25	49.99	59.81	**28.69**	59.99	63.76
Concordance	0.500	0.816	**0.817**	0.706	0.739	0.740
Non-returning AUC	0.743	0.793	0.788	0.763	**0.796**	0.794
Non-returning recall	0.000	0.246	0.461	0.000	0.538	**0.605**

4.1   Result on Performance Metrics

We report the performance metrics on the test set after training the models in Table 1. As the test dataset only contains users that were active in the activity window, the baseline model predicts that all users will return.

The CPH model uses an aggregated representation of each user's session history and additional metrics such as the absence time. The CPH model outperforms the baseline in each performance metric with the exception of the RMSE. This suggests that the improved non-returning recall rate can therefore be partially attributed to the model learning a positive bias for return times. This effect is even more pronounced for the CPHA model. However, the improvement in the concordance score demonstrates that, beyond a positive bias, a better relative ordering of predictions is achieved. Both CPH models perform particularly well in terms of the concordance score, suggesting that their predictions best reflect the relative ordering.

The RNN model cannot recall any non-returning users as its training examples only included returning users. However, the RMSE score demonstrates that the RNN model is superior in terms of predicting the return time of returning users and thus that sequential information is predictive of return time.

Finally, the recurrent neural network-based survival model (RNNSM) further improves the recall of non-returning users over the CPHA model without notable changes in the RMSE. More importantly it obtains the best performance for non-returning AUC, meaning it is the best model to discriminate between returning and non-returning users in the prediction window. Applying the absence-conditioned expectation to obtain predictions from the RNNSM

---

[6] Unlike in the cited publications, our dataset contains a large proportion of non-returning users. The score thus only reflects the performance on returning users.

[7] This incorporates the knowledge of the censored return time of non-returning users.

[8] For AUC and recall, we treat the users who did not return within the prediction window as the positive class in a binary-classification framework.

further improves the model's performance on non-returning recall. However, the concordance scores of both RNNSM models suggest that the relative ordering is not reflected as well as by the CPH model.

4.2 Prediction Error in Relation to True Return Time

To evaluate the performance of each model in more detail we group users by their true return time, rounded down to a week. We then evaluate each model's performance based on a number of error metrics. This is to assess the usefulness of each model; for example, a model which performs well on short return times but poorly on long return times would be less useful in practice than one that performs equally well on a wide distribution of return times.

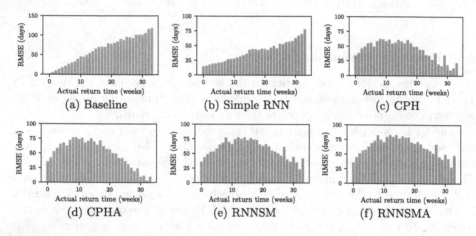

Fig. 6. Root mean squared error (RMSE) in relation to true return time compared between return time prediction models. In order to get a more detailed impression of each model's performance in relation to the true return time, we group users by their true return time rounded down to weeks. For each group, we then find the RMSE. Note the adjusted scale for the baseline model (a).

Root Mean Squared Error. The RMSE in relation to the return time in weeks is shown for each return time prediction model in Fig. 6. The majority of users in the dataset return within ten weeks; we see that for the baseline model and the RNN model the RMSE for these users is relatively low, this gives a low overall RMSE. However, for users who have longer return times both of those models perform increasingly poorly for increasing true return times.

For the models that incorporate training examples of both non-returning and returning users we see a different pattern. The performance for users with longer return times is generally better than those returning earlier. This demonstrates that these models are able to use censored observations to improve predictions for returning users. While the overall RMSE is lower for the CPH model compared the RNNSM (see Table 1), the distribution of errors is skewed, with a higher RMSE for earlier returning users.

We also see the effect of the absence-conditioned return time expectation. For the CPH model there is an increase in performance for users with very long return times, however there is a significant negative impact on users with shorter return times. This results suggest that the absence-conditioned expectation is more suitable for the RNNSM as it seems to have little effect on the RMSE distribution whilst improving the non-returning recall as can be seen in Table 1.

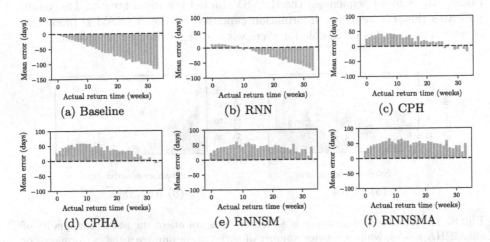

(a) Baseline   (b) RNN   (c) CPH

(d) CPHA   (e) RNNSM   (f) RNNSMA

Fig. 7. Mean error per in relation to true return time comparison for all return time prediction models. Users are grouped by their true return time to the nearest week. This allows us to determine how the prediction bias is related to the true return time. We see that models which include non-returning users in training have a positive bias on the prediction for returning users, those that don't have a negative bias. Note the adjusted range for the baseline model (a).

Mean Error. Figure 7 shows the mean error for each group and each model. The baseline model will always underestimate the true return time as by definition it can only predict a value equal to or lower than the true return time. The RNN model's performance is worse for users returning later, this is due to the restriction of predicting all users to return within a certain window leading to a negative bias for users returning later. The CPH model and the RNNSMs both overestimate the return times of the majority of users. It is possible to subtract the mean prediction error on the training set from these predictions in order to reduce the error, however this would lead to a reduction in non-returning AUC as overall return times would be reduced.

4.3   Error in Relation to Number of Active Days

In this section we group users by their number of active days, an active day is a day on which a user had a session. We plot the RMSE in days for the CPHA model and RNNSMA model in Fig. 8. These are the two best performing models which include returning users in terms of non-returning AUC and recall. We use up to 64 active days per user – we therefore group all users with 64 or more active days. We can immediately see that the RNNSM is able to make better predictions for users with a higher number of active days. This is not the case for the CPHA model. This demonstrates that for users with more active days (longer RNN input sequences) the RNNSM model improves greatly. This again indicates that the sequence information captured by the RNNSM is predictive of user return and is preferable for users with a larger number of sessions.

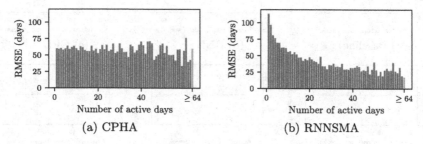

(a) CPHA   (b) RNNSMA

Fig. 8. The number of days a user is active for does not affect the prediction quality of the CPHA model, while a greater number of active days improves the performance of the RNNSMA. The last bar in both charts represents all users with 64 or more active days.

5   Discussion

We have developed the RNNSM, a novel model that overcomes the weaknesses of survival models and RNNs for user return time prediction. We have highlighted the importance of including right-censored observations in return time prediction models and extended the Cox proportional hazard model to include users' absence time. We found that for modelling recurring events, a limitation of existing survival regression models is that they only operate on aggregate representations instead of raw time-series data. We addressed this problem by using a RNN point process model, which combines the advantages of survival regression and overcomes the limitation of RNNs by including censored observations. We extended the RMTPP model to include any number of session markers and developed a method of training the model using censored observations. We further demonstrated how to include users' absence times. The RNNSM successfully learns from sequences of sessions and outperforms all other models in predicting which users are not going to return (non-returning AUC).

References

1. Bengio, Y., Simard, P., Frasconi, P.: Learning long-term dependencies with gradient descent is difficult. IEEE Trans. Neural Netw. **5**(2), 157–166 (1994)
2. Benson, A.R., Kumar, R., Tomkins, A.: Modeling user consumption sequences. In: WWW 2016, pp. 519–529 (2016)
3. Breslow, N.: Covariance analysis of censored survival data. Biometrics **30**(1), 89–99 (1974)
4. Cai, X., Zhang, N., Venayagamoorthy, G.K., Wunsch, D.C.: Time series prediction with recurrent neural networks trained by a hybrid PSO-EA algorithm. Neurocomputing **70**(13), 2342–2353 (2007)
5. Chamberlain, B.P., Cardoso, A., Liu, C.H.B., Pagliari, R., Deisenroth, M.P.: Customer lifetime value prediction using embeddings. In: KDD 2017, pp. 1753–1762. ACM (2017)
6. Chandra, R., Zhang, M.: Cooperative coevolution of Elman recurrent neural networks for chaotic time series prediction. Neurocomputing **86**, 116–123 (2012)
7. Cheng, H.T., et al.: Wide & deep learning for recommender systems. In: DLRS 2016 (RecSys 2016), pp. 7–10. ACM (2016)
8. Cho, K., van Merriënboer, B., Bahdanau, D., Bengio, Y.: On the properties of neural machine translation: encoder-decoder approaches. arXiv preprint arXiv:1409.1259 (2014)
9. Chollet, F., et al.: Keras. https://github.com/fchollet/keras (2015)
10. Covington, P., Adams, J., Sargin, E.: Deep neural networks for Youtube recommendations. In: RecSys 2016, pp. 191–198. ACM (2016)
11. Cox, D.R.: Regression models and life-tables. J. Roy. Stat. Soc. Ser. B (Methodol.) **34**(2), 187–220 (1972)
12. Cox, D.R.: Partial likelihood. Biometrika **62**(2), 269–276 (1975)
13. Cox, D.R., Oakes, D.: Analysis of Survival Data, vol. 21. CRC Press, Boca Raton (1984)
14. Davidson-Pilon, C.: Lifelines (2016). https://github.com/camdavidsonpilon/lifelines
15. Du, N., Dai, H., Trivedi, R., Upadhyay, U., Gomez-Rodriguez, M., Song, L.: Recurrent marked temporal point processes: embedding event history to vector. In: KDD 2016, pp. 1555–1564. ACM (2016)
16. Du, N., Wang, Y., He, N., Song, L.: Time-sensitive recommendation from recurrent user activities. In: NIPS 2015, pp. 3492–3500. MIT Press (2015)
17. Efron, B.: The efficiency of cox's likelihood function for censored data. J. Am. Stat. Assoc. **72**(359), 557–565 (1977)
18. Flunkert, V., Salinas, D., Gasthaus, J.: DeepAR: probabilistic forecasting with autoregressive recurrent networks. arXiv preprint arXiv:1704.04110 (2017)
19. Graves, A., Liwicki, M., Fernández, S., Bertolami, R., Bunke, H., Schmidhuber, J.: A novel connectionist system for unconstrained handwriting recognition. IEEE Trans. Pattern Anal. Mach. Intell. **31**(5), 855–868 (2009)
20. Han, M., Xi, J., Xu, S., Yin, F.L.: Prediction of chaotic time series based on the recurrent predictor neural network. IEEE Trans. Sig. Process. **52**(12), 3409–3416 (2004)
21. Harrell, F.E., Lee, K.L., Mark, D.B.: Multivariable prognostic models: issues in developing models, evaluating assumptions and adequacy, and measuring and reducing errors. Stat. Med. **15**, 361–387 (1996)

22. Hawkes, A.G.: Spectra of some self-exciting and mutually exciting point processes. Biometrika **58**(1), 83–90 (1971)
23. Hochreiter, S., Schmidhuber, J.: Long short-term memory. Neural Comput. **9**(8), 1735–1780 (1997)
24. Ishwaran, H., Kogalur, U.B., Blackstone, E.H., Lauer, M.S.: Random survival forests. Ann. Appl. Stat. **2**(3), 841–860 (2008)
25. Kalbfleisch, J.D., Prentice, R.L.: Marginal likelihoods based on cox's regression and life model. Biometrika **60**(2), 267–278 (1973)
26. Kapoor, K., Sun, M., Srivastava, J., Ye, T.: A hazard based approach to user return time prediction. In: KDD 2014, pp. 1719–1728. ACM (2014)
27. Klein, J.P., Moeschberger, M.L.: Survival Analysis: Techniques for Censored and Truncated Data. Springer, Heidelberg (2005)
28. Li, L., Jing, H., Tong, H., Yang, J., He, Q., Chen, B.C.: NEMO: next career move prediction with contextual embedding. In: WWW 2017 Companion, pp. 505–513 (2017)
29. Manzoor, E., Akoglu, L.: Rush!: Targeted time-limited coupons via purchase forecasts. In: KDD 2017, pp. 1923–1931. ACM (2017)
30. Mikolov, T., Chen, K., Corrado, G., Dean, J.: Efficient estimation of word representations in vector space. arXiv preprint arXiv:1301.3781 (2013)
31. Rajesh, R., Perotte, A., Elhadad, N., Blei, D.: Deep survival analysis. In: Proceedings of the 1st Machine Learning for Healthcare Conference, pp. 101–114 (2016)
32. Rodríguez, G.: Survival models. In: Course Notes for Generalized Linear Statistical Models (2010). http://data.princeton.edu/wws509/notes/c7.pdf
33. Sutskever, I., Vinyals, O., Le, Q.V.: Sequence to sequence learning with neural networks. In: NIPS 2014, pp. 3104–3112 (2014)
34. Vinyals, O., Toshev, A., Bengio, S., Erhan, D.: Show and tell: a neural image caption generator. In: CVPR 2015, pp. 3156–3164 (2015)
35. Wangperawong, A., Brun, C., Laudy, O., Pavasuthipaisit, R.: Churn analysis using deep convolutional neural networks and autoencoders. arXiv preprint arXiv:1604.05377 (2016)

Implicit Linking of Food Entities in Social Media

Wen-Haw Chong[✉] and Ee-Peng Lim

Singapore Management University, 80 Stamford Road, Singapore 178902, Singapore
whchong.2013@phdis.smu.edu.sg, eplim@smu.edu.sg

Abstract. Dining is an important part in people's lives and this explains why food-related microblogs and reviews are popular in social media. Identifying food entities in food-related posts is important to food lover profiling and food (or restaurant) recommendations. In this work, we conduct Implicit Entity Linking (IEL) to link food-related posts to food entities in a knowledge base. In IEL, we link posts even if they do not contain explicit entity mentions. We first show empirically that food venues are *entity-focused* and associated with a limited number of food entities each. Hence same-venue posts are likely to share common food entities. Drawing from these findings, we propose an IEL model which incorporates venue-based query expansion of test posts and venue-based prior distributions over entities. In addition, our model assigns larger weights to words that are more indicative of entities. Our experiments on Instagram captions and food reviews shows our proposed model to outperform competitive baselines.

Keywords: Entity linking · Food entities · Query expansion

1 Introduction

In social media, food-related topics are highly popular. Many users post food-related microblogs or reviews on various platforms such as Instagram, Foursquare, Yelp, etc. Such user generated content can be mined for profiling food lovers or for food and dining venue recommendations. In fact, identifying the local cuisines in posts has been justified [1] as useful for helping tourists in their dining choices. In this work, we propose to link food-related posts to a knowledge base of food entities. Given a test post that mention or *merely imply* some food entity, the task is to rank food entities in order of relevance.

We refer to this problem of linking posts as *Implicit Entity Linking* (IEL) [2,3]. In IEL, one links each test post to one or more related entities, without the need for mention extraction. This contrasts with the Explicit Entity Linking (EL) problem [4–8] which links mentions of named entities. Notably IEL circumvents the challenge of mention extraction in social media where posts are often grammatically noisy and colloquial. IEL also generalizes easily to various content scenarios. For example, consider the text snippets "XX Chicken Rice", "rice with

© Springer Nature Switzerland AG 2019
U. Brefeld et al. (Eds.): ECML PKDD 2018, LNAI 11053, pp. 169–185, 2019.
https://doi.org/10.1007/978-3-030-10997-4_11

chicken" and "having lunch". These are cases where food entities are respectively mentioned via proper nouns, improper nouns and merely implied. All snippets can be processed via IEL while EL is mention-dependent and will process only the first snippet comprising proper nouns. Lastly, IEL is also easier to conduct if one is only focused on a certain entity type, e.g. food entities. There is no need to ensure that only mentions of the right type are extracted.

Problem Setting. We formulate IEL as a ranking problem. For each post, we rank candidate food entities such that high ranking entities are more likely to be related. We assume that posts are not labeled with food entities for training, but are associated with posting venues. Both assumptions are realistic. Firstly labeled data are typically expensive to obtain. Secondly venue information is often available for platforms such as Foursquare, Instagram, review websites etc. We use Wikipedia as the knowledge base to link against.

Contributions. Our contributions are (1) an empirical analysis whereby we highlight that venues are focused around a limited set of food entities each, i.e., *entity-focused characteristic* and (2) a series of models for IEL. Our best performing model comprises the following aspects:

- **Entity-Indicative Weighting**: We propose a weighting scheme in our model to assign more weights to entity-indicative words. The intuition is that such words are more important for inferring entities than other words.
- **Query Expansion:** The entity-focused characteristic implies that a test post is likely to share common food entities as other same-venue posts. Hence we augment each test post via query expansion to include words from other same-venue posts.
- **Venue-based Prior:** Leveraging the same entity-focused characteristic, we generate venue-based prior distribution over food entities in an initial entity linking stage. This prior is used to bias the entity scores for the next stage.

By combining all above aspects, our best model EW-EWQE(v) outperforms state-of-the-art baselines that have been adapted for implicit entity linking.

2   Empirical Analysis

2.1   Datasets

In our empirical analysis and subsequent experiments, we use data from Instagram and Burpple[1]. The latter is a popular food review website in Singapore. Both datasets are generated by users from Singapore, a city well known for its wide range of food choices. Since both datasets are from Singapore users, we link their posts against a list of 76 food entities derived from the Wikipedia page on Singapore's cuisines[2]. Further details are discussed in Sect. 4.1.

---

[1] https://www.burpple.com/sg.
[2] https://en.wikipedia.org/wiki/Singaporean_cuisine.

For Instagram, we collect highly popular food-related captions from 2015 using hashtags of food e.g. '#foodporn' [3], or food entities e.g. '#chillicrab'. Following data cleaning and venue deduplication, we obtained 278,647 Instagram posts from 79,496 distinct venues. For Burpple, all its posts are food reviews and filtering by hashtags is not required. From Burpple, we obtained 297,179 posts over 13,966 venues. Table 1 illustrates four sample posts, two each from Instagram and Burpple. Clearly some posts are more informative about food entities than others. For example, the first instagram example does not reveal the food entity explicitly while the second example mentions fish ball noodle.

Table 1. Sample posts comprising Instagram captions and Burpple reviews.

Instagram	"super heavy lunch. and spicy! but its a must-try cafe! #food #foodporn #foodie #foodgasm #badoquecafe #instagood"
	"yesterday's lunch! #fishballnoodle #food #foodporn the soup was damn good"
Burpple	"*Signature Lamb Rack ($46++)* Very neat rectangular bricks of lamb, which we requested to be done medium-well.Nothing too impressive.. hurhur. Service is top -notch though"
	"*Good morning!* One of my favourite old school breakfast but he not his fav"

2.2   Analysis

A food venue typically focuses on some cuisines or food themes and is unlikely to serve an overly wide variety of dishes. For example, it is more probable for a restaurant to serve either Western or Asian cuisines, rather than both. Consequently, each food venue is likely to be associated with a limited number of food entities. We termed this as the *entity-focused characteristic*. To verify this characteristic, we compare the number of distinct food entities per venue against a null model where the characteristic is absent. We expect food venues to be associated with fewer food entities when compared against the null model.

For each venue v with multiple posts, we first compute the number of distinct entities over its posts. We then compute the expected number of distinct entities under the null model following the steps below:

- For each post from v, sample an entity e based on global entity probability i.e. entity popularity. Add to entity set $\mathbb{E}_{null}(v)$.
- Compute $|\mathbb{E}_{null}(v)|$, the distinct entity count under the null model.

---

[3] The most popular food related hashtag on our Instagram dataset.

We conduct our analysis on 2308 venues from Instagram and 362 venues from Burpple which have at least two user-labeled posts each. Such posts contain hashtags with food entity names, e.g. '#chillicrab', '#naan' etc. As sampling is required for the null model, we conduct 10 runs and take the average expected food entity count for each venue. For further analysis, we also repeat a similar procedure for users to compare their actual and expected food entity count. The intuition is that users may possess the entity-focused characteristic as well due to food preferences or constraints e.g vegetarian. The user statistics are computed over 2843 Instagram users and 218 Burpple users.

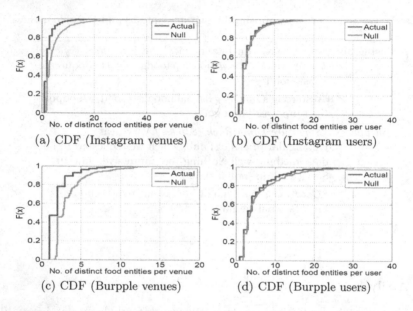

Fig. 1. CDFs of actual and expected distinct food entities for venues and users. F(x) on y-axis is probability of venues or users with $\leq x$ distinct food entities. (Color figure online)

Figure 1 plots the Cumulative Distribution Function (CDF) of distinct food entities for venues and users on both Instagram and Burpple, whereby distinct entity counts are on a per venue or user basis. In each graph, the blue line represents the actual count while the red line is for counts from the null model (averaged over 10 runs). For Figs. 1(a) and (c) venues are shown to be focused around specific food entities such that on average, each venue has fewer distinct food entities than expected under the null model. For example in Fig. 1(a), around 98% of the Instagram venues are associated with 10 distinct food entities or less in the actual data. In contrast, the null model has a corresponding proportion of around 91%. A similar trend can be observed for Burpple venues as shown in Fig. 1(c). Thus, the entity-focused characteristic is clearly evident for the venues of both datasets.

Figure 1(b) and (d) plot for Instagram and Burpple users respectively. There is much less difference between the actual and null model count, as both the blue and red lines overlap substantially in both figures. Comparing the plots for venues and users, we conclude that users are relatively less focused on food entities when compared to venues. These findings have implications for entity linking and should be considered when designing models. In particular, given a test post with both user and venue information, it may be easier to improve linking accuracy by exploiting other posts from the same venue rather than from the same user. In Sect. 3.2, we shall introduce a query expansion approach based on exploiting the entity-focused characteristic of venues.

3   Models

In this section, we present a series of models for IEL, culminating in a final best performing model. We start with the naive Bayes model. This can be regarded as a standard information retrieval baseline. Let $\mathbf{w}$ be the set of words in a post, where for notation simplicity, we assume each unique word $w \in \mathbf{w}$ occurs only once in the post. In our problem setting, we assume the entity probability $p(e)$ to be uniform as labeled posts are unavailable for estimating $p(e)$. The probability of food entity e given $\mathbf{w}$ is:

$$p(e|\mathbf{w}) \propto \prod_{w \in \mathbf{w}} p(w|e) = \prod_{w \in \mathbf{w}} \frac{f(e, w) + \gamma}{\sum_{w'} f(e, w') + W\gamma} \tag{1}$$

whereby $f(e, w)$ is the number of co-occurrences of word w with entity e, γ is the smoothing parameter and W is the vocabulary size. In the absence of labeled posts, the co-occurrences are estimated solely from the Wikipedia knowledge base. For each food entity e, we derive $f(e, w)$ by the count of w occurrences in the Wikipedia page of e and in Wikipedia text snippets around hyperlinks to e (refer Sect. 4.1). Finally entities are ranked by $p(e|\mathbf{w})$. The naive Bayes model is efficient and highly amenable to extensions.

3.1   Entity-Indicative Weighting (EW)

The naive Bayes model multiplies word probabilities without considering which words are more important for entity linking. Intuitively, some words are more indicative of food entities than others and should be assigned greater importance in entity linking models. Formally, an entity-indicative word w has relatively high $p(e|w)$ for some entity/entities in comparison with other words, e.g. 'sushi' is more entity-indicative than 'dinner'.

An entity-indicative word is different from a high probability word given an entity. For example, a food entity e may have high probability of generating the word 'rice', i.e. $p(\text{'rice'}|e)$ is high. However if many other food entities are also related to rice, then the word may not indicate e with high probability i.e. low $p(e|\text{'rice'})$. If a post $\mathbf{w}$ mentions other more entity-indicative words, e.g.

related to ingredients or cooking style, then such words should be assigned more importance when computing $p(e|\mathbf{w})$.

To capture the above intuition, we propose the entity-indicative weighting (EW) model. This assigns continuous weights to words and incorporates easily into the naive Bayes model. Let $\beta(w)$ be the *entity-indicative* weight for word w. This weight $\beta(w)$ is added as an exponent to the term $p(w|e)$ in Eq. 1. By taking the log to avoid underflow errors, we obtain the EW model:

$$\ln p(e|\mathbf{w}) \propto \sum_{w \in \mathbf{w}} \beta(w) \ln p(w|e) \tag{2}$$

Interestingly, Eq. (2) is similar in form to the weighted naive Bayes model proposed in prior work [9,10] for classification tasks. Here, we use it for IEL.

To compute the weights $\beta(w)$, we apply the vector space model and treat entities as documents. By definition, entity-indicative words are associated with fewer entities and have large weights. Weights are defined by:

$$\beta(w) = \log(1 + E/df(w)) \tag{3}$$

where E is the number of distinct food entities considered and $df(w)$ counts number of food entities with at least one occurrence of w.

3.2 Query Expansion with Same-Venue Posts

Based on the entity-focused characteristic, we expect that as a venue accumulates posts over time, its set of entities will be discussed repeatedly over different posts. This implies that for a test post discussing some entity e, there may exist other same-venue posts related to e. Hence if we augment the test post appropriately with words from other same-venue posts, we can potentially overcome information sparsity in one post and improve entity linking accuracy. This strategy is also known as query expansion.

Let test post $\mathbf{w}$ be posted from venue v. The idea is then to score candidate words w' appearing in other posts from v and whereby $w' \notin \mathbf{w}$. The expanded words w's aim to provide additional information for inferring the latent entity in $\mathbf{w}$. Among the many scoring schemes in the literature, we adopt a relatively simple cosine similarity scheme from [11]. This scheme scores each candidate word w' by its average relatedness $0 \leq \alpha_v(w', \mathbf{w}) \leq 1$ to the test post as:

$$\alpha_v(w', \mathbf{w}) = \frac{1}{|\mathbf{w}|} \sum_{w \in \mathbf{w}} \frac{d_v(w', w)}{\sqrt{d_v(w')d_v(w)}} \tag{4}$$

where $|\mathbf{w}|$ is the number of words in $\mathbf{w}$, $d_v(w', w)$ is the count of v's posts containing both w' and w; and $d_v(w)$ is the count of v's posts with w. Intuitively, if w' co-occurs more with each word from $\mathbf{w}$ on average, then average relatedness is higher. However, relatedness can be over-estimated for common words. To mitigate this, Eq. (4) includes in the denominator the product of word frequencies as the normalization term.

Following query expansion using same-venue posts, we combine two different word sets in a weighted naive Bayes model, which we refer to as QE(v):

$$\ln p(e|\{\mathbf{w}, \mathbf{w}'\}, v) \propto \sum_{w \in \mathbf{w}} \ln p(w|e) + \sum_{w' \in \mathbf{w}'} \alpha_v(w', \mathbf{w}) \ln p(w'|e) \quad (5)$$

where $\mathbf{w}'$ is the set of added words for post $\mathbf{w}$ from venue v. Since $0 \leq \alpha_v(w', \mathbf{w}) \leq 1$, Eq. (5) illustrates that the original query words $w \in \mathbf{w}$ have greatest importance in the model while the importance of newly added words $w' \in \mathbf{w}'$ vary based on how related they are to the query.

In our experiments, we shall also compared against a model variant QE(u), which selects augmenting words from same-user posts. As conjectured in Sect. 2.2, this model may be less likely to improve linking accuracy.

3.3 Fused Model (EWQE)

We now combine the EW and QE(v) models to create a new fused model called EWQE. Intuitively, we consider a word as important only when it is both entity-indicative *and* highly related to the test post. For example, if a word is not indicative of any entities, then it is less useful for entity linking even if it is present in the test post or is a highly related word based on Eq. (4). On the other hand, a non-related word may be indicative of some entity which is unrelated to the test post, such that test post augmentation with it introduces noise and lowers accuracy.

To model the discussed conjunction logic, we multiply the weights from entity-indicative weighting and query expansion to obtain the model EWQE(v):

$$\ln p(e|\{\mathbf{w}, \mathbf{w}'\}, v) \propto \sum_{w \in \mathbf{w}} \beta(w) \ln p(w|e) + \sum_{w' \in \mathbf{w}'} \beta(w')\alpha_v(w', \mathbf{w}) \ln p(w'|e) \quad (6)$$

Alternatively, one can combine entity-indicative weighting with user-based query expansion. We denote such a model as EWQE(u) and include it for experiments.

3.4 Venue-Based Prior

In our final model, we augment the probabilistic generative process in Eq. (6) with a venue-based prior distribution over entities $p(e|v)$. Let joint probability $p(e, \{\mathbf{w}, \mathbf{w}'\}, v)$ be factorized as $p(v)p(e|v)p(\{\mathbf{w}, \mathbf{w}'\}|e)$. We now need to compute $p(e|v)$ while $p(\{\mathbf{w}, \mathbf{w}'\}|e)$ can be computed as before with the EWQE(v) model. Assuming uniform venue probability $p(v)$ and incorporating a weighting term η $(0 \leq \eta \leq 1)$, we have:

$$\ln p(e|\{\mathbf{w}, \mathbf{w}'\}, v) \propto \eta \ln p(e|v) +$$
$$(1 - \eta) \left(\sum_{w \in \mathbf{w}} \beta(w) \ln p(w|e) + \sum_{w' \in \mathbf{w}'} \beta(w')\alpha_v(w', \mathbf{w}) \ln p(w'|e) \right) \quad (7)$$

Basically $p(e|v)$ bias the entity score in a venue-specific manner, rather than a post-specific manner as prescribed by query expansion. Given a set of training posts labeled with food entities, $p(e|v)$ is computed trivially. However in our setting, we assume no labeled posts for training. Hence we compute Eq. (7) in a 2-stage process as follows:

- Stage 1: With a desired IEL model, link the training posts. For each venue v, compute the aggregated entity scores $\tilde{p}(e|v)$, eg. if using the EW model, we compute $\tilde{p}(e|v) = \sum_{\mathbf{w} \in v} p(e|\mathbf{w})$. Normalize $\tilde{p}(e|v)$ to obtain $p(e|v)$.
- Stage 2: Combine $p(e|v)$ with the scores from the EWQE(v) model as detailed in Eq. (7) to derive the final entity scores for ranking.

4 Experiments

4.1 Setup

Our experiment setup is weakly supervised. Training posts are assumed to be unlabeled with respect to food entities. These training posts are used only for query expansion and for computing the venue prior over entities, but not for computing the entity profile $p(w|e)$. The entity profile $p(w|e)$ and entity-indicative weights $\beta(w)$ are computed using only Wikipedia pages. However, we retain a small validation set of entity-labeled posts for tuning model parameters with respect to the ranking metrics. Also, all posts are associated with posting venues, regardless of whether they are in the training, test or validation set.

For discussion ease, denote posts with food entity hashtags e.g., '#chillicrab' as type A posts and post without such hashtags as type B posts. Type A posts are easily associated with Wikipedia food entities, which facilitates the construction of test and validation sets. Our datasets contain a mixture of both post types. For Instagram, we have 18,333 type A vs 216,881 type B posts[4] whereas for Burpple, we have 1944 type A vs 200,293 type B posts. We conduct 10 experiment runs for each dataset, whereby in each run, we *mask the food entity hashtags* of type A posts and randomly assign 50% of them to the training set, 20% to the validation set and 30% to the test set. The type B posts are all assigned to the training set. Lastly, most of our type A posts contain only one food entity hashtag each, hence we use such single-entity posts for evaluation in our test set.

Food Entities. We consider 76 food entities that are defined by Wikipedia as local cuisines of Singapore[5], as well as associated with distinct pages/descriptions. For each entity e, we construct its profile, i.e. $p(w|e)$ from its Wikipedia description page and Wikipedia text snippets with hyperlinks to e. For example, the Wikipedia page 'Pakistani_cuisine' contains many hyperlinks to the food entity 'Naan'[6]. When building the profile for 'Naan', we include the preceding and succeeding 10 words around each hyperlink.

---

[4] Filtering by vocabulary has been applied, hence the numbers sum to less than the total food-related posts in Sect. 2.1.

[5] https://en.wikipedia.org/wiki/Singaporean_cuisine.

[6] Oven-baked flatbread.

Models to Be Evaluated. We evaluate the following models:

- NB: The naive Bayes model from Eq. (1).
- EW: Entity-indicative weighting as indicated in Eq. (2).
- QE(v): Venue-based query expansion whereby each test post is augmented with words from other same-venue posts, as indicated in Eq. (5).
- QE(u): User-based query expansion whereby each test post is augmented with words from other same-user posts.
- EWQE(v): Fusion of venue-based query expansion and entity-indicative weighting as shown in Eq. (6).
- EWQE(u): Fusion of user-based query expansion and entity-indicative weighting.
- NB-EWQE(v): In stage 1, we compute $p(e|v)$ with the NB model, which is then combined with the EWQE(v) model in stage 2. See Eq. (7).
- EW-EWQE(v): In stage 1, we use the EW model to compute $p(e|v)$. In stage 2, the computed $p(e|v)$ is combined with EWQE(v) model to score entities.

For each model, we use the validation set to tune γ, the smoothing parameter for $p(w|e)$, based on the grid $[0.01, 0.1, 1, 10]$. For NB-EWQE(v) and EW-EWQE(v), γ is jointly tuned with η whereby η is varied in steps of 0.1 from 0 to 1.

For further comparison, we introduce three other baselines. We adapt two EL models from [12, 13] such that they can be used for IEL. Without any adaptation, it is impossible for the vanilla EL models to link posts directly to entities. Our adaptations also aim to exploit the entity-focused characteristic of venues, or other related characteristics. Lastly, we include a word embedding baseline [15] that does not require any adaptation. The baselines are:

- TAGME: In the TAGME model [12,14], candidate entities for a mention are voted for by candidate entities from other mentions in the same post. Adapting the idea to IEL, candidate entities for a post are voted for by candidate entities from other posts in the same venue. Since a candidate entity gathers larger votes from the same or related entities, this voting process exploits the entity-focused characteristic of venues as well. Let $\mathbf{w}_{i,v}$ denote the i-th post from venue v. Then candidate entity e_i for $\mathbf{w}_{i,v}$ gathers a vote from $\mathbf{w}_{j,v}$ computed as

$$vote(\mathbf{w}_{j,v} \rightarrow e_i) = \frac{1}{|e_j : p(e_j|\mathbf{w}_{j,v}) > 0|} \sum_{e_j : p(e_j|\mathbf{w}_{j,v}) > 0} sr(e_i, e_j)p(e_j|\mathbf{w}_{j,v}) \quad (8)$$

where $sr(e_i, e_j)$ is the Jaccard similarity of incoming Wikipedia links [14] between e_i, e_j, and $p(e_j|\mathbf{w}_{j,v})$ can be based on any implicit entity linking models. Finally for ranking entities, we compute the final score for entity e_i as $p(e_i|\mathbf{w}_{i,v}) \sum_j vote(\mathbf{w}_{j,v} \rightarrow e_i)$.

- LOC: Locations in the form of grid cells can be entity-focused as well. To exploit this, we implement the framework from [13]. For each grid cell, the distributions over entities are inferred via EM learning and integrated with implicit entity linking models. Unlike [13], we omit the dependency on posting time as our targeted posts include food reviews which are usually posted after, rather than during meal events. We tune grid cell lengths based on grid [200 m, 500 m, 1 km, 2 km].

- PTE: This is a graph embedding method [15] that learns continuous vector representation for words, posts and entities over a heterogeneous graph. The graph consists of word nodes, post nodes and entity nodes, connected via the following edge types: word-word, post-word and entity-word. For each test post, we compute its vector representation by averaging over the representations of its constituent words. We then compute the cosine similarities to entity representations for ranking. As in [15], we use an embedding dimension of 100. We set the number of negative samples to be 200 million.

For the baselines TAGME and LOC, we integrate the implicit entity linking models NB, EW and EW-EWQE(v). For each model, we replace the relevant mention-to-entity computations with post-to-entity computations. For example, TAGME(NB) computes $p(e_j|\mathbf{w}_{j,v})$ in Eq. (8) using the NB model. Such integration leads to the following baseline variants: TAGME(NB), TAGME(EW), TAGME(EW-EWQE(v)), LOC(NB), LOC(EW) and LOC(EW-EWQE(v)).

Metrics. We use the Mean Reciprocal Rank (MRR) as our evaluation metric. Given a post $\mathbf{w}_i$, let the rank of its food entity be $r(\mathbf{w}_i)$, where $r(\mathbf{w}_i) = 0$ for the top rank. Over N test cases, MRR is defined as:

$$\text{MRR} = N^{-1} \sum_{i=1}^{N} (r(\mathbf{w}_i) + 1)^{-1} \tag{9}$$

The above MRR definition is a micro measure. In a sample of test posts, more popular food entities contribute more to MRR. For further analysis, we consider treating all entities as equally important, regardless of their popularities. Thus we introduce Macro-MRR, the macro-averaged version of MRR. For all test posts related to the same food entity, we compute the MRR of the food entity. We then average the MRRs over distinct food entities. Formally:

$$\text{Macro-MRR} = E^{-1} \sum_{i=1}^{E} \text{MRR}(e_i) \tag{10}$$

where $\text{MRR}(e_i)$ is MRR values averaged over all test posts about entity e_i and E is the number of distinct food entities.

4.2 Results

Table 2 displays the MRR and Macro-MRR values averaged over 10 runs for each dataset. In subsequent discussions, a model is said to perform better or worse than another model only when the differences are statistically significant at p-level of 0.05 based on the Wilcoxon signed rank test.

EW and QE(v) easily outperform NB, which affirms the utility of entity-indicative weighting and venue-based query expansion. EW also outperforms

Table 2. MRR and Macro-MRR values averaged over 10 runs for each dataset. The best performing model is bolded.

Model	Instagram		Burpple	
	MRR	Macro-MRR	MRR	Macro-MRR
NB	0.344	0.218	0.335	0.259
EW	0.461	0.301	0.467	0.377
QE(v)	0.403	0.236	0.389	0.252
QE(u)	0.326	0.215	0.336	0.237
EWQE(v)	0.543	0.323	0.503	0.388
EWQE(u)	0.449	0.284	0.419	0.329
NB-EWQE(v)	0.543	0.323	0.500	0.389
EW-EWQE(v)	**0.593**	**0.340**	**0.537**	**0.401**
TAGME(NB)	0.368	0.233	0.344	0.259
TAGME(EW)	0.462	0.293	0.446	0.363
TAGME(EW-EWQE(v))	0.520	0.296	0.507	0.390
LOC(NB)	0.409	0.236	0.357	0.259
LOC(EW)	0.472	0.254	0.413	0.315
LOC(EW-EWQE(v))	0.520	0.271	0.467	0.333
PTE	0.288	0.216	0.291	0.274

QE(v), e.g. EW's MRR is 0.461 on Instagram posts, higher than QE(v)'s MRR of 0.403. By combining both models together in EWQE(v), we achieve even better performance than applying EW or QE(v) alone. This supports EWQE(v)'s modeling assumption that a word is important if it is both entity-indicative and highly related to the test post.

While venue-based query expansion is useful, user-based query expansion is less promising. Over the different datasets and metrics, QE(u) is inferior or at best on par with NB. This may be due to the entity-focused characteristic being weaker in users. This observation is consistent with our earlier empirical findings that users are less focused on food entities when compared to venues. Consequently user-based query expansion may augment test posts with noisy words less related to their food entities. Combining user-based query expansion with entity-indicative weighting also leads to mixed results. Although EWQE(u) outperforms QE(u), the former still underperforms EW.

Our results also show that the venue-based prior distribution over entities is useful, but only if it is computed from a reasonably accurate linking model. Over all dataset-metric combination, the best performing model is EW-EWQE(v) which incorporates a prior computed using the EW model. Although NB-EWQE(v) incorporates a prior as well, it utilizes the less accurate NB model. For Instagram, the tuning procedure consistently indicates in each run that the optimal η is 0 for NB-EWQE(v), thus it is equivalent to the model EWQE(v). For

Burpple, the optimal η is non-zero for some runs, but NB-EWQE(v) performs only on par with EWQE in terms of statistical significance.

The TAGME variants exploit the entity-focused characteristic of venues via a voting mechanism. Performance depends on the voting mechanism as well as the underlying entity linking models. Intuitively better underlying models should lead to higher ranking accuracies in the corresponding variants. For example, TAGME(EW-EWQE(v)) outperforms TAGME(EW) while TAGME(EW) outperforms TAGME(NB). However comparing the variants against their underlying models, we note that only TAGME(NB) consistently improves over NB, while TAGME(EW) and TAGME(EW-EWQE(v)) fails to outperform EW and EW-EWQE(v) respectively. The same observation applies to the LOC variants. LOC(NB) consistently outperforms NB. LOC(EW) only outperforms EW for MRR on Instagram and is inferior in other dataset-metric combination. LOC(EW-EWQE(v)) is also inferior to EW-EWQE(v). Such mixed results of LOC variants may be due to grid cells being less entity-focused than venues. Lastly, PTE performs poorly. We note that each entity has only one Wikipedia description page and appears in a limited number of Wikipedia contexts. Hence the Wikipedia content of food entities may be overly sparse for learning good entity representations. There are also language differences between Wikipedia pages and social media posts. This may impact cross-linking if embeddings are trained on only one source, but not the other. In conclusion, our proposed model EW-EWQE(v) performs well, despite its conceptually simple design.

4.3 Case Studies

Tables 3, 4 and 5 illustrate different model aspects by comparing model pairs on Instagram posts. Comparison is based on the ranked position of the ground truth food entity (under column e) for each post. The ranked position is denoted as r_X for model X and is 0 for the top ranked. The ground truth entities can be inspected by appending the entity name to the URL 'https://en.wikipedia.org/wiki/'.

Table 3. Sample test posts to illustrate entity-indicative weighting. Words in larger fonts indicate larger weights under the EW model.

		e	r_{NB}	r_{EW}
S1	"#singapore we already ate claws ."	Chilli_crab	2	0
S2	"finally got to eat rojak !!!"	Rojak	5	0
S3	"#singapore #tourist "	Hainanese_chicken_rice	18	2

Entity-Indicative Weighting. Table 3 compares the models NB and EW. For each test post, words with larger weights under the EW model are in larger fonts. For S1 with food entity 'Chilli_crab'[7], the largest weighted word is 'claws', referring to a crab body part. This word is rarely mentioned with other food entities, but appears in the context around the 'Chilli_crab' anchor in the Wikipedia page for 'The_Amazing_Race_25', hence it is highly indicative of 'Chilli_crab'. By assigning 'claws' a larger weight, EW improves the entity ranking over NB, from a position of 2 to 0. For S2, the word 'rojak' is indicative of the food entity 'Rojak' [8]. While NB does well with a ranked position of 5, EW further improves the ranked position to 0 by weighting 'rojak' more relative to other words. For S3, the food entity 'Hainanese_chicken_rice'[9] is described in the Wikipedia page 'Singaporean_cuisine' as the most popular dish for tourists in the meat category. Thus by assigning a larger weight to 'tourist', EW improves the linking of S3.

Table 4. Sample test posts with added words (in brackets) from query expansion (QE model). The top 5 added words with largest weights are listed.

		e	r_{NB}	r_{QE}
S4	"last night dinner at #singapore #foodporn" (rice,0.25),(chicken,0.23), (late,0.21),(food,0.21), (to,0.20)	Hainanese_chicken_rice	19	3
S5	"indian feast #daal #palakpaneer #mangolassi @rebekkariis du vil elske det!" (pakistani,0.17),(cuisine,0.17) (buffet,0.17) (lunch,0.17)(team,0.17)	Naan	1	0

Query Expansion. Table 4 illustrates posts where the QE model improves over the NB model. While S4 mentions dinner, the food entity is not evident. However the word 'dinner' co-occurs with more informative words such as 'chicken' and 'rice' in other posts from the same venue. Such words are retrieved with query expansion and used to augment the post. The augmented post is then linked more accurately by the QE model. For S5, query expansion augments the post with 6 words of which 5 words share similar weights. Out of the 5 words, the word 'pakistani' is indicative of the food entity 'Naan', helping to improve the ranked position further from 1 to 0.

Venue-Based Prior. Table 5 compares EWQE(v) and EW-EWQE(v). S6 is posted from a food venue which serves 'Mee_pok' [10] as one of its food entities. This food entity is mentioned explicitly in other same-venue posts. Hence

---

[7] Crabs stir-fried in chilli-based sauce.
[8] A traditional fruit and vegetable salad dish.
[9] Roasted or steamed chicken with rice cooked in chicken stock.
[10] A Chinese noodle dish.

Table 5. Sample test posts for comparing models EWQE(v) and EW-EWQE(v). $r_{p(e|v)}$ corresponds to ranking with the venue prior $p(e|v)$.

		e	$r_{p(e|v)}$	$r_{EWQE(v)}$	$r_{EW-EWQE(v)}$
S6	"life's simple pleasures. #gastronomy"	Mee_pok	0	56	0
S7	"the black pepper sauce is robust and quite spicy, one of my favourite in singapore."	Black_pepper_crab	1	9	2

on applying the EW model, we infer this venue as having a high prior probability for this entity. In fact if we rank food entities by the venue prior $p(e|v)$ alone, 'Mee_pok' is ranked at position 0. Integrating the prior distribution with other information as done in EW-EWQE(v), the same rank position of 0 is obtained. For S7, the ingredient black pepper sauce is mentioned, which is indicative to some extent of 'Black_pepper_crab' [11]. However EWQE(v) manages only a ranked position of 9. From other same-venue posts, the venue prior is computed and indicates the food entity to be highly probable at S7's venue. Integrating this information, EW-EWQE(v) improves the ranked position to 2.

4.4   Parameter Sensitivity

For models with γ as the sole tuning parameter, we compare their sensitivity with respect to γ. Figure 2 plots the performance of NB, EW, EWQE and EWQE(v), averaged over 10 runs for different γ values. It can be seen that EWQE(v) outperforms NB over most of the applied γ values, i.e. 0.1, 1 and 10. Although EWQE(v) is simply a product combination of the EW and QE(v) models, it easily outperforms its constituent models, validating our combination approach. This trend is consistent across both metrics and datasets. We also note that in the absence of a validation set for tuning, a natural option is to use Laplace smoothing, i.e. $\gamma = 1$. In this perfectly unsupervised setting, it is reassuring that EWQE(v) remains the best performing model. Lastly when γ is very small at 0.01, EW and EWQE(v) are under-smoothed and perform worse than NB. In this setting where smoothing is limited, QE(v) outperforms all other models, possibly because augmenting each test post with additional words is analogous to additional smoothing for selected words.

---

[11] Crabs stir-fried in black pepper sauce.

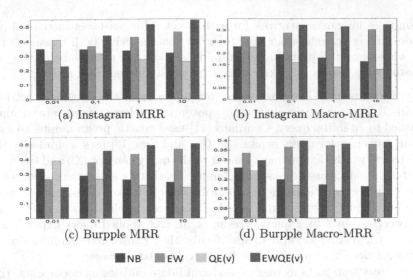

(a) Instagram MRR   (b) Instagram Macro-MRR

(c) Burpple MRR   (d) Burpple Macro-MRR

■ NB ■ EW ■ QE(v) ■ EWQE(v)

Fig. 2. Model performance (Y-axis) with different γ values (X-axis).

5 Related Work

Explicit Entity Linking. Compared to IEL, there has been more work in EL [4–6,16]. In [4], Liu et al. constructed an objective function based on mention-entity features, mention-mention features etc. When linking mentions, entities are assigned via a decoding algorithm. In [16], the objective function is defined over a graph that connects tweets close in space and time. The assumption is that such tweets are likely to mention closely-related entities. In [5], Shen et al. propagate users' interest scores over an entity graph built from inter-entity semantic-relatedness [17]. Given a test mention, candidate entities with higher interest scores are preferred. Huang et al. [6] proposed label propagation over graphs with mention-entity tuples as nodes. After label propagation, high scoring tuples provide the mention-entity assignments. Finally, our baselines include extensions of EL models [12,13]. Fang and Chang [13] learned entity distributions over time and grid cells and integrate them into a base linking system. We use their learning framework and integrate grid cell information into our model. In [12], the idea is to let candidate entities across intra-document mentions vote for each other. For each mention, top k entities with the most votes are then filtered again by entity probability. In our extension, our voting entities are candidates for posts from the same venue, not mentions from the same document.

Implicit Entity Linking. For IEL, Perera et al. [2] built information network to link entities and knowledge nodes, using factual knowledge from the knowledge base and contextual knowledge from labeled tweets. They then use graph features to rank entities. The work in [3] engineered features from labeled tweets to train decision trees for ranking entities for each tweet. This per-tweet instead of per-mention linking resembles our IEL task. In contrast with both discussed IEL work, we assume the posts in our training set are not entity-labeled, but are associated with venues. Thus our work explores a different task setting.

Query Expansion. Query expansion originates from the document retrieval problem. To improve retrieval accuracy, potentially relevant words are weighted and added to an initial query. Cummins [11] used genetic programming to learn weighting schemes for query expansion. Qiu and Frei [18] uses a similarity thesaurus to add words that are most similar to the query concept. Xu and Croft [19] compared various query expansion techniques exploiting either the corpora-wide word occurrences/relationships or exploiting the top ranked documents returned for the initial query. Query expansion has also been applied [21] to retrieve relevant tweets given a user query. Fresno et al. [20] applied query expansion to retrieve event-related keywords. Specifically they considered candidate words from tweets close in space and time to an event-related tweet.

If we treat test posts as queries and candidate entities as documents, then IEL can be viewed as a form of document retrieval. In this work, we use query expansion to exploit the entity-focused characteristics of venues.

6   Conclusion

We have proposed novel yet well principled models for implicit food entity linking in social media posts. Our best model exploits the entity-focused characteristic of food venues and the intuition that entity-indicative words are more important for IEL, in order to outperform more complex state-of-the-art models. In future work, we intend to explore IEL in non-geotagged social media posts, where posting venues are unknown. Lastly we point out that the entity-focused characteristic appears in various forms in other problems. For example, in linking tweets to posting venues [22], users may be focused in their visits, preferring venues close to their home regions. Hence potentially, our model can be generalized to other information retrieval problems.

Acknowledgments. This research is supported by the National Research Foundation, Prime Minister's Office, Singapore under its International Research Centres in Singapore Funding Initiative, and DSO National Laboratories.

References

1. Ming, Z.-Y., Chua, T.-S.: Resolving local cuisines for tourists with multi-source social media contents. Multimedia Syst. **22**, 443–453 (2016)
2. Perera, S., Mendes, P.N., Alex, A., Sheth, A.P., Thirunarayan, K.: Implicit entity linking in tweets. In: Sack, H., Blomqvist, E., d'Aquin, M., Ghidini, C., Ponzetto, S.P., Lange, C. (eds.) ESWC 2016. LNCS, vol. 9678, pp. 118–132. Springer, Cham (2016). https://doi.org/10.1007/978-3-319-34129-3_8

3. Meij, E., Weerkamp, W., de Rijke, M.: Adding semantics to microblog posts. In: WSDM 2012 (2012)
4. Liu, X., Li, Y., Wu, H., Zhou, M., Wei, F., Lu, Y.: Entity linking for tweets. In: ACL 2013 (2013)
5. Shen, W., Wang, J., Luo, P., Wang, M.: Linking named entities in tweets with knowledge base via user interest modeling. In: KDD 2013 (2013)
6. Huang, H., Cao, Y., Huang, X., Ji, H., Lin, C.-Y.: Collective tweet wikification based on semi-supervised graph regularization. In: ACL 2014 (2014)
7. Shen, W., Wang, J., Luo, P., Wang, M.: LIEGE: link entities in web lists with knowledge base. In: KDD 2012 (2012)
8. Shen, W., Wang, J., Luo, P., Wang, M.: LINDEN: linking named entities with knowledge base via semantic knowledge. In: WWW 2012 (2012)
9. Zaidi, N.A., Cerquides, J., Carman, M.J., Webb, G.I.: Alleviating naive Bayes attribute independence assumption by attribute weighting. JMLR **14**(1), 1947–1988 (2013)
10. Ferreira, J.T.A.S., Denison, D.G.T., Hand, D.J.: Weighted naive Bayes modelling for data mining. Department of Mathematics, Imperial College (2001)
11. Cummins, R.: The evolution and analysis of term-weighting schemes in information retrieval. Ph.D. thesis, National University of Ireland, Galway (2008)
12. Ferragina, P., Scaiella, U.: TAGME: on-the-fly annotation of short text fragments (by wikipedia entities). In: CIKM 2010 (2010)
13. Fang, Y., Chang, M.-W.: Entity linking on microblogs with spatial and temporal signals. In: TACL 2014 (2014)
14. Piccinno, F., Ferragina, P.: From TagME to WAT: a new entity annotator. In: ERD 2014 (2014)
15. Tang, J., Qu, M., Mei, Q.: PTE: predictive text embedding through large-scale heterogeneous text networks. In: KDD 2015 (2015)
16. Chong, W.-H., Lim, E.-P., Cohen, W.: Collective entity linking in tweets over space and time. In: Jose, J.M., Hauff, C., Altıngovde, I.S., Song, D., Albakour, D., Watt, S., Tait, J. (eds.) ECIR 2017. LNCS, vol. 10193, pp. 82–94. Springer, Cham (2017). https://doi.org/10.1007/978-3-319-56608-5_7
17. Milne, D., Witten, I.H.: An effective, low-cost measure of semantic relatedness obtained from Wikipedia links. In: AAAI 2008 (2008)
18. Qiu, Y., Frei, H.-P.: Concept based query expansion. In: SIGIR 1993 (1993)
19. Xu, J., Croft, W.B.: Query expansion using local and global document analysis. In: SIGIR 1996 (1996)
20. Fresno, V., Zubiaga, A., Ji, H., Martínez, R.: Exploiting geolocation, user and temporal information for natural hazards monitoring in Twitter. Procesamiento del Lenguaje Nat. **54**, 85–92 (2015)
21. Bandyopadhyay, A., Mitra, M., Majumder, P.: Query expansion for microblog retrieval. In: TREC 2011 (2011)
22. Chong, W.-H., Lim, E.-P.: Tweet geolocation: leveraging location, user and peer signals. In: CIKM 2017 (2017)

A Practical Deep Online Ranking System in E-commerce Recommendation

Yan Yan[1(✉)], Zitao Liu[2], Meng Zhao[1], Wentao Guo[1], Weipeng P. Yan[1], and Yongjun Bao[1]

[1] Intelligent Advertising Lab, JD.COM,
675 E Middlefield Road, Mountain View, CA, USA
{yan.yan,zhaomeng1,guowentao,paul.yan,baoyongjun}@jd.com
[2] TAL AI Lab, TAL Education Group, Beijing, China
liuzitao@100tal.com

Abstract. User online shopping experience in modern e-commerce websites critically relies on real-time personalized recommendations. However, building a productionized recommender system still remains challenging due to a massive collection of items, a huge number of online users, and requirements for recommendations to be responsive to user actions. In this work, we present our relevant, responsive, and scalable deep online ranking system (DORS) that we developed and deployed in our company. DORS is implemented in a three-level architecture which includes (1) candidate retrieval that retrieves a board set of candidates with various business rules enforced; (2) deep neural network ranking model that takes advantage of available user and item specific features and their interactions; (3) multi-arm bandits based online re-ranking that dynamically takes user real-time feedback and re-ranks the final recommended items in scale. Given a user as a query, DORS is able to precisely capture users' real-time purchasing intents and help users reach to product purchases. Both offline and online experimental results show that DORS provides more personalized online ranking results and makes more revenue.

Keywords: Recommender system · E-commerce · Deep learning Multi-arm bandits

1 Introduction

Building a relevant and responsive recommender system is the key to the success of e-commerce and online retailing companies. A desired recommender system helps people discover things they love and potentially purchase them, which not only makes companies profitable but also brings convenient online shopping experience to people's daily lives.

Electronic supplementary material The online version of this chapter (https://doi.org/10.1007/978-3-030-10997-4_12) contains supplementary material, which is available to authorized users.

U. Brefeld et al. (Eds.): ECML PKDD 2018, LNAI 11053, pp. 186–201, 2019.
https://doi.org/10.1007/978-3-030-10997-4_12

There is a huge amount of data generated on e-commerce websites everyday, such as user historical online behaviors, product inventory specifics, purchase transactions, etc. However, how to utilize them and build a practical recommender system that recommends billions of items to millions of daily active users still remains challenging due to the *relevance* and *responsiveness* trade-offs arising from online shopping scenarios. Briefly, when a recommender system is designed to serve the most relevant items, it uses every piece of available information about the users, such as demographical information, past shopping habits, favorite and disliked items, etc., to score, rank and select recommendation candidates. Because of the tremendous dimensions of information and the huge size of candidates, the entire recommendation process is time consuming and cannot be finished in real time[1]. Hence, the recommended items become stale to the most recent shopping interests of users. On the other hand, a good recommender system needs to be responsive and is able to capture user shopping intent in real time. User shopping intents drift quite a lot and users tend to shop in different categories even within a single session[2]. Responsiveness is difficult to achieve since given a limited time window (request time out window), only a few features can be used to score and rank the results. The majority of existing approaches proposed for e-commerce recommendation in the literature only focus on either improving recommendation relevance or achieving responsiveness in a computationally expensive approach which is not affordable or applicable when serving thousands of recommendation requests every second.

In this work, we address this problem by presenting our *D*eep *O*nline *R*anking *S*ystem, i.e., *DORS* which is relevant, responsive and scalable and is able to serve millions of recommendation requests everyday. DORS is designed and implemented in a three-level novel architecture, which includes (1) candidate retrieval; (2) learning-to-rank deep neural network (DNN) ranking; and (3) online re-ranking via multi-arm bandits (MAB). The candidate retrieval stage enables us to incorporate various business rules to filter out irrelevant items, such as out-of-stock items, etc. The learning-to-rank DNN stage fully utilizes all available information about users, items and their corresponding interaction features and conducts a full scoring, which provides a fine-grained high-recall set of candidates. The last stage of online re-ranking takes the output of DNN top K ranking results and incorporates user real-time online feedback, such as impressions, clicks, purchases, etc. to adjust the final recommendations.

Overall this paper makes the following contributions:

- It presents a practical three-stage recommendation system that is able to serve relevant and responsive recommendations to millions of users.
- It provides a robust and flexible production system that is easy to implement and is superior in terms of computational efficiency and system latency.
- We evaluate our approach and benefits in our real-world scenarios with millions of real traffics.

---

[1] Real time is defined as under 200 ms.
[2] Session is defined as a 30-min window in this paper.

2   Related Work

Various methodologies have been built by machine learning and data mining community to conduct better recommendations in different scenarios [14,21,35], which can be divided into the following categories: static recommendation (Sect. 2.1), time-aware recommendation (Sect. 2.2) and online recommendation (Sect. 2.3).

2.1   Static Recommendation

Static recommendation makes the assumption that both the users and items rarely change over time and it tries to optimize the relatedness between users and items given the entire history. All standard and classic recommendation algorithms such as content based recommendations [18,20], user or item based collaborative filtering [26], non-negative matrix factorization [24,36] fall into this category. Such approaches are effective and serve as the core services in lots of Internet companies [9,13,16]. However, the recommender system by nature is a dynamic process, which means user online actions or events happened in a sequential manner and user shopping interests drift over time. Moreover, with the development of fast and easy Internet access, more frequent online user behaviors are observed in real time (under 500 ms). It is challenging or even inapplicable for static recommendation approaches to be responsive and get updated once user feedback is observed.

2.2   Time-Aware Recommendation

Besides modeling the relations between users and items, various researches have been done by taking *time* as the third dimension to capture the evolution of user interest. Time-aware recommendation approaches aim at explicitly modeling user interests over time slices [15,28,29,32] and they can be combined with various classic recommendation algorithms in static recommendation (Sect. 2.1). For examples, Ding and Li introduced a personalized decay factor according to purchase behaviors and proposed time weight collaborative filtering [11]. Koren developed a collaborative filtering with temporal dynamics which each prediction is composed of a static average value and a dynamic changing factor [17]. Yin et al. proposed a user-tag-specific temporal interests model for tracking users' interests over time by maximizing the time weighted data likelihood [34]. Xiong et al. used a tensor factorization approach to model temporal factors, where the latent factors are under Markovian assumption [33]. Lu et al. improved the standard low-rank matrix factorization by using Kalman filter to infer changes of user and item factors [22]. Gao et al. modeled the correlations between user check-in timestamps and locations through an improved matrix factorization approach [12]. Although many studies above have been considered the temporal aspects of data, most of them are far away from being responsive. The majority of proposed approaches model the temporal dependence by a pre-defined time window, such as hours, days, or weeks, which is far away from serving real-time recommendations.

2.3   Online Recommendation

To capture user real-time feedback, several studies focus on improving online learning algorithms so that recommendation models get updated immediately once the user online actions happen [1,6,7,10,25]. Agarwal et al. developed an online bilinear factor model to learn item-specific factors through online regression. The online regression for each item can be performed independently and hence the procedure is fast, scalable and easily parallelizable [1]. Rendle et al. derived an online-update algorithm for regularized kernel matrix factorization models that allows to solve the cold start problem [25]. Diaz et al. presented a stream ranking matrix factorization technique, which optimizes the personalized ranking of topics. They also proposed a selective sampling strategy to perform online model updates based on active learning principles, that closely simulates the task of identifying relevant items from a pool of mostly uninteresting ones [10]. Chang et al. viewed the online recommendation problems as an explicit continuous-time random process, which generates user interested topics over time. They also provided a variational Bayesian approach to permit instantaneous online inference [6]. However, even the majority work above presents approaches to conduct parallelized online updates with user feedback there are two drawbacks. First, recommendation model parameters tend to be resistant to a single user feedback online update and hence it is not sufficient to provide personalized results. Second, the majority work mentioned above doesn't scale well when building a recommender system to suggest billions of items to millions of daily active users.

3   The Deep Online Ranking System

In this work, we develop a three-level recommender system architecture that is able to flexibly provide the most relevant and responsive recommendation to our millions of daily active users. The entire recommendation workflow in our system includes three phases: (1) candidate retrieval that scans all available item inventory by using forward and backward indices and trims out candidates violating the business rules (Sect. 3.1); (2) full features scoring and ranking via DNN that uses the pre-trained DNN to rank all candidates in order to optimize the gross merchandise volume (Sect. 3.2); (3) online re-ranking via MAB that dynamically adjusts the ranking results from DNN by using a novel multi-arm bandits algorithm. It takes user real-time feedback such as clicks and captures user shopping intents as fast as possible (Sect. 3.3). The entire workflow is illustrated in Fig. 1.

3.1   Candidate Retrieval

Candidate retrieval serves as the first phase of our recommendation system and it is triggered when a user request is received. The recommender system fetches both user historical data and item features. Then all these information are passed

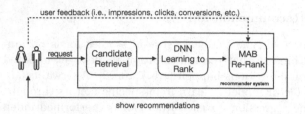

Fig. 1. Workflow overview of our recommender system.

to our retrieval system to select initial candidates for full feature scoring. Furthermore, business rule based item trimming or filtering are also done in candidate retrieval phase. In a productionized e-commerce recommender system, various business rules have to be enforced, such as out-of-stock items shouldn't be retrieved, the number of items from the same brand should be limited, etc.

3.2   Learning-to-Rank via DNN

In the ranking stage, we develop our novel *learning-to-rank* DNN model to score all candidates returned from the retrieval stage by using full sets of features and hence generate a fine-grained set of items S. Generally speaking, this is achieved by sophisticated machine learning models such as factorization machines [23], pairwise *learning-to-rank* [4,27], listwise *learning-to-rank* [5], etc. In this work, the ranking module is implemented by a pairwise *learning-to-rank* DNN.

The Ranking Problem. Let x_i be the m-dimensional feature vector of item i. Each item i belongs to its category c_i. A category in e-commerce generally represents a group of products that share the same utilities or have similar functionalities: *e.g.* apparels, cellphones, snacks, etc. Each item i also associates with its *gross merchandise volume* (GMV) gmv_i which represents the product's retail value and it in most cases corresponds to the item's final price.

In this work, the goal is that given all retrieved items, select the top K items so that the discounted cumulative gain (DCG) of GMV is optimized and the DCG_K is defined as follows $DCG_K = \sum_{s_i \in S} gmv_{s_i} \mathcal{I}(s_i) / \log_2(i+1)$ and $\mathcal{I}(s_i)$ is the indicator function such that $\mathcal{I}(s_i) = 1$ when s_i is purchased and otherwise, $\mathcal{I}(s_i) = 0$.[3]

Learning. In order to optimize the DCG objective, we develop a *learning-to-rank* DNN model [3,8]. More specifically, our DNN model is implemented by a

---

[3] The motivation for optimizing GMV is related to the e-commerce company's business model. Since investors and shareholders utilize the settled GMV from all e-commerce transactions generated to estimate the current health and future potential of the corresponding company, maximizing GMV becomes critical. In general, if each item's GMV sets to be **1**, optimizing GMV degrades to optimizing the sales conversion.

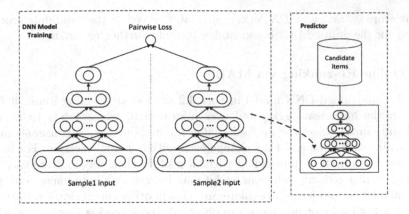

Fig. 2. Pairwise architecture of our learning-to-rank DNN.

pair of 5-layer DNNs that share the same set of model parameters. Each DNN includes 1 input layer, 3 fully-connected layers and 1 output layer. The input layer takes feature $\mathbf{x}_i$, which contains both item and user information via one-hot encoding. The high level model structure is shown in Fig. 2 (Left).

In our training, instead of using the binary labels which indicate whether items are bought, we conduct calibrations on binary labels and convert them into real-valued labels y_is, where $0 \leq y_i \leq 1$. The calibrated labels (y_is) are computed by using item s_i's GMV gmv_{s_i} that represents how much revenue will be generated if s_i is impressed comparing against the maximum gmv under the same category c_{s_i}, shown in Eq. (1).

$$y_i = \mathcal{Y}(gmv_{s_i}, \mathcal{I}(s_i)) = \frac{gmv_{s_i}}{\max_{s \in c_{s_i}} gmv_s} \times \mathcal{I}(s_i) \tag{1}$$

To train our pairwise DNN model (shown in Fig. 2), we generate training example pairs by combining a positively scored item and a randomly selected zero scored item ($y = 0.0$). Furthermore, we define the loss function of the proposed DNN as follows.

$$\mathcal{L} = \sum_j [(\hat{y}_{1j} - y_{1j})^2 + (\hat{y}_{2j} - y_{2j})^2 + \lambda \max(0, \gamma - (\hat{y}_{1j} - \hat{y}_{2j})(y_{1j} - y_{2j}))] \tag{2}$$

where $<\mathbf{x}_{1j}, y_{1j}>$ and $<\mathbf{x}_{2j}, y_{2j}>$ are the first and second examples in the jth pair. $\hat{y}_{1j}$ and $\hat{y}_{2j}$ are the outputs of the DNN predictions of the first and second sample: $\hat{y} = \mathcal{DNN}(\mathbf{x})$. λ is the weighting parameter that balances the square loss of each prediction and the hinge loss of the prediction distance for such pair. γ serves as the classification margin to make sure that separability between each pair is preferred.

Prediction. In the prediction phase, each item $\mathbf{x}_i \in \mathcal{S}$ passes through the well trained *learning-to-rank* DNN predictor (Fig. 2 (Right)), and DNN predicts the output score $\hat{y}_i$, i.e., $\hat{y}_i = \mathcal{DNN}(\mathbf{x}_i)$. $\hat{y}_i$ is later served as the candidate score of item $\mathbf{x}_i$ for the online MAB based ranker to make further re-ranking.

3.3   Online Re-ranking via MAB

Even the pre-trained DNN model in Sect. 3.2 is good at selecting items, it fully relies on the historical aggregated user information, which tends to be outdated and cannot reflect the up-to-date user intents. Furthermore, user shopping intents vary with time and they change even within a single session. For example, tens or hundreds of impressions without a single click may indicate users are bored of current recommendations and it may be worthwhile to make "broader" recommendation and let users explore products in other categories. On the other hand, if a few clicks or purchases are observed, the following recommendations probably need to be "focused" and allow users to fully exploit relevant products. It is necessary to have such responsive recommender systems that keep track of transient shopping intentions while users are browsing the product feeds.

It is very difficult to achieve such responsiveness by using the DNN model in the second phase (Sect. 3.2) since doing full scoring of all the candidates after each user click or purchase event is computationally prohibitive. Moreover, it is not easy to have an online update to incorporate most recent events into the ranking model and hence the ranking model may become stale.

In this work, we develop a dynamic online re-ranking algorithm by revising the classic Thompson sampling. We pick Thompson sampling out of other MAB approaches due to its implementation straightforwardness. Further, by careful revising, the *revised*-Thompson sampling turns out to be superior in terms of performance metrics which we will elaborate in more detail. Our MAB based re-ranking algorithm is able to

- promote recommended items that users are potentially interested in;
- demote items that users are intentionally ignored;
- explore items from different categories to diversify the ranking results.

Here, we follow the problem settings of the contextual multi-arm bandits problem in [19], and define the contextual bandits (see Definition 1) and rewards (see Definition 2) in DORS as follows.

Definition 1 (CONTEXTUAL BANDITS IN DORS). *Let each arm represent a product category c and $\mathcal{S}$ be a set of K top ranked items from DNN. Such K items are then divided into different arms based on their own product categories. The player pulls one arm c_i at round i, selects the top item $\mathbf{x}_i$ from the chosen arm which has not been displayed yet and contiguously places the item to the item selected from round $i - 1$. The reward at round i is observed as $gmv_i\mathcal{I}(\mathbf{x}_i)$. Ideally, we would like to choose arms such that the total rewards are maximized.*

Definition 2 (REWARDS IN DORS). *The rewards are defined as the total GMV generated from $\mathcal{S}$ that users place orders on. In general it can be written as: $Revenue(K, \mathcal{S}) = \sum_{s_i \in \mathcal{S}} gmv_{s_i}\mathcal{I}(s_i)$.*

The *revised*-Thompson sampling algorithm is triggered after *learning-to-rank* DNN. After taking the DNN output as the static ranking results, DORS organizes items by categories, and fine tunes the local orders of the DNN static ranking, which enables more efficient convergence by enforcing DNN scores as the warm starts and analyzing user feedback including impressions and clicks as rewards signals for the Thompson online re-ranking.

At the initialization stage, we group the *pre*-ranked items based on their categories, and define each group as a single arm, which is modeled by a unique beta distribution as follows

$$f(r; \alpha_c, \beta_c) = \frac{\Gamma(\alpha_c + \beta_c)}{\Gamma(\alpha_c)\Gamma(\beta_c)} r^{\alpha_c - 1}(1 - r)^{\beta_c - 1} \tag{3}$$

where α_c and β_c are two hyper-parameters.

Let avg_c represent the average DNN scores of all items in category c. We initialize α_c, β_c, avg_c by DNN *pre*-ranked scores which represent how likely the user be interested in those items based on recent history. At round i, the *revised*-Thompson sampling randomly draws M samples $\{r\}_M$ based on M estimated beta distributions, and then selects the item associated with the highest adjusted score $\hat{y}_i$ for exposure at round i. If the item is clicked, the algorithm updates α_{c_i} in arm c_i. Otherwise, the algorithm updates β_{c_i} in arm c_i. The details of the MAB algorithm in DORS are described in Algorithm 1. Please note that the values of $\theta_1, \theta_2, \theta_3, \delta$ in our current production are selected by grid searches in online A/B experiments.

One challenge that stops people from using the multi-arm bandits model to the real production system might be the lack of existence of researches regarding the multi-arm bandits convergence analysis. Such convergence efficiency is important since most customers in average only browse less than one handred items *per* visit. In this section, we prove that the *revised*-Thompson sampling in DORS guarantees the efficient convergence and provide the regret boundaries in two cases.

As described in earlier sections, arms are pulled based on the beta distributions $f(r; \alpha, \beta) = \frac{\Gamma(\alpha+\beta)}{\Gamma(\alpha)\Gamma(\beta)} r^{\alpha-1}(1-r)^{\beta-1}$, and for each item, the rewards are set to be $gmv_{s_i}\mathcal{I}(s_i)$, indicating whether a gmv related user action would happen at the current round. Without loss of generality, by assuming each item's gmv identity and taking the expectation, we simplify the total rewards and translate it into the modified total expected regrets as:

$$\mathbb{E}(Regret(K, \mathcal{S})) = \mathbb{E}\left[\sum_{k=1}^{K}(\mu^* - \mu_i(k))\right] = \sum_i \Delta_i \mathbb{E}[h_i(K)] \tag{4}$$

where $\Delta_i = \mu^* - \mu_i(k)$ denotes the regret for pulling the ith arm and h_i denotes the number of plays for i^{th} arm up to round $k - 1$, μ^* denotes the maximum reward (calibrated revenue) the system could possibly reach for exposing item i in round i, here in our case $\mu^* = 1.0$. Suggested in Algorithm 1, unlike traditional Thompson sampling updating strategies, we utilized DNN score $\hat{y}$s to update hyper parameter α and β for each beta distribution with smoothing factors under our GMV identity assumptions.

Algorithm 1. MAB in DORS: the *revised*-Thompson sampling.

INPUT:
δ: control the intensity of negative feedbacks
θ_1, θ_2, θ_3: control how much the $\mathcal{DNN}$ scores to be tuned
α_c, β_c: the hyper parameter of beta distribution for category c.
$\mathcal{S}$: the items that are top ranked by the $\mathcal{DNN}$ algorithm
$\mathcal{U}_c$: the items that are not selected in category c.
$\mathcal{E}_c$: the items that are impressed but not clicked in category c.
$\mathcal{A}_c$: the items that are impressed and clicked in category c.
M: the total number of categories / arms
$|\cdot|_0$: the cardinality operator.
procedure INITIALIZATION
    for each $\langle \mathbf{x}, \hat{y} \rangle \in \mathcal{S}$ **do**
        for arm c such that $\mathbf{x} \in c$ **do**
            $\alpha_c = \alpha_c + \hat{y}$
            $\beta_c = \beta_c + (1 - \hat{y})$
            $\mathcal{U}_c = \mathcal{U}_c \cup \{\langle \mathbf{x}, \hat{y} \rangle\}$
        $avg_c = \alpha_c / |\mathcal{U}_c|_0$
procedure AT ROUND-i MAB RANKING
    PULLING ARMS:
    for each arm c **do**
        sample $r \sim f(r; \alpha_c, \beta_c)$
        update all $\hat{y} = \hat{y} \times (1 + r/\theta_1)$ for $\langle \mathbf{x}, y \rangle \in c$
    pick category $c = \arg\max_c \{r_1, r_2, \ldots, r_M\}$
    pick item $\langle \mathbf{x}_i, \hat{y}_i \rangle = \arg\max_i \{\hat{y} \in c\}$
    $\mathcal{U}_c = \mathcal{U}_{c_i} - \langle \mathbf{x}_i, \hat{y}_i \rangle$
    FEEDBACK:
    if $\langle \mathbf{x}_i, \hat{y}_i \rangle$ is impressed but not clicked **then**
        $\mathcal{E}_c = \mathcal{E}_c \cup \{\langle \mathbf{x}_i, \hat{y}_i \rangle\}$
        $\beta_{c_i} = \beta_{c_i} + (1 - avg_{c_i}) \times (1 - \exp(-\frac{|\mathcal{E}_{c_i}|_0}{\delta})) \times \theta_2$
    if $\langle \mathbf{x}_i, \hat{y}_i \rangle$ is impressed and clicked **then**
        $\mathcal{A}_{c_i} = \mathcal{A}_{c_i} \cup \{\langle \mathbf{x}_i, \hat{y}_i \rangle\}$
        $\alpha_{c_i} = \alpha_{c_i} + avg_{c_i} \times (\frac{|\mathcal{A}_{c_i}|_0}{|\mathcal{E}_{c_i}|_0}) \times \theta_3$

Theorem 1. *For the 2-armed case the revised-Thompson sampling holds the expected regret:*

$$\mathbb{E}(Regret(K, \mathcal{S})) \leq \frac{40 \ln K}{\Delta} + \frac{48}{\Delta^3} + 18\Delta = \mathcal{O}(\frac{\ln K}{\Delta} + \frac{1}{\Delta^3}) \qquad (5)$$

where Δ indicates the difference in mean between the reward from the optimal arm μ^* and the reward from the suboptimal arm μ_{sub}: $\Delta = \mu^* - \mu_{sub}$.

Theorem 2. *For the M-armed case (M > 2) the revised-Thompson sampling holds the expected regret:*

$$\mathbb{E}(Regret(K, \mathcal{S})) \leq \mathcal{O}(\frac{\Delta_{\max}}{\Delta_{\min}^3}(\sum_{a=2}^{M} \frac{1}{\Delta_a^2}) \ln K) \tag{6}$$

Detailed proof can be found in supplemental materials. With convergence guarantees, we know that the *revised*-Thompson sampling helps the system achieve the overall optimized GMV performance with the expected regret no greater than $\mathcal{O}(\frac{\Delta_{\max}}{\Delta_{\min}^3}(\sum_{a=2}^{M} \frac{1}{\Delta_a^2}) \ln K)$.

4   Experiments

4.1   Case Study

We first walk through a simple case study to evaluate different bandit algorithms under the streaming recommendation use cases. We pick three *state-of-the-art* bandit algorithms: ϵ-greedy [31], Upper Confidence Bound (UCB) [2], and Thompson sampling [30]. Specifically, we simulate two versions of Thompson sampling:

- *revised*-**Thompson:** the revised Thompson sampling with *learning-to-rank* DNN initializations (Algorithm 1);
- *normal*-**Thompson:** the normal Thompson sampling without initializations.
  The random selection is also evaluated as a naïve baseline.

In the simulation, we design $M = 5$ arms. If the user clicks, we set each item's reward as 1, and 0 otherwise. The way we simulate "click"s is by presetting a thresholding probability τ. When an item is selected by different algorithms at each round, we sample another probability f_{item} based on that item's real unknown beta distribution. If $f_{\text{item}} \geq \tau$, we assume the "click" happens; otherwise we assume the user is not interested in the item which results in "no click" at that round.

We run this simulation 10 times and at each time we operate 10,000 rounds. The average performance is shown in Figs. 3 and 4. The left subfigures are the cumulative gains/regrets for different methods and the right subfigures zoom in the early stage of different algorithms' performance. As shown, ϵ-greedy remains suboptimal regarding both rewards and regrets; UCB and *normal*-Thompson perform almost equally well; while the *revised*-Thompson performs the best in terms of faster convergence comparing to UCB and *normal*-Thompson. This is because the *revised*-Thompson initializes its arms based on the DNN scores of each item and updates its hyper-parameters via the user feedback. Hence, the *revised*-Thompson converges in less steps relative to other standard MAB approaches. The random selection no-surprisingly performs the worst against other approaches. Regarding with implementation, the *revised*-Thompson is straightforward and the overall system latency remains very low (the details are reported in Sect. 4.5). From this case study, the *revised*-Thompson proves itself to be the best choice out of many other MAB approaches for the online re-ranking module in DORS.

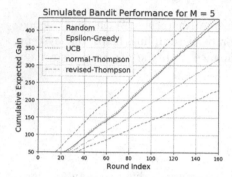
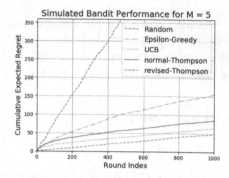

Fig. 3. MAB rewards simulation.   **Fig. 4.** MAB regrets simulation.

4.2 Experiment Setup

We conduct large-scale experiments on www.jd.com, which processes millions of requests in a daily basis. The total traffics have been divided into 11 buckets: the biggest bucket serves as the control bucket which takes about 20% of the total traffics and the rest 10 buckets gets about 8% traffics each, each of which is served as one testing bucket.

We first deploy DORS to one testing bucket for one week, and track the following metrics: GMV, order numbers, overall and page-wise (Eq. (7)) normalized discounted cumulative gains ($NDCG$).

$$DCG_{p,k} = \sum_{i=1, k \in \text{page}}^{p} \frac{gmv_{ki}\mathcal{I}(\mathbf{x}_{ki})}{\log_2(i+1)}$$

$$IDCG_{p,k} = \max_{\mathcal{H}} DCG_{p,k}$$

$$NDCG_{p,k} = DCG_{p,k}/IDCG_{p,k}$$

$$\Delta_{NDCG_{p,k}} = (NDCG_{p,k}^{\text{test}}/NDCG_{p,k}^{\text{control}} - 1.0) \times 100.0\% \qquad (7)$$

$$\mathcal{H} : \text{all possible arrangement of items in page} - k$$

Since the item lists are presented to users in a page-wise fashion and each page contains 4–20 items (front-page typically contains 8 items *per* page), page-wise $NDCG_p$ ($p = 8$) is a perfect metric for evaluating how DORS is performing in real world and how much gains observed are due to DORS online re-ranking.

Experiments include four methods and we briefly explain each of them as follows:

- **DORS:** our proposed online ranking framework which is composed by *learning-to-rank* DNN and the *revised*-Thompson sampling as online re-ranking;
- **DNN-*normal*-Thompson:** a right straight combination of *learning-to-rank* DNN and normal Thompson sampling without the specialized initialization. This model explains how the non-contextual bandits algorithm performs as the online ranker in the production system;
- **MAB-only:** is implemented via a normal Thompson sampling without *learning-to-rank* DNN and using the historical click through rate to update the beta distributions;
- **DNN-*rt-ft*:** utilizes another *learning-to-rank* DNN with the same model structure. The difference is that besides utilizing offline features used in DORS, DNN-*rt-ft* takes most recent user feedback into the DNN training process to generate its own ranking results.

We report the performance of 7-day average page-wise $NDCG$ gain in Fig. 5. At the first glance, it is clear that MAB-only performs the worst among all four algorithms, which explains the importance of the *learning-to-rank* DNN serving as the static ranking phase. Further, similar to what has been observed in the previous case study, without the specialized initialization, DNN-*normal*-Thompson fails to DNN-*rt-ft* due to the fact that by limited online signals, the *normal*-Thompson is slow in convergence. By our design, the proposed DORS beats the production baseline DNN-*rt-ft* by efficiently learning the user online intents.

4.3   Production Performance

Taking a closer look, we evaluate the DORS page-wise $NDCG$ percentage gains over DNN-*rt-ft* in Table 1 and Fig. 6. By understanding each user recent behaviors via the personalized initialization and learning the real-time signals for online ranking, although they are not quite visible at page-1 (+1.47%), the gains for DORS at page-2 (+9.96%) and page-3 (+8.90%) quickly boost up, and then gradually diminish along with more items users browse. In the end, the gap between DORS and DNN-*rt-ft* becomes closer again at page-7 (+1.54%) and page-8 (+1.34%).

In terms of overall ranking performance, we report the final ranking NDCG percentage gain between DORS and DNN-*rt-ft* in a daily basis as shown in Table 2, as one can see, DORS consistently beats DNN-*rt-ft*. In average DORS's overall NDCG is 4.37% better comparing against DNN-*rt-ft*.

We also report the daily GMV gain/loss for DORS, DNN-*normal*-Thompson and MAB-only over DNN-*rt-ft*. In average we see DORS has increased **16.69%** over DNN-*rt-ft* (Table 3). Daily GMV gains that are greater than 1.0% are typically translating into revenue in hundreds of thousands of dollars. On the one hand, DORS has proved its superiority against the current production DNN-*rt-ft* ranking algorithm in terms of operating revenue; on the other hand, without

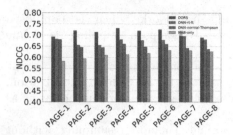

Fig. 5. 7-day page-wise NDCG for DORS, DNN-*rt-ft*, DNN-*normal*-Thompson, MAB-only.

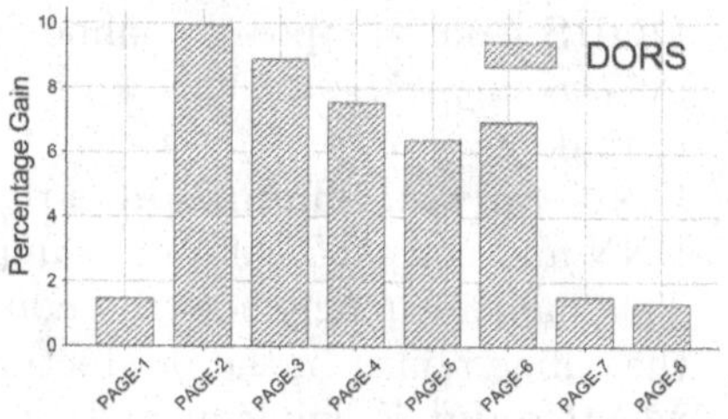

Fig. 6. 7-day page-wise NDCG percentage gain for DORS v.s. DNN-*rt-ft*.

decent *learning-to-rank* DNN static ranking or carefully designed initializations, DNN + *normal*-Thompson (Table 4) and MAB-only (Table 5) both failed to beat the baseline by losing 8.08% and 19.57% in revenue respectively.

Table 1. Page-wise NDCG gain A/B test comparison: DORS v.s. DNN-*rt-ft*.

Page	Page-1	Page-2	Page-3	Page-4	Page-5	Page-6	Page-7	Page-8
Δ_{NDCG}	+1.47%	+9.96%	+8.90%	+7.55%	+6.40%	+6.95%	+1.54%	+1.34%

Table 2. Daily NDCG gain A/B test comparison: DORS v.s. DNN-*rt-ft*.

Date	Day1	Day2	Day3	Day4	Day5	Day6	Day7	Average
Δ_{NDCG}	+5.60%	+10.23%	+2.90%	+1.93%	+1.59%	+6.96%	+2.34%	**+4.37%**

Table 3. GMV and orders comparison for DORS v.s. DNN-*rt-ft*.

Date	Day1	Day2	Day3	Day4	Day5	Day6	Day7	Summary
GMV	+22.87%	+45.45%	+20.20%	+2.73%	+0.91%	+23.15%	+1.50%	**+16.69%**
Orders	+2.14%	+1.57%	+5.18%	+0.42%	+2.79%	+4.19%	+2.20%	**+2.64%**

Table 4. GMV and orders comparison for DNN + *normal*-Thompson v.s. DNN-*rt-ft*.

Date	Day1	Day2	Day3	Day4	Day5	Day6	Day7	Summary
GMV	−12.08%	−9.33%	−4.74%	−3.24%	−18.31%	−7.49%	−1.43%	**−8.08%**
Orders	0.30%	−4.72%	−1.34%	−0.67%	−10.67%	−4.69%	−0.81%	**−3.23%**

Table 5. GMV and orders comparison for MAB-only v.s. DNN-*rt-ft*.

Date	Day1	Day2	Day3	Day4	Day5	Day6	Day7	Summary
GMV	−29.52%	−17.46%	−37.17%	−8.99%	−32.61%	−5.21%	−6.04%	**−19.57%**
Orders	−8.33%	−12.51%	−7.66%	−5.91%	−15.35%	−7.64%	−3.87%	**−8.75%**

4.4 Distribution Analysis

It is worth analyzing the page-wise click distribution in a daily basis between DORS and DNN-*rt-ft* to better understand why the GMV has been significantly driven. Figure 7 is the daily page-wise click distributions over 7 days. For the figure readability, we display y-axis in the $\log_{10}$ scale, keep three significant digits and round up all numbers that are smaller than 1.0 to 1.0 ($\log_{10}(1.0) = 0.0$). Since each page displays 8 items for recommendations, the page-wise clicks could at most reach 8. The x-axis indicates the number of clicks people make and y-axis indicates the number of people making certain clicks at that page.

At page-1, the DORS histogram resembles the DNN-*rt-ft* histogram. This is due to the fact that the online user signals have not been fed into the MAB yet, so DORS is not expected to behave differently from DNN-*rt-ft*. At page 2–7, we observe the DORS histograms consistently "flatter" than DNN-*rt-ft* histograms. Note that from page 2–7 DORS wins most cases against DNN-*rt-ft*. This could be explained by the fact that the *revised*-Thompson is better in capturing user online intents so the online ranking results are optimized and people tend to click more frequently. Finally, the DORS histogram at page-8 resembles DNN-*rt-ft* again, due to the fact that at page-8 most users either have their intents captured or they abandon the site visit.

4.5 System Specifications

Our current DORS ranking framework is maintained by hundreds of Linux servers[4]. Each machine processes 192 queries per second in average (peaking at 311), and the 99% percentile end-to-end latency is within 200.0 ms. In our current production, the feature dimension is 5.369×10^8.

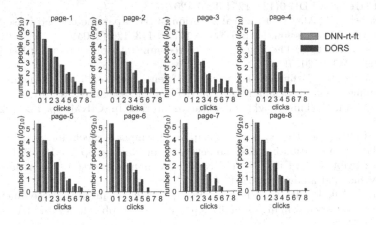

Fig. 7. Daily page-wise click distribution: DORS v.s. DNN-*rt-ft*.

---

[4] We could not release the exact number of operating servers due to the company confidentiality.

5   Conclusion

In this paper, we presented a novel three-level recommender system for e-commerce recommendation, which includes candidate retrieval, learning-to-rank DNN and MAB based online re-ranking. Compared to the traditional recommender systems, our approach is relevant, responsive, scalable and is running in production to serve millions of recommendations everyday. Our offline case studies have empirically demonstrated the efficiency for learning-to-rank DNN serving as the static ranking as well as the warm start for the revised-Thompson initializations that enable the quick convergence. Furthermore, online A/B experiments have been used to prove DORS is superior in terms of pagewise/overall NDCG as well as the operating revenue gains when serving in the real production system.

References

1. Agarwal, D., Chen, B.C., Elango, P.: Fast online learning through offline initialization for time-sensitive recommendation. In: KDD, pp. 703–712. ACM (2010)
2. Auer, P., Cesa-Bianchi, N., Fischer, P.: Finite-time analysis of the multiarmed bandit problem. Mach. Learn. **47**(2–3), 235–256 (2002)
3. Basak, D., Pal, S., Patranabis, D.C.: Support vector regression. Neural Inf. Process.-Lett. Rev. **11**(10), 203–224 (2007)
4. Burges, C., et al.: Learning to rank using gradient descent. In: ICML, pp. 89–96. ACM (2005)
5. Cao, Z., Qin, T., Liu, T.Y., Tsai, M.F., Li, H.: Learning to rank: from pairwise approach to listwise approach. In: ICML, pp. 129–136. ACM (2007)
6. Chang, S., et al.: Streaming recommender systems. In: WWW, pp. 381–389 (2017)
7. Chen, C., Yin, H., Yao, J., Cui, B.: TeRec: a temporal recommender system over tweet stream. VLDB **6**(12), 1254–1257 (2013)
8. Cherkassky, V., Ma, Y.: Practical selection of SVM parameters and noise estimation for SVM regression. Neural Netw. **17**(1), 113–126 (2004)
9. Davidson, J., et al.: The YouTube video recommendation system. In: RecSys, pp. 293–296. ACM (2010)
10. Diaz-Aviles, E., Drumond, L., Schmidt-Thieme, L., Nejdl, W.: Real-time top-n recommendation in social streams. In: RecSys, pp. 59–66. ACM (2012)
11. Ding, Y., Li, X.: Time weight collaborative filtering. In: CIKM, pp. 485–492. ACM (2005)
12. Gao, H., Tang, J., Hu, X., Liu, H.: Exploring temporal effects for location recommendation on location-based social networks. In: RecSys, pp. 93–100. ACM (2013)
13. Gomez-Uribe, C.A., Hunt, N.: The Netflix recommender system: algorithms, business value, and innovation. TMIS **6**(4), 13 (2016)
14. Guan, Z., Bu, J., Mei, Q., Chen, C., Wang, C.: Personalized tag recommendation using graph-based ranking on multi-type interrelated objects. In: SIGIR, pp. 540–547. ACM (2009)
15. Gultekin, S., Paisley, J.: A collaborative Kalman filter for time-evolving dyadic processes. In: ICDM, pp. 140–149. IEEE (2014)
16. Hannon, J., Bennett, M., Smyth, B.: Recommending Twitter users to follow using content and collaborative filtering approaches. In: RecSys, pp. 199–206. ACM (2010)

17. Koren, Y.: Collaborative filtering with temporal dynamics. Commun. ACM **53**(4), 89–97 (2010)
18. Lang, K.: NewsWeeder: learning to filter netnews. In: Proceedings of the 12th International Machine Learning Conference (ML 1995) (1995)
19. Langford, J., Zhang, T.: The epoch-greedy algorithm for multi-armed bandits with side information. In: NIPS, pp. 817–824 (2008)
20. Liu, Y., Miao, J., Zhang, M., Ma, S., Ru, L.: How do users describe their information need: query recommendation based on snippet click model. Expert Syst. Appl. **38**(11), 13847–13856 (2011)
21. Lu, D., et al.: Cross-media event extraction and recommendation. In: NAACL, pp. 72–76 (2016)
22. Lu, Z., Agarwal, D., Dhillon, I.S.: A spatio-temporal approach to collaborative filtering. In: RecSys, pp. 13–20. ACM (2009)
23. Rendle, S.: Factorization machines. In: ICDM, pp. 995–1000. IEEE (2010)
24. Rendle, S., Freudenthaler, C., Schmidt-Thieme, L.: Factorizing personalized Markov chains for next-basket recommendation. In: WWW, pp. 811–820. ACM (2010)
25. Rendle, S., Schmidt-Thieme, L.: Online-updating regularized kernel matrix factorization models for large-scale recommender systems. In: RecSys, pp. 251–258. ACM (2008)
26. Sarwar, B., Karypis, G., Konstan, J., Riedl, J.: Item-based collaborative filtering recommendation algorithms. In: WWW, pp. 285–295. ACM (2001)
27. Severyn, A., Moschitti, A.: Learning to rank short text pairs with convolutional deep neural networks. In: SIGIR, pp. 373–382. ACM (2015)
28. Tang, J., Hu, X., Gao, H., Liu, H.: Exploiting local and global social context for recommendation. In: IJCAI, vol. 13, pp. 2712–2718 (2013)
29. Tang, J., Hu, X., Liu, H.: Social recommendation: a review. Soc. Netw. Anal. Min. **3**(4), 1113–1133 (2013)
30. Thompson, W.R.: On the likelihood that one unknown probability exceeds another in view of the evidence of two samples. Biometrika **25**(3/4), 285–294 (1933)
31. Watkins, C.J.C.H.: Learning from delayed rewards. Ph.D. thesis, University of Cambridge, England (1989)
32. Xiang, L., et al.: Temporal recommendation on graphs via long-and short-term preference fusion. In: KDD, pp. 723–732. ACM (2010)
33. Xiong, L., Chen, X., Huang, T.K., Schneider, J., Carbonell, J.G.: Temporal collaborative filtering with bayesian probabilistic tensor factorization. In: SDM. pp. 211–222. SIAM (2010)
34. Yin, D., Hong, L., Xue, Z., Davison, B.D.: Temporal dynamics of user interests in tagging systems. In: AAAI (2011)
35. Yu, X., et al.: Personalized entity recommendation: A heterogeneous information network approach. In: WSDM. pp. 283–292. ACM (2014)
36. Zhang, Y., Zhang, M., Liu, Y., Ma, S., Feng, S.: Localized matrix factorization for recommendation based on matrix block diagonal forms. In: WWW (2013)

ADS Engineering and Design

Helping Your Docker Images to Spread Based on Explainable Models

Riccardo Guidotti[1,2]([✉]), Jacopo Soldani[1], Davide Neri[1], Antonio Brogi[1], and Dino Pedreschi[1]

[1] University of Pisa, Largo B. Pontecorvo, 3, Pisa, Italy
{riccardo.guidotti,jacopo.soldani,davide.neri,antonio.brogi,
dino.pedreschi}@di.unipi.it
[2] KDDLab, ISTI-CNR, via G. Moruzzi, 1, Pisa, Italy
guidotti@isti.cnr.it

Abstract. Docker is on the rise in today's enterprise IT. It permits shipping applications inside portable containers, which run from so-called Docker images. Docker images are distributed in public registries, which also monitor their popularity. The popularity of an image impacts on its actual usage, and hence on the potential revenues for its developers. In this paper, we present a solution based on interpretable decision tree and regression trees for estimating the popularity of a given Docker image, and for understanding how to improve an image to increase its popularity. The results presented in this work can provide valuable insights to Docker developers, helping them in spreading their images. Code related to this paper is available at: https://github.com/di-unipi-socc/DockerImageMiner.

Keywords: Docker images · Popularity estimation
Explainable models

1 Introduction

Container-based virtualization provides a simple yet powerful solution for running software applications in isolated virtual environments, called *containers* [30]. Containers are rapidly spreading over the spectrum of enterprise information technology, as they feature much faster start-up times and less overhead than other existing visualization approaches, e.g., virtual machines [14].

Docker is the de-facto standard for container-based virtualization [21]. It permits building, shipping and running applications inside portable containers. Docker containers run from Docker images, which are the read-only templates used to create them. A Docker image permits packaging a software together with all the dependencies needed to run it (e.g., binaries, libraries). Docker also provides the ability to distribute and search (images of) Docker containers through so-called Docker registries. Given that any developer can create and distribute its own created images, other users have at their disposal plentiful

© Springer Nature Switzerland AG 2019
U. Brefeld et al. (Eds.): ECML PKDD 2018, LNAI 11053, pp. 205–221, 2019.
https://doi.org/10.1007/978-3-030-10997-4_13

repositories of heterogeneous, ready-to-use images. In this scenario, public registries (e.g., Docker Hub) are playing a central role in the distribution of images.

DOCKERFINDER [6] permits searching for existing Docker images based on multiple attributes. These attributes include (but are not limited to) the name and size of an image, its popularity within the Docker community (measured in terms of so-called *pulls* and *stars*), the operating system distribution they are based on, and the software distributions they support (e.g., java 1.8 or python 2.7). DOCKERFINDER automatically crawls all such information from the Docker Hub and by directly inspecting the Docker containers that run from images. With this approach, DOCKERFINDER builds its own dataset of Docker images, which can be queried through a GUI or through a RESTful API.

The popularity of an image directly impacts on its usage [19]. Maximising the reputation and usage of an image is of course important, as for every other kind of open-source software. The higher is the usage of an open-source software, the higher are the chances of revenue from related products/services, as well as the self-marketing and the peer recognition for its developers [13].

In line with [9], the main objectives of this paper are (i) to exploit the features retrieved by DOCKERFINDER to understand how the features of an image impact on its popularity, and (ii) to design an approach for recommending how to update an image to increase its popularity. In this perspective, we propose:

(i) DARTER (*Decision And Regression Tree EstimatoR*), a mixed hierarchical approach based on decision tree classifiers and regression trees, which permits estimating the popularity of a given Docker image, and
(ii) DIM (*Docker Image Meliorator*), an explainable procedure for determining the smallest changes to a Docker image to improve its popularity.

It is worth highlighting that our approach is *explainable* by design [10,11]. That is, we can understand which features delineate an estimation, and we can exploit them to improve a Docker image. Besides being a useful peculiarity of the model, comprehensibility of models is becoming crucial, as the European Parliament in May 2018 adopted the GDPR for which a "right of explanation" will be required for automated decision making systems [8].

Our results show that (i) DARTER outperforms state-of-the-art estimators and that popular images are not obtained "by chance", and that (ii) DIM recommends successful improvements while minimizing the number of required changes. Thanks to the interpretability of both DARTER and DIM, we can analyze not-yet-popular images, and we can automatically determine the most recommended, minimal sets of changes allowing to improve their popularity.

The rest of the paper is organized as follows. Section 2 provides background on Docker. Sections 3 and 4 illustrate the popularity problems we aim to target and show our solutions, respectively. Section 5 presents a dataset of Docker images and some experiments evaluating our solutions. Sections 6 and 7 discuss related work and draw some conclusions, respectively.

2   Background

Docker is a platform for running applications in isolated user-space instances, called *containers*. Each Docker *container* packages the applications to run, along with all the software support they need (e.g., libraries, binaries, etc.).

Containers are built by instantiating so-called Docker *images*, which can be seen as read-only templates providing all instructions needed for creating and configuring a container (e.g., software distributions to be installed, folders/files to be created). A Docker image is made up of multiple file systems layered over each other. A new Docker image can be created by loading an existing image (called *parent* image), by performing updates to that image, and by committing the updates. The commit will create a new image, made up of all the layers of its parent image plus one, which stores the committed updates.

Existing Docker images are distributed through Docker *registries*, with the Docker Hub (hub.docker.com) being the main registry for all Docker users. Inside a registry, images are stored in *repositories*, and each repository can contain multiple Docker images. A repository is usually associated to a given software (e.g., *Java*), and the Docker images contained in such repository are different versions of such software (e.g., *jre7, jdk7, open-jdk8*, etc.). Repositories are divided in two main classes, namely *official* repositories (devoted to curated sets of images, packaging trusted software releases—e.g., *Java, NodeJS, Redis*) and *non-official* repositories, which contain software developed by Docker users.

The success and popularity of a repository in the Docker Hub can be measured twofold. The number of *pulls* associated to a repository provides information on its actual usage. This is because whenever an image is downloaded from the Docker Hub, the number of pulls of the corresponding repository is increased by one. The number of *stars* associated to a repository instead provides significant information on how much the community likes it. Each user can indeed "star" a repository, in the very same way as eBay buyers can "star" eBay sellers.

DOCKERFINDER is a tool for searching for Docker images based on a larger set of information with respect to the Docker Hub. DOCKERFINDER automatically builds the description of Docker images by retrieving the information available in the Docker Hub, and by extracting additional information by inspecting the Docker containers. The Docker image descriptions built by DOCKERFINDER are stored in a JSON format[1], and can be retrieved through its GUI or HTTP API.

Among all information retrieved by DOCKERFINDER, in this work we shall consider the size of images, the operating system and software distributions they support, the number of layers composing an image, and the number of pulls and stars associated to images. A formalization of data structure considered is provided in the next section. Moreover, in the experimental section we will also observe different results for official and non-official images.

---

[1] An example of raw Docker image data is available at https://goo.gl/hibue1.

3   Docker Images and Popularity Problems

We hereafter provide a formal representation of Docker images, and we then illustrate the popularity problems we aim to target.

A *Docker image* can be represented as a tuple indicating the operating system it supports, the number of layers forming the image, its compressed and actual size, and the set of software distributions it supports. For the sake of readability, we shall denote with $\mathbb{U}_{os}$ the finite universe of existing operating system distributions (e.g., "Alpine Linux v3.4", "Ubuntu 16.04.1 LTS"), and with $\mathbb{U}_{sw}$ the finite universe of existing software distributions (e.g., "java", "python").

Definition 1 (Image). *Let $\mathbb{U}_{os}$ be the finite universe of operating system distributions and $\mathbb{U}_{sw}$ be the finite universe of software distributions. We define a Docker image I as a tuple $I = \langle os, layers, size_d, size_a, \mathcal{S} \rangle$ where*

- *$os \in \mathbb{U}_{os}$ is the operating system distribution supported by the image I,*
- *$layers \in \mathbb{N}$ is the number of layers stacked to build the image I,*
- *$size_d \in \mathbb{R}$ is the download size of I,*
- *$size_a \in \mathbb{R}$ is the actual size[2] of I, and*
- *$\mathcal{S} \subseteq \mathbb{U}_{sw}$ is the set of software distributions supported by the image I.*

A concrete example of Docker image is the following:

$$\langle \text{Ubuntu 16.04.1 LTS}, \ 6, \ 0.78, \ 1.23, \ \{\text{python}, \text{perl}, \text{curl}, \text{wget}, \text{tar}\} \rangle.$$

A *repository* contains multiple Docker images, and it stores the amount of pulls and stars associated to the images it contains.

Definition 2 (Repository). *Let $\mathbb{U}_I$ be the universe of available Docker images. We define a repository of images as a triple $R = \langle p, s, \mathcal{I} \rangle$ where*

- *$p \in \mathbb{R}$ is the number (in millions) of pulls from the repository R,*
- *$s \in \mathbb{N}$ is the number of stars assigned to the repository R, and*
- *$\mathcal{I} \subseteq \mathbb{U}_I$ is the set of images contained in the repository R.*

For each repository, the number of pulls and stars is not directly associated with a specific image, but it refers to the overall repository. We hence define the notion of *imager*, viz., an image that can be used as a "representative image" for a repository. An imager essentially links the pulls and stars of a repository with the characteristic of an image contained in such repository.

---

[2] Images downloaded from registries are compressed. The download size of an image is hence its compressed size (in GBs), while its actual size is the disk space (in GBs) occupied after decompressing and installing it on a host.

Definition 3 (Imager). *Let $R = \langle p, s, \mathcal{I} \rangle$ be a repository, and let $I = \langle os,$ layers, $size_d, size_a, \mathcal{S} \rangle \in \mathcal{I}$ be one of the images contained in R. We define an imager I_R as a tuple directly associating the pulls and stars of R with I, viz.,*

$$I_R = \langle p, s, I \rangle = \langle p, s, \langle os, layers, size_d, size_a, \mathcal{S} \rangle \rangle.$$

A concrete example of imager is the following:

$$\langle 1.3,\ 1678,\ \langle \texttt{Ubuntu 16.04.1 LTS},\ 6,\ 0.7,\ 1.2,\ \{\texttt{python}, \texttt{perl}, \texttt{curl}, \texttt{wget}\} \rangle \rangle.$$

An imager can be obtained from any image I contained in R, provided that I can be considered a "medoid" representing the set of images contained in R.

We can now formalize the *popularity estimation problem*. As new Docker images will be released from other users, image developers may be interested in estimating the popularity of a new image in terms of pulls and stars.

Definition 4 (Popularity Estimation Problem). *Let $I_R = \langle p, s, I \rangle$ be an imager, whose values p and s of pulls and stars are unknown. Let also $\mathcal{I}_R$ be the context[3] where I_R is considered (viz., $I_R \in \mathcal{I}_R$). The* popularity estimation problem *consists in estimating the actual values p and s of pulls and stars of I_R in the context $\mathcal{I}_R$.*

Notice that, due to the currently available data, the popularity estimation problem is not considering time. For every image we only have a "flat" representation and, due to how DOCKERFINDER currently works, we cannot observe its popularity evolution in time. However, the extension of the problem that considers also the temporal dimension is a future work we plan to pursue.

Image developers may also be interested in determining the minimum changes that could improve the popularity of a new image. This is formalized by the *recommendation for improvement problem*.

Definition 5 (Recommendation for Improvement Problem). *Let $I_R = \langle p, s, I \rangle$ be an imager, whose values p and s of pulls and stars have been estimated in a context X (viz., $I_R \in X$). The* recommendation for pulls improvement problem *consists in determining a set of changes C^* such that*

- $I_R \xrightarrow{C^*} I_R^* = \langle p^*, \cdot, \cdot \rangle$, with $I_R^* \in X \wedge p^* > p$, and
- $\nexists C^\dagger$ s.t. $|C^\dagger| < |C^*|$ and $I_R \xrightarrow{C^\dagger} I_R^\dagger = \langle p^\dagger, \cdot, \cdot \rangle$, with $I_R^\dagger \in X \wedge p^\dagger > p^*$

(where $x \xrightarrow{C} y$ denotes that y is obtained by applying the set of changes C to x). The recommendation for stars improvement problem *is analogous.*

---

[3] As I_R is the representative image for the repository R, $\mathcal{I}_R$ may be (the set of imagers representing) the registry containing the repository R.

In other words, a solution of the recommendation for improvement problem is an imager I'_R obtained from I_R such that I'_R is more likely to get more pulls/stars than I_R and that the number of changes to obtain I'_R from I_R is minimum.

4   Proposed Approach

We hereby describe our approaches for solving the popularity estimation problem and the recommendation for stars/pulls improvement problem.

4.1   Estimating Popularity

We propose DARTER (*Decision And Regression Tree EstimatoR*, Algorithm 1) as a solution of the popularity estimation problem[4]. From a general point of view the problem can be seen as a regression problem [31]. However, as we show in the following, due to the fact that few imagers are considerably more popular than the others and most of the imagers are uncommon, usual regression methods struggle in providing good estimation (popularity distributions are provided in Sect. 5.1). DARTER can be used to estimate both pulls p and stars s. We present the algorithm in a generalized way by considering a popularity target u.

DARTER can be summarized in three main phases. First, it estimates a popularity threshold pt (line 2) with respect to the known imagers $\mathcal{I}_R$. Then, it labels every image as *popular* or *uncommon* (equals to 1 and 0 respectively) with respect to this threshold (line 3). Using these labels, DARTER trains a *decision tree* Ψ (line 4). This phase could be generalized using multiple threshold and labels.

In the second phase (lines 5–9), DARTER exploits the decision tree to classify both the imagers $\mathcal{I}_R$ and the images to estimate $\mathcal{X}$ as *popular* ($^+$) or *uncommon* ($^-$). In this task, DARTER exploits the function $isPopular(\Psi, I_R)$, which follows a path along the decision tree Ψ according to the imager I_R to estimate whether I_R will be popular or uncommon (see Fig. 1 *(left)*).

In the third phase (lines 10–14), DARTER trains two regression trees Λ^-, Λ^+ for uncommon and popular images, respectively. These regression trees are specialized to deal with very different types of images, which may have very different estimations on the leaves: High values for the popular regression tree, low values for the uncommon regression tree (see Fig. 1 *(center) & (right)*). Finally, DARTER exploits the two regression trees to estimate the popularity of the images in X and returns the final estimation Y.

---

[4] DARTER is designed to simultaneously solve multiple instances of the popularity estimation problem, given a set $\mathcal{X}$ of imagers whose popularity is unknown.

Algorithm 1. $DARTER(\mathcal{I}_R, \mathcal{X}, u)$

Input : $\mathcal{I}_R$ - context/set of imagers, $\mathcal{X}$ - set of images to estimate, u - popularity type (can be equal to "pulls" or to "stars")

Output: Y - popularity estimation

1 $T \leftarrow getPopularity(\mathcal{I}_R, u);$ // extract imagers popularity
2 $pt \leftarrow estimatePopularityThreshold(T);$ // estimate popularity threshold
3 $L \leftarrow \{l \mid l = 0 \text{ if } t < pt \text{ else } 1, \; \forall\, t \in T\};$ // label a target value as popular or not
4 $\Psi \leftarrow trainDecisionTree(\mathcal{I}_R, L);$ // train decision tree

5 $\mathcal{I}_R^- \leftarrow \{I_R \mid \text{if } isPopular(\Psi, I_R) = 0 \; \forall\, I_R \in \mathcal{I}_R\};$ // classify uncommon imagers
6 $\mathcal{I}_R^+ \leftarrow \{I_R \mid \text{if } isPopular(\Psi, I_R) = 1 \; \forall\, I_R \in \mathcal{I}_R\};$ // classify popular imagers
7 $\mathcal{X}^- \leftarrow \{I_R \mid \text{if } isPopular(\Psi, I_R) = 0 \; \forall\, I_R \in \mathcal{X}\};$ // classify uncommon imagers
8 $\mathcal{X}^+ \leftarrow \{I_R \mid \text{if } isPopular(\Psi, I_R) = 1 \; \forall\, I_R \in \mathcal{X}\};$ // classify popular imagers
9 $T^- \leftarrow getPopularity(\mathcal{I}_R^-, u); \; T^+ \leftarrow getPopularity(\mathcal{I}_R^+, u);$ // extract popularity

10 $\Lambda^- \leftarrow trainRegressionTree(\mathcal{I}_R^-, T^-);$ // train decision tree uncommon
11 $\Lambda^+ \leftarrow trainRegressionTree(\mathcal{I}_R^+, T^+);$ // train decision tree popular
12 $Y^- \leftarrow estimatePopularity(\Lambda^-, \mathcal{X}^-);$ // estimate popularity for uncommon
13 $Y^+ \leftarrow estimatePopularity(\Lambda^+, \mathcal{X}^+);$ // estimate popularity for popular
14 $Y \leftarrow buildResult(\mathcal{X}, \mathcal{X}^-, \mathcal{X}^+, Y^-, Y^+);$ // build final result w.r.t. original order

15 **return** Y;

In summary, DARTER builds a hierarchical estimation model. At the top of the model, there is a decision tree Ψ that permits discriminating between popular and uncommon imagers. The leaves of Ψ are associated with two regression trees Λ^+ or Λ^-. Λ^+ permits estimating the level of popularity of an imager, while Λ^- permits estimating the level of uncommonness of an imager. It is worth noting that specialized regression trees for each leaf of Ψ could be trained. However, since in each leaf there are potentially few nodes, this could lead to model overfitting [31] decreasing the overall performance. On the other hand, the two regression trees Λ^+, Λ^- result to be much more general since they are trained on all the popular/uncommon imagers.

As an example we can consider the imager $I_R = \langle ?, ?, \langle$Ubuntu 16.04.1 LTS$, 6, 0.78, 1.23, \{$python, perl, curl, wget, tar$\}\rangle\rangle$. Given $\Psi, \Lambda^+, \Lambda^-$ we can have that $isPopular(\Psi, I_R) = 0$. The latter means that I_R is uncommon, hence requiring to estimate its popularity with Λ^-, viz., the popularity of I_R is estimated as $y = estimatePopularity(\Lambda^-, I_R) = 0.56$ millions of pulls.

More in detail, we realized Algorithm 1 as follows. Function *estimate PopularityThreshold*$(\cdot)$ is implemented by the so-called "knee method" [31]. The knee method sorts the target popularity T and then, it selects the point threshold pt on the curve which has the maximum distance with the closest point on the straight line passing through the minimum and the maximum of the curve described by the sorted T (examples in Fig. 4 *(bottom)*). As models for decision and regression trees we adopted an optimized version of $CART$ [5]. We used the *Gini* criteria for the decision tree and the *Mean Absolute Error* for the regression

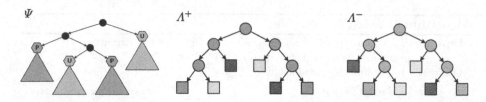

Fig. 1. Decision tree discriminating between popular and uncommon imagers (left), regression trees evaluating the popularity (center), and the degree of uncommonness (right). The regression trees are linked to corresponding leaves of Ψ: Regression trees like Λ^+ are linked to leaves marked with P (*popular*), while those like Λ^- to leaves marked with U (*uncommon*). (Best view in color)

trees [31]. We used a cost matrix for training the decision tree in order to improve the tree recall and precision in identifying popular images.

Finally, besides the good performance reported in the experimental section, the choice of decision and regression tree as estimation models lies in the fact that these models are easily interpretable [11]. Indeed, as shown in the following, we can extract from these trees an explanation of the estimation, and this explanation can be exploited to understand which are the changes that can lead to an improvement of the popularity of an image.

4.2 Recommending Improvements

To solve the recommendation for improvement problem we propose DIM (*Docker Image Meliorator*, Algorithm 2). First, DIM estimates the popularity y of the imager under analysis I_R in the context given by the decision tree Ψ (lines 1–2). The path along the tree leading to y constitutes the explanation for such estimation. In such terms, the proposed model is a *transparent box* which is both local and global explainable by design [11]. Then, it extracts the paths Q with a popularity higher than the one estimated for I_R (line 3), and it selects the shortest path sp among them (line 4). Finally, it returns an updated imager I_R^* built from the input imager I_R by applying it the changes composing the improvement shortest path sp (lines 5–6).

Functions *getPathsWithGreaterPopularity* and *getShortestPath* respectively collects all the paths in the tree ending in a leaf with a popularity higher than y, and selects the shortest path among them (see Fig. 2). When more than a shortest path with the same length is available, DIM selects the path with the highest overlap with the current path and with the highest popularity estimation.

Getting back to our example we have an estimation of $y = 0.56$ millions of pulls, viz., $I_R = \langle 0.56, ., \langle \texttt{Ubuntu 16.04.1 LTS}, 6, 0.78, 1.23, \{\texttt{python, perl, curl, wget, tar}\}\rangle\rangle$. By applying DIM on I a possible output is $I_R = \langle 0.64, ., I_R^* = \langle \texttt{Ubuntu 16.04.1 LTS}, 7, 0.78, 1.23, \{\texttt{python, perl, curl, java}\}\rangle\rangle$. That is, DIM recommends to update I_R by adding a new layer, which removes $\texttt{wget}$ and $\texttt{tar}$, and which adds the support for $\texttt{java}$.

Algorithm 2. $DIM(I_R, \Psi, \Lambda^-, \Lambda^+)$

Input : I_R - imager to improve, Ψ - decision tree, Λ^-, Λ^+ - regression trees.
Output: I_R^* - updated imager.

1 **if** $isPopular(\Psi, I_R) = 0$ **then** $y \leftarrow estimatePopularity(\Lambda^-, I_R)$;
2 **else** $y \leftarrow estimatePopularity(\Lambda^+, I_R)$;
3 $Q \leftarrow getPathsWithGreaterPopularity(I_R, y, \Psi, \Lambda^-, \Lambda^+)$;   `// get improving paths`
4 $sp \leftarrow getShortestPath(Q, I_R, y, \Psi, \Lambda^-, \Lambda^+)$;   `// get shortest path`
5 $I_R^* \leftarrow updateImager(I_R, sp)$;   `// update docker image`
6 **return** I_R^*;

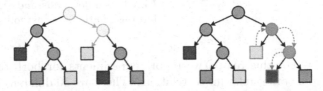

Fig. 2. The tree on the left shows the explanation of an estimation (viz., the yellow path). The tree on the right shows an example of recommendations for improvement, which is given by the shortest path leading to a leaf with a higher popularity (highlighted in blue). The latter indicates the minimum number of attribute changes that can lead to a popularity improvement. (Best view in color)

5   Experiments

5.1   Dataset

DOCKERFINDER autonomously collects information on all the images available in the Docker Hub that are contained in official repositories or in repositories that have been starred by at least three different users. The datasets collected by DOCKERFINDER[5] ranges from January 2017 to March 2018 at irregular intervals. If not differently specified in this work we refer to the most recent backup where 132,724 images are available. Since the popularity estimation problem require a notion of popularity, i.e., pulls or stars, from the available images we select 1,067 imagers considering for each repository the "latest" image (i.e., the most recent image of each repository). We leave as future work the investigation of the effect of considering other extraction of imagers. Some examples can be the smallest image, the one with more softwares, or a medoid or centroid of each repository.

Details of the imagers extracted from the principal dataset analyzed can be found in Fig. 3. $size_d$, $size_a$, p and s follow a long tailed distribution highlighted by the large difference between the median $\tilde{x}$ and the mean μ in Fig. 3. The power-law effect is stronger for *pulls* and *stars* (see Fig. 4). There is a robust Pearson correlation between pulls and stars of 0.76 (p-value 1.5e-165). However, saying that a high number of pulls implies a high number of stars could be

---

[5] Publicly available at https://goo.gl/ggvKN3.

	$size_d$	$size_a$	layers	$|S|$	pulls	stars
$\tilde{x}$	0.16	0.41	10.00	8.00	0.06	26.0
μ	0.27	0.64	12.67	7.82	6.70	134.46
σ	0.48	1.11	9.62	2.26	46.14	564.21

Fig. 3. Statistics of imagers, viz., median $\tilde{x}$, mean μ and standard deviation σ. The most frequent OSs and softwares are Debian GNU/Linux 8 (jessie), Ubuntu 14.04.5 LTS, Alpine Linux v3.4, and erl, tar, bash, respectively.

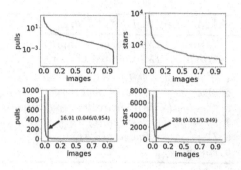

Fig. 4. Semilog pulls and stars distributions *(top)*. Knee method *(bottom)*.

a tall statement. For this reason we report experiments for both target measures. There are no other relevant correlations. There are 50 different *os* and the most common ones are Debian GNU/Linux 8 (jessie), Ubuntu 14.04.5 LTS and Alpine Linux v3.4. The most common softwares among the 28 available (without considering the version) are erl, tar and bash.

5.2 Experimental Settings

The experiments reported in the following sections are the results of a 5-fold cross validation [31] repeated ten times. We estimate the goodness of the proposed approach by using the following indicators to measure regression performance [16,28,36]: Median absolute error (*MAE*), and mean squared logarithmic error (*MSLE*). These indicators are more relevant than mean absolute error, mean squared error or explained variance because we are in the case when target values have an exponential growth. MSLE penalizes an under-predicted estimate greater than an over-predicted estimate, which is precisely what we are interested in, as there are only few popular images. Besides aggregated statistics on these measures on the ten runs, we report (a sort of) area under the curve plot [31], which better enhances the overall quality of the estimation in terms of quantification [20], i.e., how good is the method in estimating the popularity of a set of images. We do not report learning and prediction times as they are negligible (less than a second for all the methods analyzed), and also because the experiments are more focused in highlighting the quality of the results.

For the popularity estimation problem we compare DARTER against the following baselines: Regression tree (*RegTree*) [5], k-nearest-neighbor (*RegKnn*) [1], linear regression model (*LinReg*) [37], *Lasso* model [32], and *Ridge* model [33], besides the *Null* model estimating the popularity using the mean value. We selected these approaches among the existing one because *(i)* they are adopted in some of the works reported in Sect. 6, *(ii)* they are interpretable [11] differently from more recent machine learning methods. On the other hand, as (to the best of

Table 1. Mean and standard deviation of MAE and MSLE for pulls and stars.

Model	pulls		stars	
	MAE	MSLE	MAE	MSLE
Darter	**0.222 ± 0.066**	**1.606 ± 0.268**	**19.925 ± 1.904**	**2.142 ± 0.171**
RegTree	0.355 ± 0.092	1.857 ± 0.430	22.650 ± 3.223	2.416 ± 0.233
RegKnn	0.748 ± 0.084	2.251 ± 0.195	30.020 ± 1.679	3.419 ± 0.445
Lasso	7.051 ± 1.207	4.978 ± 0.813	95.423 ± 13.445	4.767 ± 0.631
LinReg	7.998 ± 1.874	84.611 ± 123.256	112.794 ± 17.435	48.180 ± 69.352
Ridge	7.575 ± 1.736	8.236 ± 1.283	107.305 ± 15.207	5.169 ± 0.599
Null	3.471 ± 0.367	6.814 ± 1.023	3.122 ± 0.236	117.969 ± 13.459

our knowledge) no method is currently available for solving the recommendation for improvement problem, we compare DIM against a random null model[6].

5.3 Estimating Image Popularity

We hereby show how DARTER outperforms state-of-the-art methods in solving the popularity estimation problem. Table 1 reports the mean and standard deviation of MAE and MSLE for pulls and stars. The Null model performs better than the linear models (Lasso, LinReg and Ridge). This is probably due to the fact that linear models fail in treating the vectorized sets of softwares, which are in the image descriptions used to train the model. DARTER has both a lower mean error than all the competitors and a lower error deviation in term of standard deviation, i.e., it is more stable when targeting *pulls* and/or *stars*. The results in Table 1 summarize the punctual estimation of each method for each image in the test sets. In Fig. 5 we observe the overall quantification of the estimation for the best methods. It reports the cumulative distribution of the estimation against the real values. The more a predicted curve is adherent to the real one, the better is the popularity estimation. All the approaches are good in the initial phase when uncommon images are predicted. Thus, the image difficult to estimate are those that somehow lay in the middle, and DARTER is better than the others is assessing this challenging task.

Further evidence on the stability of DARTER is provided by the fact that its MSLE keeps stable even when considering different datasets extracted by DOCKERFINDER in different times, and steadily better than all other estimators. These results are highlighted in Fig. 6 for both pulls and stars. Moreover, in Fig. 7 we show the performance of the estimators in terms of MAE (pulls first row, stars second row) for increasing size of the training set in order to test the so called "cold start" problem [27]. Results show that DARTER suffer less than the other approaches when using less data for the training phase[7].

---

[6] The python code is available here https://goo.gl/XnJ7yD.

[7] The non-intuitive fact that with 50% training data MAE seems to be best for some algorithms can be explained with overfitting and partial vision of the observations.

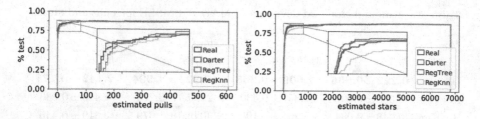

Fig. 5. Cumulative distribution of the estimations.

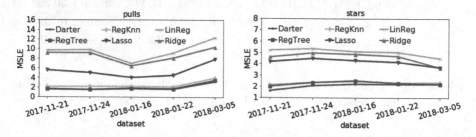

Fig. 6. Estimators stability for different datasets.

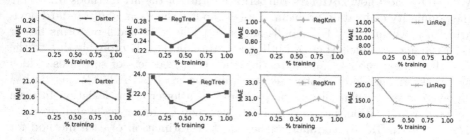

Fig. 7. MAE varying the portion of training set used: pulls first row, stars second row.

5.4   Recommending Image Improvements

We now show that the recommendation for stars/pulls improvement problem cannot be solved by simply using a null model approach. We build a random null model (RND) that, given in input an imager I_R, changes a feature of I_R (e.g., os, $layers$, $size_d$) with probability π by selecting a new value according to the distribution of these values in the dataset, hence creating a new imager I_R^*. Then we apply RND and DIM on a test set of images. For each updated imager I_R^* we keep track of the number of changes performed to transform I_R into I_R^*, and of the variation Δ between the original popularity and the estimated popularity of I_R^*. We estimate the popularity of the improved images using DARTER since it is the best approach as we observed in the previous section.

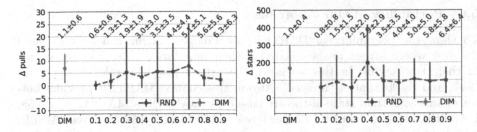

Fig. 8. Upper part: mean and standard deviation of the number of changes. Error plot: mean and standard deviation of the popularity improvement Δ.

Figure 8 reports the results of these experiment for pulls and stars. For the null model we vary $\pi \in [0.1, 0.9]$ with step 0.1. Every point in Fig. 8 represents the mean popularity improvement Δ, while the vertical lines are one fourth of the standard deviation. The numbers in the upper part are the means and standard deviations of the number of changes. We notice that a random choice of the features to change can lead to an average improvement comparable to the one of DIM ($\pi = 0.7$ for pulls, $\pi = 0.4$ for stars). However, two aspects must not be neglected. The first one is that when RND has a higher Δ it also has a higher variability. The second one is that on average DIM uses just one or two changes to improve the image, while RND requires a consistently higher number of changes. This allows us to conclude that, given an imager I_R, DIM provides can effectively suggest how to build an imager I_R^* whose estimated popularity will be higher, and which can be obtained by applying very few changes to I_R.

5.5   Explaining Improvement Features

In this section we exploit the fact that DARTER and DIM are explainable to retrieve the most important features that should be changed to obtain the improved imager. We focus on the analysis of the most uncommon imagers by analyzing the tree Λ^- of uncommon imagers. In particular, among them we signal the subsequent: *(i)* required presence of one of the following *oss*: Alpine Linux v3.7, Alpine Linux v3.2, Ubuntu 16.04.3 LTS, *(ii)* having a size $size_d$ lower than 0.0241, *(iii)* having $size_a \leq 0.319$ or $size_a > 0.541$ (depending on the other features), *(iv)* having less than six software avoiding tar but including ruby.

Since the images realized by private developers can rarely reach the popularity of official imager repositories (e.g., *java*, *python*, *mongo*, etc.) we repeated the previous analysis by excluding official imagers. Results highlights that again Alpine Linux v3.2, Ubuntu 16.04.3 LTS are the required *oss*, but it is also generally recommended by DIM of having $size_a > 0.301$, $size_d \leq 0.238$ and to support the following software distributions: gunicorn, go and ping.

5.6 Portability of Our Approach

To show the portability of DARTER, we analyzed a musical Spotify-based dataset, where artists are associated with a popularity score and to the set of their tracks [26]. In this scenario, the artists play the role of "repositories", and tracks that of "images". Also in this context DARTER provides better estimations than state-of-the-art baselines (DARTER's MAE: 12.80 ± 0.58, DARTER's MSLE: 4.36 ± 0.175, RegTree's MAE: 13.91 ± 0.57, RegTree's MSLE: 4.57 ± 0.14).

6 Related Work

The problem of estimating and analysing popularity of Docker images resembles the discovery of success performed in various other domains.

A well-known domain is related to quantifying the changes in productivity throughout a research career in science. [35] defines a model for the citation dynamics of scientific papers. The results uncover the basic mechanisms that govern scientific impact. [25] points out that, besides dependent variables, also contextual information (e.g., prestige of institutions, supervisors, teaching and mentoring activities) should be considered. The latter holds also in our context, where we can observe that official images behave differently with respect to non-official images. Sinatra et al. [29] recently designed a stochastic model that assigns an individual parameter to each scientist that accurately predicts the evolution of her impact, from her h-index to cumulative citations, and independent recognitions (e.g., prizes). The above mentioned approaches (viz., [35], [25] and [29]) model the success phenomena using the fitting of a mathematical formulation given from an assumption. In our proposal, we are not looking for just an indicator but for an explainable complex model that not only permits analyzing a population, but also to reveal suggestions for improvements.

Another domain of research where the study of success is relevant is sport. The level of competitive balance of the roles within the four major North American professional sport leagues is investigated in [2]. The evidence suggests that the significance of star power is uncovered only by multiplicative models (rather than by the commonly employed linear ones). As shown by our experiments, this holds also in our context: our complex model outperforms ordinary linear ones. Franck et al. [7] provide further evidence on contextual factors, by showing that the emergence of superstars in German soccer depends not only on their investments in physical talent, but also on the cultivation of their popularity. An analysis of impact of technical features on performances of soccer teams is provided in [22]. The authors find that draws are difficult to predict, but they obtain good results in simulating the overall championships. Instead, the authors of [23] try to understand which are the features driving human evaluation with respect to performance in soccer. Like us, they use a complex model to mimic an artificial judge which accurately reproduces human evaluation, which permits showing how humans are biased towards contextual features.

Another field of research where the study of success and popularity is quite useful is that of online social networks, like Twitter, Instagram, Youtube, Facebook, etc. In [18], the authors propose a method to predict the popularity of new hashtags on Twitter using standard classification models trained on content features extracted from the hashtag and on context features extracted from the social graph. The difference with our approach is that an explanation is not required, neither a way to produce a more popular hashtag. For understanding the ingredients of success of fashion models, the authors of [24] train machine learning methods on Instagram images to predict new popular models. Instead, Trzciński and Rokita [34] present a regression method to predict the popularity of an online video (from YouTube or Facebook) measured in terms of its number of views. Results show that, despite the visual content can be useful for popularity prediction before content publication, the social context represents a much stronger signal for predicting the popularity of a video.

Some forms of analytics have been recently applied to GitHub repositories. The authors of [15] present a first study on the main characteristics of GitHub repositories, and on how users take advantage of their main features, e.g., commits, pull requests, and issues. A deeper analysis is provided in [3], where the authors analyze various features of GitHub with respect to the impact they have on the popularity of a GitHub repository. A model for predicting such popularity is then described in [4], where multiple linear regressions are used to predict the number of stars assigned to a GitHub repository. The crucial difference between the approach in [4] and ours is that we exploit features that concretely describe a Docker image (such as the operating system and software distributions it supports, for instance), while in [4] the authors build models based only on the time series of the amounts of stars previously assigned to repositories.

Further domains where the analysis and prediction of success is a challenging task are music [26], movies [17] and school performances [12]. However, to the best of our knowledge, our approach is the first that is based on complex descriptions such as those of Docker images, and which tries to estimate success and to provide recommendations for improvements based on an explainable model.

7 Conclusion

In this paper we have proposed DARTER and DIM, two methods specifically designed to analyze the popularity of Docker images. In particular, DARTER is a mixed hierarchical model formed by a decision tree and by two regression trees able to outperform state-of-the-art approaches in understanding the degree of popularity an image will get (in terms of pulls and stars). Moreover, DARTER predictions are explainable in terms of the characteristics of Docker images. This aspect is exploited by DIM to determine how to improve the popularity of a given image performing by applying it a minimal set of changes.

It is worth noting that DARTER and DIM are focused on the technical content of images, as their ultimate objective is to provide explainable models helping developers in analyzing and improving their Docker images. Hence, other factors that can orthogonally impact on the popularity of images (e.g., the previous reputation of a developer, or external endorsements by widely known experts in the field) are outside of the scope of this paper, as they could not lead to technical updates on images geared towards improving their popularity.

Besides testing the proposed method on other domains, we would like to strengthen the experimental section by means of a real validation. The idea is to release on Docker Hub a set of images and their improved versions and to observe how good are the prediction of DARTER and the recommendation of DIM in a real case study, and how long it takes to reach the estimated values. Time is indeed another crucial component that was not considered because the current version of DOCKERFINDER is not updating the status of a repository at constant time intervals. The extension of our approach to also consider time is in the scope of our future work. Finally, another interesting direction for future work is to extend DIM by allowing users to indicate the desired popularity for an image and constraints on acceptable image updates (e.g., software that cannot be removed from an image, or its minimum/maximum acceptable size).

Acknowledgments. Work partially supported by the EU H2020 Program under the funding scheme "INFRAIA-1-2014-2015: Research Infrastructures", grant agreement 654024 *"SoBigData"* (http://www.sobigdata.eu).

References

1. Altman, N.S.: An introduction to kernel and nearest-neighbor nonparametric regression. Am. Stat. **46**(3), 175–185 (1992)
2. Berri, D.J., Schmidt, M.B., Brook, S.L.: Stars at the gate: the impact of star power on nba gate revenues. J. Sports Econ. **5**(1), 33–50 (2004)
3. Borges, H., et al.: Understanding the factors that impact the popularity of GitHub repositories. In: ICSME, pp. 334–344. IEEE (2016)
4. Borges, H., Hora, A., Valente, M.T.: Predicting the popularity of GitHub repositories. In: PROMISE, p. 9. ACM (2016)
5. Breiman, L., et al.: Classification and Regression Trees. CRC Press, Boca Raton (1984)
6. Brogi, A., Neri, D., Soldani, J.: DockerFinder: multi-attribute search of Docker images. In: IC2E, pp. 273–278. IEEE (2017)
7. Franck, E., Nüesch, S.: Mechanisms of superstar formation in german soccer: empirical evidence. Eur. Sport Manag. Q. **8**(2), 145–164 (2008)
8. Goodman, B., Flaxman, S.: EU regulations on algorithmic decision-making and a right to explanation. In: ICML (2016)
9. Guidotti, R., Davide, S.J.N., Antonio, B.: Explaining successful Docker images using pattern mining analysis. In: Mazzara, M., Ober, I., Salaün, G. (eds.) STAF 2018. LNCS, vol. 11176, pp. 98–113. Springer, Cham (2018). https://doi.org/10.1007/978-3-030-04771-9_9
10. Guidotti, R., et al.: Local rule-based explanations of black box decision systems. arXiv preprint arXiv:1805.10820 (2018)
11. Guidotti, R., Monreale, A., Ruggieri, S., Turini, F., Giannotti, F., Pedreschi, D.: A survey of methods for explaining black box models. CSUR **51**(5), 93 (2018)
12. Harackiewicz, J.M., et al.: Predicting success in college. JEP **94**(3), 562 (2002)
13. Hars, A., Ou, S.: Working for free? - motivations of participating in open source projects. IJEC **6**(3), 25–39 (2002)

14. Joy, A.: Performance comparison between Linux containers and virtual machines. In: ICACEA, pp. 342–346, March 2015
15. Kalliamvakou, E., Gousios, G., Blincoe, K., Singer, L., German, D.M., Damian, D.: The promises and perils of mining GitHub. In: MSR, pp. 92–101. ACM (2014)
16. Lehmann, E.L., Casella, G.: Theory of Point Estimation. Springer, New York (2006). https://doi.org/10.1007/b98854
17. Litman, B.R.: Predicting success of theatrical movies: an empirical study. J. Popular Cult. **16**(4), 159–175 (1983)
18. Ma, Z., Sun, A., Cong, G.: On predicting the popularity of newly emerging hashtags in twitter. JASIST **64**(7), 1399–1410 (2013)
19. Miell, I., Sayers, A.H.: Docker in Practice. Manning Publications Co., New York (2016)
20. Milli, L., Monreale, A., Rossetti, G., Giannotti, F., Pedreschi, D., Sebastiani, F.: Quantification trees. In: ICDM, pp. 528–536. IEEE (2013)
21. Pahl, C., Brogi, A., Soldani, J., Jamshidi, P.: Cloud container technologies: a state-of-the-art review. IEEE Trans. Cloud Comput. (2017, in press)
22. Pappalardo, L., Cintia, P.: Quantifying the relation between performance and success in soccer. Adv. Complex Syst. 1750014 (2017)
23. Pappalardo, L., Cintia, P., Pedreschi, D., Giannotti, F., Barabasi, A.-L.: Human perception of performance. arXiv preprint arXiv:1712.02224 (2017)
24. Park, J., et al.: Style in the age of Instagram: predicting success within the fashion industry using social media. In: CSCW, pp. 64–73. ACM (2016)
25. Penner, O., Pan, R.K., Petersen, A.M., Kaski, K., Fortunato, S.: On the predictability of future impact in science. Sci. Rep. **3**, 3052 (2013)
26. Pollacci, L., Guidotti, R., Rossetti, G., Giannotti, F., Pedreschi, D.: The fractal dimension of music: geography, popularity and sentiment analysis. In: Guidi, B., Ricci, L., Calafate, C., Gaggi, O., Marquez-Barja, J. (eds.) GOODTECHS 2017. LNICST, vol. 233, pp. 183–194. Springer, Cham (2018). https://doi.org/10.1007/978-3-319-76111-4_19
27. Resnick, P., Varian, H.R.: Recommender systems. CACM **40**(3), 56–58 (1997)
28. Shcherbakov, M.V., et al.: A survey of forecast error measures. World Appl. Sci. J. **24**, 171–176 (2013)
29. Sinatra, R., Wang, D., Deville, P., Song, C., Barabási, A.-L.: Quantifying the evolution of individual scientific impact. Science **354**(6312), aaf5239 (2016)
30. Soltesz, S., et al.: Container-based operating system virtualization: a scalable, high-performance alternative to hypervisors. SIGOPS **41**(3), 275–287 (2007)
31. Tan, P.-N., et al.: Introduction to Data Mining. Pearson Education India, Bangalore (2006)
32. Tibshirani, R.: Regression shrinkage and selection via the lasso. J. Roy. Stat. Soc. Ser. B (Methodol.) 267–288 (1996)
33. Tikhonov, A.: Solution of incorrectly formulated problems and the regularization method. Soviet Meth. Dokl. **4**, 1035–1038 (1963)
34. Trzciński, T., Rokita, P.: Predicting popularity of online videos using support vector regression. IEEE Trans. Multimedia **19**(11), 2561–2570 (2017)
35. Wang, D., Song, C., Barabási, A.-L.: Quantifying long-term scientific impact. Science **342**(6154), 127–132 (2013)
36. Willmott, C.J., Matsuura, K.: Advantages of the mean absolute error (MAE) over the root mean square error (RMSE). Clim. Res. **30**(1), 79–82 (2005)
37. Yan, X., Su, X.: Linear Regression Analysis: Theory and Computing. World Scientific, Singapore (2009)

ST-DenNetFus: A New Deep Learning Approach for Network Demand Prediction

Haytham Assem[1,2], Bora Caglayan[1], Teodora Sandra Buda[1],
and Declan O'Sullivan[2]

[1] Cognitive Computing Group, Innovation Exchange, IBM, Dublin, Ireland
{haythama,tbuda}@ie.ibm.com, bora.caglayan@ibm.com
[2] School of Computer Science and Statistics, Trinity College Dublin, Dublin, Ireland
declan.osullivan@cs.tcd.ie

Abstract. Network Demand Prediction is of great importance to network planning and dynamically allocating network resources based on the predicted demand, this can be very challenging as it is affected by many complex factors, including spatial dependencies, temporal dependencies, and external factors (such as regions' functionality and crowd patterns as it will be shown in this paper). We propose a deep learning based approach called, ST-DenNetFus, to predict network demand (i.e. uplink and downlink throughput) in every region of a city. ST-DenNetFus is an end to end architecture for capturing unique properties from spatio-temporal data. ST-DenNetFus employs various branches of dense neural networks for capturing temporal closeness, period, and trend properties. For each of these properties, dense convolutional neural units are used for capturing the spatial properties of the network demand across various regions in a city. Furthermore, ST-DenNetFus introduces extra branches for fusing external data sources that have not been considered before in the network demand prediction problem of various dimensionalities. In our case, these external factors are the crowd mobility patterns, temporal functional regions, and the day of the week. We present an extensive experimental evaluation for the proposed approach using two types of network throughput (uplink and downlink) in New York City (NYC), where ST-DenNetFus outperforms four well-known baselines.

Keywords: Spatio-temporal data · Deep learning
Convolutional neural networks · Dense networks
Network demand prediction

1 Introduction

Mobile data traffic has increased dramatically in the last few decades [8]. Besides, the increase in the number of devices accessing the cellular network, emerging social networking platforms such as Facebook and Twitter has further added to

© Springer Nature Switzerland AG 2019
U. Brefeld et al. (Eds.): ECML PKDD 2018, LNAI 11053, pp. 222–237, 2019.
https://doi.org/10.1007/978-3-030-10997-4_14

Fig. 1. Network demand prediction using the introduced ST-DenNetFus approach.

the mobile data traffic [10]. This led to the need to increase the network resources provided to end-users and consequently this has caused a huge cost increase on the operators. The mobile network operators are striving for solutions to reduce the OPEX and CAPEX costs of such emerging demand. Reducing the OPEX and CAPEX cost is not only of importance to the operators but as well to the environment. The statistics show that the total CO_2 emissions from the information and communication technology (ICT) infrastructure contributes for 2% of total CO_2 emissions across the globe in which the telecommunication industry is a major part of it [15]. There are significant spatial and temporal variations in cellular traffic [24] and the cellular system is designed using the philosophy of worst case traffic (such as to fulfil the quality of service (QoS) in case of a peak traffic). Hence, there is a growing need to have a spatio-temporal prediction based model for the Network Demand Prediction problem [26].

In this paper, two types of cellular network throughput, downlink and uplink are predicted using a new deep learning based-approach called ST-DenNetFus (see Fig. 1): (i) Downlink is the total downloaded network throughput in a region during a given time interval. (ii) Uplink denotes the total uploaded network throughput in a region during a given time interval. Both types of throughput track the overall Network Demand[1] of a region.

Deep learning has been applied successfully in many applications, and is considered one of the most cutting edge techniques in Artificial Intelligence (AI) [18]. There are two types of deep neural networks that try to capture spatial and temporal properties: (a) Convolutional Neural Networks (CNNs) for capturing spatial structure and dependencies. (b) Recurrent Neural Networks (RNNs) for learning temporal dependencies. However, (and similar to the crowd flow prediction problem in the urban computing domain presented in [37]) it is still very challenging to apply these type of techniques to the spatio-temporal Network

---

[1] In this paper terminology, we refer to both types of throughput, uplink and downlink as "Network Demand".

Demand Prediction problem due to the following reasons: **(A) Spatial dependencies: (A.1) Nearby** - The downlink throughputs of a region might be affected by the uplink throughputs of nearby regions and vice versa. In addition, the downlink throughputs of a region would affect its own uplink throughputs as well. **(A.2) Distant** - The network demand of a region can be affected by the network demand of distant regions especially if both are supported by same Base Station geographically. **(B) Temporal dependencies: (B.1) Closeness** - Intuitively, the network demand of a region is affected by recent time intervals. For instance, a high network demand occurring due to a crowded festival occurring at 9 PM will affect that of 10 PM. **(B.2) Period** - Network demand during morning or evening hours may be similar during consecutive weekdays, repeating 24 h. **(B.3) Trend** - Network demand may increase as summer approaches especially on weekends. Recent study [32] showed that the summer usage increases in the evening and early morning hours from about midnight to 4 AM, which indicates that young adults are not putting down mobile devices just because it is a summer break. **(C) External Factors: (C.1) Multiple** - Some external factors may impact the network demand such as the temporal functional regions, crowd mobility patterns and day of the week. For example, a business functional region may rely on the wireless networks more than the cellular networks. In addition, a highly crowded area has a higher chance for more cellular network usage. **(C.2) Dimensionality** - The external factors may vary in the dimensionality of the data. For example, the day of the week data will be in 1-dimensional space since it varies across time only but crowd mobility or temporal functional regions data will be in 2-dimensional space since it varies across space and time.

To tackle the above challenges, in this paper a spatio-temporal deep learning based-approach called ST-DenNetFus is proposed that collectively predict the uplink and downlink throughputs in every region in a city. The proposed contributions in this paper are five-fold:

1. ST-DenNetFus employs convolutional-based dense networks to model both nearby and distance spatial dependencies between regions in cities.
2. ST-DenNetFus employs several branches for fusing various external data sources of different dimensionality. The ST-DenNetFus architecture proposed is expandable according to the availability of the external data sources needed to be fused.
3. ST-DenNetFus uses three different dense networks to model various temporal properties consisting of temporal closeness, period, and trend.
4. The proposed approach has been evaluated on a real network data extracted from NYC and in particular Manhattan, for 6 months. The results reinforce the advantages of the new approach compared to 4 other baselines.
5. For the first time, it is shown that extracted urban patterns that have not been considered before in this particular problem (such as crowd mobility patterns and temporal functional regions) when fused as an external data sources for estimating the network demand, lead to more accurate prediction results.

2   Related Work

In this section, the state-of-the-art is reviewed from two different perspectives. First, an overview on the Network Demand Prediction related work is presented and then an overview on the recent advancements of convolutional neural networks is presented.

Cellular network throughput prediction plays an important role in network planning. To name a few works, in [9], ARIMA and exponential smoothing model are used for predicting the network demand for a single cell and whole region scenarios. ARIMA was found to outperform for a whole region scenario while the exponential smoothing model had better performance for the single cell scenario. In [35], a hybrid method using both ARMA and FARIMA is introduced to predict the cellular traffic where FARIMA found to work effectively on the time series that hold long range dependence. For long time prediction, the authors in [25] presented an approach with 12-hour granularity that allows to estimate aggregate demands up to 6 months in advance. Shorter and variable time scales are studied in [27] and [39] adopting ARIMA and GARCH techniques respectively. Recently, researchers started to exploit external sources trying to achieve more reliable and accurate predictions. In [1], the authors propose a dynamic network resources allocation framework to allocate downlink radio resources adaptively across multiple cells of 4G systems. Their introduced framework leverages three types of context information: user's location and mobility, application-related information, and radio maps. A video streaming simulated use case is used for evaluating the performance of the proposed framework. Another interesting work presented in [34] focuses on building geo-localized radio maps for a video streaming use-case in which the streaming rate is changed dynamically on the basis of the current bandwidth prediction from the bandwidth maps. *To the best our knowledge, in the field of telecommunications, for the first time end-to-end deep learning with fusing urban patterns as external data sources is considered, showing the effectiveness of the proposed deep learning approach along with the impact of some of the urban patterns in achieving higher accuracy for predicting network demand.*

In the last few years, deep learning has led to very good performance on a variety of problems, such as visual recognition, speech recognition and natural language processing [23] as well as spatio-temporal prediction problems [4] and environmental challenges [3]. Among different types of deep neural networks, convolutional neural networks (CNN) have been most extensively studied. Since 2006, various methods have been developed to overcome the limitations and challenges encountered in training deep CNNs. The most notable work started by Krizhevsky et al. when they introduced an architecture called AlexNet [17]. The overall architecture of AlexNet is similar to LeNet-5 but with deeper structure and showed significant improvements compared to LeNet-5 on the image classification task. With the success of AlexNet, several successful architectures have evolved, ZFNet [36], VGGNet [30], GoogleNet [31] and ResNet [11]. One of the main typical trends with these evolving architectures is that the networks are getting deeper. For instance, ResNet, the winner of ILSVRC 2015 competition

(a) The extracted temporal functional   (b) The extracted crowd mobility patterns
regions.

Fig. 2. External factors extraction.

got deeper 20 times more deeper than AlexNet and 8 times deeper than VGGNet. This typical trend is because networks can better approximate the target function when they are deeper. However, the deeper the network the more complex it is, which makes it more difficult to optimize and easier to suffer overfitting. Recently in 2016, a new architecture has been introduced called DenseNets [13] that exploits the potential of the network through feature reuse, yielding condensed models that are easy to train and highly parameter efficient. DenseNets obtain significant improvements over most of the state-of-the-art networks to date, whilst requiring less memory and computation to achieve high performance [13]. *Hence, in this work we rely mainly on leveraging the dense blocks as a core part of the proposed ST-DenNetFus architecture as will be described in the sections to follow. To the best our knowledge, this is the first work to show the effectiveness of DenseNet on a different domain than computer vision.*

3   External Factors Extraction

In this section, our approach for extracting the considered three external factors that are fused in the ST-DenNetFus architecture are described.

For the first external factor called *temporal functional regions* and as defined in [2,5], a temporal functional region is defined as "a set of regions that change their functionality over space and time, depending on the activity shifts across the temporal variation of the different time slots considered." The approach introduced for extracting these temporal functional regions was based on using location-based social networks (LBSNs) data and applying clustering techniques based on the time-slot. However, in this paper we propose a different approach for extracting temporal functional regions based on Point-Of-Interests (POIs) data. We argue that our proposed approach is more advantageous as accessing POIs data is easier compared to LBSNs data. We extracted the POIs in Manhattan region using the Google Places API[2] after finding various inconsistencies

---

[2] https://developers.google.com/places/.

in the openly accessible Open Street Maps data. We take into consideration the
opening and closing times of the point of interest to increase the accuracy of
the functionality mapping. If opening and closing time of a particular POI was
not available, we used the mean opening and closing times for that category.
The functionality of a point of interest can be: (1) Education, (2) Business,
(3) Shopping, (4) Nightlife, (5) Eating, (6) Social services and this information
is inferred from the POI description accessed through Google Places API. We
count the active point of interest counts for each hour and for each function
to generate this particular external factor. The output of this external factor is
six 2D-dimensional matrices where each corresponds to a different functionality
based on the POI frequency count with shape $(N * M * T)$. Figure 2a shows an
example of extracted functional regions during certain time-slot in a day across
the defined 32×32 grid map.

For the second external factor related to *crowd mobility patterns*, we calculated the crowd count in our experiments based on the unique anonymized
identifiers $(ueid)$ of the users in the mobile network sessions available in the
network dataset. The crowd count is the number of unique $ueid$ in a particular
region in an hour. A user that travels between regions in the period is assumed
to be present in multiple regions in that hour. The output of this external 2D-
dimensional external factor follows a shape of $(N * M * T)$. Figure 2b shows an
example of a crowd mobility pattern during certain time-slot during the day
across the defined 32×32 grid map visualized using heat-map.

The third and final external factor that we took into account when predicting
network demand is related to the day of the week which for each sample in the
dataset, we inferred the corresponding day of the week as well as whether this
day is weekend or not. This resulted in a 1D-dimensional external factor that
change across time.

4   Deep Spatio-Temporal Dense Network with Data Fusion (ST-DenNetFus)

Figure 3 presents the proposed architecture of ST-DenNetFus where each of the
downlink and uplink network throughputs at time t is converted to a 32×32
of 2-channel image-like matrix spanning over a region. Then the time axis is
divided into three fragments denoting recent time, near history and distant history. Further, these 2-channel image-like matrices are fed into three branches on
the right side of the diagram for capturing the trend, periodicity, and closeness
and output $\mathbf{X}_{in}$. Each of these branches starts with convolution layer followed
by L dense blocks and finally another convolution layer. These three convolutional based branches capture the spatial dependencies between nearby and
distant regions. Moreover, there are number of branches that fuse external factors based on their dimensionality. In our case, the temporal functional regions
and the crowd mobility patterns are 2-dimensional matrices $(\mathbf{X}_{Ext-2D})$ that
change across space and time but on the other side, the day of the week is
1-dimensional matrix that change across time only $(\mathbf{X}_{Ext-1D})$. At that stage

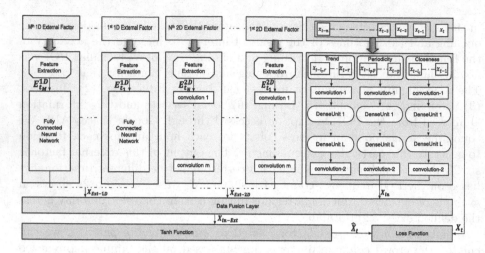

Fig. 3. ST-DenNetFus architecture. (Color figure online)

a data fusion layer is introduced that fuses the $\mathbf{X}_{in}$, $\mathbf{X}_{Ext-2D}$, and $\mathbf{X}_{Ext-1D}$. The output is $\mathbf{X}_{in-Ext}$ which is fed to *tanh* function to be mapped to $[-1, 1]$ range. This helps towards a faster convergence in the backpropagation learning compared to a standard logistic function [19].

4.1   Network Throughput Input Data

The network throughput data for both uplink and downlink are fed into the first three branches (shown in blue in Fig. 3).

Convolution Design. Since a city usually has a very large size with many regions, and intuitively the network demand may be affected by nearby as well as distant regions, convolutional neural network can handle this effectively as it captures the spatial structure through convolutions [38]. In order to capture the dependency between regions, there is a need to design many convolutional layers. Subsampling techniques have been introduced to preserve distant dependencies and avoid the loss of resolution especially in video sequence generating tasks [21]. Unlike with the common approach to CNN, we do not use subsampling but instead rely only on convolutions [14]. Support for such an approach can be found in [38], where the authors were trying to capture the spatial dependencies at a citywide scale similar to our problem here. They concluded that one convolution naturally captures spatial near dependencies, and a stack of convolutions afterwards can further capture the spatial distant citywide dependencies. The closeness, periodicity, and trend components adapt 2-channel image-like matrices according to the time interval as follows, $[\mathbf{X}_{t-l_c}, ..., \mathbf{X}_{t-1}]$, $[\mathbf{X}_{t-l_p.p}, ..., \mathbf{X}_{t-p}]$, and $[\mathbf{X}_{t-l_r.r} ..., \mathbf{X}_{t-r}]$ respectively. l_c, l_p and l_r represent the length of the dependent sequence for the closeness, period and trend while c, p and r depicts their span respectively. In our detailed implementation, the p captures the daily periodicity while the r is equal to one week that reveals the weekly trend of the network demand.

Each of these inputs is concatenated across the first axis (time interval) as tensors, $\mathbf{X}_c^{(0)}$, $\mathbf{X}_p^{(0)}$, and $\mathbf{X}_r^{(0)}$ and then followed by a convolution (convolution-1 in Fig. 3) for each branch as follows:

$$\mathbf{X}_c^{(1)} = f(\mathbf{W}_c^{(1)} * \mathbf{X}_c^{(0)} + \mathbf{b}_c^{(1)}) \tag{1}$$

$$\mathbf{X}_p^{(1)} = f(\mathbf{W}_p^{(1)} * \mathbf{X}_p^{(0)} + \mathbf{b}_p^{(1)}) \tag{2}$$

$$\mathbf{X}_r^{(1)} = f(\mathbf{W}_r^{(1)} * \mathbf{X}_r^{(0)} + \mathbf{b}_r^{(1)}) \tag{3}$$

where $*$ denotes the convolution operation, $f(.)$ is an activation rectifier function [17], and the $(\mathbf{W}_c^{(1)}, \mathbf{b}_c^{(1)})$, $(\mathbf{W}_p^{(1)}, \mathbf{b}_p^{(1)})$, and $(\mathbf{W}_r^{(1)}, \mathbf{b}_r^{(1)})$ are the learnable parameters for the first layer of the three branches.

Since our objective is to have the final output size as same as the size of the input (size of the grid map), a specific type of convolution called "same convolution" is employed which allows the filter to go outside of the border of the input padding each with a zero.

Dense Blocks Design. Since in our case, there is a need to capture large citywide dependencies for increasing the accuracy in predicting network demand, a deep network will be required. This will place both computational power and complexity burden on its implementation. To address this issue, DenseNet has been employed with some modifications that exploits the potential of the network through *feature reuse*, yielding condensed models that are easily trained and highly parameter efficient [13]. In our proposed ST-DenNetFus architecture, each of the outputs from the first convolution layer (shown as convolution-1 in Fig. 3), $\mathbf{X}_c^{(1)}$, $\mathbf{X}_p^{(1)}$ and $\mathbf{X}_r^{(1)}$ is passed through L layers, each of which implements a non-linear transformation $\mathbf{H}_l(.)$, where l depicts the layer. In our implementation, $\mathbf{H}_l(.)$ is defined as a composite function of two consecutive operations of Rectified Linear Unit (ReLU) followed by a 3×3 convolution. On top of the Lth dense block, a convolutional layer is appended (shown as convolution-2 in Fig. 3). The final outputs of each of these branches after convolution-2 are $\mathbf{X}_c^{(L+2)}$, $\mathbf{X}_p^{(L+2)}$ and $\mathbf{X}_r^{(L+2)}$.

4.2 External Factors and Fusion

Network demand can be affected by many complex external factors. As discussed earlier, one of the main contributions of this research is to explore if extracted urban patterns in cities can be of impact to one of the most important challenges in the telecommunications domain, Network Demand Prediction [29].

The reason behind this is that there might be a relation between the cellular data usage and external factors such as the functionality of the regions, crowd mobility and day of the week. Taking the functional regions as an example, thinking about a business district, then intuitively one could expect that most companies will be empowered by a WiFi network and hence people once they arrive to their work will probably rely on the WiFi network more than the cellular network. In contrast, in a shopping district, the cellular network might be used more than the WiFi network as usually people are walking in streets or in shops in which WiFi is not universally, freely available or has poor signal coverage. For capturing finer granularity of functional regions in our proposed model, we relied on what so-called *temporal functional regions*[3]. The concept of *temporal functional regions* has been recently introduced in [2] and shows the possibility of recognizing regions' functionalities that not only change spatially but also temporally.

Another external factor that intuitively could impact the network demand is the crowd mobility patterns as it is expected that the more crowded an area is, the higher network demand. In addition and as shown before in [26], the day of the week is of an impact to the network demand variation. The simple example is that people typically rely on their cellular networks in a different pattern on the weekends compared to the weekdays. To predict the network demand at time t, the prior three external factors: temporal functional regions, day of the week and the crowd mobility patterns can be already obtained. However, the challenge in embedding these external factors into a model is that they vary in their dimensionality. In other words, the temporal functional regions and the crowd mobility patterns are both 2-dimensional features that vary across time however, the day of the week is 1-dimensional feature that varies across the time. For addressing this challenge, various branches have been introduced in the ST-DenNetFus architecture to fuse the external features according to their dimensionality as shown in the yellow branches of Fig. 3. Let $[E_{t_1}^{1D}, ..., E_{t_N}^{1D}]$ and $[E_{t_1}^{2D}, ..., E_{t_M}^{2D}]$ depict the features vectors for the 1-dimensional and 2-dimensional features respectively, where N and M indicate the number of the external 1-dimensional and 2-dimensional features respectively. For the 1-dimensional features, fully-connected layers are stacked and for the 2-dimensional features, convolutional layers with 5×5 filter are stacked for capturing the spatial dependencies of these features employing the "same convolution", for preserving the final output size to be the same as the size of the input.

After the network demand data is input and output $\mathbf{X}_{in}$ is generated, and the other branches for the external data sources $\mathbf{X}_{Ext-2D}$ and $\mathbf{X}_{Ext-1D}$ for the 2-dimensional and 1-dimensional features are produced, then a fusing layer (shown in red in Fig. 3) is used. The output $\mathbf{X}_{in-Ext}$ is further fed to a *tanh* function to generate $\hat{\mathbf{X}}_t$ which denotes the predicted value at the tth time interval. These operations can be summarized with the following equations:

$$\hat{\mathbf{X}}_t = tanh(\mathbf{X}_{in-Ext}) \tag{4}$$

$$\mathbf{X}_{in-Ext} = \mathbf{X}_{in} + \mathbf{X}_{Ext-2D} + \mathbf{X}_{Ext-1D} \tag{5}$$

---

[3] For same region, it could be classified as business district in the morning, eating in the afternoon and entertainment at night.

The ST-DenNetFus architecture can then be trained to predict $\hat{\mathbf{X}}_t$ from the Network Throughput input data and the external features by minimizing mean squared error between the predicted demand and the true demand matrix:

$$\kappa(\varepsilon) = ||\mathbf{X}_t - \hat{\mathbf{X}}_t||^2 \tag{6}$$

where ε represents all the learnable parameters in the whole ST-DenNetFu architecture.

5   Experiments Setup

5.1   Dataset

The dataset used in this paper captures the application and network usage gathered from Truconnect LLC[4], a mobile service provider based in US. The raw real data contains more than 200 billion records of mobile sessions that span across 6 months starting from July 2016 until end of December 2016 in NYC. All of the mobile sessions are geo-tagged with longitude and latitude. Mobile sessions can be any type of application usage on the phone that uses mobile network. These sessions might include additional session types information such as Youtube video views, application downloads and updates, and web browsing sessions.

Each sample in the dataset is created due to one of the following: (a) every hour, (b) every change of a pixel (lat, long of 4 digits with a resolution of 10×10), (c) every application used within this pixel and hour results in a new record in the dataset. On average per mobile device, there are between 1000–1200 records per day. The features that are filtered and used within this dataset are as follows:

- **Ueid:** This feature represents a mobile device unique identifier.
- **Latitude:** This value represents the latitude of the bottom-right corner of the pixel with resolution of 0.0001°.
- **Longitude:** This value represents the longitude of the bottom-right corner of the pixel with resolution of 0.0001°.
- **MaxRxThrpt:** This value represents the maximum downlink throughput observed on the network in the current pixel (in bps).
- **MaxTxThrpt:** This value represents the maximum uplink throughput observed on the network in the current pixel (in bps).

5.2   Baselines

The proposed ST-DenNetFus approach has been compared with the following 4 baselines:

- **Naive:** A naive model [20] works by simply setting the forecast at time t to be the value of the observation at time $t - l$ where l is the lag value. Several lag values are tested considering l equals to 1, 24, and 168 that corresponds to

---

[4] https://www.truconnect.com.

hourly, daily or weekly which we refer to as Naive-1, Naive-24, and Naive-168 respectively. For instance, in the case of daily, the network demand at time t on Monday is considered the same as at time t on Sunday. In the case of weekly, the network demand at time t on Monday is considered the same as at time t on previous Monday and for hourly, the network demand at time t is considered the same as at time $t - 1$. We concluded from these comparisons that Naive-1 shows the best accuracy and following that Naive-24 and Naive-168 respectively.

- **ARIMA:** An ARIMA model [33] is a well-known model for analyzing and forecasting time series data. ARIMA is considered a generalization of the simpler AutoRegressive Moving Average and adds the notion of integration. A nonseasonal ARIMA model is titled as ARIMA(p, d, q) where p is the number of autoregressive terms, d is the number of nonseasonal differences needed for stationarity, and q is the number of lagged forecast errors in the prediction equation. In our trained model, p, d, q are set to a default value with 1 [33].

- **RNN:** Recurrent Neural Networks (RNNs) [12] are a special type of neural network designed for sequence problems. Given a standard feedforward Multilayer Perceptron network, a recurrent neural network can be thought of as the addition of loops to the architecture. For example, in a given layer, each neuron may pass its signal latterly (sideways) in addition to forward to the next layer. The output of the network may feedback as an input to the network with the next input vector. And so on. In our experiments, the length of the input sequence is fixed to one of the $\{1, 3, 6, 12, 24, 48, 168\}$ and we concluded that the best accurate model is RNN-12.

- **LSTM:** The Long Short-Term Memory (LSTM) [22] network is a recurrent neural network that is trained using back propagation through time and overcomes the vanishing gradient problem. As such it can be used to create large (stacked) recurrent networks, that in turn can be used to address difficult sequence problems in machine learning and achieve state-of-the-art results [28]. Instead of neurons, LSTM networks have memory blocks that are connected into layers. The experiments are conducted on 6 LSTM variants following the same settings of RNN, including, LSTM-1, LSTM-3, LSTM-6, LSTM-12, LSTM-24, LSTM-48, LSTM-168. We concluded that LSTM-6 is the most accurate model.

Going forward in this paper, the best performing baseline models, Naive-1, RNN-12, and LSTM-6 will be referred to as simply Naive, RNN, and LSTM in the baselines comparison respectively.

6 Results

In this section, we give an overview on the preprocessing procedures taken before performing our experiments. In addition, we illustrate our procedures for selecting the tuning parameters and their effect on the prediction accuracy. Finally, we discuss how the proposed ST-DenNetFus architecture perform compared to the baselines. ST-DenNetFus is evaluated by the Mean Absolute Error (MAE) and Root Mean Squared Error (RMSE) as they are two of the most common metrics used to measure the accuracy of continuous variables [6].

The learnable parameters of the ST-DenNetFus are initialized using a uniform distribution in Keras [7]. The convolutions of convolution-1 and convolution-2 use 24 and 2 filters of size 3×3 respectively. Convolution-2 uses 2 filters to match the desired number of outputs needed for the downlink and uplink throughput. Adam [16] is used for optimization, and the batch size is set to 15 for fitting the memory of the GPU used in the experiments. The number of dense blocks is set to 5. For p and r, they are empirically set to capture one-day and one-week respectively where l_c, l_p and $l_r \in \{1, 2, 3, 4\}$. From the training dataset, 90% is selected for training each model and the remaining 10% for the validation dataset which is used for choosing the best model as well as to early-stop the training algorithm if there is no improvement found after 5 consecutive epochs.

In the Dense Networks, if each function $\mathbf{H}$ produced k feature-maps as output, then the l_{th} layer will have $k \times (l - 1) + k_0$ input feature-maps, where k_0 is the number of channels in the input matrix. To prevent the network from growing too wide and to improve the parameters efficiency, k is limited to a bounded integer. This hyperparamater is referred to as *growth rate* [13]. We experimented the impact of increasing the growth rate from 5 to 35 on the prediction's accuracy and we observed that the accuracy improves by increasing the growth rate until reaching certain point 24 in which widening further the network starts to have a counter impact on the accuracy. Hence it was concluded that the optimum growth rate is 24. We experimented as well the impact of the network depth and we concluded that a network depth of 5 has the optimum results to sufficiently capture with the close spatial dependence as well as the distant one. In ST-DenNetFus, each of the external features are input into a separate branch unlike the traditional approach. The traditional approach of fusing external features of same dimensionality merges these features first and then fuses them into one branch. However, in our proposed approach, a separate branch for each of the external features is used and then merge their outputs in a later stage after the feedforward execution of each of the branches (shown in yellow in Fig. 3). In our case, although both the temporal functional regions and crowd mobility patterns are of same dimensionality where both are 2-dimensional matrices that vary across time (1-hour time-interval), they are each input on a separate branch and then fused later. We concluded that the impact of feeding the external features in this way performs 10% RMSE and 8% MAE better for the downlink throughput prediction and 8% RMSE and 7% MAE better for the uplink throughput prediction. In order to determine the optimum length of closeness, period and trend for the network demand dataset, the length of period and trend are set to 1 and then the length of closeness is varied from 0 to 5 where $l_c = 0$ indicates that the closeness component/branch is not employed which we concluded that l_c equals to 3 has the lowest RMSE and MAE and $l_c = 0$ has the highest error. Further, l_c is set to 3 and l_r is set to 1 and then l_p is varied from 0 to 5. Then we concluded that the best performance is when l_p equals to 3. Similarly, we did for

Table 1. Prediction accuracy comparisons with baselines

Model	Evaluation Metric (Downlink Throughput)		Evaluation Metric (Uplink Throughput)	
	RMSE	MAE	RMSE	MAE
Naive	13278936.747	7667966.397	5237542.197	2406522.277
ARIMA	12177307.197	8073921.767	4816366.977	2783808.635
RNN	11199525.956	8576942.055	4335734.302	2865784.639
LSTM	10580656.522	7660093.113	4216037.533	2713482.051
ST-DenNetFus-NoExt	9675762.836	6315907.039	3907936.380	2131071.282
ST-DenNetFus-Ext	**9600259.526**	**6206750.047**	**3847875.555**	**1933871.466**

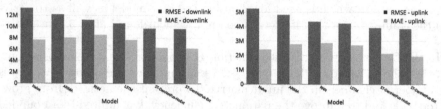

(a) Comparison for downlink throughput prediction accuracy.　(b) Comparison for uplink throughput prediction accuracy.

Fig. 4. Model ranking for network demand prediction. The smaller the better.

l_r and found that the best value is at 4. Based on this analysis, it is concluded that the best configuration for the $\{l_c, l_p, l_r\}$ is $\{3, 3, 4\}$.

Table 1 shows the ST-DenNetFus Network Demand Prediction accuracy comparisons with the baselines for both the throughput downlink and uplink. As shown, the proposed ST-DenNetFus consistently and significantly outperforms all baselines. Specifically, the results for the downlink throughput prediction demonstrate that ST-DenNetFus (with 5 dense-blocks) is relatively 30% RMSE and 20% MAE better than the Naive model, 20% RMSE and 23% MAE better than ARIMA, 15% RMSE and 30% MAE better than RNN and 10% RMSE and 20% MAE better than LSTM. For the uplink throughput prediction, ST-DenNetFus is 27% RMSE and 20% MAE better than the Naive model, 20% RMSE and 30% MAE better than ARIMA, 12% RMSE and 33% MAE better than RNN, and 10% RMSE and 30% MAE better than LSTM. ST-DenNetFus-NoExt is our proposed version of ST-DenNetFus-Ext that does not consider external factors (e.g. temporal functional regions). It can be seen that ST-DenNetFus-NoExt is worse than the ST-DenNetFus-Ext indicating that external factors and patterns fused are always beneficial. Intuitively, the models in RMSE can be ranked as illustrated in Fig. 4.

7   Conclusion

In this paper, a new deep learning based approach called ST-DenNetFus is proposed for forecasting the network demand (throughput uplink and downlink) in each and every region of a city. For the first time, it has been shown that fusing some external patterns such as temporal functional regions and crowd mobility patterns improves the accuracy of the forecasting due to their intuitive correlation with the network demand variation. Compared to other 4 baselines, the proposed approach outperforms, confirming that the proposed approach is better and more applicable to the Network Demand Prediction problem. The introduced ST-DenNetFus is capable of learning the spatial and temporal dependencies. In addition, it employs various branches for fusing external data sources of various dimensionalities. Furthermore, we argue that the introduced ST-DenNetFus architecture could be leveraged for solving other spatio-temporal prediction problems that requires fusing external data sources such as energy demand forecasting and others. In the future, we plan to consider more external data sources such as weather data and the type of applications used by individuals in each grid cell of the city which could further boost the accuracy of the network demand prediction. In addition, we would like in the future to extend our data fusion layer from all branches using more complex techniques such as the parametric-matrix-based fusion mechanisms.

Acknowledgment. This work was partly supported by the Science Foundation Ireland ADAPT centre (Grant 13/RC/2106) and by the EC project ASGARD, 700381 (H2020-ICT-2016-09, Research and Innovation action).

References

1. Abou-Zeid, H., Hassanein, H.S.: Predictive green wireless access: exploiting mobility and application information. IEEE Wirel. Commun. **20**(5), 92–99 (2013)
2. Assem, H., Buda, T.S., O'sullivan, D.: RCMC: recognizing crowd-mobility patterns in cities based on location based social networks data. ACM Trans. Intell. Syst. Technol. (TIST) **8**(5), 70 (2017)
3. Assem, H., Ghariba, S., Makrai, G., Johnston, P., Gill, L., Pilla, F.: Urban water flow and water level prediction based on deep learning. In: Altun, Y., et al. (eds.) ECML PKDD 2017. LNCS (LNAI), vol. 10536, pp. 317–329. Springer, Cham (2017). https://doi.org/10.1007/978-3-319-71273-4_26
4. Assem, H., O'Sullivan, D.: Discovering new socio-demographic regional patterns in cities. In: Proceedings of the 9th ACM SIGSPATIAL Workshop on Location-Based Social Networks, p. 1. ACM (2016)
5. Assem, H., Xu, L., Buda, T.S., O'Sullivan, D.: Spatio-temporal clustering approach for detecting functional regions in cities. In: 2016 IEEE 28th International Conference on Tools with Artificial Intelligence (ICTAI), pp. 370–377. IEEE (2016)
6. Chai, T., Draxler, R.R.: Root mean square error (RMSE) or mean absolute error (MAE)?-Arguments against avoiding RMSE in the literature. Geoscientific Model Dev. **7**(3), 1247–1250 (2014)
7. Chollet, F.: Deep learning library for python. Runs on TensorFlow, Theano or CNTK. https://github.com/fchollet/keras. Accessed 09 Aug 2017

8. Cisco, I.: Cisco visual networking index: forecast and methodology, 2011–2016. CISCO White paper, pp. 2011–2016 (2012)
9. Dong, X., Fan, W., Gu, J.: Predicting lte throughput using traffic time series. ZTE Commun. **4**, 014 (2015)
10. Hasan, Z., Boostanimehr, H., Bhargava, V.K.: Green cellular networks: a survey, some research issues and challenges. IEEE Commun. Surv. Tutorials **13**(4), 524–540 (2011)
11. He, K., Zhang, X., Ren, S., Sun, J.: Deep residual learning for image recognition. In: Proceedings of the IEEE Conference on Computer Vision and Pattern Recognition, pp. 770–778 (2016)
12. Hochreiter, S., Schmidhuber, J.: Long short-term memory. Neural Comput. **9**(8), 1735–1780 (1997)
13. Huang, G., Liu, Z., Weinberger, K.Q., van der Maaten, L.: Densely connected convolutional networks. arXiv preprint arXiv:1608.06993 (2016)
14. Jain, V., et al.: Supervised learning of image restoration with convolutional networks. In: IEEE 11th International Conference on Computer Vision, ICCV 2007, pp. 1–8. IEEE (2007)
15. Khan, L.U.: Performance comparison of prediction techniques for 3G cellular traffic. Int. J. Comput. Sci. Netw. Secur. (IJCSNS) **17**(2), 202 (2017)
16. Kingma, D., Ba, J.: Adam: a method for stochastic optimization. arXiv preprint arXiv:1412.6980 (2014)
17. Krizhevsky, A., Sutskever, I., Hinton, G.E.: Imagenet classification with deep convolutional neural networks. In: Advances in Neural Information Processing Systems, pp. 1097–1105 (2012)
18. LeCun, Y., Bengio, Y., Hinton, G.: Deep learning. Nature **521**(7553), 436–444 (2015)
19. LeCun, Y.A., Bottou, L., Orr, G.B., Müller, K.-R.: Efficient BackProp. In: Montavon, G., Orr, G.B., Müller, K.-R. (eds.) Neural Networks: Tricks of the Trade. LNCS, vol. 7700, pp. 9–48. Springer, Heidelberg (2012). https://doi.org/10.1007/978-3-642-35289-8_3
20. Makridakis, S., Wheelwright, S.C., Hyndman, R.J.: Forecasting Methods and Applications. Wiley, Hoboken (2008)
21. Mathieu, M., Couprie, C., LeCun, Y.: Deep multi-scale video prediction beyond mean square error. arXiv preprint arXiv:1511.05440 (2015)
22. Mikolov, T., Karafiát, M., Burget, L., Cernocký, J., Khudanpur, S.: Recurrent neural network based language model. In: Interspeech, vol. 2, p. 3 (2010)
23. Najafabadi, M.M., Villanustre, F., Khoshgoftaar, T.M., Seliya, N., Wald, R., Muharemagic, E.: Deep learning applications and challenges in big data analytics. J. Big Data **2**(1), 1 (2015)
24. Oh, E., Krishnamachari, B., Liu, X., Niu, Z.: Toward dynamic energy-efficient operation of cellular network infrastructure. IEEE Commun. Mag. **49**(6), 56–61 (2011)
25. Papagiannaki, K., Taft, N., Zhang, Z.L., Diot, C.: Long-term forecasting of internet backbone traffic. IEEE Trans. Neural Netw. **16**(5), 1110–1124 (2005)
26. Paul, U., Subramanian, A.P., Buddhikot, M.M., Das, S.R.: Understanding traffic dynamics in cellular data networks. In: 2011 Proceedings IEEE INFOCOM, pp. 882–890. IEEE (2011)
27. Sadek, N., Khotanzad, A.: Multi-scale high-speed network traffic prediction using k-factor Gegenbauer ARMA model. In: 2004 IEEE International Conference on Communications, vol. 4, pp. 2148–2152. IEEE (2004)

28. Sahu, A.: Survey of reasoning using neural networks. arXiv preprint arXiv:1702.06186 (2017)
29. Sayeed, Z., Liao, Q., Faucher, D., Grinshpun, E., Sharma, S.: Cloud analytics for wireless metric prediction-framework and performance. In: 2015 IEEE 8th International Conference on Cloud Computing (CLOUD), pp. 995–998. IEEE (2015)
30. Simonyan, K., Zisserman, A.: Very deep convolutional networks for large-scale image recognition. arXiv preprint arXiv:1409.1556 (2014)
31. Szegedy, C., et al.: Going deeper with convolutions. In: Proceedings of the IEEE Conference on Computer Vision and Pattern Recognition, pp. 1–9 (2015)
32. Waber, A.: The seasonality of mobile device usage (2014). https://marketingland.com/seasonality-mobile-device-usage-warmer-weather-tempers-tech-95937
33. Wu, J., Wei, S.: Time Series Analysis. Hunan Science and Technology Press, Chang-Sha (1989)
34. Yao, J., Kanhere, S.S., Hassan, M.: Improving QoS in high-speed mobility using bandwidth maps. IEEE Trans. Mob. Comput. **11**(4), 603–617 (2012)
35. Yu, Y., Song, M., Fu, Y., Song, J.: Traffic prediction in 3G mobile networks based on multifractal exploration. Tsinghua Sci. Technol. **18**(4), 398–405 (2013)
36. Zeiler, M.D., Fergus, R.: Visualizing and understanding convolutional networks. In: Fleet, D., Pajdla, T., Schiele, B., Tuytelaars, T. (eds.) ECCV 2014. LNCS, vol. 8689, pp. 818–833. Springer, Cham (2014). https://doi.org/10.1007/978-3-319-10590-1_53
37. Zhang, J., Zheng, Y., Qi, D.: Deep spatio-temporal residual networks for citywide crowd flows prediction. In: AAAI, pp. 1655–1661 (2017)
38. Zhang, J., Zheng, Y., Qi, D., Li, R., Yi, X., Li, T.: Predicting citywide crowd flows using deep spatio-temporal residual networks. arXiv preprint arXiv:1701.02543 (2017)
39. Zhou, B., He, D., Sun, Z., Ng, W.H.: Network traffic modeling and prediction with ARIMA/GARCH. In: Proceedings of HET-NETs Conference, pp. 1–10 (2005)

On Optimizing Operational Efficiency in Storage Systems via Deep Reinforcement Learning

Sunil Srinivasa[1]([⊠]), Girish Kathalagiri[1], Julu Subramanyam Varanasi[2],
Luis Carlos Quintela[1], Mohamad Charafeddine[1], and Chi-Hoon Lee[1]

[1] Samsung SDS America, 3655 North First Street, San Jose, CA 95134, USA
ssunil@gmail.com, girish.sk@gmail.com,
{l.quintela,mohamad.c,lee.chihoon}@samsung.com
[2] Samsung Semiconductors Inc., 3655 North First Street, San Jose, CA 95134, USA
sv.julu@gmail.com

Abstract. This paper deals with the application of deep reinforcement learning to optimize the operational efficiency of a solid state storage rack. Specifically, we train an on-policy and model-free policy gradient algorithm called the Advantage Actor-Critic (A2C). We deploy a dueling deep network architecture to extract features from the sensor readings off the rack and devise a novel utility function that is used to control the A2C algorithm. Experiments show performance gains greater than 30% over the default policy for deterministic as well as random data workloads.

Keywords: Data center · Storage system · Operational efficiency
Deep reinforcement learning · Actor-critic methods

1 Introduction

The influence of artificial intelligence (AI) continues to proliferate our daily lives at an ever-increasing pace. From personalized recommendations to autonomous vehicle navigation to smart personal assistants to health screening and diagnosis, AI has already proven to be effective on a day-to-day basis. But it can also be effective in tackling some of the world's most challenging control problems – such as minimizing the power usage effectiveness[1] (PUE) in a data center.

From server racks to large deployments, data centers are the backbone to delivering IT services and providing storage, communication, and networking to the growing number of users and businesses. With the emergence of technologies such as distributed cloud computing and social networking, data centers have an even bigger role to play in today's world. In fact, the Cisco® Global Cloud Index,

---

[1] PUE [1] is defined as the ratio of the total energy used by the data center to the energy delivered towards computation. A PUE of 1.0 is considered ideal.

© Springer Nature Switzerland AG 2019
U. Brefeld et al. (Eds.): ECML PKDD 2018, LNAI 11053, pp. 238–253, 2019.
https://doi.org/10.1007/978-3-030-10997-4_15

an ongoing effort to forecast the growth of data center and cloud-based traffic, estimates that the global IP traffic will grow 3-fold over the next 5 years [2].

Naturally, data centers are sinkholes of energy (required primarily for cooling). While technologies like virtualization and software-based architectures to optimize utilization and management of compute, storage and network resource are constantly advancing [4], there is still a lot of room to improve the utilization of energy on these systems using AI.

1.1 The Problem

In this paper, we tackle an instance of the aforementioned problem, specifically, optimizing the operation of a solid state drive (SSD) storage rack (see Fig. 1). The storage rack comprises 24 SSDs and a thermal management system with 5 cooling fans. It also has 2 100G ethernet ports for data Input/Output (I/O) between the client machines and the storage rack. From data I/O operations off the SSDs, the rack's temperature increases and that requires the fans to be turned on for cooling. Currently, the fan speeds are controlled simply based on a tabular (rule-based) method - thermal sensors throughout the chassis including one for each drive bay record temperatures and the fan speeds are varied based on a table that maps the temperature thresholds to desired fan speeds. In contrast, our solution uses a deep reinforcement learning[2] (DRL)-based control algorithm called the Advantage Actor-Critic to control the fans. Experimental results show significant performance gains over the rule-based current practices.

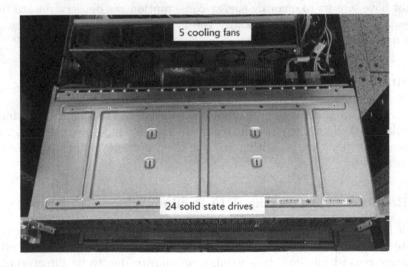

Fig. 1. The SSD storage rack we used for our experiments. It comprises 24 SSDs that are kept cool via 5 fans.

---

[2] DRL [3] is the latest paradigm for tackling AI problems.

Ideally, if we are able to measure the PUE of the storage rack, that will be the precise function that needs to be minimized. However, the rack did not contain any hooks or sensors for directly measuring energy. As a proxy for the PUE, we instead designed a utility (or reward) function that represents the *operational efficiency* of the rack. Our reward function is explained in detail in Sect. 2.2, and across the space of its arguments, comprises contours of good as well as contours of not-so-good values. The problem now becomes being able to learn a control algorithm that optimizes the operational efficiency of the rack by always driving it towards the reward contours with good values.

1.2   Related Work and Our Contributions

Prior work in the area of developing optimal control algorithms for data centers [4–6] build approximate models to study the effects of thermal, electrical and mechanical subsystem interactions in a data center. These model-based methods are sometimes inadequate and suffer from error propagation, which leads to sub-optimal control policies. Recently, Google DeepMind published a blog [7] on using AI to reduce Google's data centre cooling bill by 40%. In [8], the authors use the deep deterministic policy gradient technique on a simulation platform and achieve a low PUE as well as a 10% reduction in cooling energy costs.

Our novel contributions are three-fold. First, unlike prior model-based approaches, we formulate a model-free method that does not require any knowledge of the SSD server behavior dynamics. Second, we train our DRL algorithm on the real system[3], and do not require a simulator. Finally, since our SSD rack does not have sensors to quantify energy consumption, we devise a reward function that is used not only to quantify the system's operational efficiency, but also as a control signal for training.

2   Our DRL-Based Solution

In this section, we provide the details of our solution to the operation optimization problem. We first introduce the reader to reinforcement learning (RL), and subsequently explain the algorithm and the deep network architecture used for our experiments.

2.1   Reinforcement Learning (RL) Preliminaries

RL is a field of machine learning that deals with how an agent (or algorithm) ought to take actions in an environment (or system) so as to maximize a certain cumulative reward function. It is gaining popularity due to its direct applicability to many practical problems in decision-making, control theory and multi-dimensional optimization. RL problems are often modeled as a Markov Decision

---

[3] In fact, the SSD server rack also has a Intel Xeon processor with 44 cores to facilitate online (in-the-box) training.

Process with the following typical notation. During any time slot t, the environment is described by its *state* notated s_t. The RL agent interacts with the environment in that it observes the state s_t and takes an *action* a_t from some set of actions, according to its *policy* $\pi(a_t|s_t)$ - the policy of an agent (denoted by $\pi(a_t|s_t)$) is a probability density function that maps states to actions, and is indicative of the agent's behavior. In return, the environment provides an immediate *reward* $r_t(s_t, a_t)$ (which is evidently a function of s_t and a_t) and transitions to its next state s_{t+1}. This interaction loops in time until some terminating criterion is met (for example, say, until a time horizon H). The set of states, actions, and rewards the agent obtains while interacting (or rolling-out) with the environment, $\tau =: \{(s_0, a_0, r_0), (s_1, a_1, r_1), \ldots, (s_{H-1}, a_{H-1}, r_{H-1}), s_H\}$ forms a *trajectory*. The cumulative reward observed in a trajectory τ is called the *return*, $\mathcal{R}(\tau) = \sum_{t=0}^{H-1} \gamma^{H-1-t} r_t(s_t, a_t)$, where γ is a factor used to discount rewards over time, $0 \leq \gamma \leq 1$. Figure 2 represents an archetypal setting of a RL problem.

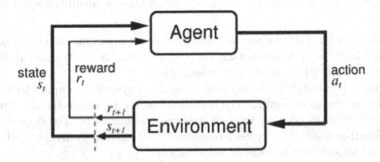

Fig. 2. The classical RL setting. Upon observing the environment state s_t, the agent takes an action a_t. This results in an instantaneous reward r_t, while the environment transitions to its next state s_{t+1}. The objective of the agent is to maximize the cumulative reward over time.

In the above setting, the goal of the agent is to optimize the policy π so as to maximize the expected return $\mathbb{E}_\tau[\mathcal{R}_\tau]$ where the expectation operation is taken across several trajectories. Two functions related to the return are (a) the *action value function* $Q^\pi(s_t, a_t)$, which is the expected return for selecting action a_t in state s_t and following the policy π, and (b) the *state value function* $V^\pi(s_t)$, which measures the expected return from state s_t upon following the policy π. The *advantage* of action a_t in state s_t is then defined as $A^\pi(s_t, a_t) = Q^\pi(s_t, a_t) - V^\pi(s_t)$.

2.2 State, Action and Reward Formulations

In order to employ RL for solving our problem, we need to formulate state, action and reward representations.

State: We use a vector of length 7 for the state representation comprising the following (averaged and further, normalized) scalars:

1. **tps** (transfers per second): the mean number of transfers per second that were issued to the SSDs. A transfer is an I/O request to the device and is of indeterminate size. Multiple logical requests can be combined into a single I/O request to the device.
2. **kB_read_per_sec**: the mean number of kilo bytes read from an SSD per second.
3. **kB_written_per_sec** (writes per second): the mean number of kilo bytes written to an SSD per second.
4. **kb_read**: the mean number of kilo bytes read from an SSD in the previous time slot.
5. **kb_written**: the mean number of kilo bytes written to an SSD in the previous time slot.
6. **temperature**: the mean temperature recorded across the 24 SSD temperature sensors on the rack.
7. **fan_speed**: the mean speed of the 5 cooling fans in revolutions per minute (rpm).

Recall there are 24 SSDs in our rack (and hence 24 different values of transfers per second, bytes read, bytes written, etc.), but we simply use the averaged (over the 24 SSDs) values for our state representation[4].

To obtain 1. through 5. above, we use the linux system command **iostat** [20] that monitors the input/output (I/O) device loading by observing the time the devices are active in relation to their average transfer rates. For 6. and 7., we use the **ipmi-sensors** [21] system command that displays current readings of sensors and sensor data repository information.

For normalizing, 1. through 7., we use the *min-max* strategy. Accordingly, for feature X, an averaged value of $\bar{x}$ is transformed to $\Gamma(\bar{x})$, where

$$\Gamma(\bar{x}) = \frac{\bar{x} - \text{minX}}{\text{maxX} - \text{minX}}, \tag{1}$$

where minX and maxX are the minimum and maximum values set for the feature X. Table 1 lists the minimum and maximum values we used for normalizing the state features, which were chosen based on empirically observed range of values.

Action: The action component of our problem is simply in setting the fan speeds. In order to keep the action space manageable[5], we use the same action

---

[4] We shall see shortly that we are constrained to use the same control setting (speed) on all the 5 fans. Hence, it is not meaningful to individualize the SSDs based on their slot location inside the server. Using averaged values is practical in this scenario.

[5] The action space grows exponentially with the number of fans, when controlling each fan independently. For example, with just 3 fan speed settings, the number of possible actions with 5 fans becomes $3^5 = 243$. DRL becomes ineffective when handling such a large problem. Instead, if we set all the fan speeds to the same value, the action space dimension reduces to just 5, and is quite manageable. Incidentally, setting separate speeds to the fans is also infeasible from a device driver standpoint - the ipmitool utility can only set all the fans to the same speed.

Table 1. Minimum and maximum values of the various state variables used for the min-max normalization.

Feature X	minX	maxX
tps	0.0	1000.0
kB_read_per_sec	0.0	4000000.0
kB_written_per_sec	0.0	4000000.0
kB_read	0.0	4000000.0
kB_written	0.0	4000000.0
temperature	27.0	60.0
fan_speed	7500.0	16000.0

(i.e., speed-setting) on all the 5 fans. For controlling the fan speeds, we use the **ipmitool** [22] command-line interface. We consider two separate scenarios:

- **raw action:** the action space is discrete with values 0 through 6, where 0 maps to 6000 rpm, while 6 refers to 18000 rpm. Accordingly, only 7 different rpm settings are allowed: 6000 through 18000 in steps of 2000 rpm. Note that consecutive actions can be very different from each other.
- **incremental action:** the action space is discrete taking on 3 values - 0, 1 or 2. An action of 1 indicates no change in the fan speed, while 0 and 2 refer to an decrement or increment of the current fan speed by 1000 rpm, respectively. This scenario allows for smoother action transitions. For this case, we allow 10 different rpm values: 9000 through 18000, in steps of 1000 rpm.

Reward: In this section, we design a reward function that functions as a proxy for the operational efficiency of the SSD rack. One of the most important components of a RL solution is *reward shaping*. Reward shaping refers to the process of incorporating domain knowledge towards engineering a reward function, so as to better guide the agent towards its optimal behavior. Being able to devise a good reward function is critical since it explicitly relates to the expected return that needs to be maximized. We now list some desired properties of a meaningful reward function that will help perform reward shaping in the context of our problem.

- Keeping both the devices' temperatures and fan speeds low should yield the highest reward, since this scenario means the device operation is most efficient. However, also note that this case is feasible only when the I/O loads are absent or are very small.
- Irrespective of the I/O load, a low temperature in conjunction with a high fan speed should yield a bad reward. This condition is undesirable since otherwise, the agent can always set the fan speed to its maximum value, which in turn will not only consume a lot of energy, but also increase the wear on the mechanical components in the system.

- A high temperature in conjunction with a low fan speed should also yield a poor reward. If not, the agent may always choose to set the fan speed to its minimum value. This may result in overheating the system and potential SSD damages, in particular when the I/O loads are high.
- Finally, for different I/O loads, the optimal rewards should be similar. Otherwise, the RL agent may learn to overfit and perform well only on certain loads.

While there are several potential candidates for our desired reward function, we used the following mathematical function:

$$R = -\max\left(\frac{\Gamma(\bar{T})}{\Gamma(\bar{F})}, \frac{\Gamma(\bar{F})}{\Gamma(\bar{T})}\right), \tag{2}$$

where $\bar{T}$ and $\bar{F}$ represent the averaged values of temperature (over the 24 SSDs) and fan speeds (over the 5 fans), respectively. $\Gamma(\cdot)$ is the normalizing transformation explained in (1), and is performed using the temperature and fan speed minimum and maximum values listed in Table 1. Note that while this reward function weighs F and T equally, we can tweak the relationship between F and T to meet other preferential tradeoffs that the system operator might find desirable. Nevertheless, the DRL algorithm will be able to optimize the policy for any designed reward function.

Figure 3 plots the reward function we use as a function of mean temperature $\bar{T}$ (in °C) and fan speed $\bar{F}$ (in rpm). Also shown on the temperature-fan speed plane are contours representing regions of similar rewards. The colorbar on the right shows that blue and green colors represent the regions with poor rewards, while (dark and light) brown shades the regions with high rewards. All the aforementioned desired properties are satisfied with this reward function - the reward is maximum when both fan speed temperature are low; when either of them becomes high, the reward drops and there are regions of similar maximal rewards for different I/O loads (across the space of temperature and fan speeds).

2.3   Algorithm: The Advantage Actor-Critic (A2C) Agent

Once we formulate the state, action and reward components, there are several methods in the RL literature to solve our problem. A rather classic approach is the policy gradient (PG) algorithm [9], that essentially uses gradient-based techniques to optimize the agent's policy. PG algorithms have lately become popular over other traditional RL approaches such as Q-learning [10] and SARSA [11] since they have better convergence properties and can be effective in high-dimensional and continuous action spaces.

While we experimented with several PG algorithms including Vanilla Policy Gradient [12] (and its variants [13,14]) and Deep Q-Learning [15], the most encouraging results were obtained with the **Advantage Actor-Critic (A2C)** agent. A2C is essentially a synchronous, deterministic variant of Asynchronous

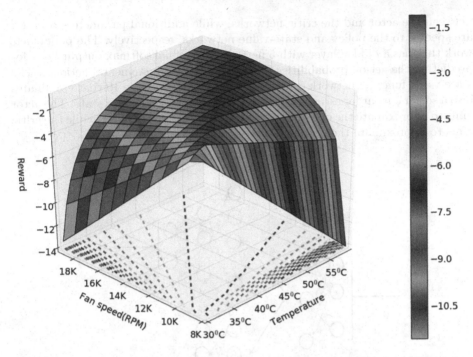

Fig. 3. Depiction of the reward (operational efficiency) versus temperature and fan speeds. The reward contours are also plotted on the temperature-fan speed surface. The brown colored contour marks the regions of optimal reward. (Color figure online)

Advantage Actor-Critic (A3C) [16], that yields state-of-the-art performance on several Atari games as well as on a wide variety of continuous motor control tasks.

As the name suggests, actor-critic algorithms comprise two components, an *actor* and a *critic*. The actor determines the best action to perform for any given state, and the critic estimates the actor's performed action. Iteratively, the actor-critic network implements generalized policy iteration [11] - alternating between a policy evaluation step and a policy improvement step. Architecturally, both the actor and the critic are best modeled via functional approximators, such as deep neural networks.

2.4   Actor-Critic Network Architecture

For our experiments, we employed a **dueling network architecture**, similar to the one proposed in [17]. The exact architecture is depicted in Fig. 4: the state of the system is a 7-length vector that is fed as input to a fully connected (FC) layer with 10 neurons represented by trainable weights θ. The output of this FC layer explicitly branches out to two separate feed-forward networks - the *policy* (actor) *network* (depicted on the upper branch in Fig. 4) and the *state-value function* (critic) *network* (depicted on the lower branch). The parameters θ are shared between the actor and the critic networks, while additional parameters α and β are specific to the policy and state-value networks, respectively. The policy network that has a hidden layer with 5 neurons and a final softmax output layer for predicting the action probabilities (for the 7 raw or 3 incremental actions). The state-value function network comprises a hidden layer of size 10 that culminates into a scalar output for estimating the value function of the input state. The actor aims to approximate the optimal policy π^*: $\pi(a|s; \theta, \alpha) \approx \pi^*(a|s)$, while the critic aims to approximate the optimal state-value function: $V(s; \theta, \beta) \approx V^*(s)$.

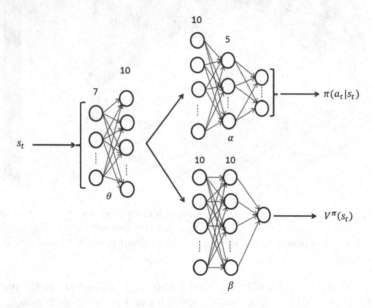

Fig. 4. The employed dueling network architecture with shared parameters θ. α and β are the actor- and critic-specific parameters, respectively. All the layers are fully connected neural networks; the numbers shown above represent the hidden layer dimensions. The policy network on the upper branch estimates the action probabilities via a softmax output layer, while the critic network on the lower branch approximates the state-value function.

Prior DRL architectures for actor-critic methods [16,18,19] employ single-stream architectures wherein the actor and critic networks do not share parameters. The advantage of our dueling network lies partly in its ability to compute both the policy and state-value functions via fewer trainable parameters vis-à-vis single-stream architectures. The sharing of parameters also helps mitigate overfitting one function over the other (among the policy and state-value functions). In other words, our dueling architecture is able to learn both the state-value and the policy estimates efficiently. With every update of the policy network parameters in the dueling architecture, the parameters θ get updated as well - this contrasts with the updates in a single-stream architecture wherein when the policy parameters are updated, the state-value function parameters remain untouched. The more frequent updating of the parameters θ mean a higher resource allocation towards the learning process, thus resulting in faster convergence in addition to obtaining better function approximations.

The pseudocode for our A2C algorithm in the context of the dueling network architecture (see Fig. 4) is described in Algorithm 1. Note that R represents the Monte Carlo return, and well-approximates the action-value function. Accordingly, we use $R - V(s; (\theta, \beta))$ as an approximation to the advantage function.

Algorithm 1. Advantage Actor-Critic (A2C) - pseudocode

// Notate shared parameters by θ and actor- and critic-specific parameters by α and β, respectively.

// Assume same learning rates η for θ, α as well as β. In general, they may all be different.

Initialize θ, α and β via uniformly distributed random variables.

repeat

    Reset gradients $d\theta = 0$, $d\alpha = 0$ and $d\beta = 0$.

    Sample N trajectories $\tau_1, \ldots, \tau_N$ under the (current) policy $\pi(\cdot; (\theta, \alpha))$.

    $i = 1$

    repeat

        $t_{\text{start}} = t$

        Obtain state s_t

        repeat

            Perform action a_t sampled from policy $\pi(a_t|s_t; (\theta, \alpha))$.

            Receive reward $r_t(s_t, a_t)$ and new state s_{t+1}

            $t \leftarrow t + 1$

        until $t - t_{\text{start}} = H$

        $i \leftarrow i + 1$

        Initialize R: $R = V(s_t; (\theta, \beta))$

        for $i \in \{t - 1, \ldots, t_{\text{start}}\}$ **do**

            $R \leftarrow r_i(s_i, a_i) + \gamma R$

            Sum gradients w.r.t θ and α: // gradient ascent on the actor parameters

            $d\theta \leftarrow d\theta + \nabla_\theta \log \pi(a_i|s_i; (\theta, \alpha)) (R - V(s_i; (\theta, \beta)))$

            $d\alpha \leftarrow d\alpha + \nabla_\alpha \log \pi(a_i|s_i; (\theta, \alpha)) (R - V(s_i; (\theta, \beta)))$

            Subtract gradients w.r.t β and θ: //gradient descent on the critic parameters

            $d\theta \leftarrow d\theta - \nabla_\theta (R - V(s_i; (\theta, \beta)))^2$

            $d\beta \leftarrow d\beta - \nabla_\beta (R - V(s_i; (\theta, \beta)))^2$

        end for

    until $i = N$

    // Optimize parameters

    Update θ, α and β: $\theta \leftarrow \theta + \eta d\theta$, $\alpha \leftarrow \alpha + \eta d\alpha$, $\beta \leftarrow \beta + \eta d\beta$.

until convergence.

---

3   Experimental Setup and Results

3.1   Timelines

Time is slotted to the duration of 25 s. At the beginning of every time slot, the agent observes the state of the system and prescribes an action. The system is then allowed to stabilize and the reward is recorded at the end of the time slot (which is also the beginning of the subsequent time slot). The system would have, by then, proceeded to its next state, when the next action is prescribed. We use a time horizon of 10 slots ($H = 250$ s), and each iteration comprises $N = 2$ horizons, i.e., the network parameters θ, α and β are updated every 500 s.

3.2   I/O Scenarios

We consider two different I/O loading scenarios for our experiments.

- **Simple periodic workload:** We assume a periodic load where within each period, there is no I/O activity for a duration of time (roughly 1000 s) followed by heavy I/O loading for the same duration of time. I/O loading is performed using a combination of 'read', 'write' and 'randread' operations with varying block sizes of data ranging from 4 KBytes to 64 KBytes. A timeline of the periodic workload is depicted in Fig. 5 (left).
- **Complex stochastic workload:** This is a realistic workload where in every time window of 1000 s, the I/O load is chosen uniformly randomly from three possibilities: no load, medium load or heavy load. A sample realization of the stochastic workload is shown in Fig. 5 (right).

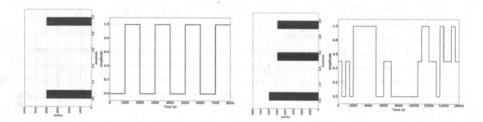

Fig. 5. The simple periodic workload (left) with a period of 2000 s, and a realization of the more realistic stochastic workload (right). Histograms of the load types are also shown for clarity. While the simple load chooses between no load and heavy load in a periodic manner, the complex load chooses uniformly randomly between the no load, medium load and heavy load scenarios.

3.3   Hyperparameters

Table 2 lists some hyperparameters used during model training.

Table 2. Table of hyperparameters.

Parameter	Value
γ: discount factor used in computing the returns	0.99
Optimizer for training θ, α and β	Stochastic gradient descent
SGD learning rate η	0.01
Entropy bonus scaling (See [23])	0.05

3.4   Results

In this section, we present our experimental results. Specifically, we consider three separate scenarios - (a) periodic load with raw actions, and the stochastic load with both (b) raw and (c) incremental actions. In each of the cases, we compare the performances of our A2C algorithm (after convergence) against the default policy (which we term the *baseline*). Recall that he baseline simply uses a tabular method to control fan speeds based on temperature thresholds.

(a) Scenario 1: Periodic Load with Raw Actions. We first experimented with the periodic I/O load shown in Fig. 5 (left). Figure 6 summarizes the results; it shows the I/O activity, normalized cumulative rewards[6], fan speeds and temperature values over time for both the baseline and our method. Compared to the baseline, the A2C algorithm provided a cumulative reward uplift of ∼33% for similar I/O activity! The higher reward was obtained primarily as a result of the A2C algorithm prescribing a higher fan speed when there was heavy I/O loading (roughly 16000 rpm versus 11000 rpm for the baseline), which resulted in

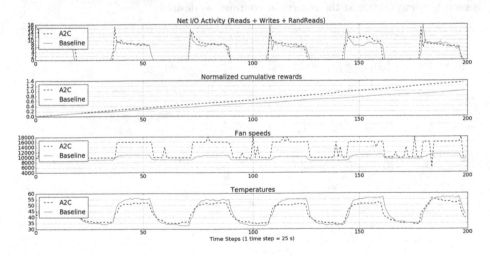

Fig. 6. Performance comparison of baseline and A2C for scenario 1.

---

[6] We normalize the cumulative baseline reward to 1.0, and correspondingly scale the cumulative A2C reward. This also helps quantify the reward uplift.

a lower temperature (52 °C versus 55 °C). To clarify this, Fig. 7 plots the contour regions of the temperature and fan speeds for the baseline (left) and the A2C algorithm, at convergence (right). The black-colored blobs essentially mark the operating points of the SSD rack for the two types of load. Evidently, the A2C method converges to the better reward contour as compared to the baseline.

(b) Scenario 2: Stochastic Load with Raw Actions. With the stochastic I/O load (see Fig. 8), the overall reward uplift obtained is smaller (only 12% after averaging over 3000 time steps) than the periodic load case. Again, the A2C algorithm benefits by increasing the fan speeds to keep the temperatures lower. Upon looking more closely at the convergence contours (Fig. 9), it is noted

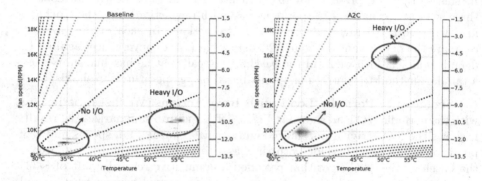

Fig. 7. For the no load scenario, the baseline policy settles to 36 °C and 9 K rpm while the A2C algorithm converges to 35 °C and 10 K rpm. With heavy I/O loading, the corresponding numbers are 55 °C and 11 K versus 52 °C and 16 K. The A2C algorithm is seen to always settle at the innermost contour, as desired.

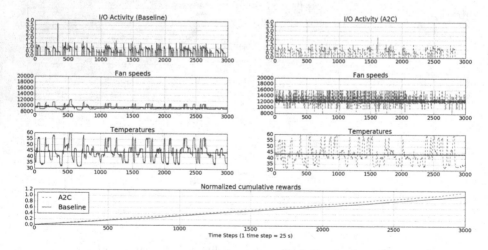

Fig. 8. Performance comparison of baseline and A2C for scenario 2. The mean values of temperatures and fan speeds are shown using the black line.

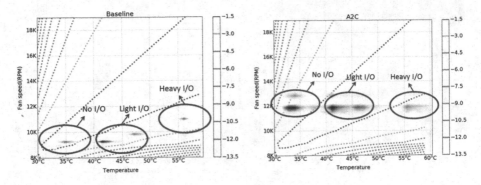

Fig. 9. Temperature and fan speed contours for the baseline (left) and the A2C method (right). With the stochastic loading with 7 actions, the A2C does not converge as well as in the periodic load case (see Fig. 9 (right)).

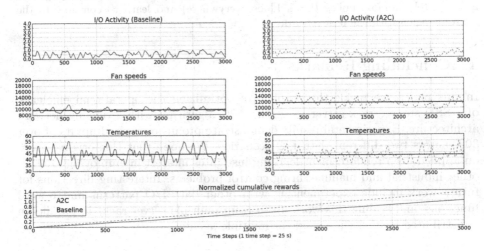

Fig. 10. Performance comparison of baseline and A2C for scenario 3.

that with the stochastic load, the A2C method sometimes settles to sub-optimal reward contours. We believe this happened due of insufficient exploration.

(c) Scenario 3: Stochastic Load with Incremental Actions. With raw actions, the action space is large to explore given the random nature of the I/O load, and this slows learning. To help the algorithm explore better, we study scenario 3. wherein actions can take on only 3 possible values (as compared to 7 values in the prior scenario). With this modification, more promising results are observed - specifically, we observed a cumulative reward uplift of 32% (see Fig. 10). In fact, the A2C algorithm is able to start from a completely random policy (Fig. 11 (left)) and learn to converge to the contour region with the best reward (Fig. 11 (right)).

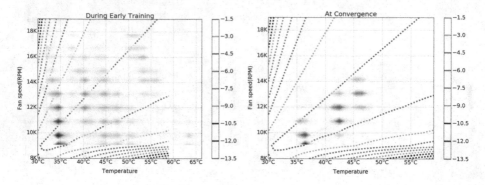

Fig. 11. Temperature and fan speed contours for the A2C method under scenario 3. The left plot is taken during early steps of training, while the plot on the right is taken at convergence. This illustrates that the A2C algorithm is able to start exploring from a completely random policy (black blobs everywhere) and learn to converge to the contour region with the best reward.

4 Concluding Remarks

In this paper, we tackle the problem of optimizing the operational efficiency of a SSD storage rack server using the A2C algorithm with a dueling network architecture. Experimental results demonstrate promising reward uplifts of over 30% across two different data I/O scenarios. We hope that this original work on applied deep reinforcement learning instigates interest in employing DRL to other industrial and manufacturing control problems. Interesting directions for future work include experimenting with other data I/O patterns and reward functions, and scaling this work up to train multiple server racks in parallel in a distributed fashion via a single or multiple agents.

Acknowledgements. We want to thank the Memory Systems lab team, Samsung Semiconductors Inc. for providing us a SSD storage rack, workload, data and fan control API for running our experiments. We also thank the software engineering team at Samsung SDS for developing a DRL framework [24] that was used extensively for model building, training and serving.

References

1. Power Usage Effectiveness. https://en.wikipedia.org/wiki/Power_usage_effective ness
2. Cisco Global Cloud Index: Forecast and Methodology, 2016–2021 White Paper, February 2018. https://www.cisco.com/c/en/us/solutions/collateral/service-prov ider/global-cloud-index-gci/white-paper-c11-738085.html
3. Arulkumaran, K., Deisenroth, M.P., Brundage, M., Bharath, A.A.: A brief survey of deep reinforcement learning. IEEE Sig. Process. Mag. **34**(6), 26–28 (2017)
4. Shuja, J., Madani, S.A., Bilal, K., Hayat, K., Khan, S.U., Sarwar, S.: Energy-efficient data centers. Computing **94**(12), 973–994 (2012)

5. Sun, J., Reddy, A.: Optimal control of building HVAC systems using complete simulation-based sequential quadratic programming (CSBSQP). Build. Environ. **40**(5), 657–669 (2005)
6. Ma, Z., Wang, S.: An optimal control strategy for complex building central chilled water systems for practical and real-time applications. Build. Environ. **44**(6), 1188–1198 (2009)
7. Evans, R., Gao, J.: DeepMind AI Reduces Google Data Centre Cooling Bill by 40%, July 2016. Blog: https://deepmind.com/blog/deepmind-ai-reduces-google-data-centre-cooling-bill-40/
8. Li, Y., Wen, Y., Guan, K., Tao, D.: Transforming Cooling Optimization for Green Data Center via Deep Reinforcement Learning (2017). https://arxiv.org/abs/1709.05077
9. Peters, J., Schaal, S.: Reinforcement learning of motor skills with policy gradients. Neural Netw. (2008 Spec. Issue) **21**(4), 682–697 (2008)
10. Watkins, C.J.C.H., Dayan, P.: Q-learning. Mach. Learn. **8**(3–4), 279–292 (1992)
11. Sutton, R.S., Barto, A.G.: Reinforcement Learning: An Introduction, 2nd edn. MIT Press, Cambridge (2017)
12. Sutton, R.S., McAllester, D., Singh, S., Mansour, Y.: Policy gradient methods for reinforcement learning with function approximation. In: Advances in Neural Information Processing Systems, vol. 12, pp. 1057–1063 (2000)
13. Williams, R.J.: Simple statistical gradient-following algorithms for connectionist reinforcement learning. Mach. Learn. **8**(3–4), 229–256 (1992)
14. Greensmith, E., Bartlett, P.L., Baxter, J.: Variance reduction techniques for gradient estimates in reinforcement learning. J. Mach. Learn. Res. **5**, 1471–1530 (2004)
15. Mnih, V., et al.: Human-level control through deep reinforcement learning. Nature **518**(7540), 529–533 (2015)
16. Mnih, V., et al.: Asynchronous methods for deep reinforcement learning. In: International Conference on Machine Learning, pp. 1928–1937 (2016)
17. Wang, Z., Schaul, T., Hessel, M., Hasselt, H.V., Lanctot, M., Freitas, N.D.: Dueling network architectures for deep reinforcement learning. In: International Conference on International Conference on Machine Learning, vol. 48 (2016)
18. Silver, D., Lever, G., Heess, N., Degris, T., Wierstra, D., Riedmiller, M.: Deterministic policy gradient algorithms. In: International Conference on Machine Learning (2014)
19. Lillicrap, T.P., et al.: Continuous control with deep reinforcement learning. US Patent Application, No. US20170024643A1. https://patents.google.com/patent/US20170024643A1/en
20. iostat Man Page. https://linux.die.net/man/1/iostat
21. ipmi-sensors Man Page. https://linux.die.net/man/8/ipmi-sensors
22. ipmitool Man Page. https://linux.die.net/man/1/ipmitool
23. O'Donoghue, B., Munos, R., Kavukcuoglu, K., Mnih, V.: Combining policy gradient and Q-learning. In: International Conference on Learning Representations (2017)
24. Parthasarathy, K., Kathalagiri, G., George, J.: Scalable implementation of machine learning algorithms for sequential decision making. In: Machine Learning Systems, ICML Workshop, June 2016

Automating Layout Synthesis
with Constructive Preference Elicitation

Luca Erculiani[1], Paolo Dragone[1,3], Stefano Teso[2(✉)], and Andrea Passerini[1]

[1] University of Trento, Trento, Italy
{luca.erculiani,paolo.dragone,andrea.passerini}@unitn.it
[2] KU Leuven, Leuven, Belgium
stefano.teso@cs.kuleuven.be
[3] TIM-SKIL, Trento, Italy

Abstract. Layout synthesis refers to the problem of arranging objects subject to design preferences and structural constraints. Applications include furniture arrangement, space partitioning (e.g. subdividing a house into rooms), urban planning, and other design tasks. Computer-aided support systems are essential tools for architects and designers to produce custom, functional layouts. Existing systems, however, do not learn the designer's preferences, and therefore fail to generalize across sessions or instances. We propose addressing layout synthesis by casting it as a *constructive preference elicitation* task. Our solution employs a coactive interaction protocol, whereby the system and the designer interact by mutually improving each other's proposals. The system iteratively recommends layouts to the user, and learns the user's preferences by observing her improvements to the recommendations. We apply our system to two design tasks, furniture arrangement and space partitioning, and report promising quantitative and qualitative results on both. Code related to this paper is available at: https://github.com/unitn-sml/constructive-layout-synthesis/tree/master/ecml18.

Keywords: Constructive learning · Preference elicitation
Layout synthesis · Furniture arrangement · Space partitioning

1 Introduction

Layout synthesis refers to the problem of arranging objects in accordance to design preferences and hard constraints. It encompasses tasks like arranging furniture within a room [8,31,32], planning entire floors [16], and designing block- or city-sized urban spaces [19]. The constraints are meant to encode functional and structural requirements, as well as any applicable human design guideline [3,18] (visibility, accessibility, *etc.*). In this paper we focus on the challenging problem of synthesizing layouts customized for a particular user.

PD is a fellow of TIM-SKIL Trento and is supported by a TIM scholarship.

© Springer Nature Switzerland AG 2019
U. Brefeld et al. (Eds.): ECML PKDD 2018, LNAI 11053, pp. 254–270, 2019.
https://doi.org/10.1007/978-3-030-10997-4_16

Striking the right balance between personal taste and hard requirements is notoriously difficult. For instance, an apartment may be furnished in a minimalist or industrial style depending on the owner's preferences. But when furnishing it, the tenants, who often have no knowledge of interior design principles, proceed intuitively, producing layouts that are not entirely functional or do not "look or feel right" [32]. Even experienced designers often rely on trial-and-error [14,32]. Understandably, the more complex the constraints (regulatory and design guidelines, engineering requirements), the harder it is for the user to produce a functional design that she actually enjoys. Layout synthesis tools assist both expert and amateur designers in this endeavor, enhancing productivity and outcome quality.

Now, consider a customer wishing to buy an apartment from a set of alternatives. To evaluate a candidate, she could use a layout synthesis tool to "fill in" the missing furniture according to her taste. Existing tools assist the user in this task [13,14]. However, they do not explicitly learn the user's preferences, and thus can not generalize across synthesis instances or design sessions. This implies that, in order to compare the alternative flats, the customer would have to furnish every one of them individually, a tedious and time consuming activity. In stark contrast, a tool capable of interactively learning her preferences could automatically furnish any flat based on the preferences estimated on the previous ones. To the best of our knowledge, this challenging learning problem has not been considered before.

Building on recent work on constructive learning [26,27], we propose a *constructive preference elicitation* approach to synthesizing custom layouts. Our method leverages Coactive Learning [24], a framework for interactive preference learning from manipulative interaction intended specifically for structured domains. As in previous work [13,14], our interaction protocol involves iteratively presenting high-scoring recommendations to the user, and then asking the latter to improve them. The user is free to perform small, large, or even multiple adjustments at each round. The algorithm automatically acquires a preference estimate from the observed improvements.

In our approach, layouts are represented as ensembles of components whose properties (position, size, *etc.*) determine the feasible space of candidates. The salient properties of the layouts are captured by features of its components (rooms, furniture) and their arrangement (e.g. the maximal distance between the bedrooms and the bathroom), whose weight is learned during the interaction process. In contrast to standard preference elicitation [20], where the set of alternatives can be easily searched over (e.g. looking up a movie in a catalogue), in layout synthesis the number of possible arrangements is large or infinite. For this reason, synthesizing a layout to be recommended to the user requires full-fledged mathematical optimization [16]. Borrowing from previous work [16,26], we cast this synthesis step as a constrained combinatorial optimization problem, which can be readily handled by off-the-shelf solvers.

We apply the proposed method to two tasks: arranging furniture in a room and planning the layout of an apartment. In the first case, the user can interactively adjust the values of features of interest (e.g. the average distance between tables in a cafè), while in the second the user modifies the layout directly by, for instance, moving or removing walls. We empirically evaluate our algorithm on instances of increasing complexity in terms of recommendation quality and computational cost, showing that it can effectively learn the user's preferences and recommend better layouts over time. We also apply it to larger synthesis instances and show how to tame the runtime of the synthesis step by employing approximation techniques. Our results show that, even in this case, the algorithm can reliably learn the user's preferences with only a minor loss in recommendation quality. Finally, we also positively evaluate the ability of the algorithm to deal with users with very different preferences and learn how to generate layouts that suit their taste.

2   Related Work

Synthesizing a custom layout requires to solve two problems: generating a layout consistent with the known requirements and preferences (*synthesis*), and biasing the synthesis process toward layouts preferred by the user (*customization*).

Broadly speaking, synthesis can be solved in two ways. One is to design a parameterized distribution over layouts (e.g. a probabilistic graphical model [31] or a probabilistic grammar [11]), whose structure encodes the set of validity constraints on objects and arrangements. Synthesis equates to sampling from the distribution via MCMC. A major downside of probabilistic approaches is that enforcing hard constraints (other than those implied by the structure of the distribution) may severely degrade the performance of the sampler, potentially compromising convergence, as discussed for instance in [27].

An alternative strategy, adopted by our method, is to define a scoring function that ranks candidate layouts based on the arrangement of their constituents. In this case, synthesis amounts to finding a high-scoring layout subject to design and feasibility constraints. This optimization problem may be solved using stochastic local search [1,32], mathematical optimization [14], or constraint programming [16]. We opted for the latter: constraint programming [22] allows to easily encode expressive local and global constraints, and is supported by many efficient off-the-shelf solvers. Further, in many cases it is easy to instruct the solver to look for (reasonably) sub-optimal solutions, allowing to trade-off solution quality for runtime, for enhanced scalability. This synergizes with our learning method, which is robust against approximations, both theoretically [23] and experimentally (see Sect. 4).

Many of the existing tools are concerned with synthesis only, and do not include a customization step: their main goal is to automate procedural generation of realistic-looking scenes or to produce concrete examples for simplifying requirement acquisition from customers [30]. Other approaches bias the underlying model (distribution or scoring function) toward "good" layouts by fitting it on sensibly furnished examples [13,31,32]. However, the generated configurations are not customized for the target user[1]. More generally, offline model estimation may be used in conjunction with our method to accelerate layout fine-tuning for the end user. We will explore this possibility in a future work.

Akase and colleagues proposed two interactive methods based on iterative evolutionary optimization [1,2]. Upon seeing a candidate furniture arrangement, the user can tweak the fitness function either directly using sliders [1] or indirectly via conjoint analysis [2]. In both works the number of customizable parameters is small and not directly related to the scene composition (e.g. illumination, furniture crowdedness). Contrary to these methods, we enable the user to graphically or physically manipulate a proposed layout to produce an improved one. This kind of interaction was successfully employed by a number of systems [13,14,28] using ideas from direct manipulation interfaces [10,25]. We stress that our method works even if user improvements are small, as shown by our empirical tests. The major difference to the work of Akase *et al.*, however, is that we leverage constraint programming rather than generic evolutionary optimization algorithms. This enables our method naturally handle arbitrary feasibility constraints on the synthesized layouts, extending its applicability to a variety of layout synthesis settings.

Among interactive methods, the one of Merrell *et al.* [13] is the closest to ours. Both methods rely on a scoring function, and both require the user and system to interact by suggesting modifications to each other. In contrast to our approach, the method of Merrel *et al.* does not learn the scoring function in response to the user suggestions, i.e., it will always suggest configurations in line with fixed design guidelines. Since no user model is learned, this method does not allow to transfer information across distinct design session.

3   Coactive Learning for Automated Layout Synthesis

In this section we frame custom layout synthesis as a *constructive preference elicitation* task. In constructive preference elicitation [7], the candidate objects are complex configurations composed of multiple components and subject to feasibility constraints. Choosing a recommendation amounts to synthesizing a novel configuration that suits the user's preferences and satisfies all the feasibility constraints. In this setting, learning can be cast as an interactive *structured-output prediction* problem [4]. The goal of the system is to learn a *utility* function u over objects y that mimics the user's preferences. The higher the utility, the better the object. In layout synthesis, the objects y may represent, for instance, the positions of all furniture pieces in a given room or the positions and shapes of all rooms on a given floor. The utility can optionally depend on extra contextual information x, e.g., the size and shape of the target apartment. More formally, the

---

[1] While the examples may be provided by the end user, it is unreasonable to expect the latter to manually select the large number of examples required for fine-grained model estimation. Through interaction, our system allows a more direct control over the end result.

```
1 Procedure PreferencePerceptron(T)
2 │   Initialize w₁
3 │   for t = 1,...,T do
4 │   │   Receive user context xₜ
5 │   │   yₜ ← argmax_{y∈𝒴} ⟨wₜ, φ(xₜ, y)⟩
6 │   │   Receive improvement ȳₜ from the user
7 │   │   wₜ₊₁ ← wₜ + φ(xₜ, ȳₜ) − φ(xₜ, yₜ)
8 │   end
9 end
```

Algorithm 1. The Preference Perceptron from Coactive Learning [24].

utility is a function $u : \mathcal{X} \times \mathcal{Y} \to \mathbb{R}$, where $\mathcal{X}$ and $\mathcal{Y}$ are the sets of all the contexts and objects respectively. Typically, the utility is assumed to be a linear model of the type: $u(x, y) = \langle w, \phi(x, y) \rangle$. Here the *feature function* $\phi : \mathcal{X} \times \mathcal{Y} \to \mathbb{R}^d$ maps an object to a vector of d features summarizing its high-level properties. For instance, in furniture arrangement the features might capture the maximum distance of the furniture pieces from the walls or the minimum distance between each other. The weights $w \in \mathbb{R}^d$ associated to the features represent the user preferences and are estimated by interacting with the user.

In our approach, the learning step is based on Coactive Learning [24], whereby the system iteratively proposes recommendations y to the user and the user returns an improvement $\bar{y}$, i.e. a new object with a (even slightly) higher utility than y. Using this information, the algorithm updates the current estimate of the parameters w in order to make better recommendations in the future.

Perhaps the simplest, yet effective, Coactive Learning algorithm is the Preference Perceptron [24], listed in Algorithm 1. The algorithm iteratively elicits the user preferences by making recommendations and getting user feedback in response. Across the iterations $t \in [1, T]$, the algorithm keeps an estimate of the parameters w_t which are continuously updated as new feedback is observed. At each iteration t, the algorithm first receives the context x_t and then produces a new object y_t by maximizing the current utility estimate $y_t = \text{argmax}_{y \in \mathcal{Y}} u_t(x_t, y) = \langle w_t, \phi(x_t, y) \rangle$ (line 5). The object y_t is recommended to the user, who then provides an improvement $\bar{y}_t$. Lastly, the algorithm obtains a new parameter vector w_{t+1} using a simple perceptron update (line 7).

Depending on the type of objects $y \in \mathcal{Y}$, inference (line 5) may be solved in different ways. In layout synthesis, the objects $y \in \mathcal{Y}$ are sets of variable assignments representing the components of the layouts. For instance, in furniture arrangement y may contain the position of each piece of furniture, its size, type, color, and so on. The space of potential configurations (i.e. possible assignments of attributes) is combinatorial or even infinite (for continuous variables) in the number of attributes. The space $\mathcal{Y}$ is also typically restricted by feasibility constraints, enforcing functional and structural requirements, e.g. non-overlap between furniture pieces. In this setting, the inference problem can be formalized in the general case as a mixed integer program (MIP). In practice, to make inference computationally feasible, we often restrict it to the mixed integer *linear* program (MILP) case by imposing that constraints and features be linear in *y*. Exploring in which cases more general constrained optimization problems (e.g. mixed integer quadratic programs) are computationally feasible is an interesting future direction (as outlined in the Conclusion). While being NP-hard in the general case, MILP problems with hundreds of variables can be quickly solved to optimality by existing off-the-shelf solvers. For problems that are too complex for exact solution strategies, approximation techniques can be used to speed-up the inference process. Reasonably sub-optimal synthesized layouts do not significantly alter the performance of Coactive Learning, as proven theoretically in [21] and shown empirically by our experiments.

The Coactive Learning paradigm can be seen as a cooperation between the learning system and the user to mutually improve each other's design proposals. This is particularly appealing for devising a layout synthesis system, since coactive feedback may be acquired through visual or physical manipulation of the recommended objects (see e.g. [13,14,28]). Such a system could be integrated into graphical design applications and used by architects and designers to automatically improve their products. Depending on the type and complexity of the design task, a visual layout recommendation system may also be used by an amateur designer.

Fig. 1. Illustration of feature- and object-level improvements. Left: recommended configuration. Right: corresponding user-provided improvement. The two top images illustrate a *feature-level* improvement where the user increased the minimum distance between tables (a feature) directly, affecting all tables. The bottom images show an *object-level* improvement where the user modified the object itself by adding a new wall. Best viewed in color.

4   Experiments

In this section we evaluate the proposed system on two layout synthesis applications. The first is a furniture arrangement problem in which the system recommends arrangements of tables in a room, for furnishing, e.g., bars versus offices. The second is a space partitioning problem in which the system suggests how to partition the area of an apartment into rooms. In our empirical analysis, we [i] perform a quantitative evaluation of the system's ability to learn the user preferences, with an emphasis on the trade-off between exact and approximate synthesis; and [ii] perform a qualitative evaluation by illustrating the layouts recommended at different stages of the learning process.

In our experiments we simulate a user's behavior according to standard feedback models [24]. Namely, we assume that the user judges objects according to her own true utility[2] $u^*(x, y) = \langle w^*, \phi(x, y) \rangle$, and measure the quality of recommendations by their *regret*. The regret at iteration t is simply the difference between the true utility of the best possible object and that of the actual recommendation y_t, that is, $\text{REG}(x_t, y_t) = (\max_{y \in \mathcal{Y}} u^*(x_t, y)) - u^*(x_t, y_t)$. Since varying contexts makes comparing regrets at different iterations impossible [24], in the experiments we measure the *average regret* $\frac{1}{T} \sum_{t=1}^{T} \text{REG}(x_t, y_t)$.

We observe that, very intuitively, good recommendations can't be improved by much, while bad recommendations offer ample room for improvement. For this reason, we assume the user's feedback to follow the α-informative feedback model [24], which states that the true utility of $\bar{y}_t$ is larger than the true utility of y_t by some fraction $\alpha \in (0, 1]$ of the regret $\text{REG}(x_t, y_t)$, that is $u^*(x_t, \bar{y}_t) - u^*(x_t, y_t) \geq \alpha \text{REG}(x_t, y_t)$. Here α represents the user "expertise". Higher values of α imply better improvements and fewer iterations for learning to suggest good (or optimal) recommendations. Indeed, if α-informative holds the average regret of the Preference Perceptron decreases at a rate of $\mathcal{O}(1/(\alpha \sqrt{T}))$ (see [24]).

Crucially, the α-informative model is very general, since it is always possible to find an α, no matter how small, that satisfies the feedback model[3]. In our experiments we assume that no user expertise is required to interact usefully with our system, and thus set $\alpha = 0.1$ to simulate a non-expert user. Furthermore, we assume that changes made by the user are small, in order to keep her effort to a minimum. We thus take a conservative approach and simulate the user behavior by selecting a "minimal" α-informative improvement (more details below).

For both settings, in the quantitative experiment we compare the average regret of the system on instances of increasing complexity. As the complexity increases, approximate solutions become necessary for guaranteeing real-time interaction. We perform approximate synthesis by setting a time cut-off on the solver and choosing the best solution found in that time. Regret bounds similar to the aforementioned one can be found also in approximate settings [21]. We

---

[2] Note that the learner can observe the user's feedback, but has no access to the true utility itself.

[3] Assuming the user makes no mistakes, i.e., $u^*(x_t, \bar{y}_t) > u^*(x_t, y_t) \forall t \in [T]$. This model can be easily extended to allow user mistakes (see [24]).

evaluate empirically the effect of using approximate inference on the quality of the recommendations in both settings. For the quantitative experiments we simulated 20 users with randomly generated true parameter vectors w^* and plotted the median performance.

We consider two types of user improvements, as exemplified in Fig. 1. In the first experiment we use a feature-based improvement, in which a user may variate the value of a feature (e.g. with a simple set of UI controllers) to generate a better configuration. In the top example of Fig. 1, the user sets the minimum distance between the tables to a higher value. The second type of improvement considered is an object-based improvement, in which the user directly shapes the configuration by adding, moving or removing parts. This is the case of the bottom example of Fig. 1, in which the user adds a wall to create a new room. The details of the user feedback simulation models are reported in the corresponding subsections. In both experimental settings, we also report a qualitative evaluation showcasing the behavior of the system in interacting with some "prototypical" type of user (e.g. a cafè owner arranging the tables in her place). We show that the system achieves the goal of finding good configurations matching the user taste.

The system is implemented in Python[4] and uses MiniZinc to model the constrained optimization problems [17], and an external MILP solver[5] for the inference and the improvement problems. All the experiments were run on a 2.8 GHz Intel Xeon CPU with 8 cores and 32 GiB of RAM.

4.1 Furniture Arrangement

In the first experimental setting, the goal of the system is to learn to arrange tables in a room according to the user preferences. Rooms are 2D spaces of different shapes. We model the rooms as squared bounding boxes, plus several inaccessible areas making up internal walls. The size of the bounding box and the inaccessible areas are given in the context x, together with the number of tables to place. The available space is discretized into unit squares of fixed size. Tables are rectangles of different shape occupying one or more unit squares. The objects y consist in the table arrangements in the given room. More precisely, tables are represented by their bottom-left coordinates (h, v) in the bounding box and their sizes (dh, dv) in horizontal and vertical directions. The object y contains the coordinates (h_t, v_t) and the sizes (dh_t, dv_t) of each table t. Several constraints are imposed to define the feasible configurations. Tables are constrained to fit all in the bounding box, to not overlap, and to not be positioned in unfeasible areas. Tables must keep a minimum "walking" distance between each other. Doors are also placed on the room walls (in the context) and tables are required to keep a minimum distance from the doors.

In our experiment the total size of the bounding box is 12×12. Tables are either 1×1 squares (occupying one unit square) or 1×2 rectangles (occupying

---

[4] See https://github.com/unitn-sml/constructive-layout-synthesis for the complete implementation.

[5] Opturion CPX: http://opturion.com.

Table 1. Summary of the structure of the objects in the two experimental settings. In the furniture arrangement task, *Tables* is the set of tables, $bbdist(t)$ is the distance of table t from the bounding box, $wdist(t)$ is the distance of table t from the inaccessible areas (walls), $dist(t_1, t_2)$ is distance between tables t_1 and t_2. In the floor planning setting, instead, *Types* is the set of room types, *Rooms* is the set of rooms, R_t the set of rooms of type t, A_r the area of room r (number of unit squares), dh_r and dv_r the horizontal and vertical size of room r, $type(r)$ the type of room r, $adj(r, s)$ is a boolean function denoting the adjacency between rooms r and s, $sdist(r)$ is the distance between r and the south edge of the bounding box. All distances considered here are Manhattan distances.

Furniture arrangement	
Context x	- Size of bounding box - Position of doors - Inaccessible areas - Number of tables
Object y	- Position (h, v) of all tables - Sizes (dh, dv) of all tables

Features $\phi(x, y)$	- Max and min distance of tables from bounding box:
	  $\max_{t \in Tables} bbdist(t)$; $\min_{t \in Tables} bbdist(t)$
	- Max and min distance of tables from inaccessible areas:
	  $\max_{t \in Tables} wdist(t)$; $\min_{t \in Tables} wdist(t)$
	- Max and min distance between tables:
	  $\max_{t_1, t_2 \in Tables} dist(t_1, t_2)$; $\min_{t_1, t_2 \in Tables} dist(t_1, t_2)$
	- Number of tables per type (1×1 and 1×2):
	  $|\{t \in Tables \mid dh_t + dv_t \leq 2\}|$; $|\{t \in Tables \mid dh_t + dv_t \geq 3\}|$

Floor planning	
Context x	- Size of bounding box - Position of entrance door - Inaccessible areas - Max and min rooms per type
Object y	- Position (h, v) of all rooms - Type t_r of each room r - Sizes (dh, dv) of all rooms

Features $\phi(x, y)$	- Ranges of occupied space (percent) per room type:
	  $\forall t \in Types \ \sum_{r \in R_t} A_r \leq 15\%$
	  $\forall t \in Types \ 15\% < \sum_{r \in R_t} A_r \leq 30\%$
	  $\forall t \in Types \ \sum_{r \in R_t} A_r > 30\%$
	- Upper bound D_r of difference of sides for each room r:
	  $\forall r \in Rooms \ D_r$ s.t. $|dh_r - dv_r| \leq D_r$
	- Number of rooms per type: $\forall t \in Types \ |R_t|$
	- Room with entrance door r_{door} is of type t:
	  $\forall t \in Types \ t = type(r_{\text{door}})$
	- Sum of pairwise difference of room areas per type:
	  $\forall t \in Types \ \sum_{i, j \in R_t} |A_i - A_j|$
	- Number of rooms adjacent to corridors:
	  $|\{r \in Rooms \mid \exists s \in R_{\text{corridor}} \ adj(r, s)\}|$
	- Distance of each room from South (bottom edge):
	  $\forall r \in Rooms \ sdist(r)$

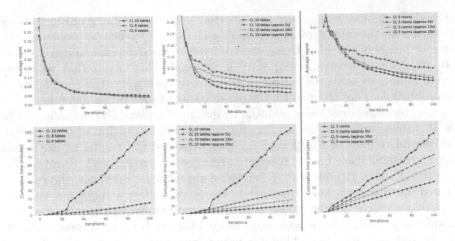

Fig. 2. Median average regret (top) and median cumulative time (bottom) in three settings. Left to right: furniture arrangement with exact inference on 6, 8 and 10 tables; comparison between exact and approximate inference in furniture arrangement with 10 tables; exact versus approximate inference on floor planning. Best viewed in color.

two unit squares). Room shapes were selected randomly at each iteration from a pool of five candidates.

The feature vector $\phi(x, y)$ is composed of several numeric properties of the configuration, such as the maximum and minimum distance between tables, the maximum and minimum distance between tables and walls, and the number of tables per type (1×1 and 1×2). The upper part of Table 1 contains a detailed summary of the structure of x, y and $\phi(x, y)$ in this setting.

As mentioned, in this setting we employ a feature-based improvement scheme to simulate the user behavior. In particular, the following constrained problem is solved to generate an α-informative improvement $\bar{y}_t$:

$$\bar{y}_t = \operatorname*{argmin}_{y \in \mathcal{Y}} \|\phi(x_t, y) - \phi(x_t, y_t)\|_0$$
$$\text{s.t.} \quad u^*(x_t, \bar{y}_t) - u^*(x_t, y_t) \geq \alpha(u^*(x_t, y_t^*) - u^*(x_t, y_t))$$

where $\|\phi(x_t, y) - \phi(x_t, y_t)\|_0$ is the ℓ_0 norm of the difference between the feature vectors, i.e. the number of different features between $\phi(x_t, y)$ and $\phi(x_t, y_t)$. This is in line with the assumption made on the minimal user effort.

In the quantitative evaluation we run the recommendation algorithm for an increasing number of tables to be placed. A high number of tables makes the inference problem more complex, as it involves more optimization variables and constraints. We test the algorithm on problems with 6, 8 and 10 tables. We compare the average regret and the running time of the system in each of these scenarios. Figure 2 shows the median results (over all users) on settings with different number of tables. The plots show the median average regret (top) and the median cumulative inference time (bottom). The first column of Fig. 2 shows the results for the table arrangement task with exact inference on problems with different numbers of tables. Using exact inference, the difference in regret decay between different levels of complexity is minimal. This means that when the system is able to solve the inference problem to optimality, the complexity of the problem does not affect much the performance of the system. Inference time, however, increases drastically with the increasing complexity. Exact inference in the 10 tables setting is already largely impractical for an interactive system. The second column of Fig. 2 shows a comparison of the results of exact and approximate inference on the furniture arrangement setting with 10 tables, for time cut-offs at 5, 10 and 20 s[6]. When using approximate inference, the running times drop to a much lower rate, while the regret suffers a slight increase but keeps decreasing at a similar pace as the exact variant. We can see that the time cut-off can be modulated to achieve the desired balance between recommendation quality and inference time. This is a promising behaviour suggesting that the method can scale with the problem size with predictable running time without compromising performance.

In order to get a visual grasp of the quality of the recommended solutions, we also evaluated our system on two prototypical arrangement problems, namely a user interested in furnishing a café and one willing to furnish an office. Cafés are usually furnished with small tables (1×1), positioned along the walls in a regular fashion. Offices, instead, contain mostly desks (1×2) positioned along the walls or in rows/columns across the room. We sampled two users according to the above criteria. Figure 3 showcases the recommendations of the system at different stages of the learning procedure. Initially, tables are randomly spread across the room. Gradually, the system learns to position tables in a more meaningful way. In the case of the café, the intermediate image shows that the algorithm has learned that a café should have mostly 1×1 tables and that they should be placed along the walls. For the office, the intermediate figure shows that the algorithm has roughly figured out the position of tables, but not their correct type. At the end of the elicitation, the final configurations match the user desiderata.

4.2   Floor Planning

Our second experimental setting is on floor planning, that is recommending partitionings of apartments into separate rooms. The outer shape of the apartment is provided by the context, while the user and the system cooperate on the placement of the inner walls defining the room boundaries. As in the previous setting, the space is discretized into unit squares. Each room is a rectangle described by four variables: (h, v) indicate its position, (dh, dv) its size. Coordinates and sizes are measured in unit squares. Rooms must fit in the apartment and must not overlap. Rooms can be of one among five types, namely kitchen, living room, bedroom, bathroom and corridor. In the context, the user can also specify an

---

[6] The time cut-off is on the solver time, but the actual inference time has some more computational overhead, taking on average 2.21 s more.

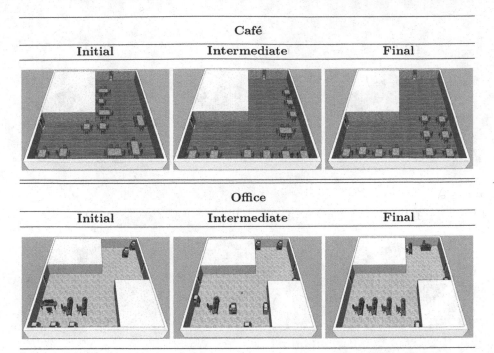

Fig. 3. Two use cases of our system. The images are 3D renderings of configurations recommended by our system when interacting with users whose goal is to furnish a café (top) and an office (bottom). Horizontally, the figures show different stages of the elicitation process. In the café, 1×1 and 1×2 tables are seen as dining tables of different sizes, whereas in the office 1×2 tables represent desks while 1×1 tables contain utilities such as printers. Best viewed in colors.

upper and lower bound on the number of rooms of each type. For instance, a user may look for an apartment with exactly one kitchen, one living room, and between one and two bathrooms and bedrooms. After placing all the rooms, the spaces left in the apartment are considered corridors. The context also specifies the position of the entrance door to the apartment.

In this experiment we consider a 10×10 bounding box. We define five different apartment shapes and generate random contexts with any combination of room types, summing to a maximum of five rooms, with random lower bounds.

While the number of generated rooms is variable, we impose a maximum. The dimension of feature vector $\phi(x, y)$ depends on the maximum number of rooms, hence it must be fixed in advance and cannot change throughout the iterations. Features are normalized in order to generalize different contexts and different numbers of rooms. The features include: [i] the percentage of space occupied by the rooms of each type, discretized in several ranges of values, each denoting a certain target size for each room type; [ii] an upper-bound on the difference between the sides dh_r and dv_r of each room r, which is used to modulate how

Flat

Initial	Intermediate	Final

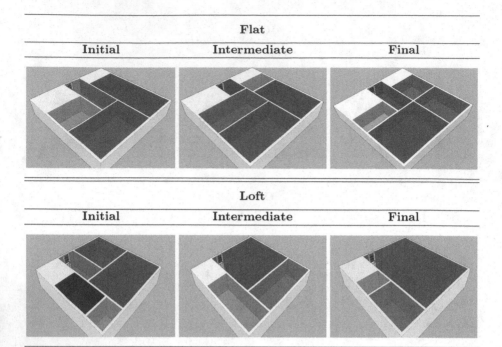

Loft

Initial	Intermediate	Final

Fig. 4. Two use cases of our system for the task of floor planning. The images are 3D renderings of configurations recommended by our system when interacting with users whose goal is to build a flat (top) and an loft (bottom). Horizontally, the figures show different stages of the elicitation process. Room colors are associated to room types: the kitchen is in red, the living room is in blue, the bathroom is in turquoise, the bedroom in green, the corridor is in violet. Best viewed in colors.

"squared" the room should be; [iii] the actual number of rooms per type; [iv] a boolean value for each room type indicating whether the entrance door is in a room of that type; [v] the sum of the pairwise difference between the areas of rooms of the same type, to modulate how similar in size rooms of a certain type should be; [vi] the number of rooms that are adjacent to corridors; [vii] the distance of each room from the south border of the apartment, as living rooms are usually made to look south and bedrooms look north for lighting purposes. A summary of all the features of this setting is listed in the lower part of Table 1.

Differently from the previous setting, here we employ an object-based improvement schema. We simulate the user feedback by solving the following optimization problem:

$$\bar{y}_t = \operatorname*{argmin}_{y \in \mathcal{Y}} \|U_y - U_{y_t}\|_0$$

$$\text{s.t.} \quad u^*(x_t, \bar{y}_t) \geq u^*(x_t, y_t) + \alpha(u^*(x_t, y_t^*) - u^*(x_t, y_t))$$

where U_y is the matrix 10×10 containing the room types per unit square. We assume a user to perform a change that involves as least rooms as possible and we simulate this behavior by minimizing the number of unit squares affected by the improvement. This is done to simulate a minimal effort for the user.

In this case, the problem complexity is mainly given by the maximum number of rooms to be placed in the apartment. Notice that this problem is more difficult than the previous one, as it has more optimization variables, more features and it has to learn from a more diverse set of possible contexts. We evaluate this setting only on a scenario with a maximum of five rooms. As in the previous experiment, we report a comparison of the results of exact inference and approximate inference. We again run approximate inference with time cut-offs at 5, 10, and 20 s. The last column of Fig. 2 shows the median average regret and the median cumulative inference time in this setting. Both regret and times follow the same trend as the ones in the previous experiment. Approximate inference allows for substantial computational savings[7] at the cost of a small reduction in recommendation quality.

In the qualitative experiment we compare two users who are interested in different kinds of apartments. In the first case, the user is interested in a "traditional" apartment (here dubbed "flat" to avoid ambiguities), which contains a corridor from the entrance door to the rooms, two separate rooms for the kitchen and living room, with the former slightly smaller that the latter, a bedroom and a bathroom. The second user is interested in a loft, which is composed by fewer rooms, usually a big living room with a kitchenette, a bedroom and a bathroom. In Fig. 4 we can see different stages of the learning process for both users. At the beginning the recommended configurations are random. The system then is able to learn that a flat should have a corridor as entrance and a smaller bathroom, and that a loft should have only a moderately large living room plus a bedroom and a bathroom of approximately equal size. Finally the system reaches good recommendations that meet the preferences of the users: an apartment with a kitchen smaller than the living room and a corridor connecting rooms, and a loft with a big living room, a bedroom and a small bathroom.

5   Conclusion

We presented an approach to layout synthesis suitable for suggesting layouts customized for a particular user. We cast layout synthesis as a constructive preference elicitation problem, where the set of potential arrangements is determined by hard constraints encoding the functional and structural requirements. Contrary to previous solutions, our approach learns the user's preferences and thus generalizes across synthesis instances and design sessions. Our interactive learning strategy pairs a very natural interactive protocol based on direct manipulation with a simple but principled learning framework for manipulative feedback [24]. We applied our system to two design tasks, namely furniture arrangement and floor planning, and evaluated it on instances of increasing complexity.

---

[7] Exact inference becomes impractical for more that five rooms.

The results show that our approach can reliably learn the user's preferences even when synthesis is (moderately) sub-optimal, for improved scalability, at the cost of a minor degradation of recommendation quality. We showcased the flexibility of our system by learning from users with radically different preferences, e.g., users that prefer lofts to highly partitioned apartments and vice-versa.

This work can be extended in several directions. First, it makes sense to bias the learner toward "known working" layouts, as done in [13,31,32], to accelerate convergence towards promising candidates and reduce the amount of user intervention. Second, although our presentation focuses on linear constraints and features (which yield a MILP synthesis problem) our learning procedure is not restricted to this setup. As solver technology for mixed integer quadratic problems matures[8], it becomes progressively more viable to employ non-linear terms to model even richer layout properties, such as surface areas, Euclidean distances, and variances. The increased computational requirements could be handled with appropriate quality-runtime trade-offs, as done in our experiments. Third, decomposition strategies like those presented in [6] offer a promising direction for aggressively reducing the cognitive and computational costs of layout synthesis. This is especially fitting for inherently modular layouts such as buildings, which can be decomposed into progressively simpler parts (floors, rooms, *etc.*). This would also facilitate the introduction of more computationally demanding features, as hinted to above, by restricting inference to portions of layouts. We are actively investigating these research directions. Finally, the proposed method can in principle be employed to automate tasks other than layout synthesis, like environmental [12] and chemical engineering [29] and synthetic biology [9].

Acknowledgments. This work has received funding from the European Research Council (ERC) under the European Unions Horizon 2020 research and innovation programme (grant agreement No. [694980] SYNTH: Synthesising Inductive Data Models).

References

1. Akase, R., Okada, Y.: Automatic 3D furniture layout based on interactive evolutionary computation. In: CISIS, pp. 726–731. IEEE (2013)
2. Akase, R., Okada, Y.: Web-based multiuser 3D room layout system using interactive evolutionary computation with conjoint analysis. In: VINCI, p. 178. ACM (2014)
3. Alexander, C.: A Pattern Language: Towns, Buildings, Construction. Oxford University Press, Oxford (1977)
4. Bakir, G.H., Hofmann, T., Schölkopf, B., Smola, A.J., Taskar, B., Vishwanathan, S.V.N.: Predicting Structured Data. MIT Press, Cambridge (2007)
5. Boyd, S., Kim, S.J., Vandenberghe, L., Hassibi, A.: A tutorial on geometric programming. Optim. Eng. **8**(1), 67 (2007)

---

[8] Especially for expressive but restricted fragments such as mixed integer second order cone programming [15,33] and mixed integer geometric programming [5].

6. Dragone, P., Teso, S., Kumar, M., Passerini, A.: Decomposition strategies for constructive preference elicitation. In: AAAI (2018)
7. Dragone, P., Teso, S., Passerini, A.: Constructive preference elicitation. Front. Robot. AI **4**, 71 (2017)
8. Fisher, M., Ritchie, D., Savva, M., Funkhouser, T., Hanrahan, P.: Example-based synthesis of 3D object arrangements. ACM TOG **31**(6), 135 (2012)
9. Galdzicki, M., et al.: The Synthetic Biology Open Language (SBOL) provides a community standard for communicating designs in synthetic biology. Nature Biotechnol. **32**(6), 545 (2014)
10. Harada, M., Witkin, A., Baraff, D.: Interactive physically-based manipulation of discrete/continuous models. In: SIGGRAPH, pp. 199–208. ACM (1995)
11. Liu, T., Chaudhuri, S., Kim, V.G., Huang, Q., Mitra, N.J., Funkhouser, T.: Creating consistent scene graphs using a probabilistic grammar. ACM Trans. Graph. (TOG) **33**(6), 211 (2014)
12. Masters, G.M., Ela, W.P.: Introduction to Environmental Engineering and Science, vol. 3. Prentice Hall, Upper Saddle River (1991)
13. Merrell, P., Schkufza, E., Li, Z., Agrawala, M., Koltun, V.: Interactive furniture layout using interior design guidelines. ACM Trans. Graph. (TOG) **30**, 87 (2011)
14. Michalek, J., Papalambros, P.: Interactive design optimization of architectural layouts. Eng. Optim. **34**(5), 485–501 (2002)
15. Misener, R., Smadbeck, J.B., Floudas, C.A.: Dynamically generated cutting planes for mixed-integer quadratically constrained quadratic programs and their incorporation into GloMIQO 2. Optim. Methods Softw. **30**(1), 215–249 (2015)
16. Mitchell, W.J., Steadman, J.P., Liggett, R.S.: Synthesis and optimization of small rectangular floor plans. Environ. Plan. B: Plan. Des. **3**(1), 37–70 (1976)
17. Nethercote, N., Stuckey, P.J., Becket, R., Brand, S., Duck, G.J., Tack, G.: MiniZinc: towards a standard CP modelling language. In: Bessière, C. (ed.) CP 2007. LNCS, vol. 4741, pp. 529–543. Springer, Heidelberg (2007). https://doi.org/10.1007/978-3-540-74970-7_38
18. Panero, J., Zelnik, M.: Human Dimension and Interior Space: A Source Book of Design Reference Standards. Watson-Guptill, New York (1979)
19. Parish, Y.I., Müller, P.: Procedural modeling of cities. In: Proceedings of the 28th Annual Conference on Computer Graphics and Interactive Techniques, pp. 301–308. ACM (2001)
20. Pigozzi, G., Tsoukiàs, A., Viappiani, P.: Preferences in artificial intelligence. Ann. Math. Artif. Intell. **77**(3–4), 361–401 (2016)
21. Raman, K., Shivaswamy, P., Joachims, T.: Online learning to diversify from implicit feedback. In: Proceedings of the 18th ACM SIGKDD International Conference on Knowledge Discovery and Data Mining, pp. 705–713. ACM (2012)
22. Rossi, F., Van Beek, P., Walsh, T.: Handbook of Constraint Programming. Elsevier, Amsterdam (2006)
23. Shivaswamy, P., Joachims, T.: Online structured prediction via coactive learning. In: ICML, pp. 1431–1438 (2012)
24. Shivaswamy, P., Joachims, T.: Coactive learning. JAIR **53**, 1–40 (2015)
25. Sutherland, I.E.: Sketchpad a man-machine graphical communication system. Trans. Soc. Comput. Simul. **2**(5), R–3 (1964)
26. Teso, S., Passerini, A., Viappiani, P.: Constructive preference elicitation by setwise max-margin learning. In: IJCAI, pp. 2067–2073 (2016)
27. Teso, S., Sebastiani, R., Passerini, A.: Structured learning modulo theories. Artif. Intell. **244**, 166–187 (2015)

28. Tidd, W.F., Rinderle, J.R., Witkin, A.: Design refinement via interactive manipulation of design parameters and behaviors. In: ASME DTM (1992)
29. Turton, R., Bailie, R.C., Whiting, W.B., Shaeiwitz, J.A.: Analysis, Synthesis and Design of Chemical Processes. Pearson Education, London (2008)
30. Xu, W., Wang, B., Yan, D.M.: Wall grid structure for interior scene synthesis. Comput. Graph. **46**, 231–243 (2015)
31. Yeh, Y.T., Yang, L., Watson, M., Goodman, N.D., Hanrahan, P.: Synthesizing open worlds with constraints using locally annealed reversible jump MCMC. ACM TOG **31**(4), 56 (2012)
32. Yu, L.F., et al.: Make it home: automatic optimization of furniture arrangement. SIGGRAPH **30**(4) (2011)
33. Zhao, Y., Liu, S.: Global optimization algorithm for mixed integer quadratically constrained quadratic program. J. Comput. Appl. Math. **319**, 159–169 (2017)

Configuration of Industrial Automation Solutions Using Multi-relational Recommender Systems

Marcel Hildebrandt[1,2(✉)], Swathi Shyam Sunder[1,3], Serghei Mogoreanu[1], Ingo Thon[1], Volker Tresp[1,2], and Thomas Runkler[1,3]

[1] Siemens AG, Corporate Technology, Munich, Germany
{marcel.hildebrandt,swathi.sunder,serghei.mogoreanu,ingo.thon,
volker.tresp,thomas.runkler}@siemens.com
[2] Ludwig Maximilian University, Munich, Germany
[3] Technical University of Munich, Munich, Germany

Abstract. Building complex automation solutions, common to process industries and building automation, requires the selection of components early on in the engineering process. Typically, recommender systems guide the user in the selection of appropriate components and, in doing so, take into account various levels of context information. Many popular shopping basket recommender systems are based on collaborative filtering. While generating personalized recommendations, these methods rely solely on observed user behavior and are usually context free. Moreover, their limited expressiveness makes them less valuable when used for setting up complex engineering solutions. Product configurators based on deterministic, handcrafted rules may better tackle these use cases. However, besides being rather static and inflexible, such systems are laborious to develop and require domain expertise. In this work, we study various approaches to generate recommendations when building complex engineering solutions. Our aim is to exploit statistical patterns in the data that contain a lot of predictive power and are considerably more flexible than strict, deterministic rules. To achieve this, we propose a generic recommendation method for complex, industrial solutions that incorporates both past user behavior and semantic information in a joint knowledge base. This results in a graph-structured, multi-relational data description – commonly referred to as a knowledge graph. In this setting, predicting user preference towards an item corresponds to predicting an edge in this graph. Despite its simplicity concerning data preparation and maintenance, our recommender system proves to be powerful, as shown in extensive experiments with real-world data where our model outperforms several state-of-the-art methods. Furthermore, once our model is trained, recommending new items can be performed efficiently. This ensures that our method can operate in real time when assisting users in configuring new solutions.

M. Hildebrandt and S. S. Sunder—Contributed equally to this work.

U. Brefeld et al. (Eds.): ECML PKDD 2018, LNAI 11053, pp. 271–287, 2019.
https://doi.org/10.1007/978-3-030-10997-4_17

Keywords: Recommender system · Cold start · Knowledge graph
Link prediction · Tensor factorization

1   Introduction

Industrial automation systems consist of a wide variety of components – in general, a combination of mechanical, hydraulic, and electric devices – described in a plan. For instance, a plan for a control cabinet might specify the model of the controller, the number of digital input ports, the number of sensors to be connected, and so on. Based on the plan, the user incrementally selects products which in combination fulfill all required functionalities. The goal of our solution is to support the user in this process by reordering product lists, such that products which most likely correspond to the user needs and preferences are on top.

A good ordering of the products may depend on the properties of the components, such as the line voltage common in the country (110 V in US, 230 V in Europe), marine certification, or whether or not the customer has a preference towards premium or budget products. While some of this information is explicitly modeled as product features – exploited by means of a knowledge graph in our work – a big part of it is implicit information. In our system, the current partial solution, which consists of the already selected components, acts as a source of information about the actual requirements of the user.

The main advantage of our proposed approach, for both the end customer and the sales team, is the reduction of time required to select the right product. Moreover, the system guides the customer towards a typical, and therefore most likely useful, combination of products. In addition, the learned information allows vendors to optimize the portfolio as the learned knowledge about item combinations makes implicit interdependencies transparent.

The problem of finding the right products to fulfill the customer needs can be addressed from two directions. The first direction is towards general recommender systems that typically find products based on customer preferences. However, in our domain, the customer preferences are not fixed, but rather depend on implicit requirements of the solution currently being built. The second direction focuses on configuration systems which find a combination of products that fulfill all requirements specified in the plan. The practical use of the latter, however, is subject to the precise specification of the requirements. This is rarely the case, especially in industry-based applications such as the one we are dealing with. The use case addressed in this paper concerns an internal R&D project at Siemens. One of the sources of data for this project is the Siemens product database,[1] which presents a diversity-rich environment – offering services related to 135,000 products and systems with more than 30 million variants. Such massive amounts of data make it increasingly difficult to specify all the requirements explicitly and directly add to the computational overhead of the

---

[1] https://mall.industry.siemens.com.

recommender system. Nevertheless, it is crucial to deliver instantaneous results without compromising quality in industrial use cases like ours.

Since our method is intended for real-world usage, it needs to satisfy the following constraints:

- Reasonable effort to set up the system in terms of data preparation, maintenance, and integration of new data. The last point is particularly important when new products are being launched.
- Computing recommendations efficiently is crucial because the system must work in real time when the user is configuring new solutions.
- The method must be able to make recommendations for items without or with only little historical customer data.

In this paper, we present RESCOM. The key idea is to employ a modified version of RESCAL [16] for the recommendation task. We not only show that RESCOM outperforms all baseline methods but also demonstrate that we achieved all of the goals stated above.

This paper is organized into six sections. Section 2 introduces the notation and reviews some of the basic principles of knowledge graphs. In Sect. 3, we proceed by investigating methods that are traditionally used for recommendation tasks and outline the challenges arising from the real-world nature of our data. Furthermore, we review methods that are commonly employed for the link prediction task on knowledge graphs. In Sect. 4, we present RESCOM, a generic recommendation method for complex, industrial solutions, that incorporates both past user behavior and semantic information in a joint knowledge base. The results of our extensive real-world experiments are presented in Sect. 5. In particular, we show that RESCOM outperforms several state-of-the-art methods in standard measures. Finally, Sect. 6 concludes our work by discussing the results and proposing some directions for future research.

2   Notation and Background

Before proceeding, we first define the mathematical notation that we use throughout this work and provide the necessary background on knowledge graphs.

Scalars are given by lower case letters ($x \in \mathbb{R}$), column vectors by bold lower case letters ($\mathbf{x} \in \mathbb{R}^n$), and matrices by upper case letters ($X \in \mathbb{R}^{n_1 \times n_2}$). Then $X_{i,:} \in \mathbb{R}^{n_2}$ and $X_{:,j} \in \mathbb{R}^{n_1}$ denote the i-th row and j-th column of X, respectively. Third-order tensors are given by bold upper case letters ($\mathbf{X} \in \mathbb{R}^{n_1 \times n_2 \times n_3}$). Further, slices of a tensor (i.e., two-dimensional sections obtained by fixing one index) are denoted by $\mathbf{X}_{i,:,:} \in \mathbb{R}^{n_2 \times n_3}$, $\mathbf{X}_{:,j,:} \in \mathbb{R}^{n_1 \times n_3}$, and $\mathbf{X}_{:,:,k} \in \mathbb{R}^{n_1 \times n_2}$, respectively.

In Sect. 4, we propose a recommender system that is based on constructing a knowledge base that contains not only historical data about configured solutions, but also descriptive features of the items available in the marketplace. This leads to a graph-structured, multi-relational data description. Loosely speaking, such kind of data is commonly referred to as a knowledge graph. The basic idea is to represent all entities under consideration as vertices in a graph and link those vertices that interact with each other via typed, directed edges.

More formally, let $\mathcal{E} = \{e_1, e_2, \ldots, e_{n_E}\}$ and $\mathcal{R} = \{r_1, r_2, \ldots, r_{n_R}\}$ denote the set of entities and the set of relations under consideration, respectively. In our setting, entities correspond to either solutions, items, or technical attributes of items. Relations specify the interconnectedness of entities. Pertaining to our use case, the *contains* relation is of particular interest. It links solutions and items by indicating which items were configured in each solution. The remaining relations connect items to their technical attributes. In this work, a knowledge graph is defined as a collection of triples $S \subset \mathcal{E} \times \mathcal{R} \times \mathcal{E}$ which are interpreted as known facts. Each member of S is of the form (e_i, r_k, e_j), where e_i is commonly referred to as subject, r_k as predicate, and e_j as object. Figure 1 depicts an example of a knowledge graph in the context of an industrial purchasing system and the kinds of entities and relations we consider in this paper.

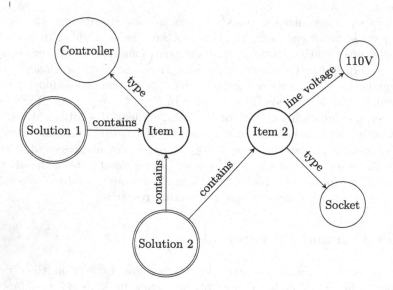

Fig. 1. A knowledge graph in the context of an industrial purchasing system. The nodes correspond either to solutions, items, or technical properties of items. The relationships between the different entities are determined by edges that come in multiple types.

A knowledge graph has a natural representation in terms of an adjacency tensor $\mathbf{X} \in \mathbb{R}^{n_E \times n_E \times n_R}$ with entries

$$\mathbf{X}_{i,j,k} = \begin{cases} 1, & \text{if the triple } (e_i, r_k, e_j) \in S, \\ 0, & \text{otherwise.} \end{cases} \tag{1}$$

Usually, a positive entry in the adjacency tensor is interpreted as a known fact. Under the so-called closed-world assumption, the absence of a triple indicates a false relationship. Alternatively, one can make the open world assumption which states that a zero entry is not interpreted as false but rather as unknown. The approach that we present in Sect. 4 is based on the local closed-world assumption. Thereby, a knowledge graph is assumed to be only locally complete in the sense that $(e_i, r_k, e_j) \notin S$ is assumed to be false if and only if there exists an entity $e_l \in \mathcal{E}$ such that $(e_i, r_k, e_l) \in S$. Otherwise, the triple is interpreted as unknown.

3   Related Methods

Here we cover both well-established general recommender systems and knowledge graph models. Given that knowledge graphs offer a natural representation for semantically structured information (see Fig. 1), they are closely related to the domain under consideration.

3.1   General Recommender Systems

Traditional approaches to recommender systems are based on collaborative filtering, where the problem is formulated as that of matrix completion pertaining to the user-item matrix. The winning solution to the Netflix challenge [2] and Amazon's item-to-item collaborative filtering method [12] are two prominent examples. A general class of collaborative filtering methods is based on factorizing the user-item matrix $X \in \mathbb{R}^{n_U \times n_I}$, where n_U and n_I correspond to the number of user and items, respectively. Further, the entries of X are only well-defined on an index set $\mathcal{I}$ that correspond to observed user-item interactions. The underlying optimization problem consists of factorizing X into the product of two low-dimensional matrices, i.e.,

$$\min_{U,V} \sum_{(i,j) \in \mathcal{I}} (X_{i,j} - U_{i,:} V_{j,:}^T)^2 + \lambda(||U||_F^2 + ||V||_F^2), \qquad (2)$$

where $|| \cdot ||_F$ denotes the Frobenius norm. Moreover, the factor matrices $U \in \mathbb{R}^{n_U \times d}$ and $V \in \mathbb{R}^{n_I \times d}$, where d denotes the number of latent factors, represent the strength of the associations between the latent features of users and items. So the product UV^T essentially performs a sum of the features weighted by the corresponding coefficients. Imposing elementwise non-negativity constraints on the factor matrices U and V in Eq. (2) leads to the non-negative matrix factorization (NMF). Similar to most factorization-based approaches, NMF is most effective when there is a high availability of user behavior information.

However, sparsity, which stems from the fact that users typically will have rated only few of the items, constitutes a challenge to collaborative filtering methods. Further, collaborative filtering recommender systems are also prone to the cold start issue, i.e., the problem of dealing with new users or novel items, while suffering from a lack of preferences or content information. This is caused by the context information not being taken into account in collaborative filtering approaches.

Although context-aware recommender systems (CARS) adapt recommendations to the specific contextual situations of users, they fail to address the issue with user/product cold start. Additionally, as described in [1], most of the existing work in this area require all of the context information to be known a priori and to be modeled explicitly. This is in contrast to our use case of industry automation, where a majority of the context information is rather implicit.

While content-based filtering does consider context, recommendations are primarily generated based on a comparison between the content of items and the user profile. This causes only those items to be recommended that are similar to the items already rated by the user, resulting in a lack of diversity and novelty. While discussing the shortcomings of content-based systems, [13] also states that this problem of over-specialization limits the range of applications where the method could be useful.

Contrary to pure approaches, hybrid recommender systems are based on combining multiple models in one of the several possible ways described in [6]. While weighted hybrid recommenders (e.g., [7] and [18]) manually assign weights to the collaborative and content-based recommendation components, switching hybrid recommender systems such as [3] choose among the different recommendation components based on some criteria. Further, there also exist feature-based techniques where a set of features is first computed and subsequently fed into the recommendation algorithm. Although hybrid methods can provide more accurate recommendations than any method on its own, they are still required to strike the right balance between collaborative approaches and knowledge-based methods. The strength of our method lies in its ability to automatically learn this balance without heavy reliance on either deterministic rules or user behavior.

Together with the multitude of practical applications trying to help users cope with information overload, the increased development of new approaches to recommender systems – including this work – may also be attributed to tackling the existing challenges discussed in this section.

3.2   Knowledge Graph Methods

Most knowledge graphs that are currently in practical use are far from being complete in the sense that they are missing many true facts about the entities at hand. Therefore, one of the most important machine learning tasks related to knowledge graphs consists of predicting new links (i.e., facts) given the remaining knowledge graph. This problem is sometimes also referred to as knowledge graph completion. In our setting, the recommendation task is equivalent to predicting new links of a certain type based on the other items that are currently configured.

Many link prediction methods belong to the class of latent feature models. These methods approach the knowledge graph completion task by modeling the score of a candidate triple as a function of learned latent representations of the involved entities and relations. The method that we propose in this paper is based on RESCAL [16]. RESCAL constitutes a collective tensor factorization model that scores triples based on a set of bilinear forms. A more detailed description of RESCAL along with our modifications to make it applicable in our setting is presented in the next section.

Another popular class of methods against which we compare our approach is called translational methods. Here, the key idea is to find embeddings for the entities in a low-dimensional vector space and model their relatedness through translations via relation-specific vectors. TransE, introduced by [4], was the first among this class of methods. It aims at learning representations such that $\mathbf{e}_i + \mathbf{r}_k \approx \mathbf{e}_j$ if $(e_i, r_k, e_j) \in S$, where the bold letters refer to the vector space embeddings of the corresponding entities and relations.

After realizing that TransE is quite limited when dealing with multi-cardinality relations, several other methods that extend the idea of TransE and add more expressiveness have been developed: TransH [20], TransR [11], and TransD [9]. All these methods learn linear mappings specific to the different relations and then map the embeddings of entities into relation-specific vector spaces where they again apply TransE.

One of the ways to apply this translation-based approach to recommender systems was introduced in [8] where personalized recommendations are generated by capturing long-term dynamics of users. However, the focus is geared more towards sequential prediction, i.e., predicting the next item for a user, knowing the item previously configured in the solution. While it does not consider possible interdependencies among all different items previously configured in the solution, it also does not deal with as much semantic information of the items, as this work.

4 Our Method

In analogy to existing collaborative filtering approaches which are based on matrix factorizations, RESCOM relies on exploiting the sparsity pattern of $\mathbf{X}$ by finding a low-rank approximation via tensor factorization.

Our method, RESCOM, is based on a modified version of RESCAL [16] which is a three-way-tensor factorization that has shown excellent results in various relational learning tasks (e.g., in [17]). The key idea is to approximate the adjacency tensor $\mathbf{X}$ by a bilinear product of the factor matrix $E \in \mathbb{R}^{d \times n_E}$ and a core tensor $\mathbf{R} \in \mathbb{R}^{d \times d \times n_R}$, where d corresponds to the number of latent dimensions. More concretely, we impose

$$\mathbf{X}_{:,:,r} \approx E^T \mathbf{R}_{:,:,r} E \quad , \forall r = 1, 2, \ldots, n_R. \tag{3}$$

Thus, after fitting the model, the columns of E denoted by $(\mathbf{e}_i)_{i=1,2,\ldots,n_E} \subset \mathbb{R}^d$ contain latent representations of the entities in $\mathcal{E}$. Similarly, each frontal slice of $\mathbf{R}$ contains the corresponding latent representations of the different relations in $\mathcal{R}$. These embeddings preserve local proximities of entities in the sense that, if a large proportion of the neighborhood of two entities overlaps, their latent representations are also similar. Hence, we obtain that, if items are similar from a technical point of view or if they are often configured within the same solution, they will have similar latent representations.

In its original form, the parameters of RESCAL are obtained by minimizing the squared distance between the observed and the predicted entries of $\mathbf{X}$. Hence, the objective function is given by

$$\text{loss}(\mathbf{X}, E, \mathbf{R}) = \sum_{r=1}^{n_R} ||\mathbf{X}_{:,:,r} - E^T \mathbf{R}_{:,:,r} E||_F^2. \tag{4}$$

While logistic extensions of RESCAL that model the entries of $\mathbf{X}$ explicitly as Bernoulli random variables were proposed (see [15]), we stick to the formulation given by Eq. (4). This is mainly due to the fact that when using the squared error loss, recommending new items for partial solutions can be performed efficiently via an orthogonal projection into the latent feature space (see below for more details). We also ran experiments with DistMult [21], which can be obtained as a special case of RESCAL when the frontal slices of the core tensor $\mathbf{R}$ are restricted to be diagonal. However, since this did not lead to a better performance, we do not report the results in this work.

Usually Eq. (4) is minimized via alternating least squares (ALS). However, the sparsity in our data (only about $6 \cdot 10^{-6}\%$ of the entries of $\mathbf{X}$ are non-zero) caused numerical instability when computing the least-squares projections for ALS. In particular, we tested two different implementations of RESCAL[2] that are based on ALS and found that they were both infeasible in our setting. Therefore, we implemented our own version of RESCAL that aims to minimize a modified loss function. In order to obtain an approximate version of Eq. (4) that does not require to take all the entries of $\mathbf{X}$ into account, we sample negative examples from the set of unknown triples $S' \subset \mathcal{E} \times \mathcal{R} \times \mathcal{E} \setminus S$. More specifically, we employ the following loss function during training

$$\text{loss}(\mathbf{X}, E, \mathbf{R}) = \sum_{(e_i, r_k, e_j) \in S} (1 - \mathbf{e}_i^T \mathbf{R}_{:,:,k} \mathbf{e}_j)^2 + (\mathbf{e}_i^T \mathbf{R}_{:,:,k} \tilde{\mathbf{e}})^2, \tag{5}$$

where $\tilde{\mathbf{e}}$ corresponds to the latent representation of a randomly sampled entity $\tilde{e}$ such that $(e_i, r_k, \tilde{e}) \in S'$. Further, we impose the additional constraint that $\tilde{e}$ appears as object in a known triple with respect to the k-th relation (i.e., there exists an entity e so that $(e, r_k, \tilde{e}) \in S$). This sampling procedure can be interpreted as an implicit type constraint which teaches the algorithm to discriminate between known triples on one side and unknown but plausible (i.e., semantically correct) triples on the other side. Conceptually, our sampling mechanism is related to the local closed-world assumption proposed in [10], where the goal is to infer the domain and range of relations based on observed triples. While [10] proposes a type-constrained ALS procedure, our training process is based on generating negative samples and is carried out by minimizing Eq. (5) using different versions of stochastic gradient descent.

In the context of industrial automation, generating recommendations is most relevant right at the time when new solutions are being configured, thus making

---

[2] https://github.com/mnick/rescal.py   and   https://github.com/nzhiltsov/Ext-RESCAL.

it necessary to tune the previously trained model to make predictions on the new data. While being time consuming, re-training the model on the freshly available partial solutions is possible. However, the need for immediate real-time updates makes model re-building an infeasible option for the recommendation task. Hence we propose to achieve this through the *projection* step. This is a special case of the more general embedding mapping for tensor factorization models proposed in [22]. To ease the notation, let $\mathcal{E}_I \subset \mathcal{E}$ with $|\mathcal{E}_I| =: n_{E_I}$ denote the collection of entities that correspond to configurable items.

We start by constructing a binary vector $\mathbf{x} \in \mathbb{R}^{n_{E_I}}$, where a value of 1 at the i-th position indicates that the corresponding item was part of the partial solution. For the purpose of recommendations, we only consider the *contains* relation which represents the information linking solutions to the items configured in them. In order to ease the notation, let r_1 denote the *contains* relation.

Then, we perform an orthogonal projection of this partial solution into latent space guided by the following equation

$$\mathcal{P}\left(\mathbf{x}\right) = \left(\mathbf{R}_{:,:,1} E_I E_I^T \mathbf{R}_{:,:,1}^T\right)^{-1} \mathbf{R}_{:,:,1} E_I \mathbf{x}, \tag{6}$$

where $E_I \in \mathbb{R}^{d \times n_{E_I}}$ contains the latent representations of all items. Thus, we have that a completion of $\mathbf{x}$ is given by

$$\hat{\mathbf{x}} = \mathcal{P}\left(\mathbf{x}\right)^T \mathbf{R}_{:,:,1} E_I. \tag{7}$$

The values of the entries of $\hat{\mathbf{x}}$ can be interpreted as scores indicating whether or not the corresponding items are likely to be configured in the particular solution. With this interpretation, the items may be reordered in decreasing order of their scores and the ranking can then be used for recommendations.

Note that the matrix representation of the linear map $\mathcal{P}(\cdot)$ can be precomputed and stored for later use. Thus the computational complexity of the whole *projection* step is given by $\mathcal{O}(dN_{E_I})$. To the best of our knowledge, there is no equivalent, simple procedure for the class of translational models. Hence, they require re-training when a novel solution is set up by the user. This constitutes a major limitation in their applicability to our use case.

5   Real-World Experimental Study

5.1   Data

The experiments have been conducted on real-world data collected from Siemens internal projects. It can be roughly divided into two categories:

1. *Historical solutions*: This data contains (anonymized) information about automation solutions that have been previously configured. Only considering this part of the data (as it is done by some of the traditional recommender systems) would result in recommending items that have been configured together most frequently in the past.

2. *Descriptive features of the items*: These features may either specify the type of an item (e.g., panel, controller) or a certain technical specification (e.g., line voltage, size). An example of an entry in this part of the dataset would be a Siemens SIMATIC S7-1500 controller (with a unique identifier 6ES7515-2FM01-0AB0) being fit for fail-safe applications.

Some pre-processing steps have been applied before conducting our experiments. These include:

1. Removing solutions that contain only one item.
2. Removing duplicate descriptive features that were named in different languages.
3. Removing descriptive features that have unreasonably long strings as possible values. In most cases, these are artifacts of an automatic documentation processing procedure for legacy items.
4. Removing items with no known descriptive features or descriptive features with just one possible value.

After applying these pre-processing steps, we obtained 426,710 facts about 25,473 solutions containing 3,003 different items that share 3,981 different technical features among themselves. Figure 2 illustrates the three-way adjacency tensor obtained from the data: both rows and columns correspond to the entities (i.e., solutions, items, as well as all possible values of the descriptive features of the items), while each slice corresponds to one of the relations (with the first one being the *contains* relation).

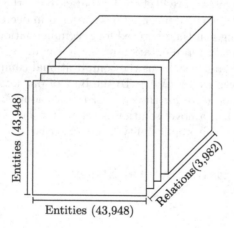

Fig. 2. The three-way adjacency tensor obtained after data pre-processing

5.2   Implementation

We compared RESCOM to state-of-the-art methods, namely TransE, TransR, TransD, and TransH, as well as NMF, considering its successful application in popular recommender systems. For experimental consistency, we re-implemented these methods within the same framework as our method, using Python and TensorFlow. All experiments were conducted on a Linux machine and the models were trained/tested on NVIDIA Tesla K80 GPUs. Training our model takes approximately 60 min for 50 epochs (∼100 batches) on one GPU.

To further facilitate a fair evaluation, we generated the training (70%), validation (20%) and test (10%) sets once and used this fixed split in all experiments. In each case, the best set of hyperparameters were chosen based on the highest mean reciprocal rank on the validation set. The results are then reported for the chosen parameters on the test set.

Having tried optimizers such as stochastic gradient descent, Adam, and AdaGrad, we found that the Adam optimizer resulted in the best convergence in our case. The dimensionality of the embeddings was tuned from the range $\{10, 15, 20, 30, 40, 50, 60, 70, 80, 90, 100\}$ and the regularization parameter was chosen among $\{10^{-1}, 10^{-2}, 10^{-3}, 10^{-4}, 10^{-5}, 10^{-6}\}$. We also investigated the influence of the number of negative triples per positive training sample by choosing among $\{1, 2, 3, 4, 5, 10, 20\}$. All experiments were executed for a maximum of 500 epochs, with a check for early stopping every 50 iterations.

5.3   Evaluation Scheme

The particular metrics used for the evaluation of recommender systems depend greatly on the characteristics of the data set and the ultimate purpose of recommendation. In the current setting, the goal is to evaluate the predictive accuracy, i.e., how well the recommender system can predict the likelihood of occurrence of items in partial solutions, in the context of the already configured items.

For evaluation of the various translational models, we followed the ranking procedure proposed by [5]. For each test triple (e_i, r_k, e_j), the object is removed and in turn replaced by each of the entities e from the set of entities $\mathcal{E}$. Dissimilarities of all those resulting triples (also referred to as corrupted triples) are first computed by the models and then sorted in ascending order; the rank of the correct entity is finally stored. Unlike [5], we do not repeat the procedure by replacing the subject entity. This is in accordance with the recommendation setting, where we are interested in obtaining a ranking for the items that only appear as object nodes in the *contains* relation.

To evaluate non-negative matrix factorization (NMF) and our own method, we employ a common setting where we construct an adjacency matrix of solutions and configured items with missing entries. We then perform an inverse transformation (in the case of NMF) or an orthogonal projection (for RESCOM - see Eqs. (6) and (7)) to obtain a completed matrix. The values in each row of this matrix are then interpreted as scores deciding whether or not the corresponding items are present in the particular solution, with a higher value implying a greater chance of the item being present. Next, the items in each row are reordered in decreasing order of their scores, yielding the ranking.

We report the mean of those predicted ranks, the mean reciprocal rank (MRR), and the hits@10%, i.e., the proportion of correct entities ranked in the top 10%. We consider the MRR owing to its robustness towards outliers, which is not the case with the mean rank. Further, a trend analysis indicating that users typically look at the top 10% of the displayed results before opting for a new search, forms the basis for choosing hits@10% for evaluation (see Sect. 5.4 for more details).

As noted by [4], these metrics can be flawed when some corrupted triples are in fact valid ones (i.e., are present in the training, validation, or test set), causing them to be ranked higher than the test triple. This however should not be counted as an error because both triples are actually true. To avoid such misleading behavior, we discard all those triples in the ranking process that appear either in the training, validation, or test set (except the test triple of interest). This ensures that none of the corrupted triples belong to the dataset and grants a clearer view of the ranking performance. We refer to this as the *filtered (filt.)* setting and the original (possibly flawed) one as the *raw* setting.

In Sect. 5.4, results are reported according to both raw and filtered settings.

Evaluation Mechanism for Cold Start. We adapt the evaluation mechanism to specifically suit the scenario of the cold start problem. Experiments are run on the best hyperparameters chosen earlier, without the need for cross-validation. We introduce a set of new items into the dataset. In doing so, we only add technical features for these items, without including them in any solutions. Additionally, we ensure that items belonging to the same category as these new items are in fact available. This should allow the model to pick up features common among similar items. Thereafter, we carry out the same workflow as before for model training and testing and report the metrics on the test set.

5.4   Results

Table 1 displays the results of the recommendation task for all methods under consideration. The filtered setting produces lower mean rank and higher values of MRR and hits@10% for all the methods, as expected. As described in Sect. 5.3, this setting offers a better evaluation of the performance. RESCOM outperforms all methods in most of the standard metrics. However, NMF has the highest value of MRR in the filtering setting, even better than that of RESCOM. This may be explained by considering that MRR favours correct entities being retrieved exactly in a few cases, against correct entities being in the top-k ranks consistently. To verify if this affects our use case involving the industrial purchasing system, we further analyzed the number of items that users are willing to consider until they arrive at the desired product. This was done by looking at the filters applied by users at the time of solution configuration and noting the positions of selected items among the results. Trends show that users typically look at the top 10% of the displayed results before opting for a new search and since RESCOM still has the best hits@10%, we consider it to be superior to all the counterparts.

Table 1. Results of all evaluated methods on the test set

Metric	Mean rank		Mean reciprocal rank		Hits@10%	
Eval. setting	Raw	Filt.	Raw	Filt.	Raw	Filt.
TransR	898.42	894.58	0.02	0.02	0.29	0.32
TransE	322.40	318.58	0.09	0.10	0.70	0.72
TransD	332.62	328.82	0.09	0.12	0.68	0.71
TransH	316.04	312.22	0.09	0.10	0.69	0.72
NMF	182.38	177.84	0.12	**0.22**	0.82	0.87
RESCOM	**81.32**	**76.76**	**0.13**	0.18	**0.93**	**0.95**

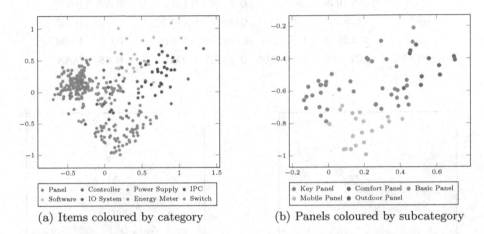

(a) Items coloured by category   (b) Panels coloured by subcategory

Fig. 3. Item embeddings obtained using RESCOM after PCA (with 33.42% variance explained) (Color figure online)

Furthermore, Fig. 3 depicts a scatter plot of the item embeddings. After training a model using RESCOM (with 15 latent dimensions), the resulting embeddings were further subjected to dimensionality reduction via principal component analysis (PCA) to arrive at the 2-dimensional embeddings shown. In Fig. 3(a), the items are coloured by *category*, which is one of the available technical features. As indicated by a few dense regions resembling clusters, it may be seen that the model learns the associations among the items for most categories. Figure 3(b) provides a more detailed view of the items in the Panel category. The items are coloured by subcategory and the values correspond to the subcategories within the Panel category. This demonstrates that our model is capable of learning the subspace embeddings as well, although we do not encode this hierarchical structure explicitly.

Cold Start: Model Evaluation Results. As explained in Sect. 5.3, we perform an explicit evaluation of the methods for the cold start scenario. Since it is not possible to test for previously unseen data in the case of NMF, we exclude it from consideration. The results are as shown in Table 2. We can clearly see that RESCOM performs better than all of the other methods, proving that it is capable of recommending even those items that were not previously configured.

Table 2. Cold start: Results of all evaluated methods on the test set

Metric	Mean rank		Mean reciprocal rank		Hits@10%	
Eval. setting	Raw	Filt.	Raw	Filt.	Raw	Filt.
TransR	766.7216	766.1959	0.0044	0.0036	0.284	0.289
TransE	486.7938	482.2165	0.0983	0.1073	0.655	0.686
TransD	533.8299	529.3041	0.0982	0.1066	0.649	0.655
TransH	473.2113	468.7423	0.0993	0.1081	0.65	0.658
RESCOM	**127.5792**	**122.3405**	**0.1423**	**0.1801**	**0.884**	**0.887**

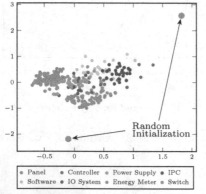

(a) Pre-trained embeddings for old items and 2 new items randomly initialized

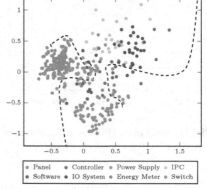

(b) After training the new items for 50 epochs

Fig. 4. Cold start: Item embeddings obtained using RESCOM after PCA (with 33.42% variance explained)

Further, we conducted an experiment aimed at testing the ability of the model to learn the embeddings for newly added items, while using the previously trained embeddings for old items. We added two new items – one belonging to the Panel category and another belonging to the Switch category. The model was then trained while keeping the embeddings of the previously trained items fixed. Figure 4(a) shows the two new randomly initialized items while the previous embeddings remain the same as in Fig. 3(a). The new items were then trained in the usual way and we observed their embeddings during the training. After each epoch, we tracked these items to see if they moved closer to the clusters formed by the other items in the same categories. Figure 4(b) illustrates the embeddings obtained after training for 50 epochs. As evident from the trajectories shown, the model learns the associations among items of the same category quite well.

6   Conclusion

We proposed RESCOM, a multi-relational recommender system, and applied it in the context of an industrial purchasing system for setting up engineering solutions. In this domain, the necessity of expressive recommendation methods arises because the functionality of each solution is highly dependent on the interplay of its components and their technical properties. RESCOM aims to take this complexity into account by embedding all past solutions, items, and technical properties into the same vector space and modelling their relatedness via a set of bilinear forms. We conducted extensive experiments based on real-world data from the Siemens product database. To sum up, the main findings of our real-world study are:

- RESCOM significantly outperforms all the considered baseline methods in most of the common performance measures.
- RESCOM offers a natural remedy to the cold start problem. Even in the absence of any historical data for particular items, our method is able to produce reasonable recommendations.
- RESCOM requires minimal effort in terms of data preparation and maintenance. Moreover, training the model and computing recommendations can be performed efficiently. Apart from sorting the items, we have shown that recommending items boils down to a sequence of matrix multiplications. This ensures that our method can operate in real time when guiding the user to set up new solutions.

So far, we have considered semantic annotations of the configurable items, but ignored contextual information about the solutions such as their area of applications or temporal aspects. We will explore this direction in future work when the relevant data becomes available. Other possibilities for future research include extending this approach from recommendations to pairwise preferences that can be represented in preference graphs [19] and incorporating semantic knowledge using reasoning mechanisms [14].

Finally, we would like to stress that this work evolved from a real-world industrial R&D project at Siemens and our method is now an integral part of one of the major solution configurators at Siemens.

Acknowledgements. We would like to thank Siemens Digital Factory Division for providing the data and helping us to get a better understanding of the application area, our colleague Martin Ringsquandl for the insightful discussions and helpful remarks on early drafts, as well as the head of our research group, Steffen Lamparter, for providing all the necessary resources.

References

1. Adomavicius, G., Tuzhilin, A.: Context-aware recommender systems. In: Ricci, F., Rokach, L., Shapira, B. (eds.) Recommender Systems Handbook, pp. 191–226. Springer, Boston (2015). https://doi.org/10.1007/978-1-4899-7637-6_6
2. Bell, R.M., Koren, Y., Volinsky, C.: The Bellkor 2008 solution to the Netflix prize. Stat. Res. Dept. AT&T Res. **1** (2008)
3. Billsus, D., Pazzani, M.J.: User modeling for adaptive news access. User Model. User-Adap. Interact. **10**(2–3), 147–180 (2000)
4. Bordes, A., Usunier, N., Garcia-Duran, A., Weston, J., Yakhnenko, O.: Translating embeddings for modeling multi-relational data. In: Advances in Neural Information Processing Systems, pp. 2787–2795 (2013)
5. Bordes, A., Weston, J., Collobert, R., Bengio, Y., et al.: Learning structured embeddings of knowledge bases. In: AAAI, vol. 6, p. 6 (2011)
6. Burke, R.: Hybrid recommender systems: survey and experiments. User Model. User-Adap. Interact. **12**(4), 331–370 (2002)
7. Claypool, M., Gokhale, A., Miranda, T., Murnikov, P., Netes, D., Sartin, M.: Combining content-based and collaborative filters in an online newspaper (1999)
8. He, R., Kang, W.C., McAuley, J.: Translation-based recommendation. In: Proceedings of the 11th ACM Conference on Recommender Systems, pp. 161–169. ACM (2017)
9. Ji, G., He, S., Xu, L., Liu, K., Zhao, J.: Knowledge graph embedding via dynamic mapping matrix. In: Proceedings of the 53rd Annual Meeting of the Association for Computational Linguistics and the 7th International Joint Conference on Natural Language Processing (Volume 1: Long Papers), vol. 1, pp. 687–696 (2015)
10. Krompaß, D., Baier, S., Tresp, V.: Type-constrained representation learning in knowledge graphs. In: Arenas, M., et al. (eds.) ISWC 2015. LNCS, vol. 9366, pp. 640–655. Springer, Cham (2015). https://doi.org/10.1007/978-3-319-25007-6_37
11. Lin, Y., Liu, Z., Sun, M., Liu, Y., Zhu, X.: Learning entity and relation embeddings for knowledge graph completion. In: AAAI, vol. 15, pp. 2181–2187 (2015)
12. Linden, G., Smith, B., York, J.: Amazon.com recommendations: item-to-item collaborative filtering. IEEE Internet Comput. **7**(1), 76–80 (2003)
13. Lops, P., de Gemmis, M., Semeraro, G.: Content-based recommender systems: state of the art and trends. In: Ricci, F., Rokach, L., Shapira, B., Kantor, P.B. (eds.) Recommender Systems Handbook, pp. 73–105. Springer, Boston, MA (2011). https://doi.org/10.1007/978-0-387-85820-3_3
14. Mehdi, G., Brandt, S., Roshchin, M., Runkler, T.: Towards semantic reasoning in knowledge management systems. In: Mercier-Laurent, E., Boulanger, D. (eds.) AI4KM 2016. IAICT, vol. 518, pp. 132–146. Springer, Cham (2018). https://doi.org/10.1007/978-3-319-92928-6_9
15. Nickel, M., Tresp, V.: Logistic tensor factorization for multi-relational data. arXiv preprint arXiv:1306.2084 (2013)
16. Nickel, M., Tresp, V., Kriegel, H.P.: A three-way model for collective learning on multi-relational data. In: ICML, vol. 11, pp. 809–816 (2011)
17. Nickel, M., Tresp, V., Kriegel, H.P.: Factorizing YAGO: scalable machine learning for linked data. In: Proceedings of the 21st International Conference on World Wide Web, pp. 271–280. ACM (2012)
18. Pazzani, M.J.: A framework for collaborative, content-based and demographic filtering. Artif. Intell. Rev. **13**(5–6), 393–408 (1999)

19. Runkler, T.A.: Mapping utilities to transitive preferences. In: Medina, J., et al. (eds.) IPMU 2018. CCIS, vol. 853, pp. 127–139. Springer, Cham (2018). https://doi.org/10.1007/978-3-319-91473-2_11
20. Wang, Z., Zhang, J., Feng, J., Chen, Z.: Knowledge graph embedding by translating on hyperplanes. In: AAAI, vol. 14, pp. 1112–1119 (2014)
21. Yang, B., Yih, W., He, X., Gao, J., Deng, L.: Embedding entities and relations for learning and inference in knowledge bases. arXiv preprint arXiv:1412.6575 (2014)
22. Yang, Y., Esteban, C., Tresp, V.: Embedding mapping approaches for tensor factorization and knowledge graph modelling. In: Sack, H., Blomqvist, E., d'Aquin, M., Ghidini, C., Ponzetto, S.P., Lange, C. (eds.) ESWC 2016. LNCS, vol. 9678, pp. 199–213. Springer, Cham (2016). https://doi.org/10.1007/978-3-319-34129-3_13

Learning Cheap and Novel Flight Itineraries

Dmytro Karamshuk[✉] and David Matthews

Skyscanner Ltd., Edinburgh, UK
{Dima.Karamshuk,David.Matthews}@skyscanner.net

Abstract. We consider the problem of efficiently constructing cheap and novel round trip flight itineraries by combining legs from different airlines. We analyse the factors that contribute towards the price of such itineraries and find that many result from the combination of just 30% of airlines and that the closer the departure of such itineraries is to the user's search date the more likely they are to be cheaper than the tickets from one airline. We use these insights to formulate the problem as a trade-off between the recall of cheap itinerary constructions and the costs associated with building them.

We propose a supervised learning solution with location embeddings which achieves an AUC $= 80.48$, a substantial improvement over simpler baselines. We discuss various practical considerations for dealing with the staleness and the stability of the model and present the design of the machine learning pipeline. Finally, we present an analysis of the model's performance in production and its impact on Skyscanner's users.

1 Introduction

Different strategies are used by airlines to price round trip tickets. Budget airlines price a complete round trip flight as the sum of the prices of the individual outbound and inbound journeys (often called flight legs). This contrasts with traditional, national carrier, airlines as their prices for round trip flights are rarely the sum of the two legs. Metasearch engines, such as Skyscanner[1], can mix outbound and inbound tickets from different airlines to create combination itineraries, e.g., flying from Miami to New York with United Airlines and returning with Delta Airlines (Fig. 1)[2]. Such combinations are, for a half of search requests, cheaper than the round trip tickets from one airline.

A naïve approach to create such combinations with traditional airlines, requires an extra two requests for prices per airline, for both the outbound and the inbound legs, on top of the prices for complete round trips. These additional requests for quotes is an extra cost for a metasearch engine. The cost, however,

---

[1] https://www.skyscanner.net/.

[2] Our constructions contrast with those built through interlining which involve two airlines combining flights on the same leg of a journey organised through a commercial agreement.

© Springer Nature Switzerland AG 2019
U. Brefeld et al. (Eds.): ECML PKDD 2018, LNAI 11053, pp. 288–304, 2019.
https://doi.org/10.1007/978-3-030-10997-4_18

can be considerably optimized by constructing only the combinations which are competitive against the round trip fares from airlines.

To this end, we aim to predict price competitive combinations of tickets from traditional airlines given a limited budget of extra quote requests. Our approach is as follows.

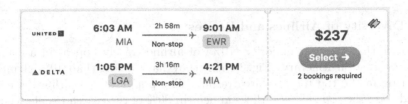

Fig. 1. Example of a combination flight itinerary in Skyscanner's search results.

Firstly, we analyse a data set of 2.3M search queries from 768K Skyscanner's users, looking for the signals which impact the competitiveness of combination itineraries in the search results. We find the that the vast majority of competitive combination itineraries are composed of only 30% of airlines and are more likely to appear in the searches for flights departing within days of the user's search.

Secondly, we formulate the problem of predictive itinerary construction as a trade-off between the computation cost and resulting coverage, where the cost is associated with the volume of quote requests the system has to make to construct combination itineraries, and the coverage represents the model's performance in finding all such itineraries that are deemed price competitive. To the best of our knowledge this is the first published attempt to formulate and solve the problem of constructing flight itineraries using machine learning.

Thirdly, we evaluate different supervised learning approaches to solve this problem and propose a solution based on neural location embeddings which outperforms simpler baselines and achieves an AUC = 80.48. We also provide an intuition on the semantics of information that such embedding methods are able to learn.

Finally, we implement and deploy the proposed model in a production environment. We provide simple guidance for achieving the right balance between the staleness and stability of the production model and present the summary of its performance.

2   Data Set

To collect a dataset for our analysis we enabled the retrieval of both outbound and inbound prices for all airlines on a sample of 2.3M Skyscanner search results for round trip flights in January 2018. We constructed all possible combination itineraries and recorded their position in the ranking of the cheapest search results, labelling them competitive, if they appeared in the cheapest ten search results, or non-competitive otherwise[3]. This resulted in a sample of 16.9M combination itineraries (both competitive and non-competitive) for our analysis, consisting of 768K users searching for flights on 147K different routes, i.e., origin and destination pairs.

Our analysis determined that the following factors contribute towards a combination itinerary being competitive.

2.1   Diversity of Airlines and Routes

We notice that the vast majority (70%) of airlines rarely appear in a competitive combination itinerary (Fig. 2), i.e., they have a less than 10% chance of appearing in the top ten of search results. The popularity of airlines is highly skewed too. The top 25% of airlines appear in 80% of the search results whereas the remaining 75% of airlines account for the remaining 20%. We found no correlation between airlines' popularity and its ability to appear in a competitive combination itinerary.

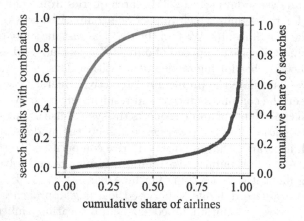

Fig. 2. Search results with competitive combinations across different airlines. The cumulative share of all search results (red) and search results with competitive combinations (blue) for top x% of airlines (x-axis). (Color figure online)

The absence of a correlation with popularity is even more vividly seen in the analysis of combination performance on different search routes (Fig. 3). The share of competitive combinations on unpopular and medium popular routes is rather stable ($\approx$45%) and big variations appear only in the tail of popular routes. In fact, some of those very popular routes have almost a 100% chance to have combination itineraries in the top ten results, whereas some other ones of a comparable popularity almost never feature a competitive combination itinerary.

---

[3] Skyscanner allows to rank search results by a variety of other parameters apart from the cheapest. The analysis of these different ranking strategies is beyond the scope of this paper.

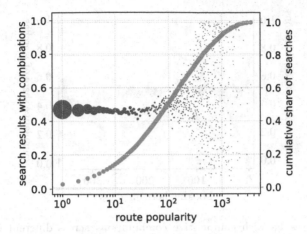

Fig. 3. Search results with competitive combinations across routes with different popularity. Red: the cumulative distribution function of the volume of searches across different origin and destination pairs (routes). Blue: the share of search results with competitive combinations (y-axis) on the routes of a given popularity (x-axis). (Color figure online)

This finding is in line with our modelling results in Sect. 3 where we observe that the popularity of a route or an airline is not an indicative feature to predict price competitiveness of combination itineraries. We therefore focus on a small number of airlines and routes which are likely to create competitive combination itineraries. We explore different supervised learning approaches to achieve this in Sect. 3.

2.2 Temporal Patterns

We also analyse how the days between search and departure (number of days before departure) affects the competitiveness of combinations in the top ten of search results (Fig. 4). We find that combination itineraries are more likely to be useful for searches with short horizons and gradually become less so as the days between search and departure increases. One possible explanation lies in the fact that traditional single flight tickets become more expensive as the departure day approaches, often unequally so across different airlines and directions. Thus, a search for a combination of airlines on different flight legs might give a much more competitive result. This observation also highlights the importance to consider the volatility of prices as the days between search and departure approaches, the fact which we explore in building a production pipeline in Sect. 4.

3 Predictive Construction of Combination Itineraries

Only 10% of all possible combination itineraries are cheap enough to appear in the top ten cheapest results and therefore be likely to be seen by the user. The

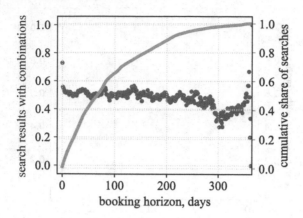

Fig. 4. Search results with competitive combinations across different days between search and departures (booking horizon). Red: the cumulative distribution function of the booking horizon. Blue: the share of search results with competitive combinations (y-axis) for a given booking horizon (x-axis). (Color figure online)

difficulty is in the fact that the cost of enabling combinations in Skyscanner search results is proportional to the volume of quote requests required to check their competitiveness. In this section we formulate the problem of predictive combination itinerary construction where we aim to train an algorithm to speculatively construct only those combinations which are likely to be competitive and thus to reduce the overall cost associated with enabling combinations in production.

3.1   Problem Formulation

We tackle the predictive combination itinerary construction as a supervised learning problem where we train a classifier $F(Q, A, F) \rightarrow \{True, False\}$ to predict whether any constructed combination itinerary in which airline A appears on the flight leg F, either outbound or inbound, will yield a competitive combination itinerary in the search results for the query Q. The current formulation is adopted to fit in Skyscanner's current pricing architecture which requires an advance decision about whether to request a quote from airline A on a leg F for a query Q. To measure the predictive performance of any such classifier $F(Q, A, F)$ we define the following metrics:

Recall or coverage is measured as a share of competitive itineraries constructed by the classifier $F(X)$, more formally:

$$Recall@10 = \frac{|L_{pred}^{@10} \cap L_{all}^{@10}|}{|L_{all}^{@10}|} \tag{1}$$

where $L_{pred}^{@10}$ is the set of competitive combination itineraries constructed by an algorithm and $L_{all}^{@10}$ is the set of all possible competitive combination itineraries.

In order to estimate the latter we need a mechanism to sample the ground truth space which we discuss in Sect. 4.

Quote requests or cost is measured in terms of all quote requests required by the algorithm to construct combination itineraries, i.e.:

$$Quote\ Requests = \frac{|L_{pred}|}{|L_{all}|} \tag{2}$$

where L_{all} - is the set of all possible combination itineraries constructed via the ground truth sampling process. Note that our definition of the cost is sometimes also named as predictive positive condition rate in the literature.

The problem of finding the optimal classifier $F(Q, A, F)$ is then one of finding the optimal balance between the recall and quote requests. Since every algorithm can yield a spectrum of all possible trade-offs between the recall and the quote requests we also use the area under the curve (AUC) as an aggregate performance metric.

3.2   Models

We tried several popular supervised learning models including logistic regression, multi-armed bandit and random forest. The first two algorithms represent rather simple models which model a linear combination of features (logistic regression) or their joint probabilities (multi-armed bandit). In contrast, random forest can model non-linear relations between individual features and exploits an idea of assembling different simple models trained on a random selection of individual features. We use the scikit-learn[4] implementation of these algorithms and benchmark them against:

Popularity Baseline. We compare the performance of the proposed models against a naïve popularity baseline computed by ranking the combinations of (origin, destination, airline) by their popularity in the training set and cutting-off the top K routes which are estimated to cumulatively account for a defined share of quote requests. We note that this is also the model which was initially implemented in the production system.

Oracle Upper-Bound. We also define an upper-bound for the prediction performance of any algorithm by considering an oracle predictor constructed with the perfect knowledge of the future, i.e., the validation data set. The aim of the oracle predictor is to estimate the upper-bound recall of competitive combinations achieved with a given budget of quote requests.

Results. From Fig. 5 we observe that all proposed supervised models achieve a superior performance in comparison to the naïve popularity baseline (AUC = 51.76%), confirming our expectations from Sect. 2 that popularity alone cannot explain competitiveness of combinations itineraries. Next, we notice that the

---

[4] http://scikit-learn.org/.

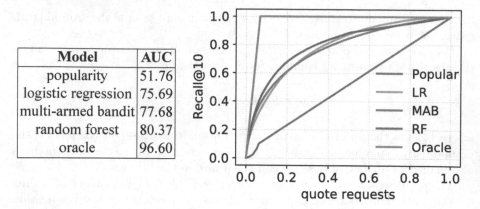

Model	AUC
popularity	51.76
logistic regression	75.69
multi-armed bandit	77.68
random forest	80.37
oracle	96.60

Fig. 5. Performance of different supervised learning models (logistic regression (LR), nearest neighbour (NN), multi-armed bandit (MAB) and random forest (RF)) benchmarked over a naïve popularity baseline (popular) and the upper-bound performance attainable with a perfect knowledge of the future (oracle).

random forest model outperforms other models and achieves an AUC = 80.37%, a large improvement from the second best performing model (AUC = 77.68%). At the same time, the results of our best performing model still lag behind the oracle predictor which achieves 100% recall with as little as 10% of total cost or AUC = 96.60%. In order to improve the performance of our best model even further in the following section we focused on experimenting with the representation of the feature space and more specifically the representation of location information identified as the most important predictor across all experiments.

3.3   Location Representations

This section describes different approaches we tried to more richly represent location information.

Trace-Based Embeddings. In this approach we collected the histories of per-user searches in the training data set and built sequences of origin and destination pairs appearing in them. For instance, if a user searched for a flight from London to Barcelona, followed by a search from London to Frankfurt, followed by another one from Frankfurt to Budapest, then we will construct a sequence of locations [London, Barcelona, London, Frankfurt, Frankfurt, Budapest] to represent the user's history. We also filter out the users who searched for less than 10 flights in our data set and remove the duplicates in consecutive searches. We feed the resulting sequences into a Word2Vec algorithm [13], treating each location as a word and each user sequence as a sentence. We end up with a representation of each origin and destination locations as vectors from the constructed space of location embeddings.

This approach is inspired by the results in mining distributed representations of categorical data, initially proposed for natural language processing [13], but recently applied also for mining graph [16] and location data [12,15,20]. Specifically, we tried the approach proposed in [15] and [20], but since the results were quite similar we only describe one of them.

Co-trained Embeddings. In this alternate approach we train a neural network with embedding layers for origin and destination features, as proposed in [8] and implemented in Keras embedding layers[5]. We use a six-layer architecture for our neural network where embedding layers are followed by four fully connected layers of 1024, 512, 256, 128 neurons with relu activation functions.

Note that the goal of this exercise is to understand whether we can learn useful representation of the location data rather than to comprehensively explore the application of deep neural networks as an alternative to our random forest algorithm which, as we discuss in Sect. 4, is currently implemented in our production pipeline. Hence, we focus on the representations we learn from the first layer of the proposed network.

Table 1. Examples of location embeddings for airports most similar to London Heathrow (left) and Beijing Capital (right) in the embedded feature space.

London Heathrow		Beijing Capital	
Airport	Similarity	Airport	Similarity
Frankfurt am Main	0.71	Chubu Centrair	0.91
Manchester	0.69	Taipei Taoyuan	0.90
Amsterdam Schipol	0.62	Seoul Incheon	0.90
Paris Charles de Gaulle	0.62	Miyazaki	0.88
London Gatwick	0.61	Shanghai Pudong	0.88

Learned Embeddings. In Table 1 we present few examples of the location embeddings we learn with these proposed approaches. Particularly, we take few example airports (London Heathrow and Beijing Capital) and find other airports which are located in vicinity in the constructed vector spaces. The results reveal two interesting insights. Firstly, the resulting location embeddings look like they are capturing the proximity between the airports. The airports most closely located to London Heathrow and Beijing Capital are located in the western Europe and south-east Asia, correspondingly. Secondly, we notice that the algorithm is able to capture that London Heathrow is semantically much closer to transatlantic hubs such as Paris Charles de Gaulle, Amsterdam Schipol and London Gatwick rather than a geographically closer London Luton or London Stansted airports which are mainly focused on low-cost flights within Europe.

---

[5] https://keras.io/layers/embeddings/.

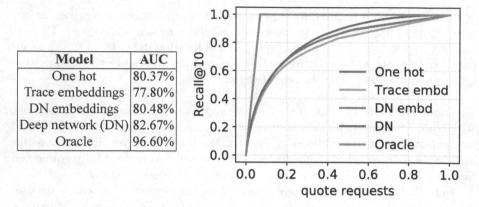

Model	AUC
One hot	80.37%
Trace embeddings	77.80%
DN embeddings	80.48%
Deep network (DN)	82.67%
Oracle	96.60%

Fig. 6. Performance of the random forest model with different representations of origin and destination data (one hot encoding, trace-based embeddings, co-trained (DN) embeddings) and a neural network with embedding layers (DN). (Color figure online)

3.4 Prediction Performance

In Fig. 6 we compare the results of applying different location representations to the random forest algorithm proposed in the previous section. We use the random forest trained with one-hot representation as a baseline and compare it with: (a) the random forest model trained with trace-based embeddings (orange curve) and (b) the random forest trained with co-trained embeddings from the deep neural network model discussed early (green curve). In this latter approach we decouple the embedding layer from the rest of the layers in the neural network and use that as an input to our random forest model. We are able to assess how the embedding learned in the neural network can effectively represent the location data. Finally, we provide the results of the deep neural network itself for comparison (red curve).

The results of the model trained from trace-based embeddings performed worse than a baseline one-hot encoding, Fig. 6. The random forest model with co-trained embeddings outperforms both results and achieves AUC = 80.48%. The characteristic curves of the random forest model with one-hot encoding (blue curve) and co-trained embeddings (green curve) overlap largely in Fig. 6, but a closer examination reveals a noticeable improvement of the latter in the area between 0 and 20% and above 50% of the quote request budget. One possible explanation behind these results might be that the embeddings we have trained from user-traces, in contrast to the co-trained embeddings, have been learning the general patterns in user-searches rather than optimising for our specific problem.

We also notice that the performance of the deep neural network surpasses that of the random forest but any such comparison should also consider the complexity of each of the models, e.g., the number and the depth of the decision trees in the random forest model versus the number and the width of the layers in the neural network.

4 Putting the Model in Production

4.1 Model Parameters

Training Data Window. To decide on how far back in time we need to look for data to train a good model we conduct an experiment where samples of an equivalent size are taken from each of the previous N days, for increasing values of N (Fig. 7). We observe that the performance of the model is initially increasing as we add more days into the training window, but slows down for N between [3..7] days and the performance even drops as we keep increasing the size of the window further. We attribute this observation to the highly volatile nature of the flight fares and use a training window of 7 days to train the model in production.

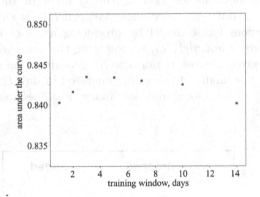

Fig. 7. The impact of the selected training window on the prediction performance of the model.

Model Staleness. To decide how frequently to retrain the model in production we measure its staleness in an experiment (Fig. 8). We consider a six day long period with two variants: when the model is trained once before the start of the experiment and when the model is retrained every single day. The results suggest, that the one-off trained model quickly stales by an average of 0.3% in AUC with every day of the experiment. The model retrained every single day, although also affected by daily fluctuations, outperforms the one-off trained model. This result motivates our decision to retrain the model every day.

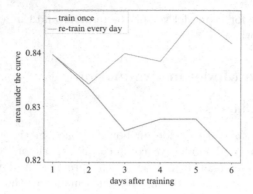

Fig. 8. Model staleness of the one-off trained model vs. the model retrained every day.

Model Stability. Frequent retraining of the model comes at a price of its stability, i.e., giving the same prediction for the same input day in day out. To explain this phenomena we look at the changes in the rules that the model is learning in different daily runs. We generate a simplified approximation of our random forest model by producing a set of decision rules of a form $(origin, destination, airline)$, representing the cases when combination itineraries with a given $airline$ perform well on a given $(origin, destination)$ route. We analyse how many of the rules generated in day T_{i-1} were dropped in the day T_i's run of the model and how many new ones were added instead (Fig. 9).

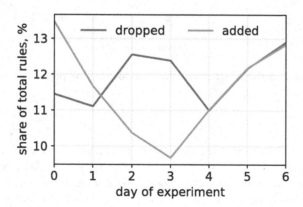

Fig. 9. Model stability, daily changes of (origin, destination, airline) rules inferred from the random forest model.

We see that around 88% of rules remain relevant between the two consecutive days the remaining ≈12% are dropped and a similar number of new ones are added. Our qualitative investigation followed from this experiment suggested that dropping a large number of rules may end up in a negative user experience. Someone who saw a combination option on day T_{i-1} might be frustrated from not seeing it on T_i even if the price went up and it is no longer in the top ten of the search results. To account for this phenomenon we have introduced a simple heuristic in production which ensures that all of the rules which were generated on day T_{i-1} will be included for another day T_i.

4.2   Architecture of the Pipeline

Equipped with the observations from the previous section we implement a machine learning pipeline summarised in Fig. 10. There are three main components in the design of the pipeline: the data collection process which samples the ground truth space to generate training data; the training component which runs daily to train and validate the model and the serving component which delivers predictions to the Skyscanner search engine.

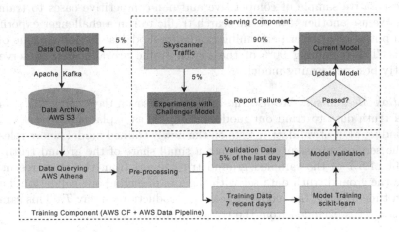

Fig. 10. The architecture of the machine learning pipeline.

Training Infrastructure: The training infrastructure is orchestrated by AWS Cloud Formation[6] and AWS Data Pipeline[7]. The data querying and preprocessing is implemented with Presto distributed computing framework[8] managed by AWS Athena[9]. The model training is done with scikit-learn library on a high-capacity virtual machine. Our decision for opting towards a single large virtual machine vs. a multitude of small distributed ones has been dictated by the following considerations:

---

[6] https://aws.amazon.com/cloudformation/.
[7] https://aws.amazon.com/datapipeline/.
[8] https://prestodb.io/.
[9] https://aws.amazon.com/athena/.

Data Volume: Once the heavy-lifting of data collection and preprocessing is done in Presto, the size of the resulting training data set becomes small enough to be processed on a single high capacity virtual machine.

Performance: By avoiding expensive IO operations characteristic of distributed frameworks, we decreased the duration of a model training cycle to less than 10 min.

Technological Risks: The proposed production environment closely resembles our offline experimentation framework, considerably reducing the risk of a performance difference between the model developed during offline experimentation and the model run in production.

Traffic Allocation. We use 5% of Skyscanner search traffic to enable ground truth sampling and prepare the data set for training using Skyscanner's logging infrastructure[10] which is built on top of Apache Kafka[11]. We enable construction of all possible combination itineraries on this selected search traffic, collecting a representative sample of competitive and non-competitive cases to train the model. We use another 5% of the search traffic to run a challenger experiment when a potentially better performing candidate model is developed using offline analysis. The remaining 90% of the search traffic are allocated to serve the currently best performing model.

Validation Mechanism. We use the most recent seven days, $T_{i-7}..T_{i-1}$, of the ground truth data to train our model on day T_i as explained in Sect. 4.1. We also conduct a set of validation tests on the newly trained model before releasing it to the serving infrastructure. We use a small share of the ground truth data (5% out of 5% of the sampled ground truth data) from the most recent day T_{i-1} in the ground truth data set with the aim of having our validation data as close in time to when the model appears in production on day T_i. This sampled validation set is excluded from the training data.

4.3   Performance in Production

When serving the model in production we allow a budget of an additional 5% of quote requests with which we expect to reconstruct 45% of all competitive combination itineraries (recall Fig. 6). From Fig. 11 we note that the recall measured in production deviates by ≈5% from expectations in our offline experiments. We attribute this to model staleness incurred from 24 h lag in the training data we use from the time when the model is pushed to serve users' searches.

Analysing the model's impact on Skyscanner users, we note that new cheap combination itineraries become available in 22% of search results. We see evidence of users finding these additional itineraries useful with a 20% relative increase in the booking transactions for combinations.

---

[10] More details here https://www.youtube.com/watch?v=8z59a2KWRIQ.
[11] https://kafka.apache.org/.

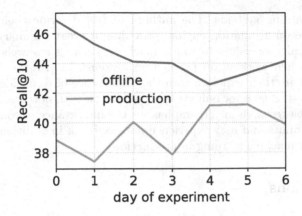

Fig. 11. Performance of the model in offline experiments vs. production expressed in terms of Recall@10 at 5% of quote requests.

5   Related Work

Mining Flights Data. The problem of airline fare prediction is discussed in detail in [2] and several data mining models were benchmarked in [5]. The authors of [1] modelled 3D trajectories of flights based on various weather and air traffic conditions. The problem of itinerary relevance ranking in one of the largest Global Distributed Systems was presented in [14]. The systematic patterns of airline delays were analysed in [7]. And the impact of airport network structure on the spread of global pandemics was weighed up in [4].

Location Representation. Traditional ways to model airline prices have been based on complex networks [4,7] or various supervised machine learning models [5,14]. A more recent trend is around incorporating neural embeddings to model location data. Embeddings have seen great success in natural language processing [13], modelling large graphs [16] and there has been a spike of enthusiasm around applying neural embedding to geographic location context with a variety of papers focusing on: (a) mining embeddings from sequences of locations [12,15,20]; (b) modelling geographic context [6,9,19] and (c) using alternative neural architectures where location representations are learned while optimising towards particular applications [19] and different approaches are mixed together in [9] and [6]. The practicalities of augmenting existing non-deep machine learning pipelines with neural embeddings are discussed in [21] and in [3].

Productionising Machine Learning Systems. The research community has recently started recognising the importance of sharing experience and learning in the way machine learning and data mining systems are implemented in production systems. In [17] the authors stress the importance of investing considerable thinking and resources in building long-lasting technological infrastructures for machine learning systems. The authors of [10] describe their experiences in building a recommendation engine, providing a great summary of business and technological constraints in which machine learning researchers and engineers operate when working on production systems. In [18] the developers of Google Drive share their experience on the importance of reconsidering UI metrics and launch strategies for online experimentation with new machine learning features. Alibaba research in [11] emphasises the importance of considering performance constraints and user experience and feedback in addition to accuracy when deploying machine learning in production.

6   Conclusions

We have presented a system that learns to build cheap and novel round trip flight itineraries by combining legs from different airlines. We collected a sample of all such combinations and found that the majority of competitive combinations were concentrated around a minority of airlines but equally spread across routes of differing popularity. We also found that the performance of these combinations in search results increases as the time between search and departure date decreases.

We formulated the problem of predicting competitive itinerary combinations as a trade-off between the coverage in the search results and the cost associated with performing the requests to airlines for the quotes needed for their construction. We considered a variety of supervised learning approaches to model the proposed prediction problem and showed that richer representations of location data improved performance.

We put forward a number of practical considerations for putting the proposed model into production. We showed the importance of considering the trade-off between the model stability and staleness, balancing keeping the model performant whilst minimising the potential negative impact on the user experience that comes with changeable website behaviour.

We also identify various considerations we took to deliver proposed model to users including technological risks, computational complexity and costs. Finally, we provided an analysis of the model's performance in production and discuss its positive impact on Skyscanner's users.

Acknowledgement. The authors would like to thank the rest of the Magpie team (Boris Mitrovic, Calum Leslie, James Eastwood, Linda Edstrand, Ronan Le Nagard, Steve Morley, Stewart McIntyre and Vitaly Khamidullin) for their help and support with this project and the following people for feedback on drafts of this paper: Bryan Dove, Craig McIntyre, Kieran McHugh, Lisa Imlach, Ruth Garcia, Sri Sri Perangur, Stuart Thomson and Tatia Engelmore.

References

1. Ayhan, S., Samet, H.: Aircraft trajectory prediction made easy with predictive analytics. In: KDD, pp. 21–30 (2016)
2. Boyd, E.: The Future of Pricing: How Airline Ticket Pricing Has Inspired a Revolution. Springer, Heidelberg (2016)
3. Chamberlain, B.P., Cardoso, A., Liu, C.H., Pagliari, R., Deisenroth, M.P.: Customer life time value prediction using embeddings. In: Proceedings of the Ninth ACM SIGKDD International Conference on Knowledge Discovery and Data Mining. ACM (2017)
4. Colizza, V., Barrat, A., Barthélemy, M., Vespignani, A.: The role of the airline transportation network in the prediction and predictability of global epidemics. Proc. Nat. Acad. Sci. U.S.A. **103**(7), 2015–2020 (2006)
5. Etzioni, O., Tuchinda, R., Knoblock, C.A., Yates, A.: To buy or not to buy: mining airfare data to minimize ticket purchase price. In: Proceedings of the Ninth ACM SIGKDD International Conference on Knowledge Discovery and Data Mining, pp. 119–128. ACM (2003)
6. Feng, S., Cong, G., An, B., Chee, Y.M.: POI2Vec: geographical latent representation for predicting future visitors. In: AAAI, pp. 102–108 (2017)
7. Fleurquin, P., Ramasco, J.J., Eguiluz, V.M.: Systemic delay propagation in the us airport network. Sci. Rep. **3**, 1159 (2013)
8. Guo, C., Berkhahn, F.: Entity embeddings of categorical variables. arXiv preprint arXiv:1604.06737 (2016)
9. Kejriwal, M., Szekely, P.: Neural embeddings for populated geonames locations. In: d'Amato, C., et al. (eds.) ISWC 2017. LNCS, vol. 10588, pp. 139–146. Springer, Cham (2017). https://doi.org/10.1007/978-3-319-68204-4_14
10. Liu, D.C., et al.: Related pins at pinterest: the evolution of a real-world recommender system. In: Proceedings of the 26th International Conference on World Wide Web Companion, pp. 583–592. International World Wide Web Conferences Steering Committee (2017)
11. Liu, S., Xiao, F., Ou, W., Si, L.: Cascade ranking for operational e-commerce search. In: Proceedings of the 23rd ACM SIGKDD International Conference on Knowledge Discovery and Data Mining, pp. 1557–1565. ACM (2017)
12. Liu, X., Liu, Y., Li, X.: Exploring the context of locations for personalized location recommendations. In: IJCAI, pp. 1188–1194 (2016)
13. Mikolov, T., Sutskever, I., Chen, K., Corrado, G.S., Dean, J.: Distributed representations of words and phrases and their compositionality. In: Advances in Neural Information Processing Systems, pp. 3111–3119 (2013)
14. Mottini, A., Acuna-Agost, R.: Deep choice model using pointer networks for airline itinerary prediction. In: Proceedings of the 23rd ACM SIGKDD International Conference on Knowledge Discovery and Data Mining, pp. 1575–1583. ACM (2017)
15. Pang, J., Zhang, Y.: DeepCity: a feature learning framework for mining location check-ins. arXiv preprint arXiv:1610.03676 (2016)
16. Perozzi, B., Al-Rfou, R., Skiena, S.: DeepWalk: online learning of social representations. In: Proceedings of the 20th ACM SIGKDD International Conference on Knowledge Discovery and Data Mining, pp. 701–710. ACM (2014)
17. Sculley, D., et al.: Hidden technical debt in machine learning systems. In: Advances in Neural Information Processing Systems, pp. 2503–2511 (2015)
18. Tata, S., et al.: Quick access: building a smart experience for Google drive. In: Proceedings of the 23rd ACM SIGKDD International Conference on Knowledge Discovery and Data Mining, pp. 1643–1651. ACM (2017)

19. Yan, B., Janowicz, K., Mai, G., Gao, S.: From ITDL to Place2Vec-reasoning about place type similarity and relatedness by learning embeddings from augmented spatial contexts. Proc. SIGSPATIAL **17**, 7–10 (2017)
20. Zhao, S., Zhao, T., King, I., Lyu, M.R.: Geo-teaser: geo-temporal sequential embedding rank for point-of-interest recommendation. In: Proceedings of the 26th International Conference on World Wide Web Companion, pp. 153–162. International World Wide Web Conferences Steering Committee (2017)
21. Zhu, J., Shan, Y., Mao, J., Yu, D., Rahmanian, H., Zhang, Y.: Deep embedding forest: forest-based serving with deep embedding features. In: Proceedings of the 23rd ACM SIGKDD International Conference on Knowledge Discovery and Data Mining, pp. 1703–1711. ACM (2017)

Towards Resource-Efficient Classifiers for Always-On Monitoring

Jonas Vlasselaer[1], Wannes Meert[2(✉)], and Marian Verhelst[1]

[1] MICAS, Department of Electrical Engineering, KU Leuven, Leuven, Belgium
[2] DTAI, Department of Computer Science, KU Leuven, Leuven, Belgium
`wannes.meert@cs.kuleuven.be`

Abstract. Emerging applications such as natural user interfaces or smart homes create a rising interest in electronic devices that have always-on sensing and monitoring capabilities. As these devices typically have limited computational resources and require battery powered operation, the challenge lies in the development of processing and classification methods that can operate under extremely scarce resource conditions. To address this challenge, we propose a two-layered computational model which enables an enhanced trade-off between computational cost and classification accuracy: The bottom layer consists of a selection of state-of-the-art classifiers, each having a different computational cost to generate the required features and to evaluate the classifier itself. For the top layer, we propose to use a Dynamic Bayesian network which allows to not only reason about the output of the various bottom-layer classifiers, but also to take into account additional information from the past to determine the present state. Furthermore, we introduce the use of the Same-Decision Probability to reason about the added value of the bottom-layer classifiers and selectively activate their computations to dynamically exploit the computational cost versus classification accuracy trade-off space. We validate our methods on the real-world SINS database, where domestic activities are recorded with an accoustic sensor network, as well as the Human Activity Recognition (HAR) benchmark dataset.

1 Introduction

There is a rising interest in electronic devices that have always-on sensing and monitoring capabilities [9,11]. Consider, for example, the task of Ambient Assisted Living (AAL), where the goal is to allow recovering patients and elderly persons to stay in their home environment instead of traveling to a hospital or living in a care center. Crucial to achieve this goal, is a continuous, always-on monitoring setup. Such a setup entails three components: First, wireless sensors and wearables equipped with various sensors are connected in a wireless network to sense and record the physical conditions of the environment. Second, the measurements are processed using signal processing techniques to extract useful features and compress the data without loss of relevant information. Third, machine learning algorithms are used to train a model for the task at

© Springer Nature Switzerland AG 2019
U. Brefeld et al. (Eds.): ECML PKDD 2018, LNAI 11053, pp. 305–321, 2019.
https://doi.org/10.1007/978-3-030-10997-4_19

hand (e.g. an activity detector). Typically, these three components are analyzed and optimized separately [13,16,20]. For the machine learning model, the input features are assumed given and a classifier is trained to maximize its accuracy. The use of battery-powered, computationally-limited embedded devices, however, poses new challenges as it requires methods that can operate under scarce resource conditions. To address these challenges, we propose to optimize these components simultaneously. Not only offline but also online, when the model is deployed, using the recently introduced *Same-Decision Probability* [3]. This allows us to dynamically minimizes the setup's resource usage, while maintaining good detection accuracy.

This work was triggered by our previous work on monitoring domestic activities based on multi-channel acoustics. The resulting dataset has been consolidated in the publicly available *SINS database* [1,6]. The data was gathered using an acoustic sensor network where various low-cost microphones were distributed in an home environment. The SINS database is different from many other publicly available datasets as the data is recorded as a continuous stream and activities were being performed in a spontaneous manner. The original paper proposes to compute the Mel-Frequency Cepstral Coefficients (MFCC) as well as its Delta and Acceleration coefficients as input features, and makes use of a Support Vector Machine based classifier. Computing these features and classifier comes with a high computational cost as it requires various transformations of the raw measurements. Executing these operations in an always-on fashion will drain the battery of the targeted embedded sensing nodes, requiring alternative solutions.

Theoretically, operating under scarce resource constraints can be avoided by pushing the computational overhead to a more powerful computing device or to the *cloud*. One can, for example, use the sensor nodes to only collect the data and then communicate the raw measurements towards another device which then performs the necessary computations and returns the result. In practice, however, this approach comes with various concerns as communication of the raw signals is often infeasible for sensors with high sampling rates due to bandwidth limitations and communicating this much data might lead to delays and packet losses. In many cases, the only feasible approach is thus to use the sensor node or embedded device to also process the measurements and to run the classification algorithm locally. As such, in order to improve the device's resource consumption, each of the components in the classification system should be made resource-aware. As the consumption of the sensors themselves are typically negligible in the classification pipeline [18], resource-aware classification implies we should obtain battery savings during feature generation as well as execution of the classifier. As there is no free lunch, battery savings in this context typically require to explore the trade-off space between computational cost and classification accuracy.

To enable this, we propose a two-layered classifier with as top layer a Bayesian Network that allows us to make decisions and as bottom layer a mixture of classifiers operating on features with diverse computational costs (e.g. Tree Augmented Naive Bayes or Convolutional Neural Networks). We rely on Bayesian networks for the decision making, i.e. the top layer, as they come with several advantages: (1) They are generative models and can deal elegantly with missing data. This allows us to dynamically decide which of the bottom-layer classifiers and features to compute or request. (2) They allow us to reason about uncertainty. In the context of activity tracking, for example, this allows us to precisely quantify the probability of transitions between activities. (3) They allow us to reason about the Same-Decision Probability. Same-Decision Probability (SDP) is a recently introduced technique to investigate the usefulness of observing additional features [3]. SDP is the probability one would make the same-decision, had one known the value of additional features. Our resource-aware method relies on SDP to dynamically evaluate the usefulness of spending additional computational resources.

The main contributions of this paper are thus twofold, both leading to increased resource-awareness and resource-efficiency of an always-on embedded classifier pipeline. Firstly, we present a two-layered classifier that dynamically chooses which features to compute and which models to evaluate. Secondly, we show how same-decision probability can be used in a resource-aware classifier to select features, both in an offline and an online manner. We apply variations of the proposed two-layered model to investigate the computational cost versus classification accuracy trade-off space. We validate our models using both the real-world SINS database and the *Human Activity Recognition Using Smartphones* (HAR) benchmark dataset [2].

2   Background and Notations

Upper-case letters (Y) denote random variables and lower case letters (y) denote their instantiations. Bold letters represent sets of variables ($\mathbf{Y}$) and their instantiations ($\mathbf{y}$).

2.1   Bayesian Classifiers

A *Bayesian network* is a directed acyclic graph where each node represents a random variable and each edge indicates a direct influence among the variables [15]. The network defines a conditional probability distribution $\Pr(X|\mathbf{Pa}(X))$ for every variable X, where $\mathbf{Pa}(X)$ are the parents of X in the graph. The joint distribution of the network can be factored into a product of conditional distributions as follows:

$$\Pr(X^1, X^2, \ldots, X^n) = \prod_{i=1}^{n} \Pr(X^i|\mathbf{Pa}(X^i))$$

A *Bayesian classifier* is a Bayesian network where one variable (or more) represents the class and each of the features are represented by variables that can be observed. For example, a Tree Augmented Naive Bayes (TAN) classifier has the class variable as its root, and all other variables represent features that have as parents the root and one other feature variable. A TAN classifier comes with less strong assumptions compared to a Naive Bayes classifier and, as a result, typically comes with a higher classification accuracy while still maintaining moderate computational complexity.

A *dynamic Bayesian network* is defined by two networks: B_1, which specifies the prior or initial state distribution $\Pr(\mathbf{Z}_1)$, and $B_\rightarrow$, a two-slice temporal BN that specifies the transition model $\Pr(\mathbf{Z}_t | \mathbf{Z}_{t-1})$. Together, they represent the distribution

$$\Pr(\mathbf{Z}_{1:T}) = \Pr(\mathbf{Z}_1) \prod_{t=2}^{T} \Pr(\mathbf{Z}_t | \mathbf{Z}_{t-1})$$

2.2 Bayesian Inference and Knowledge Compilation

Many techniques exist to perform inference in a Bayesian Network. In this paper we use knowledge compilation and weighted model counting [5]. This is a state-of-the-art inference technique for Bayesian networks and Dynamic Bayesian networks [19]. After encoding BNs into a logic knowledge base, we use the c2d knowledge compiler [4] to obtain an Arithmetic Circuit. An Arithmetic Circuit is a directed acyclic graph where leaf nodes represent variables and their weights, and where inner nodes represent additions and multiplication. Inference in the BN now boils down to performing an upward pass through the AC. Hence, the size of the obtained arithmetic circuit directly denotes the required number of computational operations.

2.3 Same-Decision Probability

Recently, the concept of Same-Decision Probability (SDP) has been introduced in the context of Bayesian Networks that are employed to support decision making in domains such as diagnosis and classification [3]. Intuitively, SDP is the probability that we would have made the same threshold-based decision, had we known the state of some hidden variables pertaining to our decision. Within the context of classification, SDP denotes the probability we would classify a data instance towards the same class had we known the value of some additional features.

Let d and $\mathbf{e}$ be instantiations of the decision variable D and evidence variables $\mathbf{E}$. Let T be a threshold. Let $\mathbf{X}$ be a set of variables distinct from D and $\mathbf{E}$. The same-decision probability (SDP) is computed as:

$$SDP_{d,T}(\mathbf{X}|\mathbf{e}) = \sum_{\mathbf{x}} \Big(\underbrace{[\Pr(d|\mathbf{x}, \mathbf{e}) =_T \Pr(d|\mathbf{e})]}_{\alpha} \cdot \Pr(\mathbf{x}|\mathbf{e}) \Big) \tag{1}$$

Here, the equality $=_T$ holds if both sides evaluate to a probability on the same side of threshold T, and $[\alpha]$ is 1 when α is true and 0 otherwise. We can compute the SDP on a given Bayesian Network using knowledge compilation and model counting [14].

3 Two-Layered Model: Building Blocks

In order to explore and enhance the accuracy versus computational cost trade-off, we propose to use a two-layered model of classifiers (see Fig. 1). Intuitively, the bottom-layer consists of various state-of-the-art classifiers and their classification result is combined by means of a top-layer network. Additionally, feature generation comes with a certain cost and, in many applications, this feature extraction cost even significantly surpasses the cost of classifier model evaluations. Hence, also this component should be taken into account in the accuracy versus resource trade-off.

To summarize, resource-aware classification provides three degrees of freedom that we will explore: (1) Generation of the required features; (2) Training and evaluation of the bottom-layer classifiers; (3) A top-layer network that dynamically combines the bottom-layer classifiers.

3.1 Feature Generation

Sensor nodes typically produce a continuous stream of measurements which are sampled at a certain rate. For microphones, as used for the SINS dataset, the sampling rate is 16 kHz while for the accelerometer and gyroscope, as used for the HAR dataset, the sampling rate is 50 Hz. Optionally, the measurements can be preprocessed with, for example, a noise filter. Next, one applies a fixed-width sliding window with a certain overlap to split the measurements into sets with a fixed number of values. Finally, these fixed-width windows of values can be used to compute various features.

While we do not aim to formally quantize this in general, we differentiate between features based on their computational cost of generating these features. We report the cost of the features used in the empirical evaluation in Sect. 5 and more detailed analysis about the cost of various features is available in other research [8,10,12]. We differentiate between three general types:

1. Raw signal: Not all classifiers require features. Convolutional Neural Networks, for example, can be trained on the raw signal as they perform implicit feature extraction in the convolutional layers of the network. This type of feature preprocessing does not require any effort and have thus no cost.

2. Cheap features: We refer to *cheap* features as those features for which the computational cost is low. For example, computing the *min*, *max* and *mean* of a window is rather cheap as it only requires a constant number of operations for each of the values in the window. Hence, resource-aware feature generation and classification have a preference for this type of features.

3. Expensive features: We refer to *expensive* features as those features for which the computational cost is high. For example, computing the Mel-Frequency Cepstral Coefficients of an audio signal is rather expensive as it requires multiple transformations. Hence, considerable battery savings could be obtained in case we can minimize the need to compute these features.

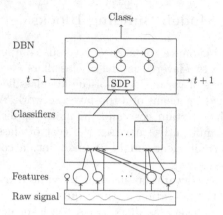

Fig. 1. Overview of the two-layered classifier. Arrows indicate flow of information.

3.2   Bottom-Layer: Mixture of Classifiers

Any type of classifier can be used in the bottom-layer, for example (Tree Augmented) Naive Bayes or Convolutional Neural Networks. We only make a distinction how they represent the classes to predict: as a one-vs-all classifier or a multi-class classifier. The former typically results in a set of rather small classifiers while the latter yields one large classifier.

A ONE-VS-ALL classifier is an ensemble of classifiers, each trained for one of the N available classes. In other words, N binary classifiers denoting whether a data instance belongs to a particular class or whether it belongs to any of the other classes. The final prediction needs to be derived from all predictions (e.g. using scoring). The advantage is that each classifier can focus on one class and each class is optimally represented. The disadvantage is the potentially large number of classifiers.

In our experimental section, we will use Tree Augmented Naive Bayes (TAN) classifiers for our one-vs-all classifiers as we observed good results but this classifier can be substituted with any other type of classifier.

We choose to only use cheap features to train the one-vs-all classifiers. As a result, the computational cost to evaluate whether a new data instance belongs to a certain class is rather low, at the expense of a potential accuracy impact. This property will be especially exploited in the models we propose in Sects. 4.6 and 4.7 and allows us to push the trade-off space towards models with a lower computational cost.

A MULTI-CLASS classifier represents all classes explicitly in one model. In other words, one classifier directly predicts the particular class a data instance belongs to. The advantage is that only one model needs to be executed. The disadvantage is that not all classes are equally represented.

In our experiments, we will use Tree Augmented Naive Bayes (TAN) classifiers and Convolutional Neural Networks (CNN) for our multi-class classifiers but both of them could be substituted by other types of classifiers depending on the application.

We choose to use all available features, i.e. cheap and expensive features to train the TAN multi-class classifier. The CNN, on the other hand, operates on the raw data and typically induces a large computational cost due to its large model size. Hence, our multi-class classifiers completely ignore the computational cost while they try to maximize the classification accuracy, allowing us to push the trade-off space towards models with a higher accuracy.

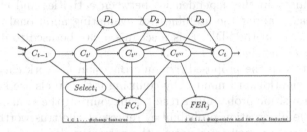

Fig. 2. Overview of our DBN approach.

3.3 Top-Layer: Dynamic Bayesian Network

The one-vs-all and the multi-class bottom-layer classifiers introduced in the previous sections make use of the available data, i.e. features or raw data, to classify a new data instance. The top-layer method, on the other hand, aims to dynamically combine the bottom-layer classifiers and reasons about their output.

As top-layer method we use a dynamic Bayesian network (DBN). Compared to a static network, a DBN additionally takes into account information from the past to reason about the present state. Within the context of monitoring and activity tracking, this leverages information that the activity a person was performing in the past will be of great help to classify the current activity: (1) There is quite a high probability the person remains doing the same activity; (2) The transition of one activity towards particular other activities might be unlikely. For example, if we classify a previous data instance as showering, then in the next time step this person is most likely still showering or drying herself. The transition of showering towards another activity is much less likely.

We present the general strategy we followed to build the DBN (see Fig. 2). The concrete various instantiations between which we compare are detailed in Sect. 4. The dynamic Bayesian Network observes the previous state (S_{t-1}), i.e. the probability distribution over the previous activities, and the result of up to three different types of bottom-layer classifiers (represented by C_t-primes) that use a selection of cheap features (FC), possibly together with expensive features and raw data (FER). This information is combined in an incremental and probabilistic manner to compute the likelihood of being in a state, i.e. the probability distribution of the current activity. Not all classifiers use all cheap features. When moving from C_{t-1} towards $C_{t'''}$ incrementally more features are being computed. This selection is encoded in the $Select_i$ variables.

The method we use in the top-layer comes with some specifications that are fulfilled by a Bayesian approach: (1) We want to dynamically choose which of the bottom-layer classifiers to evaluate. This is coordinated by the D_i variables that based on the certainty of the classifiers C_t-prime(s) decides to query more classifiers/features (i.e. go to a next C_t-prime(s)) or not and pass the current state directly to the next time step (through C_t); (2) We want to quantify the level of uncertainty in the dependencies between activities and classifiers. (3) We want to reason about the usefulness of evaluating additional bottom-layer classifiers by means of the SDP. This mechanism can be used to decide on the outcome of the D variables.

All parameters in the proposed dynamic Bayesian network classifier can be trained in a straightforward manner by counting if the labels are known during training. The transition probabilities result from counting the available transition in the data. The observation probabilities represent the trustworthiness of the classifiers and are obtained by counting the correctly classified data instances from the set of trained bottom-layer classifiers.

4   Two-Layered Model: Variations

We now propose several models that are variations of the two-layered models where the components are specified in detail. This allows us to explore and enhance the classification accuracy versus computational resources trade-off space. Each of the models makes use of one or more of the base classifiers introduced in the previous section.

The title of each of the following subsections uses a shorthand description of each of the proposed models, which we will use further on when we refer to these models. On the right side, we denote a more informative description of how the various classifiers are combined. Arrows indicate a chain of different classifiers. On top of the arrow is indicated whether we use a simple threshold to decide to evaluate the next classifier or the more advanced SDP.

4.1 MC   MC

Our first model is a static, baseline approach where we use the multi-class (MC) classifier trained over all the available classes in the dataset, without a dynamic network layer on top of it. Classifying a new data instance with our MC classifier is straightforward, as it directly denotes the most likely class the data instance belongs to.

The **advantage** if this model is that we can use any state-of-the-art method to train the multi-class classifier and, hence, this model should result in state-of-the-art classification accuracies. The **disadvantage** is that this model does not take into account the computational cost of evaluating the classifier, nor does it exploit temporal correlations.

4.2 OVA   OVA $\xrightarrow{\text{NA}}$ MC

Our second static, baseline model firstly relies on one-vs-all (OVA) classifiers to classify a new data instance, without taking into account the prediction of the previous time step. In case of agreement, i.e. exactly one of the classifiers predicts that the data instance belongs to a particular class, we pick the corresponding class as the predicted class for the data instance. In case of non-agreement (NA), we evaluate the multi-class classifier.

The **advantage** if this model is that the use of the OVA classifiers significantly reduces the computational cost. The **disadvantage** is that we do not have a way to reason about the outcome of the OVA classifiers and, in worst case, we still have to rely on the MC classifier for quite a lot of data instances, nor does this model exploits temporal correlations.

4.3 DBN$_{\text{OVA}}$ DBN$_{\text{OVA}}$

To combine the results of the OVA classifiers in a probabilistic way, we feed them as observations into a dynamic Bayesian network top-layer. Specifically, our DBN has a corresponding variable for each of the available OVA classifiers. The class variable is a multi-valued variable where the number of values is equal to the number of classes in the data. To use this model, we first evaluate each of OVA classifiers and feed their output to the DBN. Next, we compute the posterior probability of the class variable in the DBN and the value with the highest probability denotes the predicted class.

The **advantage** of this model is that the computational cost is further reduced as we completely avoided the use of the multi-class classifier and expensive features. Additionally the dynamic top-layer network allows us to exploit temporal correlations between consecutive data instances. The **disadvantage** is that the classification accuracy is expected to be quite below the state-of-the-art as we solely rely on cheap features.

4.4 DBN$_{\text{MC}}$ DBN$_{\text{MC}}$

We now combine the DBN top-layer with the multi-class classifier where the DBN contains exactly one observed variable corresponding to this MC classifier. Hence, we first evaluate the multi-class classifier and feed its output to the DBN.

The **advantage** of this model is that the DBN takes into account the probability distribution of the previously classified data instance and is expected to increase the classification accuracy compared to solely using the multi-class classifier. The **disadvantage** is that additionally evaluating the DBN classifier increases the computational cost compared to the MC alternative.

4.5 DBN$_{\text{OVA+MC}}$ DBN$_{\text{OVA}}$ $\xrightarrow{\text{TH}}$ DBN$_{\text{OVA+MC}}$

We now combine the previous two models and extend the DBN classifier with the one-vs-all as well as with the multi-class classifiers. Hence, the DBN contains a corresponding variable for each of the available classifiers. In order to reduce the computational cost, we first only compute the cheap features, evaluate each of the OVA classifiers and feed them to the DBN. In case the most likely class predicted by the DBN has a certain probability, i.e. it is above a certain threshold (TH), we proceed by picking this class. In case the most likely class comes with a rather low probability, we do additionally compute the expensive features, evaluate the MC classifier and feed it to the DBN.

The **advantage** of this model is that it nicely allows us to explore the accuracy versus computational cost trade-off space by changing the threshold (TH). A high threshold would require a considerable amount of MC classifier evaluations but most likely also increases the classification accuracy. A low threshold, on the other hand, would result in a lower classification accuracy but also significantly reduces the computational cost. The **disadvantage** is that the use of a threshold is not robust as it ignores how likely it is that the predicted class changes by evaluating additional classifiers.

4.6 $SDP_{OFFLINE}$ $DBN_{OVA\text{-}SDP} \xrightarrow{TH} DBN_{OVA} \xrightarrow{TH} DBN_{OVA+MC}$

We can now ask ourselves the question whether it is always useful to evaluate each of the OVA classifiers or to additionally evaluate the MC classifier in case a certain threshold is not reached. Consider the following scenario; Our DBN classifier predicts a data instance to belong to a particular class with a rather high probability. For the next data instance, the single OVA classifier of this particular class predicts the data instance belongs again to that class. Intuitively, there is only a small probability that observing additional classifiers will change the predicted class.

The above reasoning is exactly what we can compute with the SDP. As we will see in our experimental evaluation, the SDP in this case is indeed very high denoting there is a high probability we will still predict the same class, even after observing additional classifiers. Important to note is that, in this case, we can precompute the SDP offline and, hence, we do not need to spend any additional computational resources at run-time.

The **advantage** of this model is that it allows us to further reduce the computational cost as we now start out by only evaluating a single one-vs-all classifier. The **disadvantage** is that we still have to rely on a threshold in case the single one-vs-all classifier predicts the class to be changed.

4.7 SDP_{ONLINE} $DBN_{OVA\text{-}SDP} \xrightarrow{TH} DBN_{OVA} \xrightarrow{SDP} DBN_{OVA+MC}$

In the previous model, we computed the SDP in an offline manner. This allowed us to decide whether we do need to observe additional classifiers in case the corresponding OVA classifier predicts a new data instance to belong to the same class as the previous data instance. In this model, we now aim to further optimize the opposite case, where the class of the new data instance is most likely not the same as the previous one.

Specifically, we now use the SDP in an online manner to investigate whether it is useful to additionally evaluate the MC classifier given that we have observed all OVA classifiers. While computing the SDP online comes with a certain cost, it will be less expensive compared to generating complex features in case the cost of the latter surpasses the cost of evaluating the OVA classifiers.

The **advantage** of this model is that it allows us to fully exploit the classification accuracy versus computational cost trade-off space, where we aim to maximally avoid the need to evaluate additional classifiers. The **disadvantage** of this approach is that we should aim to also reduce the number of SDP computations and therefore we again rely on a threshold to decide whether we compute the SDP or not.

5   Experiments

The goal of our experimental evaluation is to: (1) explore the classification accuracy versus computational cost trade-off space with the various models we introduced in the previous section and (2) investigate the use of the Same-Decision Probability for resource-aware classification.

5.1   Datasets

The first dataset (SINS) [6] aims to track the daily activities performed by a person in a home environment, based on the recordings of an acoustic sensor network. The daily activities are performed in a spontaneous manner and are recorded as a continuous acoustic stream. As a result, the dataset contains also the transitions of going from one activity towards the next activity. The data is recorded with a microphone and sampled at a rate of 16 kHz. We make use of the data recorded in the bathroom where the dataset distinguishes seven activities; showering, getting dry, shaving, toothbrushing, absence, vacuum cleaner and others.

The second benchmark dataset, Human Activity Recognition dataset (HAR) [2], aims to recognize the activity a person is performing based on the accelerometer and gyroscope signals measured by a smart phone. The data is sampled at a rate of 50 Hz, preprocessed with various filters, and then sampled in fixed-width sliding windows of 2.56 s and 50% overlap, i.e. 128 readings per window. The raw data as well as 561 precomputed features are available from the authors. The data distinguishes six activities but, in contrast to the SINS dataset, the data is collected by means of a scripted experiment and does not contain the transitions of one activity towards to next activity. We did however concatenate the data in order to introduce such transitions and clearly show the effect of temporal information in a controlled manner.

5.2   Feature Generation

For the SINS data, the expensive features are the mean and standard deviation of the MFCC, the delta, and acceleration coefficients as described in the original paper [6]. These features are computed on a sliding window of 15 s with a step size of 10 s. For each window, we have 84 complex features. For the cheap features, we first down-sample the original signal to 1.6 kHz and next compute the energy, zero-crossing rate, min and max on windows of 3 s within the 15 s window. Hence, for each window we have 20 cheap features.

For the HAR data, the expensive features are the 561 precomputed features available in the dataset computed on the 2.56 s sliding window. For the multi-class Convolutional Neural Networks classifier we make use of the inertial signals. For the cheap features, we use the mean, standard deviation, max, min and energy for each of the three axes of the time domain of the body acceleration, gravity acceleration and gyroscope signal. Hence, for each window (data instance) we have 45 cheap features.

5.3 Training of the Classifiers

Preparing data for a one-vs-all classifier leads to an unbalanced dataset as the 'other' label is much more common than the current class. We balance the data by duplicating data instances of the minority class and we add a small amount of Gaussian noise (0.05 of the standard deviation).

To train the Tree Augmented Naive Bayes classifiers, we make use of Weka with the default options [7]. To compile the obtained Bayesian Networks into an Arithmetic Circuit, in order to estimate their computational cost, we make use of ACE with the default options.[1] To train the CNN for the HAR data we make use of Keras with the hyper-parameters set as in the corresponding paper [17]. To model the Dynamic Bayesian Network, we make use of Problog where we use the default options for compilation and inference.[2]

For the SINS data we use 7-fold cross validation where each of the folds corresponds to one day of the recorded data. For the HAR data, we randomly generate 7 train and test sets, out of the train and test data provided by the dataset, and report average results.

5.4 Cost Computation

The computational cost of generating the features, evaluating the base classifiers, and computing the SDP in an online manner is shown in Table 1. For the TAN classifiers as well as the DBN we report the number of edges in the compiled Arithmetic Circuit representation. This is, up to a constant, identical to the required number of operations. For the CNN classifier we computed the number of required operations after training the network with the hyper-parameters described in the corresponding paper [17]. For the features we use an approximated number of operations based on our experience when implementing these in hardware. For the cheap features, this is a cost of 5 operations for each of the values in the fixed window. For expensive features on the HAR data, we assume a cost of 10 operations for each of the values in the fixed window. For the expensive features on the SINS data, we estimate the number of required operations to be 600k to compute the MFCC, delta and acceleration coefficients per second. Following Eq. (1), computing the SDP requires to evaluate the corresponding Bayesian network $2 \cdot N + 1$ times with N the number of classes in the data. In our case, online computations of the SDP is done on the DBN. We observe that the cost of computing the SDP in an online manner is rather low compared to computing the expensive features and evaluating the multi-class classifier.

---

[1] http://reasoning.cs.ucla.edu/ace/.
[2] https://dtai.cs.kuleuven.be/problog/.

Table 1. Computational cost (required number of operations) for features generation, evaluation of the base classifiers and online computation of the SDP. We use HAR_{DL} to refer to the HAR dataset with the CNN multi-class classifier

		SINS (x1000)	HAR (x1000)	HAR_{DL} (x1000)
ONE-VS-ALL	Features	500	29	29
	Base classifier	54	204	204
MULTI-CLASS	Features	10000	690	0
	Base classifier	581	965	7000
DBN	Base classifier	4	4	4
SDP (ONLINE)		52	52	52

5.5 Cost Versus Accuracy

The total cost and obtained classification accuracy for each of the different models introduced in the previous section is reported in Fig. 3. We show the average cost of classifying one data instance and additionally report this cost relative to $SDP_{OFFLINE}$. While DBN_{OVA} comes with the lowest cost, as it does not require any of the expensive features nor the multi-class classifier, it also comes with the lowest accuracy. As expected, the MC and DBN_{MC} models have the highest computation cost as they only rely on the expensive features and multi-class classifier. OVA performs slightly better than MC because if the OVA classifiers do not agree, the decision is forwarded to the MC method. Using a DBN model and deciding incrementally what features to query offers a substantial decrease in resource usage (e.g. on SINS, DBN_{OVA+MC} needs half the number of operations of OVA). Applying SDP to make decisions instead of a simple threshold further reduces the number of operations (e.g. on SINS about 35%).

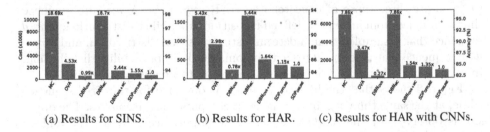

(a) Results for SINS.   (b) Results for HAR.   (c) Results for HAR with CNNs.

Fig. 3. Cost of the models (bars) versus accuracy (diamonds).

Influence of SDP. The result of the offline SDP computation, as discussed in Sect. 4.6, is depicted in Fig. 4. Such offline computation cannot take into account the exact probability distribution of the previously classified data instance.

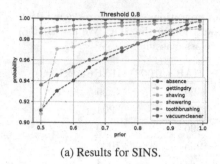
(a) Results for SINS.

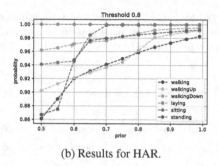
(b) Results for HAR.

Fig. 4. Offline computation of the SDP with a threshold of 0.8.

Therefore, we compute the SDP for different prior distributions (x-axis). We observe that for each of the activities, the SDP is rather high, meaning that the probability of classifying the data instance differently after observing additional features, i.e. evaluating additional classifiers, is rather low. This is also observed in Figs. 5a, 6a and 7a.

In order to investigate the SDP$_{\text{ONLINE}}$ method, we additionally introduce an *optimized* baseline model. This optimized model acts as the SDP$_{\text{OFFLINE}}$ but only picks the distribution computed after observing the MC classifier if the most likely state differs from the one before observing the MC classifier. This model represents the ideal case if we would be able to look ahead. The result for the different models for different thresholds is depicted in Figs. 5b, 6b and 7b. While the SDP$_{\text{ONLINE}}$ method is not able to completely meet with the *optimized* baseline model, it allows us to further trade off accuracy with computational resources.

Influence of the Temporal Sequence. The HAR dataset consists of separate sequences. To illustrate the effect of the DBN model that learns about sequences we generate three types of concatenated HAR sequences. A *realistic* sequence that is sampled from an expert defined transition probability distribution, a *fixed* sequence that is sampled from a deterministic transition distribution, and a *random* sequence that is sampled from a random transition probability distribution. In each of our other experiments, we made use of the *realistic* sequence.

Figure 8 shows that the highest accuracy is obtained for the *fixed* sequence while, at the same time, it requires the least computational resources. The opposite is observed for the *random* sequence, i.e. lowest accuracy for the highest computational cost. This clearly shows the benefit of the dynamic Bayesian network as it almost completely captures the deterministic transitions while it offers almost no benefit in case the order of activities is random.

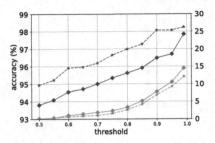

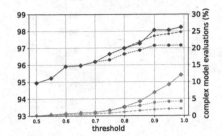

(a) Comparison of DBN$_{\text{OVA+MC}}$ (solid line) with SDP$_{\text{OFFLINE}}$ (dashed line).

(b) Comparison of SDP$_{\text{OFFLINE}}$ (solid line) with SDP$_{\text{ONLINE}}$ (dotted line) and the *optimized* model. (dashed line).

Fig. 5. Results for SINS.

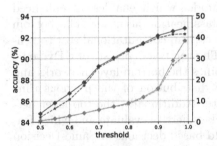

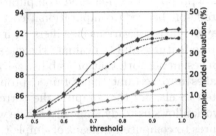

(a) Comparison of DBN$_{\text{OVA+MC}}$ (solid line) with SDP$_{\text{OFFLINE}}$ (dashed line).

(b) Comparison of SDP$_{\text{OFFLINE}}$ (solid line) with SDP$_{\text{ONLINE}}$ (dotted line) and the *optimized* model. (dashed line).

Fig. 6. Results for HAR.

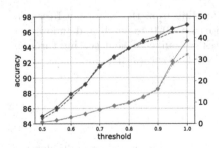

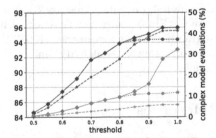

(a) Comparison of DBN$_{\text{OVA+MC}}$ (solid line) with SDP$_{\text{OFFLINE}}$ (dashed line).

(b) Comparison of SDP$_{\text{OFFLINE}}$ (solid line) with SDP$_{\text{ONLINE}}$ (dotted line) and the *optimized* model. (dashed line).

Fig. 7. Results for HAR with CNNs.

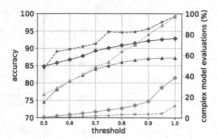

Fig. 8. Results for three different HAR sequences, being a *realistic* sequence (solid line) as also used in our other experiments, a *fixed* sequence (dashed line) and a *random* sequence (dotted line).

6   Conclusions

We propose a two-layered computational model which enables an enhanced trade-off between computational cost and classification accuracy. The bottom layer consist of various classifiers with each a certain computational cost to generate features and to evaluate the classifier. The top layer consists of a Dynamic Bayesian network that probabilistically combines the bottom layer networks and takes into account the obtained probability distribution of the previous time step to classify the compute the current state. Additionally, the top layer network allows us to dynamically decide which classifiers to evaluate and which features to compute by means of simple threshold based decisions or Same-Decision Probability. The latter is computed in an offline as well as online manner and allows us to further push the computational resources versus classification accuracy trade-off space. We validate our methods on the real-world SINS database, where domestic activities are recorded with an accoustic sensor network, as well as the Human Activity Recognition (HAR) benchmark dataset and show that resources can be dynamically minimized while maintaining a good accuracy.

Acknowledgements. This work has received funding from the European Research Council (ERC) under the European Union's Horizon 2020 research and innovation programme. Grant agreements 715037 Re-SENSE: Resource-efficient sensing through dynamic attention-scalability and 694980 SYNTH: Synthesising Inductive Data Models.

References

1. DCASE Challenge 2018: Monitoring of domestic activities based on multi-channel acoustics. http://dcase.community/challenge2018
2. Anguita, D., Ghio, A., Oneto, L., Parra, X., Reyes-Ortiz, J.L.: A public domain dataset for human activity recognition using smartphones. In: 21th European Symposium on Artificial Neural Networks, Computational Intelligence and Machine Learning. ESANN 2013 (2013)

3. Choi, A., Xue, Y., Darwiche, A.: Same-decision probability: a confidence measure for threshold-based decisions. Int. J. Approx. Reason. **53**(9), 1415–1428 (2012)
4. Darwiche, A.: New advances in compiling CNF to decomposable negation normal form. In: Proceedings of ECAI, pp. 328–332 (2004)
5. Darwiche, A.: Modeling and Reasoning with Bayesian Networks. Cambridge University Press, Cambridge (2009)
6. Dekkers, G., et al.: The SINS database for detection of daily activities in a home environment using an acoustic sensor network. In: Proceedings of the Detection and Classification of Acoustic Scenes and Events Workshop (DCASE) (2017)
7. Frank, E., Hall, M.A., Witten, I.H.: The WEKA workbench. In: Data Mining: Practical Machine Learning Tools and Techniques, 4th edn. Morgan Kaufmann, Burlington (2016)
8. Ghasemzadeh, H., Ostadabbas, S., Guenterberg, E., Pantelopoulos, A.: Wireless medical-embedded systems: a review of signal-processing techniques for classification. IEEE Sens. J. **13**, 423–437 (2013)
9. Gubbi, J., Buyya, R., Marusic, S., Palaniswami, M.: Internet of Things (IoT): a vision, architectural elements, and future directions. Future Gener. Comput. Syst. **29**(7), 1645–1660 (2013)
10. Kantorov, V., Laptev, I.: Efficient feature extraction, encoding and classification for action recognition. In: Proceedings of the IEEE Conference on Computer Vision and Pattern Recognition, pp. 2593–2600 (2014)
11. Korhonen, I., Parkka, J., Van Gils, M.: Health monitoring in the home of the future. IEEE Eng. Med. Biol. Mag. **22**(3), 66–73 (2003)
12. Mitra, J., Saha, D.: An efficient feature selection in classification of audio files. arXiv preprint arXiv:1404.1491 (2014)
13. Olascoaga, L.I.G., Meert, W., Bruyninckx, H., Verhelst, M.: Extending Naive Bayes with precision-tunable feature variables for resource-efficient sensor fusion. In: AI-IoT ECAI, pp. 23–30 (2016)
14. Oztok, U., Choi, A., Darwiche, A.: Solving PP PP-complete problems using knowledge compilation. In: Fifteenth International Conference on the Principles of Knowledge Representation and Reasoning (2016)
15. Pearl, J.: Probabilistic Reasoning in Intelligent Systems: Networks of Plausible Inference. Morgan Kaufmann, Burlington (1988)
16. Piatkowski, N., Lee, S., Morik, K.: Integer undirected graphical models for resource-constrained systems. Neurocomputing **173**, 9–23 (2016)
17. Ronao, C.A., Cho, S.B.: Human activity recognition with smartphone sensors using deep learning neural networks. Expert Syst. Appl. **59**(C), 235–244 (2016)
18. Shnayder, V., Hempstead, M., Chen, B.R., Allen, G.W., Welsh, M.: Simulating the power consumption of large-scale sensor network applications. In: Proceedings of the 2nd International Conference on Embedded Networked Sensor Systems, pp. 188–200. ACM (2004)
19. Vlasselaer, J., Meert, W., Van den Broeck, G., De Raedt, L.: Exploiting local and repeated structure in dynamic Bayesian networks. Artif. Intell. **232**, 43–53 (2016)
20. Xu, Z.E., Kusner, M.J., Weinberger, K.Q., Chen, M., Chapelle, O.: Classifier cascades and trees for minimizing feature evaluation cost. J. Mach. Learn. Res. **15**(1), 2113–2144 (2014)

ADS Financial/Security

Uncertainty Modelling in Deep Networks: Forecasting Short and Noisy Series

Axel Brando[1,3]([⊠]), Jose A. Rodríguez-Serrano[1]([⊠]), Mauricio Ciprian[2]([⊠]),
Roberto Maestre[2]([⊠]), and Jordi Vitrià[3]([⊠])

[1] BBVA Data and Analytics, Barcelona, Spain
{axel.brando,joseantonio.rodriguez.serrano}@bbvadata.com
[2] BBVA Data and Analytics, Madrid, Spain
{mauricio.ciprian,roberto.maestre}@bbvadata.com
[3] Departament de Matemàtiques i Informàtica, Universitat de Barcelona,
Barcelona, Spain
{axelbrando,jordi.vitria}@ub.edu

Abstract. Deep Learning is a consolidated, state-of-the-art Machine Learning tool to fit a function $y = f(x)$ when provided with large data sets of examples $\{(x_i, y_i)\}$. However, in regression tasks, the straightforward application of Deep Learning models provides a point estimate of the target. In addition, the model does not take into account the uncertainty of a prediction. This represents a great limitation for tasks where communicating an erroneous prediction carries a risk. In this paper we tackle a real-world problem of forecasting impending financial expenses and incomings of customers, while displaying predictable monetary amounts on a mobile app. In this context, we investigate if we would obtain an advantage by applying Deep Learning models with a Heteroscedastic model of the variance of a network's output. Experimentally, we achieve a higher accuracy than non-trivial baselines. More importantly, we introduce a mechanism to discard low-confidence predictions, which means that they will not be visible to users. This should help enhance the user experience of our product.

Keywords: Deep Learning · Uncertainty · Aleatoric models
Time-series

1 Introduction

Digital payments, mobile banking apps, and digital money management tools such as personal financial management apps now have a strong presence in the financial industry. There is an increasing demand for tools which bring higher interaction efficiency, improved user experience, allow for better browsing using different devices and take or help taking automatic decisions.

The Machine Learning community has shown early signs of interest in this domain. For instance, several public contests have been introduced since 2015.

© Springer Nature Switzerland AG 2019
U. Brefeld et al. (Eds.): ECML PKDD 2018, LNAI 11053, pp. 325–340, 2019.
https://doi.org/10.1007/978-3-030-10997-4_20

Fig. 1. Screenshots of BBVA's mobile app showing expected incomes and expenses. Global calendar view (left) and expanded view of one of the forecasts (right).

This notably includes a couple of Kaggle challenges[1,2,3] and the 2016 ECML Data Discovery Challenge. Some studies have addressed Machine Learning for financial products such as recommendation prediction [19], location prediction [22] or fraud detection [20].

Industrial Context. Our present research has been carried out within the above-mentioned context of digital money management functions offered via our organization's mobile banking app and web. This includes the latest "expense and income forecasting" tool which, by using large amounts of historical data, estimates customers' expected expenses and incomings. On a monthly basis, our algorithm detects recurrent expenses[4] (either specific operations or aggregated amounts of expenses within a specific category). We will feed the mobile app with the generated results, which enables customers to anticipate said expenses and, hence, plan for the month ahead.

This function is currently in operation on our mobile app and website. It is available to 5M customers, and generates hundreds of thousands of monthly visits. Figure 1 displays a screenshot.

The tool consists of several modules, some of which involve Machine Learning models. A specific problem solved by one of the modules is to estimate the amount of money a customer will spend in a specific financial category in the upcoming month. We have financial category labels attached to each operation. Categories include financial events, as well as everyday products and services such as ATM withdrawals, salary, or grocery shopping, among others (as we will explain hereafter). This problem can be expressed as a regression problem, where the input is a set of attributes of the customer history, while the output

---

[1] https://www.kaggle.com/c/santander-product-recommendation.

[2] https://www.kaggle.com/c/sberbank-russian-housing-market.

[3] https://www.kaggle.com/c/bnp-paribas-cardif-claims-management.

[4] From now on, we will refer to this as 'expenses' because our models treat income as "negative expenses".

is the monetary amount to be anticipated. However, the problem presents several unique challenges. Firstly, as we work with monthly data, the time series are short, and limited to a couple of years; thus, we fail to capture long-term periodicities. Secondly, while personal expenses data can exhibit certain regular patterns, in most cases series can appear erratic or can manifest spurious spikes; indeed, it is natural that not all factors for predicting a future value are captured by past values.

Indeed, preliminary tests with classical time series methods such as the Holt-Winters procedure yielded poor results. One hypothesis is that classical time series method perform inference on a per-series bases and therefore require a long history as discussed in [17].

With the aim of evolving this solution, we asked ourselves whether Deep Learning methods can offer a competitive solution. Deep Learning algorithms have become the state-of-the-art in fields like computer vision or automatic translation due to their capacity to fit very complex functions $\hat{y} = \phi(x)$ to large data sets of pairs (x, y).

Our Proposed Solution. This article recounts our experience in solving one of the main limitations of Deep Learning methods for regression: they do not take into account the variability in the prediction. In fact, one of the limitations is that the learned function $\phi(x)$ provides a point-wise estimate of the output target, while the model does not predict a probability distribution of the target or a range of possible values. In other words, these algorithms are typically incapable to assess how confident they are concerning their predictions.

This paper tackles a real-world problem of forecasting approaching customer financial expenses in certain categories based on the historical data available. We pose the problem as a regression problem, in which we build features for a user history and fit a function that estimates the most likely expense in the subsequent time period (dependent variable), given a data set of user expenses. Previous studies have shown that Deep Networks provide lower error rates than other models [8].

At any rate, Neural Networks are *forced* to take a decision for all the cases. Rather than minimising a forecast error for *all* points, we seek mechanisms to *detect* a small fraction of predictions for each user where the forecast is confident. This is done for two reasons. Since notifying a user about an upcoming expense is a value-added feature, we ask ourselves whether it is viable to reject forecasts for which we are not certain. Furthermore, as a user may have dozens of expense categories, this is a way of selecting relevant impending expenses.

To tackle the issue of prediction with confidence, potential solutions in the literature combine the good properties of Deep Learning to generically estimate functions with the probabilistic treatments including Bayesian frameworks of Deep Learning ([3, 4, 10, 18]). Inspired by these solutions, we provide a formulation of Deep regression Networks which outputs the parameters of a certain distribution that corresponds to an estimated target and an input-dependent variance (i.e. Heteroscedastic) of the estimate; we fit such a network by maximum likelihood estimation. We evaluate the network by both its accuracy and by its capability to select "predictable" cases. For our purpose involving millions of noisy time series, we performed a comprehensive benchmark and observed that it outperforms non-trivial baselines and some of the approaches cited above.

2   Method

This section presents the details of the proposed method (Fig. 2).

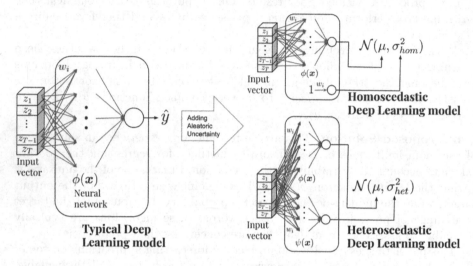

Fig. 2. Overview representation of the process to transform a typical Deep Learning model that has a single output given an input vector to a generic Aleatoric Deep Learning model with Gaussian distributions.

2.1   Deep Learning as a Point Estimation

A generic and practical solution to tackle the regression problem is Deep Learning as it is able to approximate any continuous function following the universal approximation theorem [16].

Straightforward implementations of Deep Learning models for regression do not typically consider a confidence interval (or score) of the estimated target. However, with our particular setting, it is critical that the model identifies the reliable forecasts, and can ignore forecasts where user-spending gets noisier. Furthermore, if we can find a correlation between the noisy time-series and the error of prediction, a strategy could be devised to improve the general performance of the system.

In this paper we denote $\hat{y} = \phi_{\boldsymbol{w}}(\boldsymbol{x})$ as the function computed by the Neural Network, with $\mathbf{x}$ a vector of inputs (the *attributes*) and $\hat{y}$ the output, i.e. the forecasted amount. We will denote the set of all weights by $\boldsymbol{w}$.

In particular, we will use two standard types of layers. On the one hand, we consider non-linear stacking of Dense layers. A Dense layer is a linear combination of the weights $\boldsymbol{w}$ with the input of the layer $\boldsymbol{z}$ passing through an activation function, act, i.e.

$$\boldsymbol{z}_{n+1} = \text{act}\left(\boldsymbol{w}\boldsymbol{z}_n\right)$$

On the other hand, a LSTM layer ([11,15]) which, like the other typical recurrent layers [6], maps a time-series with t steps to an output that can have t steps or fewer depending on our needs. In order to do this, for each of the steps, the input $\boldsymbol{z}_n^{(t)}$ will be combined at step t with the same weights and non-linearities for all the steps in the following way:

$$\boldsymbol{f}_t = sig\left(\boldsymbol{w}_f \cdot [\boldsymbol{h}_{t-1}, \boldsymbol{z}_n^{(t)}] + b_f\right)$$

$$\boldsymbol{i}_t = sig\left(\boldsymbol{w}_i \cdot [\boldsymbol{h}_{t-1}, \boldsymbol{z}_n^{(t)}] + b_i\right)$$

$$\boldsymbol{c}_t = \boldsymbol{f}_t * \boldsymbol{c}_{t-1} + \boldsymbol{i}_t * \tanh\left(\boldsymbol{w}_c \cdot [\boldsymbol{h}_{t-1}, \boldsymbol{z}_n^{(t)}] + b_c\right)$$

$$\boldsymbol{o}_t = sig\left(\boldsymbol{w}_o \cdot [\boldsymbol{h}_{t-1}, \boldsymbol{z}_n^{(t)}]^T + b_o\right)$$

$$\boldsymbol{h}_t = \boldsymbol{o}_t * \tanh\left(\boldsymbol{c}_t\right)$$

Where $\boldsymbol{w}_f, b_f, \boldsymbol{w}_i, b_i, \boldsymbol{w}_c, b_c, \boldsymbol{w}_o, b_o$ are the weights of the cell (shared for all the t steps) and sig are Sigmoid functions.

In the experimental section, we consider different Neural Network architectures by stacking Dense and/or LSTM layers. Fitting the network weights $\mathbf{w}$ involves minimizing a loss function, which is typically based on the standard *back-propagation* procedure. Nowadays, several optimizers are implemented in standard Deep Learning libraries which build on automatic differentiation after specifying the loss. All our Deep Learning models are implemented in Keras [7] with TensorFlow [1] backend and specific loss functions are discussed below.

2.2 Types of Uncertainty

With our system it will be critical to make highly accurate predictions, and, at the same time, assess the confidence of the predictions. We use the predictions to alert the user to potential impending expenses. Because of this, the cost of making a wrong prediction is much higher than not making any prediction at all. Considering the bias-variance trade-off between the predicted function ϕ and the real function to be predicted f, the sources of the error could be thought as:

$$\mathcal{E}(\boldsymbol{x}) = \underbrace{\left(\mathbb{E}\left[\phi(\boldsymbol{x})\right] - f(\boldsymbol{x})\right)^2}_{\text{Bias}^2} + \underbrace{\left(\mathbb{E}\left[\phi(\boldsymbol{x}) - \mathbb{E}[\phi(\boldsymbol{x})]\right]\right)^2}_{\text{Variance}} + \underbrace{\sigma_{cst}}_{\text{cst error}}$$

Where $\mathbb{E}(z)$ is the expected value of a random variable z.

To achieve a mechanism to perform rejection of predictions within the framework of Deep Learning, we will introduce the notion of variance in the prediction. According to the Bayesian viewpoint proposed by [9], it is possible to characterise the concept of uncertainty into two categories depending on the origin of the noise. On the one hand, if the noise applies to the model parameters, we will refer to *Epistemic uncertainty* (e.g. Dropout, following [10], could be seen as a way to capture the Epistemic uncertainty). On the other hand, if the noise occurs directly in the output given the input, we will refer to it as *Aleatoric uncertainty*. Additionally, Aleatoric uncertainty can further be categorised into two more categories: *Homoscedastic uncertainty*, when the noise is constant for all the outputs (thus acting as a "measurement error"), or *Heteroscedastic uncertainty* when the noise of the output also depends explicitly on the specific input (this kind of uncertainty is useful to model effects such as occlusions/superpositions of factors or variance in the prediction for an input).

2.3 A Generic Deep Learning Regression Network with Aleatoric Uncertainty Management

This section describes the proposed framework to improve the accuracy of our forecasting problem by modelling Aleatoric uncertainty in Deep Networks.

The idea is to pose Neural Network learning as a probabilistic function learning. We follow a formulation of Mixture Density Networks models [3], where we do not minimise the typical loss function (e.g. mean squares error), but rather the likelihood of the forecasts.

Following [3] or [18], they define a likelihood function over the output of a Neural Network with a Normal distribution, $\mathcal{N}(\phi(x), \sigma_{ale}^2)$, where ϕ is the Neural Network function and σ_{ale} is the variance of the Normal distribution. However, there is no restriction that does not allow us to use another distribution function if it is more convenient for our very noisy problem. So we decided to use a Laplace distribution defined as

$$LP(y \mid \phi(x), b_{ale}) = \frac{1}{2b_{ale}} \exp\left(-\frac{\mid y - \phi(x) \mid}{b_{ale}}\right), \tag{1}$$

which has similar properties and two (location and scale) parameters like the Normal one but this distribution avoids the square difference and square scale denominator of the Normal distribution with an empirically unstable behaviour in the initial points of the Neural Network weights optimisation or when the absolute error is sizeable. Furthermore, because of the monotonic behaviour of the logarithm, maximising the likelihood is equivalent to minimising the negative logarithm of the likelihood (Eq. 1), i.e. our loss function, $\mathcal{L}$, to minimise will be as follows

$$\mathcal{L}\left(w, b_{ale}; \{(x_i, y_i)\}_{i=1}^N\right) = -\sum_{i=1}^N \left[-\log(b_{ale}) - \frac{1}{b_{ale}} \mid y_i - \phi(x_i) \mid\right] \tag{2}$$

where w are the weights of the Neural Network to be optimised.

In line with the above argument, note that b_{ale} captures the Aleatoric uncertainty. Therefore, this formulation applies to both the Homoscedastic and Heteroscedastic case. In the former case, b_{ale} is a single parameter to be optimised which we will denote b_{hom}. In the latter, the $b_{ale}(x)$ is a function that depends on the input. Our assumption is that the features needed to detect the variance behaviours of the output are not directly related with the features to forecast the predicted value. Thus, $b_{ale}(x)$ can itself be the result of another input-dependent Neural Network $b_{ale}(x) = \psi(x)$, and all parameters of ϕ and ψ are optimised jointly. It is important to highlight that b_{ale} is a strictly positive scale parameter, meaning that we must restrict the output of the $\psi(x)$ or the values of the b_{hom} to positive values.

3   Experimental Settings and Results

3.1   Problem Setting

Our problem is to forecast upcoming expenses in personal financial records with a data set constructed with historical account data from the previous 24 months. The data was anonymized, e.g. customer IDs are removed from the series, and we do not deal with individual amounts, but with monthly aggregated amounts. All the experiments were carried out within our servers. The data consists of series of monetary expenses for certain customers in selected expense categories.

We cast this problem as a rolling-window regression problem. Denoting the observed values of expenses in an individual series over a window of the last T months as $z = (z_1, \ldots z_T)$, we extract L attributes $\pi(z) = (\pi_1(z), \ldots, \pi_L(z))$ from z. The problem is then to estimate the most likely value of the $(T + 1)$th month from the attributes, i.e fit a function for the problem $\hat{y} = \phi(\pi(z))$, with $y = z_{T+1}$, where we made explicit in the notation that the attributes depend implicitly on the raw inputs. To fit such a model we need a data set of N instances of pairs $(\pi(z), y)$, i.e. $\mathcal{D} = (\pi(Z), Y) = ((\pi(z_1), y_1), \ldots, (\pi(z_N), y_N))$.

To illustrate the nature and the variability of the series, in Fig. 3 we visualise the values of the raw series grouped by the clusters resulting from a k-means algorithm where $k = 16$. Prior to the clustering, the series were centered by their mean and normalised by their standard deviation so that the emergent clusters indicate scale-invariant behaviours such as periodicity.

As seen in Fig. 3, the series present clear behaviour patterns. This leads us to believe that casting the problem as a supervised learning one, using a large data set, is adequate and would capture general recurrent patterns.

Still, it is apparent that the nature of these data sets presents some challenges, as perceived from the variability of certain clusters in Fig. 3, and in some of the broad distributions of the output variable. In many individual cases, an expense may result from erratic human actions, spurious events or depending on factors not captured by the past values of the series. One of our objectives is to have an uncertainty value of the prediction to detect those series - and only communicate forecasts for those cases for which we are confident.

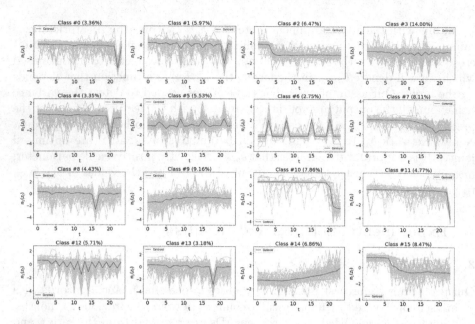

Fig. 3. Clustering of the normalised 24 points time-series by using the π_1 transformation. The grey lines are 100 samples and the blue line is the centroid for each cluster. (Color figure online)

Data Pre-processing Details. We generate monthly time series by aggregating the data over the contract id + the transaction category label given by an automatic internal categorizer which employs a set of strategies including text mining classifiers. While this process generates dozens of millions of time series in production, this study uses a random sample of 2 million time series (with $T = 24$) for the training set, and 1 million time series for the test set. In order to ensure a head-to-head comparison with the existing system, the random sample is taken only from the subset of series which enter the forecasting module (many other series are discarded by different heuristics during the production process). It is worth noticing that most of these series have non-zero target values. From these series we construct the raw data $z = (z_1, \ldots, z_T)$ and the targets $y = z_{T+1}$, from which we compute attributes.

As attributes, we use the T values of the series, normalized as follows:

$$\pi_1(z_i) = \begin{cases} \frac{z_i - \overline{z}}{std(z)} & \text{if } std(z) \geq \theta \\ z_i - \overline{z} & \text{if } std(z) < \theta \end{cases}.$$

Where θ is a threshold value. We also add the mean and standard deviation as attributes, which can be written as: $x_i \equiv (\pi_1(z_1), \ldots, \pi_1(z_T), \overline{z}, std(z))_i$.

The rationale for this choice is: (i) important financial behaviours such as periodicity tend to be scale-invariant, (ii) the mean recovers scale information (which is needed as the forecast is an real monetary value), (iii) the spread of the series could provide information for the uncertainty. We converged to these attributes after a few preliminary experiments.

3.2   Evaluation Measures

We evaluate all the methods in terms of an Error vs. Reject characteristic in order to take into account both the accuracy and the ability to reject uncertain samples. Given the point estimates $\hat{y}_i$ and uncertainty scores v_i of the test set, we can discard all cases with an uncertainty score above a threshold κ, and compute an error metric only over the set of accepted ones. Sweeping the value of κ gives rise to an Error vs. Keep trade-off curve: at any operating point, one reads the fraction of points retained and the resulting error on those – thus measuring simultaneously the precision of a method and the quality of its uncertainty estimate.

To evaluate the error of a set of points, we use the well-known mean absolute error (MAE):

$$\text{MAE}(\boldsymbol{y}, \boldsymbol{x}) = \frac{1}{N} \sum_{i=1}^{N} |\boldsymbol{y}_i - \phi(\boldsymbol{x}_i)|$$

which is the mean of absolute deviations between a forecast and its actual observed value. To specify explicitly that this MAE is computed only on the set of samples where the uncertainty score v is under κ, we use the following notation:

$$\text{MAE}_\kappa(\kappa; \boldsymbol{y}, \boldsymbol{x}) = \frac{\sum\limits_{i=1}^{N} |\boldsymbol{y}_i - \phi(\boldsymbol{x}_i)| \cdot [v_i < \kappa]}{\sum\limits_{i=1}^{N} [v_i < \kappa]} \tag{3}$$

And the kept fraction is simply $\text{Keep}(\kappa) = \sum [v_i < \theta] / N$. Note that $\text{MAE}_\kappa = $ MAE when κ is the maximum v value.

Our goal in this curve is to select a threshold such that the maximum error is controlled; since large prediction errors would affect the user experience of the product. In other words, we prefer not to predict everything than to predict erroneously.

Hence, with the best model we will observe the behaviour of the MAE error and order the points to predict according to the uncertainty score $v(\boldsymbol{x})$ for that given model.

In addition, we will add visualisations to reinforce the arguments of the advantages of using the models described in this paper that take into account Aleatoric uncertainty to tackle noisy problems.

3.3   Baselines and Methods Under Evaluation

We provide details of the methods we evaluate as well as the baselines we compare are comparing against. **Every method below outputs an estimate of the**

target along with an uncertainty score. We indicate in italics how each estimator and uncertainty score is indicated in Table 1. For methods which do not explicitly compute uncertainty scores, we will use the variance var(z) of the input series as a proxy.

Trivial Baselines. In order to validate that the problem cannot be easily solved with simple forecasting heuristics such as a moving average, we first evaluate three simple predictors: (i) the mean of the series $\hat{y} = \bar{z}$ (*mean*), (ii) $\hat{y} = 0$ (*zero*), and (ii) the value of the previous month $\hat{y} = z_T$ (*last*). In the three cases, we use the variance of the input as the uncertainty score var(z) (*var*).

Random Forest. We also compare against the forecasting method which is currently implemented in the expense forecasting tool. The tool consists of a combination of modules. The forecasting module is based on random forest (*RF**). Attributes include the values of the 24 months (z) but also dozens of other carefully hand-designed attributes. The module outputs the forecasted amount and a discrete confidence label (low/medium/high), referred to as *prec*. For the sake of completeness, we also repeat the experiment using the variance of the input (*var*) as the confidence score.

General Additive Model. We also compare against a traditional regression method able to output estimates and their distribution, specifically we use the Generalized Additive Models (*GAM*) [2] with its uncertainty score denoted as *SE*. GAM is a generalized lineal model with a lineal predictor involving a sum of smooth functions of covariates [12] that follows the following form,

$$g(\mu_i) = X_i^* \theta + f_1(x_{1i}) + f_1(x_{2i}) + f_3(x_{3i}, x_{4i}) + ...$$

where $\mu_i \equiv \mathbb{E}(Y_i)$, Y_i is the response variable that follows some exponential family distribution, X_i^* is a row of the model matrix for any strictly parametric model components, θ is the parametric vector and f_j are smooth functions of the covariates, x_j, and g is a known link function. We fit *GAM* in a simple way: $g(\mu_i) = f_1(z_1) + ... + f_N(z_N)$. The parameters in the model are estimated by penalized iteratively re-weighted least squares (P-IRLS) using Generalized Cross Validation (GCV) to control over-fitting.

Deep Learning Models. Finally, we will compare a number of Deep Learning models with various strategies to obtain the forecasted amount and an uncertainty score.

Dense Networks. We start with a plain Dense Network architecture (referred to as *Dense*). In particular, our final selected Dense model is composed by 1 layer of 128 neurons followed by another of 64 neurons, each of the layers with a ReLu activation, and finally this goes to a single output. We use the Mean Absolute Error (MAE) as loss function. Since a Dense regression Network does not provide a principled uncertainty estimate, again we use the variance of the series, var(z), as the uncertainty score.

Epistemic Models. We also evaluate Bayesian Deep Learning alternatives, which offer a principled estimate of the uncertainty score. These carry out an *Epistemic* treatment of the uncertainty (see Sect. 2.2), rather than our *Aleatoric* treatment. We consider two methods: (i) adding Dropout [10] (with a 0.5 probability parameter of Dropout) (referred to as *DenseDrop*), and (ii) Bayes-By-Backprop variation [4], denoted as *DenseBBB*. Both techniques model the distribution of the network weights, rather than weight values. This allows us to take samples from the network weights. Therefore, given a new input, we obtain a set of output samples rather than a single output; we can take the mean of those samples as the final target estimate and its standard deviation as our uncertainty score, i.e. $v(\boldsymbol{x}) = std\left(\{\phi_{w_i}(\boldsymbol{x})\}_i\right)$. In Table 1, we denote the uncertainty score of Dropout and BBB as *drop* and *BBB*, respectively. For the sake of completeness, we also consider the simple case of the variance (*var*).

Homoscedastic and Heteroscedastic Models. Here we consider fitting a network under the Aleatoric model (see 2), both for the Homoscedastic and Heteroscedastic cases. It is important to highlight that, in both cases, the way used to restrict the values of b_{ale} to positive values was by applying a function $g(b_{ale})$ to the values of the b_{hom} or output of $\psi(\boldsymbol{x})$ defined as the translation of the ELU function plus 1,

$$g(x) = ELU(\alpha, x) + 1 = \begin{cases} \alpha(e^x - 1) + 1 & \text{for } x < 0 \\ x + 1 & \text{for } x \geq 0 \end{cases}$$

In addition, in the Heteroscedastic case, we used the same Neural Network architecture for the "variance" approximation $\psi(\boldsymbol{x})$ than for the "mean" approximation $\phi(\boldsymbol{x})$.

In the Homoscedastic case, since the variance is constant for all the samples, we have to resort to the variance of the series as the uncertainty scores. In contrast, the Heteroscedastic variance estimate is input-dependent and can thus be used as the uncertainty score (denoted as b_{het}) in Table 1.

All the Deep Learning models were trained during 800 epochs with Early Stopping by using the 10% of the training set as a validation set.

LSTM Networks. To conclude, we will repeat the experiment by replacing the Dense Networks with long short-term memory (*LSTM*) networks, which are more suited to sequential tasks. Here we consider both types of aletoric uncertainty (i.e. *LSTMHom* and *LSTMHet*). In this particular case, the architecture uses two LSTM layers of 128 neurons followed by a 128 neurons Dense layer, and finally a single output.

3.4 Results

Models Comparison. In Fig. 4 we present a visual comparison between the Error-Keep curves (Eq. 3) of RF* (indicating the points of low/medium/high

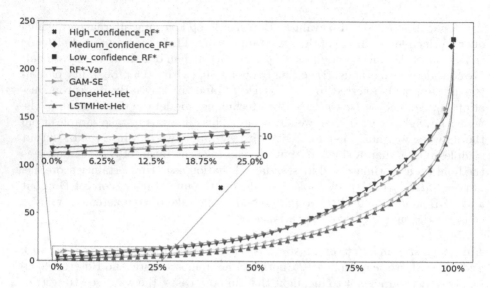

Fig. 4. Error-keep curve: MAE versus fraction of the samples kept obtained by selected forecasting methods, when cutting at different thresholds of their uncertainty scores.

confidence), RF* using *var* as confidence score, the GAM (which uses its own uncertainty estimate), and the best Dense and LSTM models (which, as will be detailed later, use the Heteroscedastic uncertainty estimates).

First we notice that the Heteroscedastic solutions are better than all the other previous solutions. This confirms our assumption that, for our given noisy problem, taking into account the variability of the output provides better accuracy at given retention rates.

For a more detailed analysis, Table 1 shows the MAE values of all the compared methods for several cut-off points of the Error-Keep curve, namely for keep = 25%, 50%, 75%, 100%, and additionally for 41% and 99.5% (for direct comparison to the currently implemented RF* method, because these are the cut-off values obtained by the RF* method with confidence = high and confidence = medium). *Note that we are more interested in low values of keep (typically below 50%)*, as they correspond to "selecting" the most confident samples; but we show all the percentages for completeness.

For Deep Learning models, to avoid a known issue of sensitivity with respect to the random seeds, we repeated all experiments 6 times with different random initialisations and report the mean and standard deviation over the 6 runs.

Table 1 shows different interesting aspects. First, we confirm that the best performing model is the LSTM network with the proposed Heteroscedastic uncertainty treatment (except in the points where keep >99.5%. However, these are not of practical importance because there is virtually no rejection, and the total number of errors are high for all the methods). Had we not considered LSTM networks as an option, the best performing model would still be a Dense Network with Heteroscedastic uncertainty. We also observed that the

Table 1. Errors (MAE) of each method at points of the Error-Keep curve corresponding to Keep = K. Each row corresponds to the combination of a predictor and an uncertainty score (described in the main text).

Predictor + Uncertainty	K = 25%	K = 41%	K = 50%	K = 75%	K = 99.5%	K = 100%
mean + var	36.93	54.09	67.54	152.84	417.93	491.68
Zeros + var	36.95	54.21	67.75	154.16	429.29	504.45
Last + var	50.25	75.19	94.53	203.27	474.12	606.58
GAM + var	16.37	23.41	29.66	68.83	191.73	7539.91
GAM + SE	13.19	20.97	27.30	62.62	191.86	7539.91
RF* + prec	N/A	76.08	N/A	N/A	224.95	232.11
RF* + var	11.69	18.68	25.21	66.90	177.15	232.11
Dense + var	12.3 ± .24	18.1 ± .26	23.5 ± .26	56.3 ± .28	146. ± .48	193. ± .49
DenseDrop + var	14.9 ± .19	20.8 ± .13	26.4 ± .07	61.1 ± .30	160. ± .80	208. ± 1.0
DenseDrop + Drop	16.1 ± 1.3	23.2 ± .91	29.5 ± .76	64.9 ± .62	162. ± 1.4	208. ± 1.0
DenseBBB + var	16.1 ± 1.6	23.5 ± 1.9	30.6 ± 2.2	79.8 ± 4.4	198. ± 5.2	248. ± 5.6
DenseBBB + BBB	19.9 ± 1.1	28.5 ± 1.2	36.0 ± 1.7	89.4 ± 4.6	199. ± 4.7	248. ± 5.6
DenseHom + var	12.3 ± .23	18.1 ± .22	23.5 ± .22	56.3 ± .30	145. ± .89	193. ± .84
DenseHet + var	10.7 ± .08	16.4 ± .10	21.7 ± .11	55.0 ± .30	150. ± .97	199. ± .84
DenseHet + b_{het}	8.24 ± 1.3	12.9 ± .90	18.1 ± .59	45.3 ± .67	153. ± 1.1	199. ± .84
LSTM + var	11.6 ± .13	17.6 ± .12	23.1 ± .13	55.7 ± .43	146. ± .81	205. ± 1.2
LSTMHom + var	11.9 ± .37	18.0 ± .59	23.5 ± .76	56.6 ± 2.0	150. ± 9.1	210. ± 14.
LSTMHet + var	10.5 ± .03	16.2 ± .07	21.6 ± .10	53.9 ± .17	147. ± 1.3	218. ± 1.6
LSTMHet + b_{het}	5.04 ± .23	10.7 ± .41	15.2 ± .24	41.1 ± .54	149. ± 2.9	218. ± 1.6

performance ranking of the incremental experiments between Dense and LSTM is consistent.

We also note that, for this particular task, the Aleatoric methods (both Homoscedastic and Heteroscedastic) perform better than Epistemic models such as Dropout or BBB. We believe this to be specific to the task at hand where we have millions of short and noisy time series. We conclude that the variability of human spending, with complex patterns, erratic behaviors and intermittent spendings, is captured more accurately by directly modelling an input-dependent variance with a complex function, than by considering model invariance (which may be better suited for cases where the data is more scarce or the noise smoother).

Another interesting observation is that the Random Forest baseline used more attributes than the proposed solutions based on Deep Learning, which just used T values of the historical time series. This confirms that fitting a Deep Network on a large data set is a preferred solution; even more when being able to model the uncertainty of the data, as is the case here.

Last but not least, comparing the same uncertainty score with a certain model and its Heteroscedastic version we also observe that there is an accuracy improvement. This means that taking into account uncertainty in the training process, in our noisy problem, not only gives to us an uncertainty score to reject uncertain samples but also it helps in order to predict better.

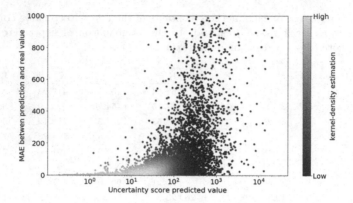

Fig. 5. Correlation between MAE of prediction and the real $T + 1$ value and its uncertainty score of random selection of $10,000$ time series of the test set. The colours represent the density of the zones by using Gaussian kernels. (Color figure online)

Error-Uncertainty Score Correlation. An interesting question is whether or not there exists any relationship between the errors of a model and the uncertainty scores v it provides. In Fig. 5 we show the correlation between the MAE of 10.000 randomly selected points of the test set and the uncertainty score of the best Heteroscedastic model expressed in the logarithm scale. To reduce the effect of clutter, we used a colour-map for each point that represents the density of the different parts of the Figure as resulting from a Gaussian Kernel Density Estimate.

This figure exhibits two main regimes. On the one hand, for high uncertainty scores (>100), we observe scattered samples (considering their purple colour) where the errors range from very low to very high values and do not seem to align with the uncertainly scores. Upon inspection, many of these series correspond to what humans would consider "unpredictable": e.g. an expense in month $T + 1$ which was much larger than any of the observed T expenses, or series without a clear pattern. On the other hand, we observe a prominent high-concentration area (the yellowish one) of low-error and low-uncertainty values; indicating that, by setting the threshold under a specific value of the uncertainty score, the system mostly selects low-error samples.

4   Related Work

In the case of using Deep Learning for classification, it is common to use a Softmax activation function in the last layer [5]. This yields a non-calibrated probability score which can be used heuristically as the confidence score or calibrated to a true probability [13]. However, in the case of regression, the output variables are not class labels and it is not possible to obtain such scores. On the other hand, [18] introduces the idea of combining different kinds of Aleatoric and Epistemic uncertainties. Nevertheless, as we saw in Table 1, the use of Epistemic uncertainty for our problem worsens the accuracy and the general performance of the model.

While there have been several proposals to deal with uncertainties in Deep Learning, they boil down to two main families. On the one hand, some approaches consider the uncertainty of the output. Typically, they construct a network architecture so that its output is not a point estimate, but a distribution [3]. On the other hand, some approaches consider the uncertainty of the model. These apply a Bayesian treatment to the optimisation of the parameters of a network ([4,10,14,21]). In the present work we observe that if a highly noisy problem can change the loss function in order to minimise a likelihood function as in [3], and by introducing some changes explained above, we are provided with a significant improvement which is crucial to identifying forecasts with a high degree of confidence and even improve the accuracy.

We are dealing with a problem that contains high levels of noise. To make predictions is therefore risky. These are the two reasons why it was our goal to find a solution that improves the mean-variance solution that can be a "challenging" solution. In order to do that, we grouped several theories proposed in [18] and [3] to create a Deep Learning model that takes into account the uncertainty and provides a better performance.

5   Conclusion

We explore a new solution for an industrial problem of forecasting real expenses of customers. Our solution is based on Deep Learning models for effectiveness and solve the challenge of uncertainty estimation by learning both a target output and its variance, and performing maximum likelihood estimation of the resulting model that contains one network for the target output and another for its variance. We show that this solution obtains better error-reject characteristics than other (traditional and Deep) principled models for regression uncertainty estimation, and outperforms the characteristic that would be obtained by the current industrial system in place. While Epistemic models such as Dropout or BBB did not improve the performance in this specific task, we are already working in combining them with our Aleatoric treatment to consider both types of uncertainty in the same model. This is considered future work. We also highlight that, while the present model seems to be able to detect confident predictions, it still lacks mechanisms to deal with the "unknown unknowns" problem; and believe that incorporating ideas such as those in [13] may help in future work.

Acknowledgements. We gratefully acknowledge the Industrial Doctorates Plan of Generalitat de Catalunya for funding part of this research. The UB acknowledges the support of NVIDIA Corporation with the donation of a Titan X Pascal GPU and recognizes that part of the research described in this chapter was partially funded by TIN2015-66951-C2, SGR 1219. We also thank Alberto Rúbio and César de Pablo for insightful comments as well as BBVA Data and Analytics for sponsoring the industrial PhD.

References

1. Abadi, M., et al.: Tensorflow: a system for large-scale machine learning. In OSDI, vol. 16, pp. 265–283, November 2016
2. Anderson-Cook, C.M.: Generalized Additive Models: An Introduction with R. American Statistical Association, UK (2007)
3. Bishop, C.M.: Mixture density networks, p. 7. Technical report NCRG/4288. Aston University, Birmingham, UK (1994)
4. Blundell, C., Cornebise, J., Kavukcuoglu, K., Wierstra, D.: Weight uncertainty in neural networks. In: Proceedings of the 32nd International Conference on International Conference on Machine Learning, vol. 37, pp. 1613–1622, July 2015
5. Bridle, J.S.: Probabilistic interpretation of feedforward classification network outputs, with relationships to statistical pattern recognition. In: Soulié, F.F., Hérault, J. (eds.) Neurocomputing. NATO ASI Series, vol. 68, pp. 227–236. Springer, Heidelberg (1990). https://doi.org/10.1007/978-3-642-76153-9_28
6. Cho, K., et al.: Learning phrase representations using RNN encoder-decoder for statistical machine translation. arXiv preprint arXiv:1406.1078 (2014)
7. Chollet, F.: Keras: deep learning library for Theano and TensorFlow (2015)
8. Ciprian, M., et al.: Evaluating uncertainty scores for deep regression networks in financial short time series forecasting. In: NIPS Workshop on Machine Learning for Spatiotemporal Forecasting (2016)
9. Der Kiureghian, A., Ditlevsen, O.: Aleatory or epistemic? Does it matter? Struct. Saf. 31(2), 105–112 (2009)
10. Gal, Y., Ghahramani, Z.: Dropout as a Bayesian approximation: representing model uncertainty in deep learning. In: ICML, June 2016
11. Gers, F.A., Schmidhuber, J., Cummins, F.: Learning to forget: continual prediction with LSTM. IET Digital Library (1999)
12. Marra, G., Wood, S.N.: Coverage properties of confidence intervals for generalized additive model components. SJS 39(1), 53–74 (2012)
13. Hendrycks, D., Gimpel, K.: A baseline for detecting misclassified and out-of-distribution examples in neural networks. arXiv preprint arXiv:1610.02136 (2016)
14. Hernández-Lobato, J.M., Adams, R.: Probabilistic backpropagation for scalable learning of Bayesian neural networks. In ICML, pp. 1861–1869, June 2015
15. Hochreiter, S., Schmidhuber, J.: Long short-term memory. Neural Comput. 9(8), 1735–1780 (1997)
16. Hornik, K., Stinchcombe, M., White, H.: Multilayer feedforward networks are universal approximators. Neural Netw. 2(5), 359–366 (1989)
17. Hyndman, R.J., Koehler, A.B.: Another look at measures of forecast accuracy. Int. J. Forecast. 22(4), 679–688 (2006)
18. Kendall, A., Gal, Y.: What uncertainties do we need in Bayesian deep learning for computer vision? In: NIPS, pp. 5574–5584 (2017)
19. Mitrović, S., Singh, G.: Predicting branch visits and credit card up-selling using temporal banking data. arXiv preprint arXiv:1607.06123 (2016)
20. Mutanen, T., Ahola, J., Nousiainen, S.: Customer churn prediction-a case study in retail banking. In: Proceedings of ECML/PKDD Workshop on Practical Data Mining, pp. 13–19, September 2006
21. Rasmussen, C.E.: A practical Monte Carlo implementation of Bayesian learning. In: Advances in Neural Information Processing Systems, pp. 598–604 (1996)
22. Wistuba, M., Duong-Trung, N., Schilling, N., Schmidt-Thieme, L.: Bank card usage prediction exploiting geolocation information. arXiv preprint arXiv:1610.03996 (2016)

Using Reinforcement Learning to Conceal Honeypot Functionality

Seamus Dowling[1]($\boxtimes$) ⓘ, Michael Schukat[2], and Enda Barrett[2]

[1] Galway Mayo Institute of Technology, Castlebar, Mayo, Ireland
seamus.dowling@gmit.ie
[2] National University of Ireland Galway, Galway, Ireland

Abstract. Automated malware employ honeypot detecting mechanisms within its code. Once honeypot functionality has been exposed, malware such as botnets will cease the attempted compromise. Subsequent malware variants employ similar techniques to evade detection by known honeypots. This reduces the potential size of a captured dataset and subsequent analysis. This paper presents findings on the deployment of a honeypot using reinforcement learning, to conceal functionality. The adaptive honeypot learns the best responses to overcome initial detection attempts by implementing a reward function with the goal of maximising attacker command transitions. The paper demonstrates that the honeypot quickly identifies the best response to overcome initial detection and subsequently increases attack command transitions. It also examines the structure of a captured botnet and charts the learning evolution of the honeypot for repetitive automated malware. Finally it suggests changes to an existing taxonomy governing honeypot development, based on the learning evolution of the adaptive honeypot. Code related to this paper is available at: https://github.com/sosdow/RLHPot.

Keywords: Reinforcement learning · Honeypot · Adaptive

1 Introduction

Honeypots have evolved to match emerging threats. From "packets found on the internet" in 1993 [1] to targeting and capturing IoT attacks [2], honeypot development has become a cyclic process. Malware captured on a honeypot is retrospectively analysed. This analysis informs defence hardening and subsequent honeypot redevelopment. Because of this, honeypot contribution to security is considered a reactive process. The value of honeypot deployment comes from the captured dataset. The longer an attack interaction can be maintained, the larger the dataset and subsequent analysis. Global honeypot projects track emerging threats [3]. Virtualisation technologies provide honeypot operators with the means of abstracting deployments from production networks and bare metal infrastructure [4]. To counter honeypot popularity, honeypot detection tools were developed and detection techniques were incorporated into malware deployments [5]. These have the effect of ending an attempted compromise

© Springer Nature Switzerland AG 2019
U. Brefeld et al. (Eds.): ECML PKDD 2018, LNAI 11053, pp. 341–355, 2019.
https://doi.org/10.1007/978-3-030-10997-4_21

once it is discovered to be a honeypot. The only solution is to modify the honeypot or incorporate a fix into a new release or patch. This is a cyclic process. The rapid mutation of malware variants often makes this process redundant. For example, the recent Mirai bot [6] exploited vulnerabilities in Internet of Things (IoT) deployments. IoT devices are often constrained reduced function devices (RFD). Security measures can be restricted and provide an opportunity for compromise. When the Mirai bot source code was made public, it spawned multiple variants targeting different attack vectors. Bots and their variants are highly automated. Human social engineering may play a part in how the end devices are compromised. Botmasters communicate with command and control (C&C) to send instructions to the compromised hosts. But to generate and communicate with a global botnet of compromised hosts, malware employs highly automated methods. Coupled with honeypot detection techniques, it becomes an impossible task for honeypot developers to release newer versions and patches to cater for automated malware and their variants. This paper describes a reinforcement learning approach to creating adaptive honeypots that can learn the best responses to attack commands. In doing so it overcomes honeypot detection techniques employed by malware. The adaptive honeypot learns from repetitive, automated compromise attempts. We implement a state action space formalism designed to reward the learner for prolonging attack interaction. We present findings on a live deployment of the adaptive honeypot on the Internet. We demonstrate that after an initial learning period, the honeypot captures a larger dataset with four times more attack command transitions when compared with a standard high interaction honeypot. Thus we propose the following contributions over existing work:

- Whilst the application of reinforcement learning to this domain is not new, previous developments facilitated human attack interactions which is not relevant for automated malware. Our proposed state action space formalism is new, targeting automated malware and forms a core contribution of this work.
- Presents findings on a live deployment with a state action space formalism demonstrating intelligent decision making, overcoming initial honeypot detection techniques.
- Demonstrates that using the proposed state action space formalism for automated malware produces a larger dataset with four times more command transitions compared to a standard high interaction honeypot.
- Demonstrates that our proposed state action space formalism is more effective than similar deployments. Previous adaptive honeypots using reinforcement learning collected three times more command transitions when compared to a high interaction honeypot.
- Demonstrates that the collected dataset can be used in a controlled environment for policy evaluation, which results in a more effective honeypot deployment.

This article is organised as follows:

Section 2 presents the evolution of honeypots and the taxonomy governing their development. This section also addresses honeypot detection measures used by malware and the use of reinforcement learning to create adaptive honeypots capable of learning from attack interactions.

Section 3 outlines the implementation of reinforcement learning within a honeypot, creating an adaptive honeypot. It rewards the learning agent for overcoming detection attempts and prolonging interaction with automated malware.

Section 4 presents findings from two experiments. The first experiment is a live deployment of the adaptive honeypot on the Internet, which captures automated attack interactions. The second is a controlled experiment, which evaluates the learning policies used by the adaptive honeypot.

Sections 5 and 6 discuss the limitations of the adaptive honeypot model that provides for future work in this area and our conclusions.

2   Related Work

2.1   Honeypot Evolution

In 2003, Spitzner defined a honeypot as a "security resource whose value lies in being probed, attacked or compromised" [7]. After a decade of monitoring malicious traffic [1], honeypots were seen as a key component in cybersecurity. Capturing a successful compromise and analysing their methods, security companies can harden subsequent defences. This cycle of capture-analyse-harden-capture, is seen as a reactive process. Virtualisation allowed for the easy deployment of honeypots and gave rise to popular deployments such as Honeyd [4]. This increase in honeypot popularity raised the question of their role. To address this, Zhang [8] introduced honeypot taxonomy. This taxonomy identifies security as the role or class of a honeypot, and could have prevention, detection, reaction or research as values. Deflecting an attacker away from a production network, by luring them into a honeypot, has a prevention value. Unauthorized activity on a honeypot is red flagged immediately providing a detection value. Designing a honeypot to maintain an attackers interest, by offering choices or tokens [9] has a reaction value. Finally the research value gives insights into the behaviour and motivation of an attacker. As more devices connected and threats evolved on the Internet, Seifert [10] introduced a new taxonomy, a summery of which is displayed in Table 1. The possible combinations for class/value from Table 1 informed the development of complex honeypots. The purpose of honeypots developed under this taxonomy is to collect data for the retrospective analysis of malware methods. However, firewalls and intrusion prevention and detection systems (IPS, IDS) became standard IT infrastructure for proactively limiting malware infections. Therefore honeypots were briefly considered as alternatives to IDS and IPS [11, 12].

Table 1. Seifert's taxonomy for honeypot development [10]

Class	Value	Note
Interaction level	High	High degree of functionality
	Low	Low degree of functionality
Data capture	Events	Collect data about changes in state
	Attacks	Collect data about malicious activity
	Intrusions	Collect data about security compromises
	None	Do not collect data
Containment	Block	Identify and block malicious activity
	Diffuse	Identify and mitigate against malicious activity
	Slow down	Identify and hinder malicious activity
	None	No action taken
Distribution appearance	Distributed	Honeypot is or appears to be composed of multiple systems
	Standalone	Honeypot is or appears to be one system
Communications interface	Network interface	Directly communicated with via a NIC
	Non-network interface	Directly communicated with via interface other than NIC (USB etc.)
	Software API	Honeypot can be interacted with via a software API (SSH, HTTP etc.)
Role in multi-tiered	Client	Honeypot acts as a server
Architecture	Server	Honeypot acts as a client

2.2 Anti-honeypot and Anti-detection

Honeypots can be deployed simply and quickly with virtualization tools and cloud services. Seiferts taxonomy from Table 1 provides the framework to create modern, relevant honeypots. For example, ConPot is a SCADA honeypot developed for critical industrial IoT architectures [13]; IoTPot is a bespoke honeypot designed to analyze malware attacks targeting IoT devices [2]. Botnets provide a mechanism for global propagation of cyber attack infection and control. They are under the control of a single C&C [14]. A typical botnet attack will consist of 2 sets of IP addresses. The first set of IPs is the compromised hosts. These are everyday compromised machines that are inadvertently participating in an attack. The second set of IPs is the C&C reporters and software loaders. These are the sources from which the desired malware is downloaded. There is a multitude of communication channels opened between C&C, report and loader servers, bot victims and target. The main methods of communication are IRC (Internet Relay Chat), HTTP and P2P based models where bots use peer-to-peer communications. Botmasters obfuscate visibility by changing the C&C connection channel. They use Dynamic DNS (DDNS) for botnet communication, allowing them to shut down a C&C server on discovery and start up a new server for uninterrupted attack service. Malware developers became aware of the existence honeypots, capturing and analyzing attack activity. To counter this, dedicated honeypot detection tools were developed and evasion techniques we designed into malware [15]. Tools such as Honeypot Hunter performed tests to identify a honeypot [5]. False services were created and connected to by the anti-honeypot tool. Honeypots are predominately designed to prolong attacker interaction and therefore pretend to facilitate the creation and execution of these false services. This immediately tags the system as a honeypot. With the use of virtualization for honeypot deployments [4], anti-detection techniques issued simple kernel commands to identify the presence of virtual infrastructure instead of bare metal [16]. New versions of honeypots are redesigned and redeployed continuously to counter new malware methods. More recently, the Mirai botnet spawned multiple variants targeting IoT devices through various attack vectors [6]. Honeypots played an active part in capturing and analyzing the Mirai structure [17]. Variants were captured on Cowrie, a newer version of the popular Kippo honeypot [18]. Analysis of the Mirai variants found that Cowrie failed initial anti-honeypot tests before honeypot functionality was manually amended [19]. Examining the structure of the Mirai variant [20], when a "mount" command is issued, the honeypot returns it's standard response at which point the attack ends. It indicates that the variant is implementing anti-detection techniques. Virtual machine (VM) aware botnets, such as Conficker and Spybot scan for the presence of a virtualized environment and can refuse to continue or modify its methods [21].

2.3   Adaptive Honeypots

Machine learning techniques have previously been applied to honeypots. These techniques have analysed attacker behaviour retrospectively on a captured dataset. Supervised and unsupervised methods are used to model malware interaction and to classify attacks [22–24]. This analysis is a valuable source of information for security groups. However, by proactively engaging with an attacker, a honeypot can prolong interaction and capture larger datasets potentially leading to better analysis. To this end, the creation of intelligent, adaptive honeypots has been explored. Wagener [25] uses reinforcement learning to extract as much information as possible from the intruder. A honeypot called Heliza was developed to use reinforcement learning when engaging an attacker. The honeypot implemented behavioural strategies such as blocking commands, returning error messages and issuing insults. Previously, he had proposed the use of game theory [26] to define the reactive action of a honeypot towards attacker's behaviour. Pauna [27] also presents an adaptive honeypot using the same reinforced learning algorithms and parameters as Heliza. He proposes a honeypot named RASSH that provides improvements to scalability, localisation and learning capabilities. It does this by using newer libraries with a Kippo honeypot and delaying responses to frustrate human attackers. Both Heliza and RASSH use PyBrain [28] for their implementation of reinforcement learning and use actions, such as insult and delay targeting human interaction. Whilst Heliza and RASSH implement reinforcement learning to increase command transitions from human interactions, our work pursues the goal of increasing command transitions to overcome automated honeypot detection techniques.

2.4   Reinforcement Learning in Honeypots

Reinforcement learning is a machine learning technique in which a learning agent learns from its environment, through trial and error interactions. Rather than being instructed as to which action it should take given a specific set of inputs, it instead learns based on previous experiences as to which action it should take in the current circumstance.

Markov Decision Processes. Reinforcement learning problems can generally be modelled using Markov Decision Processes (MDPs). In fact reinforcement learning methods facilitate solutions to MDPs in the absence of a complete environmental model. This is particularly useful when dealing with real world problems such as honeypots, as the model can often be unknown or difficult to approximate. MDPs are a particular mathematical framework suited to modelling decision making under uncertainty. A MDP can typically be represented as a four tuple consisting of states, actions, transition probabilities and rewards.

- S, represents the environmental state space;
- A, represents the total action space;
- $p(.|s, a)$, defines a probability distribution governing state transitions $s_{t+1} \sim p(.|s_t, a_t)$;
- $q(.|s, a)$, defines a probability distribution governing the rewards received $R(s_t, a_t) \sim q(.|s_t, a_t)$;

S the set of all possible states represents the agent's observable world. At the end of each time period t the agent occupies state $s_t \in S$. The agent must then choose an action $a_t \in A(s_t)$, where $A(s_t)$ is the set of all possible actions within state s_t. The execution of the chosen action, results in a state transition to s_{t+1} and an immediate numerical reward $R(s_t, a_t)$. Formula (1) represents the reward function, defining the environmental distribution of rewards. The learning agent's objective is to optimize its expected long-term discounted reward.

$$R_a s, s' = E\{r_{t+1} | s_t = s, a_t = a, s_{t+1} = s'\} \tag{1}$$

The state transition probability $p(s_{t+1}|s_t, a_t)$ governs the likelihood that the agent will transition to state s_{t+1} as a result of choosing a_t in s_t.

$$P_a s, s' = Pr\{s_{t+1} = s' | s_t = s, a_t = a\} \tag{2}$$

The numerical reward received upon arrival at the next state is governed by a probability distribution $q(s_{t+1}|s_t, a_t)$ and is indicative as to the benefit of choosing a_t whilst in s_t. In the specific case where a complete environmental model is known, i.e. (S, A, p, q) are fully observable, the problem reduces to a planning problem and can be solved using traditional dynamic programming techniques such as value iteration. However if there is no complete model available, then reinforcement learning methods have proven efficacy in solving MDPs.

SARSA Learning. During the reinforcement learning process the agent can select an action which exploits its current knowledge or it can decide to use further exploration. Reinforcement learning provides parameters to help the learning environment decide on the reward and exploration values. Figure 1 models a basic reinforcement learning interaction process. We have to consider how the model in Fig. 1, applies to adaptive honeypot development. Throughout its deployment, the honeypot is considered to be an environment with integrated reinforcement learning. We are using SSH as an access point and a simulated Linux server as a vulnerable environment. SSH is responsible for 62% of all compromise attempts [29]. Within this environment, the server has states that are examined and changed with bash scripts. Examples are iptables, wget, sudo, etc. The reinforcement learning agent can perform actions on these states such as to allow, block or substitute the execution of the scripts. The environment issues a reward to the agent for performing that action. The agent learns from this process as the honeypot is attacked and over time learns the optimum policy π^*, mapping the optimal action to be taken each time, for each state s. The learning process will eventually converge as the honeypot is rewarded for each attack episode. This temporal difference method for on-policy learning uses the transition from one state/action pair to the next state/action pair, to derive the reward. State, Action, Reward, State, Action also known as SARSA, is a common implementation of on-policy reinforcement learning (3). The reward policy Q is estimated for a given state s_t and a given action a_t. The environment is

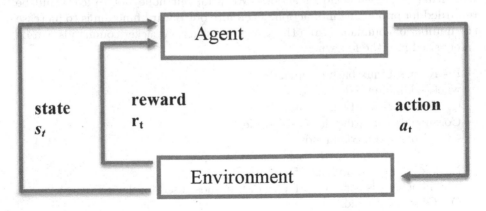

Fig. 1. Reinforcement learning model.

explored using a random component ϵ or exploited using learned Q values. The estimated Q value is expanded with a received reward r_t plus an estimated future reward $Q(s_{t+1}, a_{t+1})$, that is discounted (γ). A learning rate parameter is also applied (α).

$$Q(s_t, a_t) \leftarrow Q(s_t, a_t) + \alpha[r_{t+1} + \gamma Q(s_{t+1}, a_{t+1}) - Q(s_t, a_t)] \qquad (3)$$

As each attack is considered an episode, the policy is evaluated at the end of each episode. SARSA is implemented with the following parameters:

- epsilon-greedy policy
  The honeypot environment is unknown to the learning agent. We want it to learn without prior knowledge and eventually converge. To this end, we set our explorer to ϵ-greedy.
- Discount factor $\gamma = 1$
  γ is applied when a future reward is estimated. For our honeypot no discount factor is applied, as attack episodes are readily defined commands between SSH open and close events. This allows for retrospective reward calculations.
- Step size $\alpha = 0.5$
  A step size parameter is applied for the creation of the state/action space.

3   Reinforcement Learning Honeypot Implementation

Previous contributions demonstrate that there is a relationship between honeypot and malware development and evolution. Botnet traffic is highly automated as it attacks, compromises and reports without any intervention. From a malware developer's perspective, there is very little human interaction, post launch. Honeypot evasion techniques can be implemented, as new honeypots are uncovered. This paper uses reinforcement learning to prolong interaction with an attack sequence. Initial bot commands will attempt to return known honeypot responses, or positives to false requests. Depending on the response, the bot will either cease or amend its actions. We want our honeypot to learn and be rewarded for prolonging interaction. Therefore our reward function is to increase the number of commands from the attack sequence. Attacker commands can be categorised into the following:

- L - Known Linux bash commands
  wget, cd, mount, chmod, etc.
- C - Customised attack commands
  Commands executing downloaded files
- CC - Compound commands
  Multiple commands with bash separators/operators
- NF - known commands not facilitated by honeypot
  The honeypot is configured for 75 of the 'most used' bash commands.
- O - Other commands
  Unhandled keystrokes such as ENTER and unknown commands.

We propose a transition reward function whereby the learning agent is rewarded if a bot command is an input string i, comprised of bash commands (L), customised commands (C) or compound commands (CC). Therefore $Y = C \cup L \cup CC$. For example 'iptables stop' is an input string i that transitions the state (s) of iptables on a Linux system. Other commands or commands not facilitated by the honeypot are not given any reward. We propose an action set a = allow, block, substitute. Allow and Block are realistic responses to malware behaviour. Botnets can use complex if-else structures to determine next steps during compromise [30]. Using Substitute to return an alternative response to an attack command potentially increases the number of attack transitions to newer commands. This action set coupled with state set Y, creates a discrete state/action space. The reward function is as follows:

$$r_t(s_i, a) = \begin{cases} 1, \text{if i } \in Y \\ 0, \text{otherwise} \end{cases} \tag{4}$$

$$\text{where } Y = C \cup L \cup CC$$

The formula for transition reward r_t, based on state/action (s,a), is as follows:

- If the input string i is a customised attacker command C or a known bash command L or a compound command CC, then reward $r_t(s_i, a) = 1$
- otherwise the reward $r_t(s_i, a) = 0$

A model of the adaptive learning process from the reinforcement learning model in Fig. 1, is shown in Fig. 2.

The adaptive honeypot has the following elements:

- Modified honeypot. Cowrie is a widely used honeypot distribution that logs all SSH interactions in a MySQL database. This has been modified to generate the parameters to pass to the learning agent. Depending on the action selected by the learning agent, the honeypot will allow, block or substitute attack commands.
- SARSA agent. This module receives the required parameters from the adaptive honeypot and calculates $Q(s_t, a_t)$ as per Eq. (3). It determines the responses chosen by the adaptive honeypot and learns over time which actions yield the greatest amount of reward.

More detailed learning and reward functionality for this adaptive honeypot is available [31].

4   Results

We deployed two Cowrie honeypots at the same time; one adaptive as detailed in the previous section, and one standard high interaction. The adaptive honeypot was developed to generate rewards on 75 states and is freely available [32]. This was facilitated within PyBrain, which allows us to define the state/action

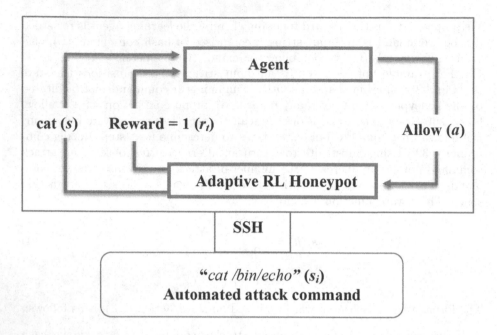

Fig. 2. Adaptive honeypot with reinforcement learning process.

space. We used Amazon Web Services (AWS) EC2 to facilitate an Internet facing honeypots. Cowrie, PyBrain, MySQL and other dependencies were installed on the adaptive honeypot EC2 instance. Both were accessible through SSH and immediately started to record malware activity. Initially it logged dictionary and bruteforce attempts. To compare performance, we undertake to extract all commands executed on the honeypots. Therefore events such as failed attempts, dictionary and bruteforce attacks are excluded as they represent pre-compromise interactions. Thereafter it captured other malware traffic including a Mirai-like bot [20]. These commands all represent interactions post-compromise. This bot became the dominant attacking tool over a 30-day period, until over 100 distinct attacks were recorded on the honeypot. Other SSH malware interacted with the honeypot. But these were too infrequent to excessively modify the rewards. Prior to presenting the results it is important to discuss the format of the bot, a sample of which is displayed in Table 2. In reality it has a sequence of 44 commands. The high interaction honeypot only ever experienced the first 8 commands in the sequence before the attack terminated. During 100 captures of the same bot, it never captured more than 8 transitions in the attack sequence. The adaptive honeypot initially experienced only the first 8 commands but then the sequence count started to increase as the honeypot learned from its state actions and rewards. Examining the entire bot structure, a mount command appears in the initial command sequence. A standard high-interaction honeypot has preconfigured parameters, and the response of mount is a known anti-evasion technique [19]. Figure 3 presents a comparison of the cumulative transitions for

Table 2. Sample of bot commands and categories

Sequence	Bot command	Category
38	/gweerwe323f	Other
39	cat/bin/echo	L
40	cd/	L
41	wget http:// <IP >/bins/usb_bus.x86 -O - >usb_bus; chmod 777 usb_bus	CC
42	./usb_bus	C

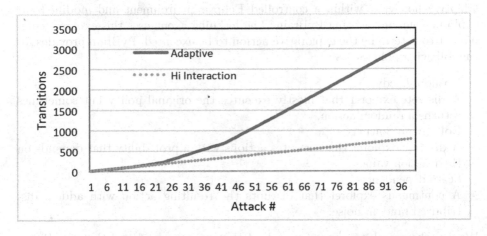

Fig. 3. Cumulative transitions for all commands.

both honeypots. This includes all command transitions for Y. At attack 16, the adaptive honeypot increased the number of commands in the sequence to 12. Examining the logs, we find that for the first time, the adaptive honeypot substituted a result for cat command and blocked a mount command. At attack 12 the honeypot blocked a compound echo command and continued to do so until attack 42. This is a very interesting result when compared to the high interaction honeypot's linear accumulation. The honeypot continued to learn until attack 42 when it allowed all compound commands in the sequence to be executed. Investigating the machinations of the bot, through sandboxing and code exploration, could explain this learning evolution. This is however beyond the scope of this article. The adaptive honeypot subsequently collected four times more command transitions than a standard high interaction honeypot. This task demonstrates that the adaptive honeypot was rewarded to increase the command transitions captured. By doing so it overcame initial honeypot identification techniques and continued to reward the learning agent in the event of further evasion measures.

Previous research has used reinforcement learning to create adaptive honeypots provisioning human interaction. Heliza presented findings from an adaptive honeypot deployed on the Internet. When compared to a high interaction honeypot, it produced three times more transitions to attack commands after 347 successful attacks. We point out that malware is an automated process and providing actions such as insult and delay are not relevant. After 100 successful attacks, our adaptive honeypot overcame initial honeypot detection techniques and subsequently captured four times more attack commands. It demonstrates that our state action space formalism for automated malware in more effective and produces a larger dataset for analysis. The captured dataset and bot analysis are valuable elements for further honeypot research. As an example, we used the captured dataset as an input stream and considered how the honeypot can be optimised by performing policy evaluation. We implemented the adaptive honeypot within a controlled Eclipse environment and modified the explorer component using PyBrain. The learning agent uses this explorer component to determine the explorative action to be executed. PyBrain provides for the following policies [28]:

- Epsilon greedy
  A discrete explorer that mostly executes the original policy but sometimes returns a random action.
- Boltzmann Softmax
  A discrete explorer that executes actions with a probability that depends on their action values.
- State dependent
  A continuous explorer that disrupts the resulting action with added, distributed random noise.

We ran the attack sequence for each of the policies and found that the Boltzmann explorer performed better in the controlled environment, resulting in more command transitions (Fig. 4). Boltzmann learned the best action to overcome detection at attack 11. State dependent learned the best action to overcome detection at attack 52. Epsilon greedy was the least effective as it learned the best action to overcome detection at attack 67. This policy evaluation demonstrates that parameters within the learning environment can be optimised to inform the development and deployment of more effective adaptive honeypots.

5   Limitations and Future Work

This paper improves upon an existing RL implementation by reducing the action set and simplifying the reward. By doing so it targets automated attack methods using SSH as a popular attack vector [23]. It is important to note that all interactions captured on both the high-interaction and adaptive honeypots were automated. There were no interactions showing obvious human cognition such as delayed responses, typing errors etc. This was verified by the timestamps of an entire attack sequence being executed in seconds. However, there are many honeypots deployed to capture malware using known vulnerabilities at different application and protocol levels [33]. There also exists tools to detect honeypots deployed on various attack vectors [5]. Therefore our research can be considered to have a narrow focus. We have presented previous and current adaptive

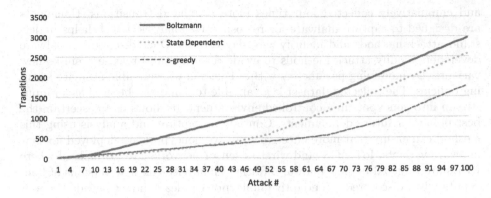

Fig. 4. Cumulative transitions for policy evaluation

honeypots targeting human and automated malware. We have also demonstrated that reinforcement learning can overcome initial detection honeypot techniques and prolong attack interaction. It is not unreasonable to consider a similar approach for application and protocol dependent honeypots. Further research into adaptive honeypot deployment could incorporate machine learning libraries such as PyBrain into these application and protocol dependent honeypots. Prior to pursuing this future work, research is required to determine if malware targeting these other honeypots is predominately human or automated. Honeypot operators have the tools to target human [25] or automated [31] attack traffic. Thereafter, adaptive honeypots using reinforcement learning can facilitate detection evasion. This ultimately leads to prolonged attack engagement and the capture of more valuable datasets.

Table 3. Modified Seifert's taxonomy

Class	Value	Note
Interaction level	High	High degree of functionality
	Low	Low degree of functionality
	Adaptive	*Learns from attack interaction*

6 Conclusions

This paper presents a state action space formalism designed to overcome anti detection techniques used by malware to evade honeypots. The reinforcement learning function is configured to reward the honeypot for increasing malware interaction. After an initial period, the honeypot learned the best method to avoid an attack command known to discover the presence of a honeypot. A standard high interaction honeypot always failed to maintain an attack beyond 8 commands. This indicated that the bot detected a honeypot and ceased operations. Our adaptive honeypot learned how to overcome this initial detection, and cumulatively collected four times more command transitions. Honeypots are designed to capture malware for retrospective analysis, which helps understand attack methods and identify zero day attacks. Once discovered, malware developers modify attack methods to evade detection and release them as new variants. By modifying parameters within PyBrain we were able evaluate learning policies. The collected dataset is a valuable tool and can be used as an input stream into honeypots in a controlled environment. It allows us to ascertain the best policy for future deployments. Continuous capture and analysis using this method can ensure that more effective adaptive honeypots are deployed for new malware variants. IoT, ICS, and wireless sensor networks are example of where RFDs are targeted by new malware. Rather than releasing new versions of honeypots when discovered, an adaptive honeypot can learn how to evade the anti detection methods used. It gives security developers an advantage in the cat-and-mouse game of malware development and discovery. For future work it is incumbent on honeypot developers to revisit Seifert's taxonomy in Table 1 and consider adding to interaction level class and value, a sample of which is shown in Table 3.

References

1. Bellovin, S.M.: Packets found on an internet. ACM SIGCOMM Comput. Commun. Rev. **23**(3), 26–31 (1993)
2. Pa, Y.M.P., Suzuki, S., Yoshioka, K., Matsumoto, T., Kasama, T., Rossow, C.: IoTPOT: analysing the rise of IoT compromises. EMU **9**, 1 (2015)
3. Watson, D., Riden, J.: The honeynet project: data collection tools, infrastructure, archives and analysis. In: WOMBAT Workshop on 2008 IEEE Information Security Threats Data Collection and Sharing. WISTDCS 2008, pp. 24–30 (2008)
4. Provos, N., et al.: A virtual honeypot framework. In: USENIX Security Symposium, vol. 173, pp. 1–14 (2004)
5. Krawetz, N.: Anti-honeypot technology. IEEE Secur. Priv. **2**(1), 76–79 (2004)
6. Kolias, C., Kambourakis, G., Stavrou, A., Voas, J.: DDoS in the IoT: Mirai and other botnets. Computer **50**(7), 80–84 (2017)
7. Spitzner, L.: Honeypots: Tracking Hackers, vol. 1. Addison-Wesley, Reading (2003)
8. Zhang, F., et al.: Honeypot: a supplemented active defense system for network security. In: 2003 Proceedings of the Fourth International Conference on Parallel and Distributed Computing, Applications and Technologies. PDCAT-2003, pp. 231–235. IEEE (2003)
9. Spitzner, L.: Honeytokens: the other honeypot (2003). https://www.symantec.com/connect/articles/honeytokens-other-honeypots. Accessed 17 Feb 2014
10. Seifert, C., Welch, I., Komisarczuk, P.: Taxonomy of honeypots. Technical report CS-TR-06/12, School of Mathematical and Computing Sciences, Victoria University of Wellington, June 2006
11. Kuwatly, I., et al.: A dynamic honeypot design for intrusion detection. In: 2004 Proceedings of The IEEE/ACS International Conference on Pervasive Services. ICPS 2004, pp. 95–104. IEEE (2004)
12. Prasad, R., Abraham, A.: Hybrid framework for behavioral prediction of network attack using honeypot and dynamic rule creation with different context for dynamic blacklisting. In: 2010 Second International Conference on Communication Software and Networks. ICCSN 2010, pp. 471–476. IEEE (2010)

13. Jicha, A., Patton, M., Chen, H.: SCADA honeypots: an indepth analysis of Conpot. In: 2016 IEEE Conference on Intelligence and Security Informatics (ISI) 2016
14. Vormayr, G., Zseby, T., Fabini, J.: Botnet communication patterns. IEEE Commun. Surv. Tutor. **19**(4), 2768–2796 (2017)
15. Wang, P., et al.: Honeypot detection in advanced botnet attacks. Int. J. Inf. Comput. Secur. **4**(1), 30–51 (2010)
16. Holz, T., Raynal, F.: Detecting honeypots and other suspicious environments. In: 2005 Proceedings from the Sixth Annual IEEE SMC Information Assurance Workshop. IAW 2005, pp. 29–36. IEEE (2005)
17. Antonakakis, M., et al.: Understanding the Mirai botnet. In: USENIX Security Symposium, pp. 1092–1110 (2017)
18. Valli, C., Rabadia, P., Woodward, A.: Patterns and patter-an investigation into SSH activity using kippo honeypots (2013)
19. Not capturing any Mirai samples. https://github.com/micheloosterhof/cowrie/issues/411. Accessed 02 Feb 2018
20. SSH Mirai-like bot. https://pastebin.com/NdUbbL8H. Accessed 28 Nov 2017
21. Khattak, S., et al.: A taxonomy of botnet behavior, detection, and defense. IEEE Commun. Surv. Tutor. **16**(2), 898–924 (2014)
22. Hayatle, O., Otrok, H., Youssef, A.: A Markov decision process model for high interaction honeypots. Inf. Secur. J.: A Global Perspect. **22**(4), 159–170 (2013)
23. Ghourabi, A., Abbes, T., Bouhoula, A.: Characterization of attacks collected from the deployment of Web service honeypot. Secur. Commun. Netw. **7**(2), 338–351 (2014)
24. Goseva-Popstojanova, K., Anastasovski, G., Pantev, R.: Using multiclass machine learning methods to classify malicious behaviors aimed at web systems. In: 2012 IEEE 23rd International Symposium on Software Reliability Engineering (ISSRE), pp. 81–90. IEEE (2012)
25. Wagener, G., Dulaunoy, A., Engel, T., et al.: Heliza: talking dirty to the attackers. J. Comput. Virol. **7**(3), 221–232 (2011)
26. Wagener, G., State, R., Dulaunoy, A., Engel, T.: Self adaptive high interaction honeypots driven by game theory. In: Guerraoui, R., Petit, F. (eds.) SSS 2009. LNCS, vol. 5873, pp. 741–755. Springer, Heidelberg (2009). https://doi.org/10.1007/978-3-642-05118-0_51
27. Pauna, A., Bica, I.: RASSH-reinforced adaptive SSH honeypot. In: 2014 10th International Conference on Communications (COMM), pp. 1–6. IEEE (2014)
28. Schaul, T., et al.: PyBrain. J. Mach. Learn. Res. **11**(Feb), 743–746 (2010)
29. Initial analysis of four million login attempts. http://www.honeynet.org/node/1328. Accessed 17 Nov 2017
30. Dowling, S., Schukat, M., Melvin, H.: A ZigBee honeypot to assess IoT cyberattack behaviour. In: 2017 28th Irish Signals and Systems Conference (ISSC), pp. 1–6. IEEE (2017)
31. Dowling, S., Schukat, M., Barrett, E.: Improving adaptive honeypot functionality with efficient reinforcement learning parameters for automated malware. J. Cyber Secur. Technol. 1–17 (2018) https://doi.org/10.1080/23742917.2018.1495375
32. An adaptive honeypot using reinforcement learning implementation. https://github.com/sosdow/RLHPot. Accessed 19 Dec 2017
33. Bringer, M.L., Chelmecki, C.A., Fujinoki, H.: A survey: recent advances and future trends in honeypot research. Int. J. Comput. Netw. Inf. Secur. **4**(10), 63 (2012)

Flexible Inference for Cyberbully Incident Detection

Haoti Zhong[1], David J. Miller[1(✉)], and Anna Squicciarini[2]

[1] Electrical Engineering Department, Pennsylvania State University,
State College, PA 16802, USA
hzz113@psu.edu, djmiller@engr.psu.edu
[2] College of Information Sciences and Technology, Pennsylvania State University,
State College, PA 16802, USA
acs20@psu.edu

Abstract. We study detection of cyberbully incidents in online social networks, focusing on session level analysis. We propose several variants of a customized convolutional neural networks (CNN) approach, which processes users' comments largely independently in the front-end layers, but while also accounting for possible conversational patterns. The front-end layer's outputs are then combined by one of our designed output layers – namely by either a max layer or by a novel sorting layer, proposed here. Our CNN models outperform existing baselines and are able to achieve classification accuracy of up to 84.29% for cyberbullying and 83.08% for cyberaggression.

1 Introduction

Cyberbullying, along with other forms of online harassment such as cyberaggression and trolling [22] are increasingly common in recent years, in light of the growing adoption of social network media by younger demographic groups. Typically, a cyberaggression incident refers to a negative comment with rude, vulgar or aggressive content. Cyberbullying is considered a severe form of cyberaggression, with repeated cyberaggression incidents targeting a person who cannot easily self-defend [5,24].

To date, researchers from various disciplines have addressed cyberbullying (e.g. [10,28]) via detection and warning mechanisms. A growing body of work has proposed approaches to detect instances of bullying by analysis of individual posts, focused on detection of rude, vulgar, and offensive words. Recently, acknowledging that offensive messages are not solely (or always) defined by the presence of a few selected words, studies have also considered lexical features and sentence construction to better identify more subtle forms of offensive content [4,7]. Yet, despite some promising results, previous works typically ignore other characteristics of bullying, such as its repetitive and targeted nature [22]. As such, previous work is typically unable to distinguish between bullying and mere isolated aggressive or offensive messages, oversimplifying the cyberbullying

U. Brefeld et al. (Eds.): ECML PKDD 2018, LNAI 11053, pp. 356–371, 2019.
https://doi.org/10.1007/978-3-030-10997-4_22

detection problem. We believe a better way to detect both cyberbullying *and* cyberaggression is to consider the contextual cues surrounding these incidents, as they are exposed in a conversation. Accordingly, we focus on session-level detection of cyberbullying and cyberaggression, particularly sessions generated in response to posted media, e.g. images.

Our work aims at answering the following questions:

– Can we detect both cyberaggresion and cyberbullying based on a common model structure?
– Can we detect cyberbully incidents at the session level, rather than simply identifying individual aggressive comments?
– Can session-specific elements (e.g. the image, or the caption of the image) help improve inference of session-level bullying episodes?

Note that our research questions are intentionally focused on *session-level* inference. While in some cases bullying may be tied to a person (e.g. a single account) rather than the content of their messages, we here omit observations related to the specific user's history and patterns within a network (i.e. we do not consider social network features or users' features). We investigate the above questions by developing customized Convolutional Neural Network (CNN) models, with a single convolutional layer, that can be trained to attempt to fulfill the above requirements. The model performs a "session-level" analysis and takes all comments in a session as input. Here, a session is an Instagram-like session, with a thread of replies created after an initial post of an image+caption.

Our CNNs process individual users' comments, while also accounting for possible conversational patterns (a comment and its referrant). The CNN outputs, one per comment, are combined by one of our designed output layers – namely by either a max layer or by a novel sorting layer, proposed here. The sorting layer is a generalization of a max layer, which takes all comments into account according to their probabilities of being aggressive.

We test multiple variants of our proposed CNN architecture, training it *both* for bullying and aggression detection. Compared to prior work, our proposed approach is more powerful and comprehensive. Unlike previous works [16], our approach provides flexible inferences – it can distinguish sessions affected by cyberbullying from those affected by cyberaggression and also possibly identify the victim of cyberbullying in a session (be it the poster *or* one of the commenters). Further, it is truly designed to detect cyberbullying as it unfolds during a conversation (it can be applied to make progressive detection decisions, as more and more comments are made), unlike a simple offensive speech detector.

The paper is organized as follows. The next section summarizes related work. Section 3 presents a detailed description of our network. Section 4 describes all the datasets we used for our experiments. Section 5 compares the performance of our models with baselines, then gives insights into the nature of cyberaggression and cyberbullying. Lastly, Sect. 6 discusses our findings and potential extensions of this work.

2 Related Work

Studies in psychology and sociology have investigated the dynamics of cyber-bullying, bullies' motives and interactions [2,14,18,30,31,37]. In particular, a number of methods have been proposed for detecting instances of cyberbullying, most focused on offensive textual features [4,9,21,36] (e.g., URLs, part-of-speech, n-grams, Bag of Words as well as sentiment) [6,10,29]. Several recent studies have also begun to take user features into consideration to help detect bullies themselves [3,6,12]. [17] conducted a simple study hinting at the importance of social network features for bully detection. They considered a corpus of Twitter messages and associated local *ego-networks* to formalize the local neighborhood around each user. Here, however, we focus on inference in the absence of such information, *i.e.*, purely based on the snapshot offered by one session.

[16] provides an initial effort on the detection of bullying and aggressive comments. The authors tested several text and social features and found that the counts of 10 predefined words offered the best results. Our work not only significantly increases the overall performance compared to this prior work, but also naturally gives the capability to identify the aggressive comments within a cyberbullying incident.

Our methodology is related to the recent body of work on CNN and deep neural net (DNN) models for textual analysis. CNNs were first designed for image-based classification [25], and have more recently been applied to various domains in NLP including document classification and sentence modeling [19,23]. Many efforts have focused on learning word or sentence vector representations [1,8,27], converting a word or sentence into a low dimensional vector space in order to overcome the curse of dimensionality and to allow calculation of similarities between words. Also many works use DNNs to classify a sentence or document, e.g. [13,33]. Specifically, Kim et al. [20] use a CNN-based model for sentence classification tasks. CNNs use a filtering (convolutional) layer to extract spatially invariant low-level feature "primitives". Subsequent layers perform spatial pooling of primitive information, and also pooling across primitive types. This is followed by several fully connected layers, leading to an output layer that predicts the class for the sentence/word (e.g., its sentiment).

Here, we propose a CNN approach (without a great number of layers), rather than a DNN approach. This is because CNNs (without many layers) are naturally *parsimonious* in that there is weight sharing and therefore they are more suitable for applications like ours, where there is a limited amount of available training data. The scarcity of labeled cyberbullying data for training does not support use of DNNs with many layers and a huge number of free parameters to learn.

3 Approach

3.1 CNN for Session-Level Bully Incident Detection

In designing our approach, we rely on commonly accepted definitions of cyber-bullying and cyberaggression, consistent with recent literature in the field [5,16].

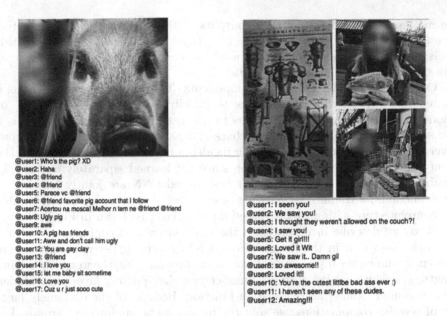

Fig. 1. Examples of sessions with cyberbullying (left) and without (right).

We view cyberbullying as repeated acts of explicit targeted aggression, whereas cyberaggression refers to occasional vulgar or rude attacks in isolation. Accordingly, we exploit the fact that cyberbullying events are considered a subset of cyberaggression.[1] In Fig. 1, we report examples of two sessions, one with instances of bullying, and one that is instead *not* affected by bullying.

Therefore, in order to identify a cyberbullying incident, we should observe *multiple* negative comments within a conversation, all related to the same victim (i.e. a single account denoting an individual or group). We consider this in the context of a *session*, a collection of temporally sorted comments related to the same topic/discussion. Since there are multiple comments in a session, if we simply train a detector of offensive or aggressive comments, the false positive rate may be close to 100% since in this case the probability of a session being bullied is the probability of at least one offensive comment, i.e. $P(bully) = 1 - \Pi_i(1 - p_i(bully))$ – even a small overestimated probability for comment detection may lead to a significant error rate at the session level. Also, this approach is inconsistent with the very definition of cyberbullying, as it ignores the targeted and repeated nature of these incidents.

Our model aims at detecting cyberaggresion first, treating this as a multi-instance learning problem. The assumption is that a bag of instances (i.e. a session) can be labeled as positive (i.e. inclusive of an aggressive comment) if there is at least one positive instance. Specifically, to detect aggressive incidents,

---

[1] As we discuss in Sect. 4, our experimental evaluation is based on a dataset labeled consistently with these definitions.

we maximize over all (probability) outputs of the aggressive comment detection CNN: $P_{incident}(agg) = Max_i(P_{sentence_i}(agg))$. By suitably generalizing this decision rule, our model can also be used to detect cyberbullying incidents, as will be explained in the following sections.

Our model includes two main components. The front-end component is a shared CNN model which learns how to classify sentences as "aggressive" or "non-aggressive". Here, "shared" refers to the fact that all sentences go through the same (a common) network. The back-end part of the model is the output layer, learning to detect cyberbullying incidents based on the outputs from the front-end model. These two components are not learned separately and simply combined together – all layers of our customized CNN are jointly learned, so as to maximize a training set measure of cyberbullying detection accuracy. Our CNN architecture with a sorting-based decision layer is shown in Fig. 2.

Next, we describe in more depth the main layers and configuration of our network. Note that in the design of our CNN, several hyper-parameters were chosen, including the top number of words, the dropout rate, the word embedding length, convolutional layer filters' number and size, pooling layer size, hidden layer's neuron number and activation function. Because of the extremely large set of possible combinations, we split the hyperparameters into two groups. For each group, we chose the optimized hyper-parameter combination by means of nested cross-validation (CV). We selected these hyperparameters once, and used the same configuration across all extensions and variants of our model.

3.2   Input Layer

The input to the comment-level network is a vector of indices where each index uniquely identifies one of the top 20,000 words (in terms of frequency of occurrence in the training set). Each such index is mapped to a vector via word2vec. We first preprocessed all the words by removing all non ASCII characters, then applied tokenizing and stemming. Since for any comment, we also know to which comment it replied (or if it is a direct reply to the post itself), this contextual information should also be exploited, in modelling each comment. To achieve this, we concatenate to the "responding" comment's index vector the index vector of the "referrant" comment. If there is no such comment, we simply concatenate a vector of 0s; if it is a response to the original posting, the referrant is the representation of the image's caption.

3.3   Shared CNN Model for Sentence Classification

Inspired by prior work [20], we first define a shared comment level aggression detection model. The output of this shared CNN will be a comment-level probability of cyberaggression, for each comment in a session.

The CNN is constructed as follows:

– *Dropout layer* [32]: Downsamples some of the features during training to increase the robustness of the model; the dropout rate is a hyper-parameter (not shown in Fig. 2 for simplicity).

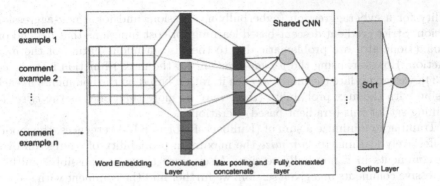

Fig. 2. Our CNN architecture, consisting of a shared CNN model applied to each comment, and an output (fusion) layer. In this case the fusion is based on sorting shared CNN outputs that represent the aggression probabilities for all sentences and feeding these sorted probabilities to a sigmoidal output activation layer. The decision layer is the sorting layer.

- *Embedding layer:* Applies a pre-trained word2vec [27] model to embed each word into a vector of length 100, so now each comment is a (25, 100) matrix, since we use a fixed number (25) of words for each comment – longer comments are truncated. Shorter comments are padded with repetition of a "null" word to fill out to length 25.
- *Convolutional layer:* Uses several filters (1, 3, 5, 8) to get a convolution response from the original matrix; ReLU activation functions (linear response if positive; zero if negative) are used to process these responses. A filter with length one encompasses only individual words, whereas if the length is increased to encompass multiple words what is learned is word-order dependent.
- *Max pool layer:* Downsamples the responses of the convolutional layer, and concatenates them into a single vector. We use a large pooling size (e.g. each sentence will pool to 1 output). This allows us to capture cases where a single offensive word or an aggressive phrase expresses a strong negative meaning, and also may reduce potential overfitting (since smaller pooling sizes require more model parameters).
- *Concatenation layer:* Concatenates all channel's outputs into a single vector.
- *Fully connected layer:* A hidden layer like a traditional MLP hidden layer with 100 neurons using ReLU activations.
- *Output layer:* 1 unit output using a sigmoid activation function, which evaluates the probability that a comment is aggressive.

3.4 Decision Layers for Bully Incident Detection

Max Layer: The main idea of the max layer is to identify the comment with the maximum aggression probability in each session; on the training set, we essentially learn a good "threshold" for discriminating between the maximum probability for a cyberaggression (cyberbullying) session, and for a non-aggression session. Strict gradient descent-based learning for cost functions that depend on a max() operator are problematic due to the discontinuous nature of the max function. However, using the *pseudo-gradient* for the max() function introduced in [34], we can update the model at each iteration based on the comments in each session with the max probability (there may be more than one) as the *effective* training set for this (gradient-based) iteration.

Training to minimize a sum of (Kullback-Leibler) KL divergences, the model is effectively learning to *minimize* the maximum probability of aggression over the comments in a non-bully session. Also, since there are possibly multiple aggressive comments in a given session, we further find the comment with second highest probability if its value is higher than a certain ratio compared with the maximum probability and we take these two top comments into account in the training process. This additional information is expected to boost the performance of the CNN. The learning procedure is shown in Algorithm 1. The final model was chosen by maximizing the average validation accuracy. In the experiment we found the model tends to overfit after a few iterations; thus we set the maximum iteration number as 10.

During the inference phase, the model's output for each session can be obtained by calculating the max probability over all comments in the session, and this probability is treated as the probability of a cyberaggressive incident in a session. Moreover, our trained cyberaggresion model can be used as a cyber-bully incident detector by changing the classification rule to adhere more strictly to the definition of cyberbullying: i.e. at least two comments with a probability greater than 0.5 must target the same user (either the original poster *or* a commenter).

Algorithm 1: Train the cyberaggression incident detector CNN

---

initialize parameter θ of CNN f(x);
iter $= 0$;
while *iter $\leq$ MaxIter* **do**
    iter += 1;
    training_data = [];
    training_label = [];
    for *each session s* **do**
        $idx1 = \arg\max_{c \in s} f(x_c|\theta)$
        $idx2 = \arg\max_{c \in s, c \neq idx1} f(x_c|\theta)$
        training_data. append(comment idx1);
        training_label. append(label$_s$)
        if $f(x_{idx2}|\theta) \leq 0.9 * f(x_{idx1}|\theta)$ **then**
            training_data. append(comment idx2);
            training_label. append(label$_s$)
        end
    end
    save model if it achieves the best validation accuracy;
    update θ with ADAGRAD[11];
end

Sorting Layer: Generalizing beyond the "top two" comments in a session, we propose a novel *sorting layer* which exploits the *top K* comments (those with the K largest probabilities of aggression) in learning a multi-instance CNN. Each comment goes through a shared (front-end) model (the same as the model we described before), which produces the comment's probability of aggression. These probabilities are then sorted, with only the top K retained (K a hyperparameter) and fed into a final sigmoid output layer. The learned sorting layer will tend to give the comment more "weight" if it is an aggressive comment, and zero weight if it is *uninformative*– e.g., if $K = 10$, the learning for this layer may determine that only the top four weights to the output layer are non-zero. In such case, only the top 4 comments (in terms of aggression) are found useful for discriminating bullying from non-bullying sessions. Likewise, if e.g. the 7-th ranked comment has a low probability of being labeled as aggressive, this information is likely neutral as to whether the session involves bullying or not, and the weight on this comment may thus be learned to be zero. The learning phase for this model (i.e. the CNN with sorting layer) is a natural generalization of the learning approach described above involving max() functions, just as a sorting function itself is a natural generalization of a max() function. Similar to the previous (max-based) model, this sorting-based model can be trained to predict either cyberbullying (versus negation) or cyberaggression (versus negation), depending on the supervising labels that are chosen. The pseudo-code for the algorithm is reported in Algorithm 2. Note that we set the maximum iteration number as 50, larger than for Algorithm 1, in order to allow all the parameters (more than for Algorithm 1) to be reasonably learned.

Algorithm 2: Train the bully incident detector sorted-CNN

---

initialize parameter θ of shared CNN f(x);
iter = 0;
while *iter* $\leq$ *MaxIter* **do**
    training_data = [];
    training_label = [];
    for *each session s* **do**
        keys=[];
        for *each comment c* **do**
          | *keys.append(f($x_c|\theta$)*
        end
        indexs=argsort(keys);
        new_session=[];
        for *i in indexs and i $\leq$ 30* **do**
          | *new_session.append(c_i)*
        end
        training_data.append(new_session);
        training_label.append($label_s$)
    end
    save model if it achieves the best validation accuracy;
    update full model with ADAGRAD;
end

---

4 Datasets

The dataset we used for our experiments is taken from [16], which was collected from Instagram posts. The authors collected Instagram sessions with more than 15 comments so that labelers could adequately assess the frequency or repetition of aggression. Labelers were asked to label the whole post, whether or not there is (1) a cyberaggression incident and/or (2) a cyberbullying incident. For each Instagram post, the dataset includes images, captions, poster's id, the number of followers, all comments and all commenters' ids and post times. In this paper, we call each post a session. Most comments are short, with few words. The dataset has 1540 non-bully sessions and 678 bully sessions. It likewise has 929 aggression sessions and 1289 non-aggression sessions (consistent with cyberbullying being a subset of cyberaggression).

To validate our claim that a comment level aggressive detector is not suitable for detecting cyberbullying incidents, we trained a CNN baseline with a comment level aggression dataset, which was taken from [38]. (We will report results for this method in the next section). The dataset was crawled from publicly visible accounts on the popular Instagram social platform through the site's official API, and had a subset of the comments labeled for aggression. Labelers had access to the image, the image's commentary, and indicated whether or not each comment represented aggression. Overall, the training dataset for this model includes 1483 non aggressive comments and 666 aggressive comments.

5 Experiments

In order to validate our model, we carried out a number of experiments. We validate the accuracy of our CNN frameworks, and investigate whether additional inference power can be gained by exploiting the session's uploaded image (which triggered the session). We compare accuracy of our approach with two baselines, discussed next.

5.1 Baselines

We considered two main baselines for comparative purposes.

First, we replicated [16]. In this work, several text and image features were tested to find out the best features for detecting evidence of cyberbullying within a session. The authors eventually claimed that using the count for 10 specific (mostly vulgar) words as the features input to a logistic regression model provided the best performance. The model is directly trained with cyberbully labels applied at the session level.

As a second baseline, we used the same CNN architecture described in Sect. 3.3, but trained it with an Instagram dataset labeled at the comment level for aggression [38] (see Sect. 4 for a description). We call this the COMMENT CNN. In this case, given a comment label {aggressive,non-aggressive}, we assess the accuracy of detection for a *cyberbully* incident. After training as a comment-level detector, we refine the classification rule during testing so that, only if there is more than one aggressive comment targeting the same user, the session is declared "bullied". Testing is performed on [16]'s dataset.

5.2   Variants of Our Model

In addition to comparing with state-of-the art baseline models, we experiment with several variants of our proposed model. For all the model variants (besides the CNN) each comment is concatenated with the comment to which it replied, as input to the shared CNN structure.

- *CNN:* The shared CNN is coupled with a max function in the decision layer and takes all comments as input. Each comment is separately input to the shared CNN structure. The model is trained with session level bully labels.
- *CNN-R:* The shared CNN (using comments and their referrant comments) is coupled with a max function in the decision layer. The model is trained with session level bully labels.
- *SORT CNN-R:* The shared CNN is coupled with a sorting function in the decision layer followed by a sigmoidal nonlinearity. The front-end model parameters are initialized using the trained CNN-R model. This model is trained with session level bully labels.
- *ACNN-R:* The shared CNN is coupled with a max function in the decision layer. The model is trained with session level *aggression* labels. We can use this model directly to detect cyberaggression. Alternatively, for cyberbully detection, during testing, we can replace the max layer by a layer that checks whether there are at least two comments with a front-end output greater than 0.5 which responded to the same user.
- *SORT ACNN-R:* The shared CNN is coupled with a sorting function followed by a sigmoidal nonlinearity in the decision layer. Front-end model parameters are initialized using the trained ACNN-R model. This model is trained with session level aggression labels.

5.3   Results for Session-Level Cyberbullying Detection

Results for cyberbully detection of our models are reported in Table 1. Cross validation was used; the dataset was divided into 10 (outer) folds, and 8 of these folds were used for training, one for validation, and the last fold for testing. As shown, the Hosseinmardi et al. baseline approach achieves a True Positive Rate (TPR) of 67.98% and True Negative Rate (TNR) of 75.01%. The COMMENT CNN gives a TNR of 27.39% and TPR of 97.63% – it performs as expected, since even a small overestimated probability of comment-level aggression will lead to a large false positive rate (low false negative rate). All variants of our CNN models consistently outperform Hosseinmardi et al.'s method, as indicated next.

CNN achieves a TPR of 73.06% and a TNR of 86.44%. The biggest advantage over the baseline is that our model captures not only sensitive words indicative

Table 1. Performance for cyberbully detection CNNs on [15] dataset.

Classifier	Overall accuracy	TNR	TPR	F1-measure
Baseline [16]	71.5%	75.01%	67.98%	0.7132
COMMENT CNN	62.51%	27.39%	97.63%	0.4278
CNN	79.75%	86.44%	73.06%	0.7919
CNN-R	81.25%	89.83%	72.67%	0.8034
SORT CNN-R	82.05%	88.89%	75.2%	0.8147
ACNN-R	82.92%	82.24%	83.59%	0.829
SORT ACNN-R	84.29%	83.24%	85.33%	0.8427

of a cyberaggression attack, such as the 10 words used by the baseline, but also information from the whole sentence.

As mentioned in Sect. 3.3, in order to capture contextual cues about the ongoing discussion, we concatenated the target comment with the comment to which the target comment replied and trained CNN-R. This simple change increases overall accuracy by about 1.5%, showing the importance of leveraging insights from users' conversational patterns. SORT CNN-R is more effective in detecting bully incidents than directly choosing the max aggressive comment probability as the session level output. SORT CNN-R increases accuracy by about 1% compared with CNN-R (81.25% vs. 82.05% for SORT CNN-R). However, likely due to an insufficient number of training posts, SORT CNN-R tends to overfit if we use all the comments in a session. Thus, we only use the top 30 comments, with the highest aggressive probabilities, and also stop model training once training accuracy exceeds validation accuracy. We also note that if we train a CNN with aggression labels rather than cyberbullying labels, using either max layer (ACNN-R) or sorting layer (SORT ACNN-R), we consistently obtain an overall accuracy gain of about 2% for *cyberbullying* detection. One of the reasons is that the aggression domain is more class-balanced than the bullying domain. This is likely helping the trained model to better learn the characteristics of negative comments since there are more negative incidents from which the model can learn, compared with the case of cyberbullying labels.

5.4 Applying a CNN Model for Cyberaggression Detection

We also explored whether our CNN model could help detect cyberaggression. We compare the performance of two variants of our model: ACNN-R with a modified logistic classification rule (three classes: no detection, cyberaggression, cyberbullying), and SORT ACNN-R.

Note that ACNN-R is able to detect *both* cyberaggression and cyberbully sessions, simultaneously. We achieve a TPR of 85.13% and a TNR of 80.64% for cyberaggression detection with ACNN-R, and the overall accuracy for cyberbullying detection is 82.92%. With SORT ACNN-R for cyberaggression detection, we achieve a TPR of 82.44% and a TNR of 83.71%. This result shows our model's ability to detect incidents of cyberaggression. Since cyberaggression occurs when there is even just one aggressive comment, the sorting layer, as expected, is not very necessary for this task (Sort ACNN-R does not outperform ACNN-R here).

5.5   Image-Related Information for Cyberbully Detection

We also investigated several approaches for integrating image (i.e. visual content) information within the CNN model. We tested both a model that included the image features concatenated in our model's hidden layer, and a model version wherein we let the model's convolutional layers predict the image content as an additional task. However, we did not find image information to be helpful in improving our detection accuracy. We validated these two approaches over 1142 available full Instagram sessions (i.e. images and comments were available), which include 763 non-bully and 379 bully sessions. We applied these two ideas into our ACNN-R model. The image features concatenated in the hidden layer yields to a decrease of TNR and TPR to 73.68% and 58.83%, respectively. A multi-task approach instead achieves a TNR of 74.11% and TPR of 88.09% and still shows no significant gain or harm compared to the original inference power of the convolutional layers for the text-only CNN.

5.6   Insights on Cyberbullied Sessions and Their Context

In order to gather additional insights on the correlation between images and comments, we carried out two additional experiments. First, we generated an image caption according to Xu's recent approach [35]: images' visual features are extracted from a CNN model, then a Recurrent Neural Network (RNN) converts these features into word sequences. We then calculated sentence similarity between the detected aggressive comments and the image's generated caption[2]. We analyze similarity distribution using sentence similarity based on semantic nets and corpus statistics [26].

Second, we calculated the similarity distribution between the aggressive comment and the caption written by the image poster. As shown in Fig. 3, left, both similarity distributions have a low average (0.2595 and 0.1900, respectively), indicating that there is no strong relationship between aggressive comments and the posted content in general. This may be due to the (skewed) nature of the dataset – as discussed in Sect. 4, only sessions with a sufficiently large number of comments were crawled. This may have generated a dataset of sessions mainly from popular users and celebrities. In these cases, it is possible that users posting aggressive comments simply target the poster, not the posted content - or that responders are intrinsically aggressive.

To further gain insights on the conversations triggering bullying, we estimated how cohesive the posted comments are within a session (in terms of the content being discussed). To do so, we first applied the Twitter-to-vector model [8] to generate a fixed length vector for every comment. The Twitter-to-vector

---

[2] Two comments are considered similar if the model's output is greater than 0.5.

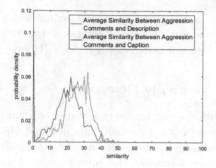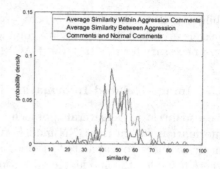

Fig. 3. Left: Similarity between aggressive comment and image captions. Right: Comment similarity comparison

model finds vector space representations of whole tweets by learning complex, non-local dependencies in character sequences. We calculated the average cosine similarity among all aggressive comments in a session (similarity is calculated pair-wise between all the aggressive comments), and compared with the average cosine similarity between aggressive comments and normal comments in the same session. As shown in Fig. 3, right, pair-wise comment similarity among aggressive comments is larger than pair-wise similarity between normal and aggressive comments. This partially supports the hypothesis that aggressive comments share similar content for a given post or bullies simply attack users in a similar and repeated fashion.

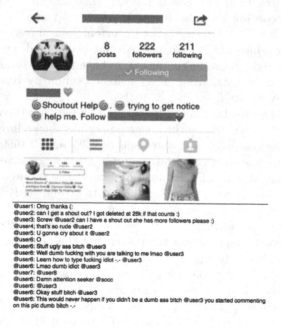

Fig. 4. Example of highlighted aggressive comments in a session

6 Conclusions and Future Work

In this work, we presented a session-level approach for detection of sessions affected by bullying or aggressive comments. We present several CNN-based models and demonstrate that our proposed models increase the average accuracy by about 13% compared with baselines. All of our models achieve similar performance in detecting cyberbullying and cyberaggression.

Our model lends itself to several interesting applications and extensions. One potential application is to explicitly infer which comment(s) are the likely triggers of bullying incidents. This is a natural extension of the current approach, as we already determine the probabilities of aggression for each of the top K comments. Moreover, each such comment has an earlier *referrant* comment. Thus, likely candidates for the trigger include: (1) the comment with greatest probability; (2) the referrant comment for the comment with greatest probability, if the referrant comment's probability itself is above a threshold; (3) the first comment with probability exceeding a given threshold.

We carried out an initial experiment with our CNN with max layer output. Figure 4 shows the outcome for a sample session. We highlighted the detected aggressive comments in red, and the comments with a probability greater than 0.25 as orange. As shown, our model correctly identifies most of the bully comments. We will continue along this direction and provide a systematic way to rank comments within conversations, not only with respect to their aggressive or bullying nature but also with respect to their specific "role" in the conversation.

An additional possible refinement includes further studying how to effectively leverage information contained in the images included in the analyzed sessions. Finally, future work also includes extending our model to support a more thorough distinction among aggressive comments (e.g. trolling vs harrassment) in a semi-automated manner.

Acknowledgements. Work from Dr. Squicciarini and Haoti Zhong was partly supported by the National Science Foundation under Grant 1453080 and Grant 1421776.

References

1. Bengio, Y., Ducharme, R., Vincent, P., Jauvin, C.: A neural probabilistic language model. JMLR **3**, 1137–1155 (2003)
2. Berson, I.R., Berson, M.J., Berson, M.J.: Emerging risks of violence in the digital age: lessons for educators from an online study of adolescent girls in the united states. J. Sch. Violence **1**(2), 51–71 (2002)
3. Chatzakou, D., Kourtellis, N., Blackburn, J., De Cristofaro, E., Stringhini, G., Vakali, A.: Mean birds: detecting aggression and bullying on twitter. arXiv preprint arXiv:1702.06877 (2017)
4. Chen, Y., Zhou, Y., Zhu, S., Xu, H.: Detecting offensive language in social media to protect adolescent online safety. In: PASSAT, pp. 71–80. IEEE (2012)

5. Corcoran, L., Guckin, C.M., Prentice, G.: Cyberbullying or cyber aggression?: A review of existing definitions of cyber-based peer-to-peer aggression. Societies 5(2), 245–255 (2015)
6. Dadvar, M., Trieschnigg, D., de Jong, F.: Experts and machines against bullies: a hybrid approach to detect cyberbullies. In: Sokolova, M., van Beek, P. (eds.) AI 2014. LNCS (LNAI), vol. 8436, pp. 275–281. Springer, Cham (2014). https://doi.org/10.1007/978-3-319-06483-3_25
7. Dadvar, M., Trieschnigg, D., Ordelman, R., de Jong, F.: Improving cyberbullying detection with user context. In: Serdyukov, P., et al. (eds.) ECIR 2013. LNCS, vol. 7814, pp. 693–696. Springer, Heidelberg (2013). https://doi.org/10.1007/978-3-642-36973-5_62
8. Dhingra, B., Zhou, Z., Fitzpatrick, D., Muehl, M., Cohen, W.W.: Tweet2Vec: character-based distributed representations for social media. arXiv preprint arXiv:1605.03481 (2016)
9. Dinakar, K., Reichart, R., Lieberman, H.: Modeling the detection of textual cyberbullying. In: The Social Mobile Web (2011)
10. Djuric, N., Zhou, J., Grbovic, M., et al.: Hate speech detection with comment embeddings. In: WWW, pp. 29–30. ACM (2015)
11. Duchi, J., Hazan, E., Singer, Y.: Adaptive subgradient methods for online learning and stochastic optimization. JMLR 12, 2121–2159 (2011)
12. Galán-García, P., de la Puerta, J.G., Gómez, C.L., Santos, I., Bringas, P.G.: Supervised machine learning for the detection of troll profiles in Twitter social network: application to a real case of cyberbullying. In: Herrero, Á., et al. (eds.) SOCO 2013-CISIS 2013-ICEUTE 2013, vol. 239, pp. 419–428. Springer, Cham (2014). https://doi.org/10.1007/978-3-319-01854-6_43
13. Glorot, X., Bordes, A., Bengio, Y.: Domain adaptation for large-scale sentiment classification: a deep learning approach. In: ICML, pp. 513–520 (2011)
14. Hinduja, S., Patchin, J.W.: Social influences on cyberbullying behaviors among middle and high school students. J. Youth Adolesc. 42(5), 711–722 (2013)
15. Hosseinmardi, H., Mattson, S.A., Rafiq, R.I., Han, R., Lv, Q., Mishra, S.: Detection of cyberbullying incidents on the Instagram social network. CoRR abs/1503.03909 (2015)
16. Hosseinmardi, H., Rafiq, R.I., Han, R., Lv, Q., Mishra, S.: Prediction of cyberbullying incidents in a media-based social network. In: ASONAM, pp. 186–192. IEEE (2016)
17. Huang, Q., Singh, V.K., Atrey, P.K.: Cyber bullying detection using social and textual analysis. In: SAM, pp. 3–6. ACM (2014)
18. Juvonen, J., Graham, S.: Bullying in schools: the power of bullies and the plight of victims. Annu. Rev. Psychol. 65, 159–185 (2014)
19. Kalchbrenner, N., Grefenstette, E., Blunsom, P.: A convolutional neural network for modelling sentences. arXiv preprint arXiv:1404.2188 (2014)
20. Kim, Y.: Convolutional neural networks for sentence classification. arXiv preprint arXiv:1408.5882 (2014)
21. Kontostathis, A., Reynolds, K., Garron, A., Edwards, L.: Detecting cyberbullying: query terms and techniques. In: WebSci, pp. 195–204. ACM (2013)
22. Kowalski, R.M., Limber, S.P., Limber, S., Agatston, P.W.: Cyberbullying: Bullying in the Digital Age. Wiley, Hoboken (2012)
23. Lai, S., Xu, L., Liu, K., Zhao, J.: Recurrent convolutional neural networks for text classification. In: AAAI, vol. 333, pp. 2267–2273 (2015)
24. Langos, C.: Cyberbullying: the challenge to define. Cyberpsychol. Behav. Soc. Netw. 15(6), 285–289 (2012)

25. LeCun, Y., et al.: Backpropagation applied to handwritten zip code recognition. Neural Comput. **1**(4), 541–551 (1989)
26. Li, Y., McLean, D., Bandar, Z.A., O'shea, J.D., Crockett, K.: Sentence similarity based on semantic nets and corpus statistics. IEEE Trans. Knowl. Data Eng. **18**(8), 1138–1150 (2006)
27. Mikolov, T., Chen, K., Corrado, G., Dean, J.: Efficient estimation of word representations in vector space. arXiv preprint arXiv:1301.3781 (2013)
28. Nahar, V., Unankard, S., Li, X., Pang, C.: Sentiment analysis for effective detection of cyber bullying. In: Sheng, Q.Z., Wang, G., Jensen, C.S., Xu, G. (eds.) APWeb 2012. LNCS, vol. 7235, pp. 767–774. Springer, Heidelberg (2012). https://doi.org/10.1007/978-3-642-29253-8_75
29. Nobata, C., Tetreault, J., Thomas, A., Mehdad, Y., Chang, Y.: Abusive language detection in online user content. In: WWW, Republic and Canton of Geneva, Switzerland, pp. 145–153. International World Wide Web Conferences Steering Committee (2016)
30. Rodkin, P.C., Farmer, T.W., Pearl, R., Acker, R.V.: They're cool: social status and peer group supports for aggressive boys and girls. Soc. Dev. **15**(2), 175–204 (2006)
31. Salmivalli, C., Isaacs, J.: Prospective relations among victimization, rejection, friendlessness, and children's self-and peer-perceptions. Child Dev. **76**(6), 1161–1171 (2005)
32. Srivastava, N., Hinton, G.E., Krizhevsky, A., Sutskever, I., Salakhutdinov, R.: Dropout: a simple way to prevent neural networks from overfitting. JMLR **15**(1), 1929–1958 (2014)
33. Tang, D., Qin, B., Liu, T.: Document modeling with gated recurrent neural network for sentiment classification. In: EMNLP, pp. 1422–1432 (2015)
34. Teow, L.-N., Loe, K.-F.: An effective learning method for max-min neural networks. In: IJCAI, pp. 1134–1139 (1997)
35. Xu, K., et al.: Show, attend and tell: neural image caption generation with visual attention. In: ICML, pp. 2048–2057 (2015)
36. Yin, D., Xue, Z., Hong, L., Davison, B.D., Kontostathis, A., Edwards, L.: Detection of harassment on web 2.0. In: Proceedings of the Content Analysis in the WEB, vol. 2, pp. 1–7 (2009)
37. Zalaquett, C.P., Chatters, S.J.: Cyberbullying in college. Sage Open **4**(1), 1–8 (2014). https://doi.org/10.1177/2158244014526721
38. Zhong, H., et al.: Content-driven detection of cyberbullying on the Instagram social network. In: IJCAI, pp. 3952–3958 (2016)

Solving the False Positives Problem in Fraud Prediction Using Automated Feature Engineering

Roy Wedge[1], James Max Kanter[1(✉)], Kalyan Veeramachaneni[1], Santiago Moral Rubio[2], and Sergio Iglesias Perez[2]

[1] Data to AI Lab, LIDS, MIT, Cambridge, MA 02139, USA
kanter@mit.edu
[2] Banco Bilbao Vizcaya Argentaria (BBVA), Madrid, Spain

Abstract. In this paper, we present an automated feature engineering based approach to dramatically reduce false positives in fraud prediction. False positives plague the fraud prediction industry. It is estimated that only 1 in 5 declared as fraud are actually fraud and roughly 1 in every 6 customers have had a valid transaction declined in the past year. To address this problem, we use the Deep Feature Synthesis algorithm to automatically derive behavioral features based on the historical data of the card associated with a transaction. We generate 237 features (>100 behavioral patterns) for each transaction, and use a random forest to learn a classifier. We tested our machine learning model on data from a large multinational bank and compared it to their existing solution. On an unseen data of 1.852 million transactions, we were able to reduce the false positives by 54% and provide a savings of 190K euros. We also assess how to deploy this solution, and whether it necessitates streaming computation for real time scoring. We found that our solution can maintain similar benefits even when historical features are computed once every 7 days.

1 Introduction

Fraud detection problems are well-defined supervised learning problems, and data scientists have long been applying machine learning to help solve them [2,7]. However, false positives still plague the industry [10] with rates as high as 10–15%. Only 1 in 5 transactions declared as fraud be truly fraud [10]. Analysts have pointed out that these high false positives may be costing merchants more then fraud itself[1].

To mitigate this, most enterprises have adopted a multi-step process that combines work by human analysts and machine learning models. This process usually starts with a machine learning model generating a risk score and combining it with expert-driven rules to sift out potentially fraudulent transactions.

---

[1] https://blog.riskified.com/true-cost-declined-orders/.

© Springer Nature Switzerland AG 2019
U. Brefeld et al. (Eds.): ECML PKDD 2018, LNAI 11053, pp. 372–388, 2019.
https://doi.org/10.1007/978-3-030-10997-4_23

The resulting alerts pop up in a 24/7 monitoring center, where they are examined and diagnosed by human analysts. This process can potentially reduce the false positive rate by 5% – but this improvement comes only with high (and very costly) levels of human involvement. Even with such systems in place, a large number of false positives remain.

In this paper, we present an improved machine learning solution to drastically reduce the "false positives" in the fraud prediction industry. Such a solution will not only have financial implications, but also reduce the alerts at the 24/7 control center, enabling security analysts to use their time more effectively, and thus delivering the true promise of machine learning/artificial intelligence technologies.

We use a large, multi-year dataset from BBVA, containing 900 million transactions. We were also given fraud reports that identified a very small subset of transactions as fraudulent. Our task was to develop a machine learning solution that: (a) uses this rich transactional data in a transparent manner (no black box approaches), (b) competes with the solution currently in use by BBVA, and (c) is deployable, keeping in mind the real-time requirements placed on the prediction system.

We would be remiss not to acknowledge the numerous machine learning solutions achieved by researchers and industry alike (more on this in Sect. 2). However, the value of extracting behavioral features from historical data has been only recently recognized as an important factor in developing these solutions – instead, the focus has generally been on finding the best possible model given a set of features, and even after this, few studies focused on extracting a handful of features. Recognizing the importance of feature engineering [5] – in this paper, we use an automated feature engineering approach to generate hundreds of features to dramatically reduce the false positives.

Key to Success is Automated Feature Engineering: Having access to rich information about cards and customers exponentially increases the number of possible features we can generate. However, coming up with ideas, manually writing software and extracting features can be time-consuming, and may require customization each time a new bank dataset is encountered. In this paper, we use an automated method called *deep feature synthesis* (DFS) to rapidly generate a rich set of features that represent the patterns of use for a particular account/card. Examples of features generated by this approach are presented in Table 4.

As per our assessment, because we were able to perform feature engineering automatically *via* Featuretools and machine learning tools, we were able to focus our efforts and time on understanding the domain, evaluating the machine learning solution for financial metrics (>60% of our time), and communicating our results. We imagine tools like these will also enable others to focus on the real problems at hand, rather than becoming caught up in the mechanics of generating a machine learning solution.

Deep Feature Synthesis Obviates the Need for Streaming Computing: While the deep feature synthesis algorithm can generate rich and complex features using historical information, and these features achieve superior accuracy when put through machine learning, it still needs to be able to do this in real time in order to feed them to the model. In the commercial space, this has prompted the development of streaming computing solutions.

But, what if we could compute these features only once every t days instead? During the training phase, the abstractions in deep feature synthesis allow features to be computed with such a *"delay"*, and for their accuracy to be tested, all by setting a single parameter. For example, for a transaction that happened on August 24th, we could use features that had been generated on August 10th. If accuracy is maintained, the implication is that aggregate features need to be only computed once every few days, obviating the need for streaming computing.

Table 1. A transaction, represented by a number of attributes that detail every aspect of it. In this table, we are showing *only* a fraction of what is being recorded in addition to the *amount, timestamp* and *currency* for a transaction. These range from whether the customer was present physically for the transaction to whether the terminal where the transaction happened was serviced recently or not. We categorize the available information into several categories.

	Information type	Attribute recorded
Verification results	Card	Captures information about unique situations during card verification
	Terminal	Captures information about unique situations during verification at a terminal
About the location	Terminal	Can print/display messages
		Can change data on the card
		Maximum pin length it can accept
		Serviced or not
		How data is input into the terminal
	Authentication mode	Device type
About the merchant		Unique id
		Bank of the merchant
		Type of merchant
		Country
About the card		Authorizer
About the transaction		Amount
		Timestamp
		Currency
		Presence of a customer

What did we Achieve?

DFS Achieves a 91.4% Increase in Precision Compared to BBVA's Current Solution. This comes out to a reduction of 155,870 false positives in our dataset – a 54% reduction.

Table 2. Overview of the data we use in this paper

Item	Number
Cards	7,114,018
Transaction log entries	903,696,131
Total fraud reports	172,410
Fraudulent use of card number reports	122,913
Fraudulent card reports matched to transaction	111,897

The DFS-Based Solution Saves 190K Euros over 1.852 Million Transactions - A Tiny Fraction of the Total Transactional Volume. These savings are over 1.852 million transactions, only a tiny fraction of BBVA's yearly total, meaning that the true annual savings will be much larger.

We can Compute Features, Once Every 35 Days and Still Generate Value. Even when DFS features are only calculated once every 35 days, we are still able to achieve an improvement of 91.4% in precision. However, we do lose 67K euros due to approximation, thus only saving 123K total euros. This unique capability makes DFS is a practically viable solution.

2   Related Work

Fraud detection systems have existed since the late 1990s. Initially, a limited ability to capture, store and process data meant that these systems almost always relied on expert-driven rules. These rules generally checked for some basic attributes pertaining to the transaction – for example, "Is the transaction amount greater then a threshold?" or "Is the transaction happening in a different country?" They were used to block transactions, and to seek confirmations from customers as to whether or not their accounts were being used correctly.

Next, machine learning systems were developed to enhance the accuracy of these systems [2,7]. Most of the work done in this area emphasized the modeling aspect of the data science endeavor – that is, learning a *classifier*. For example, [4,12] present multiple classifiers and their accuracy. Citing the non-disclosure agreement, they do not reveal the fields in the data or the features they created. Additionally, [4] present a solution using only transactional features, as information about their data is unavailable.

Starting with [11], researchers have started to create small sets of handcrafted features, aggregating historical transactional information [1,9]. [13] emphasize the importance of aggregate features in improving accuracy. In most of these studies, aggregate features are generated by aggregating transactional information from the immediate past of the transaction under consideration. These are features like "number of transactions that happened on the same day", or "amount of time elapsed since the last transaction".

Fraud detection systems require instantaneous responses in order to be effective. This places limits on real-time computation, as well as on the amount of data that can be processed. To enable predictions within these limitations, the aggregate features used in these systems necessitate a streaming computational paradigm in production[2,3] [3]. As we will show in this paper, however, aggregate summaries of transactions that are as old as 35 days can provide similar precision to those generated from the most recent transactions, up to the night before. This poses an important question: When is streaming computing necessary for predictive systems? Could a comprehensive, automatic feature engineering method answer this question?

3   Dataset Preparation

Looking at a set of multiyear transactional data provided to us – a snapshot of which is shown in Table 1 – a few characteristics stand out:

Rich, Extremely Granular Information: Logs now contain not only information about a transaction's *amount, type, time stamp* and *location*, but also tangentially related material, such as the attributes of the terminal used to make the transaction. In addition, each of these attributes is divided into various subcategories that also give detailed information. Take, for example, the attribute that tells *"whether a terminal can print/display messages"*. Instead of a binary *"yes"* or *"no,"* this attribute is further divided into multiple subcategories: *"can print"*, *"can print and display"*, *"can display"*, *"cannot print or display"*, and *"unknown"*. It takes a 59-page dictionary to describe each transaction attribute and all of its possible values.

Historical Information About Card Use: Detailed, transaction-level information for each card and/or account is captured and stored at a central location, starting the moment the account is activated and stopping only when it is closed. This adds up quickly: for example, the dataset we received, which spanned roughly three years, contained 900 million transactions. Transactions from multiple cards or accounts belonging to the same user are now linked, providing a full profile of each customer's financial transactions.

Table 2 presents an overview of the data we used in this paper – a total of 900 million transactions that took place over a period of 3 years. A typical transactional dataset is organized into a three-level hierarchy: *Customers* ← *Cards* ← *Transactions*. That is, a transaction belongs to a card, which belongs to a customer. Conversely, a card may have several transactions, and a customer may have multiple cards. This relational structure plays an important role in identifying subsamples and developing features.

Before developing predictive models from the data, we took several preparative steps typical to any data-driven endeavor. Below, we present two data preparation challenges that we expect to be present across industry.

Identifying a Data Subsample: Out of the 900 million transactions in the dataset, only 122,000 were fraudulent. Thus, this data presents a challenge that is

---

[2] https://mapr.com/blog/real-time-credit-card-fraud-detection-apache-spark-and-event-streaming/.

[3] https://www.research.ibm.com/foiling-financial-fraud.shtml.

Table 3. The representative sample data set we extracted for training.

	Original fraud	Non-fraud
# of Cards	34378	36848
# of fraudulent transactions	111,897	0
# of non-fraudulent transactions	4,731,718	4,662,741
# of transactions	4,843,615	4,662,741

very common in fraud detection problems – less then 0.002% of the transactions are fraudulent. To identify patterns pertaining to fraudulent transactions, we have to identify a subsample. Since we have only few examples of fraud, each transaction is an important training example, and so we choose to keep every transaction that is annotated as fraud.

However, our training set must also include a reasonable representation of the non-fraudulent transactions. We could begin by sampling randomly – but the types of features we are attempting to extract also require historical information about the card and the customer to which a given transaction belongs. To enable the transfer of this information, we have to sample in the following manner:

1. Identify the cards associated with the fraudulent transactions,
   - Extract all transactions from these cards,
2. Randomly sample a set of cards that had no fraudulent transactions and,
   - Extract all transactions from these cards.

Table 3 presents the sampled subset. We formed a training subset that has roughly 9.5 million transactions, out of which only 111,897 are fraudulent. These transactions give a complete view of roughly 72K cards (Fig. 1).

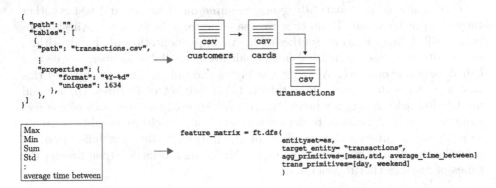

Fig. 1. The process of automatic feature generation. User specifies a `metadata` file that describes the relationships between multiple `csv`s, the path to the data files and several properties for each of the fields in each of the `csv`s. The three files for this problem are `customers` that has the list of all unique customers, `cards` that has the list of all unique cards, and `transactions`. The arrows represent one-to-many relationships. Given these two pieces of information, a user can select primitives in the featuretools library and compute the features. The library is available as open source at: https:// github.com/featuretools/featuretools/

4 Automated Feature Generation

Given the numerous attributes collected during every transaction, we can generate hypotheses/features in two ways:

- **By using only transaction information:** Each recorded transaction has a number of attributes that describe it, and we can extract multiple features from this information alone. Most features are binary, and can be thought of as answers to yes-or-no questions, along the lines of *"Was the customer physically present at the time of transaction?"*. These features are generated by converting categorical variables using `one-hot-encoding`. Additionally, all the numeric attributes of the transaction are taken as-is.
- **By aggregating historical information:** Any given transaction is associated with a *card*, and we have access to all the historical transactions associated with that *card*. We can generate features by aggregating this information. These features are mostly numeric – one example is, *"What is the average amount of transactions for this card?"*. Extracting these features is complicated by the fact that, when generating features that describe a transaction at time t, one can only use aggregates generated about the *card* using the transactions that took place *before t*. This makes this process computationally expensive during the model training process, as well as when these features are put to use.

Broadly, this divides the features we can generate into two types: (a) so-called "transactional features", which are generated from transactional attributes alone, and (b) features generated using historical data along with transactional features. Given the number of attributes and aggregation functions that could be applied, there are numerous potential options for both of these feature types.

Our goal is to automatically generate numerous features and test whether they can predict fraud. To do this, we use an automatic feature synthesis algorithm called Deep Feature Synthesis [8]. An implementation of the algorithm, along with numerous additional functionalities, is available as open source tool called `featuretools` [6]. We exploit many of the unique functionalities of this tool in order to to achieve three things: (a) a rich set of features, (b) a fraud model that achieves higher precision, and (c) approximate versions of the features that make it possible to deploy this solution, which we are able to create using a unique functionality provided by the library. In the next subsection, we describe the algorithm and its fundamental building blocks. We then present the types of features that it generated.

Deep Feature Synthesis. The purpose of Deep Feature Synthesis (DFS) is to automatically create new features for machine learning using the relational structure of the dataset. The relational structure of the data is exposed to DFS as *entities* and *relationships*.

An entity is a list of instances, and a collection of attributes that describe each one – not unlike a table in a database. A transaction entity would consist of a set of transactions, along with the features that describe each transaction, such as the transaction amount, the time of transaction, etc.

A relationship describes how instances in two entities can be connected. For example, the point of sale (POS) data and the historical data can be thought of as a "Transactions" entity and a "Cards" entity. Because each card can have many transactions, the relationship between Cards and Transactions can be described as a "parent and child" relationship, in which each parent (Card) has one or more children (Transactions).

Given the relational structure, DFS searches a built-in set of *primitive feature functions*, or simply called "primitives", for the best ways to synthesize new features. Each primitive in the system is annotated with the data types it accepts as inputs and the data type it outputs. Using this information, DFS can stack multiple primitives to find *deep features* that have the best predictive accuracy for a given problems.

The primitive functions in DFS take two forms.

- Transform primitives: This type of primitive creates a new feature by applying a function to an existing column in a table. For example, the Weekend primitive could accept the transaction date column as input and output a columns indicating whether the transaction occurred on a weekend.
- Aggregation primitives: This type of primitive uses the relations between rows in a table. In this dataset, the transactions are related by the id of the card that made them. To use this relationship, we might apply the Sum primitive to calculate the total amount spent to date by the card involved in the transaction.

Synthesizing Deep Features: For high value prediction problems, it is crucial to explore a large space of potentially meaningful features. DFS accomplishes this by applying a second primitive to the output of the first. For example, we might first apply the Hour transform primitive to determine when during the day a transaction was placed. Then we can apply Mean aggregation primitive to determine average hour of the day the card placed transactions. This would then read like cards.Mean(Hour(transactions.date)) when it is auto-generated. If the card used in the transaction is typically only used at one time of the day, but the transaction under consideration was at a very different time, that might be a signal of fraud.

Following this process of stacking primitives, DFS enumerates many potential features that can be used for solving the problem of predicting credit card fraud. In the next section, we describe the features that DFS discovered and their impact on predictive accuracy.

Table 4. Features generated using DFS primitives. Each feature aggregates data pertaining to past transactions from the card. The *left* column shows how the feature is computed via. an expression. The *right* column describes the feature in English. These features capture patterns in the transactions that belong to a particular card. For example, what was the mean value of the amount.

Features aggregating information from all the past transactions	
Expression	Description
cards.MEAN(transactions.amount)	Mean of transaction amount
cards.STD(transactions.amount)	Standard deviation of the transaction amount
cards.AVG_TIME_BETWEEN(transactions.date)	Average time between subsequent transactions
cards.NUM_UNIQUE(transactions.DAY(date))	Number of unique days
cards.NUM_UNIQUE(transactions.tradeid)	Number of unique merchants
cards.NUM_UNIQUE(transactions.mcc)	Number of unique merchant categories
cards.NUM_UNIQUE(transactions.acquirerid)	Number of unique acquirers
cards.NUM_UNIQUE(transactions.country)	Number of unique countries
cards.NUM_UNIQUE(transactions.currency)	Number of unique currencies

5 Modeling

After the feature engineering step, we have 236 features for 4,843,615 transactions. Out of these transactions, only 111,897 are labeled as fraudulent. With machine learning, our goal is to (a) learn a model that, given the features, can predict which transactions have this label, (b) evaluate the model and estimate its generalizable accuracy metric, and (c) identify the features most important for prediction. To achieve these three goals, we utilize a random forest classifier, which uses subsampling to learn multiple decision trees from the same data. We used `scikit-learn`'s classifier with 100 trees by setting `n_estimators=100`, and used `class_weight = 'balanced'`.

5.1 Evaluating the Model

To enable comparison in terms of "false positives", we assess the model comprehensively. Our framework involves (a) meticulously splitting the data into multiple exclusive subsets, (b) evaluating the model for machine learning metrics, and (c) comparing it to two different baselines. Later, we evaluate the model in terms of the financial gains it will achieve (in Sect. 6).

Machine Learning Metric. To evaluate the model, we assessed several metrics, including the area under the receiver operating curve (AUC-ROC). Since non-fraudulent transactions outnumber fraudulent transactions 1000:1, we first pick the operating point on the ROC curve (and the corresponding threshold) such that the true positive rate for fraud detection is $> 89\%$, and then assess the model's `precision`, which measures how many of the blocked transactions were in fact fraudulent. For the given true positive rate, the precision reveals what losses we will incur due to false positives.

Data Splits: We first experiment with all the cards that had one or more fraudulent transactions. To evaluate the model, we split it into mutually exclusive subsets, while making sure that fraudulent transactions are proportionally represented each time we split. We do the following:

- we first split the data into training and testing sets. We use 55% of the data for training the model, called D_{train}, which amounts to approximately 2.663 million transactions,
- we use an additional 326K, called D_{tune}, to identify the threshold - which is part of the training process,
- the remaining 1.852 million transactions are used for testing, noted as D_{test}.

Baselines: We compare our model with two baselines.

- **Transactional features baseline:** In this baseline, we only use the fields that were available at the time of the transaction, and that are associated with it. We do not use any features that were generated using historical data via DFS. We use one-hot-encoding for categorical fields. A total of 93 features are generated in this way. We use a random forest classifier, with the same parameters as we laid out in the previous section.
- **Current machine learning system at BBVA:** For this baseline, we acquired risk scores that were generated by the existing system that BBVA is currently using for fraud detection. We do not know the exact composition of the features involved, or the machine learning model. However, we know that the method uses only transactional data, and probably uses neural networks for classification.

Evaluation Process:

- Step 1: Train the model using the training data - D_{train}.
- Step 2: Use the trained model to generate prediction probabilities, P_{tu} for D_{tune}.
- Step 3: Use these prediction probabilities, and true labels L_{tu} for D_{tune} to identify the threshold. The threshold γ is given by:

$$\gamma = \underset{\gamma}{\arg\max} \; \text{precision}_\gamma \times u\left(\text{tpr}_\gamma - 0.89\right) \tag{1}$$

where tpr_γ is the true positive rate that can be achieved at threshold γ and u is a unit step function whose value is 1 when $\text{tpr}_\gamma \geq 0.89$. The true positive rate (tpr) when threshold γ is applied is given by:

$$\text{tpr}_\gamma = \frac{\sum_i \delta(P_{tu}^i \geq \gamma)}{\sum_i L_{tu}^i}, \forall i, \quad where \quad L_{tu}^i = 1 \tag{2}$$

where $\delta(.) = 1$ when $P_{tu}^i \geq \gamma$ and 0 otherwise. Similarly, we can calculate fpr_γ (false positive rate) and $precision_\gamma$.

– Step 4: Use the trained model to generate predictions for D_{test}. Apply the threshold γ and generate predictions. Evaluate `precision`, `recall` and `f-score`. Report these metrics (Table 5).

Table 5. `Precision` and `f-score` achieved in detecting non-fraudulent transactions at the fixed `recall` (a.k.a true positive rate) of $>= 0.89$. We compare the performance of features generated using the deep feature synthesis algorithm to those generated by ``one-hot-encoding`` of transactional attributes, and those generated by the baseline system currently being used. These baselines are described above. A

Metric	Transactional	Current system	DFS
Precision	0.187	0.1166	**0.41**
F-Score	0.30	0.20	**0.56**

Results and Discussion DFS Solution: In this solution we use the features generated by the DFS algorithm as implemented in `featuretools`. A total of 236 features are generated, which include those generated from the fields associated with the transaction itself. We then used a random forest classifier with the hyperparameter set described in the previous section.

In our case study, the transactional features baseline system has a false positive rate of 8.9%, while the machine learning system with DFS features has a false positive rate of 2.96%, a reduction of 6%.

When we fixed the true positive rate at >89%, our precision for the transactional features baseline was 0.187. For the model that used DFS features, we got a precision of 0.41, a >2x increase over the baseline. When compared to the current system being used in practice, we got a >3x improvement in precision. The current system has a precision of only about 0.1166.

6 Financial Evaluation of the Model

To assess the financial benefit of reducing false positives, we first detail the impact of false positives, and then evaluate the three solutions. When a false positive occurs, there is the possibility of losing a sale, as the customer is likely to abandon the item s/he was trying to buy. A compelling report published by Javelin Strategy & Research reports that these blocked sales add up to $118 billion, while the cost of real card fraud only amounts to $9 billion [10]. Additionally, the same [10] study reports that 26% of shoppers whose cards were declined reduced their shopping at that merchant following the decline, and 32% stopped entirely. There are numerous other costs for the merchant when a customer is falsely declined (See footnote 1).

From a card issuer perspective, when possibly authentic sales are blocked, two things can happen: the customer may try again, or may switch to a different issuer (different card). Thus issuers also lose out on millions in interchange fees, which are assessed at 1.75% of every transaction.[4] Additionally, it also may cause customer retention problems. Hence, banks actively try to reduce the number of cards affected by false positives.

Table 6. Losses incurred due to false positives and false negatives. This table shows the results when `threshold` is tuned to achieve $tpr \geq 0.89$. **Method:** We aggregate the `amount` for each false positive and false negative. False negatives are the frauds that are not detected by the system. We assume the issuer fully reimburses this to the client. For false positives, we assume that 50% of transactions will not happen using the card and apply a factor of 1.75% for interchange fee to calculate losses. These are estimates for the validation dataset which contained approximately 1.852 million transactions.

Method	False positives		False negatives		Total cost (€)
	Number	Cost (€)	Number	Cost (€)	
Current system	289,124	319,421.93	**4741**	**125,138.24**	**444,560**
Transactional features only	162,302	96,139.09	5061	818,989.95	915,129.05
DFS	**53,592**	**39,341.88**	5247	638,940.89	678,282.77

To evaluate the financial implications of increasing the precision of fraud detection from 0.1166 to 0.41, we do the following:

- We first predict the *label* for the 1.852 million transactions in our test dataset using the model and the threshold derived in Step 3 of the "evaluation process". Given the true *label*, we then identify the transactions that are falsely labeled as frauds.
- We assess the financial value of the false positives by summing up the amount of each of the transactions (in Euros).
- Assuming that 50% of these sales may successfully go through after the second try, we estimate the loss in sales using the issuer's card by multiplying the total sum by 0.5.
- Finally, we assess the loss in interchange fees for the issuer at 1.75% of the number in the previous step. This is the cost due to false positives - $cost_{fp}$
- Throughout our analysis, we fixed the true positive rate at 89%. To assess the losses incurred due to the remaining 10%, we sum up the total amount across all transactions that our model failed to detect as fraud. This is the cost due to false negatives - $cost_{fn}$
- The total cost is given by

$$totalcost = cost_{fp} + cost_{fn}$$

---

[4] "Interchange fee" is a term used in the payment card industry to describe a fee paid between banks for the acceptance of card-based transactions. For sales/services transactions, the merchant's bank (the "acquiring bank") pays the fee to a customer's bank (the "issuing bank").

By doing the simple analysis as above, we found that our model generated using the DFS features was able to reduce the false positives significantly and was able to reduce the $cost_{fp}$ when compared to BBVA's solution (€39,341.88 vs. €319,421.93). But it did not perform better then BBVA overall in terms of the total $cost$, even though there was *not* a significant difference in the number of false negatives between DFS based and BBVA's system. Table 6 presents the detailed results when we used our current model as if. This meant that BBVA's current system does really well in detecting high valued fraud. To achieve similar effect in detection, we decided to re-tune the threshold.

Retuning the Threshold: To tune the threshold, we follow the similar procedure described in Sect. 5.1, under subsection titled "Evaluation process", except for one change. In Step 2 we weight the probabilities generated by the model for a transaction by multiplying the amount of the transaction to it. Thus,

$$P_{tu}^i \leftarrow P_{tu}^i \times amount^i \tag{3}$$

We then find the `threshold` in this new space. For test data, to make a decision we do a similar transformation of the probabilities predicted by a classifier for a transaction. We then apply the threshold to this new transformed values and make a decision. This weighting essentially reorders the transactions. Two transactions both with the same prediction probability from the classifier, but vastly different amounts, can have different predictions.

Table 7 presents the results when this new `threshold` is used. A few points are noteworthy:

- **DFS model reduces the total cost BBVA would incur by atleast 190K euros.** It should be noted that these set of transactions, 1.852 million, only represent a tiny fraction of overall volume of transactions in a year. We further intend to apply this model to larger dataset to fully evaluate its efficacy.
- **When threshold is tuned considering financial implications, precision drops.** Compared to the precision we were able to achieve previously, when we did not tune it for high valued transactions, we get less precision (that is more false positives). In order to save from high value fraud, our threshold gave up some false positives.
- **"Transactional features only" solution has better precision than existing model, but smaller financial impact:** After tuning the threshold to weigh high valued transactions, the baseline that generates features only using attributes of the transaction (and no historical information) still has a higher precision than the existing model. However, it performs worse on high value transaction so the overall financial impact is the worse than BBVA's existing model.
- **54% reduction in the number of false positives.** Compared to the current BBVA solution, DFS based solution cuts the number of false positives by more than a half. Thus reduction in number of false positives reduces the number of cards that are false blocked - potentially improving customer satisfaction with BBVA cards.

Table 7. Losses incurred due to false positives and false negatives. This table shows the results when the `threshold` is tuned to consider high valued transactions. **Method**: We aggregate the `amount` for each false positive and false negative. False negatives are the frauds that are not detected by the system. We assume the issuer fully reimburses this to the client. For false positives, we assume that 50% of transactions will not happen using the card and apply a factor of 1.75% for interchange fee to calculate losses. These are estimates for the validation dataset which contained approximately 1.852 million transactions.

Method	False positives		False negatives		Total cost (€)
	Number	Cost (€)	Number	Cost (€)	
Current system	289,124	319,421.93	4741	125,138.24	444,560
Transactional features only	214,705	190,821.51	**4607**	686,626.40	877,447
DFS	**133,254**	**183,502.64**	4729	**71,563.75**	**255,066**

7   Real-Time Deployment Considerations

So far, we have shown how we can utilize complex features generated by DFS to improve predictive accuracy. Compared to the baseline and the current system, DFS-based features that utilize historical data improve the precision by 52% while maintaining the recall at 90%.

However, if the predictive model is to be useful in real life, one important consideration is: how long does it take to compute these features in real time, so that they are calculated right when the transaction happens? This requires thinking about two important aspects:

- Throughput: This is the number of predictions sought per second, which varies according to the size of the client. It is not unusual for a large bank to request anywhere between 10–100 predictions per second from disparate locations.
- Latency: This is the time between when a prediction is requested and when it is provided. Latency must be low, on the order of milliseconds. Delays cause annoyance for both the merchant and the end customer.

While throughput is a function of how many requests can be executed in parallel as well as the time each request takes (the latency), latency is strictly a function of how much time it takes to do the necessary computation, make a prediction, and communicate the prediction to the end-point (either the terminal or an online or digital payment system). When compared to the previous system in practice, using the complex features computed with DFS adds the additional cost of computing features from historical data, on top of the existing costs of creating transactional features and executing the model. Features that capture the aggregate statistics can be computed in two different ways:

- **Use aggregates up to the point of transaction:** This requires having infrastructure in place to query and compute the features in near-real time, and would necessitate streaming computation.

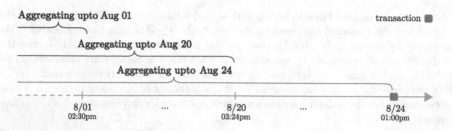

Fig. 2. The process of approximating of feature values. For a transaction that happens at 1 PM on August 24, we can extract features by aggregating from transactions up to that time point, or by aggregating up to midnight of August 20, or midnight of August 1, and so on. Not shown here, these approximations implicitly impact how frequently the features need to be computed. In the first case, one has to compute the features in real time, but as we move from left to right, we go from computing on daily basis to once a month.

- **Use aggregates computed a few time steps earlier:** We call these approximate features – that is, they are the features that were current a few time steps t ago. Thus, for a transaction happening at 1PM, August 24, we could use features generated on August 1 (24 days old). This enables feature computation in a batch mode: we can compute features once every month, and store them in a database for every card. When making a prediction for a card, we query for the features for the corresponding card. Thus, the real-time latency is only affected by the query time.

It is possible that using old aggregates could lead to a loss of accuracy. To see whether this would affect the quality of the predictions, we can simulate this type of feature extraction during the training process. Featuretools includes an option called *approximate*, which allows us to specify the intervals at which features should be extracted before they are fed into the model. We can choose approximate = ''1 day'', specifying that Featuretools should only aggregate features based on historical transactions on a daily basis, rather than all the way up to the time of transaction. We can change this to approximate = ''21 days'' or approximate = ''35 days'' Fig. 2 illustrates the process of feature approximation. To test how different metrics of accuracy are effected – in this case, the precision and the f1-score – we tested for 4 different settings: {1 day, 7 days, 21 days, and 35 days}.

Using this functionality greatly affects feature computation time during the model training process. By specifying a higher number of days, we can dramatically reduce the computation time needed for feature extraction. This enables data scientists to test their features quickly, to see whether they are predictive of the outcome.

Table 8 presents the results of the approximation when threshold has been tuned to simply achieve > 0.89 tpr. In this case, there is a loss of 0.05 in precision when we calculate features every 35 days.

In Table 9 presents the `precision` and `f1-score` for different levels of approximation, when `threshold` is tuned taking the financial value of the transaction into account. Surprisingly, we note that even when we compute features once every 35 days, we do not loose any `precision`. However, we loose approximately $67K$ euros in money.

Implications: This result has powerful implications for our ability to deploy a highly precise predictive model generated using a rich set of features. It implies that the bank can compute the features for all cards once every 35 days, and still be able to achieve better accuracy then the baseline method that uses only transactional features. Arguably, a 0.05 increase in `precision` as per Table 8 and €67K benefit as per Table 9 is worthwhile in some cases, but this should be considered alongside the costs it would incur to extract features on a daily basis. (It is also important to note that this alternative still only requires feature extraction on a daily basis, which is much less costly than real time.)

Table 8. `Precision` and `f-score` achieved in detecting non-fraudulent transactions at the fixed `recall` (a.k.a true positive rate) of $>= 0.89$, when feature approximation is applied and `threshold` is tuned only to achieve a *tpr* $>= 0.89$. A loss of 0.05 in precision is observed. No significant loss in financial value is noticed.

Metric	DFS with feature approximation			
	1	7	21	35
Precision	0.41	0.374	0.359	0.36
F1-score	0.56	0.524	0.511	0.512
Total-cost	678,282.77	735,229.05	716,157.54	675,854.12

Table 9. `Precision` and `f-score` achieved in detecting non-fraudulent transactions at the fixed `recall` (a.k.a true positive rate) of $>= 0.89$, when feature approximation is applied and `threshold` is tuned to weigh high valued transactions more. No significant loss in `precision` is found, but an additional cost of approximately 67K euros is incurred.

Metric	DFS with feature approximation			
	1	7	21	35
Precision	0.22	0.223	0.23	0.236
F1-score	0.35	0.356	0.366	0.373
Total-cost	255,066	305,282.26	314,590.34	322,250.67

References

1. Bhattacharyya, S., Jha, S., Tharakunnel, K., Westland, J.C.: Data mining for credit card fraud: a comparative study. Decis. Support Syst. **50**(3), 602–613 (2011)
2. Brause, R., Langsdorf, T., Hepp, M.: Neural data mining for credit card fraud detection. In: Proceedings of 11th IEEE International Conference on Tools with Artificial Intelligence, pp. 103–106. IEEE (1999)
3. Carcillo, F., Dal Pozzolo, A., Le Borgne, Y.A., Caelen, O., Mazzer, Y., Bontempi, G.: SCARFF: a scalable framework for streaming credit card fraud detection with spark. Inf. Fusion **41**, 182–194 (2017)
4. Chan, P.K., Fan, W., Prodromidis, A.L., Stolfo, S.J.: Distributed data mining in credit card fraud detection. IEEE Intell. Syst. Appl. **14**(6), 67–74 (1999)
5. Domingos, P.: A few useful things to know about machine learning. Commun. ACM **55**(10), 78–87 (2012)
6. Feature Labs, I.: Featuretools: automated feature engineering (2017)
7. Ghosh, S., Reilly, D.L.: Credit card fraud detection with a neural-network. In: Proceedings of the Twenty-Seventh Hawaii International Conference on System Science, vol. 3, pp. 621–630. IEEE (1994)
8. Kanter, J.M., Veeramachaneni, K.: Deep feature synthesis: towards automating data science endeavors. In: IEEE International Conference on Data Science and Advanced Analytics (DSAA), 36678 2015, pp. 1–10. IEEE (2015)
9. Panigrahi, S., Kundu, A., Sural, S., Majumdar, A.K.: Credit card fraud detection: a fusion approach using dempster-shafer theory and Bayesian learning. Inf. Fusion **10**(4), 354–363 (2009)
10. Pascual, A., Marchini, K., Van Dyke, A.: Overcoming false positives: saving the sale and the customer relationship. In: Javelin Strategy and Research Reports (2015)
11. Shen, A., Tong, R., Deng, Y.: Application of classification models on credit card fraud detection. In: 2007 International Conference on Service Systems and Service Management, pp. 1–4. IEEE (2007)
12. Stolfo, S., Fan, D.W., Lee, W., Prodromidis, A., Chan, P.: Credit card fraud detection using meta-learning: issues and initial results. In: AAAI-97 Workshop on Fraud Detection and Risk Management (1997)
13. Whitrow, C., Hand, D.J., Juszczak, P., Weston, D., Adams, N.M.: Transaction aggregation as a strategy for credit card fraud detection. Data Min. Knowl. Discov. **18**(1), 30–55 (2009)

Learning Tensor-Based Representations from Brain-Computer Interface Data for Cybersecurity

Md. Lutfor Rahman$^{(\boxtimes)}$, Sharmistha Bardhan, Ajaya Neupane, Evangelos Papalexakis, and Chengyu Song

University of California Riverside, Riverside, USA
{mrahm011,sbard002,ajaya}@ucr.edu, {epapalex,csong}@cs.ucr.edu

Abstract. Understanding, modeling, and explaining neural data is a challenging task. In this paper, we learn tensor-based representations of electroencephalography (EEG) data to classify and analyze the underlying neural patterns related to phishing detection tasks. Specifically, we conduct a phishing detection experiment to collect the data, and apply tensor factorization to it for feature extraction and interpretation. Traditional feature extraction techniques, like power spectral density, autoregressive models, and Fast Fourier transform, can only represent data either in spatial or temporal dimension; however, our tensor modeling leverages both spatial and temporal traits in the input data. We perform a comprehensive analysis of the neural data and show the practicality of multi-way neural data analysis. We demonstrate that using tensor-based representations, we can classify real and phishing websites with accuracy as high as 97%, which outperforms state-of-the-art approaches in the same task by 21%. Furthermore, the extracted latent factors are interpretable, and provide insights with respect to the brain's response to real and phishing websites.

1 Introduction

Phishing is a type of social engineering attacks, where attackers create fake websites with the look and feel similar to the real ones, and lure users to these websites with the intention of stealing their private credentials (e.g., password, credit card information, and social security numbers) for malicious purposes. Because phishing attacks are a big threat to cybersecurity, many studies have been conducted to understand why users are susceptible to phishing attacks [10,34,35,41], and to design automated detection mechanisms, e.g., by utilizing image processing [39], URL processing [7,37], or blacklisting [36]. Recently, Neupane et al. [24–26] introduced a new detection methodology based on the

---

Md. L. Rahman and S. Bardhan—Contributed equally.

© Springer Nature Switzerland AG 2019
U. Brefeld et al. (Eds.): ECML PKDD 2018, LNAI 11053, pp. 389–404, 2019.
https://doi.org/10.1007/978-3-030-10997-4_24

differences in the neural activity levels when users are visiting real and phishing websites. In this paper, we advance this line of work by introducing tensor decomposition to represent phishing detection related brain-computer interface data.

With the emergence of the Brain-Computer Interface (BCI), electroencephalography (EEG) devices have become commercially available and have been popularly used in gaming, meditation, and entertainment sectors. Thus, in this study, we used an EEG-based BCI device to collect the neural activities of users when performing a phishing detection task. The EEG data are often analyzed with methodologies like time-series analysis, power spectral analysis, and matrix decomposition, which consider either the temporal or spatial spectrum to represent the data. However, in this study, we take advantage of the multi-dimensional structure of the EEG data and perform tensor analysis, which takes into account spatial, temporal and spectral information, to understand the neural activities related to phishing detection and extract related features.

In this paper, we show that the tensor representation of the EEG data helps better understanding of the activated brain areas during the phishing detection task. We also show that the tensor decomposition of the EEG data reduces the dimension of the feature vector and achieves higher accuracy compared to the state-of-the-art feature extraction methodologies utilized by previous research [25].

Our Contributions: In this paper, we learned tensor representations of brain data related to phishing detection task. Our contributions are three-fold:

- We show that the multi-way nature of tensors is a powerful tool for the analysis and discovery of the underlying hidden patterns in neural data. To the best of our knowledge, this is the first study which employs the tensor representations to understand human performance in security tasks.
- We perform a comprehensive tensor analysis of the neural data and identify the level of activation in the channels or brain areas related to the users' decision making process with respect to the real and the fake websites based on the latent factors extracted.
- We extract features relevant to real and fake websites, perform cross-validation using different machine learning algorithms and show that using tensor-based representations can achieve the accuracy of above 94% consistently across all classifiers. We also reduce the dimension of the feature vector keeping the features related to the highly activated channels, and show that we can achieve better accuracy (97%) with the dimension-reduced feature vector.

The tensor representations of the data collected in our study provided several interesting insights and results. We observed that the users have higher component values for the channels located in the right frontal and parietal areas, which meant the areas were highly activated during the phishing detection task. These areas have been found to be involved in decision-making, working memory, and memory recall. Higher activation in these areas shows that the users were trying hard to infer the legitimacy of the websites, and may be recalling the properties of the website from their memory. The results of our study are consistent with the findings of the previous phishing detection studies [25, 26]. Unlike these studies, our study demonstrates a tool to obtain the active brain areas or channels involved in the phishing detection task without performing multiple statistical comparisons. On top of that, our methodology effectively derives more predictive features from these channels to build highly accurate machine-learning based automated phishing detection mechanism.

2   Data Collection Experiments

In this section, we describe details on data collection and preprocessing.

2.1   Data Collection

The motivation of our study is to learn tensor based representations from the BCI measured data for a phishing detection task, where users had to identify the phishing websites presented to them. We designed and developed a phishing detection experiment that measured the neural activities when users were viewing the real and fake websites. We designed our phishing detection study inline with the prior studies [10, 24–26]. Our phishing websites were created by obfuscating the URL either by inserting an extra similar looking string in the URL, or by replacing certain characters of the legitimate URL. The visual appearances of the fake websites were kept intact and similar to the real websites. We designed our fake webpages based on the samples of phishing websites and URLs available at PhishTank [32] and OpenPhish [28]. We choose twenty websites from the list of top 100 popular websites ranked by Alexa [3] and created fake versions of the 17 websites applying the URL obfuscation methodology. We also used the real versions of these 17 websites in the study. We collected data in multiple sessions and followed the EEG experiments like prior studies [21, 40]. In each session of the experiment, the participants were presented with 34 webpages in total.

We recruited fifteen healthy computer science students after getting the Institutional Review Board (IRB) approval and gave them $10 Amazon gift-card for participating in our study. We had ten (66.66%) male participants, and five (33.33%) female participants with the age-range of 20–32 years. The participants were instructed to look at the webpage on the screen and give response by pressing a 'Yes'/'No' button using a computer mouse. We used commercially available, light-weight EEG headset [1] to simulate a near real-world browsing experience. EmotivPro software package was used to collect raw EEG data. We presented with the same set of (randomized) trials to all the participants. All participants performed the same tasks for four different sessions. There was a break of approximate 5 min between two consecutive sessions. We collected all sessions data in the same day and same room. We have total 2040 (Participants (15) × Number of sessions (4) × Number of events per session (34)) responses. We discarded 187 wrong responses and only considered 1853 responses for our analysis.

2.2   Data Preprocessing

The EEG signals can be contaminated by eye blink, eyeball movement, breath, heart beats, and muscles movement. They can overwhelm the neural signals and may eventually degrade the performance of the classifiers. So, we preprocess the data to reduce the noise before modeling the data for tensor decomposition. Electrooculogram (EOG) produced by eye movements and Electromyography (EMG) produced by muscles movement are the common noise sources contaminating the EEG data. We used the AAR (Automatic Artifact Removal) toolbox [13] to remove both EOG and EMG [17]. After removing the EOG and EMG artifacts, EEG data were band pass filtered with the eighth-order Butterworth filter with the pass-band 3 to 60 Hz to remove other high frequency noises. The band pass filter keeps signals within the specified frequency range and rejects the rest. The electrical activities in the brain are generated by billions of neuron and the raw EEG signals we collected using sensors of Emotiv Epoc+ device had received signals from a mixture of sources. So we applied the Independent Component Analysis (ICA) [16], a powerful technique to separate independent sources linearly mixed in several sensors, to segregate the electrical signals related to each sensor. Our EEG data pre-processing methodology is similar to the process reported in [23].

3   Problem Formulation and Proposed Data Analysis

Tensor decomposition method is useful to capture the underlying structure of the analyzed data. In this experiment, the tensor decomposition method is applied to the EEG brain data measured for a phishing detection task.

One of the most popular tensor decomposition is the so-called PARAFAC decomposition [14]. In PARAFAC, by following an Alternating Least Square (ALS) method we decompose the tensor into 3 factor matrices. The PARAFAC decomposition decomposes the tensor into a sum of component rank-one tensors. Therefore, for a 3-mode tensor where $X \in R^{I \times J \times K}$, the decomposition will be,

$$X = \sum_{r=1}^{R} a_r \circ b_r \circ c_r \tag{1}$$

Here, R is a positive integer and $a_r \in R^I$, $b_r \in R^J$ and $c_r \in R^K$ are the factor vectors which we combine over all the modes and get the factor matrices. Figure 1 is showing the graphical representation of PARAFAC decomposition. However, PARAFAC model assumes that, for a set of variables the observations are naturally aligned. Since, in our phishing experiments, this is not guaranteed, we switched to PARAFAC2 model which is a variation of PARAFAC model.

The dimension of the feature matrix varies in dimension $68 \times N$, where 68 is for the number of event and N indicates the number of components or features. We have selected different number of features for our experiment to test what number of features trains a better model.

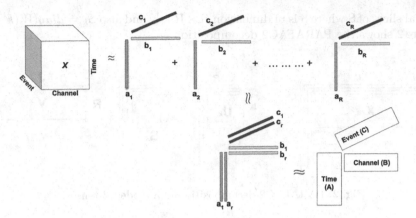

Fig. 1. PARAFAC decomposition with 3 factor matrices (Time, Channel and Event). Event matrix (blue colored) is used as features. (Color figure online)

3.1   PARAFAC2 Decomposition

In real life applications, a common problem is the dataset is not completely aligned in all modes. This situation occurs for different problems for example, clinical records for different patients where patients had different health problems and depending on that the duration of treatments varied over time [31]. Moreover, participants response record for phishing detection where each of them took a variable amount of time to select and decide whether the website presented is a real one or phishing one. In these examples, the number of samples per participant does not align naturally. The traditional models (e.g., PARAFAC and Tucker) assume that, the data is completely aligned. Moreover, if further preprocessing is applied in the data to make it completely aligned it might be unable to represent actual representation of the data [15,38]. Therefore, in order to model unaligned data, the traditional tensor models need changes. The PARAFAC2 model is designed to handle such data.

The PARAFAC2 model is the flexible version of the PARAFAC model. It also follows the uniqueness property of PARAFAC. However, the only difference is that the way it computes the factor matrices. It allows the other factor matrix to vary while applying the same factor in one mode. Suppose, the dataset contains data for K subjects. For each of these subjects (1, 2,..., K) there are J variables across which I_k observations are recorded. The I_k observations are not necessarily of equal length. The PARAFAC2 decomposition can be expressed as,

$$X_k \approx U_k S_k V^T \tag{2}$$

This is an equivalence relation of Eq. 1. It only represents the frontal slices X_k of the input tensor X. Where, for subject k and rank R, U_k is the factor matrix in the first mode with dimension $I_k \times R$, S_k is a diagonal matrix with dimension $R \times R$ and V is the factor matrix with dimension $J \times R$. The S_k is the frontal slices of S where S is of dimension R × R × K and also $S_k = diag(W(k, :))$. Figure 2 shows the PARAFAC2 decomposition.

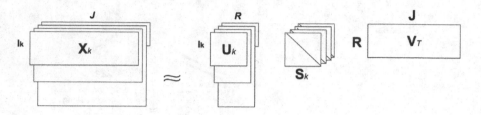

Fig. 2. PARAFAC2 decomposition of a mode - 3 tensor.

PARAFAC2 can naturally handle sparse data or dense data [18]. However, this statement was true only for a small number of subject [6]. The SPARTan algorithm is used for PARAFAC2 decomposition when the dataset is large and sparse [31].

3.2   Formulating Our Problem Using PARAFAC2

In order to apply different tensor decomposition method, at first we need to form the tensor. We form the initial tensor by considering all participants phishing detection brain data. The tensor for this experiment is of three dimensions, time × channel × events.

In this experiment, the participants were given the option to take the necessary time to decide whether the current website is phishing or not. Since, the participants were not restricted to take a decision within a particular time-frame, it has been found that for each event different participants took variable amount of time. Therefore, it is not possible to apply general tensor decomposition algorithm and even form a general tensor.

In order to solve the above problem, the PARAFAC2 model is used in this experiment. The SPARTan [31] algorithm is used to compute the PARAFAC2 decomposition. This algorithm has used the Matricized-Tensor-Times-Khatri-Rao-Product (MTTKRP) kernel. The major benefit of SPARTan is that it can handle large and sparse dataset properly. Moreover, it is more scalable and faster than existing PARAFAC2 decomposition algorithms.

3.3   Phishing Detection and Tensor

In this project, each participant was shown the real and phishing website and during that time, the brain EEG signal was captured. The participants were given the flexibility to take the required amount of time to select whether the website is real or not. Therefore, the observations for a set of variables do not align properly and the PARAFAC2 model is used to meaningfully align the data.

In order to create the PARAFAC2 model, the EEG brain data for all user for both real/phishing website was merged. The 3-mode tensor was then formed as Time × Channel × Events. In events, both the real and the phishing website are considered. Therefore, the tensor formed from this dataset consists of 1853 events, 14 channels (variables) and a maximum of 3753 observations (time in seconds). Figure 3 shows the PARAFAC2 model of the phishing experiment.

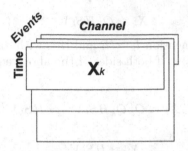

Fig. 3. PARAFAC2 model representing the brain EEG data across different events.

The 3 factor matrices obtained from the decomposition are U, V and W. These factor matrices representing the mode Time, Channel and Events respectively. In this experiment, we analyzed the V and W factor matrices to see which channels capture the high activity of brain regions and also distinguish between real and phishing events respectively.

In the SPARTAN algorithm [31], a modified version of the Matricized-Tensor-Times-Khatri-Rao-Product (MTTKRP) kernel has been used. It computes a tensor that is required in the PARAFAC2 decomposition algorithm. For a PARAFAC2 model, if our factor matrices are H, V and W and of dimension RXR, JXR, and KXR respectively, then for mode 1 with respect to K MTTKRP is computed as,

$$M^{(1)} = Y_{(1)}(W \odot V) \tag{3}$$

The computation here is then parallelized by computing the matrix multiplication as the sum of outer products for each block of $(W \odot V)$. The efficient way to compute the specialized MTTKRP is, first computing $Y_k V$ for each row of the intermediate result and then computing the Hadamard product with W(k, :). Since Y_k is column sparse, it reduces the computation of redundant operations. For this project, we have computed the factor matrices in Channel mode and Events mode using the above method.

Brain Data vs Tensor Rank. In exploratory data mining problems, it is really important to determine the quality of the results. In order to ensure a good quality of the decomposition, it is important to select a right number of components as the rank of the decomposition. In this experiment, we used the

AutoTen [29] algorithm to assess the performance of the decomposition with different ranks.

The application of AutoTen algorithm is not straightforward for the phishing experiment, since the observations for a set of variables do not align properly. Therefore, a number of additional operations are performed to bring the tensor of the whole dataset into a naturally aligned form. From Eq. (2), if we decompose U_k as $Q_k H$, then we can rewrite Eq. 2 as,

$$X_k \approx Q_k H S_k V^T \tag{4}$$

Where Q_k is with dimension $I_k \times R$ and H is with dimension $R \times R$. Q_k has orthonormal columns. Now, if both sides of the above equation is multiplied by Q_k^T, then we get,

$$Q_k^T X_k \approx Q_k^T Q_k H S_k V^T \approx H S_k V^T \tag{5}$$

Therefore, we can write,

$$Y_k \approx H S_k V^T \tag{6}$$

Where Y_k is the outer product of Q_k^T and X_k. The above equation is now same as the PARAFAC decomposition with consistency in all the modes. Y_k is also a tensor and is used in the AutoTen algorithm as input. The AutoTen algorithm was run for maximum rank 20 and it has been found that 3 is the rank for which the model can perform better. Therefore, for the PARAFAC2 decomposition using SPARTan, rank 3 is used.

4   Classification Performance

In this section, we discuss our classification performance for detecting the real and phishing page based on neural data. We merge all the data across all the sessions and across all the users. We extracted features from brain data using tensor decomposing with rank 3 computed by our modification of AutoTen as discussed in Sect. 3.3. We then applied the different type of machine learning algorithms for distinguishing the real and fake website based on brain data and checked their performance. We tested with Bayesian type BayesNet (BN), Function type Logistic Regression and MultilayerPerceptron, Rules type JRip and DecisionTable, Lazy type KStar and IB1 and Tree type J48, RandomTree, Logistic Model Tree (LMT), and RandomForest (RF). We present the best one (BayesNet, Logistic Regression, JRip, IB1, RandomForest) from each type of machine learning algorithms. We use 15-fold cross validation because we have 15 users data in our dataset. Here, the dataset is divided into 15 subsets where 14 subsets will be in training set and rest one subset will be in the testing subset.

We tested our model using several metrics: accuracy, precision, recall, F1 score and Area Under the Curve (AUC). We compared our classification performance in two different cases.

- **All Channels:** In this setting, we consider all 14 channel's data as feature vectors.
- **Top 6 Channels:** In this setting, we consider only top 6 highly activated channel's data as feature vectors. Details discussion for this can be found in Sect. 5.

Table 1. Classification Performance: In this table, we present the classification results of the five classifiers. Here, we have classification results for two scenarios. One for considering all channels for features extraction and another for considering only top 6 channels based on their activation. We have highlighted the accuracy of the best performing classifier in grey.

Metric	Accuracy		Recall		Precision		F-measure	
Algorithm	All	Top 6	All	Top 6	All	Top 6	All	Top 6
BayesNet	84.83	92.49	84.83	92.49	85.86	92.92	84.74	92.48
Logistic Regression	94.98	95.08	94.98	95.08	95.00	95.16	94.98	95.08
JRip	91.90	97.07	91.90	97.03	91.92	97.05	91.90	97.03
IB1	94.44	97.57	94.44	97.57	94.44	97.57	94.44	97.57
RandomForest	93.41	97.62	93.43	97.63	93.41	97.63	93.41	97.62

The summary of classification performance for different metrics (Accuracy, Recall, Precision, and F-measure) can be found in Table 1. We have seen that for considering all channels logistic regression algorithm gives 94% accuracy. We get 97% accuracy for considering top 6 highly activated channels using Random Forest algorithm. We achieved improved performance than the prior study which reported 76% accuracy of their phishing detection model built using neural signals when the participants were asked to identify real and fake websites under fNIRS scanning [25].

We also validated our classification performance by plotting the ROC curve in Fig. 4 using the Random Forest algorithm which gives the best accuracy among all the algorithms. In an ideal scenario, the AUC should be 100%. The baseline for AUC is 50%, which can be achieved through purely random guessing. Our model achieved 97.32% AUC for when considering all channels data and 99.22% when considering only top 6 highly activated channels data. We have seen that our True Positive Rate is 79.04 in case of all channels data and True Positive Rate is 94.91 in case of top 6 channels data while keeping False Positive Rate less than 1%. Reducing the channels gives us better phishing detection accuracy.

5   Discussion

In this section, we answer why we are getting good accuracy in classifying real and fake websites using brain data. We highlight the several key points for getting the good accuracy. First, we show that certain brain areas are highly activated during the phishing detection task. Second, we show that there is a statistically significant difference between the real and fake components.

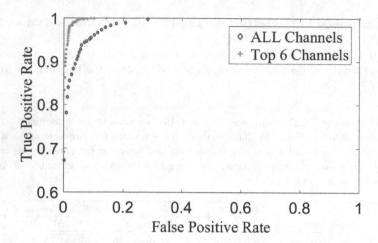

Fig. 4. AUC curve for all channels vs top 6 channels using the Random Forest algorithm. Here, we observed that TPR for all channels is 79.04% and 94.91% for top 6 channels when FPR is <1%

5.1 Phishing Detection vs Brain Areas

In this section, we provide a concise neuro-scientific insight of the brain data measured for the phishing detection. We discuss the relationship between the brain activities and phishing detection task. In our experiments, we collected brain data from human scalp using a commercially available non-invasive brain computer interface device. The data we collected using Emotiv Epoc+ device come from fourteen (AF3, F7, F3, FC5, T7, P7, O1, O2, P8, T8, FC6, F4, F8, AF4) different sensors as shown in Fig. 5. These sensors are placed on different regions according to the International 10−20 system. Two sensors positioned above the participant's ears (CMS/DRL) are used as references. Sensors location and functionality of each region is given below:

- *Frontal Lobe*, located at the front of the brain and associated with reasoning, attention, short memory, planning, and expressive language. The sensors that are placed in those area are AF3, F7, F3, FC5, FC6, F4, F8, and AF4.
- *Parietal Lobe*, located in the middle section of the brain and associated with perception, making sense of the world, and arithmetic. The censors P7 and P8 belongs to this area.
- *Occipital Lobe*, located in the back portion of the brain and associated with vision. The sensors from this location are O1 and O2.
- *Temporal Lobe*, located on the bottom section of the brain and associated with sensory input processing, language comprehension, and visual memory retention. The sensors of this location are T7 and T8.

Based on the factor analysis in channel dimension, we observed that mostly Frontal lobe and Parietal lobe sensors (AF3, F3, FC5, F7, P7, and P8) are highly activated for the phishing detection task. In Fig. 5(a), we present the channel activity based on channel factor data. Here, we consider all phishing detection events and get the factor matrix data in channel dimension using rank 3. We consider the first component data for drawing this graph. We have found that same subset of channels while considering the second and the third component data. In Fig. 5(b) we show the corresponding brain mapping for phishing detection task. Higher the red is the higher brain activity for phishing detection task. Our findings are aligned with the prior fMRI [26] and fNIRS [25] studies.

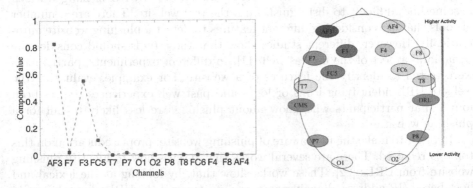

Fig. 5. (a) Shows the channel activity after the application of SPARTan decomposition on the tensor. The channel data for the first component is plotted in this figure to determine which channels have high activity. (b) Shows the corresponding brain region activation. (Color figure online)

5.2   Statistical Analysis: Real vs Fake Events

In this subsection, we present the statistical analysis of the components obtained from the tensor analysis. First, we performed the Kolmogorov-Smirnov (KS) test to determine the statistical distribution of the first component values of the real and fake factor matrix. In KS test we observed that the distribution of the real and fake samples was non-normal (p < .0005). We then applied Wilcoxon Singed-Rank Test, a non-parametric test comparing two sets of scores that come from the same participants, to measure the difference between real and fake components. We observed that there was statistically significantly high differences between the real and fake components (Z = 6.8, p < .0005).

5.3   Feature Space Reduction

One of the primary application of our study is the reduction of the dimension of the feature vector by keeping the features related to highly activated frontal and parietal channels. We observed that the prediction accuracy of the machine learning model trained on the features belonging to the top 6 highly activated channels was better than the prediction accuracy of the models better trained on features related to all channels. Our model achieved 97% of accuracy while applying reduced features vector. From the ROC curve in Fig. 4, we can see that our true positive rate increases from 79% to 94% when we use reduced feature vector in classification while keeping false positive rate <1%.

6 Related Works

Phishing attacks usually come in different forms or structures. In the case of the phishing website, the front-end structure of the website or URL is changed which is sometimes difficult to distinguish from the real website. There are a number of tools that are considering different features to detect a phishing website automatically. However, different studies show that these tools should consider the behavioral aspect of the user as well [11]. In different experiments, participants were tested to identify the features of a website. For example, evaluating the website URL, identifying icons or logos and past web experiences. It has been found that participants who know about phishing are less likely to fall for a phishing website.

In order to make the user aware of phishing website, proper education on this topic is required. There are several works that discuss how to identify phishing website from URLs [22]. These works show that, by looking at the lexical and host-based (IP-address, domain name, etc.) features of the URL, it can be easily found out whether the website is phishing or not. In this work, the accuracy obtained in classifying the phishing and the real webpage is 95–99%. Furthermore, it has been found that if appropriate education is provided, the user will be more efficient in avoiding phishing website [4]. Moreover, it has also been studied that what type of browser phishing warnings works better for the user and the performance of active warnings outperform the passive ones [12].

Apart from understanding user behavior while browsing the internet, it is also possible to prevent phishing by focusing on tracking the hacker's behavior. The hybrid feature selection method is applied to capture the phishing attacker's behavior from email header [2] and they achieved an accuracy of 94%. In these methods, both the content of email header and behavioral basis of it is considered for feature selection.

Automated Phishing Detection Method: In order to automatically detect phishing website, the pattern of the URL is considered as the primary method, and with the aid of machine learning algorithms it can protect the user from a phishing attack. However, these models do not perform well due to the lack in the number of features. Moreover, the domain top-page similarity based method is also used for phishing detection [33] where they obtained maximum AUC of 93%.

There are few more automated phishing detection system that use density based spatial clustering techniques to distinguish phishing and real website [20] with the accuracy of 91.44%. Linear classifiers are also used for phishing detection problem, and phishing domain ontology is also used for this task [42]. The content of a webpage is analyzed and based on their linguistic feature, an accuracy of 97% is achieved.

Tensor Decomposition and Phishing Detection: Tensor is useful for EEG brain data representation and visualization as well. It provides a compact representation of the brain network data. Moreover, it is useful to use tensor decomposition method to capture the underlying structure of the brain data. In Cichocki et al. [8], a brain computer interface system is used where tensor decomposition is applied in EEG signals. Tensor decomposition has already been applied for feature extraction in different problems involving EEG data. In P300 based BCIs, tensor decomposition is used to extract hidden features because of its multi-linear structures [27]. Unlike the general Event-related Potentials (ERP) based BCI problems, tensor can consider both temporal and spatial structure for feature extraction instead of only temporal structure which ensures better accuracy [8,9]. Tensor decomposition method has also been used for the classification of Mild and Severe Alzheimer's Disease using brain EEG data [19].

Tensor decomposition has been used for brain data analysis as well. GEBM is an algorithm that models the brain activity effectively [30]. SEMIBAT is a semi-supervised Brain network analysis approach based on constrained Tensor factorization [5]. The optimization objective is solved using the Alternating Direction Method of Multipliers (ADMM) framework. The proposed SEMIBAT method showed 31.60% improved results over plain vanilla tensor factorization for graph classification problem in EEG brain network.

Tensor decomposition methods have been applied for a variety of problems related to the analysis of brain signal. However, the idea of applying tensor decomposition methods in an automated system where the main task is to classify phishing and real websites based on brain EEG data is novel. In our case, we achieved the classification accuracy of real and phishing websites as high as 97% using neural signatures.

7   Conclusion

In this paper, we show that the tensor representation of brain data helps better understanding of the brain activation during the phishing detection task. In this scheme, owing to tensor representation on multi-modes of channel, time, and event, different characteristics of EEG signals can be presented simultaneously. We observed that right frontal and parietal areas are highly activated for participants during the phishing website detection task. These areas are involved in decision making, reasoning, and attention. We use the AutoTen algorithm to measure the quality of the result and also to choose a proper rank for the decomposition. We reduce the dimension of feature vectors and achieve a maximum 97% of classification accuracy while considering only highly activated brain area sensor's data. Our results show that the proposed methodology can be used in the cybersecurity domain for detecting phishing attacks using human brain data.

References

1. Emotiv EEG headset (2017). https://www.emotiv.com. Accessed 17 May 2017
2. Hamid, I.R.A., Abawajy, J.: Hybrid feature selection for phishing email detection. In: Xiang, Y., Cuzzocrea, A., Hobbs, M., Zhou, W. (eds.) ICA3PP 2011. LNCS, vol. 7017, pp. 266–275. Springer, Heidelberg (2011). https://doi.org/10.1007/978-3-642-24669-2_26
3. Amazon.com Inc.: Alexa skill kit (2027). https://developer.amazon.com/alexa-skills-kit
4. Arachchilage, N.A.G., Love, S.: Security awareness of computer users: a phishing threat avoidance perspective. Comput. Hum. Behav. **38**, 304–312 (2014)
5. Cao, B., Lu, C.-T., Wei, X., Yu, P.S., Leow, A.D.: Semi-supervised tensor factorization for brain network analysis. In: Frasconi, P., Landwehr, N., Manco, G., Vreeken, J. (eds.) ECML PKDD 2016. LNCS, vol. 9851, pp. 17–32. Springer, Cham (2016). https://doi.org/10.1007/978-3-319-46128-1_2
6. Chew, P.A., Bader, B.W., Kolda, T.G., Abdelali, A.: Cross-language information retrieval using PARAFAC2. In: Proceedings of the 13th ACM SIGKDD International Conference on Knowledge Discovery and Data Mining, KDD 2007, pp. 143–152. ACM (2007)
7. Chu, W., Zhu, B.B., Xue, F., Guan, X., Cai, Z.: Protect sensitive sites from phishing attacks using features extractable from inaccessible phishing URLs. In: 2013 IEEE International Conference on Communications, ICC, pp. 1990–1994. IEEE (2013)
8. Cichocki, A., et al.: Noninvasive BCIs: multiway signal-processing array decompositions. Computer **41**(10), 34–42 (2008)
9. Cong, F., Lin, Q.H., Kuang, L.D., Gong, X.F., Astikainen, P., Ristaniemi, T.: Tensor decomposition of EEG signals: a brief review. J. Neurosci. Methods **248**, 59–69 (2015)
10. Dhamija, R., Tygar, J.D., Hearst, M.: Why phishing works. In: Proceedings of the SIGCHI Conference on Human Factors in Computing Systems, pp. 581–590. ACM (2006)
11. Downs, J.S., Holbrook, M., Cranor, L.F.: Behavioral response to phishing risk. In: Proceedings of the Anti-phishing Working Groups 2nd Annual eCrime Researchers Summit, eCrime 2007, pp. 37–44. ACM (2007)
12. Egelman, S., Cranor, L.F., Hong, J.: You've been warned: an empirical study of the effectiveness of web browser phishing warnings. In: Proceedings of the SIGCHI Conference on Human Factors in Computing Systems, CHI 2008, pp. 1065–1074. ACM (2008)
13. Gómez-Herrero, G., et al.: Automatic removal of ocular artifacts in the EEG without an EOG reference channel. In: NORSIG, Signal Processing Symposium, pp. 130–133. IEEE (2006)
14. Harshman, R.A.: Foundations of the PARAFAC procedure: models and conditions for an "explanatory" multimodal factor analysis (1970)
15. Ho, J.C., Ghosh, J., Sun, J.: Marble: high-throughput phenotyping from electronic health records via sparse nonnegative tensor factorization. In: Proceedings of the 20th ACM SIGKDD International Conference on Knowledge Discovery and Data Mining, KDD 2014, pp. 115–124. ACM (2014)
16. Hyvärinen, A., Oja, E.: Independent component analysis: algorithms and applications. Neural Netw. **13**(4), 411–430 (2000)
17. Joyce, C.A., Gorodnitsky, I.F., Kutas, M.: Automatic removal of eye movement and blink artifacts from EEG data using blind component separation. Psychophysiology **41**(2), 313–325 (2004)

18. Kiers, H.A.L., Ten Berge, J.M.F., Bro, R.: PARAFAC2 - Part I. A direct fitting algorithm for the PARAFAC2 model. J. Chemometr. **13**, 275–294 (1999)
19. Latchoumane, C.F.V., Vialatte, F.B., Jeong, J., Cichocki, A.: EEG classification of mild and severe Alzheimer's disease using parallel factor analysis method. In: Ao, S.I., Gelman, L. (eds.) Advances in Electrical Engineering and Computational Science. LNEE, vol. 39, pp. 705–715. Springer, Dordrecht (2009). https://doi.org/10.1007/978-90-481-2311-7_60
20. Liu, G., Qiu, B., Wenyin, L.: Automatic detection of phishing target from phishing webpage. In: 2010 20th International Conference on Pattern Recognition, pp. 4153–4156, August 2010
21. Luck, S.J.: Ten simple rules for designing ERP experiments. In: Event-Related Potentials: A Methods Handbook 262083337 (2005)
22. Ma, J., Saul, L.K., Savage, S., Voelker, G.M.: Beyond blacklists: learning to detect malicious web sites from suspicious URLs. In: Proceedings of the 15th ACM SIGKDD International Conference on Knowledge Discovery and Data Mining, KDD 2009, pp. 1245–1254. ACM (2009)
23. Neupane, A., Rahman, M.L., Saxena, N.: PEEP: passively eavesdropping private input via brainwave signals. In: Kiayias, A. (ed.) FC 2017. LNCS, vol. 10322, pp. 227–246. Springer, Cham (2017). https://doi.org/10.1007/978-3-319-70972-7_12
24. Neupane, A., Rahman, M.L., Saxena, N., Hirshfield, L.: A multi-modal neurophysiological study of phishing detection and malware warnings. In: Proceedings of the 22nd ACM SIGSAC Conference on Computer and Communications Security, pp. 479–491. ACM (2015)
25. Neupane, A., Saxena, N., Hirshfield, L.: Neural underpinnings of website legitimacy and familiarity detection: an fNIRS study. In: Proceedings of the 26th International Conference on World Wide Web, pp. 1571–1580. International World Wide Web Conferences Steering Committee (2017)
26. Neupane, A., Saxena, N., Kuruvilla, K., Georgescu, M., Kana, R.: Neural signatures of user-centered security: an fMRI study of phishing, and malware warnings. In: Proceedings of the Network and Distributed System Security Symposium, NDSS, pp. 1–16 (2014)
27. Onishi, A., Phan, A.H., Matsuoka, K., Cichocki, A.: Tensor classification for P300-based brain computer interface. In: 2012 IEEE International Conference on Acoustics, Speech and Signal Processing, ICASSP, pp. 581–584. IEEE (2012)
28. OpenPhish: Phishing url (2017). https://openphish.com/feed.txt. Accessed 10 May 2017
29. Papalexakis, E.E.: Automatic unsupervised tensor mining with quality assessment. ArXiv e-prints, March 2015
30. Papalexakis, E.E., Fyshe, A., Sidiropoulos, N.D., Talukdar, P.P., Mitchell, T.M., Faloutsos, C.: Good-enough brain model: challenges, algorithms and discoveries in multi-subject experiments. In: Proceedings of the 20th ACM SIGKDD International Conference on Knowledge Discovery and Data Mining, KDD 2014, pp. 95–104. ACM, New York (2014)
31. Perros, I., et al.: SPARTan: scalable PARAFAC2 for large & sparse data. In: Proceedings of the 23rd ACM SIGKDD International Conference on Knowledge Discovery and Data Mining, KDD 2017, pp. 375–384. ACM (2017)
32. PhishTank: Join the fight against phishing (2017). https://www.phishtank.com/. Accessed 10 May 2017
33. Sanglerdsinlapachai, N., Rungsawang, A.: Using domain top-page similarity feature in machine learning-based web phishing detection. In: 2010 Third International Conference on Knowledge Discovery and Data Mining, pp. 187–190, January 2010

34. Sheng, S., Holbrook, M., Kumaraguru, P., Cranor, L.F., Downs, J.: Who falls for phish?: a demographic analysis of phishing susceptibility and effectiveness of interventions. In: Proceedings of the SIGCHI Conference on Human Factors in Computing Systems, pp. 373–382. ACM (2010)

35. Sheng, S., et al.: Anti-phishing phil: the design and evaluation of a game that teaches people not to fall for phish. In: Proceedings of the 3rd Symposium on Usable Privacy and Security, pp. 88–99. ACM (2007)

36. Sheng, S., Wardman, B., Warner, G., Cranor, L.F., Hong, J., Zhang, C.: An empirical analysis of phishing blacklists (2009)

37. Thomas, K., Grier, C., Ma, J., Paxson, V., Song, D.: Design and evaluation of a real-time URL spam filtering service. In: 2011 IEEE Symposium on Security and Privacy, SP, pp. 447–462. IEEE (2011)

38. Wang, Y., et al.: Rubik: knowledge guided tensor factorization and completion for health data analytics. In: Proceedings of the 21th ACM SIGKDD International Conference on Knowledge Discovery and Data Mining, KDD 2015, pp. 1265–1274. ACM (2015)

39. Whittaker, C., Ryner, B., Nazif, M.: Large-scale automatic classification of phishing pages. In: Proceedings of the Network and Distributed System Security Symposium, NDSS (2015)

40. Woodman, G.F.: A brief introduction to the use of event-related potentials in studies of perception and attention. Attention Percept. Psychophys. **72**(8), 2031–2046 (2010)

41. Wu, M., Miller, R.C., Garfinkel, S.L.: Do security toolbars actually prevent phishing attacks? In: Proceedings of the SIGCHI Conference on Human Factors in Computing Systems, pp. 601–610. ACM (2006)

42. Zhang, J., Li, Q., Wang, Q., Geng, T., Ouyang, X., Xin, Y.: Parsing and detecting phishing pages based on semantic understanding of text. J. Inf. Comput. Sci. **9**, 1521–1534 (2012)

ADS Health

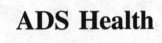

Can We Assess Mental Health Through Social Media and Smart Devices? Addressing Bias in Methodology and Evaluation

Adam Tsakalidis[1,2]([✉]), Maria Liakata[1,2], Theo Damoulas[1,2],
and Alexandra I. Cristea[1,3]

[1] Department of Computer Science, University of Warwick, Coventry, UK
{a.tsakalidis,m.liakata,t.damoulas,a.i.cristea}@warwick.ac.uk
[2] The Alan Turing Institute, London, UK
[3] Department of Computer Science, Durham University, Durham, UK

Abstract. Predicting mental health from smartphone and social media data on a longitudinal basis has recently attracted great interest, with very promising results being reported across many studies [3,9,13,26]. Such approaches have the potential to revolutionise mental health assessment, if their development and evaluation follows a real world deployment setting. In this work we take a closer look at state-of-the-art approaches, using different mental health datasets and indicators, different feature sources and multiple simulations, in order to assess their ability to generalise. We demonstrate that under a pragmatic evaluation framework, none of the approaches deliver or even approach the reported performances. In fact, we show that current state-of-the-art approaches can barely outperform the most naïve baselines in the real-world setting, posing serious questions not only about their deployment ability, but also about the contribution of the derived features for the mental health assessment task and how to make better use of such data in the future.

Keywords: Mental health · Bias · Evaluation · Wellbeing
Natural language processing · Smartphones · Sensors · Social media
Challenges

1 Introduction

Establishing the right indicators of mental well-being is a grand challenge posed by the World Health Organisation [7]. Poor mental health is highly correlated with low motivation, lack of satisfaction, low productivity and a negative economic impact [20]. The current approach is to combine census data at the population level [19], thus failing to capture well-being on an individual basis. The latter is only possible via self-reporting on the basis of established psychological scales, which are hard to acquire consistently on a longitudinal basis, and they capture long-term aggregates instead of the current state of the individual.

© Springer Nature Switzerland AG 2019
U. Brefeld et al. (Eds.): ECML PKDD 2018, LNAI 11053, pp. 407–423, 2019.
https://doi.org/10.1007/978-3-030-10997-4_25

The widespread use of smart-phones and social media offers new ways of assessing mental well-being, and recent research [1–3,5,9,10,13,14,22,23,26] has started exploring the effectiveness of these modalities for automatically assessing the mental health of a subject, reporting very high accuracy. What is typically done in these studies is to use features based on the subjects' smart phone logs and social media, to predict some self-reported mental health index (e.g., "wellbeing", "depression" and others), which is provided either on a Likert scale or on the basis of a psychological questionnaire (e.g., PHQ-8 [12], PANAS [29], WEMWBS [25] and others).

Most of these studies are longitudinal, where data about individuals is collected over a period of time and predictions of mental health are made over a sliding time window. Having such longitudinal studies is highly desirable, as it can allow fine-grained monitoring of mental health. However, a crucial question is *what constitutes an appropriate evaluation framework*, in order for such approaches to be employable in a real world setting. Generalisation to previously unobserved users can only be assessed via leave-N-users-out cross-validation setups, where typically, N is equal to one (**LOUOCV**, see Table 1). However, due to the small number of subjects that are available, such generalisation is hard to achieve by any approach [13]. Alternatively, personalised models [3,13] for every individual can be evaluated via a within-subject, leave-N-instances-out cross-validation (for $N = 1$, **LOIOCV**), where an instance for a user u at time i is defined as a $\{X_{ui}, y_{ui}\}$ tuple of {features(u, i), mental-health-score(u, i)}. In a real world setting, a *LOIOCV* model is trained on some user-specific instances, aiming to predict her mental health state at some future time points. Again however, the limited number of instances for every user make such models unable to generalize well. In order to overcome these issues, previous work [2,5,9,10,22,26] has combined the instances $\{X_{u_j i}, y_{u_j i}\}$ from different individuals u_j and performed evaluation using randomised cross validation (**MIXED**). While such approaches can attain optimistic performance, the corresponding models fail to generalise to the general population and also fail to ensure effective personalised assessment of the mental health state of a single individual.

Table 1. Summary of the three evaluation frameworks.

	LOUOCV	LOIOCV	MIXED
Real world aim	Build a model m that generalises to a previously unseen user u	Build a personalised model m_u per user u that generalises on u, given some manual input by u	Build a model m that generalises to new instances of a specific pool of previously seen users
Train	$\{\{X_{u'i}, y_{u'i}\}\}$	$\{\{X_{ui'}, y_{ui'}\}\}$	$\{\{X_{u_0 i'}, y_{u_0 i'}\}, ..., \{X_{u_n i'}, y_{u_n i'}\}\}$
Test	$\{X_{ui}, y_{ui}\}$	$\{X_{ui}, y_{ui}\}$	$\{\{X_{u_0 i}, y_{u_0 i}\}, ..., \{X_{u_n i}, y_{u_n i}\}\}$
Limits	Few users for training and evaluation	Few instances per user for training and evaluation	Cannot ensure generalisation neither over new users nor in a personalised way

Table 2. Works on predicting mental health in a longitudinal manner.

Work	Target	Modalities	Type	Size	Eval
Ma et al. [14]	Tiredness, Tensity, Displeasure (1–5)	Location, accelerometer, sms, calls	Class.	15	N/A
Bogomolov et al. [2]	Happiness (1–7)	Weather, calls, bluetooth, sms, Big Five	Class.	117	MIXED
LiKamWa et al. [13]	Activeness, Pleasure (1–5)	Email/phone/sms contacts, location, apps, websites	Regr.	32	LOIOCV LOUOCV
Bogomolov et al. [1]	Stress (1–7)	Weather, calls, bluetooth, sms, Big Five	Class.	117	MIXED
Canzian and Musolesi [3]	PHQ-8	GPS	Class.	48	LOIOCV
Jaques et al. [9,10]	Happiness [9], {Happiness, Health, Stress, Energy, Alertness} (0–100) [10]	Electrodermal activity, calls, accelerometer, sms, surveys, phone usage, locations	Class.	68	MIXED
Tsakalidis et al. [26]	PANAS, WEMWBS	Social media, calls, sms, locations, headphones, charger, screen/ringer mode, wifi	Regr.	19	MIXED
Farhan et al. [5]	PHQ-9	GPS, PHQ-9 scores	Class.	79	MIXED
Wang et al. [27]	Positive, Negative (0–15), Positive-Negative	GPS, calls, accelerometer, microphone, light sensor, sms, apps, phone locked	Regr.	21	LOIOCV LOUOCV
Servia-Rodriguez et al. [22]	Positive/Negative, Alert/Sleepy	microphone, accelerometer, calls, sms	Class.	726	MIXED
Suhara et al. [23]	Mood (binary)	Daily surveys	Class.	2,382	LNUOCV

In this paper we demonstrate the challenges that current state-of-the-art models face, when tested in a real-world setting. We work on two longitudinal datasets with four mental health targets, using different features derived from a wide range of heterogeneous sources. Following the state-of-the-art experimental methods and evaluation settings, we achieve very promising results, regardless of the features we employ and the mental health target we aim to predict. However, when tested under a pragmatic setting, the performance of these models drops heavily, *failing to outperform the most naïve – from a modelling perspective – baselines*: majority voting, random classifiers, models trained on the identity of the user, etc. This poses serious questions about the contribution of the features derived from social media, smartphones and sensors for the task of automatically assessing well-being on a longitudinal basis. Our goal is to flesh out, study and discuss such limitations through extensive experimentation across multiple settings, and to propose a pragmatic evaluation and model-building framework for future research in this domain.

2   Related Work

Research in *assessing mental health on a longitudinal basis* aims to make use of relevant features extracted from various modalities, in order to train models for automatically predicting a user's mental state (target), either in a classification or a regression manner [1–3, 9, 10, 13, 26]. Examples of state-of-the-art work in this domain are listed in Table 2, along with the number of subjects that was used and the method upon which evaluation took place. Most approaches have used the *MIXED* approach to evaluate models [1, 2, 5, 9, 10, 22, 26], which, as we will show, is vulnerable to bias, due to the danger of recognising the user in the test set and thus simply inferring her average mood score. *LOIOCV* approaches that have not ensured that their train/test sets are independent are also vulnerable to bias in a realistic setting [3, 13]. From the works listed in Table 2, only Suhara et al. [23] achieves unbiased results with respect to model generalisability; however, the features employed for their prediction task are derived from self-reported questionnaires of the subjects and not by automatic means.

3   Problem Statement

We first describe three major problems stemming from unrealistic construction and evaluation of mental health assessment models and then we briefly present the state-of-the-art in each case, which we followed in our experiments.

P1 Training on past values of the target variable: This issue arises when the past N mood scores of a user are required to predict his/her next mood score in an autoregressive manner. Since such an approach would require the previous N scores of past mood forms, it would limit its ability to generalise without the need of manual user input in a continuous basis. This makes it impractical for a real-world scenario. Most importantly, it is difficult to measure the contribution of the features towards the prediction task, unless the model is evaluated using target feature ablation. For demonstration purposes, we have followed the experimental setup by LiKamWa et al. [13], which is one of the leading works in this field.

P2 Inferring test set labels: When training personalised models (*LOIOCV*) in a longitudinal study, it is important to make sure that there are no overlapping instances across consecutive time windows. Some past works have extracted features $\{f(t - N), ..., f(t)\}$ over N days, in order to predict the $score_t$ on day $N + 1$ [3, 13]. Such approaches are biased if there are overlapping days of train/test data. To illustrate this problem we have followed the approach by Canzian and Musolesi [3], as one of the pioneering works on predicting depression with GPS traces, on a longitudinal basis.

P3 Predicting users instead of mood scores: Most approaches merge all the instances from different subjects, in an attempt to build user-agnostic models in a randomised cross-validation framework [2, 9, 10, 26]. This is problematic, especially when dealing with a small number of subjects, whose behaviour (as captured through their data) and mental health scores differ on an individual basis. Such approaches are in danger of "predicting" the user in the test set, since her (test set) features might be highly correlated with her features in the training set, and thus infer her average well-being score, based on the corresponding observations of the training set. Such approaches cannot guarantee that they will generalise on either a population-wide ($LOUOCV$) or a personalised ($LOIOCV$) level. In order to examine this effect in both a regression and a classification setting, we have followed the experimental framework by Tsakalidis et al. [26] and Jaques et al. [9].

3.1   P1: Training on Past Values of the Target (LOIOCV, LOUOCV)

LiKamWa et al. [13] collected smartphone data from 32 subjects over a period of two months. The subjects were asked to self-report their "pleasure" and "active-ness" scores at least four times a day, following a Likert scale (1 to 5), and the average daily scores served as the two targets. The authors aggregated various features on social interactions (e.g., number of emails sent to frequently interact-ing contacts) and routine activities (e.g., browsing and location history) derived from the smartphones of the participants. These features were extracted over a period of three days, along with the two most recent scores on activeness and pleasure. The issue that naturally arises is that such a method cannot generalise to new subjects in the $LOUOCV$ setup, as it requires their last two days of self-assessed scores. Moreover, in the $LOIOCV$ setup, the approach is limited in a real world setting, since it requires the previous mental health scores by the subject to provide an estimate of her current state. Even in this case though, the feature extraction should be based on past information only – under $LOIOCV$ in [13], the current mood score we aim at predicting is also used as a feature in the (time-wise) subsequent two instances of the training data.

Experiments in [13] are conducted under $LOIOCV$ and $LOUOCV$, using Mul-tiple Linear Regression (LR) with Sequential Feature Selection (in $LOUOCV$, the past two pairs of target labels of the test user are still used as features). In order to better examine the effectiveness of the features for the task, the same model can be tested without any ground-truth data as input. Nevertheless, a simplistic model predicting the per-subject average outperforms their LR in the $LOUOCV$ approach, which poses the question of whether the smartphone-derived features can be used effectively to create a generalisable model that can assess the men-tal health of unobserved users. Finally, the same model tested in the $LOIOCV$ setup achieves the lowest error; however, this is trained not only on target scores overlapping with the test set, but also on features derived over a period of three days, introducing further potential bias, as discussed in the following.

3.2   P2: Inferring Test Labels (LOIOCV)

Canzian and Musolesi [3] extracted mobility metrics from 28 subjects to predict their depressive state, as derived from their daily self-reported PHQ-8 question-naires. A 14-day moving average filter is first applied to the PHQ-8 scores and the mean value of the same day (e.g., Monday) is subtracted from the normalised scores, to avoid cyclic trends. This normalisation results into making the target score s_t on day t dependent on the past $\{s_{t-14}, ..., s_{t-1}\}$ scores. The normalised PHQ-8 scores are then converted into two classes, with the instances deviat-ing more than one standard deviation above the mean score of a subject being assigned to the class "1" ("0", otherwise). The features are extracted over various time windows (looking at $T_{HIST} = \{0, ..., 14\}$ days before the completion of a mood form) and personalised model learning and evaluation are performed for every T_{HIST} separately, using a $LOIOCV$ framework.

What is notable is that the results improve significantly when features are extracted from a wider T_{HIST} window. This could imply that the depressive state of an individual can be detected with a high accuracy if we look back at her history. However, by training and testing a model on instances whose features are derived from the same days, there is a high risk of over-fitting the model to the timestamp of the day in which the mood form was completed. In the worst-case scenario, there will be an instance in the train set whose features (e.g., total covered distance) are derived from the 14 days, 13 of which will also be used for the instance in the test set. Additionally, the target values of these two instances will also be highly correlated due to the moving average filter, making the task artificially easy for large T_{HIST} and not applicable in a real-world setting.

While we focus on the approach in [3], a similar approach with respect to feature extraction was also followed in LiKamWa et al. [13] and Bogomolov et al. [2], extracting features from the past 2 and 2 to 5 days, respectively.

3.3 P3: Predicting Users (LOUOCV)

Tsakalidis et al. [26] monitored the behaviour of 19 individuals over four months. The subjects were asked to complete two psychological scales [25, 29] on a daily basis, leading to three target scores (positive, negative, mental well-being); various features from smartphones (e.g., time spent on the preferred locations) and textual features (e.g., ngrams) were extracted passively over the 24 h preceding a mood form timestamp. Model training and evaluation was performed in a randomised ($MIXED$) cross-validation setup, leading to high accuracy ($R^2 = 0.76$). However, a case demonstrating the potential user bias is when the models are trained on the textual sources: initially the highest R^2 (0.22) is achieved when a model is applied to the mental-wellbeing target; by normalising the textual features on a per-user basis, the R^2 increases to 0.65. While this is likely to happen because the vocabulary used by different users is normalised, there is also the danger of over-fitting the trained model to the identity of the user. To examine this potential, the $LOIOCV/LOUOCV$ setups need to be studied alongside the $MIXED$ validation approach, with and without the per-user feature normalisation step.

A similar issue is encountered in Jaques et al. [9] who monitored 68 subjects over a period of a month. Four types of features were extracted from survey and smart devices carried by subjects. Self-reported scores on a daily basis served as the ground truth. The authors labelled the instances with the top 30% of all the scores as "happy" and the lowest 30% as "sad" and randomly separated them into training, validation and test sets, leading to the same user bias issue. Since different users exhibit different mood scores on average [26], by selecting instances from the top and bottom scores, one might end up separating users and convert the mood prediction task into a user identification one. A more suitable task could have been to try to predict the highest and lowest scores of every individual separately, either in a $LOIOCV$ or in a $LOUOCV$ setup.

While we focus on the works of Tsakalidis et al. [26] and Jaques et al. [9], similar experimental setups were also followed in [10], using the median of scores to separate the instances and performing five-fold cross-validation, and by Bogomolov et al. in [2], working on a user-agnostic validation setting on 117 subjects to predict their happiness levels, and in [1], for the stress level classification task.

4   Experiments

4.1   Datasets

By definition, the aforementioned issues are feature-, dataset- and target-independent (albeit the magnitude of the effects may vary). To illustrate this, we run a series of experiments employing two datasets, with different feature sources and four different mental health targets.

Dataset 1: We employed the dataset obtained by Tsakalidis et al. [26], a pioneering dataset which contains a mix of longitudinal textual and mobile phone usage data for 30 subjects. From a textual perspective, this dataset consists of social media posts (1,854/5,167 facebook/twitter posts) and private messages (64,221/132/47,043 facebook/twitter/SMS messages) sent by the subjects. For our ground truth, we use the {positive, negative, mental well-being} mood scores (in the ranges of [10–50], [10–50], [14–70], respectively) derived from self-assessed psychological scales during the study period.

Dataset 2: We employed the StudentLife dataset [28], which contains a wealth of information derived from the smartphones of 48 students during a 10-week period. Such information includes samples of the detected activity of the subject, timestamps of detected conversations, audio mode of the smartphone, status of the smartphone (e.g., charging, locked), etc. For our target, we used the self-reported stress levels of the students (range [0–4]), which were provided several times a day. For the approach in LiKamWa et al. [13], we considered the average daily stress level of a student as our ground-truth, as in the original paper; for the rest, we used all of the stress scores and extracted features based on some time interval preceding their completion, as described next, in 4.3[1].

---

[1] For P3, this creates the P2 cross-correlation issue in the $MIXED/LOIOCV$ settings. For this reason, we ran the experiments by considering only the last entered score in a given day as our target. We did not witness any major differences that would alter our conclusions.

4.2   Task Description

We studied the major issues in the following experimental settings (see Table 3):

P1: Using Past Labels: We followed the experimental setting in [13] (see Sect. 3.1): we treated our task as a regression problem and used Mean Squared Error (MSE) and classification accuracy[2] for evaluation. We trained a Linear Regression (LR) model and performed feature selection using Sequential Feature Selection under the $LOIOCV$ and $LOUOCV$ setups; feature extraction is performed over the previous 3 days preceding the completion of a mood form. For comparison, we use the same baselines as in [13]: Model A always predicts the average mood score for a certain user (`AVG`); Model B predicts the last entered scores (`LAST`); Model C makes a prediction using the LR model trained on the ground-truth features only (`-feat`). We also include Model D, trained on non-target features only (`-mood`) in an unbiased $LOUOCV$ setting.

P2: Inferring Test Labels: We followed the experimental setting presented in [3]. We process our ground-truth in the same way as the original paper (see Sect. 3.2) and thus treat our task as a binary classification problem. We use an SVM_{RBF} classifier, using grid search for parameter optimisation, and perform evaluation using specificity and sensitivity. We run experiments in the $LOIOCV$ and $LOUOCV$ settings, performing feature extraction at different time windows ($T_{HIST} = \{1, ..., 14\}$). In order to better demonstrate the problem that arises here, we use the previous label classifier (`LAST`) and the SVM classifier to which we feed only the mood timestamp as a feature (`DATE`) for comparison. Finally, we replace our features with completely random data and train the same SVM with $T_{HIST} = 14$ by keeping the same ground truth, performing 100 experiments and reporting averages of sensitivity and specificity (`RAND`).

P3: Predicting Users: We followed the evaluation settings of two past works (see Sect. 3.3), with the only difference being the use of 5-fold CV instead of a train/dev/test split that was used in [9]. The features of every instance are extracted from the past day before the completion of a mood form. In **Experiment 1** we follow the setup in [26]: we perform 5-fold CV ($MIXED$) using SVM (SVR_{RBF}) and evaluate performance based on R^2 and $RMSE$. We compare the performance when tested under the $LOIOCV/LOUOCV$ setups, with and without the per-user feature normalisation step. We also compare the performance of the $MIXED$ setting, when our model is trained on the one-hot-encoded user id only. In **Experiment 2** we follow the setup in [9]: we label the instances as "high" ("low"), if they belong to the top-30% (bottom-30%) of mood score values ("UNIQ" – for "unique" – setup). We train an SVM classifier in 5-fold CV

---

[2] Accuracy is defined in [13] as follows: 5 classes are assumed (e.g., [0, ..., 4]) and the squared error e between the centre of a class halfway towards the next class is calculated (e.g., 0.25). If the squared error of a test instance is smaller than e, then it is considered as having been classified correctly.

using accuracy for evaluation and compare performance in the *LOIOCV* and *LOUOCV* settings. In order to further examine user bias, we perform the same experiments, this time by labelling the instances on a per-user basis ("PERS" – for "personalised" – setup), aiming to predict the per-user high/low mood days[3].

Table 3. Summary of experiments. The highlighted settings indicate the settings used in the original papers; "Period" indicates the period before each mood form completion during which the features were extracted.

Issue	P1: Training on past labels	P2: Inferring test labels	P3: Predicting users
Setting	**LOIOCV, LOUOCV**	**LOIOCV**, LOUOCV	**MIXED**, LOIOCV, LOUOCV
Task	Regr.	Class.	Regr. (E1); Class. (E2)
Metrics	MSE, accuracy	Sensitivity, specificity	R^2, RMSE (E1); accuracy (E2)
Period	Past 3 days	Past $\{1,...,14\}$ days	Past day
Model	LR_{sfs}	SVM_{rbf}	SVR_{rbf}; SVM_{rbf}
Baselines	AVG, LAST, -feat, -mood	LAST, DATE, RAND	Model trained on user id

4.3 Features

For *Dataset 1*, we first defined a "user snippet" as the concatenation of all texts generated by a user within a set time interval, such that the maximum time difference between two consecutive document timestamps is less than 20 min. We performed some standard noise reduction steps (converted text to lower-case, replaced URLs/user mentions and performed language identification[4] and tokenisation [6]). Given a mood form and a set of snippets produced by a user before the completion of a mood form, we extracted some commonly used feature sets for every snippet written in English [26], which were used in all experiments. To ensure sufficient data density, we excluded users for whom we had overall fewer than 25 snippets on the days before the completion of the mood form or fewer than 40 mood forms overall, leading to 27 users and 2,368 mood forms. For *Dataset 2*, we extracted the features presented in Table 4. We only kept the users that had at least 10 self-reported stress questionnaires, leading to 44 users and 2,146 instances. For our random experiments used in P2, in Dataset 1 we replaced the text representation of every snippet with random noise ($\mu = 0, \sigma = 1$) of the same feature dimensionality; in Dataset 2, we replaced the actual inferred value of every activity/audio sample with a random inference class; we also replaced each of the detected conversation samples and samples detected in a dark environment/locked/charging, with a random number (<100, uniformly distributed) indicating the number of pseudo-detected samples.

---

[3] In cases where the lowest of the top-30% scores (s) was equal to the highest of the bottom-30% scores, we excluded the instances with score s.

[4] https://pypi.python.org/pypi/langid.

Table 4. Features that were used in our experiments in Datasets 1, 2 (left, right).

(a) **duration** of the snippet; *(b)* binary **ngrams** ($n = 1, 2$); *(c)* cosine similarity between the words of the document and the 200 **topics** obtained by [21]; *(d)* functions over **word embeddings** dimensions [24] (mean, max, min, median, stdev, 1st/3rd quartile); *(e)* **lexicons** [8,11,16–18,30]: for lexicons providing binary values (pos/neg), we counted the number of ngrams matching each class and for those with score values, we used the counts and the total summation of the corresponding scores. *(f)* **number** of Facebook posts/messages/images, Twitter posts/messages, SMS, number of tokens/messages/posts in the snippet	*(a)* percentage of collected samples for each **activity** (stationary, walking, running, unknown) and *(b)* **audio** mode (silence, voice, noise, unknown); *(c)* number and total duration of detected **conversations**; number of samples and total duration of the time during which the phone was *(d)* in a **dark environment**, *(e)* **locked** and *(f)* **charging**

5 Results

5.1 **P1**: Using Past Labels

Table 5 presents the results on the basis of the methodology by LiKamWa et al. [13], along with the average scores reported in [13] – note that the range of the mood scores varies on a per-target basis; hence, the reported results of different models should be compared among each other when tested on the *same* target.

Table 5. P1: Results following the approach in [13].

	Positive		Negative		Wellbeing		Stress		[13]	
	MSE	acc	MSE	acc	MSE	acc	MSE	acc	MSE	acc
LOIOCV	15.96	84.5	11.64	87.1	20.94	89.0	1.07	47.3	0.08	93.0
LOUOCV	36.77	63.4	31.99	68.3	51.08	72.8	0.81	45.4	0.29	66.5
A (AVG)	29.89	71.8	27.80	73.1	41.14	78.9	0.70	51.6	0.24	73.5
B (LAST)	43.44	60.4	38.22	63.2	55.73	71.6	1.15	51.5	0.34	63.0
C (-feat)	33.40	67.2	28.60	72.3	45.66	76.6	0.81	49.8	0.27	70.5
D (-mood)	113.30	30.9	75.27	44.5	138.67	42.5	1.08	44.4	N/A	N/A

As in [13], always predicting the average score (AVG) for an unseen user performs better than applying a LR model trained on other users in a *LOUOCV* setting. If the same LR model used in *LOUOCV* is trained without using the previously self-reported ground-truth scores (Model D, -mood), its performance drops further. This showcases that personalised models are needed for more accurate mental health assessment (note that the AVG baseline is, in fact, a personalised baseline) and that there is no evidence that we can employ effective models in real-world applications to predict the mental health of previously unseen individuals, based on this setting.

The accuracy of LR under *LOIOCV* is higher, except for the "stress" target, where the performance is comparable to *LOUOCV* and lower than the AVG baseline. However, the problem in *LOIOCV* is the fact that the features are extracted based on the past three days, thus creating a temporal cross-correlation in our input space. If a similar correlation exists in the output space (target), then we end up in danger of overfitting our model to the training examples that are temporally close to the test instance. This type of bias is essentially present if we force a temporal correlation in the output space, as studied next.

5.2 P2: Inferring Test Labels

The charts in Fig. 1 (top) show the results by following the $LOIOCV$ approach from Canzian and Musolesi [3]. The pattern that these metrics take is consistent and quite similar to the original paper: specificity remains at high values, while sensitivity increases as we increase the time window from which we extract our features. The charts on the bottom in Fig. 1 show the corresponding results in the $LOUOCV$ setting. Here, such a generalisation is not feasible, since the increases in sensitivity are accompanied by sharp drops in the specificity scores.

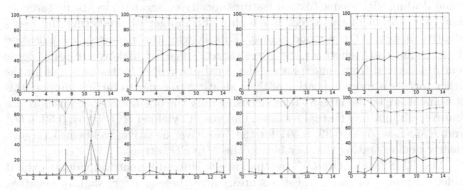

Fig. 1. P2: Sensitivity/specificity (blue/red) scores over the {positive, negative, well-being, stress} targets by training on different time windows on the $LOIOCV$ (top) and $LOUOCV$ (bottom) setups, similar to [3]. (Color figure online)

The arising issue though lies in the $LOIOCV$ setting. By training and testing on the same days (for $T_{HIST} > 1$), the kernel matrix takes high values for cells which are highly correlated with respect to time, making the evaluation of the contribution of the features difficult. To support this statement, we train the same model under $LOIOCV$, using only the mood form completion date (Unix epoch) as a feature. The results are very similar to those achieved by training on $T_{HIST} = 14$ (see Table 6). We also include the results of another naïve classifier (LAST), predicting always the last observed score in the training set, which again achieves similar results. The clearest demonstration of the problem though is by comparing the results of the RAND against the FEAT classifier, which shows that under the proposed evaluation setup we can achieve similar performance if we replace our inputs with random data, clearly demonstrating the temporal bias that can lead to over-optimistic results, even in the $LOIOCV$ setting.

Table 6. P2 : Performance (sensitivity/specificity) of the SVM classifier trained over 14 days of smartphone/social media features (**FEAT**) compared against 3 naïve baselines.

	Positive		Negative		Wellbeing		Stress	
	Sens	Spec	Sens	Spec	Sens	Spec	Sens	Spec
FEAT	64.02	95.23	60.03	95.07	65.06	94.97	45.86	95.32
DATE	59.68	95.92	62.75	95.19	63.29	95.47	46.99	95.17
LAST	67.37	94.12	69.08	94.09	66.05	93.42	58.20	93.83
RAND	64.22	95.17	60.88	95.60	64.87	95.09	45.79	95.41

5.3   P3 : Predicting Users

Experiment 1: Table 7 shows the results based on the evaluation setup of Tsakalidis et al. [26]. In the *MIXED* cases, the pattern is consistent with [26], indicating that normalising the features on a per-user basis yields better results, when dealing with sparse textual features (positive, negative, wellbeing targets). The explanation of this effect lies within the danger of predicting the user's identity instead of her mood scores. This is why the per-user normalisation does not have any effect for the stress target, since for that we are using dense features derived from smartphones: the vocabulary used by the subjects for the other targets is more indicative of their identity. In order to further support this statement, we trained the SVR model using only the one-hot encoded user id as a feature, without any textual features. Our results yielded $R^2 = \{0.64, 0.50, 0.66\}$ and $RMSE = \{5.50, 5.32, 6.50\}$ for the {positive, negative, wellbeing} targets, clearly demonstrating the user bias in the *MIXED* setting.

The RMSEs in *LOIOCV* are the lowest, since different individuals exhibit different ranges of mental health scores. Nevertheless, R^2 is slightly negative, implying again that the average predictor for a single user provides a better estimate for her mental health score. Note that while the predictions across all individuals seem to be very accurate (see Fig. 2), by separating them on a per-user basis, we end up with a negative R^2.

In the unbiased *LOUOCV* setting the results are, again, very poor. The reason for the high differences observed between the three settings is provided by the R^2 formula itself $(1 - (\sum_i (pred_i - y_i)^2)/(\sum_i (y_i - \bar{y})^2))$. In the *MIXED* case, we train and test on the same users, while $\bar{y}$ is calculated as the mean of the mood scores across all users, whereas in the *LOIOCV/LOUOCV* cases, $\bar{y}$ is calculated for every user separately. In *MIXED*, by identifying who the user is, we have a rough estimate of her mood score, which is by itself a good predictor, if it is compared with the average predictor across all mood scores of all users. Thus, the effect of the features in this setting cannot be assessed with certainty.

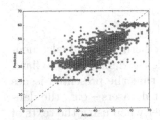

Fig. 2. P3: Actual vs predicted chart for the "wellbeing" target in *LOIOCV*. The across-subjects R^2 is negative.

Table 7. P3: Results following the evaluation setup in [26] (*MIXED*), along with the results obtained in the *LOIOCV* and *LOUOCV* settings with (+) and without (−) per-user input normalisation.

	Positive		Negative		Wellbeing		Stress	
	R^2	RMSE	R^2	RMSE	R^2	RMSE	R^2	RMSE
MIXED+	0.43	6.91	0.25	6.49	0.48	8.04	0.02	1.03
MIXED−	0.13	8.50	0.00	7.52	0.13	10.33	0.03	1.03
LOIOCV+	−0.03	5.20	−0.04	5.05	−0.03	6.03	−0.08	0.91
LOIOCV−	−0.03	5.20	−0.04	5.05	−0.03	6.03	−0.08	0.91
LOUOCV+	−4.19	8.98	−1.09	7.24	−4.66	10.61	−0.67	1.01
LOUOCV−	−4.38	8.98	−1.41	7.23	−4.62	10.62	−0.69	1.02

Experiment 2: Table 8 displays our results based on Jaques et al. [9] (see Sect. 3.3). The average accuracy on the *"UNIQ"* setup is higher by 14% compared to the majority classifier in *MIXED*. The *LOIOCV* setting also yields very promising results (mean accuracy: 81.17%). As in all previous cases, in *LOUOCV* our models fail to outperform the majority classifier. A closer look at the *LOIOCV* and *MIXED* results though reveals the user bias issue that is responsible for the high accuracy. For example, 33% of the users had all of their "positive" scores binned into one class, as these subjects were exhibiting higher (or lower) mental health scores throughout the experiment, whereas another 33% of the subjects had 85% of their instances classified into one class. By recognising the user, we can achieve high accuracy in the *MIXED* setting; in the *LOIOCV*, the majority classifier can also achieve at least 85% accuracy for 18/27 users.

In the *"PERS"* setup, we removed the user bias, by separating the two classes on a per-user basis. The results now drop heavily even in the two previously well-performing settings and can barely outperform the majority classifier. Note that the task in Experiment 2 is relatively easier, since we are trying to classify instances into two classes which are well-distinguished from each other from a psychological point of view. However, by removing the user bias, the contribution of the user-generated features to this task becomes once again unclear.

Table 8. P3: Accuracy by following the evaluation setup in [9] (*MIXED*), along with the results obtained in *LOIOCV* & *LOUOCV*.

	Positive		Negative		Wellbeing		Stress	
	UNIQ	PERS	UNIQ	PERS	UNIQ	PERS	UNIQ	PERS
MIXED	65.69	51.54	60.68	55.79	68.14	51.00	61.75	56.44
LOIOCV	78.22	51.79	84.86	53.63	88.06	52.89	73.54	55.35
LOUOCV	47.36	50.74	42.41	52.45	45.57	50.10	49.77	55.11

6 Proposal for Future Directions

Our results emphasize the difficulty of automatically predicting individuals' mental health scores in a real-world setting and demonstrate the dangers due to flaws in the experimental setup. Our findings do not imply that the presented issues will manifest themselves to the same degree in different datasets – e.g., the danger of predicting the user in the *MIXED* setting is higher when using the texts of 27 users rather than sensor-based features of more users [1, 2, 9, 22]. Nevertheless, it is crucial to establish appropriate evaluation settings to avoid providing false alarms to users, if our aim is to build systems that can be deployed in practice. To this end, we propose model building and evaluation under the following:

- **LOUOCV:** By definition, training should be performed strictly on features and target data derived from a sample of users and tested on a completely new user, since using target data from the unseen user as features violates the independence hypothesis. A model trained in this setting should achieve consistently better results on the unseen user compared to the naïve (from a modelling perspective) model that always predicts his/her average score.
- **LOIOCV:** By definition, the models trained under this setting should not violate the iid hypothesis. We have demonstrated that the temporal dependence between instances in the train and test set can provide over-optimistic results. A model trained on this setting should consistently outperform naïve, yet competitive, baseline methods, such as the last-entered mood score predictor, the user's average mood predictor and the auto-regressive model.

Models that can be effectively applied in any of the above settings could revolutionise the mental health assessment process while providing us in an unbiased setting with great insights on the types of behaviour that affect our mental well-being. On the other hand, positive results in the *MIXED* setting cannot guarantee model performance in a real-world setting in either *LOUOCV* or *LOIOCV*, even if they are compared against the user average baseline [4].

Transfer learning approaches can provide significant help in the *LOUOCV* setting. However, these assume that single-domain models have been effectively learned beforehand – but all of our single-user (*LOIOCV*) experiments provided negative results. Better feature engineering through **latent feature representations** may prove to be beneficial. While different users exhibit different behaviours, these behaviours may follow similar patterns in a latent space. Such representations have seen great success in recent years in the field of natural language processing [15], where the aim is to capture latent similarities between seemingly diverse concepts and represent every feature based on its context. Finally, working with **larger datasets** can help in providing more data to train on, but also in assessing the model's ability to generalise in a more realistic setting.

7   Conclusion

Assessing mental health with digital media is a task which could have great impact on monitoring of mental well-being and personalised health. In the current paper, we have followed past experimental settings to evaluate the contribution of various features to the task of automatically predicting different mental health indices of an individual. We find that under an unbiased, real-world setting, the performance of state-of-the-art models drops significantly, making the contribution of the features impossible to assess. Crucially, this holds for both cases of creating a model that can be applied in previously unobserved users ($LOUOCV$) and a personalised model that is learned for every user individually ($LOIOCV$).

Our major goal for the future is to achieve positive results in the $LOUOCV$ setting. To overcome the problem of having only few instances from a diversely behaving small group of subjects, transfer learning techniques on latent feature representations could be beneficial. A successful model in this setting would not only provide us with insights on what types of behaviour affect mental state, but could also be employed in a real-world system without the danger of providing false alarms to its users.

Acknowledgements. The current work was supported by the EPSRC through the University of Warwick's CDT in Urban Science and Progress (grant EP/L016400/1) and through The Alan Turing Institute (grant EP/N510129/1). We would like to thank the anonymous reviewers for their detailed feedback and the authors of the works that were analysed in our paper (N. Jaques, R. LiKamWa, M. Musolesi) for the fruitful discussions over several aspects of the presented challenges.

References

1. Bogomolov, A., Lepri, B., Ferron, M., Pianesi, F., Pentland, A.S.: Pervasive stress recognition for sustainable living. In: 2014 IEEE International Conference on Pervasive Computing and Communications Workshops (PERCOM Workshops), pp. 345–350. IEEE (2014)
2. Bogomolov, A., Lepri, B., Pianesi, F.: Happiness recognition from mobile phone data. In: 2013 International Conference on Social Computing (SocialCom), pp. 790–795. IEEE (2013)
3. Canzian, L., Musolesi, M.: Trajectories of depression: unobtrusive monitoring of depressive states by means of smartphone mobility traces analysis. In: Proceedings of the 2015 ACM International Joint Conference on Pervasive and Ubiquitous Computing, pp. 1293–1304. ACM (2015)
4. DeMasi, O., Kording, K., Recht, B.: Meaningless comparisons lead to false optimism in medical machine learning. PLoS One **12**(9), e0184604 (2017)
5. Farhan, A.A., et al.: Behavior vs. Introspection: refining prediction of clinical depression via smartphone sensing data. In: Wireless Health, pp. 30–37 (2016)
6. Gimpel, K., et al.: Part-of-speech tagging for Twitter: annotation, features, and experiments. In: Proceedings of the 49th Annual Meeting of the Association for Computational Linguistics: Human Language Technologies: Short Papers, vol. 2, pp. 42–47. Association for Computational Linguistics (2011)

7. Herrman, H., Saxena, S., Moodie, R., et al.: Promoting mental health: concepts, emerging evidence, practice: a report of the world health organization, Department of Mental Health and Substance Abuse in Collaboration with the Victorian Health Promotion Foundation and the University of Melbourne. World Health Organization (2005)

8. Hu, M., Liu, B.: Mining and summarizing customer reviews. In: Proceedings of the 10th ACM SIGKDD International Conference on Knowledge Discovery and Data Mining, pp. 168–177. ACM (2004)

9. Jaques, N., Taylor, S., Azaria, A., Ghandeharioun, A., Sano, A., Picard, R.: Predicting students' happiness from physiology, phone, mobility, and behavioral data. In: 2015 International Conference on Affective Computing and Intelligent Interaction (ACII), pp. 222–228. IEEE (2015)

10. Jaques, N., Taylor, S., Sano, A., Picard, R.: Multi-task, multi-kernel learning for estimating individual wellbeing. In: Proceedings NIPS Workshop on Multimodal Machine Learning, Montreal, Quebec (2015)

11. Kiritchenko, S., Zhu, X., Mohammad, S.M.: Sentiment analysis of short informal texts. J. Artif. Intell. Res. **50**, 723–762 (2014)

12. Kroenke, K., Strine, T.W., Spitzer, R.L., Williams, J.B., Berry, J.T., Mokdad, A.H.: The PHQ-8 as a measure of current depression in the general population. J. Affect. Disord. **114**(1), 163–173 (2009)

13. LiKamWa, R., Liu, Y., Lane, N.D., Zhong, L.: MoodScope: building a mood sensor from smartphone usage patterns. In: Proceeding of the 11th Annual International Conference on Mobile Systems, Applications, and Services, pp. 389–402. ACM (2013)

14. Ma, Y., Xu, B., Bai, Y., Sun, G., Zhu, R.: Daily mood assessment based on mobile phone sensing. In: 2012 9th International Conference on Wearable and Implantable Body Sensor Networks (BSN), pp. 142–147. IEEE (2012)

15. Mikolov, T., Sutskever, I., Chen, K., Corrado, G.S., Dean, J.: Distributed representations of words and phrases and their compositionality. In: Advances in Neural Information Processing Systems, pp. 3111–3119 (2013)

16. Mohammad, S.: #Emotional Tweets. In: *SEM 2012: The 1st Joint Conference on Lexical and Computational Semantics - Proceedings of the Main Conference and the Shared Task, and Proceedings of the 6th International Workshop on Semantic Evaluation (SemEval 2012), vols. 1 and 2, pp. 246–255. Association for Computational Linguistics (2012)

17. Mohammad, S., Dunne, C., Dorr, B.: Generating high-coverage semantic orientation lexicons from overtly marked words and a thesaurus. In: Proceedings of the 2009 Conference on Empirical Methods in Natural Language Processing, vol. 2, pp. 599–608. Association for Computational Linguistics (2009)

18. Nielsen, F.Å.: A new ANEW: evaluation of a word list for sentiment analysis in microblogs. In: Workshop on 'Making Sense of Microposts': Big Things Come in Small Packages, pp. 93–98 (2011)

19. OECD: How's Life? 2013: Measuring Well-being (2013). https://doi.org/10.1787/9789264201392-en

20. Olesen, J., Gustavsson, A., Svensson, M., Wittchen, H.U., Jönsson, B.: The economic cost of brain disorders in Europe. Eur. J. Neurol. **19**(1), 155–162 (2012)

21. Preoţiuc-Pietro, D., Volkova, S., Lampos, V., Bachrach, Y., Aletras, N.: Studying user income through language, behaviour and affect in social media. PloS One **10**(9), e0138717 (2015)

22. Servia-Rodríguez, S., Rachuri, K.K., Mascolo, C., Rentfrow, P.J., Lathia, N., Sandstrom, G.M.: Mobile sensing at the service of mental well-being: a large-scale longitudinal study. In: Proceedings of the 26th International Conference on World Wide Web, pp. 103–112. International World Wide Web Conferences Steering Committee (2017)
23. Suhara, Y., Xu, Y., Pentland, A.: DeepMood: forecasting depressed mood based on self-reported histories via recurrent neural networks. In: Proceedings of the 26th International Conference on World Wide Web, pp. 715–724. International World Wide Web Conferences Steering Committee (2017)
24. Tang, D., Wei, F., Yang, N., Zhou, M., Liu, T., Qin, B.: Learning sentiment-specific word embedding for Twitter sentiment classification. In: Proceedings of the 52nd Annual Meeting of the Association for Computational Linguistics (Volume 1: Long Papers), vol. 1, pp. 1555–1565 (2014)
25. Tennant, R., et al.: The Warwick-Edinburgh mental well-being scale (WEMWBS): development and UK validation. Health Qual. Life Outcomes 5(1), 63 (2007)
26. Tsakalidis, A., Liakata, M., Damoulas, T., Jellinek, B., Guo, W., Cristea, A.I.: Combining heterogeneous user generated data to sense well-being. In: Proceedings of the 26th International Conference on Computational Linguistics, pp. 3007–3018 (2016)
27. Wang, R., et al.: CrossCheck: toward passive sensing and detection of mental health changes in people with schizophrenia. In: Proceedings of the 2016 ACM International Joint Conference on Pervasive and Ubiquitous Computing, pp. 886–897. ACM (2016)
28. Wang, R., et al.: StudentLife: assessing mental health, academic performance and behavioral trends of college students using smartphones. In: Proceedings of the 2014 ACM International Joint Conference on Pervasive and Ubiquitous Computing, pp. 3–14. ACM (2014)
29. Watson, D., Clark, L.A., Tellegen, A.: Development and validation of brief measures of positive and negative affect: the PANAS scales. J. Pers. Soc. Psychol. 54(6), 1063 (1988)
30. Zhu, X., Kiritchenko, S., Mohammad, S.M.: NRC-Canada-2014: recent improvements in the sentiment analysis of Tweets. In: Proceedings of the 8th International Workshop on Semantic Evaluation (SemEval 2014), pp. 443–447. Citeseer (2014)

AMIE: Automatic Monitoring of Indoor Exercises

Tom Decroos[1]([✉]), Kurt Schütte[2], Tim Op De Beéck[1], Benedicte Vanwanseele[2], and Jesse Davis[1]

[1] Department of Computer Science, KU Leuven,
Celestijnenlaan 200A, Leuven, Belgium
{tom.decroos,tim.opdebeeck,jesse.davis}@cs.kuleuven.be
[2] Department of Movement Sciences, KU Leuven,
Tervuursevest 101, Leuven, Belgium
{kurt.schutte,benedicte.vanwanseele}@kuleuven.be

Abstract. Patients with sports-related injuries need to learn to perform rehabilitative exercises with correct movement patterns. Unfortunately, the feedback a physiotherapist can provide is limited by the number of physical therapy appointments. We study the feasibility of a system that automatically provides feedback on correct movement patterns to patients using a Microsoft Kinect camera and Machine Learning techniques. We discuss several challenges related to the Kinect's proprietary software, the Kinect data's heterogeneity, and the Kinect data's temporal component. We introduce AMIE, a machine learning pipeline that detects the exercise being performed, the exercise's correctness, and if applicable, the mistake that was made. To evaluate AMIE, ten participants were instructed to perform three types of typical rehabilitation exercises (squats, forward lunges and side lunges) demonstrating both correct movement patterns and frequent types of mistakes, while being recorded with a Kinect. AMIE detects the type of exercise almost perfectly with 99% accuracy and the type of mistake with 73% accuracy. Code related to this paper is available at: https://dtai.cs.kuleuven.be/software/amie.

1 Introduction

Being active is crucial to a healthy lifestyle. Initiatives such as *Start to Run* [3] in Belgium and *Let's Move* in the USA [28] encourage people to become more active. These initiatives are paying off, as in the USA, almost every generation is becoming more active, according to a report made by the Physical Activity Council [9]. However, this increase in activity inevitably also leads to an increase in sports-related injuries [12,24]. Besides the short and long term physical discomforts, there are substantial costs associated with injuries. A significant portion of these costs are allocated to rehabilitation [13,18]. People with injuries usually need to visit a physiotherapist. The physiotherapist will then prescribe a program of rehabilitation exercises that the injured patient must follow at home.

U. Brefeld et al. (Eds.): ECML PKDD 2018, LNAI 11053, pp. 424–439, 2019.
https://doi.org/10.1007/978-3-030-10997-4_26

This current rehabilitation paradigm has several drawbacks. First, due to time constraints of the patients, and the cost of physiotherapy sessions, the interaction between the patient and physiotherapist is necessarily limited. Second, many patients simply do not do their exercises [4], with research reporting adherence rates to home exercise programs of only 15–40% [5,15]. Third, it is hard for a patient to learn how to correctly perform the exercise due to the limited feedback by a physical therapist. These drawbacks can cause problems such as prolonged recovery time, medical complications, and increased costs of care [22].

One possible way to address these drawbacks is to exploit technological advances to develop an automated system to monitor exercises performed at home. Patients have expressed a willingness to use such a system because it allows them to perform exercises in the comfort of their own home while having fast access to feedback [19]. Such a home monitoring system could provide three important benefits, by:

1. Motivating the patient to do his exercises;
2. Showing the patient the correct movement patterns for his exercises; and
3. Monitoring the quality of the performed exercises and giving feedback in case an exercise is poorly executed.

Currently, most effort has been devoted towards addressing the first two tasks. First, researchers have shown that home-systems can successfully motivate patients to adhere to their home exercise programs by applying tools such as gamification and social media [14,17,27]. Second, several approaches have demonstrated the ability to show the correct movement patterns of exercises in a clear way such that people are able to understand and reproduce these movement patterns [7,19,25]. There has been less work on the third task: monitoring the correctness of exercises. The current approaches typically make unrealistic assumptions such as the availability of perfect tracking data [26], fail to describe how the system determines if an exercise is performed correctly [7,15], or do not quantitatively evaluate their systems [1,10,26,29,30].

In this paper, we propose AMIE (Automatic Monitoring of Indoor Exercises), a machine learning pipeline that uses the Microsoft Kinect 3D camera to monitor and assess the correctness of physiotherapy exercise performed by a patient independently in his home. At a high-level, AMIE works as follows. First, it identifies an individual repetition of an exercise from the Kinect's data that tracks the absolute location of multiple joints over time. Second, in order to capture the movement of the patient, AMIE rerepresents the time-series data for an exercise with a set of simple statistical features about the angles between interconnected joints. Finally, it detects an exercise's type, correctness, and, if applicable, mistake type. We evaluated AMIE on a data set of 1790 exercise repetitions comprising ten different test subjects performing three commonly used rehabilitation exercises (i.e., squat, forward lunge and side lunge). AMIE detects what exercise is being performed with 99.0% accuracy. In terms of predicting the correctness of an exercise and which mistake was made, AMIE achieves accuracies of 73.4% and 73.8% respectively.

To summarize, this paper makes the following contributions:

1. Details the data collected in this study comprehensively;
2. Discusses a number of challenges related to representing the Kinect data;
3. Describes the entire pipeline for classifying exercises correctly, including how to automatically detect an individual exercise repetition and how to predict if an exercise is performed correctly;
4. Assesses AMIE's ability to (a) detect the exercise being performed, (b) determine if the exercise was performed correctly, and (c) identify the type of mistake that was made; and
5. Releases both the collected data set and the source code of AMIE at http:// dtai.cs.kuleuven.be/software/amie, as a resource to the research community.

2   Data Collection

We describe the characteristics of the subjects who participated in this study and the collected data.

2.1   Subjects

Data of 7 male and 3 female subjects (26.7 ± 3.95 years, 1.76 ± 0.12 m, 73.5 ± 13.3 kg, 23.38 ± 2.61 BMI) were collected. All subjects were free of injuries and cardiovascular, pulmonary and neurological conditions that impeded the ability to perform daily activities or physical therapy exercises. The study was conducted according to the requirements of the Declaration of Helsinki and was approved by the KU Leuven ethics committee (file number: s59354).

2.2   Exercises

The subjects were instructed to perform three types of exercises, which are illustrated in Fig. 1:

Squat. The subject stands with his feet slightly wider than hip-width apart, back straight, shoulders down, toes pointed slightly out. Keeping his back straight, the subject lowers his body down and back as if the subject is sitting down into a chair, until his thighs are parallel to the ground (or as close as parallel as possible). Next, the subjects rises back up slowly.

Forward lunge. The subject stands with his feet shoulder's width apart, his back long and straight, his shoulders back and his gaze forward. Next, the subject steps forward with his left (or right) leg into a wide stance (about one leg's distance between feet) while maintaining back alignment. The subject lowers his hips until his forward knee is bent at approximately a 90° angle. Keeping his weight on his heels, the subject pushes back up to his starting position.

Side lunge. The subject stands with his feet shoulder's width apart, his back long and straight, his shoulders back and his gaze forward. Next, the subject steps sideways with his right leg into a wide stance while maintaining back alignment. Keeping his weight on his heels, the subject pushes back up to his starting position.

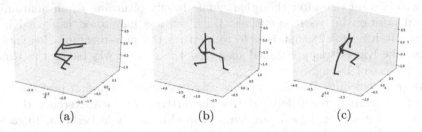

(a)   (b)   (c)

Fig. 1. Figures recorded by the Kinect of a person doing (a) a squat, (b) a forward lunge and (c) a side lunge.

2.3   Mistake Types

In addition to correct repetitions of each exercise, the subjects were instructed to perform repetitions that illustrate common incorrect ways to perform each exercise. Specifically, we consider the following types of mistakes:

Squat Knees Over Toes (KOT). The subject executes a squat, but while lowering his back, the subject goes beyond alignment so that the knees go far over the toes.

Squat Knock Knees (KK). The subject executes a squat, but while lowering his back, the subject collapses his knees inward.

Squat Forward Trunk Lean (FTL). The subject executes a squat, but while lowering his back, the subject tilts his trunk forward, so that his back is no longer straight or perpendicular to the ground.

Forward lunge KOT. The subject executes a forward lunge, but while stepping forward, the subject goes beyond alignment so that the knees go far over the toes and the forward knee is bent further than a 90° angle.

Forward lunge KK. The subject executes a forward lunge, but while stepping forward, the subject collapses his forward knee inward.

Forward lunge FTL. The subject executes a forward lunge, but while stepping forward, the subject tilts his trunk forward, so that his back is no longer straight or perpendicular to the ground.

Side lunge FTL. The subject executes a side lunge, but while stepping sideways with his right leg, the subject tilts his trunk forward, so that his back is no longer straight or perpendicular to the ground.

For side lunges, only the forward trunk lean was performed because the other two mistakes are not applicable to this exercise.

2.4 Protocol

The study protocol was designed in collaboration with biomechanics researchers with extensive expertise in collecting and analyzing data of rehabilitation exercises. Each subject was instructed to perform six sets of ten repetitions of each type of exercise (squat, forward lunge, and side lunge). Given sufficient rest in between the sets, the subjects were asked to perform ten repetitions of the exercise within a set one after the other, while briefly returning to an anatomical neutral position between repetitions. Each set was monitored using a Kinect (Microsoft Kinect V2) that is able to capture the movement of 25 joints at 30 Hz. The Kinect was positioned such that the subject was facing the Kinect at a distance of 1–2 m.

More specifically, the subjects were asked per exercise to first perform three sets of ten correct repetitions. Before the first set of each exercise, the correct execution was explained and demonstrated by a physiotherapist. In case of the forward lunge, the subjects were asked to alternate between stepping forward with their left and right leg between executions. Next, the physiotherapist demonstrated mistakes that are often made by patients while executing these three exercises. For squats and forward lunges, the *KOT*, *KK*, and *FTL* mistakes were demonstrated and the subjects performed one set of ten repetitions of each. For side lunges, only the *FTL* mistake was demonstrated as the other two mistakes are not applicable. The subjects performed three sets of ten repetitions of the *FTL* mistake in case of the side lunges, because we wanted to collect the same number of recorded repetitions per exercise.

3 The AMIE System

Our goal is to develop a Kinect-based system that provides automatic feedback to patients. Such a system requires performing the following steps:

1. Extracting the raw data from the Kinect;
2. Partitioning the stream of data into individual examples;
3. Rerepresenting the data into a format that is suitable for machine learning;
4. Learning a model to predict if an exercise was done correctly or not; and
5. Providing feedback to the user about his/her exercise execution using the learned model.

In this paper, we study whether detecting if an exercise was performed correctly or not is feasible. We establish a proof of concept called AMIE (Automatic Monitoring of Indoor Exercises) that currently addresses only the first four tasks.

3.1 Extracting the Kinect Data

The Kinect records a set of exercise repetitions as a video, which consists of a sequence of depth frames. A depth frame has a resolution of 512 × 424 pixels where each pixel represents the distance (in millimeters) of the closest object seen by that pixel. Using the Kinect's built-in algorithms, each depth frame can be processed into a stick figure (Fig. 2).

Each set of repetitions is stored as a XEF file (eXtended Event Files), which is a format native to the Kinect SDK that can only be interpreted by applications developed in the closed software system of the Kinect. It cannot be directly examined by conventional data analysis tools such as Excel, R, and Python. Through manual examination of the Kinect SDK and some reverse engineering, we have developed a tool that takes as input a Kinect video of n depth frames stored as a XEF file and outputs a sequence of n stick figures in JSON format. This tool is freely available at http://dtai.cs.kuleuven.be/software/amie.

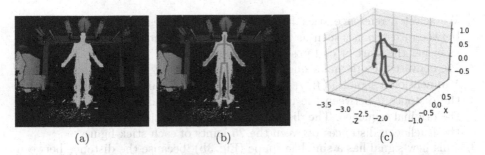

(a)   (b)   (c)

Fig. 2. (a) A depth frame as shown in KinectStudio, an application for recording and viewing Kinect videos built on the Kinect SDK (b) the stick figure built by the Kinect's algorithms as shown in KinectStudio (c) the same stick figure extracted from the native Kinect file format and plotted using Python, a popular data analysis tool.

All ten subjects performed roughly six sets of ten repetitions for three exercises. Our data set D contains exactly 186 videos $v \in D$. We sometimes have more than 18 videos (3 exercises × 6 execution sets) per subject, because a video recording of ten repetitions could get broken up in two separate videos due to technical issues.[1] Hence, not every video contains exactly ten executions.

Each video $v \in D$ can be represented as a tuple

$$v = ([f_i]_{i=0}^n, s, e, m)$$

where $[f_i]_{i=0}^n$ is a sequence of n stick figures $f_0, \ldots, f_n$, s is the identifier of the subject performing the exercise, e is the exercise type (squat, forward lunge or side lunge), and m is the mistake type. The mistake type m is *KOT*, *KK*, *FTL* or *None*. *None* means the exercise was performed correctly. A stick figure f_i is a vector of 25 joints. Each joint is represented by (x, y, z) coordinates, where z represents the distance to the Kinect camera and x and y represent respectively

---

[1] The subject were, in addition to the Kinect, also tracked with a Vicon camera system using reflective markers attached to the body. Due to sweat and movement, these markers sometimes fell off and the exercise repetition set was interrupted to reattach a marker. The collected Vicon data is not used in this paper.

horizontal and vertical positions in space. Examples of joints are the left ankle, the right knee, the left elbow, the spine base and middle, the left shoulder, etc.

3.2 Partitioning a Set of Exercises into a Single Repetition

Each video v in our data set D contains a sequence of stick figures $[f_i]_{i=0}^n$ that represents multiple repetitions of an exercise. This is problematic because we need to work on the level of an individual repetition in order to ascertain if it was performed correctly or not. Therefore, a sequence containing one set of k executions needs to be subdivided into k subsequences, with one subsequence for each repetition. We employed the following semi-supervised approach to accomplish this:

1. We select a reference stick figure f_{ref} that captures the pose of being in-between executions, such as in Fig. 3a. Typically, such a pose can be found as either the start or end position in the sequence.
2. We convert the original sequence of stick figures $[f_i]_{i=0}^n$ into an equal length 1-dimensional signal $[d(f_{ref}, f_i)]_{i=0}^n$, where the ith value of the new signal is the distance d between the reference stick figure and the ith stick figure in the original sequence. The distance d between two stick figures is the sum of the Euclidean distances between the 25 joints of each stick figure.
3. This new signal has a sine-like shape (Fig. 3b), because the distance between a stick figure in-between repetitions and f_{ref} is small whereas the distance between a stick figure in mid-exercise and f_{ref} is high. The valleys in the signal (i.e., the negative peaks) represent likely points in time when the subject is in between repetitions. To detect the valleys, we employ a modified version of the peak-finding algorithm in the signal processing toolbox of Matlab.[2] These valleys are used to subdivide the original sequence series into a number of subsequences, where one subsequence encompasses one repetition.
4. Depending on the quality of the resulting subsequences in the previous step, we do some additional manual modifications, such as inserting an extra splitting point or removing some stick figures from the start or the end of the original sequence.

Using our approach, we transformed the dataset D of 186 videos into a new dataset D' that contains 1790 repetitions $r \in D'$. The last step in our semi-supervised approach was necessary only for 15 out of 186 videos.[3]

Each repetition r is represented by a 4-tuple $([f_i]_{i=0}^n, s, e, m)$ just like a video v. The difference is that the sequence of stick figures $[f_i]_{i=0}^n$ of each repetition $r \in D'$ now only contains one repetition of the performed exercise instead of multiple repetitions. The length of the stick figure sequence per repetition ranges from 40 to 308 (136 ± 37).

---

[2] Our peak-finding algorithm takes the minimal peak distance as input, which we estimate from the data using the length of the sequence and the dominating frequency in the Fourier transform.

[3] Manual modifications were typically needed if the video recording was cut off too late, adding extra noise at the end.

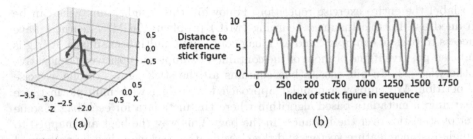

Fig. 3. (a) The reference stick figure. (b) The blue signal shows the distance between the reference stick figure and each stick figure in a sequence containing ten repetitions of an exercise. The automatically generated split points are highlighted with red vertical lines. (Color figure online)

3.3   Feature Construction

Two main challenges prevent us from applying an off-the-shelf machine learning algorithm directly to our data, its heterogeneity (e.g., examples are recorded from different subjects) and temporal nature (e.g., examples are sequences of varying length). We detail two consecutive transformations that address these challenges and construct a feature vector for each repetition r. We refer to these transformations as the heterogeneity transformation and the temporal transformation.

Heterogeneity Transformation: Not only do the subjects differ in height and weight, but also their relative position and orientation to the Kinect camera vary from exercise to exercise. All these variations affect the absolute coordinates recorded by the Kinect, but are independent to the executed exercise and its correctness. Therefore, we aim to remove these variations from the data, by using the geometrical angles in the interconnected joint triplets instead of the absolute coordinates. For example, *(left hip, left knee, left ankle)* is an interconnected joint triplet because the knee connects the hip and the ankle. Its geometrical angle is the angle formed at the *left knee* joint enclosed by the *(left hip, left knee)* and *(left knee, left ankle)* segments. For each stick figure f_i, the angles of all 30 such interconnected joint triplets are used as features. An additional advantage is that our new representation mimics the language physiotherapists often use to describe whether an exercise is performed correct or wrong (e.g., a good forward lunge has the forward knee bent at 90°).

Temporal Transformation: This transformation maps the representation of an exercise repetition from the variable-length sequence of highly self-correlated stick figures (represented by angles) to a fixed length feature vector. We observe that the temporal relationship between stick figures is not important to our tasks, because the exercises and mistakes can be recognized by specific poses. For example, the exercises are recognizable by the poses in Fig. 1, and a *KOT* mistake is made if a stick figure's knees go far over its toes. Moreover, only a subset of the stick figures need to be specific to an exercise or mistake type to label the entire exercise repetition. Following this insight, our tasks can be framed as Multiple Instance Learning (MIL) problems. In MIL, learners have access to bags of examples, and each bag containing multiple examples. A bag is labeled positive if one or more of the contained examples is positive and negative otherwise [23]. In our case, a bag contains all the stick figures of one exercise repetition. We employ a popular approach for dealing with multiple instance learning: a metadata-based algorithm where the metadata for each bag is some set of statistics over the instances in the bag. This way the bags are mapped to single-instance feature vectors and the classification task can then be performed by an arbitrary single-instance machine learning algorithm [11,23]. Our set of statistics aims to describe the value distribution for each interconnected joint triplet angle using the following five summary statistics: minimum, maximum, mean, median and standard deviation. Each exercise is therefore mapped to a fixed length feature vector of 150 summary statistics (30 angles × 5 statistics).

3.4 Model Learning

AMIE learns three separate models, one for each task we consider:

T1: Identifying which exercise the patient is performing.
T2: Predicting whether the exercise was performed correctly or not.
T3: Detecting the type of mistake that was made when performing the exercise.

The first task may not strictly be necessary as a home monitoring system could ask the patient to perform the exercises in a set order. However, the ability to detect automatically which exercise is being performed would give the patient more autonomy when conducting his rehabilitation and would allow him to dynamically decide on and update his exercise routine.

Given that we have represented our data in a fixed-length feature format, it is possible to solve each of these learning problems using standard, off-the-shelf machine learning techniques. We tested five popular algorithms and found XGBoost [8] to be the most suitable. We provide further details on the process in the following section.

4 Experiments

The goal of the empirical evaluation is to address the following six research questions:

Q1: Can we accurately detect what exercise is being performed?
Q2: Can we accurately detect whether the exercise was performed correctly?
Q3: Can we accurately detect what type of mistake was made?
Q4: How does our classification approach compare to using hand-crafted rules and a nearest-neighbor approach?
Q5: Can our pipeline provide feedback to a patient in real-time?
Q6: Is the accuracy of AMIE dependent on the type of mistake?

4.1   Evaluation Methodology

When evaluating how well our learned models will generalize to unseen data, special care has to be taken in our setting to account for two types of dependencies that appear in our data. The first dependency arises due to the temporal nature of the data. An individual example is a single repetition of an exercise, but that repetition is done in a set of ten consecutive repetitions. Hence it will be correlated to the other examples in that set. The second dependency arises because one subject performs multiple repetition sets. Consequently, standard cross-validation is unsuitable as a repetition from the same set (or subject) may appear in both the train set and the test set, which could lead to over-optimistic accuracy estimates.[4] Therefore, we see two possibilities for performing cross-validation.

Leave-one-set-out cross-validation. In this setting, the data of one repetition set appears in the test set and the data of all other 185 repetition sets appears in the training set. Practically, this setting estimates the accuracy of a system that would only be deployed to monitor patients if it had examples of them performing the specific exercises that they must complete at home.

Leave-one-subject-out cross-validation. In this setting, the data for nine subjects appears in the training set and the data for the remaining subject appears in the test set. Practically, this setting estimates the accuracy of a system that can be deployed without collecting any data about the new patient performing his exercises. In other words, a system that is trained based on a fixed set of subjects and then deployed on new (i.e., previously unseen) subjects.

For each of our research questions, we consider both setups.

4.2   Results for Q1 Through Q3

Research questions Q1 through Q3 correspond to evaluating our accuracy on tasks T1 through T3. The learners we considered are Logistic Regression, Naive Bayes, Decision Tree, Random Forest, and XGBoost [8]. For all learners, we performed no hyperparameter tuning to avoid overfitting due to our limited amount of data. That is, we used the standard parameter settings from the Scikit-learn [20] and XGBoost [8] Python packages.

We trained models for each learner on all three classification tasks using both cross-validation schemes (Table 1). On our data, we can almost perfectly determine which exercise is being performed with all models. However, further investigation is needed to determine if this result holds when confronted with a wider range of exercise types, particularly for exercises that exhibit highly similar movement patterns.[5] When determining the correctness and mistake of

---

[4] Preliminary research suggests that the standard cross-validation setting indeed leads to an over-optimistic accuracy estimate.

[5] For example, a normal squat and a single-leg-squat exhibit similar movement patterns which could confuse our learner.

an exercise, XGBoost performs best under both cross-validation settings with an accuracy of at least 73%.[6] While we perform significantly better in T3 than random guessing (25%) or predicting the majority class (50%), we deem AMIE's current accuracy insufficient to be used as an autonomous system without the supervision of a physiotherapist.

Table 1. Accuracy of AMIE and baselines for different tasks, learners and cross-validation settings. We can almost perfectly identify the type of exercise (T1) with every learner. XGBoost generally performs the best at detecting correctness (T2) and mistake type (T3).

Task		T1		T2		T3	
Cross-validation setting		Set	Subject	Set	Subject	Set	Subject
AMIE	Decision Tree	0.992	0.973	0.731	0.671	0.642	0.555
	Logistic Regression	0.999	0.989	0.772	0.708	0.726	0.672
	Naive Bayes	0.982	0.972	0.633	0.646	0.478	0.547
	Random Forest	0.997	0.987	0.762	0.700	0.705	0.675
	XGBoost	0.997	**0.990**	0.790	**0.734**	**0.741**	**0.738**
Baselines	NN-DTW (absolute coord.)	**1.000**	0.965	**0.840**	0.623	0.627	0.555
	NN-DTW (angles)	0.997	**0.990**	0.713	0.648	0.576	0.549
	Handcrafted Rule Set	X	X	0.634	0.634	0.590	0.590

4.3 Results for Q4

We compared AMIE against two popular approaches in the literature: a nearest-neighbor approach using Dynamic Time Warping as a distance measure (NN-DTW), and a rule set handcrafted by a biomechanics researcher.

NN-DTW: This baseline is based on the work of Su et al. [25], who provide feedback on rehabilitation exercises using the distance of the executed exercise to a library of reference exercises. We employ the NN-DTW baseline using two different representations of our stick figures: the initial representation with the absolute (x, y, z)-coordinates of 25 joints and the representation using the geometrical angles in the interconnected joint triplets, which is obtained after applying the heterogeneity transformation as detailed in Sect. 3.3.

Handcrafted Rule Set: This baseline is inspired by Zhao et al. [29, 30], who introduce a system that allows physiotherapists to express 'correctness rules'. Our rule set consists of three rules, one for each mistake. For example, the *KOT* rule states that if both the left and right knee joints have a z-coordinate that is closer to the camera than the z-coordinates of the left and right toes, then the subject is performing a *KOT* mistake. The *KK* and *FTL* mistakes are encoded in a similar way. To reduce the effect of noise in the data, we only predict an

---

[6] All learners employ only one model to detect mistakes for all three exercises. We tried learning one model per exercise, but noticed no difference in accuracy.

exercise repetition to have a specific mistake if at least ten stick figures in the repetition show that mistake. If this is not the case, the repetition is predicted to be correct. If multiple mistakes are detected in the exercise repetition, then the mistake with the most supporting stick figures is predicted.

The results of our baselines are shown in the lower half of Table 1. Except for one occurrence, AMIE (using the XGBoost classifier) always outperforms both the NN-DTW baselines and the handcrafted rule set on T2 and T3. This suggests that to provide accurate feedback, we cannot rely purely on domain experts, as a more flexible approach than a handcrafted rule set is necessary. However, we also cannot blindly apply machine learning techniques; NN-DTW is the most popular way to classify time series [2], yet it performs significantly worse than AMIE.

4.4   Results for Q5

The machine learning pipeline AMIE consists of four steps: (1) extracting the raw data from the Kinect, (2) partitioning the stream of data into individual examples, (3) constructing a feature vector for each example, and (4) detecting the examples' correctness using the trained models. Extracting the raw data from the Kinect into the JSON format and loading it in Python takes 0.15 s for one repetition set. Partitioning one repetition set into individual examples takes 0.03 s on average. Constructing the feature vectors takes 0.05 s on average per repetition set and detecting correctness (i.e., predicting the labels for $T1$ through $T3$ using our trained models) takes 0.0001 s on average per repetition set. In summary, a patient receives feedback from AMIE within 0.28 s after performing his exercises, which is almost instantaneous for human subjects. In addition, the largest fraction of our processing time is due to unnecessary disk I/O that could be avoided in a commercial implementation of AMIE. In this way, a patient can immediately adapt his or her incorrect movement patterns to the correct movement patterns, therefore accelerating the learning process of rehabilitation and mimicking a real-life scenario where a focused physiotherapist typically provides expert feedback after a few repetitions in practice.

4.5   Results for Q6

To check whether the accuracy of AMIE is dependent on the type of mistake, we inspect the confusion matrices for detecting mistake type for both our cross-validation schemes for XGBoost, our best performing learner (Table 2). We observe a higher accuracy, precision, and recall on the FTL mistake than on the KOT and KK mistake types. We can think of two hypotheses as to why this is the case. First, based on preliminary research and manual examination of the data, we hypothesize that the Kinect tracks the upper body more accurately than the lower body. This would naturally explain why we can accurately detect FTL, which is a mistake related to the upper body, and not KOT nor KK, which are mistakes related to the lower body. However, further research is needed to confirm this hypothesis. A second hypothesis is that our representation is imperfect in that it contains the information necessary to detect the *FTL* mistake, but lacks the necessary features to detect other mistakes. For example, *KK* is a type of mistake which will show almost no notable difference in angles of interconnected joint triplets, as the geometrical angle between the left and right *(hip, knee, ankle)* joint triplets will be unaffected. We conclude that the accuracy of AMIE depends on the type of mistake that was made during an exercise. To discern the exact reason as to why this is the case, further research is needed.

Table 2. Confusion matrices for detecting the type of mistake for (a) leave-one-set-out cross-validation and (b) leave-one-subject-out cross-validation.

Predicted / Actual	None	KOT	KK	FTL
None	816	30	19	30
KOT	161	10	20	5
KK	110	26	52	6
FTL	40	8	8	449

(a)

Predicted / Actual	None	KOT	KK	FTL
None	724	56	66	49
KOT	121	38	28	9
KK	75	10	101	8
FTL	37	3	6	459

(b)

5   Related Work

Previous work on the topic of home monitoring systems for rehabilitation exercises can be roughly divided in three categories: (1) work that qualitatively evaluates whether patients are willing to use a home monitoring system and what their expectations are of such a system [7,16]; (2) work that investigates the readiness and accuracy of several tracking systems to be used in a home monitoring system [10,21,26]; and (3) work that investigates home monitoring systems that can provide feedback using tracking data [1,6,10,15,25,26,29,30].

One of the hypotheses in our paper relevant to the second category is that the Kinect tracking system is not accurate enough to detect lower body mistakes. Pfister et al. [21] and Tang et al. [26] partially confirm this hypothesis by comparing the tracking capabilities of the Kinect to that of a Vicon camera system, which is considered to be the gold standard for tracking movements of the human body in biomechanics research. This hypothesis also explains why a large portion of the related work that incorporates the Kinect in a home monitoring systems focuses on upper body exercises [6,7,16,25,26].

Each paper in the third category contains one or more of three contributions: (a) describing the model in technical depth, (b) describing the used data set and experimental setup, and (c) outlining a clear vision on how the system should be implemented in practice. Typically, papers in this category contain either (a, c) or (b, c), but rarely (a, b, c). We consider our paper to be an (a, b) paper. We did not outline a vision on how the system should be implemented in practice both for brevity and due to the fact that our paper is mostly a feasibility study.

Examples of (a, c) work are Anton et al. who introduce KiReS [1], Tang et al. who showcase Phyio@Home [26], and Zhao et al. who introduce a rule-based approach for real-time exercise quality assessment and feedback [29,30].

An example of (b, c) work is Komatireddy et al. who introduce the VERA system [15]. In terms of experimental setup, it is the most similar work to our paper; they collected data of ten healthy subjects within age 18–36 and asked them to perform ten correct repetitions of four different exercises (sitting knee extension, standing knee flexion, deep lunge, and squat). However, they provide no description on how the correctness of an exercise is determined and do not discuss the accuracy of the system compared to a physiotherapist in-depth.

Su et al. [25] wrote one of the few (a, b, c) papers. They introduce a distance-based approach to provide feedback on rehabilitation exercises using previous recordings. A physiotherapist first recorded correct executions of the exercises together with the patient. Feedback was then provided on new exercise executions at home using the distance to those reference executions. The task they consider is simpler than the one addressed in this paper however, because they evaluate their approach on shoulder exercises, which exhibit less complex movement patterns than the exercises we consider and are more accurately tracked by the Kinect. They also construct a per-subject model, which is easier than constructing a global model that can generalize over unseen subjects. A final note is that we could not find any information on the size of the employed test and training data, so it is unknown how reliable the estimated accuracy of their approach is.

6   Conclusion

We presented AMIE, a machine learning approach for automatically monitoring the execution of commonly used rehabilitation exercises using the Kinect, a low-cost and portable 3D-camera system. This paper contributes with respect to existing work by being one of the first to comprehensively detail the collected data set, describe the used classification system in depth, report quantitative results about our performance, and publicly release both the collected data set and used software tools.

We evaluated AMIE on a data set of ten test subjects who each performed six sets of ten repetitions of three commonly used rehabilitation exercises (i.e., squat, forward lunge and side lunge). AMIE detects the type of exercise with 99% accuracy and the type of mistake that was made with 73% accuracy. It does this almost in real-time. An important limitation of AMIE is that it can accurately detect movement mistakes of the upper body, but struggles with movement mistakes related to the lower body. We hypothesize that some non-trivial technical improvements (i.e., more accurate extraction of stick figures from depth frames and a better representation of our data) are necessary to solve the remainder of our task and implement the system in practice.

Acknowledgements. Tom Decroos is supported by the Research Foundation-Flanders (FWO-Vlaanderen). Kurt Schütte and Benedicte Vanwanseele are partially supported by the KU Leuven Research Fund (C22/15/015) and imec.icon research funding. Tim Op De Beéck and Jesse Davis are partially supported by the KU Leuven Research Fund (C22/15/015, C32/17/036).

References

1. Antón, D., Goñi, A., Illarramendi, A., Torres-Unda, J.J., Seco, J.: KiRes: a kinect-based telerehabilitation system. In: 2013 IEEE 15th International Conference on e-Health Networking, Applications & Services (Healthcom), pp. 444–448 (2013)
2. Bagnall, A., Lines, J., Bostrom, A., Large, J., Keogh, E.: The great time series classification bake off: a review and experimental evaluation of recent algorithmic advances. Data Min. Knowl. Discov. **31**(3), 606–660 (2017)
3. Borgers, J., Vos, S., Scheerder, J.: Belgium (Flanders). In: Running Across Europe, pp. 28–58. Palgrave Macmillan, London (2015)
4. Campbell, R., Evans, M., Tucker, M., Quilty, B., Dieppe, P., Donovan, J.: Why don't patients do their exercises? Understanding non-compliance with physiotherapy in patients with osteoarthritis of the knee. J. Epidemiol. Commun. Health **55**(2), 132–138 (2001)
5. Chan, D.K., Lonsdale, C., Ho, P.Y., Yung, P.S., Chan, K.M.: Patient motivation and adherence to postsurgery rehabilitation exercise recommendations: the influence of physiotherapists' autonomy-supportive behaviors. Arch. Phys. Med. Rehabil. **90**(12), 1977–1982 (2009)
6. Chang, C.Y., et al.: Towards pervasive physical rehabilitation using Microsoft Kinect. In: 2012 6th International Conference on Pervasive Computing Technologies for Healthcare (PervasiveHealth), pp. 159–162. IEEE (2012)
7. Chang, Y.J., Chen, S.F., Huang, J.D.: A kinect-based system for physical rehabilitation: a pilot study for young adults with motor disabilities. Res. Dev. Disabil. **32**(6), 2566–2570 (2011)
8. Chen, T., Guestrin, C.: XGBoost: a scalable tree boosting system. In: Proceedings of the 22nd ACM SIGKDD International Conference on Knowledge Discovery and Data Mining, pp. 785–794. ACM (2016)
9. Council, Physical Activity: 2018 participation report: the Physical Activity Council's annual study tracking sports, fitness, and recreation participation in the US (2017)
10. Fernández-Baena, A., Susín, A., Lligadas, X.: Biomechanical validation of upper-body and lower-body joint movements of kinect motion capture data for rehabilitation treatments. In: 2012 4th International Conference on Intelligent Networking and Collaborative Systems (INCoS), pp. 656–661. IEEE (2012)
11. Foulds, J., Frank, E.: A review of multi-instance learning assumptions. Knowl. Eng. Rev. **25**(1), 1–25 (2010)
12. Stanford Children's Health: Sports Injury Statistics (2010). http://www.stanfordchildrens.org/en/topic/default?id=sports-injury-statistics-90-P02787
13. Hootman, J.M., Dick, R., Agel, J.: Epidemiology of collegiate injuries for 15 sports: summary and recommendations for injury prevention initiatives. J. Athl. Train. **42**(2), 311–319 (2007)
14. Knight, E., Werstine, R.J., Rasmussen-Pennington, D.M., Fitzsimmons, D., Petrella, R.J.: Physical therapy 2.0: leveraging social media to engage patients in rehabilitation and health promotion. Phys. Ther. **95**(3), 389–396 (2015)

15. Komatireddy, R., Chokshi, A., Basnett, J., Casale, M., Goble, D., Shubert, T.: Quality and quantity of rehabilitation exercises delivered by a 3D motion controlled camera. Int. J. Phys. Med. Rehabil. **2**(4) (2014)
16. Lange, B., Chang, C.Y., Suma, E., Newman, B., Rizzo, A.S., Bolas, M.: Development and evaluation of low cost game-based balance rehabilitation tool using the Microsoft Kinect sensor. In: 2011 Annual International Conference of the IEEE Engineering in Medicine and Biology Society, EMBC, pp. 1831–1834. IEEE (2011)
17. Levac, D.E., Miller, P.A.: Integrating virtual reality video games into practice: clinicians' experiences. Physiother. Theory Pract. **29**(7), 504–512 (2013)
18. de Loes, M., Dahlstedt, L., Thomee, R.: A 7-year study on risks and costs of knee injuries in male and female youth participants in 12 sports. Scand. J. Med. Sci. Sports **10**(2), 90–97 (2000)
19. Palazzo, C., et al.: Barriers to home-based exercise program adherence with chronic low back pain: patient expectations regarding new technologies. Ann. Phys. Rehabil. Med. **59**(2), 107–113 (2016)
20. Pedregosa, F., et al.: Scikit-learn: machine learning in python. Journal of machine learning research **12**(Oct)
21. Pfister, A., West, A.M., Bronner, S., Noah, J.A.: Comparative abilities of Microsoft Kinect and Vicon 3D motion capture for gait analysis. J. Med. Eng. Technol. **38**(5), 274–280 (2014)
22. Pisters, M.F., et al.: Long-term effectiveness of exercise therapy in patients with osteoarthritis of the hip or knee: a systematic review. Arthritis Care Res. **57**(7), 1245–1253 (2007)
23. Ray, S., Scott, S., Blockeel, H.: Multi-instance learning. In: Sammut, C., Webb, G.I. (eds.) Encyclopedia of Machine Learning, pp. 701–710. Springer, Boston (2011). https://doi.org/10.1007/978-0-387-30164-8_569
24. Sheu, Y., Chen, L.H., Hedegaard, H.: Sports-and recreation-related injury episodes in the united states, 2011–2014. Natl. Health Stat. Rep. (99), 1–12 (2016)
25. Su, C.J., Chiang, C.Y., Huang, J.Y.: Kinect-enabled home-based rehabilitation system using dynamic time warping and fuzzy logic. Appl. Soft Comput. **22**, 652–666 (2014)
26. Tang, R., Yang, X.D., Bateman, S., Jorge, J., Tang, A.: Physio@home: exploring visual guidance and feedback techniques for physiotherapy exercises. In: Proceedings of the 33rd Annual ACM Conference on Human Factors in Computing Systems, pp. 4123–4132. ACM (2015)
27. Taylor, M.J., McCormick, D., Shawis, T., Impson, R., Griffin, M.: Activity-promoting gaming systems in exercise and rehabilitation. J. Rehabil. Res. Dev. **48**(10), 1171–1186 (2011)
28. Wojcicki, J.M., Heyman, M.B.: Let's move - childhood obesity prevention from pregnancy and infancy onward. N. Engl. J. Med. **362**(16), 1457–1459 (2010)
29. Zhao, W.: On automatic assessment of rehabilitation exercises with realtime feedback. In: 2016 IEEE International Conference on Electro Information Technology (EIT), pp. 0376–0381. IEEE (2016)
30. Zhao, W., Feng, H., Lun, R., Espy, D.D., Reinthal, M.A.: A kinect-based rehabilitation exercise monitoring and guidance system. In: 2014 5th IEEE International Conference on Software Engineering and Service Science (ICSESS), pp. 762–765. IEEE (2014)

Rough Set Theory as a Data Mining Technique: A Case Study in Epidemiology and Cancer Incidence Prediction

Zaineb Chelly Dagdia[1,2](✉), Christine Zarges[1], Benjamin Schannes[3],
Martin Micalef[4], Lino Galiana[5], Benoît Rolland[6], Olivier de Fresnoye[7],
and Mehdi Benchoufi[7,8,9]

[1] Department of Computer Science, Aberystwyth University, Aberystwyth, UK
{zaineb.chelly,c.zarges}@aber.ac.uk
[2] LARODEC, Institut Supérieur de Gestion de Tunis, Tunis, Tunisia
chelly.zaineb@gmail.com
[3] Department of Statistics, ENSAE, 5 avenue Henry Le Chatelier,
91120 Palaiseau, France
benjamin.schannes@gmail.com
[4] Actuaris, 13/15 boulevard de la Madeleine, 75001 Paris, France
martin.micalef@gmail.com
[5] ENS Lyon, 15 parvis René Descartes, 69342 Lyon Cedex 07, France
lino.galiana@ens-lyon.fr
[6] Altran Technologies S.A., 96 rue Charles de Gaulle, 92200 Neuilly-sur-Seine, France
benoit.rolland@free.fr
[7] Coordinateur Scientifique Programme Épidemium, Paris, France
olivier@epidemium.cc
[8] Centre d'Épidémiologie Clinique,
Hôpital Hôtel Dieu, Assistance Publique-Hôpitaux de Paris, Paris, France
mehdi.benchoufi@aphp.fr
[9] Faculté de Médecine, Université Paris Descartes and INSERM UMR1153,
Paris, France

Abstract. A big challenge in epidemiology is to perform data pre-processing, specifically feature selection, on large scale data sets with a high dimensional feature set. In this paper, this challenge is tackled by using a recently established distributed and scalable version of Rough Set Theory (RST. It considers epidemiological data that has been collected from three international institutions for the purpose of cancer incidence prediction. The concrete data set used aggregates about 5 495 risk factors (features), spanning 32 years and 38 countries. Detailed experiments demonstrate that RST is relevant to real world big data applications as

This work is part of a project that has received funding from the European Union's Horizon 2020 research and innovation programme under the Marie Skłodowska-Curie grant agreement No. 702527. This work was based on a first version of a database provided by the OpenCancer organization, part of Épidemium—a data challenge oriented and community—based open science program. Additional thanks go to the Épidemium group, Roche, La Paillasse and to the Supercomputing Wales project, which is part-funded by the European Regional Development Fund via the Welsh Government.

© Springer Nature Switzerland AG 2019
U. Brefeld et al. (Eds.): ECML PKDD 2018, LNAI 11053, pp. 440–455, 2019.
https://doi.org/10.1007/978-3-030-10997-4_27

it can offer insights into the selected risk factors, speed up the learning process, ensure the performance of the cancer incidence prediction model without huge information loss, and simplify the learned model for epidemiologists. Code related to this paper is available at: https://github.com/zeinebchelly/Sp-RST.

Keywords: Big data · Rough set theory · Feature selection Epidemiology · Cancer incidence prediction · Application

1   Introduction

Epidemiology is a sub-field of public health that looks to determine where and how often disease occur and why. It is more formally defined as the study of distributions (patterns) and determinants (causes) of health related states or events within a specified human population, and the application of this study to managing health problems [4]. The ultimate goal of epidemiology is to apply this knowledge to the control of disease through prevention and treatment, resulting in the preservation of public health.

In this context, epidemiologists study chronic diseases such as arthritis, cardiovascular disease such as heart attacks and stroke, cancer such as breast and colon cancer, diabetes, epilepsy and obesity problems. To conduct such studies, one of the most important considerations is the source and content of data, as this will often determine the quality of the results. As a general rule, the larger the data, the more accurate the results, since a larger sample is less likely to, by chance, generate an estimate different from the truth in the full population. This leads epidemiologists to deal with large amounts of data, big data, which is however not a feasible task for them [3]. Hence, to assist epidemiologists in dealing with such large amounts of data, data analysis has become one of the major research focuses in epidemiology and specifically for the epidemiology of cancer, colon cancer, which is our main focus. More precisely, data analysis assists epidemiologists to investigate and describe the determinants and distribution of disease, disability, and other health outcomes and develop the means for prevention and control. From a technical perspective, data analysis generally comprises a number of processes that may include data collection, data (pre)-processing and feature reduction, data cleansing, and data transformation and modeling with the goal of discovering useful information, suggesting conclusions, and supporting decision making; all these tasks can be achieved via the use of adequate machine learning techniques.

Meanwhile, in epidemiology, feature reduction is a main point of interest across the various steps of data analysis and focusing on this phase is crucial as it often presents a source of potential information loss. Many techniques were proposed in the literature [2] to achieve the task of feature reduction and they can be categorized into two main categories: techniques that transform the original meaning of the features, called "transformation-based approaches" or "feature extraction approaches", and semantic-preserving techniques that attempt to retain the meaning of the original feature set, known as "selection-based approaches" [11]. Within the latter category a further partitioning can be defined where the techniques are classified into filter approaches and wrapper approaches. The main difference between the two branches is that wrapper approaches include a learning algorithm in the feature subset evaluation, and hence they are tied to a particular induction algorithm. In this work, we mainly focus on the use of a feature selection technique, specifically a filter technique, instead of a feature extraction technique. This is crucial to preserve the semantics of the features in the context of cancer incidence prediction as results should be interpretable and understandable by epidemiologists.

Yet, the adaptation of feature selection techniques for big data problems may require the redesign of these algorithms and their inclusion in parallel and distributed environments. Among the possible alternatives is the MapReduce paradigm [13] introduced by Google which offers a robust and efficient framework to address the analysis of big data. Several recent works have focused on the parallelization of machine learning tools using the MapReduce approach [12,14–16]. Recently, new and more flexible workflows have appeared to extend the standard MapReduce approach, mainly Apache Spark [18], which has been successfully applied over various data mining and machine learning problems [18]. With the aim of choosing the most relevant and pertinent subset of features, a variety of feature selection techniques were proposed to deal with big data in a distributed way [20]. Nevertheless, most of these techniques suffer from some shortcomings. For instance, they usually require expert knowledge for the task of algorithm parameterization or noise levels to be specified beforehand and some simply rank features leaving the user to choose their own subset. There are some techniques that need the user to specify how many features are to be chosen, or they must supply a threshold that determines when the algorithm should terminate. All of these require the expert or the user to make a decision based on their own (possibly faulty) judgment. To overcome the limitations of the state-of-the-art methods, it is interesting to look for a filter method that does not require any external or supplementary information to function properly. Rough Set Theory (RST) can be used as such a technique [6]. RST, as a powerful feature selection technique, has made many achievements in many applications such as in decision support, engineering, environment, banking, medicine and others [19]. In this study, we focus on the use of RST as a data mining technique within a case study in epidemiology and cancer incidence prediction.

The rest of this paper is structured as follows. Section 2 reviews the fundamentals of epidemiology. Section 3 introduces the basic concepts of rough set theory for feature selection. Section 4 details the application in epidemiology and cancer incidence prediction via the use of a distributed algorithm based on rough sets for large-scale data pre-processing. The experimental setup is introduced in Sect. 5. The results of the performance analysis are discussed in Sect. 6 and conclusions are presented in Sect. 7.

2 Epidemiology: Concepts and Context Design

2.1 Distribution and Determinants

Epidemiology is concerned with the study of the distribution of a disease based on a set of "frequency" and "pattern" of health events in a population. The frequency refers on one hand to the number of health events such as the number of cases of diabetes or cancer in a population, and on the other hand to the link of that number to the size of the human population. The resulting ratio permits epidemiologists to compare disease occurrence across diverse populations. Pattern denotes the occurrence of health-related events by person, place, and time. Personal patterns comprise demographic factors that may be tied to risk of sickness, injury, or disability such as age, sex, marital status, social class, racial group, occupation, as well as behaviors and environmental exposures. Place patterns include geographic disparity, urban/suburban/rural variances, and location of work sites or schools. Time patterns can be annual, seasonal, weekly, daily, hourly, or any other breakdown of time that may effect disease or injury occurrence [10]. Moreover, epidemiology is concerned with the search for determinants. These are the factors that precipitate disease. Formally, determinants can be defined as any factor, whether event, characteristic, or other definable entity, that brings about a change in a health condition or other defined characteristic [9]. Epidemiologists assume that a disease does not arise haphazardly in a population, but it occurs when a set of accumulation of risk factors or determinants subsists in an individual. To look for these determinants, epidemiologists use epidemiological studies to understand and answer the "Why" and "How" of such events. For instance, they assess whether groups with dissimilar rates of disease diverge in their demographic characteristics, genetic or immunologic make-up, or any other so-called potential risk factors. Ideally, the findings provide sufficient evidence to direct prompt and effective public health control and prevention measures [10].

2.2 Population and Samples

An epidemiological study involves the collection, analysis and interpretation of data from a human population. The population about which epidemiologists wish to draw conclusions is called the "target population". In many cases, this is defined according to geographical criteria or some political boundaries. The specific population from which data are collected is called the "study population". It is a question of judgment whether results of the study population may be used to draw accurate conclusions about the target population. Most of the epidemiological studies use study populations that are based on geographical, institutional or occupational definitions. Another way of classifying the study of population is by the stage of the disease, i.e., a population that is diseased, disease-free or a mixture [4]. On the other hand, a sample is any part of the fully defined population. A syringe full of blood drawn from the vein of a patient is a sample of all the blood in the patient's circulation at the moment. Similarly,

100 patients suffering from colon cancer is a sample of the population of all the patients suffering from colon cancer. To make accurate inferences, the sample has to be properly chosen, representative, and the inclusion and exclusion criteria should be well defined as well. A representative sample is one in which each and every member of the population has an equal and mutually exclusive chance of being selected [8].

2.3   Incidence and Prevalence

Epidemiology often focuses on measuring the occurrence of disease in populations. The basic measures of disease frequency in epidemiology are "incidence" and "prevalence". Incidence is the number of new cases of disease in a population occurring over a defined period of time. Another important measure of disease incidence is incidence rate, which gauges how fast disease occurs in the population by measuring the number of new cases emerging as a function of time. Prevalence, on the other hand, measures the number of existing cases, both new cases and cases that have been diagnosed in the past, in a population at any given point in time. By using these measures, epidemiologists can determine the frequency of disease within populations, and compare differences in disease risk among populations [4].

3   Rough Set Theory

Rough Set Theory (RST) [1,17] is considered to be a formal approximation of the conventional set theory, which supports approximations in decision making. It provides a filter-based technique by which knowledge may be extracted from a domain in a concise way, retaining the information content whilst reducing the amount of knowledge involved [6]. This section focuses mainly on highlighting the fundamentals of rough set theory for feature selection.

3.1   Preliminaries of Rough Set Theory

In rough set theory, an *information table* is defined as a tuple $T = (U, A)$ where U and A are two finite, non-empty sets with U the *universe* of primitive objects and A the set of attributes. Each attribute or feature $a \in A$ is associated with a set V_a of its value, called the *domain* of a. We may partition the attribute set A into two subsets C and D, called *condition* and *decision* attributes, respectively.

Let $P \subset A$ be a subset of attributes. The indiscernibility relation, denoted by $IND(P)$, is the central concept of RST and it is an equivalence relation, which is defined as: $IND(P) = \{(x, y) \in U \times U : \forall a \in P, a(x) = a(y)\}$, where $a(x)$ denotes the value of feature a of object x. If $(x, y) \in IND(P)$, x and y are said to be *indiscernible* with respect to P. The family of all equivalence classes of $IND(P)$, referring to a partition of U determined by P, is denoted by $U/IND(P)$. Each element in $U/IND(P)$ is a set of indiscernible objects with respect to P. The equivalence classes $U/IND(C)$ and

$U/IND(D)$ are called *condition* and *decision* classes, respectively. For any concept $X \subseteq U$ and attribute subset $R \subseteq A$, X could be approximated by the R-*lower* approximation and R-*upper* approximation using the knowledge of R. The lower approximation of X is the set of objects of U that are surely in X, defined as: $\underline{R}(X) = \bigcup\{E \in U/IND(R) : E \subseteq X\}$. The upper approximation of X is the set of objects of U that are possibly in X, defined as: $\overline{R}(X) = \bigcup\{E \in U/IND(R) : E \cap X \neq \emptyset\}$. The concept defining the set of objects that can possibly, but not certainly, be classified in a specific way is called the *boundary region*, which is defined as: $BND_R(X) = \overline{R}(X) - \underline{R}(X)$. If the boundary region is empty, that is $\overline{R}(X) = \underline{R}(X)$, concept X is said to be R-*definable*; otherwise X is a *rough set* with respect to R. The *positive region* of decision classes $U/IND(D)$ with respect to condition attributes C is denoted by $POS_c(D)$ where $POS_c(D) = \bigcup \overline{R}(X)$. The positive region $POS_c(D)$ is a set of objects of U that can be classified with certainty to classes $U/IND(D)$ employing attributes of C. In other words, the positive region $POS_c(D)$ indicates the union of all the equivalence classes defined by $IND(P)$ that each for sure can induce the decision class D. Based on the positive region, the *dependency of attributes* measuring the degree k of the dependency of an attribute c_i on a set of attributes C is defined as: $k = \gamma(C, c_i) = |POS_C(c_i)|/|U|$. Based on these basics, RST defines two important concepts for feature selection, which are the *Core* and the *Reduct*.

3.2 Reduction Process

RST aims at choosing the smallest subset of the conditional feature set so that the resulting reduced data set remains consistent with respect to the decision feature. To do so, RST defines the Reduct and the Core concepts. In rough set theory, a subset $R \subseteq C$ is said to be a D-*reduct* of C if $\gamma(C, R) = \gamma(C)$ and there is no $R' \subset R$ such that $\gamma(C, R') = \gamma(C, R)$. In other words, the *Reduct* is the minimal set of selected attributes preserving the same dependency degree as the whole set of attributes. Meanwhile, rough set theory may generate a set of reducts, $RED_D^F(C)$, from the given information table. In this case, any reduct from $RED_D^F(C)$ can be chosen to replace the initial information table. The second concept, the *Core*, is the set of attributes that are contained by all reducts, defined as $CORE_D(C) = \bigcap RED_D(C)$ where $RED_D(C)$ is the D-reduct of C. Specifically, the *Core* is the set of attributes that cannot be removed from the information system without causing collapse of the equivalence-class structure. This means that all attributes present in the *Core* are indispensable.

4 Application

4.1 Data Sources

The OpenCancer[1] organization gathers people working on cancer prediction issues. Their aim is to provide tools aimed at helping health authorities to take

---

[1] https://github.com/orgs/EpidemiumOpenCancer/.

public policy decisions in terms of cancer prevention. OpenCancer has linked and merged data from the World Health Organization (WHO)[2], World Bank (WB)[3], the International Labour Organization (ILO)[4] and the Food and Agriculture Organization (FAO)[5] of the United Nations to build a large data set covering 38 countries and many regions within these countries between 1970 and 2002. For this application, OpenCancer provided a first version of the database restricted to the WHO, WB and FAO sources. Each row is characterized by a 5-tuple (cancer type, country, gender, ethnicity, age group) and 5 495 features. For this application the single cancer type, which has been considered, is the colon cancer.

4.2 Data Pre-processing

Data Cleaning. The first version of this sub-database suffers from a vast number of missing cells due to the lack of information in the available repositories. To fix this issue prior to running any learning model, OpenCancer had discarded every feature exhibiting a missing data ratio higher than 50% and imputed other missing data with a standard mean strategy. The resulting database—merged from both FAO and WB, including the incidence provided from WHO—includes 3 365 risk factors (features) and 45 888 records. Each record, seen as a population, is identified via a 6-tuple defined as {Sex, Age group, Country, Region, Ethnicity, Year}. To measure the occurrence of the colon cancer disease in the population, the number of new cases of the disease within a population occurring over 1970 and 2002 is used, referring to the incidence measure.

Feature Selection. Once the consistent database is ready for use, a feature selection step is performed. To deal with the large amount of the epidemiological data, a distributed version of rough set theory for feature selection [5], named Sp-RST, is used. Sp-RST is based on a parallel programming design that allows to tackle big data sets over a cluster of machines independently from the underlying hardware and/or software. To select the most important risk factors from the input consistent database, and for the purpose of colon cancer incidence prediction, Sp-RST proceeds as follows:

Problem Formalization. Technically, the epidemiological database is first stored in an associated Distributed File System (DFS) that is accessible from any computer of the used cluster. To operate on the given DFS in a parallel way, a Resilient Distributed Data set (RDD) is created. We may formalize the latter as a given information table defined as T_{RDD}, where the universe $U = \{x_1, \ldots, x_N\}$ is the set of data items reflecting the population and is identified as a 6-tuple defined as {Sex, Age group, Country, Region, Ethnicity, Year}. The conditional

---

[2] http://www.who.int/en/.
[3] http://www.worldbank.org/.
[4] http://www.ilo.org/global/lang-en/index.htm.
[5] http://www.fao.org/home/fr/.

attribute set $C = \{c_1, \ldots, c_V\}$ contains every single feature of the T_{RDD} information table, and presents the risk factors. The decision attribute D of our learning problem corresponds to the class (label) of each T_{RDD} sample. It has continuous values d and refers to the incidence of the colon cancer. The condition attribute feature D is defined as follows: $D = \{\text{Typology}_1, \ldots, \text{Typology}_I\}$. The conditional attribute set C presents the pool from where the most convenient risk factors will be selected.

Feature Selection Process. For feature selection, the given T_{RDD} information table is partitioned first into m data blocks based on splits from the conditional attribute set C. Hence, $T_{RDD} = \bigcup_{i=1}^{m}(C_r)T_{RDD_{(i)}}$, where $r \in \{1, \ldots, V\}$. Each $T_{RDD_{(i)}}$ is constructed based on r random features selected from C, where $\forall T_{RDD_{(i)}} : \#\{c_r\} = \bigcap_{i=1}^{m} T_{RDD_{(i)}}$.

Within a distributed implementation, Sp-RST is applied to every single $T_{RDD_{(i)}}$ so that at the end all the intermediate results will be gathered from the different m partitions. Specifically, Sp-RST will first compute the indiscernibility relation for the decision class defined as $IND(D) : IND(d_i)$. More precisely, Sp-RST will calculate the indiscernibility relation for every decision class d_i by gathering the same T_{RDD} data items, which are defined in the universe $U = \{x_1, \ldots, x_N\}$ and which belong to the same class d_i. This task is independent from the m generated partitions and, as the result, depends on the data items class and not on the features. Once achieved, the algorithm generates the m random $T_{RDD_{(i)}}$ as previously explained. Then, and within a specific partition, Sp-RST creates all the possible combinations of the C_r set of features, computes the indiscernibility relation for every generated combination $IND(AllComb_{(C_r)})$ and calculates the dependency degrees $\gamma(C_r, AllComb_{(C_r)})$ of each feature combination. Then, Sp-RST looks for the maximum dependency value among all $\gamma(C_r, AllComb_{(C_r)})$. The maximum dependency reflects on one hand the dependency of the whole feature set (C_r) representing the $T_{RDD_{(i)}}$ and on the other hand the dependency of all the possible feature combinations satisfying the constraint $\gamma(C_r, AllComb_{(C_r)}) = \gamma(C_r)$. The maximum dependency is the baseline value for feature selection. Then, Sp-RST keeps the set of all combinations having the same dependency degrees as the selected baseline. In fact, at this stage Sp-RST removes in each computation level the unnecessary features that may affect negatively the performance of any learning algorithm.

Finally, Sp-RST keeps the set of combinations having the minimum number of features by satisfying the full reduct constraints discussed in Sect. 3: $\gamma(C_r, AllComb_{(C_r)}) = \gamma(C_r)$ while there is no $AllComb'_{(C_r)} \subset AllComb_{(C_r)}$ such that $\gamma(C_r, AllComb'_{(C_r)}) = \gamma(C_r, AllComb_{(C_r)})$. Each combination satisfying this condition is considered as a viable minimum reduct set. The attributes of the reduct set describe all concepts in the original training data set $T_{RDD_{(i)}}$.

The output of each partition is either a single reduct $RED_{i_{(D)}}(C_r)$ or a family of reducts $RED_{i_{(D)}}^{F}(C_r)$. Based on the RST preliminaries previously mentioned in Sect. 3, any reduct of $RED_{i_{(D)}}^{F}(C_r)$ can be used to represent the $T_{RDD_{(i)}}$ information table. Consequently, if Sp-RST generates only one reduct, for a specific $T_{RDD_{(i)}}$ block, then the output of this feature selection phase is the set of the $RED_{i_{(D)}}(C_r)$ features. These features reflect the most informative ones among the C_r attributes resulting a new reduced $T_{RDD_{(i)}}$, $T_{RDD_{(i)}}(RED)$, which preserves nearly the same data quality as its corresponding $T_{RDD_{(i)}}(C_r)$ that is based on the whole feature set C_r. On the other hand, if Sp-RST generates a family of reducts then the algorithm randomly selects one reduct among $RED^F_{i_{(D)}}(C_r)$ to represent the corresponding $T_{RDD_{(i)}}$. This random choice is justified by the same priority of all the reducts in $RED^F_{i_{(D)}}(C_r)$. In other words, any reduct included in $RED^F_{i_{(D)}}(C_r)$ can be used to replace the $T_{RDD_{(i)}}(C_r)$ features. At this stage, each i data block has its output $RED_{i_{(D)}}(C_r)$ referring to the selected features. However, since each $T_{RDD_{(i)}}$ is based on distinct features and with respect to $T_{RDD} = \bigcup_{i=1}^m (C_r) T_{RDD_{(i)}}$ a union of the selected feature sets is required to represent the initial T_{RDD}; defined as $Reduct_m = \bigcup_{i=1}^m RED_{i_{(D)}}(C_r)$. In order to ensure the performance of Sp-RST while avoiding considerable information loss, the algorithm runs over N iterations on the T_{RDD} m data blocks and thus generates N $Reduct_m$. Hence, at the end an intersection of all the obtained $Reduct_m$ is needed; defined as $Reduct = \bigcap_{n=1}^N Reduct_m$.

By removing irrelevant and redundant features, Sp-RST can reduce the dimensionality of the data from $T_{RDD}(C)$ to $T_{RDD}(Reduct)$. More precisely, Sp-RST was able to reduce the considered epidemiological database from 3 364 risk factors to only around 840 features. The pseudo-code of Sp-RST as well as details related to each of its distributed tasks can be found in [5].

4.3   Predictive Modeling

Accurately evaluating colon cancer risk in average and high-risk populations or individuals and determining colon cancer prognosis in patients are essential for controlling the suffering and death due to colon cancer. From a general perspective, cancer prediction models offer a significant approach to assessing risk and prognosis by detecting populations and individuals at high-risk, easing the design and planning of clinical cancer trials, fostering the development of benefit-risk indices, and supporting estimates of the population burden and cost of cancer. Models also may aid in the evaluation of treatments and interventions, and help epidemiologists make decisions about treatment and long-term follow-up care [7]. In this concern, for colon cancer incidence prediction, the distributed version of the Random Forest Regression model[6] is used.

5   Experimental Setup

5.1   Experimental Plan, Testbed and Tools

Our experiments are performed on the High Performance Computing Wales platform (HPC Wales), which provides a distributed computing facility. Under

---

[6] org.apache.spark.ml.regression.{RandomForestRegressionModel, RandomForestRegressor}.

this testbed, we used 12 dual-core Intel Westmere Xeon X5650 2.67 GHz CPUs and 36GB of memory to test the performance of Sp-RST, which is implemented in Scala 2.11 within Spark 2.1.1. The main aim of our experimentation is to demonstrate that RST is relevant to real world big data applications as it can offer insights into the selected risk factors, speed up the learning process, ensure the performance of the colon cancer incidence prediction model without huge information loss, and simplify the learned model for epidemiologists.

5.2 Parameters Settings

As previously mentioned, we use the Random Forest Regression implementation provided in the Spark framework with the following parameters: maxDepth = 5, numTrees = 20, featureSubsetStrategy = 'all' and impurity = 'variance'. The algorithm automatically identifies categorical features and indexes them. The database is split into training and test sets where 30% of the database is held out for testing. Meanwhile, for the Sp-RST settings, we set the number of partitions to 841 partitions; generating 4 features per partition (based on preliminary experiments). We run the settings on 8 nodes on HPC Wales. For the purpose of this study we set the number of iterations of Sp-RST to 10.

6 Results and Discussion

6.1 Categories of Selected Risk Factors

Recall that Sp-RST runs over 10 iterations and that at its last algorithmic stage an intersection of the generated reducts at each iteration is made. However, for the considered data set, this intersection is empty. Hence, we have modified the algorithm to return all 10 different reducts to be presented to the epidemiologists. In the following, we present two different types of results: averages accumulated over the 10 reduced datasets and separate numbers for each of the iterations performed.

Both parts of the data set (FAO, World Bank) contain 9 different categories of risk factors as shown in Table 1. Here, we list the number of different factors in each category, the average number of factors selected over the 10 iterations of Sp-RST and the corresponding percentages. From Table 1, and based on the WB database, we notice that there is only a small variation in the distribution of the selected risk factors. Exceptions are the gender and poverty risk factors, which are not selected by the algorithm. The same comments can be made for the FAO database where the food security risk factor does not appear in any of the selected feature sets. Epidemiologists confirm that these results are quite expected. This demonstrates that our method is able to select the most interesting features to keep—the key risk factors.

We depict the ratios of each category within the set of selected risk factors for each database (FAO & WB) in Fig. 1, and for an overall view, the distribution of the categories of the combined data sets is presented in Fig. 2. Based on these figures, epidemiologists confirmed again that the selected risk factors are

Table 1. Overview of the data set and the selected risk factors.

	#Risk factors	#selected (average)	% selected
WB: Education	31	7.4	23.87%
WB: Environment	78	19.7	25.26%
WB: Health	62	14.3	23.06%
WB: Infrastructure	19	4.9	25.79%
WB: Economy	141	34.1	24.18%
WB: Public Sector	41	8.5	20.73%
WB: Gender	0	0	-
WB: Social Protection & Labor	40	9.6	24.00%
WB: Poverty	0	0	-
WB: TOTAL	412	98.5	23.91%
FAO: Production	378	90.7	23.99%
FAO: Emissions	803	201.1	25.04%
FAO: Employment	6	1.4	23.33%
FAO: Environment	17	4.4	25.88%
FAO: Commodity	938	234.8	25.03%
FAO: Inputs	153	36	23.53%
FAO: Food Balance	70	20	28.57%
FAO: Food Supply	587	153.7	26.18%
FAO: Food Security	0	0	-
FAO: TOTAL	2952	742.1	25.14%
TOTAL	3364	840.6	24.99%

expected to appear in each of their corresponding databases (though potentially with a different overall distribution). This again supports that Sp-RST can determine the key risk factors among a large set of features. Meanwhile, epidemiologists highlighted that a higher average or proportion does not necessarily mean that a risk factor is more important than another. Indeed, no firm conclusions on the influence of one factor on the colon incidence prediction can be drawn based on this information, only. Thus, from an epidemiological perspective, the risk factors selected by Sp-RST should be further coupled with other sources of data to complete the analysis and to be able to draw specific conclusions.

We now investigate the selected risk factors per iteration (for all the 10 Sp-RST iterations). Each iteration of SP-RST reflects a possible reduced set of risk factors on which the prediction of the colon cancer incidence can be made. The categories of the selected risk factors in the FAO data set, in the WB database and for the combined database, separately for each of the 10 iterations are presented in Figs. 3, 4 and 5, respectively.

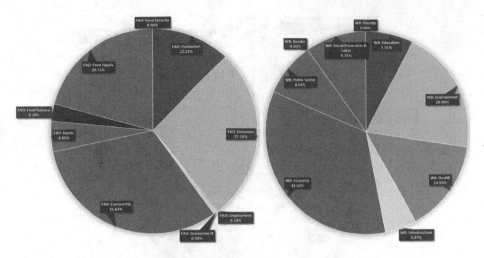

Fig. 1. Distribution of the categories of risk factors selected by our proposed method; split by data set: FAO (left) and World Bank (right).

Based on Figs. 3, 4 and 5, we can see that the risk factors partially overlap within the 10 iterations. This might be interesting from an epidemiological point of view as it can influence the consideration of other possible risk factors, which appear with different distributions. Indeed, the overlap between the selected risk factors from one iteration to another may call the attention of the epidemiologist in cases where a firm decision is taken with respect to a specific risk factor. These results are considered to be very important for the epidemiologists as they help them in the decision making process.

6.2   Evaluation of Regression

We use four different metrics to compare the obtained random forest regression models for the original data set and the reduced data sets produced by Sp-RST. Let p_i denote the predicted value of the i-th data item in the test data and v_i its actual value. We call the difference $e_i = v_i - p_i$ the sample error. We consider:

- Mean Absolute Error: $\sum_{i=0}^{n} |e_i|/n$
- Mean Squared Error: $\sum_{i=0}^{n} (e_i)^2|/n$
- Root Mean Squared Error, the square root of the mean squared error
- Coefficient of Determination (R^2): $1 - \sum_{i=0}^{n} (e_i)^2 / \sum_{i=0}^{n} (v_i - \bar{v}_i)^2$, where $\bar{v}_i$ denotes the average of the v_i.

For the first three metrics, smaller values indicate a better model. For the R2 metric values between 0 and 1 are obtained, where 1 indicates a perfect model and 0 indicates a trivial model that always predicts the average of the training samples.

Our results for all four metrics are summarized in Table 2. We see that the results are very similar, but slightly better for the original data set.

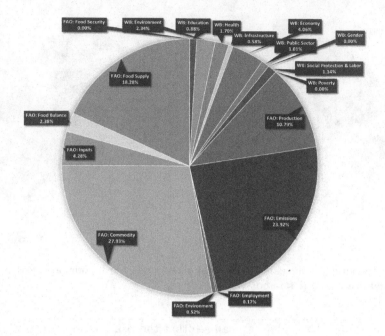

Fig. 2. Distribution of the categories of the combined data set.

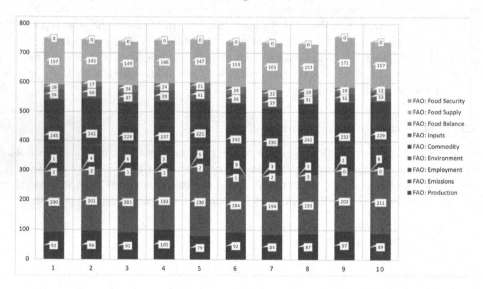

Fig. 3. Categories of the selected risk factors in the FAO data set for each of the 10 iterations.

Wilcoxon rank sum tests did not reveal any statistical significance at standard confidence level 0.05 as indicated by the p-values in Table 2. We conclude that the quality of the obtained regression models is comparable. However, the reduced

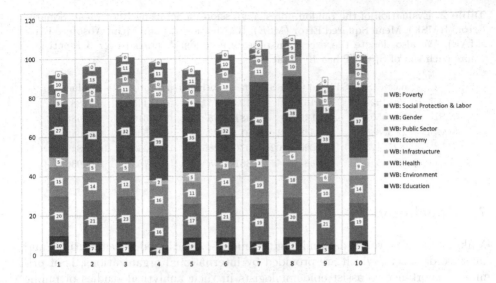

Fig. 4. Categories of the selected risk factors in the World Bank data set for each of the 10 iterations.

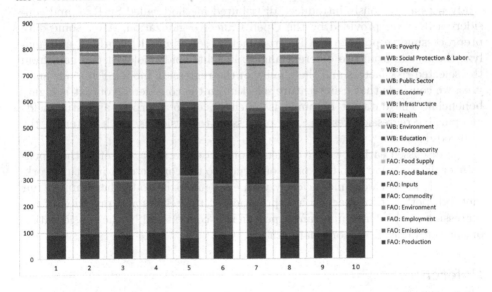

Fig. 5. Categories of the selected risk factors in the combined data set for each of the 10 iterations.

data set improves the execution time to determine the regression model considerably (by almost a factor of 5). Moreover, a data set with only around 840 risk factors is much easier to interpret and handle by epidemiologists as discussed in the previous section. We therefore argue that the reduction process is appropriate in the considered context.

Table 2. Evaluation of the random forest regression model using Root Mean Squared Error (RMSE), Mean Squared Error (MSE), R2 Metric (R2) and Mean Absolute Error (MAE). We also denote the execution time in seconds. Averaged over 3 repetitions where each run of Sp-RST has 10 iterations.

	mean (Sp-RST)	std (Sp-RST)	mean (original)	sd (original)	p-value
MAE	12.51929	0.167719	12.38475	0.174458	0.1488
MSE	501.4551	20.19929	481.8668	18.7318	0.08688
RMSE	22.38885	0.4488141	21.94872	0.4250894	0.08688
R2	0.2531238	0.02098402	0.2718285	0.01256834	0.1884
Time (s)	78.84557	10.81764	381.9735	5.647821	0.0003666

7   Conclusion

Making use of powerful data mining techniques, distributed infrastructures and massive data sets, which are provided by international organizations, is of primary importance to assist epidemiologists in their analytical studies of public interest. In this paper, we have presented a case study for using a Rough Set theory approach as a data mining technique in the context of epidemiology. Our study uses a previously introduced distributed method called Sp-RST and considers a data set provided by the Open Cancer organization. After some data preprocessing we perform feature (risk factor) selection with Sp-RST and analyze the results from two different angles: insights epidemiologist can gain from the selected risk factors and the quality of the regression model. From our analyses, we conclude that using feature selection in the considered context is highly beneficial. The data set obtained is much easier to interpret and still yields comparable regression results. The process of regression is much faster on the reduced data set.

As discussed earlier, we are currently only considering a subset of the Open-Cancer data set. We plan to expand our study to the complete set in future work. Moreover, we will work more closely with epidemiologist to further improve our method, both with respect to interpretation of the results and precision of the regression model. One important aspect in this context will be the consideration of missing values in the original data set.

References

1. Pawlak, Z., Skowron, A.: Rudiments of rough sets. Inf. Sci. **177**(1), 3–27 (2007)
2. Bagherzadeh-Khiabani, F., Ramezankhani, A., Azizi, F., Hadaegh, F., Steyerberg, E.W., Khalili, D.: A tutorial on variable selection for clinical prediction models: feature selection methods in data mining could improve the results. J. Clin. Epidemiol. **71**, 76–85 (2016)
3. Mooney, S.J., Westreich, D.J., El-Sayed, A.M.: Epidemiology in the era of big data. Epidemiology **26**(3), 390 (2015)
4. Woodward, M.: Epidemiology: Study Design and Data Analysis. CRC Press, Boca Raton (2013)

5. Dagdia, Z.C., Zarges, C., Beck, G., Lebbah, M.: A distributed rough set theory based algorithm for an efficient big data pre-processing under the spark framework. In: Proceedings of the 2017 IEEE International Conference on Big Data, pp. 911–916. IEEE, Boston (2017)
6. Thangavel, K., Pethalakshmi, A.: Dimensionality reduction based on rough set theory: a review. Appl. Soft Comput. **9**(1), 1–12 (2009)
7. Amersi, F., Agustin, M., Ko, C.Y.: Colorectal cancer: epidemiology, risk factors, and health services. Clin. Colon Rectal Surg. **18**(3), 133 (2005)
8. Banerjee, A., Chaudhury, S.: Statistics without tears: populations and samples. Ind. Psychiatry J. **19**(1), 60 (2010)
9. Porta, M.: A Dictionary of Epidemiology. Oxford University Press, Oxford (2008)
10. Dicker, R.C., Coronado, F., Koo, D., Parrish, R.G.: Principles of epidemiology in public health practice; an introduction to applied epidemiology and biostatistics. U.S. Department of Health and Human Services, Centers for Disease Control and Prevention (CDC) (2006)
11. Liu, H., Motoda, H., Setiono, R., Zhao, Z.: Feature selection: an ever evolving frontier in data mining. In: Feature Selection in Data Mining, pp. 4–13 (2013)
12. Schneider, J., Vlachos, M.: Scalable density-based clustering with quality guarantees using random projections. Data Min. Knowl. Discov. **31**, 1–34 (2017)
13. Dean, J., Ghemawat, S.: MapReduce: a flexible data processing tool. Commun. ACM **53**(1), 72–77 (2010)
14. Zhai, T., Gao, Y., Wang, H., Cao, L.: Classification of high-dimensional evolving data streams via a resource-efficient online ensemble. Data Min. Knowl. Discov. **31**, 1–24 (2017)
15. Vinh, N.X., et al.: Discovering outlying aspects in large datasets. Data Min. Knowl. Discov. **30**(6), 1520–1555 (2016)
16. Zhang, J., Wang, S., Chen, L., Gallinari, P.: Multiple Bayesian discriminant functions for high-dimensional massive data classification. Data Min. Knowl. Discov. **31**(2), 465–501 (2017)
17. Pawlak, Z.: Rough Sets: Theoretical Aspects of Reasoning About Data. Springer, Heidelberg (2012)
18. Shanahan, J.G., Dai, L.: Large scale distributed data science using apache spark. In: Proceedings of the 21th ACM SIGKDD International Conference on Knowledge Discovery and Data Mining, pp. 2323–2324. ACM (2015)
19. Polkowski, L., Tsumoto, S., Lin, T.Y.: Rough Set Methods and Applications: New Developments in Knowledge Discovery in Information Systems, vol. 56. Physica, Heidelberg (2012)
20. Guller, M.: Big Data Analytics with Spark: A Practitioner's Guide to Using Spark for Large Scale Data Analysis. Springer, Heidelberg (2015)

Bayesian Best-Arm Identification for Selecting Influenza Mitigation Strategies

Pieter J. K. Libin[1,2](✉), Timothy Verstraeten[1], Diederik M. Roijers[1], Jelena Grujic[1], Kristof Theys[2], Philippe Lemey[2], and Ann Nowé[1]

[1] Artificial Intelligence Lab, Department of Computer Science,
Vrije Universiteit Brussel, Brussels, Belgium
{pieter.libin,timothy.verstraeten,jelena.grujic,ann.nowe}@vub.be
[2] Rega Institute for Medical Research, Clinical and Epidemiological Virology,
KU Leuven - University of Leuven, Leuven, Belgium
{pieter.libin,kristof.theys,philippe.lemey}@kuleuven.be

Abstract. Pandemic influenza has the epidemic potential to kill millions of people. While various preventive measures exist (i.a., vaccination and school closures), deciding on strategies that lead to their most effective and efficient use remains challenging. To this end, individual-based epidemiological models are essential to assist decision makers in determining the best strategy to curb epidemic spread. However, individual-based models are computationally intensive and it is therefore pivotal to identify the optimal strategy using a minimal amount of model evaluations. Additionally, as epidemiological modeling experiments need to be planned, a computational budget needs to be specified a priori. Consequently, we present a new sampling technique to optimize the evaluation of preventive strategies using fixed budget best-arm identification algorithms. We use epidemiological modeling theory to derive knowledge about the reward distribution which we exploit using Bayesian best-arm identification algorithms (i.e., Top-two Thompson sampling and BayesGap). We evaluate these algorithms in a realistic experimental setting and demonstrate that it is possible to identify the optimal strategy using only a limited number of model evaluations, i.e., 2-to-3 times faster compared to the uniform sampling method, the predominant technique used for epidemiological decision making in the literature. Finally, we contribute and evaluate a statistic for Top-two Thompson sampling to inform the decision makers about the confidence of an arm recommendation. Code related to this paper is available at: https://plibin-vub.github.io/epidemic-bandits.

Keywords: Pandemic influenza · Multi-armed bandits
Fixed budget best-arm identification · Preventive strategies
Individual-based models

Electronic supplementary material The online version of this chapter (https://doi.org/10.1007/978-3-030-10997-4_28) contains supplementary material, which is available to authorized users.

© Springer Nature Switzerland AG 2019
U. Brefeld et al. (Eds.): ECML PKDD 2018, LNAI 11053, pp. 456–471, 2019.
https://doi.org/10.1007/978-3-030-10997-4_28

1   Introduction

The influenza virus is responsible for the deaths of half of a million people each year. In addition, seasonal influenza epidemics cause a significant economic burden. While transmission is primarily local, a newly emerging variant may spread to pandemic proportions in a fully susceptible host population [29]. Pandemic influenza occurs less frequently than seasonal influenza but the outcome with respect to morbidity and mortality can be much more severe, potentially killing millions of people worldwide [29]. Therefore, it is essential to study mitigation strategies to control influenza pandemics.

For influenza, different preventive measures exist: i.a., vaccination, social measures (e.g., school closures and travel restrictions) and antiviral drugs. However, the efficiency of strategies greatly depends on the availability of preventive compounds, as well as on the characteristics of the targeted epidemic. Furthermore, governments typically have limited resources to implement such measures. Therefore, it remains challenging to formulate public health strategies that make effective and efficient use of these preventive measures within the existing resource constraints.

Epidemiological models (i.e., compartment models and individual-based models) are essential to study the effects of preventive measures *in silico* [2,17]. While individual-based models are usually associated with a greater model complexity and computational cost than compartment models, they allow for a more accurate evaluation of preventive strategies [11]. To capitalize on these advantages and make it feasible to employ individual-based models, it is essential to use the available computational resources as efficiently as possible.

In the literature, a set of possible preventive strategies is typically evaluated by simulating each of the strategies an equal number of times [7,13,15]. However, this approach is inefficient to identify the optimal preventive strategy, as a large proportion of computational resources will be used to explore suboptimal strategies. Furthermore, a consensus on the required number of model evaluations per strategy is currently lacking [34] and we show that this number depends on the *hardness* of the evaluation problem. Additionally, we recognize that epidemiological modeling experiments need to be planned and that a computational budget needs to be specified a priori. Therefore, we present a novel approach where we formulate the evaluation of preventive strategies as a *best-arm identification* problem using a *fixed budget* of model evaluations. In this work, the budget choice is left to the discretion of the decision maker, as would be the case for any uniform evaluation.

As running an individual-based model is computationally intensive (i.e., minutes to hours, depending on the complexity of the model), minimizing the number of required model evaluations reduces the total time required to evaluate a given set of preventive strategies. This renders the use of individual-based models attainable in studies where it would otherwise not be computationally feasible. Additionally, reducing the number of model evaluations will free up computational resources in studies that already use individual-based models, capacitating researchers to explore a larger set of model scenarios. This is important, as considering a wider range of scenarios increases the confidence about the overall utility of preventive strategies [35].

In this paper, we contribute a novel technique to evaluate preventive strategies as a fixed budget best-arm identification problem. We employ epidemiological modeling theory to derive assumptions about the reward distribution and exploit this knowledge using Bayesian algorithms. This new technique enables decision makers to obtain recommendations in a reduced number of model evaluations. We evaluate the technique in an experimental setting, where we aim to find the best vaccine allocation strategy in a realistic simulation environment that models an influenza pandemic on a large social network. Finally, we contribute and evaluate a statistic to inform the decision makers about the confidence of a particular recommendation.

2   Background

2.1   Pandemic Influenza and Vaccine Production

The primary preventive strategy to mitigate seasonal influenza is to produce vaccine prior to the epidemic, anticipating the virus strains that are expected to circulate. This vaccine pool is used to inoculate the population before the start of the epidemic. While it is possible to stockpile vaccines to prepare for seasonal influenza, this is not the case for influenza pandemics, as the vaccine should be specifically tailored to the virus that is the source of the pandemic. Therefore, before an appropriate vaccine can be produced, the responsible virus needs to be identified. Hence, vaccines will be available only in limited supply at the beginning of the pandemic [33]. In addition, production problems can result in vaccine shortages [10]. When the number of vaccine doses is limited, it is imperative to identify an optimal vaccine allocation strategy [28].

2.2   Modeling Influenza

There is a long tradition of using individual-based models to study influenza epidemics [2,15,17], as they allow for a more accurate evaluation of preventive strategies. A state-of-the-art individual-based model that has been the driver for many high impact research efforts [2,17,18], is FluTE [6]. FluTE implements a contact model where the population is divided into communities of households [6]. The population is organized in a hierarchy of social mixing groups where the contact intensity is inversely proportional with the size of the group (e.g., closer contact between members of a household than between colleagues). Additionally, FluTE implements an individual disease progression model that associates different disease stages with different levels of infectiousness. FluTE supports the evaluation of preventive strategies through the simulation of therapeutic interventions (i.e., vaccines, antiviral compounds) and non-therapeutic interventions (i.e., school closure, case isolation, household quarantine).

2.3   Bandits and Best-Arm Identification

The *multi-armed bandit game* [1] involves a K-armed bandit (i.e., a slot machine with K levers), where each arm A_k returns a reward r_k when it is pulled (i.e., r_k represents a sample from A_k's reward distribution). A common use of the bandit game is to pull a sequence of arms such that the cumulative regret is minimized [20]. To fulfill this goal, the player needs to carefully balance between exploitation and exploration.

In this paper, the objective is to recommend the best arm A^* (i.e., the arm with the highest average reward μ^*), after a fixed number of arm pulls. This is referred to as the fixed budget best-arm identification problem [1], an instance of the pure-exploration problem [4]. For a given budget T, the objective is to minimize the *simple regret* $\mu^* - \mu_J$, where μ_J is the average reward of the recommended arm A_J, at time T [5]. Simple regret is inversely proportional to the probability of recommending the correct arm A^* [24].

3   Related Work

As we established that a computational budget needs to be specified a priori, our problem setting matches the fixed budget best-arm identification setting. This differs from settings that attempt to identify the best arm with a predefined confidence: i.e., racing strategies [12], strategies that exploit the confidence bound of the arms' means [25] and more recently fixed confidence best-arm identification algorithms [16]. We selected Bayesian fixed budget best-arm identification algorithms, as we aim to incorporate prior knowledge about the arms' reward distributions and use the arms' posteriors to define a statistic to support policy makers with their decisions. We refer to [21,24], for a broader overview of the state of the art with respect to (Bayesian) best-arm identification algorithms.

Best-arm identification algorithms have been used in a large set of application domains: i.a., evaluation of response surfaces, the initialization of hyperparameters and traffic congestion.

While other algorithms exist to rank or select bandit arms, e.g. [30], best-arm identification is best approached using adaptive sampling methods [23], as the ones we study in this paper.

In preliminary work, we explored the potential of multi-armed bandits to evaluate prevention strategies in a regret minimization setting, using default strategies (i.e., ϵ-greedy and UCB1). We presented this work at the 'Adaptive Learning Agents' workshop hosted by the AAMAS conference [26]. This setting is however inadequate to evaluate prevention strategies *in silico*, as minimizing cumulative regret is sub-optimal to identify the best arm. Additionally, in this workshop paper, the experiments considered a small and less realistic population, and only analyzed a limited range of R_0 values that is not representative for influenza pandemics.

4   Methods

We formulate the evaluation of preventive strategies as a multi-armed bandit game with the aim of identifying the best arm using a fixed budget of model evaluations. The presented method is generic with respect to the type of epidemic that is modeled (i.e., pathogen, contact network, preventive strategies). The method is evaluated in the context of pandemic influenza in the next section[1].

4.1   Evaluating Preventive Strategies with Bandits

A *stochastic epidemiological model* E is defined in terms of a model configuration $c \in C$ and can be used to evaluate a preventive strategy $p \in \mathcal{P}$. The result of a model evaluation is referred to as the *model outcome* (e.g., prevalence, proportion of symptomatic individuals, morbidity, mortality, societal cost). Evaluating the model E thus results in a sample of the model's *outcome distribution*:

$$\text{outcome} \sim E(c, p), \text{ where } c \in C \text{ and } p \in \mathcal{P} \tag{1}$$

Our objective is to find the optimal preventive strategy (i.e., the strategy that minimizes the expected outcome) from a set of alternative strategies $\{p_1, ..., p_K\} \subset \mathcal{P}$ for a particular configuration $c_0 \in C$ of a stochastic epidemiological model, where c_0 corresponds to the studied epidemic. To this end, we consider a multi-armed bandit with $K = |\{p_1, ..., p_K\}|$ arms. Pulling arm p_k corresponds to evaluating p_k by running a simulation in the epidemiological model $E(c_0, p_k)$. The bandit thus has preventive strategies as arms with reward distributions corresponding to the outcome distribution of a stochastic epidemiological model $E(c_0, p_k)$. While the parameters of the reward distribution are known (i.e., the parameters of the epidemiological model), it is intractable to determine the optimal reward analytically. Hence, we must learn about the outcome distribution via interaction with the epidemiological model. In this work, we consider prevention strategies of equal financial cost, which is a realistic assumption, as governments typically operate within budget constraints.

4.2   Outcome Distribution

As previously defined, the reward distribution associated with a bandit's arm corresponds to the outcome distribution of the epidemiological model that is evaluated when pulling that arm. Therefore, we are able to specify prior knowledge about the reward distribution using epidemiological modeling theory.

It is well known that a disease outbreak has two possible outcomes: either it is able to spread beyond a local context and becomes a fully established epidemic or it fades out [32]. Most stochastic epidemiological models reflect this reality and hence its epidemic size distribution is bimodal [32]. When evaluating preventive strategies, the objective is to determine the preventive strategy that is most

---

[1] Code is available at https://github.com/plibin-vub/bandits.

suitable to mitigate an established epidemic. As in practice we can only observe and act on established epidemics, epidemics that faded out in simulation would bias this evaluation. Consequently, it is necessary to focus on the mode of the distribution that is associated with the established epidemic. Therefore we censor (i.e., discard) the epidemic sizes that correspond to the faded epidemic. The size distribution that remains (i.e., the one that corresponds with the established epidemic) is approximately Gaussian [3].

In this study, we consider a scaled epidemic size distribution, i.e., the proportion of symptomatic infections. Hence we can assume bimodality of the full size distribution and an approximately Gaussian size distribution of the established epidemic. We verified experimentally that these assumptions hold for all the reward distributions that we observed in our experiments (see Sect. 5).

To censor the size distribution, we use a threshold that represents the number of infectious individuals that are required to ensure an outbreak will only fade out with a low probability.

4.3   Epidemic Fade-Out Threshold

For heterogeneous host populations (i.e., a population with a significant variance among individual transmission rates, as is the case for influenza epidemics [9,14]), the number of secondary infections can be accurately modeled using a negative binomial *offspring distribution* $NB(R_0, \gamma)$ [27], where R_0 is the basic reproductive number (i.e., the number of infections that is, by average, generated by one single infection) and γ is a dispersion parameter that specifies the extent of heterogeneity. The probability of epidemic extinction p_{ext} can be computed by solving $g(s) = s$, where $g(s)$ is the probability generating function (pgf) of the offspring distribution [27]. For an epidemic where individuals are targeted with preventive measures (e.g., vaccination), we obtain the following pgf

$$g(s) = pop_c + (1 - pop_c)\big(1 + \frac{R_0}{\gamma}(1 - s)\big)^{-\gamma} \tag{2}$$

where pop_c signifies the random proportion of controlled individuals [27]. From p_{ext} we can compute a threshold T_0 to limit the probability of extinction to a cutoff ℓ [19].

4.4   Best-Arm Identification with a Fixed Budget

Our objective is to identify the best preventive strategy (i.e., the strategy that minimizes the expected outcome) out of a set of preventive strategies, for a particular configuration $c_0 \in \mathcal{C}$ using a fixed budget T of model evaluations. To find the best prevention strategy, it suffices to focus on the mean of the outcome distribution, as it is approximately Gaussian with an unknown yet small variance [3], as we confirm in our experiments (see Fig. 1).

Successive Rejects was the first algorithm to solve the best-arm identification in a fixed budget setting [1]. For a K-armed bandit, Successive Rejects operates in $(K - 1)$ phases. At the end of each phase, the arm with the lowest average reward is discarded. Thus, at the end of phase $(K - 1)$ only one arm survives, and this arm is recommended.

Successive Rejects serves as a useful baseline, however, it has no support to incorporate any prior knowledge. Bayesian best-arm identification algorithms on the other hand, are able to take into account such knowledge by defining an appropriate prior and posterior on the arms' reward distribution. As we will show, such prior knowledge can increase the best-arm identification accuracy. Additionally, at the time an arm is recommended, the posteriors contain valuable information that can be used to formulate a variety of statistics helpful to assist decision makers. We consider two state-of-the-art Bayesian algorithms: BayesGap [21] and Top-two Thompson sampling [31]. For Top-two Thompson sampling, we derive a statistic based on the posteriors to inform decision makers about the confidence of an arm recommendation: the probability of success.

As we established in the previous section, each arm of our bandit has a reward distribution that is approximately Gaussian with unknown mean and variance. For the purpose of genericity, we assume an uninformative Jeffreys prior $(\sigma_k)^{-3}$ on (μ_k, σ_k^2), which leads to the following posterior on μ_k at the n_k^{th} pull [22]:

$$\sqrt{\frac{n_k^2}{S_{k,n_k}}}(\mu_k - \overline{x}_{k,n_k}) \mid \overline{x}_{k,n_k}, S_{k,n_k} \sim \mathcal{T}_{n_k} \tag{3}$$

where $\overline{x}_{k,n_k}$ is the reward mean, S_{k,n_k} is the total sum of squares and $\mathcal{T}_{n_k}$ is the standard student t-distribution with n_k degrees of freedom.

BayesGap is a gap-based Bayesian algorithm [21]. The algorithm requires that for each arm A_k, a high-probability upper bound $U_k(t)$ and lower bound $L_k(t)$ is defined on the posterior of μ_k at each time step t. Using these bounds, the gap quantity

$$B_k(t) = \max_{l \neq k} U_l(t) - L_k(t) \tag{4}$$

is defined for each arm A_k. $B_k(t)$ represents an upper bound on the simple regret (as defined in Sect. 2.3). At each step t of the algorithm, the arm $J(t)$ that minimizes the gap quantity $B_k(t)$ is compared to the arm $j(t)$ that maximizes the upper bound $U_k(t)$. From $J(t)$ and $j(t)$, the arm with the highest confidence diameter $U_k(t) - L_k(t)$ is pulled. The reward that results from this pull is observed and used to update A_k's posterior. When the budget is consumed, the arm

$$J(\operatorname*{argmin}_{t \leq T} B_{J(t)}(t)) \tag{5}$$

is recommended. This is the arm that minimizes the simple regret bound over all times $t \leq T$.

In order to use BayesGap to evaluate preventive strategies, we contribute problem-specific bounds. Given our posteriors (Eq. 3), we define

$$\begin{aligned} U_k(t) &= \hat{\mu}_k(t) + \beta \hat{\sigma}_k(t) \\ L_k(t) &= \hat{\mu}_k(t) - \beta \hat{\sigma}_k(t) \end{aligned} \tag{6}$$

where $\hat{\mu}_k(t)$ and $\hat{\sigma}_k(t)$ are the respective mean and standard deviation of the posterior of arm A_k at time step t, and β is the exploration coefficient.

The amount of exploration that is feasible given a particular bandit game, is proportional to the available budget, and inversely proportional to the game's complexity [21]. This complexity can be modeled taking into account the game's hardness [1] and the variance of the rewards. We use the hardness quantity defined in [21]:

$$H_\epsilon = \sum_k H_{k,\epsilon}^{-2} \tag{7}$$

with arm-dependent hardness

$$H_{k,\epsilon} = \max(\frac{1}{2}(\Delta_k + \epsilon), \epsilon), \text{ where } \Delta_k = \max_{l \neq k}(\mu_l) - \mu_k \tag{8}$$

Considering the budget T, hardness H_ϵ and a generalized reward variance σ_G^2 over all arms, we define

$$\beta = \sqrt{\frac{T - 3K}{4H_\epsilon \sigma_G^2}} \tag{9}$$

Theorem 1 in the Supplementary Information (Sect. 2) formally proves that using these bounds results in a probability of simple regret that asymptotically reaches the exponential lower bound of [21].

As both H_ϵ and σ_G^2 are unknown, in order to compute β, these quantities need to be estimated. Firstly, we estimate H_ϵ's upper bound $\hat{H}_\epsilon$ by estimating Δ_k as follows

$$\hat{\Delta}_k = \max_{1 \leq l < K; l \neq k} (\hat{\mu}_l(t) + 3\hat{\sigma}_l(t)) - (\hat{\mu}_k(t) - 3\hat{\sigma}_k(t)) \tag{10}$$

as in [21], where $\hat{\mu}_k(t)$ and $\hat{\sigma}_k(t)$ are the respective mean and standard deviation of the posterior of arm A_k at time step t. Secondly, for σ_G^2 we need a measure of variance that is representative for the reward distribution of all arms. To this end, when the arms are initialized, we observe their sample variance s_k^2, and compute their average $\bar{s}_G^2$.

As our bounds depend on the standard deviation $\hat{\sigma}_k(t)$ of the t-distributed posterior, each arm's posterior needs to be initialized 3 times (i.e., by pulling the arm) to ensure that $\hat{\sigma}_k(t)$ is defined, this initialization also ensures proper posteriors [22].

Top-two Thompson sampling is a reformulation of the Thompson sampling algorithm, such that it can be used in a pure-exploration context [31]. Thompson sampling operates directly on the arms' posterior of the reward distribution's mean μ_k. At each time step, Thompson sampling obtains one sample for each arm's posterior. The arm with the highest sample is pulled, and its reward is subsequently used to update that arm's posterior. While this approach has been proven highly successful to minimize cumulative regret [8,22], as it balances the exploration-exploitation trade-off, it is sub-optimal to identify the best arm [4]. To adapt Thompson sampling to minimize simple regret, Top-two Thompson sampling increases the amount of exploration. To this end, an exploration probability ω needs to be specified. At each time step, one sample is obtained for each arm's posterior. The arm A_{top} with the highest sample is only pulled with probability ω. With probability $1 - \omega$ we repeat sampling from the posteriors until we find an arm $A_{\text{top-2}}$ that has the highest posterior sample and where $A_{\text{top}} \neq A_{\text{top-2}}$. When the arm $A_{\text{top-2}}$ is found, it is pulled and the observed reward is used to update the posterior of the pulled arm. When the available budget is consumed, the arm with the highest average reward is recommended.

As Top-two Thompson sampling only requires samples from the arms' posteriors, we can use the t-distributed posteriors from Eq. 3 as is. To avoid improper posteriors, each arm needs to be initialized 2 times [22].

As specified in the previous subsection, the reward distribution is censored. We observe each reward, but only consider it to update the arm's value when it exceeds the threshold T_0 (i.e., when we receive a sample from the mode of the epidemic that represents the established epidemic).

4.5   Probability of Success

The probability that an arm recommendation is correct presents a useful confidence statistic to support policy makers with their decisions. As Top-two Thompson sampling recommends the arm with the highest average reward, and we assume that the arm's reward distributions are independent, the probability of success is

$$P(\mu_J = \max_{1 \leq k \leq K} \mu_k) = \int_{x \in \mathbb{R}} \Big[\prod_{\substack{k \neq J}}^{K} F_{\mu_k}(x) \Big] f_{\mu_J}(x) dx \tag{11}$$

where μ_J is the random variable that represents the mean of the recommended arm's reward distribution, f_{μ_J} is the recommended arm's posterior probability density function and F_{μ_k} is the other arms' cumulative density function (full derivation in Supplementary Information, Sect. 3). As this integral cannot be computed analytically, we estimate it using Gaussian quadrature.

It is important to note that, while aiming for generality, we made some conservative assumptions: the reward distributions are approximated as Gaussian and the uninformative Jeffreys prior is used. These assumptions imply that the derived probability of success will be an under-estimator for the actual recommendation success, which is confirmed in our experiments.

5   Experiments

We composed and performed an experiment in the context of pandemic influenza, where we analyze the mitigation strategy to vaccinate a population when only a limited number of vaccine doses is available (details about the rationale behind this scenario in Sect. 2.1). In our experiments, we accommodate a realistic setting to evaluate vaccine allocation, where we consider a large and realistic social network and a wide range of R_0 values.

We consider the scenario when a pandemic is emerging in a particular geographical region and vaccines becomes available, albeit in a limited number of doses. When the number of vaccine doses is limited, it is imperative to identify an optimal vaccine allocation strategy [28]. In our experiment, we explore the allocation of vaccines over five different age groups, that can be easily approached by health policy officials: pre-school children, school-age children, young adults, older adults and the elderly, as proposed in [6].

5.1   Influenza Model and Configuration

The epidemiological model used in the experiments is the FluTE stochastic individual-based model. In our experiment we consider the population of Seattle (United States) that includes 560,000 individuals [6]. This population is realistic both with respect to the number of individuals and its community structure, and provides an adequate setting for the validation of vaccine strategies [34] (more detail about the model choice in the Supplementary Information, Sect. 4).

At the first day of the simulated epidemic, 10 random individuals are seeded with an infection (more detail about the seeding choice in the Supplementary Information, Sect. 5). The epidemic is simulated for 180 days, during which no more infections are seeded. Thus, all new infections established during the run time of the simulation, result from the mixing between infectious and susceptible individuals. We assume no pre-existing immunity towards the circulating virus variant. We choose the number of vaccine doses to allocate to be approximately 4.5% of the population size [28].

We perform our experiment for a set of R_0 values within the range of 1.4 to 2.4, in steps of 0.2. This range is considered representative for the epidemic potential of influenza pandemics [2,28]. We refer to this set of R_0 values as $\mathcal{R}_0$.

Note that the setting described in this subsection, in conjunction with a particular R_0 value, corresponds to a model configuration (i.e., $c_0 \in \mathcal{C}$).

The computational complexity of FluTE simulations depends both on the size of the susceptible population and the proportion of the population that becomes infected. For the population of Seattle, the simulation run time was up to $11\frac{1}{2}$ min (median of $10\frac{1}{2}$ min, standard deviation of 6 s), on state-of-the-art hardware (details in Supplementary Information, Sect. 6).

5.2   Formulating Vaccine Allocation Strategies

We consider 5 age groups to which vaccine doses can be allocated: pre-school children (i.e., 0–4 years old), school-age children (i.e., 5–18 years old), young adults (i.e., 19–29 years old), older adults (i.e., 30–64 years old) and the elderly (i.e., >65 years old) [6]. An allocation scheme can be encoded as a Boolean 5-tuple, where each position in the tuple corresponds to the respective age group. The Boolean value at a particular position in the tuple denotes whether vaccines should be allocated to the respective age group. When vaccines are to be allocated to a particular age group, this is done proportional to the size of the population that is part of this age group [28]. To decide on the best vaccine allocation strategy, we enumerate all possible combinations of this tuple.

5.3   Outcome Distributions

To establish a proxy for the ground truth concerning the outcome distributions of the 32 considered preventive strategies, all strategies were evaluated 1000 times, for each of the R_0 values in $\mathcal{R}_0$. We will use this ground truth as a reference to validate the correctness of the recommendations obtained throughout our experiments.

$\mathcal{R}_0$ presents us with an interesting evaluation problem. To demonstrate this, we visualize the outcome distribution for $R_0 = 1.4$ and for $R_0 = 2.4$ in Fig. 1 (the outcome distributions for the other R_0 values are shown in Sect. 7 of the Supplementary Information). Firstly, we observe that for different values of R_0, the distances between top arms' means differ. Additionally, outcome distribution variances vary over the set of R_0 values in $\mathcal{R}_0$. These differences produce distinct levels of evaluation hardness (see Sect. 4.4), and demonstrate the setting's usefulness as benchmark to evaluate preventive strategies. While we discuss the hardness of the experimental settings under consideration, it is important to state that our best-arm identification framework requires no prior knowledge on the problem's hardness. Secondly, we expect the outcome distribution to be bimodal. However, the probability to sample from the mode of the outcome distribution that represents the non-established epidemic decreases as R_0 increases [27]. This expectation is confirmed when we inspect Fig. 1, the left panel shows a bimodal distribution for $R_0 = 1.4$, while the right panel shows a unimodal outcome distribution for $R_0 = 2.4$, as only samples from the established epidemic were obtained.

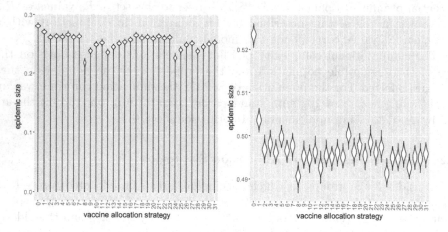

Fig. 1. Violin plot that depicts the density of the outcome distribution (i.e., epidemic size) for 32 vaccine allocation strategies (left panel $R_o = 1.4$, right panel $R_o = 2.4$).

Our analysis identified that the best vaccine allocation strategy was $\langle 0, 1, 0, 0, 0 \rangle$ (i.e., allocate vaccine to school children, strategy 8) for all R_0 values in $\mathcal{R}_0$.

5.4   Best-Arm Identification Experiment

To assess the performance of the different best-arm identification algorithms (i.e., Successive Rejects, BayesGap and Top-two Thompson sampling) we run each algorithm for all budgets in the range of 32 to 500. This evaluation is performed on the influenza bandit game that we defined earlier. For each budget, we run the algorithms 100 times, and report the recommendation success rate. In the previous section, the optimal vaccine allocation strategy was identified to be $\langle 0, 1, 0, 0, 0 \rangle$ for all R_0 in $\mathcal{R}_0$. We thus consider a recommendation to be correct when it equals this vaccine allocation strategy.

We evaluate the algorithm's performance with respect to each other and with respect to uniform sampling, the current state-of-the art to evaluate preventive strategies. The uniform sampling method pulls arm A_u for each step t of the given budget T, where A_u's index u is sampled from the uniform distribution $\mathcal{U}(1, K)$. To consider different levels of hardness, we perform this analysis for each R_0 value in $\mathcal{R}_0$.

For the Bayesian best-arm identification algorithms, the prior specifications are detailed in Sect. 4.4. BayesGap requires an upper and lower bound that is defined in terms of the used posteriors. In our experiments, we use upper bound $U_k(t)$ and lower bound $L_k(t)$ that were established in Sect. 4.4. Top-two Thompson sampling requires a parameter that modulates the amount of exploration ω. As it is important for best-arm identification algorithms to differentiate between the top two arms, we choose $\omega = 0.5$, such that, in the limit, Top-two Thompson sampling will explore the top two arms uniformly.

We censor the reward distribution based on the threshold T_0 we defined in Sect. 4.3. This threshold depends on basic reproductive number R_0 and dispersion parameter γ. R_0 is defined explicitly for each of our experiments. For the dispersion parameter we choose $\gamma = 0.5$, which is a conservative choice according to the literature [9, 14]. We define the probability cutoff $\ell = 10^{-10}$.

Figure 2 shows recommendation success rate for each of the best-arm identification algorithms for $R_0 = 1.4$ (left panel) and $R_0 = 2.4$ (right panel). The results for the other R_0 values are visualized in Sect. 8 of the Supplementary Information. To complement these results, we show the recommendation success rate with confidence intervals in Sect. 9 of the Supplementary Information. The results for different values of R_0 clearly indicate that our selection of best-arm identification algorithms significantly outperforms the uniform sampling method. Overall, the uniform sampling method requires more than double the amount of evaluations to achieve a similar recommendation performance. For the harder problems (e.g., setting with $R_0 = 2.4$), recommendation uncertainty remains considerable even after consuming 3 times the budget required by Top-two Thompson sampling.

All best-arm identification algorithms require an initialization phase in order to output a well-defined recommendation. Successive Rejects needs to pull each arm at least once, while Top-two Thompson sampling and BayesGap need to pull each arm respectively 2 and 3 times (details in Sect. 4.4). For this reason, these algorithms' performance can only be evaluated after this initialization phase. BayesGap's performance is on par with Successive Rejects, except for the hardest setting we studied (i.e., $R_0 = 2.4$). In comparison, Top-two Thompson sampling consistently outperforms Successive Rejects 30 pulls after the initialization phase. Top-two Thompson sampling needs to initialize each arm's posterior with 2 pulls, i.e., double the amount of uniform sampling and Successive Rejects. However, our experiments clearly show that none of the other algorithms reach any acceptable recommendation rate using less than 64 pulls.

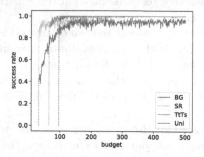 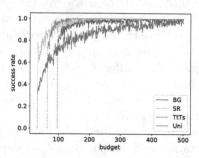

Fig. 2. In this figure, we present the results for the experiment with $R_0 = 1.4$ (left panel) and $R_0 = 2.4$ (right panel). Each curve represents the rate of successful arm recommendations (y-axis) for a range of budgets (x-axis). A curve is shown for each of the considered algorithms: BayesGap (legend: BG), Successive Rejects (legend: SR), Top-two Thompson sampling (legend: TtTs) and uniform sampling (legend: Uni).

In Sect. 4 we derived a statistic to express the probability of success (P_s) concerning a recommendation made by Top-two Thompson sampling. We analyzed this probability for all the Top-two Thompson sampling recommendations that were obtained in the experiment described above. To provide some insights on how this statistic can be used to support policy makers, we show the P_s values of all Top-two Thompson sampling recommendations for $R_0 = 2.4$ in the left panel of Fig. 3 (Figures for the other R_0 values in Sect. 10 of the Supplementary Information). This Figure indicates that P_s closely follows recommendation correctness and that the uncertainty of P_s is inversely proportional to the size of the available budget. Additionally, in the right panel of Fig. 3 (Figures for the other R_0 values in Sect. 11 of the Supplementary Information) we confirm that P_s underestimates recommendation correctness. These observations show that P_s has the potential to serve as a conservative statistic to inform policy makers about the confidence of a particular recommendation, and thus can be used to define meaningful cutoffs to guide policy makers in their interpretation of the recommendation of preventive strategies.

In this work, we define uninformed priors to ensure a generic framework. This does not exclude decision makers to use priors that include more domain knowledge (e.g., dependence between arms), if this is available. We do however show in our experiments that the use of these uninformed priors lead to a significant performance increase.

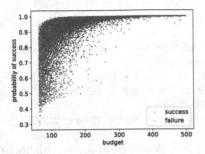 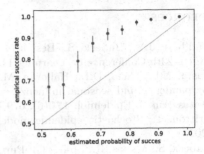

Fig. 3. Top-two Thompson sampling was run 100 times for each budget for the experiment with $R_0 = 2.4$. For each of the recommendations, P_s was computed. In the left panel, these P_s values are shown as a scatter plot, where each point's color reflects the correctness of the recommendation (see legend). In the right panel, the P_s values were binned (i.e., 0.5 to 1 in steps of 0.05). Per bin, we thus have a set of Bernoulli trials, for which we show the empirical success rate (blue scatter) and the Clopper-Pearson confidence interval (blue confidence bounds). The orange reference line denotes perfect correlation between the empirical success rate and the estimated probability of success. (Color figure online)

6 Conclusion

We formulate the objective to select the best preventive strategy in an individual-based model as a fixed budget best-arm identification problem. We set up an experiment to evaluate this setting in the context of a realistic influenza pandemic. To assess the best arm recommendation performance of the preventive bandit, we report a success rate over 100 independent bandit runs.

We demonstrate that it is possible to efficiently identify the optimal preventive strategy using only a limited number of model evaluations, even if there is a large number of preventive strategies to consider. Compared to uniform sampling, our technique is able to recommend the best preventive strategy reducing the number of required model evaluations 2-to-3 times, when using Top-two Thompson sampling. Additionally, we defined a statistic to support policy makers with their decisions, based on the posterior information obtained during Top-two Thompson sampling. As such, we present a decision support tool to assist policy makers to mitigate epidemics. Our framework will enable the use of individual-based models in studies where it would otherwise be computationally too prohibitive, and allow researchers to explore a wider variety of model scenarios.

Acknowledgments. Pieter Libin and Timothy Verstraeten were supported by a PhD grant of the FWO (Fonds Wetenschappelijk Onderzoek - Vlaanderen). Kristof Theys, Jelena Grujic and Diederik Roijers were supported by a postdoctoral grant of the FWO. The computational resources were provided by an EWI-FWO grant (Theys, KAN2012 1.5.249.12.). We thank the anonymous reviewers for their insightful comments that allowed us to improve this work.

References

1. Audibert, J.Y., Bubeck, S.: Best arm identification in multi-armed bandits. In: COLT-23th Conference on Learning Theory (2010)
2. Basta, N.E., Chao, D.L., Halloran, M.E., Matrajt, L., Longini, I.M.: Strategies for pandemic and seasonal influenza vaccination of schoolchildren in the United States. Am. J. Epidemiol. **170**(6), 679–686 (2009)
3. Britton, T.: Stochastic epidemic models: a survey. Math. Biosci. **225**(1), 24–35 (2010)
4. Bubeck, S., Munos, R., Stoltz, G.: Pure exploration in multi-armed bandits problems. In: Gavaldà, R., Lugosi, G., Zeugmann, T., Zilles, S. (eds.) ALT 2009. LNCS (LNAI), vol. 5809, pp. 23–37. Springer, Heidelberg (2009). https://doi.org/10.1007/978-3-642-04414-4_7
5. Bubeck, S., Munos, R., Stoltz, G.: Pure exploration in finitely-armed and continuous-armed bandits. Theor. Comput. Sci. **412**(19), 1832–1852 (2011)
6. Chao, D.L., Halloran, M.E., Obenchain, V.J., Longini Jr., I.M.: FluTE, a publicly available stochastic influenza epidemic simulation model. PLoS Comput. Biol. **6**(1), e1000656 (2010)
7. Chao, D.L., Halstead, S.B., Halloran, M.E., Longini, I.M.: Controlling Dengue with Vaccines in Thailand. PLoS Negl. Trop. Dis. **6**(10), e1876 (2012)
8. Chapelle, O., Li, L.: An empirical evaluation of Thompson sampling. In: Advances in Neural Information Processing Systems, pp. 2249–2257 (2011)
9. Dorigatti, I., Cauchemez, S., Pugliese, A., Ferguson, N.M.: A new approach to characterising infectious disease transmission dynamics from sentinel surveillance: application to the Italian 2009/2010 A/H1N1 influenza pandemic. Epidemics **4**(1), 9–21 (2012)
10. Enserink, M.: Crisis underscores fragility of vaccine production system. Science **306**(5695), 385 (2004)
11. Eubank, S., Kumar, V., Marathe, M., Srinivasan, A., Wang, N.: Structure of social contact networks and their impact on epidemics. DIMACS Ser. Discrete Math. Theor. Comput. Sci **70**(0208005), 181 (2006)
12. Even-Dar, E., Mannor, S., Mansour, Y.: Action elimination and stopping conditions for the multi-armed bandit and reinforcement learning problems. J. Mach. Learn. Res. **7**(Jun), 1079–1105 (2006)
13. Ferguson, N.M., Cummings, D.A.T., Cauchemez, S., Fraser, C.: Others: strategies for containing an emerging influenza pandemic in Southeast Asia. Nature **437**(7056), 209 (2005)
14. Fraser, C., Cummings, D.A.T., Klinkenberg, D., Burke, D.S., Ferguson, N.M.: Influenza transmission in households during the 1918 pandemic. Am. J. Epidemiol. **174**(5), 505–514 (2011)
15. Fumanelli, L., Ajelli, M., Merler, S., Ferguson, N.M., Cauchemez, S.: Model-based comprehensive analysis of school closure policies for mitigating influenza epidemics and pandemics. PLoS Comput. Biol. **12**(1), e1004681 (2016)
16. Garivier, A., Kaufmann, E.: Optimal best arm identification with fixed confidence. In: Conference on Learning Theory, pp. 998–1027 (2016)
17. Germann, T.C., Kadau, K., Longini, I.M., Macken, C.A.: Mitigation strategies for pandemic influenza in the United States. Proc. Nat. Acad. Sci. U.S.A. **103**(15), 5935–5940 (2006)
18. Halloran, M.E., Longini, I.M., Nizam, A., Yang, Y.: Containing bioterrorist smallpox. Science (New York, N.Y.) **298**(5597), 1428–1432 (2002)

19. Hartfield, M., Alizon, S.: Introducing the outbreak threshold in epidemiology. PLoS Pathog **9**(6), e1003277 (2013)
20. Herbert, R.: Some aspects of the sequential design of experiments. Bull. Am. Math. Soc. **58**(5), 527–535 (1952)
21. Hoffman, M., Shahriari, B., Freitas, N.: On correlation and budget constraints in model-based bandit optimization with application to automatic machine learning. In: Artificial Intelligence and Statistics, pp. 365–374 (2014)
22. Honda, J., Takemura, A.: Optimality of Thompson sampling for Gaussian bandits depends on priors. In: AISTATS, pp. 375–383 (2014)
23. Jennison, C., Johnstone, I.M., Turnbull, B.W.: Asymptotically optimal procedures for sequential adaptive selection of the best of several normal means. Stat. Decis. Theory Relat. Top. III **2**, 55–86 (1982)
24. Kaufmann, E., Cappé, O., Garivier, A.: On the complexity of best arm identification in multi-armed bandit models. J. Mach. Learn. Res. **17**(1), 1–42 (2016)
25. Kaufmann, E., Kalyanakrishnan, S.: Information complexity in bandit subset selection. In: Conference on Learning Theory, pp. 228–251 (2013)
26. Libin, P., Verstraeten, T., Theys, K., Roijers, D.M., Vrancx, P., Nowé, A.: Efficient evaluation of influenza mitigation strategies using preventive bandits. In: Sukthankar, G., Rodriguez-Aguilar, J.A. (eds.) AAMAS 2017. LNCS (LNAI), vol. 10643, pp. 67–85. Springer, Cham (2017). https://doi.org/10.1007/978-3-319-71679-4_5
27. Lloyd-Smith, J.O., Schreiber, S.J., Kopp, P.E., Getz, W.M.: Superspreading and the effect of individual variation on disease emergence. Nature **438**(7066), 355–359 (2005)
28. Medlock, J., Galvani, A.P.: Optimizing influenza vaccine distribution. Science **325**(5948), 1705–1708 (2009)
29. Paules, C., Subbarao, K.: Influenza. The Lancet (2017)
30. Powell, W.B., Ryzhov, I.O.: Optimal Learning, vol. 841. Wiley, Hoboken (2012)
31. Russo, D.: Simple Bayesian algorithms for best arm identification. In: Conference on Learning Theory, pp. 1417–1418 (2016)
32. Watts, D.J., Muhamad, R., Medina, D.C., Dodds, P.S.: Multiscale, resurgent epidemics in a hierarchical metapopulation model. Proc. Nat. Acad. Sci. U.S.A. **102**(32), 11157–11162 (2005)
33. WHO: WHO guidelines on the use of vaccines and antivirals during influenza pandemics (2004)
34. Willem, L., Stijven, S., Vladislavleva, E., Broeckhove, J., Beutels, P., Hens, N.: Active learning to understand infectious disease models and improve policy making. PLoS Comput. Biol. **10**(4), e1003563 (2014)
35. Wu, J.T., Riley, S., Fraser, C., Leung, G.M.: Reducing the impact of the next influenza pandemic using household-based public health interventions. PLoS Med. **3**(9), e361 (2006)

Hypotensive Episode Prediction in ICUs via Observation Window Splitting

Elad Tsur[1], Mark Last[1(✉)], Victor F. Garcia[2], Raphael Udassin[3], Moti Klein[4], and Evgeni Brotfain[4]

[1] Department of Software and Information Systems Engineering,
Ben-Gurion University of the Negev, 84105 Beer-Sheva, Israel
eladtsur@gmail.com, mlast@bgu.ac.il
[2] Division of Pediatric Surgery, MLC 2023, Children's Hospital Medical Center,
3333 Burnet Avenue, Cincinnati, OH 45229, USA
victor.garcia@cchmc.org
[3] Pediatric Surgery Department, Hadassah University Hospital,
Ein-Karem, 9112001 Jerusalem, Israel
raphaelu@ekmd.huji.ac.il
[4] General Intensive Care Unit, Soroka Medical Center, Beer Sheva, Israel
{MotiK,EvgeniBr}@clalit.org.il

Abstract. Hypotension, defined as dangerously low blood pressure, is a significant risk factor in intensive care units (ICUs), which requires a prompt therapeutic intervention. The goal of our research is to predict an impending Hypotensive Episode (HE) by time series analysis of continuously monitored physiological vital signs. Our prognostic model is based on the last Observation Window (OW) at the prediction time. Existing clinical episode prediction studies used a single OW of 5–120 min to extract predictive features, with no significant improvement reported when longer OWs were used. In this work we have developed the *In-Window Segmentation* (InWiSe) method for time series prediction, which splits a single OW into several sub-windows of equal size. The resulting feature set combines the features extracted from each observation sub-window and then this combined set is used by the Extreme Gradient Boosting (XGBoost) binary classifier to produce an episode prediction model. We evaluate the proposed approach on three retrospective ICU datasets (extracted from MIMIC II, Soroka and Hadassah databases) using cross-validation on each dataset separately, as well as by cross-dataset validation. The results show that InWiSe is superior to existing methods in terms of the area under the ROC curve (AUC).

Keywords: Time series analysis · Clinical episode prediction
Feature extraction · Intensive care · Patient monitoring

Partially supported by the Cincinnati Children's Hospital Medical Center; In collaboration with Soroka Medical Center in Beer-Sheva and Hadassah University Hospital, Ein Karem, Jerusalem

U. Brefeld et al. (Eds.): ECML PKDD 2018, LNAI 11053, pp. 472–487, 2019.
https://doi.org/10.1007/978-3-030-10997-4_29

1   Introduction

Hypotension is defined as dangerously low blood pressure. It is a major hemodynamic instability symptom, as well as a significant risk factor in hospital mortality at intensive care units (ICUs) [1]. As a critical condition, which may result
in a fatal deterioration, an impending Hypotensive Episode (HE) requires a
prompt therapeutic intervention [2] by ICU clinicians. However, HE prediction
is a challenging task [3]. While the clinical staff time is limited, the amount
of accumulated physiologic data per patient is massive in terms of both data
variety (multi-channel waveforms, laboratory results, medication records, nurs-
ing notes, etc.) and data volume (length of waveform time series). Even with
sufficient time, resources, and data, it is very hard to accurately estimate the
likelihood of clinical deterioration with bare-eye analysis alone.

HE may be detectable in advance by automatic analysis of continuously
monitored physiologic data; more specifically, the analysis of vital signs (multi-
parameter temporal vital data), may inform on the underlying dynamics of
organs and cardiovascular system functioning. Particularly, vital signs may contain subtle patterns which point to an impending instability [4]. Such pattern
identification is a suitable task for machine learning algorithms. Smart patient
monitoring software that could predict the clinical deterioration of high risk
patients well before there are changes in the parameters displayed by the current ICU monitors would save lives, reduce hospitalization costs, and contribute
to better patient outcomes [5].

Our research goal is to give the physicians an early warning of an impending
HE by building a prediction model, which utilizes the maximal amount of information from the currently available patient monitoring data and outperforms
state-of-the-art HE prediction systems. We present and evaluate the *In-Window
Segmentation* (InWiSe) algorithm for HE prediction, which extracts predictive
features from a set of multiple observation sub-windows rather than from a single
long observation window.

This paper is organized as follows. Section 2 surveys the previous works in
several related areas, elaborates on the limitations of these works and introduces
the contributions of our method. Section 3 describes the studied problem and
proposed methods in detail and Sect. 4 covers the results of an empirical evaluation. Finally, Sect. 5 presents the conclusions along with possible directions for
future research.

2   Related Work and Original Contributions

Several works studied the problem of clinical deterioration prediction in ICUs.
This section reviews their problem definitions, feature extraction methods, sliding window constellations, and prediction methodologies. Finally, a discussion
of the limitations of existing methods is followed by a presentation of the contributions of this study.

2.1   Clinical Episode Definitions

Previous works on clinical deterioration prediction vary mainly in two aspects [6]. The first one is an episode definition, which may be based on the recorded clinical treatment or on the behavior of vital signs within a specific time interval. The second one is the warning time, a.k.a. the *Gap Window*, which will be called in brief the *gap* in this study.

The objective in [3] was to predict the hemodynamic instability start time with a 2-h gap. The episode start time was defined by a clinical intervention recorded in the ICU clinical record of a patient. In [7], instability was also defined by some given medications and gaps of 15 min to 12 h were explored.

The 10th annual PhysioNet/Computers in Cardiology Challenge [4] conducted a competition to study an Acute Hypotensive Episode (AHE). They defined AHE as an interval, in which at least 90% of the time the Mean Arterial blood Pressure (MAP) is under 60 mmHg during any 30-min window within the interval. Their goal was to predict whether an AHE will start in the next 60 min. In [1,8], the HE and AHE definitions were identical to [4], but a lower MAP bound of 10 mmHg was added to prevent noise effects from outliers. Their goal was to predict the patient condition in a *Target Window* (called herein *target*) of 30 min, which occurs within a gap of 1–2 h (See Fig. 1a), and label it as hypotensive or normotensive (normal blood pressure). As expected, and as concluded in [8], the problem is more challenging when predicting further into the future, thus resulting in poorer performance. Note that, as indicated in [8], the accepted HE definitions for adults vary in the range of 60–80 mmHg MAP for 30+ min, where the lowest case of 60 mmHg is sometimes excluded under the definition of AHE [1,4].

2.2   Predictive Feature Types

In most works, the future episode predictive features are usually extracted from a sliding *Observation Window* (OW) over a record, which is a collection of vital sign time series of one patient in a single ICU admission. A minute-by-minute vital signs time series, like blood pressure and Heart Rate (HR) are usually used to extract features, while a few studies used the clinical information (age and temporal medications data) as well. A typically used benchmark database is MIMIC II [9], a multi-parameter ICU waveforms and clinical database.

Statistical features are the most obvious source for the extraction of predictive features, also called patterns, from intervals like OWs. In [5], the authors calculate extremes, moments, percentiles and inter-percentile ranges for every vital sign, whereas in [8] interquartile ranges and slope are extracted as well. In a more pragmatic statistical approach [10], several episode predictive indices were used, derived from the blood pressure signals only. These indices were six types of averages from Systolic Blood Pressure (SBP), Diastolic Blood Pressure (DBP), and MAP, each taken as a single feature. Another statistical approach derives *cross-correlation* features which capture the coupling between two time series by computing the sum of products of their values [8], or by estimating their variance and covariance [5].

A more recent and widely accepted feature extraction approach is the use of *wavelets*, which captures the relative energies in different spectral bands that are localized in both time and frequency. Wavelets were proven to perform well as episode predictors [11] as well as vital sign similarity detectors [12]. In [5] and [8], Daubechies (DB) and Meyer wavelet types were used, respectively, noting that the DB type dominates the basic Haar type wavelets [13] in terms of vital sign time series, which are non-stationary [5].

Apart from vital signs, patient age and vasopressors (blood pressure medications) given during OWs are added as features by Lee and Mark [8] but found to have low correlation with the target. In their other work [15], they achieve similar results without those features. Moreover, Saeed [11] mentions the low reliability of vasopressor medication timestamps, which are very important for the episode prediction task.

2.3 Observation Window Constellations

The sliding OW plays an important role in the episode prediction task. In this section, we survey the different approaches to constructing and collecting OWs.

The first important attribute of an OW is its duration. In [1,3,5,8,10,14], various OW sizes were applied (5, 10, 30, 60, 90, and 120 min). Having implemented a 60-min OW, it is claimed in [1,8] that extracting features from a longer window does not result in improvement of prediction performance.

In [15], Lee and Mark extended their previous work by implementing a weighted decision classifier that consists of four base classifiers, each predicting an HE in the same target but within different gaps (1, 2, 3 and 4 h) using a different corresponding 30-min OW. The final decision is made by weighting the four posterior probabilities from each classifier. They report insignificant improvement in prediction performance as well as independency of predictions from the 3rd and the 4th past hours (the earliest hours).

A second matter is how to collect OWs for training the prediction model. One simple approach, applied by Cao et al. [3], is to compile an OW ending gap-minutes before the start time of the first episode in every unstable record (having one or more HEs). For each stable record, one or more OWs are then sampled randomly. According to [8], collecting multiple OWs from both stable and unstable records in a random fashion, which does not collect windows exactly gap-minutes before an episode start, is proved to outperform the first method of Cao [3]. A sliding target window (with no overlap) traverses each record (Fig. 1a), and as many OWs as possible are compiled. However, they note that collecting OWs all the time, even when no HE is impending, and doing it from both stable and unstable patients will result in an extremely imbalanced dataset. Having two OW classes (hypotensive or normotensive), one way to solve this issue is by undersampling the majority class (normotensive) [5,8].

2.4   Prediction Methodology

The problem of episode prediction is most logically approached by calculating the probability of a future episode at a given time point, using features extracted from the current OW, and then classifying the target window, which starts within some gap-minutes, as hypotensive or not, based on a pre-defined probability threshold. Multiple works tackled this problem by using numerous supervised machine learning algorithms, particularly binary classifiers, with some exceptions such as in [16], where a Recurrent Neural Networks approach is used to forecast the target window MAP values, followed by a straightforward binary decision based on the episode definition.

The classifiers used by some other papers are Logistic Regression [3], Artificial Neural Network [8,15], Majority Vote [1], and Random Forest [5]. To the best of our knowledge, the most accurate HE prediction so far is reported in [8,15], which we reproduce and use for comparison in Sect. 4.

2.5   Limitations of Published Methods

Advanced methods and evaluation schemes such as in [1,5,8,14,15], solved some of the problems found in the early works [3,10], yet left some open issues, including low precision (14% in [8]) and a strict episode definition that is still far from the practical definitions used in ICUs. Moreover, a machine learning solution to a high precision HE prediction will probably need much more training data, while the current MIMIC II [9] contains only several thousands[1] of vital sign records that are matched with the clinical data.

Unfortunately, there is a lack of comprehensive public sources of ICU monitored vital signs beyond the existing MIMIC database. Consequently, the current episode prediction studies miss the crucial cross-dataset validation, which is needed to find a generic model that should work for any ICU, disregarding availability of retrospective patient records for training.

Recent papers [1,5,8] include predictions of future episodes even if the patient is going through an ongoing episode. These predictions may be less relevant to the physicians and possibly excluded from the evaluation metrics.

Finally, studies conducted over the last decade show no improvement in utilizing OWs greater than 120 min (and usually even 60 min), implying there are no additional predictive patterns to be found in the near past. On the contrary, the results from [1,5,7,8,14] show an accuracy decrease of only 1–2.5% when switching from a 60-min gap window to a 120-min one, which may imply that earlier observations may have a just a slightly lower correlation to the target. Thus, there may be additional predictive patterns, which are not utilized properly by the existing methods.

---

[1] Recently, The MIMIC III waveform database *Matched Subset*, four times larger than the MIMIC II subset, was published

2.6   Original Contributions

The main contribution of this paper is the In-Window Segmentation (*InWiSe*) method, which aims to utilize the local predictive patterns in long OWs. The method, presented in Sect. 3.2 and Fig. 1b, differs from previous methods by the following: (i) it extracts hidden local features by splitting the OW into multiple sub-windows, which improves the model predictive performance; (ii) it is flexible in terms of OW definition - if a complete sub-window set is not valid for use at the prediction time, a single OW option is used instead.

As mentioned in Sect. 2.3, a step towards multiple OW utilization was taken in [15] by combining weighted predicted posteriors of four OWs, each making an independent prediction with a distinct gap. Their approach is different from ours mainly in that we let the classifier learn the association between cross-window features, which is not possible in a weighted posterior decision. Another very recent work (DC-Prophet) [19], published while writing this paper, combines features from consecutive time series intervals (lags) to make early predictions of server failures. Their approach is similar to ours, but it has neither been applied to clinical episode prediction, nor it has handled invalid lags.

A further contribution of our work is evaluation of both our method and earlier episode prediction methods in a *cross-dataset* setting, in addition to the in-dataset cross-validation. Finally, our experiments are extended by a new evaluation approach, which excludes the clinically irrelevant in-episode predictions.

3   Methodology

This section starts with the problem definition, continues with introducing InWiSe, and concludes with the description of the data compilation, feature extraction, and classification methods used in this study.

3.1   Problem Definition and Prediction Modes

This study explores the problem of predicting a patient condition (hypotensive or normotensive) within a 60-min gap. Following the work in [1,8], we define a hypotensive episode (HE) as a 30-min target window where at least 90% of MAP values are below 60 mmHg. Any valid target (see the validity definition in Sect. 4.2) not meeting this criterion is labeled as normotensive. At the prediction time, each sub-window set is labeled with respect to its corresponding target.

Considering the implementation of the proposed method in the clinical setting, we distinguish between two alternative prediction modes: (i) *all-time prediction*, where the assumption (found in previous papers) is that episode prediction is needed continuously regardless of the clinical condition at the prediction time; (ii) *exclusive prediction*, where episode prediction is needed only when the patient is not in a currently recognized HE (the last 30 min of the ICU stay are not an HE by definition).

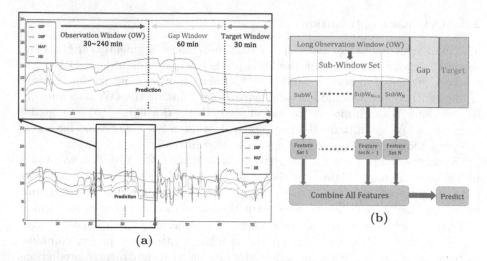

Fig. 1. (a) Basic Method: traversing over a patient record with an impending HE is demonstrated by the observation, gap and target windows with respect to the prediction time. (b) InWiSe: a given OW is split into a sub-window set of size N, followed by a prediction that is based on the combined feature set of the N sub-windows ($SubW$s).

3.2 Splitting Obervation Windows with InWiSe

In our study, which was developed based on the observation-gap-target windows scheme demonstrated in Fig. 1a, we hypothesized that taking longer OWs, splitting them into several equally sized sub-windows, also called the *sub-window set*, and combining all their features together (see Fig. 1b) would improve the predictive accuracy of the induced model versus using a smaller feature set of a single long or short OW. For example, a set of the mean MAP values from four, three, two and one hours before the same target window may be more informative for predicting the target label than the mean MAP value of a single 4-h OW.

The InWiSe method does not use a classifier based on a combined set of features if one of the current sub-windows in the set is invalid (see Sect. 4.2). In that case, the prediction is made by a simpler classification model using only the features extracted from the latest sub-window ($SubW_N$ in Fig. 1b) unless that window is invalid. Consequently, InWiSe misses less prediction points than the single OW method (more details are in Sect. 4.4, in-dataset paragraph).

3.3 Feature Extraction

We use three basic vital signs (SBP, DBP, and HR) to derive two additional vital signs for each record: Pulse Pressure calculated by $PP = SBP - DBP$, and Relative Cardiac Output calculated by $CO = HR \times PP$. Next, we extract the following three groups of features from each sub-window.

Statistical Features: mean, median, standard deviation (Std), variance, interquartile range (Iqr), skewness, kurtosis and linear regression slope are calculated for each of the vital signs. Missing values are ignored.

Wavelet Features: Similarly to [5], multi-level discrete decomposition of each vital sign can be conducted with DB wavelets. The decomposition of a single time series (signal) X is denoted by $W_X = [a_n \ d_n \ d_{n-1} \ \cdots \ d_1]$, where n is the decomposition level (window size depended), a_n is the signal approximation, and d_k is the detail signal of level k. The elements in W_X are then utilized as features by calculating the relative energy for each of them as in [8,15]. Missing values are interpolated.

Cross-Correlation Features: the cross correlation of two time series $X = \{x_1, x_2, \ldots, x_n\}$ and $Y = \{y_1, y_2, \ldots, y_n\}$ is defined as $\rho_{XY} = \frac{1}{n}\Sigma x_i y_i$ and calculated for each pair of vital signs.

The total amount of features extracted from a sub-window set is equal to the number of sub-windows N multiplied by the feature set size.

3.4 Classification

Each instance in the training dataset is composed of a sub-window set feature vector and a class label which is positive or negative (the target is either hypotensive or normotensive, respectively). Before training a binary classifier, we both normalize the training dataset (to zero mean and unit standard deviation) and undersample it to overcome the imbalance issue (Sect. 4.3).

Our classifier produces a posterior probability of the positive class, which may lead to an HE alert depending on the probability threshold determined from the Receiver Operating Characteristic (ROC) curve built on the target (testing) dataset. Following [3,8,15], the following selection criterion for the optimal threshold can be used: $Th_{selected} = \text{argmax}_{Th}\{sensitivity(Th) + specificity(Th)\}$.

4 Experiments

The experimental setup of this study includes multiple prediction modes, methods, and model configurations. We first perform an in-dataset evaluation for each prediction mode (all-time and exclusive) and for each method (single OW and InWiSe). Next, we proceed with a cross-dataset validation for each dataset pair. This section describes the datasets and their compiled OW and window-set statistics, followed by the experiments and analysis of results.

4.1 Data Description

Three databases of adult ICU admission records were prepared for this study: *Soroka* Medical Center in Beer Sheva (4,757 records), *Hadassah* Hospital, Ein-Karem, Jerusalem (8,366 records), and *MIMIC II* [9] (downloaded from [17,

18] and comprising 5, 266 records). All time-series sampling rates are minute-by-minute (some second-by-second MIMIC II records were undersampled by taking each minute median). The common-shared vital signs among the three databases are HR, SBP, DBP, MAP, peripheral capillary oxygen saturation and respiration. Similarly to Lee and Mark [8,15], we included only the HR, SBP, DBP and MAP vital signs in our data.

4.2   Data Compilation

As a pre-processing step, any outlier (out of the rage 10–200 for any vital sign) is considered as a 'missing value'. When compiling OWs from each admission record we used the observation-gap-target window scheme (Sect. 3.2) called the *single OW method*, as well as a first step for InWiSe. The window sizes of the single OW were 30, 60, 120 or 240 min, while the gap and target sizes were constant at 60 and 30 min, respectively. Furthermore, we followed Lee and Mark [8] who claimed that a prediction rate of every 30 min should represent the performance of a real-time system. Therefore, a 30-min. sliding target window was traversed with no overlaps over each record and as many OWs as possible were compiled, depending on the prediction mode. Following [8], targets with more than 10% missing MAP values were excluded from this study, as their true labels are unknown. Turning to OW validity, to prevent a classifier from learning outlier instances, more than 5% missing values for any vital sign made the window invalid and, consequently, excluded it from our work as well.

Five InWiSe configurations of window splitting were selected for this study: $60[m] \rightarrow 2 \times 30[m]$, $120 \rightarrow 2 \times 60$, $120 \rightarrow 4 \times 30$, $240 \rightarrow 2 \times 120$ and $240 \rightarrow 4 \times 60$. For each configuration and for every record in the dataset, at the prediction time, we combine a sub-window set ($[SubW_1, \ldots, SubW_N]$, Fig. 1b) if all N sub-windows are valid. The label of a complete sub-window set is the same as of its latest sub-window, which is labeled according to its corresponding target window.

Following the window label counts, the imbalance ratio for the all-time compilation was found to be 1:20 to 1:40 in favor of normotensive (negative) windows (increasing with the observation window size), as opposed to the exclusive compilation (no in-episode predictions), which was two times more imbalanced. As for the window set method, the bigger the set size N the less positive and negative examples are available. Like in a single OW, we observed an increase in the window-set labeling imbalance with an increase in the window size that reached 1:100 in the exclusive mode.

The reduction of sub-window sets availability with increasing N varied over datasets and was caused by differences in the amount of missing values (i.e., MIMIC II misses more values than Soroka). Moreover, the reason behind cross-dataset differences in terms of total OW count with respect to record count was the variance in ICU stay duration, which was higher in Soroka than in other datasets. Last, we note that using the exclusive mode resulted in a decrease of over 50% in the positive window count, probably because the average HE duration was much longer than 30 min (i.e., Hadassah average HE duration was

98 min), increasing the time intervals where we do not make a prediction under this mode.

4.3   Experimental Setup

In-dataset Evaluation: For each dataset, mode and algorithm a 5-fold cross-validation (CV) was performed. To allow the classifier to successfully learn the imbalanced data, training folds were undersampled (5 times without replacement) to attain equal counts of stable and unstable admission records within each training fold (test folds were left unbalanced). Moreover, for each record all OWs or sub-window sets were either in training or test dataset to prevent record characteristics from leaking into the test dataset. In total, the classifier produced 25 outputs (5 folds × 5 samples) which were evaluated by five metrics: area under the ROC curve (AUC), accuracy, sensitivity, specificity and precision. Furthermore, to compare between the two methods fairly, we optimized the hyper-parameters of each classifier: an inner 5-fold CV was utilized in each undersampled training fold of the outer CV and the best chosen hyper-parameters found by a grid search were used to train the outer fold (Nested CV).

To choose the prediction model, three classifiers were evaluated using a 60-min OW (single OW method) and a 4 × 60-min set (InWiSe) on all datasets combined and in the all-time prediction mode (with hyper parameters optimization). The AUCs were (0.932, 0.936) for Random Forest, (0.937, 0.940) for Artificial Neural Network (ANN) with a single hidden layer of 100 neurons, and (0.939, 0.943) for Extreme Gradient Boosting (XGBoost) [20], where each tuple represents a <single OW, sub-window set> pair. Since XGBoost outperformed Random Forest and ANN with p-values = 0.03 and 0.08, respectively (using t-test), we chose XGBoost for inducing prediction models in this study. Still, ANN was used as a baseline of [8].

XGBoost is a scalable implementation of the Gradient Boosting ensemble method that affords some additional capabilities like feature sampling (in addition to instance sampling) for each tree in the ensemble, making it even more robust to feature dimensionality and helping to avoid overfitting. Moreover, considering our minimum training dataset size of approximately 2k instances together with the maximal feature vector size of 392 features, the built-in feature selection capability of XGBoost is important.

The XGBoost classifier was grid-search optimized for each dataset or mode and for each OW size or sub-window set configuration C, where the best hyper-parameters were reproduced for all datasets, in most of the CV folds. The optimized hyper-parameters were: number of trees (500, **1000**, 1500), maximum tree depth (**3**, 5), learning rate (0.001, **0.01**, 0.1), instance sample rate (0.8, **0.4**), and feature sample rate (0.9, **0.6**). The best choices are shown above in bold.

Finally, each algorithm was tried with several OW sizes and sub-window sets Cs: four OW sizes for the single OW method and five Cs for InWiSe sub-window sets (see Sect. 4.2). As a result, a total of 54 in-dataset CVs were conducted (3 datasets × 2 modes × 9 window-set Cs and single OW sizes).

Cross-Dataset Validation: The model induced from each dataset was evaluated on other datasets using the all-time mode. XGBoost was trained on one full dataset and tested on the two other datasets separately. The source dataset was undersampled only once, justified by a mostly very low variance of AUC ($<0.1\%$) between undersamples, in each fold of the in-dataset CV. Both the single OW size and the window-set C were chosen to optimize the AUC performance in the in-dataset evaluation: 120/240-min sized OW for the single OW method and 4×60-min sub-window set for InWiSe. The hyper-parameters of the classifier (XGBoost) of each method were chosen by a majority vote over the folds in the in-dataset evaluation. A total of 18 experiments were performed (3 training datasets × 2 test datasets × 3 window-set Cs and single OW sizes).

4.4   Analysis of Results

This subsection presents the results, followed by the feature importance analysis. The reader should recall that all sub-window set results include some test instances which were classified using the latest sub-window ($SubW_N$ in Fig. 1b), if valid, in case that the sub-window set was invalid.

In-dataset: As a baseline, we reproduced the single OW method results of Lee and Mark [8] on MIMIC II with ANN. In Fig. 2, we use MIMIC II to compare the single OW method with ANN, using two more methods: single OW with XGBoost (single OW method) and sub-window set best split using XGBoost as well. In comparison with the baseline, the AUC of the 4×60 sub-window set (XGBoost) was significantly higher than for the single OW method with ANN (60, 120 or 240-min OW size) with p-values 0.009 and 0.05 for all-time and exclusive modes, respectively.

From Figs. 2 and 3, we first conclude that in terms of AUC, splitting a single OW into several sub-windows is usually better than using a single OW; we note that, the advantage of OW splitting grows with an increase in the OW duration, which emphasizes the benefit from splitting a single OW that is longer than the longest OWs used by current methods ($240 \rightarrow 4 \times 60$ versus 120 min).

Turning to XGBoost-only comparison on MIMIC II, InWiSe outperformed the single OW in the all-time prediction mode, but only with a p-value of 0.13, while performing only slightly better than the single OW in the exclusive mode (in terms of AUC). However, while these all-time prediction trends are similar in the Soroka dataset where InWiSe is better with $p = 0.15$ (Fig. 3), in the Hadassah dataset InWiSe significantly outperforms the single OW method with a p-value of 0.03. In addition, in Table 1 that shows the in-dataset best results on its diagonal, we see that, although not always statistically significant, the sub-window set method is better than the single OW alternatives in each dataset and in all metrics (in bold) when evaluating in the all-time prediction mode. Note that our significance test comparisons were between the best results of each method rather than sub-window set versus its matching long window and they were calculated with four degrees of freedom (due to five folds).

As for the exclusive mode, the AUC was lower, as expected, since less positive OW instances were available, making the prediction task harder in general.

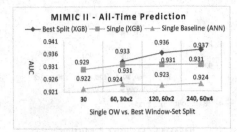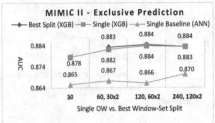

Fig. 2. The single OW method compared with its InWiSe best split, and with the ANN baseline on MIMIC II (all-time mode at the left and exclusive mode to the right).

The smaller *improvement* of InWiSe in comparison with a single OW is probably related to the considerable decrease in available positive sub-window sets relatively to the single OW count, compared to a small decrease in the case of the all-time mode. For example, Soroka HE labeled OW count decreases by 5-7% for the all-time mode, but by 35-45% for the exclusive mode, between the 4-sub-window set and the single OW methods. Table 2 shows further in-dataset metric results as Table 1 did for the all-time mode. In contrast to the all-time results, in the exclusive mode the AUC is still better with InWiSe (relatively to single OW), whereas other metrics domination depends on the posterior probability threshold.

Finally, we observed an average increase of 2.5% in valid prediction times when using InWiSe in comparison with a single OW in the size of N x $SubW_{size}$. This was mainly caused by the relaxed validation conditions in terms of missing values when splitting the windows, as well as by being able to use a single sub-window instead of a sub-window set at the beginning of a record when the available OWs are too short to be valid.

Cross-Datasets: The results of the cross-dataset experiments for the all-time prediction mode are shown in Table 1. First, one can observe the expected, but relatively small, drop in performance when training with one dataset and testing with another (0.1–0.5% in AUC). Nevertheless, we see that InWiSe outperforms other methods in terms of the AUC metric, even when applying the model to a new dataset. However, similarly to the in-dataset exclusive mode case, the other metrics domination in the cross-dataset validation (all-time mode) is threshold dependent, but this time with dependence on the source dataset. For example, the Soroka dataset sensitivity of the single OW method is higher than the sub-window set one (0.897 vs. 0.868, respectively) in the case where the model was trained on MIMIC II, while the opposite is true when it was trained on Hasassah (0.920 vs. 0.940). The reason for these results is probably the difference between the optimal threshold values in the source and the target datasets.

Feature Importance: The goal of splitting OWs was to let the classifier learn feature correlations with the target window from each sub-window separately, as well as their association and cross-window correlation. Table 3 presents the top

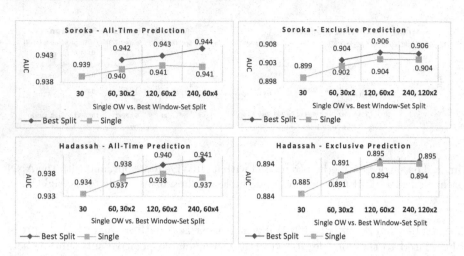

Fig. 3. In-dataset comparison between the single OW method and its InWiSe best split on the Hadassah and Soroka datasets (both prediction modes, XGBoost only).

Table 1. In-dataset and cross-dataset methods comparison (InWiSe best configuration vs. single OW best size vs. window-set matching single OW size) using all-time prediction. Metrics from top to bottom: AUC, accuracy, sensitivity, specificity and precision

Target		Source Dataset (Train)								
		MIMIC II			Soroka			Hadassah		
		4x60	120	240	4x60	120	240	4x60	120	240
MIMIC II	AUC	**0.937±0.003**	0.931±0.006	0.931±0.007	**0.933**	0.929	0.932	**0.937**	0.933	0.934
	Acc.	**0.878±0.007**	0.864±0.013	0.866±0.011	0.846	0.828	0.860	0.898	0.903	0.902
	Sens.	**0.849±0.011**	0.847±0.025	0.843±0.023	0.869	0.873	0.849	0.829	0.807	0.803
	Spec.	**0.879±0.008**	0.865±0.013	0.866±0.011	0.846	0.826	0.861	0.900	0.906	0.905
	Prec.	**0.168±0.007**	0.150±0.011	0.151±0.016	0.136	0.124	0.146	0.187	0.194	0.192
Soroka	AUC	**0.939**	0.935	0.935	**0.944±0.003**	0.941±0.003	0.941±0.003	**0.942**	0.938	0.939
	Acc.	0.867	0.822	0.858	**0.870±0.007**	0.861±0.005	0.856±0.004	0.761	0.800	0.793
	Sens.	0.868	0.897	0.865	**0.875±0.014**	0.875±0.011	0.875±0.014	0.940	0.915	0.920
	Spec.	0.867	0.820	0.857	**0.869±0.008**	0.860±0.006	0.856±0.007	0.754	0.796	0.789
	Prec.	0.187	0.150	0.178	**0.191±0.015**	0.182±0.016	0.177±0.017	0.119	0.137	0.134
Hadassah	AUC	**0.936**	0.933	0.932	**0.938**	0.935	0.935	**0.941±0.002**	0.938±0.002	0.937±0.002
	Acc.	0.842	0.816	0.815	0.838	0.814	0.834	**0.870±0.006**	0.867±0.007	0.861±0.006
	Sens.	0.889	0.899	0.896	0.894	0.904	0.888	**0.871±0.006**	0.866±0.005	0.866±0.006
	Spec.	0.839	0.812	0.811	0.835	0.809	0.831	**0.870±0.007**	0.867±0.007	0.860±0.007
	Prec.	0.207	0.184	0.182	0.204	0.183	0.198	**0.241±0.011**	0.235±0.011	0.226±0.010

Table 2. In-dataset methods comparison (InWiSe best configuration vs. single OW best size) in the exclusive prediction mode

	MIMIC II		Soroka		Hadassah	
	2 × 60	120	2 × 60	120	2 × 60	120
AUC	**0.884 ± 0.011**	0.884 ± 0.006	**0.906 ± 0.002**	0.904 ± 0.002	**0.895 ± 0.003**	0.894 ± 0.003
Accuracy	0.790 ± 0.020	**0.816 ± 0.013**	0.787 ± 0.011	**0.809 ± 0.010**	0.783 ± 0.009	**0.807 ± 0.007**
Sensitivity	**0.826 ± 0.021**	0.793 ± 0.023	**0.875 ± 0.009**	0.851 ± 0.012	**0.855 ± 0.009**	0.829 ± 0.007
Specificity	0.789 ± 0.021	**0.817 ± 0.014**	0.785 ± 0.010	**0.808 ± 0.010**	0.782 ± 0.009	**0.807 ± 0.007**
Precision	0.054 ± 0.006	**0.060 ± 0.004**	0.069 ± 0.004	**0.075 ± 0.004**	0.082 ± 0.003	**0.089 ± 0.003**

Table 3. Top 10 features rank in terms of frequency over trees in the XGBoost ensemble

	All-Time Prediction						Exclusive Prediction			
	60-min	240-min	4x60-min				60-min	120-min	2x60-min	
			$SubW_1$	$SubW_2$	$SubW_3$	$SubW_4$			$SubW_1$	$SubW_2$
SBP Slope	10	10	–	–	–	9	–	10	–	9
SBP Skewness	–	–	–	–	–	–	9	–	–	–
DBP Slope	–	–	–	–	–	–	–	–	–	10
MAP Mean	1	1	5	–	6	1	1	1	4	1
MAP Std	6	8	–	–	–	–	8	9	–	–
MAP Median	2	3	7	–	–	2	3	3	7	5
MAP Iqr	9	–	–	–	–	–	10	–	–	–
MAP Skewness	8	7	–	–	–	–	6	6	–	8
MAP Slope	4	2	–	–	–	3	5	4	–	3
HR Slope	5	6	–	–	–	8	4	5	–	6
SBP Cross Correlation w/ DBP	7	5	–	–	–	10	7	7	–	–
PP Cross Correlation w/ RCO	3	4	–	–	–	4	2	2	–	2
MAP Wavelet Detail Level-3	–	–	–	–	–	–	–	8	–	–
HR Wavelet Approximation	–	9	–	–	–	–	–	–	–	–

ten important features of XGBoost (most frequent over the ensemble trees) for the best InWiSe configuration compared with its sub-window sized OW as well as the corresponding single OW (in two prediction modes). The sub-window set columns are divided into their sub-windows, where $SubW_N$ is the sub-window ending at the prediction time and $SubW_1$ is the earliest in the set (Fig. 1b).

We first see that the Mean Arterial blood Pressure (MAP) mean is clearly dominant in all cases, which makes sense since MAP values are the ones that define an HE. Next, we observe that features from all three types (statistical, cross-correlation and wavelets) are top-ten-ranked, with the statistical ones (especially of MAP) used more frequently. Moreover, the two derived parameters, Pulse Pressure (PP) and Relative Cardiac Output (RCO), are proved to contribute particularly in their cross correlation with the target. Turning to sub-window sets, while $SubW_N$ has obviously more weight, the algorithm repeatedly chooses to combine the MAP mean and median from early sub-windows as well, with a surprisingly high rank. The early sub-window features are in favor of other higher ranked features in the single OWs (i.e., HR Slope and SBP cross-correlation with DBP). These findings support and explain our hypothesis that the model may be improved by using local sub-window features instead of extracting features from a single long OW.

5   Conclusions and Future Work

The current study presented and explored InWiSe, an enhanced feature extraction algorithm for clinical episode prediction, where physiological features are extracted from a set of observation sub-windows instead of a single OW. Our evaluation experiments have shown that the prediction performance may be improved by combining local sub-window features instead of extracting the same features from a single OW (of any size), observing an increased improvement when splitting longer OWs than used in existing methods (i.e., 240-min OW). The importance of sub-window features is confirmed by a feature importance analysis. Moreover, in the all-time prediction mode, used by the recent works, we show an improvement in comparison with the single OW method over all three experimental datasets w.r.t. all evaluated metrics[2] (up to 1% in accuracy and specificity and up to 10% in precision, while maintaining the sensitivity equal or better). We particularly focus on the AUC metric that was improved by up to 0.6%, with a statistically significant improvement in AUC performance in the case of the Hadassah dataset.

In addition to the above, we successfully evaluated the methods in a cross-dataset fashion, showing that the induced models are capable of predicting episode on a new dataset, with just a little degradation in the performance. Moreover, the AUC metric repeatedly favors the InWiSe method, even when testing the model on a new dataset. Furthermore, we explored a new prediction mode (exclusive) which may better reflect ICU needs.

With regard to InWiSe future improvements, better accuracy results may be achieved in the case of an invalid sub-window set, especially in the exclusive prediction mode. One may also evaluate alternative approaches to multiple OW utilization such as the weighted ensemble method of [15]. Another possible approach to episode prediction may be built on predicting the Mean Arterial blood Pressure (MAP) values in the target window with multivariate time series forecasting models.

From the dataset perspective, any future analysis should use the recently published MIMIC III dataset mentioned in Sect. 2.5. Applying an existing model on a new dataset should be further investigated in terms of determining a dataset-specific optimal classification threshold. Finally, the proposed methodology can be extended to other episode prediction domains.

References

1. Ghosh, S., et al.: Hypotension risk prediction via sequential contrast patterns of ICU blood pressure. IEEE J. Biomed. Health Inform. **20**(5), 1416–1426 (2016)
2. Sebat, F., et al.: Effect of a rapid response system for patients in shock on time to treatment and mortality during 5 years. Crit. Care Med. **35**(11), 2568–2575 (2007)
3. Cao, H., et al.: Predicting ICU hemodynamic instability using continuous multi-parameter trends. In: Engineering in Medicine and Biology Society (EMBS), pp. 3803–3806. IEEE (2008)
4. Moody, G.B., Lehman, L.W.H.: Predicting acute hypotensive episodes: the 10th annual physionet/computers in cardiology challenge. In: Computers in Cardiology, pp. 541–544. IEEE (2009)
5. Forkan, A.R.M., et al.: ViSiBiD: a learning model for early discovery and real-time prediction of severe clinical events using vital signs as big data. Comput. Netw. **113**, 244–257 (2017)

---

[2] All improvement percentages are in terms of a ratio between the two measures

6. Kamio, T., et al.: Use of machine-learning approaches to predict clinical deterioration in critically Ill patients: a systematic review. Int. J. Med. Res. Health Sci. **6**(6), 1–7 (2017)
7. Eshelman, L.J., et al.: Development and evaluation of predictive alerts for hemodynamic instability in ICU patients. In: 2008 AMIA Annual Symposium Proceedings, p. 379. American Medical Informatics Association (2008)
8. Lee, J., Mark, R.G.: An investigation of patterns in hemodynamic data indicative of impending hypotension in intensive care. Biomed. Eng. Online **9**(1), 62 (2010)
9. Saeed, M., et al.: Multiparameter intelligent monitoring in intensive care II (MIMIC-II): a public-access ICU database. Crit. Car. Med. **39**(5), 952 (2011)
10. Chen, X., et al.: Forecasting acute hypotensive episodes in intensive care patients based on a peripheral arterial blood pressure waveform. In: 2009 Computers in Cardiology, pp. 545–548. IEEE (2009)
11. Saeed, M.: Temporal pattern recognition in multiparameter ICU data, Doctoral dissertation, Massachusetts Institute of Technology (2007)
12. Saeed, M., Mark, R.: A novel method for the efficient retrieval of similar multiparameter physiologic time series using wavelet-based symbolic representations. In: AMIA Annual Symposium Proceedings, p. 679. American Medical Information Association (2006)
13. Rocha, T., et al.: Wavelet based time series forecast with application to acute hypotensive episodes prediction. In: Engineering in medicine and biology society (EMBC), pp. 2403–2406. IEEE (2010)
14. Ghosh, S., et al.: Septic shock prediction for ICU patients via coupled HMM walking on sequential contrast patterns. J. Biomed. Info. **66**, 19–31 (2017)
15. Lee, J., Mark, R.G.: A hypotensive episode predictor for intensive care based on heart rate and blood pressure time series. In: Computing in Cardiology, pp. 81–84. IEEE (2010)
16. Rocha, T., et al.: Prediction of acute hypotensive episodes by means of neural network multi-models. Comp. Biol. Med. **41**(10), 881–890 (2011)
17. Goldberger, A.L., et al.: PhysioBank, PhysioToolkit, and PhysioNet: components of a new research resource for complex physiologic signals. Circulation **101**(23), e215–e220 (2000)
18. The MIMIC II Waveform Database Matched Subset (Physionet Database). https://physionet.org/physiobank/database/mimic2wdb/matched/
19. Lee, Y.-L., Juan, D.-C., Tseng, X.-A., Chen, Y.-T., Chang, S.-C.: DC-Prophet: predicting catastrophic machine failures in DataCentre. In: Altun, Y., et al. (eds.) Machine Learning and Knowledge Discovery in Databases. LNCS, vol. 10536, pp. 64–76. Springer, Cham (2017). https://doi.org/10.1007/978-3-319-71273-4_6
20. Chen, T., Guestrin, C.: XGBoost: a scalable tree boosting system. In: 22nd ACM SIGKDD International Conference, pp. 785–794. ACM (2016)

Equipment Health Indicator Learning Using Deep Reinforcement Learning

Chi Zhang[✉], Chetan Gupta, Ahmed Farahat, Kosta Ristovski,
and Dipanjan Ghosh

Industrial AI Laboratory, Hitachi America Ltd., Santa Clara, CA, USA
{chi.zhang,chetan.gupta,ahmed.farahat,kosta.ristovski,
dipanjan.ghosh}@hal.hitachi.com

Abstract. Predictive Maintenance (PdM) is gaining popularity in industrial operations as it leverages the power of Machine Learning and Internet of Things (IoT) to predict the future health status of equipment. Health Indicator Learning (HIL) plays an important role in PdM as it learns a health curve representing the health conditions of equipment over time, so that health degradation is visually monitored and optimal planning can be performed accordingly to minimize the equipment downtime. However, HIL is a hard problem due to the fact that there is usually no way to access the actual health of the equipment during most of its operation. Traditionally, HIL is addressed by hand-crafting domain-specific performance indicators or through physical modeling, which is expensive and inapplicable for some industries. In this paper, we propose a purely data-driven approach for solving the HIL problem based on Deep Reinforcement Learning (DRL). Our key insight is that the HIL problem can be mapped to a credit assignment problem. Then DRL learns from failures by naturally backpropagating the credit of failures into intermediate states. In particular, given the observed time series of sensor, operating and event (failure) data, we learn a sequence of health indicators that represent the underlying health conditions of physical equipment. We demonstrate that the proposed methods significantly outperform the state-of-the-art methods for HIL and provide explainable insights about the equipment health. In addition, we propose the use of the learned health indicators to predict when the equipment is going to reach its end-of-life, and demonstrate how an explainable health curve is way more useful for a decision maker than a single-number prediction by a black-box model. The proposed approach has a great potential in a broader range of systems (e.g., economical and biological) as a general framework for the automatic learning of the underlying performance of complex systems.

Keywords: Health indicator learning
Deep Reinforcement Learning · Predictive Maintenance

Electronic supplementary material The online version of this chapter (https://doi.org/10.1007/978-3-030-10997-4_30) contains supplementary material, which is available to authorized users.

U. Brefeld et al. (Eds.): ECML PKDD 2018, LNAI 11053, pp. 488–504, 2019.
https://doi.org/10.1007/978-3-030-10997-4_30

1   Introduction

One of the important objectives in industrial operations is to minimize unexpected equipment failure. Unexpected downtime due to equipment failure can be very expensive – for example, an unexpected failure of a mining truck in the field can have a cost of up to $5000 per hour on operations. Besides the cost, unexpected failures can cause safety and environmental hazards and lead to loss of life and property. Traditionally industries have tried to address this problem through time-based maintenance. However, time-based maintenance often causes *over maintenance* [1], while still not fully being able to address the problem of unplanned downtime. With the advent of IoT, *Predictive Maintenance (PdM)* is gaining popularity. One of the key applications of PdM implies predicting the future health status of equipment (e.g., failure prediction), and then taking proactive action based on the predictions.

In the Machine Learning (ML) community, Predictive Maintenance (PdM) is typically modeled either as a problem of *Failure Prediction (FP)* [2] problem or the problem of estimating the *Remaining Useful Life (RUL)* [3], whereas in the Prognostics and Health Management (PHM) community, the problem is modeled as that of *Health Indicator Learning (HIL)*. FP answers the question (e.g., yes, no or in a probability) about whether a failure will occur in the next k days and RUL estimates the number of days l remaining before the equipment fails. HIL is the problem of estimating the future "health" $H(t)$ of the equipment, as a function of time t.

One reason for the popularity of RUL and FP in the ML community is that they are amenable to the traditional machine learning methods - FP is often modeled as a classification problem and RUL is modeled as a regression problem. RUL and FP modeled this way though useful, present operationalization challenges since ML produces black-box models and explainability is extremely desirable in industrial applications. Most domain experts are used to working with degradation curves and expect machine learning methods to produce something similar. Moreover, FP and RUL often do not provide enough information for optimal planning. For example, even if an operations manager is told that the equipment will fail in the next k days, they do not know whether to take the equipment offline tomorrow or on the kth day, since the prediction provides no visibility into the health of the equipment during the k day period (i.e., explanatory power is missing). From a practical standpoint, solving these two problems simultaneously and independently often leads to inconsistent results: FP predicts a failure will happen in k days while RUL predicts a residual life of $l > k$ days.

HIL addresses most of these concerns. Since the output of HIL is a health curve $H(t)$, one can estimate the health of the equipment at any time t and observe the degradation over time. Moreover, once the health curve is obtained, both RUL and FP can be solved using the health curve in a mutually consistent manner. However, from a ML perspective, HIL is a hard problem. The problem is essentially that of function learning, with no ground truth - i.e., there is no way to observe the actual health of the equipment during most of the operation.

We observe the health only when the equipment fails, and typically most modern industrial equipment are reliable and do not fail very often.

Researchers in the PHM community address the HIL problem by either hand-crafting a Key Performance Indicator (KPI) based on domain knowledge or try to identify some function of sensors that captures the health of the equipment through physical modeling. There are several challenges with these methods. Industrial equipment are complex (e.g., complicated and nested systems), and it is difficult and time-consuming for experts to come up with such KPIs. Additionally, domain-specific KPIs are not applicable across industries, so that developing a general method becomes infeasible. Furthermore, equipment is usually operated under varying operating conditions leading to significantly different health conditions and failure modes, making it difficult for a single manually-crafted KPI to capture health.

In this paper, we propose a machine learning based method to solve for HIL. Our key insight is that HIL can be modeled as a credit assignment problem which can be solved using Deep Reinforcement Learning (DRL). The life of equipment can be thought as a series of state transitions from a state that is "healthy" at the beginning to a state that is "completely unhealthy" when it fails. RL learns from failures by naturally backpropagating the credit of failures into intermediate states. In particular, given the observed time series of sensor, operating and event (failure) data, we learn a sequence of health indicators that represent the health conditions of equipment. The learned health indicators are then used to solve the RUL problem.

The contributions of this paper are summarized as follows:

- We propose to formalize the health indicator learning problem as a credit assignment problem and solve it with DRL-based approaches. To our knowledge, this is the first work formalizing and solving this problems within an RL framework. Additionally, the label sparsity problem (i.e., too few failures) is addressed due to the nature of RL.
- We propose a simple yet effective approach to automatically learn hyper parameters that are best for approximating health degradation behaviors. Therefore, the proposed method does not require domain-specific KPIs and are generic enough to be applied to equipment across industries.
- We use health indicators as compact representations (i.e., features) to solve the RUL problem, which is one of the most challenging problem in PdM. Therefore, we not only provide the explanation of health conditions for observed data, but can also predict the future health status of equipment.

The rest of the paper is organized as follows. Section 2 describes the details of the proposed method. Section 3 qualitatively analyzes the performance of the proposed methods on a synthetic problem. Section 4 shows experimental results applying our methods to a benchmark. Section 5 discusses the differences and relations between our method and other Markov chain based methods. Section 6 gives an overview of related work on HIL. Section 7 concludes the paper.

Table 1. Summary of notation

Notation	Meaning
$t \in \mathbb{N}$	Time step $t = 1, 2, ..., T$, where T is the last time step
$i \in \mathbb{N}$	Sequence index $i = 1, 2, ..., I$, where I is the total number of sequences
$x_i^t \in \mathbb{R}$	Sensor data x at time t from sequence i
$y_i^t \in \mathbb{R}$	Operating condition data y at time t from sequence i
$s \in \mathcal{S}$	State derived by discretizing $\{x_i^t\}$ into N bins: $s = Discretizing(x)$, where $Discretizing(\cdot)$ is a discretizing operation. $\mathcal{S} = \{s_n\}$ $n = 1, 2, ..., N$ is a finite set (i.e., state space)
$a \in \mathcal{A}$	Action defined as the change of operating conditions in two time steps: $a^t = \overrightarrow{\boldsymbol{u}^t} = c_y^{t+1} - c_y^t$, where $c_y = Clustering(y^t)$ is a clustering operation, $\boldsymbol{u}^t$ is a vector. $\mathcal{A} = \{a_m\}, m = 1, 2, ..., M$ is a finite set (i.e., action space)
$R^t \in \mathbb{R}$	Immediate reward at time t
$\pi(a|s)$	Policy function of the probability of choosing action a given state s
$v^\pi(s)$	Value function of policy π
$U(s) \in \{0, 1\}$	Failure/non-failure labels where 1 denotes failure occurrence. (Failures indicate the end-of-life of equipment)
$\mathcal{P}_{s,a}(s')$	State transition probability function: when take action a in state s, the probability of transiting to state s'
$\mathcal{R}_{s,a}$	Reward function of the expected reward received after taking action a in state s
$\mathcal{M}$	MDP model $<\mathcal{S}, \mathcal{A}, \mathcal{P}_{\cdot,\cdot}(\cdot), \mathcal{R}_{\cdot,\cdot}, \gamma>$, where $\gamma \in [0, 1]$ is a discount factor

2   Methodology

In this section, we first formalize the health indicator learning as a credit assignment problem and propose to address it with RL in two ways: in the model-based method, a Markov Decision Process (MDP) is learned and used to derive the value function in a tabular form; in the model-free method, a bootstrapping algorithm with function approximation is used to learn a continuous value function without implicitly modeling the MDP. Consequently, the learned value function maps observation data of equipment to health indicators. The notations used in the paper are presented in Table 1.

2.1   Problem Formulation

The problem of HIL is to discover a health function $H(t) = v(s|s = s^t)$ such that the following two properties [5] are satisfied:

- Property 1: Once an initial fault occurs, the trend of the health indicators should be monotonic: $H(t) \geq H(t + \Delta), t = 1, 2, ..., T - \Delta$

– *Property 2: Despite of the operating conditions and failure modes, the variance of health indicator values at failures $\sigma^2(H(T_i)), i = 1, 2, ..., I$ should be minimized.*

where Δ is a small positive integer (e.g., $\Delta = 1$ means strictly monotonic), $v(s)$ is a function mapping the equipment state s at any time t to a health indicator value, and σ^2 represents the variance.

By assuming the health degradation process is a MDP, we can transform the HIL problem to a credit assignment problem and represent $v(s)$ as the value function (a.k.a., Bellman Equation) $v^\pi(s)$ in RL [6]:

$$v^\pi(s) = E^\pi[(R^{t+1}|S^t = s) + \gamma v^\pi(S^{t+1}|S^t = s)] \tag{1}$$

where γ is a discount factor, R^t is the reward function.

2.2   Model-Based Health Indicator Learning

According to [6], we assume conditional independence between state transitions and rewards in the model-based health indicator learning approach:

$$P[s^{t+1}, R^{t+1}|s^t, a^t] = P[s^{t+1}|s^t, a^t]P[R^{t+1}|s^t, a^t] \tag{2}$$

Learning the MDP Model: According to the definition of $\mathcal{M} = <\mathcal{S}, \mathcal{A}, \mathcal{P}_{.,.}(\cdot), \mathcal{R}_{.,.}, \gamma>$, now MDP learning can be transformed to a supervised learning problem, in which learning $\mathcal{R}_{s,a} =: s, a \rightarrow R$ is a regression problem and learning $\mathcal{P}_{s,a}(s') =: s, a \rightarrow s'$ is a density estimation problem. We use a Table Lookup method to learn $\mathcal{P}, \mathcal{R}$ directly, as presented in Algorithm 1.

Algorithm 1. MDP Learning

1: **procedure** DISCRETIZATION(Process sensor and operating condition data into states and actions)

2:    $c_x \leftarrow Discretizing(\{x_i^t\})$

3:    $c_y \leftarrow Clustering(\{y_i^t\})$

4:    $\mathcal{S} = \{c_x\} = \{s\}$

5:    $\mathcal{A} = \{c_y^t - c_y^{t+1}\} = \{a\}$

6: **procedure** LEARN $\mathcal{M}$ BY COUNTING VISITS $N(s,a)$ TO EACH (s,a) PAIR

7:    $\mathcal{P}_{s,a}(s') =$

$$\frac{1}{N(s,a)} \sum_{i=1}^{I} \sum_{t=1}^{T_i} 1(s^t, a^t, s^{t+1} = s, a, s')$$

8:    $\mathcal{R}_{s,a} = \frac{1}{N(s,a)} \sum_{i=1}^{I} \sum_{t=1}^{T_i} 1(s^t, a^t = s, a)R^t$

A policy is defined as the probability of taking an action for a given state $\pi(a|s)$. Given that the objective of the proposed method is policy evaluation, the policy can be given or learned from data:

$$\pi(a|s) = \frac{1}{N(s)} \sum_{i=1}^{I} \sum_{t=1}^{T_i} 1(s^t, a^t = s, a) \tag{3}$$

where $N(s)$ is the number of occurrence of s. Note that stationary policies are assumed.

To define the immediate reward R^t, three scenarios (i.e., non-failure to non-failure, non-failure to failure, and failure to failure) are considered:

$$R^t = \begin{cases} 0 & U(s^t) = U(s^{t+1}) = 0 \\ -1.0 & U(s^t) = 0, U(s^{t+1}) = 1 \\ R_{ff} & U(s^t) = 1 \end{cases} \tag{4}$$

R_{ff} is a hyper-parameter that can be tuned and $R_{ff} > -1.0$ to impose more penalty for failure-to-failure transitions. We also propose an alternative reward function to handle faults and never-fail equipment in Appendix[1].

Learning Health Indicator Directly (HID). We rewrite Eq. 1 as:

$$v^\pi(s) = E^\pi[R^{t+1}|S^t = s] + \gamma E^\pi[v(S^{t+1})|S^t = s]$$
$$= \mathcal{R}(s) + \gamma \sum_{s' \in \mathcal{S}} \mathcal{P}_s(s')v(s') \tag{5}$$

where $\mathcal{R}(s)$ and $\mathcal{P}_s(s')$ can be calculated as:

$$\mathcal{R}(s) = \sum_{a \in \mathcal{A}} \pi(a|s)\mathcal{R}_{s,a} \tag{6}$$

$$\mathcal{P}_s(s') = \sum_{a \in \mathcal{A}} \pi(a|s)\mathcal{P}_{s,a}(s') \tag{7}$$

Equation 5 can be expressed using matrices as:

$$\boldsymbol{v} = \boldsymbol{\mathcal{R}} + \gamma \boldsymbol{P}\boldsymbol{v} \tag{8}$$

and solved as:

$$\boldsymbol{v} = (\boldsymbol{\mathcal{I}} - \gamma \boldsymbol{P})^{-1}\boldsymbol{\mathcal{R}} \tag{9}$$

where $\boldsymbol{\mathcal{I}}$ is the identity matrix. For simplicity, we omit the notation π by using $v(s)$ to represent $v^\pi(s)$ throughout the paper without losing the meaning that $v(s)$ is the value function of the policy π. To deal with a large state space, we also propose an iterative approach to learn health indicator using Dynamic Programming (HIDP), as presented in the aforementioned Appendix.

2.3   Model-Free Methods

In real-world scenarios the equipment can go through a very large number of states, which makes the model-based method less efficient and inapplicable. Thus, we propose a model free method to learn the health indicators of states for a given policy *without* explicitly modeling π and $\mathcal{P}_{s,a}(s')$.

---

[1] https://tinyurl.com/yaguzvuc.

Algorithm 2. Temporal Difference (0) with Function Approximation

---

1: Initialize value function v with random weight θ
2: Build a memory of experience $e_i^t = (x_i^t, R_i^{t+1}, x_i^{t+1})$ into a data set $D = \{e_i^t\}, i = 1, ..., I, t = 1, ..., T_i$ and randomize.
3: **do**
4:　　Sample a minibatch of m experiences $(x, R, x') \sim U(D)$
5:　　**for** $i = 1, ..., m$ **do**
6:　　　$z_i = R_i^{t+1} + \gamma \hat{v}_\theta(x_i^{t+1})$
7:　　$\theta := \arg\min_\theta (z_i - \hat{v}_\theta(x_i^t))$
8: **while** True

Health Indicator Temporal-Difference Learning with Function Approximation (HITDFA). To further extend the health indicator learning into the continuous space, we define $\hat{v}_\theta(x)$ parameterized by θ as a function to approximate the continuous state space, where θ can be a deep neural network. Given the objective function in Eq. 1, we use real experience instead of the expectation to update the value function, as shown in Algorithm 2. In the HITDFA method, memory replay [7] is used to remove correlations in the observation sequence and smoothing over changes in the data distribution.

Note that the proposed HITDFA method learns the health indicators using value function $v(s)$, but it can be easily extended to learn the action value function $q(s, a)$ for each state-action pair. Consequently, we can replace $\hat{v}_\theta(x_i^t)$ with $\hat{q}_\theta(x_i^t, a_i^t)$ and $\hat{v}_\theta(x_i^{t+1})$ with $\hat{q}_\theta(x_i^{t+1}, a_i^{t+1})$. Similarly, such extension can also be applied to HID method.

In the proposed RL-based methods, there are two hyper-parameters that need to be learned: γ and R_{ff} (Eq. 4). We use a simple yet effective grid search approach to find the hyper parameter settings that (1) make $H(t)$ monotonic (i.e., the first constraint in Sect. 2.1), and (2) obtain the minimum variance of $H(T_i), i = 1, 2, ..., I$ (i.e., second constraint).

3　Qualitative Analysis of RL-Based HIL

In this section, we first study an ideal equipment which has only three states $\{s_1, s_2, s_3\}$, representing health, intermediate, and failure states respectively. The initial state $s^0 = s_1$ and the state probability transition matrix is given as:

$$\mathcal{P} = \begin{bmatrix} 0 & 1 & 0 \\ 0 & 0 & 1 \\ 0 & 0 & 1 \end{bmatrix} \tag{10}$$

Now we use HID to learn the health indicators. By selecting the reward function $R_{ff} = -2$ and varying γ, various v can be calculated according to Eq. 9. Since the objective is to study the degradation behaviors, all the learned value functions are rescaled to $[-1, 0]$ with 0 indicating healthy and -1 indicating failed, as presented in Fig. 1(a). When we fix $\gamma = 0.9$ and vary R_{ff}, significantly different degradation behaviors are obtained. It can be observed in Fig. 1(b) that a larger γ leads to the concave property, which is because γ as the discount factor trades-off the importance of earlier versus later rewards. Therefore, a larger γ tends to emphasize the affect of failures happening at the end of the sequence. The opposite trend is observed when the immediate reward of staying in a failure state is penalized much more than transferring to a failure state (i.e., more negative R_{ff}). This is because the reward shapes the underlying relationship between failure and non-failure states. A more negative reward implies the larger difference between health indicator values at failure state and non-failure state, which makes the curve in Fig. 1(b) shift from concave to convex.

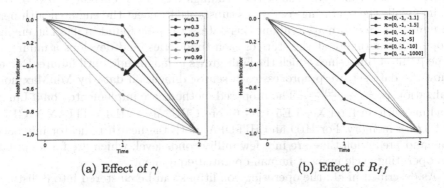

(a) Effect of γ (b) Effect of R_{ff}

Fig. 1. Using γ and R_{ff} to characterize different health degradation behaviors.

Therefore, the proposed method is capable of approximating different health degradation behaviors by tuning hyper parameters to represent different relationships between failure/non-failure states for a given problem (i.e., MDP). This also differentiates our methods from previous works such as [4] that can only model one type of degradation behavior (e.g., exponential). Potentially, the proposed methods are applicable to a wide range of physical equipment.

4   Quantitative Analysis of RL-Based HIL

To evaluate the performance of the proposed learning methods, a benchmark is used to demonstrate that the learned health indicators satisfy essential properties in health degradation development (i.e., Properties 1 and 2 in Sect. 2.1). Then, we conduct experiments to validate the effectiveness of the proposed method by examining if testing health indicator curves in run-to-failure data fall into the same failure threshold estimated from training curves. Lastly, our approaches are combined with a regression model to predict RULs to demonstrate that different PdM tasks can be achieved using the proposed methods.

4.1   Experimental Setup

To quantitatively investigate the effectiveness of the proposed methods, we conduct experiments with the NASA C-MAPSS (Commercial Modular Aero-Propulsion System Simulation) data set, as it is a standard benchmark widely used in PHM community [9]. This is a Turbofan Engine Degradation Simulation Data Set contains measurements that simulate the degradation of several turbofan engines under different operating conditions. This benchmark has four data sets (FD001 ∼ FD004), consisting of sensor data, operating condition data, and models a number of failure modes. At each operating time cycle, a snapshot of sensor data (i.e., 21 sensors) and operating condition data (i.e., 3 operating conditions) are collected. The sensor readings reflect the current status of the engine, while the operating conditions substantially effect the engine health performance. More information about this data set can be found in [9]. The engine is operating normally at the start of each time series, and develops a failure at some point in time, since which the fault grows in magnitude until failure. Before using the data set, we preprocess each sensor dimension data by MinMax normalization $x_i = \frac{x_i - \min(x_i)}{\max(x_i) - x_i}$. The proposed methods are implemented on a Linux machine with an Intel Xeon E5 1.7G 6-core CPU, an NVIDIA TITAN X GPU, and 32 Gb memory. For HID and HITDFA, both the health indicator inference time and prediction time are in a few milliseconds level, which are far less than one operating cycle in most human-operated equipment.

As described in [9], the operating conditions can be clustered into 6 distinct clusters. We apply K-means to cluster all operating condition data and construct the action set $\{a\} = \mathcal{A}$ as defined in Sect. 2. The sensor data is clustered into 400 distinct states. For each data set, we assume a unique policy and learn the corresponding value function.

Using the C-MAPSS data set, we quantitatively evaluate our methods in the following tasks:

- **Health Degradation Behavior Analysis:** Given run-to-failure data, validate the soundness of the proposed methods by examining $H(t)$ with the two essential properties and an additional property that is given in the data set, and compare the results with baselines.
- **HIL Performance Validation:** Given run-to-failure data, split the set into training and testing sets, then learn $H_{train}(t)$ on the training set to derive the distribution of the failure threshold, and finally validate on the testing set that $H_{test}(T_i)$ falls into the same threshold.
- **RUL Estimation:** Given run-to-failure data and prior-to-failure data, train a regression model based on $H_{train}(t)$ in the run-to-failure data, and use the learned regression model to predict $H_{test}(t)$ in the prior-to-failure set to estimate RULs.

4.2   Health Degradation Behavior Analysis

Using the proposed approach in Sect. 2.3, the hyper parameters satisfying the constraints for each policy (i.e., data set) are found. The health indicators of

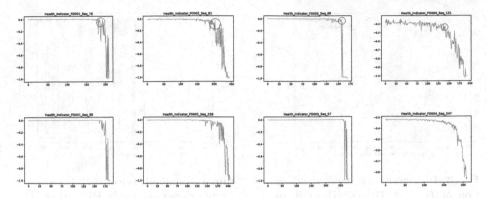

Fig. 2. Health degradation of randomly selected sequences learned by HID. First row: training, second row: testing. (Color figure online)

randomly selected training and testing sequences from each sub data set is presented in Fig. 2. It can be observed that even though there are fluctuations (due to nosies and can be smoothed by postprocessing), the monotonic property still holds (i.e., the first constraint in Sect. 2.1 is satisfied). For different $H(t)$ curves, the critical points (as highlighted by the red circles in the first row in Fig. 2) after which the value decreases sharply are also observed. A plausible explanation is that the proposed method learns the health value of each state, and the state transition in the sequence decides the shape of the curve, as well as the critical points in the curve.

To validate the second constraint in Sect. 2.1, we compare results of the learned model with a baseline health indicator learning method [4] in Table 2. Our proposed methods significantly outperform the composite health indicator and sensor-wise indicators in orders of magnitude. Note that [4] only studies FD001 set, so that we compare variances on the same data set.

Even thought the ground truth health values are not available, Saxena et al. [9] reveals that the degradation is exponential: $H(t) = 1 - d - exp\{at^b\}$, where d is the non-zero initial degradation and a, b are coefficients. It can be clearly observed from Fig. 2 that our model is able to characterize the exponential form of degradation.

Table 2. Comparison of variances between proposed methods and baselines. From T24 to W32, and CompHI are health indicators in [4], and the variance results are given by using the corresponding sensor.

Hea. Ind.	T24	T50	P30	Nf	Ps30	Phi	NRf
Variance	0.0274	0.014	0.0264	0.0683	0.0154	0.0206	0.0580
Hea. Ind.	BPR	htBleed	W31	W32	CompHI	HID	HITDFA
Variance	0.0225	0.0435	0.0220	0.0317	0.0101	7.1×10^{-5}	1.05×10^{-5}

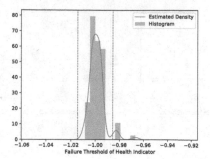

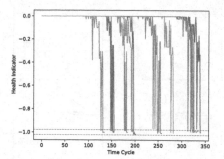

Fig. 3. Histogram and density estimation of $H(T_i)$ (FD001). (Color figure online)

Fig. 4. Randomly selected $H(t)$ for different engines learned by HID (FD001). (Color figure online)

4.3 HIL Generalization Performance Validation

It is important to validate that when apply the learned model to unseen (i.e., testing) run-to-failure sequences, it is still capable of rendering consistent health degradation behavior and ends up with consistent health indicator values at the failures. In this section, we conduct such experiments to validate the generalization performance of the proposed methods. The run-to-failure data in each data set is split into training (90%) and testing (10%) sets. In training, we apply HID method to learn $v(s)$ for each state, and obtain $H_{train}(t)$. The corresponding health indicator values at failures (i.e., $\{h_T\}_{train} = H_{train}(T_i), i = 1, 2, ...I$) are obtained as well. Note that we assume in each data set, testing data and training data are generated under the same policy.

We estimate the density function $F(h_T)$ of h_T using Gaussian kernel (as shown in Fig. 3). Then the failure threshold h_f is defined as $h_f = argmax(F(h_T))$. As suggested by [9], health scores are normalized to a fixed scale, we also normalize health indicators by h_f: $H(t) = \frac{H(t)}{|h_f|}$, so that $h_f = -1$. $H(t)$ is rescaled to a region close to $[-1, 0]$, where 0 represents the perfect health status and -1 represents the failed status. In Fig. 3, the area between the red dash lines (with the right line indicates $h_{f_{max}}$ and left line indicates $h_{f_{min}}$) indicating the confidence level c (e.g., 95%): $c = \int_{h_{f_{min}}}^{h_{f_{max}}} F(h_T)dh_T$.

In testing, we apply the learned model to obtain $H_{test}(t)$, and $\{h_T\}_{test} = H_{test}(T_i)$. By finding the health indicator that is closest to h_f, we can obtain the estimated failure time: $T' = H^{-1}(\arg\max(F(h_T)))$, where $h_{f_{min}} \leq h_T \leq h_{f_{max}}$. Figure 4 presents $H(t)$ of randomly selected testing data and the learned failure threshold region, where the red dash lines correspond to the confidence area in Fig. 3. We use Root-Mean-Square-Error (RMSE) for performance evaluation: $RMSE = \sqrt{\frac{1}{I}\sum_{i \in I}(T'_i - T_i)^2}$.

It can be observed in Fig. 5 that our approach achieves better performance on FD001 and FD003 compared to FD002 and FD004, due to the fact that FD001 and FD003 have a simpler policy (i.e., $\pi(a_0|s_i) = 1, i = 1, 2, ..., N$, where a_0 is the action that is always taken), and FD002 and FD004 involve complicated policies

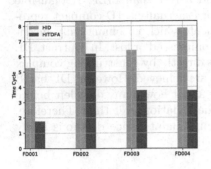

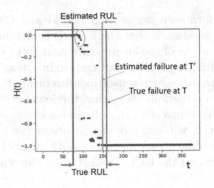

Fig. 5. Average error between actual and estimated time of failures in time cycles in the testing data.

Fig. 6. RUL estimation. (Color figure online)

that are difficult to evaluate. It can also be observed that HITDFA has a lower variance than HID, due to the fact that (1) model-free method can yield better performance since its performance does not depend on the quality of a MDP model; (2) function approximation is adopted in HITDFA so that information lost in state discretization in the HID is avoided.

The objective of this experiment is to validate that the proposed methods is powerful in characterizing the health curves when new observations are given. Our proposed health indicator learning methods can be easily combined with regression models for prediction, as presented in the next section.

4.4   Remaining Useful Life Estimation

In this experiment, we use the health indicators learned from our proposed methods to solve the RUL problem. The objective is to validate that our methods are capable of learning a good representation of health conditions, which can be used as a feature for prediction models to solve PdM problems.

We use all run-to-failure data as a training set, and prior-to-failure data as a testing set. First, we learn the value function from the training set, and derive the health indicator sequences $H_{train}(t)$ and $H_{test}(t)$. Then $H_{train}(t)$ are used to learn a Seq2Seq deep neural network [8], which can predict health indicator sequence $\hat{H}(t)$ given $H_{test}(t)$ derived from prior-to-failure data. Finally, the failure threshold in Sect. 4.3 is applied to find the failure time T' in $\hat{H}(t)$. Figure 6 shows the given health indicators (red dots), the predicted health indicators (blue dots), and the true and predicted failures. The predicted health degradation curve provides rich information that can be easily interpreted by operators. For example, the critical points (in orange circle) can be used to decide when to perform preventive maintenance on the equipment to avoid downtime and maximize utilization.

In Table 3, we compare *RMSE* of our methods with state-of-the-art Deep Learning (DL) approaches. It can be observed that our methods achieve performance worse than LSTM and CNN, and better than MLP. A plausible explanation is that the learned health indicator is actually a 1D compact representation of sensor measurement from equipment, which is semantically meaningful (i.e., health), but certain information is compressed or lost. In contrast, DL approaches map sensor measurements to RULs directly by constructing complex representations [3] so that the lower error rate are achieved. However, DL-based RUL estimation can only give a single number about how many days left before failure, without any insight about the health degradation over time. Therefore, it is difficult to make the optimal operation and maintenance decisions solely based on the RULs predicted by these approaches.

Table 3. RMSE (%) comparison of RUL with DL methods on C-MAPSS data

Data Set	#1	#2	#3	#4	Average
MLP (Multi Layer Perceptron) [12]	37.56	80.03	37.39	77.37	58.09
SVR (Support Vector Regression) [12]	20.96	42.00	21.05	45.35	32.34
RVR (Relevance Vector Regression) [12]	23.80	31.30	22.37	34.34	27.95
CNN (Convolutional Neural Network) [12]	18.45	30.29	19.82	29.16	24.43
Deep LSTM (Long-short Term Memory) [3]	16.14	24.49	16.18	28.17	21.25
LSTMBS [13]	14.89	26.86	15.11	27.11	20.99
HID	23.76	38.80	37.11	58.16	39.46
HITDFA	23.01	39.00	39.69	58.10	39.95

Besides of the comparison with DL, we are more interested in comparing our methods with other HIL approaches. In these approaches, Mean-Absolute-Percentage-Errors ($MAPE$) is the most popular metric as it measures the relative - rather than the absolute - magnitude of the prediction error in sequences. Therefore, we use the same $MAPE = \frac{1}{I} \sum_{i \in I} \left| \frac{T_i' - T_i}{T_i + O_i} \right|$ as [4], where O_i is the length of observed sequence i. From Table 4, it can be observed that our methods outperform state-of-the-art HIL approaches. The improvement is calculated as $Improvement = 1 - (MAPE_{HID}/MAPE_{state-of-the-art})$. It is noteworthy that our methods achieve a good performance even in FD002 and FD004, which are known to be difficult to learn due to the complex operating conditions and failure modes in the data. A plausible explanation is that we regard each data set independently to learn and evaluate the corresponding policy, so that the interferences in various policies are excluded. Ideally, if the environment used to generate the data (e.g., a simulator) is available, a random policy can be applied to acquire infinite experience to learn $v(s)$, despite of multiple or single operating conditions and failure modes.

Table 4. Comparison of the proposed methods with other HIL methods on C-MAPSS data based on MAPE (%) (table adapted from [18])

Data Set	#1	#2	#3	#4	Average
CompHI [4]	9	-	-	-	-
ANN [10]	43	48	44	49	46
FS-LSSVR [10]	38	44	40	47	42.5
RULCLIPPER [11]	20	32	23	34	27.25
HID	9.87	**18.05**	**13.08**	**31.00**	**18.00**
HITDFA	**8.85**	18.10	14.10	32.00	18.27
Improvement	1.67%	43.59%	43.13%	8.82%	33.95%

5 Discussion

In this section, we discuss relationship of proposed methods with Markov Chain and Hidden Markov Model. For equipment without explicitly defined controllable parts (i.e., no actions), the proposed MDP model can be easily transformed to Markov Chain by removing a^t and the proposed methods can still be applied. Our methods are also different from Hidden Markov Models (HMMs). First, HMM is supervised and requires the actual health indicator values for training, while our RL-based method does not. In HMM-based approaches, it usually creates X labels artificially by defining different health levels. The coarse definition of health levels leads to a very rough HIL results. Moreover, the definition and segmentation of health levels require a lot of domain knowledge. In contrast, the proposed method discovers the evolution of health indicators automatically by learning values for $N (>> X)$ discrete sensory states. Second, our method models actions separately from state, while HMM considers actions as some additional dimensions in the state space. This leads to $N \times M$ states (M is the number of actions) and X hidden states and hence, makes the computation expensive. In contrast, our method only has N states and no hidden states.

When generalize the proposed methods to other domains where the feature spaces are large (as opposed to 21 sensor data and 3 operating conditions in C-MAPSS), feature selection approaches can be used to preprocess the data in HIT. Automatic feature learning mechanism such as convolutional layers can be used as the first part in a deep network (i.e., θ) in HITDFA.

6 Related Work

In this section, we first give an overview of methods formalizing the health indicator learning problem as a classification problem and learning coarse-grained health indicators. Then we review methods learning fine-grained health values by mapping sensor measurement to continuous health values. Lastly, methods that map the learned health indicators to RULs to perform prognostics are reviewed.

To detect anomaly, assess health, or identify failures, a known target (i.e., ground truth) is required to evaluate the performance of the learned health indicators. Some methods [14,15] label data with pre-defined degradation levels: normal, knee corresponding, accelerated degradation, and failure. Then the learning can be converted to a classification problem that can be solved using supervised approaches such as HMM [14]. This type of methods require hand-crafted segmentations in order to label the data with the levels, which heavily depends on domain knowledge, and these labels are only approximation to ground truth, which all make it less applicable in general health indicator learning. Different from these methods, our approach is capable of automatically learning health indicators from data, without relying on hand-picked labels.

Due to that classification approaches can only learn coarse health levels, alternative approaches are proposed to map sensor measurements to continuous health indicator values. By leveraging dimension reduction methods, the work of [16] finds the most informative individual sensor and learns a health index. In contrast, in this paper we proposed to learn a mapping from all sensor measurement to health indicators, without losing any information in dimension reduction as in [16]. Similar to our methods, a composite health index is modeled by fusing data from all sensors [4]. However, the work of [4] can only deal with equipment operated under a single operating condition and falls into a single failure mode, while our proposed methods can handle more complicated situations.

Based on the learned health index, the prognostics problem is addressed by learning a second mapping links health index values to the RUL. Le Son et al. [17] models the health degradation process as a gamma process, then finds the hidden degradation states by Gibbs algorithm, and estimates RUL as a random variables with a probability distribution. As discussed in [18], the performance for RUL prediction depends on both the health indicator learning and prediction.

7 Conclusion

In the emerging area of predictive maintenance, there is a crucial demand from the users to abstract the complexity of physical systems behind explainable indicators that reflect the true status of these systems and justify the very costly actions that the users might take in response to predictions. This paper takes a major step forward toward achieving this goal by providing the ability to learn health indicators from sensor data given information about a few failure incidents. To the best of our knowledge, this is the first work to formulate this Health Indicator Learning (HIL) problem as a credit assignment problem and model the health indicator as the output of a state value function. The paper proposed both model-based and model-free RL methods to solve the value function. In particular, we proposed an automatic hyperparameter learning approach by using simple physical properties as constraints, which makes our method widely applicable across different domains and industries. We demonstrated the effectiveness of our method on synthetic data as well as well-known benchmark data sets. We showed that the method can learn the health indicators of equipment operating under various conditions even in the presence of data from various failure modes. Experiments also demonstrated that the proposed methods achieve 33.95% improvement in predicting Remaining Useful Life (RUL) in comparison to state-of-the-art methods for the HIL problem. Our method keeps the distinct quality of HIL methods in providing explainable predictions in the format of a degradation curve. This is in contrast to black-box regression models such as LSTM which directly predict RUL as a single number and provide a complex feature map that cannot be explained to decision makers.

References

1. Ahmad, R., Kamaruddin, S.: An overview of time-based and condition-based maintenance in industrial application. CAIE **63**(1), 135–149 (2012)
2. Susto, G.A., Schirru, A., Pampuri, S., McLoone, S., Beghi, A.: Machine learning for predictive maintenance: a multiple classifier approach. IINF **11**(3), 812–820 (2015)
3. Zheng, S., Ristovski, K., Farahat, A., Gupta, C.: Long short-term memory network for remaining useful life estimation. In: IEEE PHM (2017)
4. Liu, K., Gebraeel, N.Z., Shi, J.: A data-level fusion model for developing composite health indices for degradation modeling and prognostic analysis. ASE **10**, 652–664 (2013)
5. Saxena, A., et al.: Metrics for evaluating performance of prognostic techniques. In: IEEE PHM (2008)
6. Sutton, R.S., Barto, A.G.: Reinforcement Learning: An Introduction. MIT Press, Cambridge (1998)
7. Mnih, V., et al.: Human-level control through deep reinforcement learning. Nature **518**(7540), 529 (2015)
8. Sutskever, I., Vinyals, O., Le, Q.V.: Sequence to sequence learning with neural networks. In: NIPS (2014)
9. Saxena, A., Goebel, K., Simon, D., Eklund, N.: Damage propagation modeling for aircraft engine run-to-failure simulation. In: IEEE PHM (2008)
10. Li, X., Qian, J., Wang, G.G.: Fault prognostic based on hybrid method of state judgment and regression. AIME **5**, 149562 (2013)
11. Ramasso, E.: Investigating computational geometry for failure prognostics in presence of imprecise health indicator: results and comparisons on C-MAPSS datasets. PHM **5**(4), 005 (2014)
12. Sateesh Babu, G., Zhao, P., Li, X.-L.: Deep convolutional neural network based regression approach for estimation of remaining useful life. In: Navathe, S.B., Wu, W., Shekhar, S., Du, X., Wang, X.S., Xiong, H. (eds.) DASFAA 2016. LNCS, vol. 9642, pp. 214–228. Springer, Cham (2016). https://doi.org/10.1007/978-3-319-32025-0_14
13. Liao, Y., Zhang, L., Liu, C.: Uncertainty prediction of remaining useful life using long short-term memory network based on bootstrap method. In: IEEE PIIM (2018)
14. Ramasso, E.: Contribution of belief functions to Hidden Markov models with an application to fault diagnosis. In: MLSP (2009)
15. Ramasso, E., Denoeux, T.: Making use of partial knowledge about hidden states in HMMs: an approach based on belief functions. FUZZ **22**(2), 395–405 (2014)

16. El-Koujok, M., Gouriveau, R., Zerhouni, N.: Reducing arbitrary choices in model building for prognostics: an approach by applying parsimony principle on an evolving neuro-fuzzy system. Microelectron. Reliab. **51**(2), 310–320 (2011)
17. Le Son, K., Fouladirad, M., Barros, A.: Remaining useful life estimation on the non-homogenous gamma with noise deterioration based on gibbs filtering: a case study. In: IEEE PHM (2012)
18. Ramasso, E., Saxena, A.: Performance benchmarking and analysis of prognostic methods for CMAPSS datasets. PHM **5**(2), 1–15 (2014)

ADS Sensing/Positioning

PBE: Driver Behavior Assessment Beyond Trajectory Profiling

Bing He[2], Xiaolin Chen[1], Dian Zhang[1(✉)], Siyuan Liu[3], Dawei Han[4],
and Lionel M. Ni[2]

[1] College of Computer Science and Software Engineering, Shenzhen University,
Shenzhen, China
zhangd@szu.edu.cn
[2] Department of Computer and Information Science, University of Macau,
Macau SAR, China
[3] Smeal College of Business, Pennsylvania State University,
State College, PA, USA
[4] Auto Insurance Department , China Pacific Insurance Company, Shenzhen, China

Abstract. Nowadays, the increasing car accidents ask for the better
driver behavior analysis and risk assessment for travel safety, auto insurance pricing and smart city applications. Traditional approaches largely
use GPS data to assess drivers. However, it is difficult to fine-grained
assess the time-varying driving behaviors. In this paper, we employ
the increasingly popular On-Board Diagnostic (OBD) equipment, which
measures semantic-rich vehicle information, to extract detailed trajectory and behavior data for analysis. We propose **PBE** system, which consists of Trajectory Profiling Model (**P**M), Driver Behavior Model (**B**M)
and Risk Evaluation Model (**E**M). PM profiles trajectories for reminding
drivers of danger in real-time. The labeled trajectories can be utilized to
boost the training of BM and EM for driver risk assessment when data
is incomplete. BM evaluates the driving risk using fine-grained driving
behaviors on a trajectory level. Its output incorporated with the time-
varying pattern, is combined with the driver-level demographic information for the final driver risk assessment in EM. Meanwhile, the whole **PBE**
system also considers the real-world cost-sensitive application scenarios.
Extensive experiments on the real-world dataset demonstrate that the
performance of **PBE** in risk assessment outperforms the traditional systems by at least 21%.

Keywords: Driver behavior analysis · On-Board Diagnostic (OBD)

1 Introduction

Nowadays, the number of traffic accidents increases rapidly every year [6,16].
Meanwhile, researchers have found that the driver behavioral errors caused more
than 90% of the crash accidents [13], served as the most critical factor leading
to the crash accidents. Therefore, how to effectively analyze the driver behavior

© Springer Nature Switzerland AG 2019
U. Brefeld et al. (Eds.): ECML PKDD 2018, LNAI 11053, pp. 507–523, 2019.
https://doi.org/10.1007/978-3-030-10997-4_31

and assess the driver risk plays a significant role in travel safety, auto insurance pricing and smart city applications.

In the last decades, the significance of this task has led to numerous research efforts [15,16]. Most of the previous work used GPS from vehicle [27], various sensors (e.g., magnetic and accelerometer sensors) from smartphone [5] and cameras [21] to collect data for analysis. Generally, when dealing with high-dimension and heterogeneous data, these work usually fails to take the fine-grained driver actions into consideration. Therefore, the prediction and evaluation of the driver behavior is limited. Besides, most traditional work does not consider the time-varying driving behaviors, making the driver risk assessment not sufficient.

To overcome the above drawbacks, we develop **PBE** system, which is able to fine-grained analyze the driving behavior based on the increasingly popular On-Board Diagnostic (OBD) equipments[1]. Each vehicle in our experiment is integrated with such an OBD device. So we have not only GPS-related information, but also semantic-rich vehicle information including engine speed and so on. Some recent work [6,20] also explores OBD. But, they usually focus on each OBD data tuple not from the trajectory perspective, which can consider the relationship among tuples and analyze from a global view for a better assessment.

Our **PBE** system aims to build a 3-tier model: Trajectory Profiling Model (PM), Driver Behavior Model (BM) and Risk Evaluation Model (EM). PM utilizes our insight from the data (the alarm information of OBD) to predict the trajectory class for profiling. It is able to remind drivers of danger in real-time. Besides, the labeled trajectories can be utilized to boost the training of BM and EM, when partial data is missing. BM evaluates the driving risk by fine-grained behavioral information from the trajectory perspective. EM combines the driver-level demographic information and BM's trajectory-level evaluation, to provide a comprehensive assessment for each driver to denote his/her risk. Besides, the time-varying driving pattern is also incorporated in EM. Meanwhile, **PBE** fully employs a cost-sensitive setting to satisfy the real-world application requirements, e.g., to lower the cost of misclassifying high risk as low risk in the real-time alarming system and auto insurance pricing scenario.

Overall, the main contributions are listed as follows. (1) **PBE** builds a real-time system via OBD device to remind drivers of danger. (2) Beyond fine-grained trajectory profiling results, **PBE** integrates the time-varying patterns and driver-level demographic information, to provide comprehensive evaluation scores for drivers. (3) We deploy the cost-sensitive setting to provide the practical analysis of drivers in the real-world application scenarios. (4) We perform extensive experiments using real-world OBD data. The performance of **PBE** system in risk assessment much better outperforms the traditional systems, by at least 21%.

---

[1] https://en.wikipedia.org/wiki/On-board_diagnostics.

2   Related Work

The existing work usually used GPS [27] records of a vehicle to generate the trajectory and mobility pattern for driver behavior analysis, due to the easy accessibility of GPS [27]. However, it is hard for these work to capture fine-grained driving actions. Besides, other work utilized smartphones (embedded with GPS and inertial sensors) [14,23] and camera image information [21]. But, some require the installation of external cameras in vehicles, which brings concerns on cost and privacy. Alternatively, Chen et al. [6] used OBD data tuples to judge the driving state. Furthermore, incorporated with OBD, Ruta et al. [20] also used other kinds of data like map and weather information to infer the potential risk factors. However, they mainly only emphasize on each data tuple. Different from the previous work, via OBD, we extract the fine-grained driving-action-related features to analyze drivers on a trajectory level.

Concerning the driver behavior analysis techniques, fuzzy logic [5,20] and statistical approaches [18] were explored. But, they need to manually set the rules. Besides, Bayesian Network [26] and its variants (Hidden Markov Model (HMM) [8,21], Dynamical Bayesian Network [1]) were used to find the inner relationship between the driving style and sensor data for the driving behavior inference. However, they have practical challenges due to the model complexity and the required large amount of data. Additionally, some work used AdaBoost [6] and Support Vector Machine [26] classifiers to determine the driving state. Although they can achieve high precision sometimes, these work fails to consider the cost-sensitive setting with the real-world application requirement. Meanwhile, traditional trajectory classification methods [2] mainly utilized HMM-based model, which are difficult to capture the driver behaviors when encountering the fine-grained multi-dimension data. On the other hand, time-series classification can also be used to classify driving trajectories for the behavior pattern analysis, e.g., the 1-nearest neighbor classifier [24,25]. But, the trajectories in our applications are quite different from time series with each point having multi-dimension points rather than only real values. Unlike the mentioned approaches, PBE considers the cost-sensitive setting and time-varying pattern, and analyzes comprehensively from multiple perspectives of the trajectory and driver level.

3   Preliminary

3.1   Data Description

OBD is an advanced on-board equipment in vehicles to record data. Each OBD data tuple x is defined as $<u_x, t_x, lon_x, lat_x, \phi_x, \psi_x>$ where: (1) u_x is the driver identification; (2) t_x is the data recording timestamp (in second); (3) lon_x, lat_x are the longitude and latitude location record where x is created; (4) ϕ_x ($\phi_x = [v_x, a_x, \omega_x, \Omega_x]$) is a four-dimensional vector representing the real-time **physical driving state** of *speed, acceleration, engine speed* (Round Per Minutes (RPM))

Table 1. Data description (the semantic driving state is set by domain experts).

Physical driving state	Semantic driving state	Description
Speed	Vehicle speeding (vs)	Speed higher than the road speed limit after matching with road types by GPS
Acceleration	Abnormal acceleration (aa)	Acceleration value $\geq 1.8\,\text{m/s}^2$
	Abnormal deceleration (ad)	Acceleration value $\leq -1.8\,\text{m/s}^2$
Engine speed	Engine high RPM warning ($ehrw$)	Engine speed higher than the default upper engine speed of a vehicle
	Abnormal engine speed increase ($aesi$)	Engine speed increases sharply in a short time
Vehicle angular velocity	Sharp turn (st)	Vehicle angular velocity $\geq 30\,\text{Rad/s}$
	Lane change (lc)	$10\,\text{Rad/s} <$ vehicle angular velocity $< 30\,\text{Rad/s}$

and *vehicle angular velocity* (Radian per second (Rad/s)); (5) ψ_x is a seven-dimensional vector representing the **semantic driving state** to denote the real-time *warning message* about the vehicle. It is derived from physical driving state, where $\psi_x = [vs_x, aa_x, ad_x, ehrw_x, aesi_x, st_x, lc_x]$ (More details are in Table 1 and the value type is binary (i.e., 1 means driver u is in this driving state at time t, and vice versa.). Besides, OBD can offer the **crash alarm** message to denote whether the car is highly likely to have a crash accident or not[2]. Each data tuple z is defined as $<u_z, t_z, c_z>$ where: (1) u_z, t_z are similar to aforementioned identification u_x and timestamp t_x; (2) c_z is the crash alarm. Like mentioned ψ_x, c_z also uses the binary value to denote the state.

Then, given a driver's massive OBD driving data, we analyze a driver's behavior by *trajectory* [27] with the following definition:

Definition 1 (TRAJECTORY). Given a driver's physical driving state record sequence $S = x_1 x_2 \ldots x_n$ and a time gap[3] Δt, a subsequence $S' = x_i x_{i+1} \ldots x_{i+k}$ is a trajectory of S if S' satisfies: (1) $v_{x_{i-1}} = 0, v_{x_i} > 0, v_{x_{i+k}} > 0, v_{x_{i+k+1}} = 0$; (2) if there exists a subsequence $S'' = x_j x_{j+1} \ldots x_{j+g} \in S'$, where for $\forall 0 \leq q \leq g, v_{x_{j+q}} = 0$, then, $t_{x_{j+g}} - t_{x_j} \leq \Delta t$; (3) there is no longer subsequences in S that contain S', and satisfy condition (1)(2).

A trajectory essentially leverages the speed and time constraints to extract reliable record sequences in the huge-volume driving records for effective studies.

---

[2] The OBD equipment made by JUWU IoT Technology Company (www.szjuwu.com) is incorporated with multiple processing units and domain experts' knowledge to decide the crash alarm state upon receiving sensor data.

[3] Four minutes is usually selected by domain experts for practical applications.

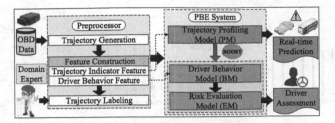

Fig. 1. Architecture of PBE system.

3.2 Problem Formulation

Given a driver set $U = \{u_i\}$ and their historical OBD data $X = \{x_i\}, Z = \{z_i\}$:

(1) How to profile driving trajectories?
(2) Based on trajectories, how to model drivers' driving risk?
(3) How to assess drivers as safe, risky or dangerous?

3.3 System Overview

In this subsection, we present PBE system as our solution to the problem. As shown in Fig. 1, it consists of four major components: **Preprocessor** performs *preprocessing* of OBD data, including generating trajectories, features and labels. The labels come from the claimed insurance data and domain experts. Concerning the *correlation* and *causality* to crash accidents, the generated features are divided into *trajectory indicator features* and *driver behavior features*. Trajectory indicator features are those trajectory variables (e.g., crash alarms) which indicate a vehicle is apt to crash accidents. They have no interpretation for the driving behavior. While, driver behavior features (e.g., abnormal accelerations/decelerations) denote driving actions during a trajectory, served as the possible reasons for crash accidents. **Trajectory Profiling Model (PM)** leverages trajectory indicator features to predict the trajectory class quickly for real-time alarming systems. When data is incomplete, PM's predicted label is able to give a boost for the latter training of BM and EM as a trajectory's pseudo label. **Driver Behavior Model (BM)** utilizes driver behavior features to model driving risk for behavior analysis from the trajectory level. Finally, **Risk Evaluation Model (EM)** computes drivers' risk evaluation scores considering both the time-varying trajectory-level pattern and demographic information.

4 Preprocessor

Preprocessor takes OBD data as the input and performs the following tasks to prepare the data for future processing:

Trajectory Generation reads a driver's OBD records to generate trajectories according to Definition 1. The filtering of noisy data tuple is conducted with a heuristic-based outlier detection method through speed information [27].

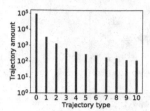
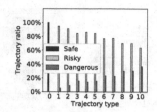
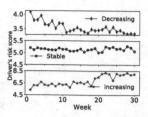

Fig. 2. Trajectory amount distribution.

Fig. 3. Trajectory ratio distribution.

Fig. 4. Time-varying pattern of driver's risk score.

Feature Construction computes features with a trajectory S ($S = x_1 x_2 \ldots x_n$):

(i) **Trajectory Indicator Features:** First, we utilize trajectory beginning time t_{x_1} and ending time t_{x_n} to query which crash alarm records exist during the trajectory period, and construct the crash alarm record sequence $\mathcal{Z} = z_1 z_2 \ldots z_m$. Then, we compute features: (1) *trajectory's running time* ($t_{x_n} - t_{x_1}$); (2) *trajectory's distance* ($\sum_{1 \le i \le n-1} \frac{1}{2}(v_{x_i} + v_{x_{i+1}})(t_{x_{i+1}} - t_{x_i})$); (3) *Crash Alarm Counts per trajectory (cac)* $\boldsymbol{cac} = \sum_{1 \le i \le m} 1(c_{z_i} = 1)$ where $1(\cdot)$ is an indicator function.

(ii) **Driver Behavior Features:** Driver behavior features (π_S, a eleven-dimensional vector) is defined as $<\overline{k}, dsc_q>$ where:

(1) $\overline{k} = \dfrac{\sum_{1 \le i \le n-1} \frac{1}{2}(k_{x_i} + k_{x_{i+1}})(t_{x_{i+1}} - t_{x_i})}{t_{x_n} - t_{x_1}}$, $k \in \{v, a, \omega, \Omega\}$, represents the *average speed/acceleration/engine speed/vehicle angular velocity*;

(2) $\boldsymbol{dsc_q} = \dfrac{\sum_{1 \le i \le n} 1(q_{x_i} = 1)}{t_{x_n} - t_{x_1}}$, $q \in \{vs, aa, ad, erhw, aesi, st, lc\}$, is the *Driving State Count per unit of time* for different semantic driving state q.

Trajectory Labeling sets the real-world ground truth label y_S of trajectory S by domain experts from insurance and transportation companies. There are three-class labels: **Safe Class (SC)**, **Risky Class (RC)** and **Dangerous Class (DC)**. Dangerous Class means the vehicle has crash accidents during the trajectory period according to auto insurance accident records. For the remaining trajectories with no accidents, domain experts judge them into Safe and Risky Class according to the driving smoothness of each trajectory.

5   Trajectory Profiling Model (PM)

In this section, we first conduct data statistics about generated trajectories to find data insights. Then based on the discovered insights, we develop two PMs of decision stump and decision tree to predict a trajectory label for profiling. Finally, we explain the PM boosting when data is incomplete.

5.1   Data Insight

After preprocessing, we count the trajectory amount distribution and the safe, risky and dangerous trajectory ratio with different trajectory types (i.e., different

Table 2. Cost matrix C.

	Predict SC	Predict RC	Predict DC
Actual SC	0	$C(SC, RC)$	$C(SC, DC)$
Actual RC	$C(RC, SC)$	0	$C(RC, DC)$
Actual DC	$C(DC, SC)$	$C(DC, RC)$	0

crash alarm counts per trajectory). As shown in Figs. 2 and 3, when crash alarm counts increase, the amount of the corresponding trajectory type decreases and the dangerous ratio increases. Interestingly, we find that the trajectories in the zero-crash-alarm-count type are all Safe Class. Furthermore, if a trajectory has more than one crash alarm, it can only be Risky or Dangerous Class. The reason may be that during the trajectory period, zero crash alarm means that the driver is driving smoothly without any risk or danger, leading to Safe Class. While, the generated crash alarm indicates the driver's aggressive driving, which results in a high probability of having crash accidents, lying in Risky or Dangerous Class. Thus, based on theses observations, we develop an insight that crash alarm can be a critical factor to predict a trajectory's label, so that it can be utilized for profiling trajectories in real-time.

5.2 Decision-Stump-Based Model

We first profile trajectories with only focusing on crash alarms (i.e., crash alarm counts per trajectory feature cac). To this end, we develop *Decision-stump-based* model to predict a trajectory S's label. Detailedly, we set two thresholds θ_1, θ_2 to generate a trajectory S's predicted label $\hat{y}_S$ as: $\hat{y}_S = \{SC : \text{if } cac \leq \theta_1; RC : \text{if } \theta_1 < cac \leq \theta_2; DC : \text{Otherwise}\}$.

To learn the parameters, we minimize a cost-sensitive objective function $C(\theta_1, \theta_2)$ with predicted label $\hat{y}_S$ and actual label y_S (ground truth label) by:

$$C(\theta_1, \theta_2) = \sum_{S \in \text{ all S}} 1(y_S \neq \hat{y}_S) \cdot C(y_S, \hat{y}_S), \tag{1}$$

where $C(i, j)$ is a cost matrix C (designed in Table 2). It means the cost that class i is mislabeled as class j and $i, j \in \{SC, RC, DC\}$. The value of the cost matrix is discussed in Sect. 8.6.

5.3 Decision-Tree-Based Model

Besides crash alarms, we also consider other trajectory indicator features (i.e., a trajectory's running time and distance) for profiling. Due to multiple features, we utilize decision tree rather than decision stump. The *Classification and Regression Tree* with *Gini index* is selected [4]. To achieve the cost-sensitive setting, we do not prune the decision tree with the max depth [9].

5.4 PM Boosting

Considering the real-world scenarios, the collected data sometimes is incomplete, e.g., labeled data missing in transmission or non-access to the private insurance claim data. Under such condition of lacking ground truth label, we use PM's predicted label instead as a pseudo label to boost the training of BM and PM.

6 Driver Behavior Model (BM)

6.1 Problem Formulation

In this part, we start to model each driver's driving behaviors from trajectories. This problem aims to predict the probability (P_S^k) of a trajectory lying in safe, risky or dangerous class given a trajectory S's driver behavior feature π_S as:

$$P_S^k = P(y_S = k | \pi_S), k \in \{SC, RC, DC\}. \tag{2}$$

This formulates the problem to a typical multi-class classification problem. Then, we employ the popular Gradient Boost Machine (GBM) with tree classifier [7] as the multi-class classifier (Open for other classifiers to plug in). However, different from traditional GBM, we include the cost-sensitive setting for the practical applications.

6.2 Cost-Sensitive Setting Design

Given the whole trajectory set $S_{all} = \{S_1, S_2, \ldots S_N\}$, we first build $N * 3$ basic regression tree classifiers. Each N classifiers classify Safe/Risky/Dangerous Class respectively through *One-vs-All* strategy and output J_S^k to denote the score of trajectory S belonging to class k. Then, with *Softmax Function*, we have trajectory S's risk probability $P_S^k = e^{J_S^k}/(e^{J_S^{SC}} + e^{J_S^{RC}} + e^{J_S^{DC}}), k \in \{SC, RC, DC\}$.

Importantly, during the process, to learn the parameters and achieve a cost-sensitive purpose, we design and minimize the following objective function,

$$\Psi = \lambda - \sum_{S \in S_{all}} \sum_{k \in \{SC, RC, DC\}} l_S^k w_k \ln P_S^k, \tag{3}$$

where λ is the regularized parameter for all tree classifiers, l_S^k is a binary value for selecting which P_S^k to compute. Detailedly, if ground truth label $y_S = k$, $l_S^k = 1$. Otherwise, $l_S^k = 0$. w_k is class k's weight for the cost-sensitive setting, achieved by multiplying different w_k values (different priorities) with the corresponding class k's cross entropy. By default, we set weights by ratio as $w_{SC} : w_{RC} : w_{DC} = C(SC, RC) : C(RC, SC) : C(DC, SC)$ from cost matrix C.

Then, for the iterative gradient tree boosting processing [7], the first and second order approximations (i.e., gradient $gred_s$ and Hessian matrix $hess_S$) are used to quickly optimize Ψ:

$$gred_S = \sum_{k \in \{SC, RC, DC\}} \frac{\partial \Psi}{\partial J_S^k} = \sum_{k \in \{SC, RC, DC\}} (P_S^k - l_S^k) w_k,$$

$$hess_S = \sum_{k \in \{SC, RC, DC\}} \frac{\partial^2 \Psi}{\partial J_S^{k^2}} = \sum_{k \in \{SC, RC, DC\}} 2(1 - P_S^k) P_S^k w_k. \tag{4}$$

7   Risk Evaluation Model (EM)

In this part, we first evaluate drivers from two perspectives: *Mobility-aware* from trajectories and *Demographic-aware* from driving habits (driver level). Then, we comprehensively consider the two evaluation scores and deploy the percentile ranking for the driver assessment.

7.1   Mobility-Aware Evaluation

For a trajectory S, after BM processing, we have the probability $P_S^{SC}, P_S^{RC}, P_S^{DC}$. Then, we compute a trajectory S's risk score, $risk_S$ by:

$$risk_S = P_S^{SC} d_{SC} + P_S^{RC} d_{RC} + P_S^{DC} d_{DC}, \tag{5}$$

where $d_{SC/RC/DC}$ is the *risk level* of probability $P_S^{SC/RC/DC}$. By default for the cost-sensitive goal, we set $d_{SC} : d_{RC} : d_{DC} = C(SC, RC) : C(RC, SC) : C(DC, SC)$ by ratio. Next, we generate driver u's m-th week's risk score $risk_u^m$ with this week's whole trajectory set ($\{S_1, S_2, \ldots, S_N\}$) by:

$$risk_u^m = \frac{1}{N} \sum_{S \in \{S_1, S_2, \ldots, S_N\}} risk_S. \tag{6}$$

Thus, with driver u's M-week OBD data, $\boldsymbol{r}_u$ ($\boldsymbol{r}_u = [risk_u^1, risk_u^2, \ldots, risk_u^M]$) denotes the risk score sequence. By generating and plotting the whole drivers' risk score sequences in Fig. 4, we find three typical time-varying patterns over time (i.e., increasing/stable/decreasing). Therefore, when rating drivers, it is

Table 3. Demographic-aware variable (habit) description. (Set by domain experts)

Group	Variable (per month)	Variable description
Time of day	Daytime hours ratio	Fraction between 8 a.m. to 8 p.m.
	Nighttime hours ratio	Fraction between 8 p.m. to 8 a.m.
Day of week	Weekday hours ratio	Fraction between Monday to Friday
	Weekend hours ratio	Fraction between Saturday to Sunday
Road type	Urban roads hours ratio	Fraction on urban roads
	Highway hours ratio	Fraction on highways
	Extra-urban hours ratio	Fraction on extra-urbans (countryside)
Mileage	Mileage per month	Overall exposure 30-day average

necessary to pay more attention to the present than the past. Then, we employ a *Linear Weight Vector* $\boldsymbol{w}$ ($|\boldsymbol{w}| = M, w_i = i$) to compute driver u's time-varying Mobility-aware evaluation score $Eval_u^{Mob}$ by: (Note that, concerning different time-varying patterns, open for other *weight vectors* to plug in)

$$Eval_u^{Mob} = \frac{1}{|\boldsymbol{w}|}\boldsymbol{w}^T \boldsymbol{r}_u. \tag{7}$$

7.2 Demographic-Aware Evaluation

We can also evaluate drivers by driving habits like the nighttime/daytime driving hours fraction per month (More driving habits in Table 3). On the other hand, viewing drivers' past τ-month trajectory data, according to domain experts ($\tau = 6$ for half of a year), there are three types of drivers: **Accident-Involved (AI)** (having more than two dangerous trajectories), **Accident-Related (AR)** (having less than one dangerous trajectory but more than fifteen risky trajectories), **Accident-Free (AF)** (the remaining). Based on these, our Demographic-aware evaluation problem aims to predict driver u's probability ($P_u^k, k \in \{AI, AR, AF\}$) of lying in AI, AR and AF by utilizing the driving habit variables.

Similar to BM, this problem also leads to a cost-sensitive multi-class classification task. After employing similar cost-sensitive solutions in BM (Due to space limit, we omit to present again), for driver u, we generate the probability of $P_u^{AI}, P_u^{AR}, P_u^{AF}$. Then, we have Demographic-aware evaluation score $Eval_u^{Dem}$ as:

$$Eval_u^{Dem} = P_u^{AI}m_{AI} + P_u^{AR}m_{AR} + P_u^{AF}m_{AF}, \tag{8}$$

where $m_{AI/AR/AF}$ is the *risk level* of probability $P_u^{AI/AR/AF}$ and we set $m_{AF} : m_{AR} : m_{AI} = C(SC, RC) : C(RC, SC) : C(DC, SC)$, similar to Eq. 5.

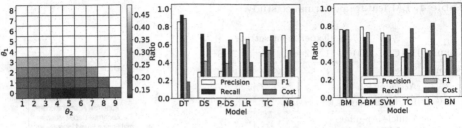

Fig. 5. PM grid search cost.

Fig. 6. PM comparison.

Fig. 7. BM comparison.

7.3 Driver Evaluation

Finally, for driver u, we sum the two scores as a driver evaluation score $Eval_u$:

$$Eval_u = \alpha Eval_u^{Mob} + \beta Eval_u^{Dem}, \tag{9}$$

where $Eval_u^{Mob}, Eval_u^{Dem}$ are normalized due to their different value ranges and α, β are weight parameters to indicate the significance/priority of corresponding evaluation score. The higher value leads to the higher importance. Generally, we set $\alpha = \beta = \frac{1}{2}$ to denote the equal significance in evaluating a driver (Flexible for other preferences to plug in for different user requirements). Finally, $Eval_u$ suggests a comprehensive risk score of driver u. The higher the score is, the more risky the driver is.

After generating all drivers' evaluation scores, we deploy the percentile ranking[4] to assess the drivers. According to domain experts' knowledge, 20% drivers can cause 80% crash accidents. Then, we set the percentile 80% drivers as the Dangerous Drivers. Among the rest drivers, usually 20% drivers are risky. Motivated by this, we set 80% percentile as the Risky Drivers and the final remaining as Safe Drivers (Available for other percentiles to plug in). Finally, by this setting, we can obtain two evaluation scores as thresholds to quickly assess a driver as safe, risky or dangerous.

8   Experiment

8.1   Setting and Dataset

During the experiments, we generated the equal number of trajectories in each class by resampling to balance the data. Besides, the 10-fold cross validation was conducted to present robust results. The real-world dataset collected drivers' OBD data from $August\ 22, 2016$ to $March\ 27, 2017$ for nearly 30 weeks, provided by a major OBD product company in China. After preprocessing, we have basic data statistics: 198 drivers, 98,218 trajectories ($Safe$ 91,687, $Risky$ 5,853 and $Dangerous$ 678), average trajectory time of 25.22 min and distance of 12.95 km.

---

[4] https://en.wikipedia.org/wiki/Percentile_rank.

Table 4. BM feature performance study.

		Experiment setting							
Column ID (COL. ID)		1	2	3	4	5	6	7	8
Feature*	Speed+	✓	✓	✓	✓	✓	✓	✓	✓
	Acceleration+		✓			✓	✓		✓
	Engine Speed+			✓		✓		✓	✓
	Vehicle Angular Velocity+				✓		✓	✓	✓
Metrics	Precision (%)	49	61	53	60	62	64	62	69
	Recall (%)	48	58	53	57	61	62	62	70
	F1 score (%)	49	57	53	56	61	61	62	69
	Cost ratio (%)	100	91	69	99	73	86	75	63

*Feature+ means the physical driving state feature plus its corresponding semantic driving state features (E.g., Speed+ means Speed plus Vehicle Speeding in Table 1).

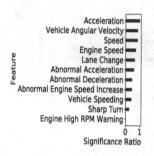

Fig. 8. BM feature significance.

8.2 Trajectory Profiling Model (PM) Evaluation

In this part, we first employ the grid search to find the optimal θ_1^*, θ_2^* in Decision-Stump-based model (DS) with the minimal cost-sensitive objective in Sect. 5.2. Then, we utilize the optimal DS and Decision-Tree-based model (DT) to profile trajectories, by predicting a trajectory's label through trajectory indicator features (i.e., crash alarm counts, distance and running time information).

After the grid search, as shown in Fig. 5 (Where cost is divided by its maximal cost as a ratio), we have the optimal thresholds of $\theta_1^* = 0, \theta_2^* = 5$. Explicitly, $\theta_1^* = 0$ means if a trajectory has no crash alarm records, it is Safe. It is consistent with the trajectory data statistics (Sect. 5.1). Besides, $\theta_2^* = 5$ denotes that if OBD generates crash alarm records during a trajectory, the crash alarm count of five is used to quickly judge whether the trajectory is Risky or Dangerous, for sending timely messages to drivers and reminding them of danger.

Figure 6 shows results for profiling trajectories in metrics of Precision, Recall, F1 score, as well as the cost (i.e., the misclassification cost in Eq. 1). We see that DT gets Precision, Recall and F1 score, close to 0.9, with the lowest cost. It outperforms the compared methods of multi-class Logistic Regression (LR) [10], Trajectory Clustering (TC) [11,12] and robust cost-sensitive Naive Bayes (NB) [9]. Furthermore, we also compare our two profiling models of DT and DS. As shown, DT is much better than DS. The reason may be that DT uses more rules (i.e., the higher depth in tree) and more features to sufficiently judge a trajectory's label, even in complex conditions. However, DS can quickly judge by only using one feature. According to the positive feedbacks from domain experts, DS is much easier to be implemented in current OBD device and has a great potential in real-time driving alarming systems with the easiest portability. Therefore, both DT and DS have their own advantages and suitable application scenarios. When users prefer the better performance, they may choose DT. For example, in PM boosting, by default, the predicted label is generated from DT.

Otherwise, if they want to quickly obtain results, DS is a good choice. *The impact of the cost-sensitive setting.* By examining the results of DS and Pure Decision-Stump-based model (P-DS), which removes the cost-sensitive setting when learning parameters, we find DS performs better with higher Recall and F1 score than P-DS. It validates the effectiveness of the cost-sensitive setting in the real-world application scenario to retrieve more risky/dangerous trajectories for higher Recall and F1 score.

8.3   Driver Behavior Model (BM) Evaluation

To evaluate BM, alternatively, we perform the formulated classification task (Sect. 6.1). We set the following compared methods: Logistic Regression (LR), Trajectory Clustering (TC), Support Vector Machine (SVM) [17], Bayesian Network (BN) [22] and Pure-BM (P-BM) where we remove BM's cost-sensitive setting.

The experiment result is shown in Fig. 7. It is observed that, BM has high Precision, Recall and F1 score close to 80% and the lowest cost. It beats SVM, LR, BN and TC in all metrics. This means that in our application, BM is more suitable to process the high-dimension trajectory feature data for the multi-class classification task. But, as mentioned before in Sect. 6.1, BM is open for other classifiers to plug in. *The impact of the cost-sensitive setting.* Compared to P-BM, BM outperforms in Recall and F1 score about 8% improvement with 27.70% lower cost. The reason may be that BM's cost-sensitive setting guides BM to give more priority to more risky classes like Risky and Dangerous class. This leads to the final prediction of Risky and Dangerous class more accurate than Safe class with higher Recall/F1 and lower cost. *The impact of feature (feature performance study).* We also test the effects of the driving behavior features under various feature combinations. The result is shown in Table 4.

It is observed that: (1) Compared to traditional GPS-related Speed+ features, adding any unique OBD-related feature like Acceleration+, Engine Speed+ and Vehicle Angular Velocity+ can improve the performance with higher Precision, Recall, F1 score and lower cost (See COL. 1 vs. COL. 2−8). It is natural to understand it because with more driving features, we can get larger feature space to describe the trajectory, leading to better predictions. Then, it effectively suggests the advantage of OBD for its involvement of more fine-grained driving features. (2) Seeing two comparing pairs: (i) (COL. 1 vs. COL. 2) vs. (COL. 1 vs. COL. 4) and (ii) (COL. 3 vs. COL. 5) vs. (COL. 3 vs. COL. 7), Acceleration+ and Vehicle Angular Velocity+ seem to have similar improving effects. The reason may be that both Acceleration+ and Vehicle Angular Velocity+ directly manifest the driving actions. Then, the improvement of adding either one feature is almost the same. However, adding Engine Speed+ leads to lower improvement in Precision, Recall, F1 score (see COL. 1 vs. COL. 2−4) probably because Engine Speed+ indicates the putting state of the drivers' feet on the oil pedal. It may not directly reflect the driving behaviors in the road like Acceleration+ and Vehicle Angular Velocity+. But viewing COL. 8, all the features together lead to the best performance. (3) Furthermore, in experiment setting 8, we investigate the significance of the whole features by measuring how many times a feature

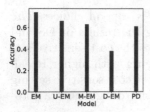
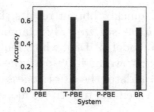
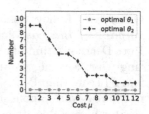

Fig. 9. EM comparison.

Fig. 10. System comparison.

Fig. 11. Effect of parameter.

is used in BM's tree classifiers to split a subtree. As shown in Fig. 8, the top two features of Acceleration and Vehicle Angular Velocity validate our previous analysis of their similar high improvements in performance compared to Speed.

8.4 Risk Evaluation Model (EM) Evaluation

In this subsection, we evaluate EM under the following enterprise scenario: with the whole driver set and the first 20-week data, after EM, we get a dangerous driver set with 80% percentile. Then, in the following 10-week dataset, we check whether these dangerous drivers have crash accident records or not. If Yes, our predictions are accurate and vice versa. We choose *Accuracy* as the metric. The compared methods are: (1) Pay as you Drive model (PD) is a state-of-the-art technique to evaluate drivers by conducting the vehicle classification task [19]. The generated classification probability is used for the evaluation and the parameters are carefully tuned to give the best performance. (2) M-EM/D-EM only contains the Mobility-aware/Demographic-aware evaluation score. (3) Unified-weight EM (U-EM) ignores the time-dependent pattern (Fig. 4) and utilizes the unified weight week vector to evaluate drivers (i.e., $w_i = 1$ in Eq. 7).

As shown in Fig. 9, EM has the highest accuracy. EM outperforms PD with 21% improvement. It may be caused from PD's vehicle classification part, which fails to consider the mobility-aware perspective. *The impact of the time-dependent pattern*. By examining the performance of EM and U-EM, we find that EM is more effective and improves the accuracy by about 13%. As aforementioned, we should give more priority to the latest driver behaviors rather than the very early, due to the changes of driving proficiency over time. *The impact of the Mobility-aware and Demographic-aware evaluation*. Compared with M-EM and D-EM, the accuracy of EM is about 47% better on average. It suggests that multiple perspectives lead to a more comprehensive evaluation for better performance. If only focusing on Mobility-aware or Demographic-aware information, we lose something for the evaluation. By viewing M-EM and D-EM only, one interesting finding is that M-EM is better. The reason is that M-EM evaluates drivers by the fine-grained driving behaviors from the trajectory perspective. It tells the dynamic mobility pattern so that it better describes and distinguishes drivers for the risk assessment. But, D-EM's rating is mainly based on general, less-distinctive and static variables like the per-month traveled mileage.

8.5 PBE System Evaluation

We investigate PBE system's performance by the same task described in EM evaluation in Sect. 8.4. To fully test the boosting effect, we consider the extreme condition with no ground truth labels, by setting PM's predicted labels as the whole trajectories' ground truth labels. The compared methods are: (1) Two-class PBE (T-PBE), which considers the two-class setting rather than PBE's three-class setting. Specially, it regards Dangerous class as one class while Risky and Safe class as another class. (2) Pure-PBE (P-PBE), one removes the cost-sensitive setting in the whole system. (3) Behavior-centric Risk-level model (BR) is a state-of-the-art method to evaluate drivers [3]. It is incorporated in an insurance pricing model to rate drivers' risk for evaluation.

The result is shown in Fig. 10. It can be found that: (1) PBE outperforms T-PBE with 9% improvement. The reason may be that PBE additionally utilizes the Risky Class trajectory to develop more semantic-rich descriptions of a trajectory and a driver (i.e., more probabilities to describe). This result suggests the advantage of the multi-class fine-grained analysis. Furthermore, PBE system is open for other multi-class settings to plug in not limited to current three classes. (2) By viewing the result of PBE and P-PBE, we observe that PBE has higher accuracy than P-PBE by 15%. It suggests that the cost-sensitive setting is effective in the whole system. The reason is aforementioned that through the cost-sensitive guidance in the system, Risky and Dangerous Classes get more priority to retrieve more risky trajectories/drivers in the real-world enterprise scenario like auto insurance. (3) Examining PBE and BR's performances, we note that PBE beats BR by 28%. Different from PBE, BR's evaluation from classification fails to consider trajectories' fine-grained driving behaviors, which leads to the lower performance. (4) Comparing the results of PBE's predicted label and the ground truth label (in Sect. 8.4), current PBE is only slightly worse by 7%. Such slight difference is acceptable in the real-world applications when the ground truth label is hard to assess and data is incomplete. It justifies the effectiveness of PM for boosting the training of BM and EM.

8.6 Parameter Tuning

In PBE, the major parameter is the cost matrix C in Table 2. Considering the trade-off between the cost-sensitive requirement and the scalable training, we set $C(SC, RC) : C(SC, DC) : C(RC, DC) : C(RC, SC) : C(DC, RC) : C(DC, SC) = 1 : 1 : 1 : \mu : \mu : \mu^2$. Through studying cost μ by Decision-Stump-based model's cost-sensitive objective $C(\theta_1, \theta_2)$ (in Sect. 5.2), where C is used for the first time in PBE, we can examine the impact of C. As shown in Fig. 11, μ greatly influences θ_2^* without affecting θ_1^*. Specifically, μ increases with θ_2^* decreasing. When μ is too low ($\mu < 4$)/high ($\mu > 6$), the optimal θ_2^* is just around θ_2's max/min value, leading to improper results. Thus, we select the middle value $\mu = 5$, resulting in a middle threshold value $\theta_2^* = 5$ in the experiment.

Table 5. The running time of PBE system.

Week #	1	5	10	15	20	25	30
Time (in second)	7	43	88	138	189	227	285

8.7 Efficiency Study

We report the running time of PBE when the week number increases (i.e., when more data is collected). As shown in Table 5, the time cost increases with more data. The reason is that with more data, more trajectories are generated and more tree classifiers are built, which result in more running time.

9 Conclusion

In this paper, we proposed PBE system, including PM, BM and EM, to assess the driver behaviors. PM utilizes the insight from the collected data for real-time alarming. BM assesses the driver behavior risk by fine-grained analyzing the trajectory data. EM evaluates drivers from multiple perspectives and gives comprehensive scores to reflect different risky scores. PBE is evaluated via extensive experiments and outperforms the traditional systems by at least 21%. In the future, we will consider more spatial factors like location/road type for analysis.

Acknowledgements. This research was supported by Shenzhen Peacock Talent Grant 827-000175, Guangdong Pre-national Project 2014GKXM054, the University of Macau Start-up Research Grant (SRG2015-00050-FST) and Research & Development Grant for Chair Professor (CPG2015-00017-FST), and Natural Science Foundation of China: 61572488 and 61673241.

References

1. Al-Sultan, S., Al-Bayatti, A.H., Zedan, H.: Context-aware driver behavior detection system in intelligent transportation systems. VT **62**(9), 4264–4275 (2013)
2. Bashir, F.I., Khokhar, A.A., Schonfeld, D.: Object trajectory-based activity classification and recognition using hidden Markov models. TIP **16**(7), 1912–1919 (2007)
3. Bian, Y., Yang, C., Zhao, J.L., Liang, L.: Good drivers pay less: a study of usage-based vehicle insurance models. Transp. Res. Part A: Policy Pract. **107**, 20–34 (2018)
4. Breiman, L.: Classification and Regression Trees. Routledge, New York (2017)
5. Castignani, G., Derrmann, T., Frank, R., Engel, T.: Driver behavior profiling using smartphones: a low-cost platform for driver monitoring. ITSM **7**(1), 91–102 (2015)
6. Chen, S.H., Pan, J.S., Lu, K.: Driving behavior analysis based on vehicle OBD information and AdaBoost algorithms. In: IMECS, vol. 1, pp. 18–20 (2015)
7. Chen, T., Guestrin, C.: XGBoost: a scalable tree boosting system. In: SIGKDD, pp. 785–794. ACM (2016)

8. Daptardar, S., Lakshminarayanan, V., Reddy, S., Nair, S., Sahoo, S., Sinha, P.: Hidden Markov model based driving event detection and driver profiling from mobile inertial sensor data. In: SENSORS, pp. 1–4. IEEE (2015)

9. Elkan, C.: The foundations of cost-sensitive learning. In: IJCAI, pp. 973–978. Lawrence Erlbaum Associates Ltd. (2001)

10. Friedman, J., Hastie, T., Tibshirani, R.: The Elements of Statistical Learning. Springer Series in Statistics, vol. 1. Springer, New York (2001). https://doi.org/10.1007/978-0-387-21606-5

11. Gaffney, S., Smyth, P.: Trajectory clustering with mixtures of regression models. In: SIGKDD, pp. 63–72. ACM (1999)

12. Hartigan, J.A., Wong, M.A.: Algorithm as 136: a k-means clustering algorithm. J. R. Stat. Society. Ser. C (Appl. Stat.) 28(1), 100–108 (1979)

13. Hendricks, D.L., Fell, J.C., Freedman, M.: The relative frequency of unsafe driving acts in serious traffic crashes. In: AAAM (2000)

14. Hong, J.H., Margines, B., Dey, A.K.: A smartphone-based sensing platform to model aggressive driving behaviors. In: SIGCHI, pp. 4047–4056. ACM (2014)

15. Kaplan, S., Guvensan, M.A., Yavuz, A.G., Karalurt, Y.: Driver behavior analysis for safe driving: a survey. ITS 16(6), 3017–3032 (2015)

16. Kumar, M.K., Prasad, V.K.: Driver behavior analysis and prediction models: a survey. IJCSIT 6(4), 3328–3333 (2015)

17. Nilsson, R., Peña, J.M., Björkegren, J., Tegnér, J.: Evaluating feature selection for SVMs in high dimensions. In: Fürnkranz, J., Scheffer, T., Spiliopoulou, M. (eds.) ECML 2006. LNCS (LNAI), vol. 4212, pp. 719–726. Springer, Heidelberg (2006). https://doi.org/10.1007/11871842_72

18. Osafune, T., Takahashi, T., Kiyama, N., Sobue, T., Yamaguchi, H., Higashino, T.: Analysis of accident risks from driving behaviors. Int. J. ITS Res. 15(3), 192–202 (2017)

19. Paefgen, J., Staake, T., Fleisch, E.: Multivariate exposure modeling of accident risk: insights from pay-as-you-drive insurance data. Transp. Res. Part A: Policy Pract. 61, 27–40 (2014)

20. Ruta, M., Scioscia, F., Gramegna, F., Di Sciascio, E.: A mobile knowledge-based system for on-board diagnostics and car driving assistance. In: UBICOMM, pp. 91–96 (2010)

21. Tran, C., Doshi, A., Trivedi, M.M.: Modeling and prediction of driver behavior by foot gesture analysis. CVIU 116(3), 435–445 (2012)

22. Tschiatschek, S., Reinprecht, P., Mücke, M., Pernkopf, F.: Bayesian network classifiers with reduced precision parameters. In: Flach, P.A., De Bie, T., Cristianini, N. (eds.) ECML PKDD 2012. LNCS (LNAI), vol. 7523, pp. 74–89. Springer, Heidelberg (2012). https://doi.org/10.1007/978-3-642-33460-3_10

23. Van Ly, M., Martin, S., Trivedi, M.M.: Driver classification and driving style recognition using inertial sensors. In: IEEE Intelligent Vehicles Symposium IV, pp. 1040–1045. IEEE (2013)

24. Wei, L., Keogh, E.: Semi-supervised time series classification. In: SIGKDD, pp. 748–753. ACM (2006)

25. Xi, X., Keogh, E., Shelton, C., Wei, L., Ratanamahatana, C.A.: Fast time series classification using numerosity reduction. In: ICML, pp. 1033–1040. ACM (2006)

26. Xu, G., Liu, L., Song, Z.: Driver behavior analysis based on Bayesian network and multiple classifiers. In: ICIS, vol. 3, pp. 663–668. IEEE (2010)

27. Zheng, Y.: Trajectory data mining: an overview. ACM TIST 6(3), 29 (2015)

Accurate WiFi-Based Indoor Positioning with Continuous Location Sampling

J. E. van Engelen[1]([⊠]), J. J. van Lier[2], F. W. Takes[1], and H. Trautmann[3]

[1] Department of Computer Science (LIACS), Leiden University,
Leiden, The Netherlands
jesper.van.engelen@gmail.com, takes@liacs.nl
[2] Big Data & Analytics, KPMG Advisory N.V., Amstelveen, The Netherlands
jvlier@gmail.com
[3] Department of Information Systems, University of Münster, Münster, Germany
trautmann@wi.uni-muenster.de

Abstract. The ubiquity of WiFi access points and the sharp increase in WiFi-enabled devices carried by humans have paved the way for WiFi-based indoor positioning and location analysis. Locating people in indoor environments has numerous applications in robotics, crowd control, indoor facility optimization, and automated environment mapping. However, existing WiFi-based positioning systems suffer from two major problems: (1) their accuracy and precision is limited due to inherent noise induced by indoor obstacles, and (2) they only occasionally provide location estimates, namely when a WiFi-equipped device emits a signal. To mitigate these two issues, we propose a novel Gaussian process (GP) model for WiFi signal strength measurements. It allows for simultaneous smoothing (increasing accuracy and precision of estimators) and interpolation (enabling continuous sampling of location estimates). Furthermore, simple and efficient smoothing methods for location estimates are introduced to improve localization performance in real-time settings. Experiments are conducted on two data sets from a large real-world commercial indoor retail environment. Results demonstrate that our approach provides significant improvements in terms of precision and accuracy with respect to unfiltered data. Ultimately, the GP model realizes continuous location sampling with consistently high quality location estimates.

Keywords: Indoor positioning · Gaussian processes
Crowd flow analysis · Machine learning · WiFi

1 Introduction

The increasing popularity of wireless networks (WiFi) has greatly boosted both commercial and academic interest in indoor positioning systems. Requiring no specialized hardware, WiFi-based positioning systems utilize off-the-shelf wireless access points to determine the location of objects and people in indoor environments using wireless signals, emitted by a multitude of electronic devices.

© Springer Nature Switzerland AG 2019
U. Brefeld et al. (Eds.): ECML PKDD 2018, LNAI 11053, pp. 524–540, 2019.
https://doi.org/10.1007/978-3-030-10997-4_32

This location information finds a broad range of applications, including in robotics, indoor navigation systems, and facility management and planning in retail stores, universities, airports, public buildings, etc. [12,13].

WiFi-based positioning systems use sensors to capture the signals transmitted by WiFi-equipped devices such as smartphones and laptops. Since the signals are attenuated (i.e., reduced in strength) as they travel through physical space, a sensor close to a device emitting a signal will measure a higher signal strength than a sensor farther away from the device. By combining measurements of the same signal by different sensors, indoor positioning systems can approximate the signal's origin. Two broad groups of localization techniques exist: fingerprinting-based and model-based methods. The former can be regarded as a form of supervised learning, where a model is trained to predict a device's location based on grouped signal strength measurements by passing it pairs of measurement groups (sample features) and known locations (labels). Consequently, fingerprinting-based methods require *training data* [5,24]. Model-based methods, on the other hand, require no training data and rely on physical models for the propagation of signals through space to determine the most likely location of the device [20]. They include lateration-based methods, which estimate the distance to multiple sensors based on the received signal strengths and use regression methods to determine the most likely position of the device.

The first concern with all of these approaches is the unpredictability of signal propagation through indoor environments. Static obstacles such as walls, ceilings, and furniture attenuate the transmitted signal and prohibit the construction of an accurate model for the received signal strength. Furthermore, the refraction of signals by these obstacles leads to multiple observations of the same signal by a single sensor, but at different signal strengths (the *multipath* phenomenon, see [16]). Dynamic obstacles such as people further complicate the process, as these cannot be modeled offline. The resulting observation noise results in significant variance of location estimates in all existing approaches [12]. The second problem encountered using these methods is that they can only estimate the location of a device when it transmits a signal. In practice, signals are transmitted infrequently and at irregular intervals, thus only allowing for intermittent positioning.

Both of these problems form major hurdles in the application of WiFi-based location analysis to real-world issues. With intermittent, inaccurate, and imprecise location estimates, any further data mining and analysis becomes exceedingly difficult. In this paper, we propose a novel method to mitigate these issues. We employ a nonparametric approach to approximate sensors' received signal strength distributions over time. For this, we use a Gaussian process (GP) model to obtain an a posterior estimate and associated variance information of signal strengths, which simultaneously allows us to resample from the distribution at any timestamp, and to sample only measurements with sufficiently low variance. The proposed approach can greatly enhance any model based on signal strengths. In addition to the Gaussian process model, we also compare several smoothing methods for location estimates. Ultimately, our main contribution is an accurate model that addresses each of the problems outlined above, reducing localization noise and providing accurate location estimates at arbitrary time intervals.

The rest of the paper is structured as follows. First, related work is outlined in Sect. 2. Section 3 gives a formal problem statement. Methods for smoothing and interpolation and the proposed Gaussian process model are discussed in Sects. 4 and 5. Experiments are covered in Sect. 6. Section 7 concludes.

2   Background and Related Work

Driven by the proliferation of WiFi-equipped devices, WiFi sensors have become a popular choice for use in indoor positioning systems. Requiring no specialized hardware, the accuracy of these approaches is generally in the range of 2 to 4 m [12]. This makes them a good candidate for indoor localization purposes in many areas, including, among others, facility planning and indoor navigation. In robotics, localization of robots in indoor spaces forms a key challenge [22].

Both physical models for signal propagation and machine learning methods are broadly applied. The latter include methods based on traditional machine learning techniques such as support vector machines and k-nearest neighbours [12]. Notably, a Gaussian process model for generating a likelihood model of signal strengths from location data is proposed in [5]. More recently, a neural network approach was suggested by Zou et al. [24]. Such fingerprinting-based models require training data to construct a prediction model, which has two major drawbacks [14]. First, training data can be expensive to obtain: to construct a proper localization model, training data consisting of signal strength measurements and known device coordinates is required from locations throughout the input space. Second, when extending a localization system to new environments or with additional sensors, the system needs to be re-trained with additional training data.

Approaches requiring no calibration data instead rely on physical models for the propagation of signals through space, only requiring the locations of the sensors to be known. They include approaches based on the measured angle of the incoming signal (*angle of arrival*) and approaches based on the time a signal takes to reach the sensor (*time of flight*) [8,15]. Furthermore, distance-based models, localizing devices using a model of the propagation of signals through space, are often used [10,23]. They exploit the fact that radio signals reduce in strength as they propagate through space, utilizing lateration-based techniques to find the most likely position of the device based on signal strength measurements at multiple sensors. Lastly, approaches exist that require neither training data nor knowledge about the environment (such as sensors locations). These approaches, known as *simultaneous localization and mapping* (SLAM [22]) methods, simultaneously infer environment information and device locations. Applications of this method using WiFi sensors include *WiFi SLAM* [4] and *distributed particle SLAM* [2]. For an extensive review of WiFi-based positioning systems and applications, we refer the reader to [3,8,12].

No single performance measure exists for evaluating localization quality, but consensus exists in literature that both accuracy and precision are important.

In [12] and [3], different performance criteria are discussed; we will use a subset of these in our work. The main contribution of this paper is the proposition of a lateration-based model using Gaussian processes for signal strength measurements, addressing the most important performance criteria of accuracy, precision, and responsiveness simultaneously. Our approach enables high-quality location estimates at arbitrary timestamps. To the best of our knowledge, this is the first generic model for continuously estimating device locations.

3   Problem Statement

Our aim is to construct a (1) precise, (2) accurate, and (3) responsive (i.e., able to provide continuous location estimates) positioning system for indoor environments where relatively few signal observations are present. Furthermore, the method should be robust and cost-effective.

Assume we are provided with a set of n signal strength measurements $X = (\mathbf{x}_1, \ldots, \mathbf{x}_n)$ for a single device. Each measurement is generated by a device at some unknown position. The vector $\mathbf{x}_i$ consists of the observation timestamp, an identifier of the sensor receiving the signal, the signal strength, and information to uniquely identify the package transmitted by the device. Let $\mathbf{d}_t$ denote the position of the device at time t. Our objective, then, is to obtain a position estimate $\hat{\mathbf{d}}_t$ for the device at any timestamp t based on the measurements X. We wish to maximize the accuracy, i.e., minimize the distance between the expected estimated location and the actual location, $||\mathbb{E}[\hat{\mathbf{d}}_t] - \mathbf{d}_t|||$, where $|| \cdot ||$ denotes the Euclidean norm. We also wish to maximize the precision, i.e., minimize the expected squared distance between the estimated location and the mean estimated location, $\mathbb{E}\left[||\hat{\mathbf{d}}_t - \mathbb{E}[\hat{\mathbf{d}}_t]||^2\right]$.

Of course, we cannot evaluate the expected accuracy and precision. Instead, we optimize the empirical accuracy and precision. Assume our calibration data consists of observations of a device at c different known locations $(\mathbf{p}_1, \ldots, \mathbf{p}_c)$. Each set C_j then consists of the estimated device positions when the device was at position $\mathbf{p}_j$. Let $\hat{\mathbf{p}}_j$ be the mean of the estimated positions for calibration point j, i.e., $\hat{\mathbf{p}}_j = \frac{1}{|C_j|} \sum_{\hat{\mathbf{d}} \in C_i} \hat{\mathbf{d}}$. The accuracy and precision are then calculated as an average over the accuracy and precision at all calibration points.

The empirical accuracy (Acc) is calculated as the mean localization error,

$$Acc = \frac{1}{c} \sum_{j=1}^{c} ||\hat{\mathbf{p}}_j - \mathbf{p}_j||. \tag{1}$$

The empirical precision ($Prec$) per calibration point is calculated as the mean squared distance between the estimated location and the mean estimated location. This yields, averaged over all calibration points,

$$Prec = \frac{1}{c} \sum_{j=1}^{c} \left(\frac{1}{|C_j|} \sum_{\hat{\mathbf{d}} \in C_j} ||\hat{\mathbf{d}} - \hat{\mathbf{p}}_j||^2 \right). \tag{2}$$

4   Lateration-Based Positioning

WiFi-equipped devices emit radio signals when connected to an access point, and when scanning for a known access point. Sensors can be utilized to listen for these signals, recording the received signal strength. We henceforth refer to this received package and the associated signal strength as a *measurement*. Our approach combines measurements of a signal by different sensors to determine its spatial origin. It requires no prior knowledge besides the sensor locations.

To localize a device, we need measurements of the same signal from multiple sensors. Each transmitted package contains information, including the package's sequence number and a device identifier, by which we can identify unique packages received by multiple sensors. We henceforth refer to these groups of measurements of the same package by different sensors as *co-occurrences*.

4.1   Localization Model

Having identified a co-occurence, we wish to translate this set of the package's signal strength measurements into an estimate of the transmitting device's location. We do so by modeling the propagation of the signal through space, and finding the device coordinates and parameters that minimize the difference between the modeled signal strengths and the observed signal strengths.

To model the propagation of a signal in space from a transmitting antenna to a receiving antenna, we make use of the Friis transmission equation [6]. It defines the relation between the transmitted signal strength P_t (from an antenna with gain G_t), the received signal strength P_r (at an antenna with gain G_r), the wavelength λ of the signal, and the distance R between the antennas. Antenna gain is a measure for the efficiency of the antenna, and is constant for each antenna. Using the *dBm* unit for the signal strengths, we calculate P_r by

$$P_r = P_t + G_t + G_r + 20 \cdot \log\left(\frac{\lambda}{4\pi R}\right), \qquad (3)$$

where $\log(\cdot)$ denotes the logarithm with base 10. Equation 3 is premised on the assumption that the path between the two antennas is unobstructed (the *free space* assumption). This assumption is captured in the last term of the equation, which is the logarithm of the inverse of what is known as the free-space path loss $(\frac{4\pi R}{\lambda})^2$ (also known as the free-space loss factor [1]). However, in most real-world indoor environments, we cannot assume free space. To combat this problem, we make use of an empirically derived formula for modeling propagation loss from [9]. It introduces the *path loss exponent* n as the exponent in the fraction of the path loss equation. For free-space environments, $n = 2$, yielding the original transmission equation. For environments with obstructions, generally $n > 2$ [18]. We introduce a method for estimating n in Sect. 4.2. We note that, by using a physical model for estimating device locations based on measurements, our lateration-based method does not require an expensive training phase. Further motivation for using lateration-based methods over fingerprinting-based methods can be found in the fact that, unlike in most machine learning problems, the true model generating our observations is known (up to corrections for the multipath phenomenon). This valuable prior knowledge is discarded in most fingerprinting-based methods.

Given the signal strength model, we formulate the localization problem as a regression problem, minimizing the sum of squared differences between the measured signal strength, and the signal strength obtained when calculating P_r from the Friis transmission equation using our estimated location parameters. Assuming i.i.d. measurements with additive Gaussian noise, this corresponds to the maximum likelihood estimator. We define the model for the transmission strength based on the Friis transmission equation from Eq. 3 and incorporate sensor i's path loss exponent n_i, rewriting it for a single sensor as:

$$P_r = P_t + G_t + G_r + n_i \cdot 10 \cdot \log\left(\frac{\lambda}{4\pi R_i}\right)$$

$$= P_t + G_t + G_r + n_i \cdot 10 \cdot \log\left(\frac{\lambda}{4\pi}\right) - n_i \cdot 10 \cdot \log R_i$$

$$= \rho - n_i \cdot 10 \cdot \log R_i,$$

where $\rho = P_t + G_t + G_r + n_i \cdot 10 \cdot \log\left(\frac{\lambda}{4\pi}\right)$ is the bias term to be estimated, and the path loss exponent n_i is assumed known and constant. The model corresponds to the transmission strength model defined in [9]. The resulting system of equations is underdetermined in the general case where G_r is dependent on the sensor. Thus, we assume that G_r is constant across all sensors, making the system overdetermined. In our case, as all WiFi sensors are of the same type, this is reasonable. Expressing R_i in terms of the sensor location vector $\mathbf{s}^{(i)} = \left(s_x^{(i)}, s_y^{(i)}\right)^T$ and the device location estimate vector $\mathbf{d} = (d_x, d_y)^T$, we obtain our model:

$$f_i(\theta) \equiv f_i(\theta | \mathbf{s}^{(i)}, n_i) = \rho - n_i \cdot 10 \cdot \log \|\mathbf{s}^{(i)} - \mathbf{d}\|, \tag{4}$$

where $\theta \equiv (\rho, d_x, d_y)^T$ are the parameters to be estimated. We are now in place to define our loss function J:

$$J(\theta) \equiv J(\theta | \mathbf{s}^{(1)}, \dots, \mathbf{s}^{(N)}, \mathbf{n}) = \sum_{i=1}^{N} (P_r^{(i)} - f_i(\theta))^2, \tag{5}$$

where $P_r^{(i)}$ is the measured signal strength at sensor i. We wish to minimize this function, i.e., we want to find $\hat{\theta} \in \arg\min_\theta J(\theta)$. The loss function is nonconvex, and has no closed-form solution. Therefore, we make use of Newton's method, which iteratively minimizes the loss function using the update rule $\theta_{t+1} = \theta_t - \frac{f'(\theta_t)}{f''(\theta_t)}$ for determining the set of parameters θ at iteration $(t + 1)$. The initial state can be chosen in several ways, e.g., by taking the weighted mean position of all sensors that received the signal.

The positioning procedure described thus far localizes each co-occurrence independently, without taking into account previous or future estimated device positions. In other words, it assumes the locations over time for a single device are independent. Furthermore, it is premised on the assumption that the signal propagates through free space, which does not generally hold. In the remainder of this section, we propose several methods to overcome these shortcomings and improve the quality of fits in the sense of the outlined performance criteria.

4.2   Estimating the Path Loss Exponent

The general Friis transmission equation is premised on the assumption that the radio signal propagates through free space, which is generally not the case in indoor environments. To combat this, [9] proposes an empirical adjustment to the model, introducing path loss exponent n in the Friis transmission equation, where n grows as the free-space assumption is relaxed.

Considering Eq. 4, we see that the received signal strength can be rewritten as $P_r = \rho - nx$, where $x = 10 \cdot \log(R)$. As all parameters in ρ are assumed to be constant with respect to R, P_r is linear in $\log(R)$. Now, using calibration data for which R, the distance between device and sensor, and P_r, the received signal strength at the sensor, are known, we can apply linear regression to estimate ρ and n. Having estimated path loss exponent n, we can use it in our model to account for attenuation effects induced by the surroundings.

4.3   Smoothing and Filtering Fits

Another improvement on fit quality, in the sense of in particular precision, but also accuracy, can be achieved by exploiting the knowledge that, during short periods of time, the position of a device is not expected to change significantly. This opens up the possibility of simply smoothing the estimated locations through time. Here, we outline the most common smoothing and filtering methods, which can be applied to the estimated x- and y-coordinates individually.

First, we consider using the *exponential moving average* (EMA) [7]. Due to its $O(1)$ complexity in smoothing a single data point, and because it only depends on previous observations, it is a good candidate for systems requiring large-scale and real-time localization. In its simplest form, the EMA assumes evenly spaced events, and calculates the smoothed value x_i' at step i as a weighted average of x_{i-1}', the smoothed value at step $i - 1$, and x_i, the unsmoothed input value. Using α to control the amount of smoothing, we obtain $x_i' = x_i \cdot \alpha + x_{i-1}' \cdot (1 - \alpha)$.

Second, we consider *Gaussian smoothing* [21], which can be seen as the convolution of the signal with a Gaussian kernel. Like EMA, the filter generally assumes evenly spaced observations. However, by adjusting the weighting of each sample based on its observation timestamp, we can apply it to non-evenly-spaced observations as well. In discrete space, we calculate the smoothed value x_i' as the weighted average of all observations, where the weight of x_j is dependent on the time difference between observations j and i. Denoting n as the number of observations and t_i as the observation timestamp of sample i, we write

$$x_i' = \frac{\sum_{j=1}^n w_j x_j}{\sum_{j=1}^n w_j} \quad , \text{ where } \quad w_j = \frac{1}{\sqrt{2\pi\sigma^2}} \exp\left(-\frac{1}{2\sigma^2}(t_j - t_i)^2\right). \quad (6)$$

Thus, the filter simply smooths the locations' x- and y-coordinates. Theoretically, the filter has time complexity $O(n)$ for a single observation, but this can be reduced to $O(1)$ for observations sorted by timestamp. As a large part of a Gaussian's density is concentrated close to its mean, we can simply use a small central section of the Gaussian without expecting significant changes in results.

Third, we consider a more sophisticated smoothing approach: the *Savitzky-Golay* filter [19], which smooths values by fitting a low-degree polynomial (centered around the observation to be smoothed) to the observed values in their vicinity. It then predicts the smoothed value at time t by evaluating the fitted polynomial function at its center. This corresponds to the bias term in the fitted polynomial. Each observation is smoothed independently, and makes use of the observations within some pre-specified window around the observation to be smoothed. For evenly spaced observations, an analytical solution exists; numerical methods are required for our non-evenly-spaced observations.

5   Gaussian Processes for Measurement Resampling

Smoothing and filtering approaches address the first of the two most significant problems of WiFi-based indoor localization: they improve accuracy and precision. However, they do not tackle the second problem, concerning the scarcity of measurements. We introduce a method to address both of the issues simultaneously, allowing arbitrary resampling of measurements while limiting variance.

Our method generates a model for the signal strengths measured by the sensors for a single device, and then resamples signals from this model. The model is constructed by means of a Gaussian process, making use of the fact that the signal of a device as measured by a sensor is expected to vary only slightly over small time intervals. Resampling facilitates the construction of new measurements at arbitrary timestamps, and reduces variance in signal strengths at the same time, by interpolating between signals received from the device around the requested timestamp. Before continuing to the implementation of this method, we provide a brief explanation of Gaussian processes; for details on prediction with Gaussian processes, we refer the reader to [17].

5.1   Gaussian Processes

Assume that we have a data set consisting of n observations, and that each observation is in $\mathbb{R}^d$. We denote $\mathcal{D} = (\mathbf{x}_i, y_i)_{i=1}^n$, where $\mathbf{x}_i \in \mathbb{R}^d$ and $y_i \in \mathbb{R}$ denote the feature vector and the target value, respectively, for the ith data point. The observations are drawn i.i.d. from some unknown distribution specified by

$$y_i = f(\mathbf{x}_i) + \epsilon_i, \tag{7}$$

where ϵ_i is a Gaussian distributed noise variable with 0 mean and variance σ_i^2.

Since we want to predict a target value for previously unseen inputs, our objective is to find a function $\hat{f}$ that models f. A Gaussian process estimates the posterior distribution over f based on the data. A model is sought that finds a compromise between fitting the input data well, and adhering to some prior preference about the shape of the model. This prior preference touches on a fundamental concept in Gaussian processes: they are based on the assumption that some similarity measure between two inputs exists that defines the correlation between their target values based on their input values. More formally, it requires a kernel function $k(\mathbf{x}, \mathbf{x}')$ to be specified that defines the correlation between any pair of inputs $\mathbf{x}$ and $\mathbf{x}'$. Furthermore, a Gaussian process requires a prior mean μ (usually 0) and variance σ^2 (chosen to reflect the uncertainty in the observations).

5.2   A Gaussian Process Model for Signal Strength Measurements

We propose a method for modeling the signal strength over time for a single device and a single sensor. We assume that each measurement consists of at least a timestamp, the measured signal strength, and information to uniquely identify the co-occurence (see Sect. 4), the device, and the sensor. In our model, we instantiate a single Gaussian process with one-dimensional feature vectors for each sensor to model the development of the signal over time. We will later elaborate on possibilities to extend this model to include additional information.

For our GP models, we propose using the exponential kernel or a similar kernel that exploits the fact that signal strengths vary relatively little over time. The measurements variance σ^2 should be chosen based on prior knowledge from the calibration data, and should reflect the observed variance in signal strength measurements. An important aspect of the GP is that it includes approximations of the variance on the posterior distribution. Consequently, we can reason about the certainty of our prediction at a specific timestamp. We exploit this knowledge in sampling from our signal strength distributions, discarding timestamps where the variance exceeds a predetermined upper bound. By sampling measurements for multiple sensors at the same timestamp, we can generate new co-occurences at arbitrary timestamps, provided that the variance is below our prespecified threshold. The latter reveals the true power of the GP model: resampling means that we are much less affected by the intermittence in the transmittal of signals by devices. Furthermore, we have estimates of the quality of the measurements, and we can vary in the trade-off between the number of measurements to generate, and their quality. We note that this capability is not natively present in existing techniques, even in those that are based on Gaussian processes: their models output device locations independently based on given measurements.

Figure 1 displays an example, where we apply the Gaussian process model to measurements of a device from a single sensor. We use data of a person carrying a device along a predefined path in a store over a time period of 15 min. We fit a Gaussian process model with an exponential kernel to these measurements, optimizing the hyperparameters ℓ and σ^2 using the Limited-memory BFGS algorithm [11]. The resulting mean estimate and 95% confidence interval are plotted in Fig. 1. As can be seen from this image, the posterior variance decreases in regions where many observations are present. This makes sense, considering we

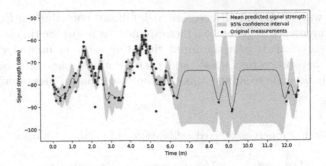

Fig. 1. Gaussian process model applied to a single sensor's signal measurements.

assume additive Gaussian noise: when multiple measurements are made in a small time range, we can more accurately estimate the target value for that time. Resampling the mean, i.e., maximum-likelihood prediction of the obtained distribution, we can generate arbitrarily many new measurements. Using a fixed upper bound on the variance at which to resample points allows us to quantify the quality of the obtained measurements.

An additional advantage of Gaussian processes is that we can incorporate additional information about the measurements in our kernel function k. For instance, we can amend the feature vectors to include the sensors' x- and y-coordinates, and make use of the fact that nearby sensors should obtain similar signal strength measurements for the same device at nearby timestamps.

6 Experiments and Results

We are now ready to evaluate the performance of our localization algorithm and the various suggested improvements. Our two experimental setups, data sets, and methods of comparison to baseline approaches are described in Sect. 6.1. We first optimize the path loss exponent hyperparameter in Sect. 6.2. Based on these experiments, we evaluate the smoothing approaches and our proposed Gaussian process model in Sects. 6.3 and 6.4, respectively.

6.1 Experimental Setup

Our experiments were conducted in an 84×110 m indoor retail environment. A total of 85 sensors were positioned throughout a single floor of the store at irregular locations, at around 3 m above ground level. The sensors, off-the-shelf WiFi access points, report integer dBm signal strengths. A multitude of obstacles were present, including furniture and walls.

The Gaussian process model generates new measurements by sampling from the modeled signal strength distributions for all sensors at regular intervals, discarding measurements with variance exceeding a certain threshold. We set this threshold such that it is just below the observation noise passed to the model. This way, we essentially require a significant contribution from multiple nearby measurements to generate a measurement at a specific timestamp: if only a single measurement is near the input timestamp, the estimated variance will be close to the observation noise, and thereby above our variance threshold. However, it is possible to adjust this threshold to balance measurement quality and quantity.

Table 1. Properties of fixed calibration and visitor path data sets.

Data set	# Measurements	# Fits	Time span
Fixed	107 751	3170	90.3 min
Visitor path	202 528	9004	248.8 min

Two data sets, listed in Table 1, were used. The *fixed calibration data set* was constructed by positioning a device at several fixed, known locations throughout the store, and emitting packages for a known amount of time. The path travelled was approximately 500 m long. A person carrying the device moved at roughly $1.4\,\mathrm{ms}^{-1}$, and traversed the path a total of 22 times. Close to 6,000 packages were received from this device during the evaluated time period. The positioning system's accuracy and precision can be evaluated by comparing the estimated locations to the real locations. This corresponds with most well-known experimental setups from literature [12]. The second *visitor path data set* was constructed by moving a device along the same known path through the environment several times in a row. The path travelled by the device is unique and known, but the associated timestamps (where the device was at what time) are not. Thus, we cannot evaluate accuracy and precision in the same way as with fixed calibration data. Instead, we rely on the shortest distance to the path, i.e., the length of the shortest line segment from the estimated location to the known path, referred to as *distance of closest approach*, or DOCA.

As data sets used in indoor localization and implementations of indoor localization methods are generally not open-source, we compare our methods to a baseline, namely the lateration-based model applied to the raw co-occurrences of measurements. The performance of this baseline corresponds with empirical results from other studies using similar methods [3].

6.2   Results: Path Loss Exponent

We estimate the path loss exponent (see Sect. 4.2) using the fixed calibration data. The true distance between device and sensor is combined with the received signal strength in a linear regression model. We apply the approach to estimate the path loss exponent globally, using 65 000 measurements with 81 unique sensors. The regression yields a path loss exponent of approximately 2.48 with an R^2 coefficient of 0.38. We validate this result by comparing predictor accuracy and precision using different path loss exponents. Figure 2a shows the mean of the distances to the calibration points, and the distance variance, for different path loss exponent values. Path loss exponent 2.0 (solid, black line) corresponds to the original localization model (which assumes free space), whereas the dashed black line corresponds to the path loss exponent calculated based on the linear regression model. The figure shows that the accuracy of the calculated exponent (2.48) is very close to that of the empirical optimum; the variance, however, is slightly higher. Precision is relatively constant for $n \in [2.0, 2.5]$, and accuracy only marginally improves with an adjusted path loss exponent.

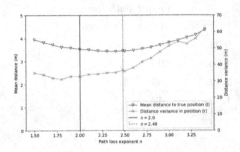 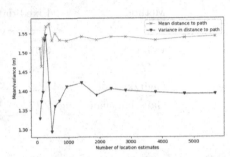

(a) Results of path loss exponent experiments.

(b) DOCA and DOCA variance for varying sample counts in GP model.

Fig. 2. Result diagrams for experiments in Sects. 6.2 and 6.4.

6.3   Results: Smoothing and Filtering

Our smoothing and filtering approaches, introduced in Sect. 4.3, attempt to improve localization accuracy and precision by averaging the estimated x- and y-coordinates over small time periods. All experiments were conducted using the path loss exponent estimated in Sect. 6.2. Based on empirical experiments with multiple hyperparameter combinations, hyperparameters of the smoothing and filtering algorithms were set:

- **Exponential Moving Average:** a constant smoothing factor α was used, with $\alpha = 0.5$. Experiments were conducted with $\alpha \in [0.1, 0.9]$.
- **Gaussian smoothing:** a Gaussian with standard deviation $\sigma = 5000$ ms was used. We experimented with $\sigma \in [1000, 20000]$.
- **Savitzky-Golay filtering:** a third-degree polynomial was fitted to a window of 50 s centered around the data point to be smoothed. Experiments with polynomials of degree 2 and 4 demonstrated inferior performance.

Results are shown in Table 2, demonstrating how, on the fixed position calibration data, all smoothing methods vastly outperformed the baseline. The average distance between the mean estimated location and the true location was reduced by 37%. Gaussian smoothing outperformed both EMA and Savitzky-Golay filtering, but the differences in accuracy were minimal. All smoothing methods show a significant improvement in precision (variance was reduced by 75% for Gaussian smoothing), which is expected as they generally shift estimates towards the mean.

Table 2. Smoothing and filtering results on fixed and visitor path data set. Results are in meters.

Method	Fixed data		Visitor path data	
	Accuracy	Precision	DOCA	Variance
Unfiltered	3.44	6.01	2.08	5.65
EMA	2.26	1.75	1.90	1.81
Savitzky-Golay	2.29	2.12	1.66	1.44
Gaussian smoothing	2.17	1.49	1.42	1.18

For the visitor path data set, we note from Table 2 that the DOCA (see Sect. 6.1) forms a lower bound on the distance between the estimated location and the actual location. This results from the fact that each actual location is on the path, but it is unclear where on the path the device was located at a given timestamp. Especially for methods with a higher deviation from the actual location, this improves observed performance, as a significantly misestimated location can still have a small DOCA when it is close to another path segment.

In general, the visitor path test results are in conformity with the fixed calibration test results. Every smoothing and filtering approach substantially improves on the baseline accuracy, and Gaussian smoothing outperforms the other two smoothing approaches. A decrease of 32% in mean DOCA relative to the baseline method was attained, and variance was significantly reduced. To visualize the effect smoothing has on our original location estimates, we include a visualization of the location estimates of a single traversal of the path. The actual path, the originally estimated path, and the smoothed path (using Gaussian smoothing) are depicted in Fig. 3a. This visualization shows the noisy nature of the original estimates, and the extent to which smoothing ensures that the estimated path corresponds with the actual path followed.

6.4   Results: Gaussian Process Measurement Resampling

Lastly, we consider the Gaussian process measurement resampling model introduced in Sect. 5, operating on the raw measurements obtained by the sensors. Because it outputs location estimates based on newly generated measurements sampled from this model, both the number of location estimates and their timestamps differ between the Gaussian process model and the smoothing approaches. To still allow for a comparison between the results of the other approaches and

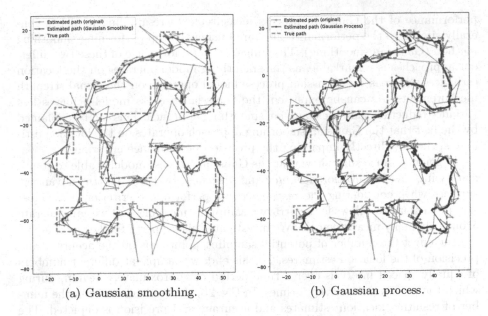

(a) Gaussian smoothing.   (b) Gaussian process.

Fig. 3. Result diagrams for experiments in Sect. 6.

the measurement resampling model, we choose the sampling interval (the time between potential measurements to be sampled) such that the number of resulting location estimates roughly corresponds to the number of location estimates for the raw measurements. The results of the Gaussian process model on the fixed and visitor path data sets are listed in comparison to the original model and the Gaussian smoothing model in Table 3. The number of location estimates is listed alongside the performance metrics.

Table 3. Gaussian process measurement resampling results on fixed and visitor path data set. Results are in meters.

Method	Fixed data			Path data		
	# Fits	Accuracy	Precision	# Fits	DOCA	Variance
Unfiltered	3170	3.44	6.01	9004	2.08	5.65
Gaussian smoothing	3170	2.17	1.49	9004	1.42	1.18
Gaussian process	3463	2.16	1.76	9095	1.54	1.50

The model is able to greatly improve the quality of location estimates when compared to the unfiltered estimates: on the fixed calibration data, accuracy improved from 3.44 m to 2.16 m, and precision improved from 6.01 m to 1.76 m. On the visitor path data, DOCA decreased from 2.08 m to 1.54 m, and the variance in DOCA decreased from 5.65 m to 1.50 m. As such, the observed performance of the Gaussian process measurements resampling model is generally similar to the performance measured using the top-performing smoothing method (Gaussian smoothing). The similarity in performance of these two different approaches is remarkable, as the smoothing model operates on the location estimates, whereas the Gaussian process model operates on the signal strength measurements. As can be expected, the Gaussian process model achieves less variance reduction than the Gaussian smoothing approach. This can be explained by the fact that the Gaussian smoothing approach operates on the location estimates, thereby directly impacting the precision performance criterion.

In general, our results show that the Gaussian process model is able to significantly improve localization *accuracy* and *precision*. However, its true advantage surfaces when considering the *responsiveness* performance criterion: the Gaussian process model allows for arbitrary resampling, meaning that we can generate arbitrarily many measurements. We investigate this property further by evaluating the effect the number of potential sampling points has on the accuracy and precision of the location estimates. To this end, we sample at different numbers of equally spaced intervals over a time period of approximately 45 min, during which the path was traversed 5 times. In Fig. 2b, the tradeoff between the number of resulting location estimates and accuracy and precision is depicted. The accuracy remains constant, and the precision converges as the number of samples increases, substantiating the claim that the number of location estimates can be increased arbitrarily, without significantly impacting performance.

A visualization of the path estimated by the GP model when using an exceedingly large number of location estimates is provided in Fig. 3b. Here, the baseline model from Fig. 3a is compared with a Gaussian process model where approximately 3 times as many points were sampled. The figure highlights the ability of the Gaussian process to provide location estimates at points in time where, previously, no location estimates were possible.

7   Conclusion and Future Work

In this paper, we have presented a novel method for continuously estimating device positions in indoor environments based on intermittent WiFi signals. Using a Gaussian process model for signal strength measurements at WiFi sensors, we realized continuous location estimation, while also significantly improving estimation accuracy and precision. Moreover, we have investigated several smoothing approaches for indoor positioning systems, which improve accuracy and precision to a similar degree, bypassing the computational costs of Gaussian processes. On our validation set of known, fixed device positions, our algorithms improved localization accuracy from 3.44 m to 2.17 m and precision from 6.01 m to 1.49 m. Performance on the visitor path data set of movements along a known path also significantly improved: the mean distance to the true path was reduced from 2.08 m to 1.42 m, and the distance variance was reduced from 5.65 m to 1.18 m. These results accomplish the goals set out in Sect. 1: an accurate, precise location estimator that is able to sample locations at arbitrary timestamps.

Further research opportunities and novel applications are manifold. The results pave the way for significant improvements in dwell-time analysis, visitor tracking, and other major application areas. In future work we aim to derive movement patterns of visitors through time. Such approaches could also be used to automatically infer layouts of indoor environments, identifying obstacles, paths, and open space based on the shape of the estimated movement distributions.

Acknowledgements. Authors acknowledge support from the European Research Center for Information Systems (ERCIS). The third author was supported by the European Research Council (ERC), EU Horizon 2020 grant agreement number 638946. Authors thank people who volunteered to generate calibration data sets.

References

1. Balanis, C.A.: Antenna Theory: Analysis and Design. Wiley, Hoboken (2016)
2. Faragher, R., Sarno, C., Newman, M.: Opportunistic radio SLAM for indoor navigation using smartphone sensors. In: IEEE PLANS, pp. 120–128 (2012)
3. Farid, Z., Nordin, R., Ismail, M.: Recent advances in wireless indoor localization techniques and system. JCNC **13**, 1–12 (2013)
4. Ferris, B., Fox, D., Lawrence, N.D.: WiFi-SLAM using Gaussian process latent variable models. In: IJCAI, pp. 2480–2485 (2007)
5. Ferris, B., Hähnel, D., Fox, D.: Gaussian processes for signal strength-based location estimation. In: Robotics: Science and Systems, vol. 2, pp. 303–310 (2006)
6. Friis, H.T.: A note on a simple transmission formula. Proc. IRE **34**(5), 254–256 (1946)
7. Gardner, E.S.: Exponential smoothing: the state of the art–Part II. Int. J. Forecast. **22**(4), 637–666 (2006)
8. Gu, Y., Lo, A., Niemegeers, I.: A survey of indoor positioning systems for wireless personal networks. IEEE Commun. Surv. Tutor. **11**(1), 13–32 (2009)
9. Hata, M.: Empirical formula for propagation loss in land mobile radio services. IEEE Trans. Veh. Technol. **29**(3), 317–325 (1980)
10. Langendoen, K., Reijers, N.: Distributed localization in wireless sensor networks: a quantitative comparison. Comput. Netw. **43**(4), 499–518 (2003)
11. Liu, D.C., Nocedal, J.: On the limited memory BFGS method for large scale optimization. Math. Program. **45**(1), 503–528 (1989)
12. Liu, H., Darabi, H., Banerjee, P., Liu, J.: Survey of wireless indoor positioning techniques and systems. IEEE Trans. Syst., Man, Cybern., Part C (Appl. Rev.) **37**(6), 1067–1080 (2007)
13. Lymberopoulos, D., Liu, J., Yang, X., Choudhury, R.R., Sen, S., Handziski, V.: Microsoft indoor localization competition: experiences and lessons learned. GetMobile: Mob. Comput. Commun. **18**(4), 24–31 (2015)
14. Madigan, D., Einahrawy, E., Martin, R.P., Ju, W.H., Krishnan, P., Krishnakumar, A.: Bayesian indoor positioning systems. IEEE INFOCOM **2**, 1217–1227 (2005)
15. Niculescu, D., Nath, B.: Ad hoc positioning system (APS) using AOA. IEEE INFOCOM **3**, 1734–1743 (2003)
16. Pahlavan, K., Li, X., Makela, J.P.: Indoor geolocation science and technology. IEEE Commun. Mag. **40**(2), 112–118 (2002)

17. Rasmussen, C.E.: Gaussian Processes for Machine Learning. The MIT Press, Cambridge (2006)
18. Sarkar, T.K., Ji, Z., Kim, K., Medouri, A., Salazar-Palma, M.: A survey of various propagation models for mobile communication. IEEE Antennas Propag. Mag. **45**(3), 51–82 (2003)
19. Savitzky, A., Golay, M.J.: Smoothing and differentiation of data by simplified least squares procedures. Anal. Chem. **36**(8), 1627–1639 (1964)
20. Seco, F., Jiménez, A.R., Prieto, C., Roa, J., Koutsou, K.: A survey of mathematical methods for indoor localization. In: IEEE WISP, pp. 9–14. IEEE (2009)
21. Smith, S.W.: The Scientist and Engineer's Guide to Digital Signal Processing. California Technical Publishing, Poway (1997)
22. Thrun, S., Burgard, W., Fox, D.: Probabilistic Robotics. The MIT Press, Cambridge (2005)
23. Yang, J., Chen, Y.: Indoor localization using improved RSS-based lateration methods. In: IEEE GLOBECOM, pp. 1–6 (2009)
24. Zou, H., Jiang, H., Lu, X., Xie, L.: An online sequential extreme learning machine approach to WiFi based indoor positioning. In: IEEE WF-IoT, pp. 111–116 (2014)

Human Activity Recognition with Convolutional Neural Networks

Antonio Bevilacqua[1]([✉]), Kyle MacDonald[2], Aamina Rangarej[2],
Venessa Widjaya[2], Brian Caulfield[1], and Tahar Kechadi[1]

[1] Insight Centre for Data Analytics, UCD, Dublin, Ireland
antonio.bevilacqua@insight-centre.org
[2] School of Public Health, Physiotherapy and Sports Science,
UCD, Dublin, Ireland

Abstract. The problem of automatic identification of physical activities performed by human subjects is referred to as Human Activity Recognition (HAR). There exist several techniques to measure motion characteristics during these physical activities, such as Inertial Measurement Units (IMUs). IMUs have a cornerstone position in this context, and are characterized by usage flexibility, low cost, and reduced privacy impact. With the use of inertial sensors, it is possible to sample some measures such as acceleration and angular velocity of a body, and use them to learn models that are capable of correctly classifying activities to their corresponding classes. In this paper, we propose to use Convolutional Neural Networks (CNNs) to classify human activities. Our models use raw data obtained from a set of inertial sensors. We explore several combinations of activities and sensors, showing how motion signals can be adapted to be fed into CNNs by using different network architectures. We also compare the performance of different groups of sensors, investigating the classification potential of single, double and triple sensor systems. The experimental results obtained on a dataset of 16 lower-limb activities, collected from a group of participants with the use of five different sensors, are very promising.

Keywords: Human activity recognition · CNN · Deep learning
Classification · IMU

1 Introduction

Human activity recognition (HAR) is a well-known research topic, that involves the correct identification of different activities, sampled in a number of ways. In particular, sensor-based HAR makes use of inertial sensors, such as accelerometers and gyroscopes, to sample acceleration and angular velocity of a body. Sensor-based techniques are generally considered superior when compared with

---

This study was approved by the UCD Office of Research Ethics, with authorization reference LS-17-107.

© Springer Nature Switzerland AG 2019
U. Brefeld et al. (Eds.): ECML PKDD 2018, LNAI 11053, pp. 541–552, 2019.
https://doi.org/10.1007/978-3-030-10997-4_33

other methods, such as vision-based, which use cameras and microphones to record the movements of a body: they are not intrusive for the users, as they do not involve video recording in private and domestic context, less sensitive to environmental noise, cheap and efficient in terms of power consumption [8,13]. Moreover, the wide diffusion of embedded sensors in smartphones makes these devices ubiquitous.

One of the main challenges in sensor-based HAR is the information representation. Traditional classification methods are based on features that are engineered and extracted from the kinetic signals. However, these features are mainly picked on a heuristic base, in accordance with the task at hand. Often, the feature extraction process requires a deep knowledge of the application domain, or human experience, and still results in shallow features only [5]. Moreover, typical HAR methods do not scale for complex motion patterns, and in most cases do not perform well on dynamic data, that is, data picked from continuous streams.

On this regard, automatic and deep methods are gaining momentum in the field of HAR. With the adoption of data-driven approaches for signal classification, the process of selecting meaningful features from the data is deferred to the learning model. In particular, CNNs have the ability to detect both spatial and temporal dependencies among signals, and can effectively model scale-invariant features [15].

In this paper, we apply convolutional neural networks for the HAR problem. The dataset we collected is composed of 16 activities from the Otago exercise program [12]. We train several CNNs with signals coming from different sensors, and we compare the results in order to detect the most informative sensor placement for lower-limb activities. Our findings show that, in most scenarios, the performance of a single sensor is comparable to the performance of multiple sensors, but the usage of multiple sensor configurations yields slightly better results. This suggests that collinearities exist among the signals sampled with sensors on different placements.

The rest of the paper is organized as follows: Sect. 2 gives a brief overview of the state of the art of deep learning models for activity recognition. Section 3 presents our dataset, the architecture of our neural network, and the methodology adopted in this study. The experimental results are discussed in Sect. 4. Some concluding remarks and future extensions for this study are provided in Sect. 5.

2   Related Works

Extensive literature has been produced about sensor-based activity recognition. Bulling et al. [6] give a broad introduction to the problem, highlighting the capabilities and limitations of the classification models based on static and shallow features. Alsheikh et al. [2] introduce a first approach to HAR based on deep learning models. They generate a spectrogram image from an inertial signal, in order to feed real images to a convolutional neural network. This approach overcomes the need for reshaping the signals in a suitable format for a CNN, however, the spectrogram generation step simply replaces the process of feature extraction, adding initial overhead to the network training. Zeng *et al.* [15] use raw acceleration signals as input for a convolutional network, applying 1-D convolution to each signal component. This approach may result in loss of spatial dependencies among different components of the same sensor. They focus on public datasets, obtained mainly from embedded sensors (like smartphones), or worn sensors placed on the arm. A similar technique is suggested by Yang *et al.* [14]. In their work, they use the same public datasets, however, they apply 2-D convolution over a single-channel representation of the kinetic signals. This particular application of CNNs for the activity recognition problem is further elaborated by Ha *et al.* [10], with a multi-channel convolutional network that leverages both acceleration and angular velocity signals to classify daily activities from a public dataset of upper-limb movements. The classification task they perform is personalized, so the signals gathered from each participant are used to train individual learning models.

One of the missing elements in all the previously described contributions about deep learning models is a comparison of the classification performance of individual sensors or group of sensors. Our aim in this paper is to implement a deep CNN that can properly address the task of activity recognition, and then compare the results obtained with the adoption of different sensor combinations. We also focus on a set of exercise activities that are part of the Otago exercise program. To the best of our knowledge, this group of activities has never been explored before in the context of activity recognition.

3   Data and Methodology

The purpose of this paper is to assess the classification performance of different groups of IMU sensors for different activities. We group the target activities into four categories, and, for each category, we aim at identifying the best placement for the inertial units, as well as the most efficient combination of sensors, with respect to the activity classification task.

3.1   Sensors and Data Acquisition

Five sensors were used for the data collection phase. Each sensor is held in place by a neoprene sleeve. For this study, we set the placement points as follows:

- two sensors placed on the distal third of each shank (left and right), superior to the lateral malleolus;
- two sensors centred on both left and right feet, in line with the head of the fifth metatarsal;
- one sensor placed on the lumbar region, at the fourth lumbar vertebrae.

We explore three main sensor configurations: with a **single device** setup, we classify the activity signal coming from each individual sensor. In the **double**

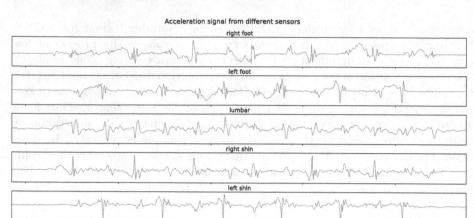

Fig. 1. This acceleration x components correspond to roughly 10 s of activity, acquired with the five sensors used in this study. Signals sampled by different sensors may show very discordant patterns and characteristics.

device setup, four combinations of two sensors are tested: shin sensors (right and left), foot sensors (right and left), right sensors and left sensors (foot and shin). When testing the **triple device** setups, the lumbar sensor is included in each one of the double sensor configurations (Fig. 1).

The chosen device for this study is Shimmer3 [7]. The Shimmer3 IMU contains a wide set of kinetic sensors, but we are interested in sampling acceleration and angular velocity only. Both these quantities are captured by triaxial sensors, so each Shimmer3 device returns a set of six signals (three acceleration components, over the axes x, y and z, and three angular velocity components, over the same axes). The sampling rate is set to 102.4 Hz for all sensors. The accelerometer is configured to have a range of $\pm 2g$, while the gyroscope range is set to 500 dps. We adopted this particular configuration in order to avoid aliasing when sampling the target activities, as gait-related human motion usually locates in the frequency range of 0.6–5.0 Hz [9].

3.2 Target Activities and Population

The physical activities targeted in this paper are part of the Otago Exercise Programme (OEP), a programme of activities designed to reduce the risk of falling among the elderlies [12]. In particular, we grouped 16 different activities into four categories: **walk**, **walking balance**, **standing balance**, and **strength**. None of the Otago warm-up activities is included in this study.

Walk: it is composed of backwards walking (bawk), sideways walking (sdwk), walking and turning around (wktrn). These three activities have all wide and diverse range of movements, especially for the foot sensors.

Walking Balance: it is composed of heel to toe walking backwards (hetowkbk), heel walking (hewk), tandem walking (tdwk), toe walking (towk). These activities are based on similar ranges of movements.

Standing Balance: it is composed of single leg stance (sls), and tandem stance (tdst). The signals sampled from these two activities are mainly flat, as they require the subject to move only once in order to change the standing leg from left to right.

Strength: it is composed of knee extension (knex), knee flexion (knfx), hip abduction (hpabd), calf raise (cars), toe raise (tors), knee bend (knbn), and sit to stand (std). As some of these activities are performed by using each individual leg separately, all the sensor configurations involving both right and left sides are not applicable to this group.

A standard operating procedure defines the execution setup for all the target activities, in terms of holding and pausing times, overall duration, starting and ending body configurations. The same operating procedure is applied to all the subjects involved in the study.

The group of 19 participants consists of 7 males and 12 females. Participants have a mean age of 22.94 ± 2.39, a mean height of 164.34 ± 7.07 cm, and a mean weight of 66.78 ± 11.92 kg.

3.3 Dataset

Once the signal is acquired from the activity, it is segmented into small overlapping windows of 204 points, corresponding to roughly 2 s of movements, with a stride of 5 points. A reduced size of the windows is generally associated with a better classification performance [3], and in the context of CNNs, it facilitates the training process as the network input has a contained shape. Therefore, each window comes in the form of a matrix of values, of shape $6N \times 204$, where N is the number of sensors used to sample the window. The dense overlapping among windows guarantees high numerosity of training and testing samples. As the activities have different execution times, and different subjects may execute the same activity at different paces, the resulting dataset is not balanced. The distributions of the example windows over the activity classes for the five target groups are listed in Table 1.

For assessing the performance of our classification system, we use a classic 5-fold cross-validation approach. We partition the available datasets based on the subjects rather than on the windows. This prevents overfitting over the subjects and helps to achieve better generalisation results. In this regard, 4 participants out of 19 are always kept isolated for testing purposes, so each fold is generated with an 80/20 split.

3.4 Input Adaptation and Network Architecture

The shape of the input examples that is fed to the network depends on the sensor configuration, as each sensor samples 6 signal components that are then

Table 1. Label distributions for the activity groups

Activity	Windows	Percentage	Activity group	Total
bawk	19204	39.88	walk	48143
sdwk	22077	45.85		
wktrn	6925	14.38		
hetowkbk	4130	9.44	walk balance	43754
hewk	17796	40.67		
tdwk	4578	10.46		
towk	17250	39.42		
sls	20006	65.05	stand balance	30759
tdst	10753	34.95		
knex	7500	12.14	strength	76854
knfx	6398	10.42		
hpabd	5954	9.62		
cars	6188	10.11		
tors	5815	9.41		
knbn	26452	34.42		
sts	8533	13.86		

arranged into a 6×204 single-channel, image-like matrix, as described in Sect. 3.3. Therefore, the input of the network has shape $6 \times 204 \times N$, where N is the number of channels and is equal to the number of sensors used for the sampling. This input adaptation is known as model-driven [13], and it is effective in detecting both spatial and temporal features among the signal components [10]. Figure 2 shows how the signal components are stacked together and form the input image for the network.

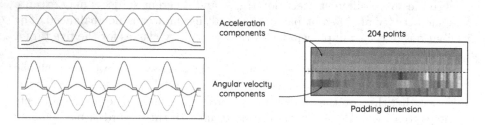

Fig. 2. The signal components are stacked on top of each other to form a bidimensional matrix of values. Additional sensors would generate new channels of the same shape.

The full structure of our convolutional model is shown in Fig. 3. After the input layer, three convolutional layers interleave with three max-pooling layers.

The depthwise convolution operation generates multiple feature maps for every input channel, with kernels of size 3×5, 2×4 and 2×2 in the first, second and third convolutional layer respectively. The input of every convolutional layer is properly padded so that no loss of resolution is determined from the convolution operation. Batch normalization is applied after each convolutional layer. The three max-pooling layers use kernels of size 3×3, 2×2 and 3×2 respectively. A fully connected network follows, composed of three dense layers of 500, 250 and 125 units. The dense layers are regularized with dropout during the training phase, with a 0.5 probability of keeping each neuron. The ReLU function is used as activation function within the whole network, while the loss is calculated with the cross entropy function. The Adam optimizer is used as stochastic optimization method [11]. The output layer is composed ok m units, where m corresponds to the number of activities in each group. The softmax function will return the most likely class of the input windows in the multi-class classification task.

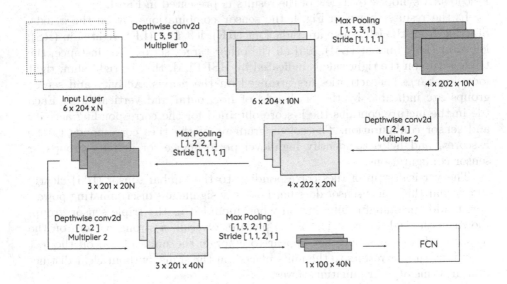

Fig. 3. Our CNN architecture, where N represents the number of sensors used during the activity sampling. Regardless of the value of N, the network structure does not change, as depthwise convolution applies different filters to each one of the input channels.

We select a set of hyperparameters that are kept constant for all the activity groups and sensor configurations, based on literature best practices [4] and empirical observations. We use a batch size of 1024, as we find this value to speed up the learning process when compared with smaller sizes, without being computationally too complex to manage. The number of training epochs varies from 150 to up to 300, according to the behaviour of individual configurations.

The initial learning rate is fixed to 0.005. The network is implemented with the TensorFlow framework [1], version 1.7. Our objective is not to build the most performant network for the task, but it is rather to compare the classification potential of different sensors. The rationale behind out architectural choices relies therefore on a rather standard network configuration, based on small kernels, standard regularization methods, and a compact set of hyperparameters. In our experience, three convolutional layers will lead to overfitting when no regularization method is applied. However, introducing dropout stabilizes the learning, and the network performance does not benefit from the inclusion of further convolutional or dense layers.

4   Experimental Results

In order to evaluate our classifier, we collect individual precision and recall scores for every combination of sensors and activities, and we then compute the F-scores. A synoptic overview of the results is presented in Fig. 4.

In the results shown in Fig. 4, the sensor combinations lay on the x axis. Starting from the left, there are right foot (RF), left foot (LF), right shin (RS), left shin (LS), lumbar (LM), and all the other target setups (for instance, the triple setup on the right side is indicated by RSRFLM, that is, right shin, right foot, lumbar). The activities are arranged on the y axis. Activity and sensor groups are indicated by the black grid of horizontal and vertical lines. Each tile in the picture contains the F-score obtained for the corresponding activity and sensor configuration. The colour gradient of the tiles corresponds to the F-scores, and helps to identify high-level performance for activity groups or sensor configurations.

The vertical strip of tiles corresponding to the lumbar sensor (LM) clearly shows that this single sensor does not hold any significant discriminating power, nor it adds meaningful information when included in the triple sensor group, shown in the rightmost region of the picture. Overall, a strong pattern on the sensor configurations does not appear to emerge: the units placed on the feet show very similar results to the units placed on the shins, without clear distinction in terms of discriminating power.

The confusion matrices resulting from the evaluation process over the test datasets are shown in Figs. 5, 6, 7 and 8, for the walk group, the walking balance group, the standing balance group, and the strength group respectively. The colour gradient for each matrix is normalized on the rows.

The walk group scores interesting results for every sensor configuration. Single sensors perform slightly worse than multiple sensor setups, however, there seems to be no difference between two sensors and three sensors. From the confusion matrix in Fig. 5, we observe that the two majority classes, bawk and sdwk, determined a reasonably limited amount of misclassified instances, while the minority class, wktrn, only recorded 4% of false negatives.

The same behaviour is shown for the walking balance group. In this case, the hetowk and tdwk activities, which represent the 9.44% and 10.46% of the

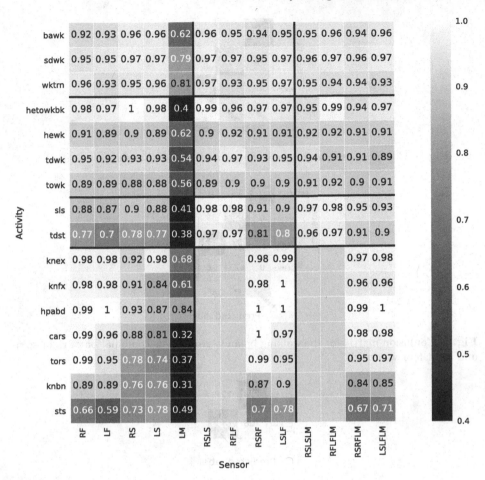

Fig. 4. F-scores for the sensor configurations applied to the activity groups. Empty tiles for the strength group correspond to regions where particular sensor configurations were not applicable (asymmetric activities cannot be classified by using sensors on both the right and the left sides at the same time). The colour scheme spans from dark tones of black for low values of F-score to lighter tones for high F-score values. Lighter tiles denote better results than darker tiles. In order to emphasize small differences in the matrix, the minimum value for the tile colour gradient is set to be 0.4 (approximatively 0.1 higher than the smallest F-score value), so scores below this value will be marked with a black tile.

entire group dataset respectively, performed remarkably well. For the first activity, only 3% of the instances were incorrectly classified, while the proportion of misclassification for the second activity is 8%. The confusion matrix in Fig. 6 indicates that the towk and hewk labels, the majority classes, got a rate of false positives of 5% and 17% respectively, in favour of one each other. From the global confusion matrix in Fig. 4, these two classes correspond to slightly darker bands

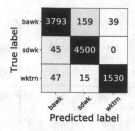

Fig. 5. Confusion matrix for the walk group. The left shin sensor was used.

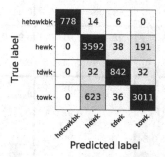

Fig. 6. Confusion matrix for the walking balance group. The combination of right shin and right foot was used.

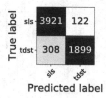

Fig. 7. Confusion matrix for the standing balance group. The combination of right shin and right foot was used.

within the activity group. As defined in the exercise program, heel walking and toe walking present some similarities.

The standing balance group, whose confusion matrix is reported in Fig. 7, was not properly classified with a single sensor. The heavy class imbalance, in conjunction with the monotonicity of the signals sampled from these two activities, skewed most of the misclassified instances towards the majority class, sls, as indicated by the confusion matrix in Fig. 7. Nonetheless, the F-scores indicate that symmetric combinations of sensors (right and left foot, right and left shin) were able to discriminate between the two better than the asymmetric ones (right side, left side).

As for the strength group, multiple sensor configurations increased the classification score remarkably when compared with single sensor configurations, in some cases reaching perfect classification for classes such as hpabd, cars or knfx.

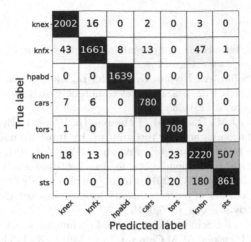

Fig. 8. Confusion matrix for the strength group. The combination of left shin, left foot and lumbar sensors was used.

The two classes that lowered the overall group performance are knbn and sts, as shown in Fig. 8. They are based on very similar movements, so weak values of precision and recall are somehow expected.

5   Conclusions and Future Works

In this paper, we presented a CNN model for the HAR problem. We focused on a set of activities extracted from a common exercise program for fall prevention, training our model data sampled from different sensors, in order to explore the classification capabilities of each individual unit, as well as groups of units. Our experimental results indicate that convolutional models can be used to address the problem of activity recognition in the context of exercise programs. In most cases, combinations of two or three sensors lead to better results compared to the adoption of single inertial units.

Further work on the application of convolutional models to real-world data is recommended. More activities could be included in the workflow, and different aggregations on the activities can be tested. In particular, it is recommended to diversify the population of participants, in order to validate the classification mechanism to wider age groups. A proper campaign of hyperparameter tuning should be carried over the same set of activities and inertial units, in order to boost the classification performance and reduce the complexity of both the training and inference phases. The very same network structure could be redesigned in an optimized fashion for the task at hand, with particular emphasis on the input adaptation step. As an example, shaping the input in a single $6N \times 204$ could lead to interesting results, as more complex kernels would allow the inclusion of features involving multiple sensors.

References

1. Abadi, M., et al.: TensorFlow: large-scale machine learning on heterogeneous systems (2015). https://www.tensorflow.org/
2. Alsheikh, M.A., Selim, A., Niyato, D., Doyle, L., Lin, S., Tan, H.P.: Deep activity recognition models with triaxial accelerometers. CoRR abs/1511.04664 (2015). http://arxiv.org/abs/1511.04664
3. Banos, O., Galvez, J.M., Damas, M., Pomares, H., Rojas, I.: Evaluating the effects of signal segmentation on activity recognition. In: International Work-Conference on Bioinformatics and Biomedical Engineering. IWBBIO 2014, pp. 759–765 (2014)
4. Bengio, Y.: Practical recommendations for gradient-based training of deep architectures. CoRR abs/1206.5533 (2012). http://arxiv.org/abs/1206.5533
5. Bengio, Y.: Deep learning of representations: looking forward. CoRR abs/1305.0445 (2013). http://arxiv.org/abs/1305.0445
6. Bulling, A., Blanke, U., Schiele, B.: A tutorial on human activity recognition using body-worn inertial sensors. ACM Comput. Surv. **46**(3), 33:1–33:33 (2014). https://doi.org/10.1145/2499621
7. Burns, A., et al.: Shimmer^{TM} a wireless sensor platform for noninvasive biomedical research. IEEE Sens. J. **10**(9), 1527–1534 (2010). https://doi.org/10.1109/JSEN.2010.2045498
8. Cook, D., Feuz, K.D., Krishnan, N.C.: Transfer learning for activity recognition: a survey. Knowl. Inf. Syst. **36**(3), 537–556 (2013). https://doi.org/10.1007/s10115-013-0665-3
9. Godfrey, A., Conway, R., Meagher, D., ÓLaighin, G.: Direct measurement of human movement by accelerometry. Med. Eng. Phys. **30**, 1364–1386 (2009)
10. Ha, S., Choi, S.: Convolutional neural networks for human activity recognition using multiple accelerometer and gyroscope sensors. In: 2016 International Joint Conference on Neural Networks (IJCNN), pp. 381–388, July 2016. https://doi.org/10.1109/IJCNN.2016.7727224
11. Kingma, D.P., Ba, J.: Adam: a method for stochastic optimization. CoRR abs/1412.6980 (2014). http://arxiv.org/abs/1412.6980
12. Thomas, S., Mackintosh, S., Halbert, J.: Does the 'otago exercise programme' reduce mortality and falls in older adults?: a systematic review and meta-analysis. Age Ageing **39**(6), 681–687 (2010). https://doi.org/10.1093/ageing/afq102
13. Wang, J., Chen, Y., Hao, S., Peng, X., Hu, L.: Deep learning for sensor-based activity recognition: a survey. CoRR abs/1707.03502 (2017). http://arxiv.org/abs/1707.03502
14. Yang, J.B., Nguyen, M.N., San, P.P., Li, X.L., Krishnaswamy, S.: Deep convolutional neural networks on multichannel time series for human activity recognition. In: Proceedings of the 24th International Conference on Artificial Intelligence. IJCAI 2015, pp. 3995–4001. AAAI Press (2015). http://dl.acm.org/citation.cfm?id=2832747.2832806
15. Zeng, M., et al.: Convolutional neural networks for human activity recognition using mobile sensors. In: 6th International Conference on Mobile Computing, Applications and Services, pp. 197–205, November 2014. https://doi.org/10.4108/icst.mobicase.2014.257786

Urban Sensing for Anomalous Event Detection:
Distinguishing Between Legitimate Traffic Changes and Abnormal Traffic Variability

Masoomeh Zameni[1]([⊠]), Mengyi He[2], Masud Moshtaghi[3], Zahra Ghafoori[1],
Christopher Leckie[1], James C. Bezdek[1], and Kotagiri Ramamohanarao[1]

[1] The University of Melbourne, Melbourne, Australia
mzameni@student.unimelb.edu.au,
{zahra.ghafoori,caleckie,jbezdek,kotagiri}@unimelb.edu.au
[2] Kepler Analytics, Melbourne, Australia
miley.he@kepleranalytics.com.au
[3] Amazon, Manhattan Beach, CA, USA
mmasud@amazon.com

Abstract. Sensors deployed in different parts of a city continuously record traffic data, such as vehicle flows and pedestrian counts. We define an unexpected change in the traffic counts as an *anomalous local event*. Reliable discovery of such events is very important in real-world applications such as real-time crash detection or traffic congestion detection. One of the main challenges to detecting anomalous local events is to distinguish them from legitimate global traffic changes, which happen due to seasonal effects, weather and holidays. Existing anomaly detection techniques often raise many false alarms for these legitimate traffic changes, making such techniques less reliable. To address this issue, we introduce an unsupervised anomaly detection system that represents relationships between different locations in a city. Our method uses training data to estimate the traffic count at each sensor location given the traffic counts at the other locations. The estimation error is then used to calculate the anomaly score at any given time and location in the network. We test our method on two real traffic datasets collected in the city of Melbourne, Australia, for detecting anomalous local events. Empirical results show the greater robustness of our method to legitimate global changes in traffic count than four benchmark anomaly detection methods examined in this paper. Data related to this paper are available at: https://vicroadsopendata-vicroadsmaps. opendata.arcgis.com/datasets/147696bb47544a209e0a5e79e165d1b0_0.

Keywords: Pedestrian event detection
Vehicle traffic event detection · Anomaly detection
Urban sensing · Smart cities

© Springer Nature Switzerland AG 2019
U. Brefeld et al. (Eds.): ECML PKDD 2018, LNAI 11053, pp. 553–568, 2019.
https://doi.org/10.1007/978-3-030-10997-4_34

1 Introduction

With the advent of the *Internet of Things* (IoT), fine-grained urban information can be continuously recorded. Many cities are equipped with such sensor devices to measure traffic counts in different locations [10]. Analyzing this data can discover anomalous traffic changes that are caused by events such as accidents, protests, sports events, celebrations, disasters and road works. For example, real-time crash detection can increase survival rates by reducing emergency response time. As another example, automatic real-time traffic congestion alarms can reduce energy consumption and increase productivity by providing timely advice to drivers [15]. Anomaly detection also plays an important role in city management by reducing costs and identifying problems with critical infrastructure.

Definition 1. *Anomalous local events:* *Events that occur in a local area of a city and cause an unexpected change in the traffic measurements are called anomalous local events in this paper. Local events can occur in a single location or a small set of spatially close neighbor locations.*

City festivals, such as the White Night event or the Queen Victoria night market (QVM) in the *Central Business District* (CBD) of Melbourne, Australia, are examples of anomalous local events. The White Night event causes a significant decrease in the vehicle counts in some local areas in the CBD due to road closures. The QVM night market causes a significant increase in the pedestrian traffic in a market in the CBD.

Definition 2. *Legitimate global traffic changes:* *Global traffic changes that occur in almost all locations of the city are called legitimate global traffic changes in this paper.*

Global changes to traffic counts due to seasonal effects, weather and holidays are examples of legitimate changes, and these changes should not be considered as anomalies. Most existing anomaly detection techniques raise many false alarms for these legitimate global traffic changes, making such anomaly detection techniques unreliable for use in real-world applications. In this paper, we propose a *City Traffic Event Detection* (CTED) method that is able to detect *anomalous local events* while ignoring *legitimate global traffic changes* as anomalous.

Consider the case study of pedestrian and road vehicle traffic counts in the Melbourne CBD. Pedestrian and vehicle count data is continuously measured by sensors at different locations. For example, pedestrian counts are recorded at hourly intervals at 32 locations, while vehicle traffic counts are recorded at 15 min intervals at 105 locations. This data has been made publicly available [1,2]. Figure 1 shows the map of the locations of the pedestrian count sensors in the CBD of Melbourne. Figure 2 shows some examples of legitimate global vehicle traffic changes including two weekends and a weekday public holiday (Australia Day), and also an anomaly that occurred due to a road closure at a location in the vehicle traffic dataset in the Melbourne CBD on 16 January 2014. Our goal is to detect *when* and *where* an anomaly occurs in the pedestrian and vehicle traffic data when a relatively small amount of training data exists.

Fig. 1. Pedestrian counting locations in Melbourne, Australia.

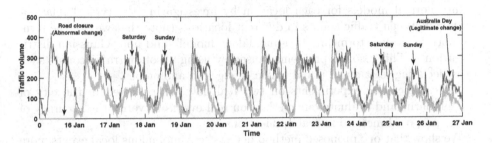

Fig. 2. Road vehicle traffic counts at two Melbourne locations (red and blue) over 12 days. (Color figure online)

Existing work [5,6,13,18] has several limitations for use in real-world applications. Specifically:

- **Unsupervised:** A key challenge in detecting anomalous local events is the lack of labeled (ground truth) data. Our proposed method, CTED, is unsupervised, so it circumvents this problem. Moreover, insufficient training data limits the use of anomaly detection methods that require a large number of observations for training. For example, techniques based on *One Class Support Vector Machines* (OCSVMs) [19] and Deep Learning [7] methods are limited by this requirement. CTED is able to work with a relatively small amount of training data.
- **Detecting both spatial and temporal anomalies:** Most anomaly detection methods for data streams [7,10,14] can only identify the time but not the location of anomalous traffic events. In contrast, CTED can detect when and where an unexpected traffic change occurs.
- **Independence of prior data distributional knowledge:** Many existing anomaly detection methods rely on prior distributional knowledge about the data [6,7,18]. In contrast, CTED is based on a simple linear regression technique that avoids this requirement.

- **Robustness to legitimate global traffic changes:** Existing anomaly detection methods often misclassify legitimate global traffic changes as anomalies. CTED offers greater robustness to legitimate global traffic changes by using linear regression and modeling the relationships between traffic counts at different locations. The main contributions of our paper are as follows:

- To the best of our knowledge, we develop CTED, which is the first unsupervised anomaly event detection method focused on legitimate global traffic changes that identifies not only the time but also the location of anomalous traffic changes in a city environment.
- Our distribution-free approach builds relative normal models instead of absolute normal models for each location by investigating the typical relationships between traffic counts in different locations (note that we use the term "normal model" to mean "non-anomalous model" and not "Gaussian distribution"). This resolves problems caused by using absolute traffic counts such as declaring legitimate global traffic changes as anomalous.
- We conduct our experiments on two real datasets collected in the city of Melbourne and evaluate our method on real events to verify the accuracy of the proposed method in real applications.
- We show that our proposed method detects real anomalous local events more accurately than the comparison methods used in this paper, while being more robust to legitimate global traffic changes.

2   Related Work

Anomaly detection (AD) methods that can identify both the location and time of anomalies [4–6,13,17] (Temporal and Spatial AD) use two different approaches to compute an anomaly score at a specific location. The first approach, the *single-profile* approach, relies solely on the traffic counts at the location itself. The second approach, the *cluster-profile* approach, combines the traffic counts at other locations when determining the anomaly score at a particular location.

Methods in [5,6,16,22–24] are single-profile anomaly detection approaches. For example, in [24], anomalies in urban traffic are detected using *Stable Principal Component Pursuit* (SPCP). The *Global Positioning System* (GPS) data from mobile phone users in Japan is used in [23] to detect anomalous events using 53 *Hidden Markov Models* (HMMs). In [5], anomalies in pedestrian flows are detected using a frequent item set mining approach, which was improved in [6] by using a window-based dynamic ensemble clustering approach.

Approaches reported in [3,9,13,18] allow traffic counts at other locations to influence AD at a specified location. For example, in [9], anomalies that are detected at a specific location are partially influenced by the traffic at locations with similar behavior found using the k-means algorithm. Rajasegarar et al. [18] detect anomalies in resource-constrained Wireless Sensor Networks (WSNs) using multiple hyperellipsoidal clusters and calculating the relative remoteness between neighbors. The method in [3] analyzes WiFi access point utilization patterns on a university campus to detect special events in physical spaces. This method has many false alarms at the beginning of each working day.

The technique that is most similar to CTED is [13], where a framework for temporal outlier detection in vehicle traffic networks, called *Temporal Outlier Discovery* (TOD), is proposed. TOD is based on updating historical similarity values using a reward/punishment rule.

None of the methods reviewed is robust to *legitimate global traffic changes*, since these methods either do not consider other sensors while calculating the anomaly score at a sensor, or only consider spatially close neighbor sensors to a specific sensor in calculating its anomaly score. This study introduces a method for detecting *anomalous local events* in traffic count data that is highly robust to *legitimate global traffic changes*. Table 1 compares properties of several methods that have been used for AD in sensor networks to those of CTED.

Table 1. Comparison of related methods for anomaly detection in sensor networks

Category	Reference	Technique	Spatial AD	Temporal AD	Unsupervised	Relative count based	Robust to global changes	Distribution-free	#parameters[1]
Single-profile	[7]	Cluster-based	×	✓	✓	×	×	×	6
	[10]	Classification-based	×	✓	×	×	×	✓	3
	[22]	Statistic-based	✓	✓	✓	×	×	✓	2
	[16]	Statistic-based	✓	✓	✓	×	×	×	5
	[24]	Classification-based	✓	✓	✓	×	×	✓	3
	[23]	HMM-based	×	✓	×	×	×	✓	4
	[5]	Frequent item set-based	✓	✓	✓	×	×	✓	3
	[6]	Window-based	✓	✓	✓	×	×	×	5
Cluster-profile	[9]	Cluster-based	✓	✓	×	×	×	✓	2
	[18]	Cluster-based	✓	✓	✓	×	×	×	4
	[3]	Window-based	✓	✓	✓	✓	×	✓	2
	[13]	Historical similarity-based	✓	✓	✓	×	×	✓	5
	CTED	Linear regression-based	✓	✓	✓	✓	✓	✓	1

[7] Number of input parameters required to implement the anomaly detection method.

3   Problem Statement

Suppose there are m different locations in a city, $L = \{1, ..., m\}$. We consider the traffic counts in an hourly basis, $H = \{1, ..., 24\}$. Assume $n_i^{(h)}(q)$ is the traffic count for location i at hour $h \in H$ of day q, denoted by (i, h, q), and let $TD = \left\{ n_i^{(h)}(q) : 1 \leqslant i \leqslant m, h \in H, 1 \leqslant q \leqslant N \right\}$ be the training traffic data collected during N days. Our goal is to detect the location and time of *unexpected traffic counts* (Definition 3) for $q > N$, which we regard as *anomalous local events* (Definition 1). The proposed method should distinguish the anomalous local events from *legitimate global traffic changes* (Definition 2). We assume that the majority of the data in TD corresponds to normal traffic.

Definition 3. Unexpected traffic count: *An Unexpected traffic count occurs at (i, h, q) if its observed traffic count, $n_i^{(h)}(q)$, is significantly different from its expected traffic count, $\hat{n}_{ij}^{(h)}(q)$.*

4   Proposed Method - CTED

4.1   Overview of the Method

The basic research questions that need to be addressed are: (a) How can we find the expected traffic count at location i? (b) Do the traffic counts at other locations affect the traffic count at location i? (c) How to reduce the false alarm rate of anomaly detection on legitimate global traffic changes?

To address these research problems, CTED consists of two phases: an *offline phase*, which builds a series of models to estimate the normal traffic for each location i at hour h of the day, and weights the normal models, and an *online phase*, which uses the accumulated weighted error of the normal models for each location i at hour h and the current traffic measurements at the other locations to compute the anomaly score at (i, h, q) for $q > N$.

Figure 3 presents the main steps of our system. Next, we explain each of these phases in detail.

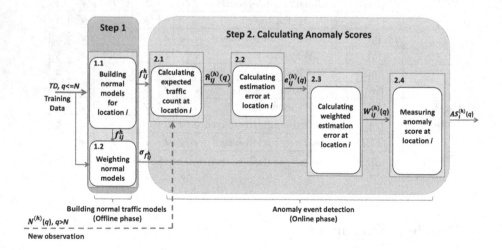

Fig. 3. Main steps of CTED for detecting anomalous events at location i.

4.2   Step 1: Build and Weight Normal Models

In a legitimate global traffic change, traffic counts change with almost the same ratio in all locations. This fact encouraged us to make our event detection model insensitive to this ratio change. To this end, we investigated the relative traffic counts between different locations. For example, we found that the traffic counts at Elizabeth Street in the Melbourne CBD are usually two times higher than the traffic at Spring Street. So, in a legitimate global traffic change, when the traffic at the Elizabeth Street increases/decreases by 1.5 times, we expect the traffic at the Spring Street to increase/decrease by almost 1.5 times. This suggested that we deploy a linear regression model.

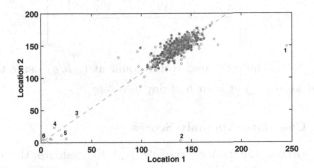

Fig. 4. Linear regression and legitimate global traffic changes for the vehicle count dataset. Linear regression considers observations affected by legitimate global traffic changes (observation numbers 3 to 6) as normal observations for the vehicle count dataset. (Color figure online)

Figure 4 shows an example where the traffic counts change almost linearly in a legitimate global traffic change (weekday holidays in this example represented by numbers 3 to 6) in two linearly correlated locations. The green linear regression line models the normal behavior between two locations at 1 pm for the vehicle traffic dataset. The blue star points (*) are training data observations and the red circle points (O) are upcoming test observations. This figure shows how using a simple linear regression for modeling the relative normal behaviours increases the robustness of CTED to legitimate global traffic changes (see observation numbers 3 to 6). This figure also shows that a linear regression models the normal behavior of data better than other methods such as clustering. Clustering techniques detect one normal cluster for the training observations and do not generalize well to the legitimate traffic changes (observation numbers 3 to 6 in the bottom left of the figure, which are affected by holidays in this example).

Let functions $f_{ij}^{(h)}, j \neq i$ represent our normal models that estimate the traffic counts at location i at hour h given the traffic counts at location j at the same hour h. We learn linear regression models of the form $f_{ij}^{(h)}(x) = a_{ij}^{(h)} x + b_{ij}^{(h)}$ for the traffic data, where $a_{ij}^{(h)}$ and $b_{ij}^{(h)}$ are the coefficients of the model learnt using the training data $\left\{ n_i^{(h)}(q), n_j^{(h)}(q) \right\}, q = 1...N$, where $n_k^{(h)}(q)$ is the traffic count at location k at hour h of day q. We train a model $f_{ij}^{(h)}$ for each hour h of the day using the training data TD. Each trained model is then used to evaluate new observations for the same hour h of the day.

Although the observed traffic counts at other locations are used to estimate the traffic counts at location i at hour h, the locations that have the highest linear correlation with location i are more important. Therefore, we assign higher weights to the locations that have the largest linear correlation with location i (step 1.2). To this end, we weight $f_{ij}^{(h)}$ models by their standard errors, $\sigma_{f_{ij}^{(h)}}$ using the training data (Eq. 1).

$$\sigma_{f_{ij}^{(h)}} = \left[\frac{\sum_{q=1}^{N} \left(n_i^{(h)}(q) - f_{ij}^{(h)}(n_j^{(h)}(q)) \right)^2}{N} \right]^{\frac{1}{2}} \tag{1}$$

where $f_{ij}^{(h)}(n_j^{(h)}(q)$ is the estimated traffic count at (i, h, q) using the observed traffic count at location j at hour h of day q, $n_j^{(h)}(q)$.

4.3   Step 2: Calculate Anomaly Scores

In this step, we use the trained models $f_{ij}^{(h)}$ to evaluate the observations (i, h, q), $q > N$. The expected traffic count at $(i, h, q > N)$, $\hat{n}_{ij}^{(h)}(q)$, based on the current traffic counts at location j, $n_j^{(h)}(q)$, is calculated in Eq. 2 (see also step 2.1 in Fig. 3).

$$\hat{n}_{ij}^{(h)}(q) = f_{ij}^{(h)}(n_j^{(h)}(q)), \quad q > N \tag{2}$$

For each upcoming observation at location i, the absolute estimation error based on the traffic counts at $j \in CK_i^{(h)}$, for (h, q) is calculated in Eq. 3.

$$e_{ij}^{(h)}(q) = \left| n_i^{(h)}(q) - \hat{n}_{ij}^{(h)}(q) \right| \tag{3}$$

To give more importance to the locations that have high linear correlation with location i, we weight the estimation error of the traffic count at (i, h) as shown in Eq. 4 using $\sigma_{f_{ij}^{(h)}}$.

$$W_{ij}^{(h)}(q) = \frac{e_{ij}^{(h)}(q)}{\sigma_{f_{ij}^{(h)}}} \tag{4}$$

where $\sigma_{f_{ij}^{(h)}}$ is the standard error of the trained normal model $f_{ij}^{(h)}$ from Eq. 1.

Measuring Anomaly Scores. The anomaly score at (i, h, q) is calculated in Eq. 5.

$$AS_i^{(h)}(q) = \sum_{j \in CK_i^{(h)}, j \neq i} W_{ij}^{(h)}(q) \tag{5}$$

$AS_i^{(h)}(q)$ is the sum of the weighted estimation errors of the traffic counts at (i, h, q). An anomalous traffic event is declared at (i, h, q) if its anomaly score, $AS_i^{(h)}(q)$, exceeds a pre-specified threshold, thr_{CTED}. We discuss the selection of this threshold in Sect. 5.

Why do we ignore lag effects? In building normal models and calculating anomaly scores, we do not consider lag effects. Usually, vehicles move in "waves", i.e., a high traffic count at location i at time h is expected to correspond to a high traffic count at a spatially close neighbor correlated location j at the next time $h + 1$. However, we do not consider lag effects because of the properties of our datasets, i.e., the low sampling rate (15-min for the vehicle traffic data and one-hour for the pedestrian data) and the small distances between locations (many of the locations are just one block apart). We considered lag effects in our studies, and noticed a small reduction in the accuracy of our anomaly event detection method. When distances between sensors and/or sampling times increase, accounting for lag becomes more effective.

4.4 DBSCAN for Removing Outliers from the Training Data

We examined the linear correlation between different locations using the *Pearson Correlation Coefficient* (PCC) and we found that in some pairs of locations, the linear regressions are affected by outliers in the training data, resulting in a low PCC (see Fig. 5(a)).

To prevent the effect of outliers, we use the *Density-Based Spatial Clustering of Applications with Noise* (DBSCAN) method [8] to remove outliers before building linear regression models. We chose DBSCAN as it is a high performance unsupervised outlier detection method used in many recent research papers [12, 20, 21]. Figure 5(b) shows the linear regression model after using DBSCAN for removing outliers from the training data, which confirms that the linear regression model after removing outliers is a more reliable fit.

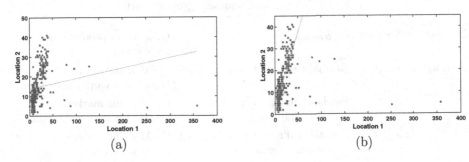

(a)   (b)

Fig. 5. Linear regression at two locations. (a) Before removing outliers, (b) After removing outliers using DBSCAN.

For the pedestrian data, removing outliers reduced the proportion of locations with low PCC values (lower than 0.6) from 6% to 0.6%. For the vehicle traffic data, this value reduced from 15% to 4%. When we used DBSCAN, we assumed that 20% of the training data are outliers. We changed this assumed outlier percentage in the range from 5 to 30 and there was little variation in the results. The effect of outliers on the linear correlations in the training vehicle traffic data is less than their effect on the coefficients in pedestrian data because the time resolution of the training data is lower in the vehicle traffic data than for the pedestrian traffic data. The smaller the sample size, the greater the effect of outliers on the normal models.

5   Experiments

5.1   Datasets and Ground Truth

We use the real datasets of the pedestrian count data [2] and the vehicle traffic count data [1] as described in Sect. 1. We ignore missing values in the training data. In the vehicle traffic data, only 3% of the training data is missing.

The performance of our method is compared to the benchmark algorithms based on some known real anomalous local events and real legitimate global traffic changes (see Table 2). For anomalous local events, we choose winter fireworks in Waterfront City (Docklands) and night market festivals in the *Queen Victoria Market* (QVM) in 2015 for the pedestrian data where an uncharacteristic pedestrian flow is experienced. We also consider the White Night event that happened in some parts of the city in 2014 as an anomalous local event for the vehicle traffic data as only some road segments in the CBD are blocked or partially affected by this event. Normal traffic patterns in April are the normal ground truth for both datasets. We also consider weekday holidays in Melbourne such as New Year's Eve, Good Friday and the Queen's Birthday as legitimate global traffic changes for both datasets (see Table 2).

Table 2. Normal and anomaly ground truth

Dataset	Scenario	GT_{Normal}	$GT_{Anomaly}$ (anomalous local events)
Pedestrian	1	Normal traffic patterns in April	QVM night market, Docklands winter fireworks
	2	Weekday holidays (Legitimate global traffic changes)	QVM night market, Docklands winter fireworks
	3	Normal traffic patterns in April and weekday holidays	QVM night market, Docklands winter fireworks
Vehicle	4	Normal traffic patterns in April	White night
	5	Weekday holidays (Legitimate global traffic changes)	White night
	6	Normal traffic patterns in April and weekday holidays	White night

5.2   Comparison Methods

We compare our method with four other methods: OCSVM [11], TOD [13], Boxplot [22], and k-sigma rule [16]. OCSVM models a training set comprising only the normal data as a single class. Dense subsets of the input space are labelled as normal whereas observations from other subsets of the input space are labelled as anomalies. We train an OCSVM model with the *Radial Basis Function* (RBF) in an unsupervised manner given the technique proposed in [11].

TOD proposed in [13] for detecting anomalies in vehicle traffic data, and is similar to CTED. The main differences are that TOD considers all locations to have the *same importance* (*weight*) when determining the anomaly score of each location and uses absolute traffic counts. TOD uses a reinforcement technique and expects two historically similar locations to remain similar and two historically dissimilar locations to stay dissimilar.

The last two comparison methods are the extended versions of the standard Boxplot [22] and the 3-sigma [16] anomaly detection methods. Boxplot constructs a box whose length is the *Inter Quartile Range* (IQR). Observations that lie outside $1.5 * IQR$ are defined as anomalies. In this paper, we learn IQR for each hour h using the training data, and then we consider different ratios of IQR as the threshold for anomalies. We define observations that are outside $k_{thr_B} * IQR$ as anomalies and investigate the overall performance of Boxplot for different threshold values, k_{thr_B}. The 3-sigma rule calculates the mean (μ) and standard deviation (σ) of the training data and then declares the current observation $(n_i^{(h)}(q))$ anomalous if $\left| n_i^{(h)}(q) - \mu \right| > 3\sigma$. We learn μ and σ for each hour h of the day using the training data. We then use a general version, k-sigma rule, where observations that are outside $\left| n_i^{(h)}(q) - \mu \right| > k_{thr_S}\sigma$ are defined as the anomalies.

5.3   Experimental Setup

We evaluate the performance of CTED by computing the *Area Under the ROC Curve* (AUC) metric. Table 2 shows the ground truth (GT) that we use for both real datasets under different experimental scenarios discussed in Sect. 5.1. In calculating the AUC metric, true positives are anomalous local events, (*location*, *hour*, *day*) triples, belonging to $GT_{Anomaly}$ that are declared anomalous, and false positives are normal traffic changes (including legitimate global ones), (*location*, *hour*, *day*) triples, that belong to GT_{Normal} but are misdetected as anomalous events.

Setting Parameters. The threshold that is compared to anomaly score values is the parameter that defines anomalies in all the comparison methods and the proposed method: thr_{AS} in CTED, thr_{SVM} in OCSVM, k_{thr_B} in Boxplot and k_{thr_S} in k-sigma. This parameter is the only parameter required by CTED.

In OCSVM, to extract features for training the models, n_c correlated sensors are identified for each sensor. The features of OCSVM are the ratio of the counts in two correlated sensors. On our experiments, we change $n_c \in \{5, 10, 15, 20\}$ for the number of correlated locations to each location and we report the best results. Note that by increasing n_c from 20, we observed reduction in the accuracy. We train a model for each hour of the day using the training data. These models are used for evaluating the current observations.

In TOD, in addition to the anomaly score threshold, a similarity threshold and three other parameters, $\alpha 1 < 1$, $\alpha 2 \geq 0$ and $\beta > 1$, must also be determined. Setting appropriate values for these parameters is difficult and is best done using prior knowledge about the dataset. In our experiments, we changed $\alpha 1$ in the range $[0-1)$, $\alpha 2$ in the range $[0-10]$ and β in the range $(1-10]$ for the test data. We found that the values of 0.7, 2 and 1 respectively for $\alpha 1$, $\alpha 2$ and β lead to the highest AUC value for the test data in the pedestrian traffic data and the values of 0.99, 0 and 1.1 respectively for $\alpha 1$, $\alpha 2$ and β lead to the highest AUC value for the test data in the vehicle traffic data. We used these parameters for TOD in our experiments. In practice, the necessity to tune the TOD parameters using test data makes this approach difficult and time consuming.

In the Boxplot and k-sigma rule methods, we estimate IQR, the observation mean μ and the observation standard error of the estimate σ for each hour h of the day using training data.

Table 3. AUC values for CTED and the benchmarks

Dataset	Scenario	OCSVM [11]	TOD [13]	Boxplot [22]	k-sigma [16]	CTED Proposed method
Pedestrian	1	0.86	0.71	0.85	0.84	**0.86**
	2	0.78	0.71	0.78	0.71	**0.84**
	3	0.84	0.71	0.83	0.82	**0.86**
Vehicle	4	0.82	0.8	0.79	0.79	**0.87**
	5	0.77	0.79	0.64	0.65	**0.88**
	6	0.82	0.79	0.77	0.78	**0.88**

5.4   Results

Figure 6 and Table 3 compare the resulting ROC curves and AUC values for the comparison methods against CTED produced by changing the threshold of anomaly scores for three different experimental scenarios in Table 2 as discussed in Sect. 5.1. In OCSVM, we set $n_c \in \{5, 10, 15, 20\}$ for the number of correlated locations to each location and we found that OCSVM is sensitive to the choice of this parameter. Specifically, increasing n_c from 20 resulted in a large reduction in the accuracy of OCSVM. We changed the number of correlated locations to each location for CTED and we found that CTED has a low sensitivity to n_c. In OCSVM, best results for the pedestrian dataset was achieved at $n_c = 15$, while we got the best results for the vehicle dataset when we set $n_c = 10$.

The bolded results in Table 3 show that the AUC values of CTED are higher than all the benchmark approaches for all the above-mentioned scenarios in both the pedestrian counts and the vehicle traffic datasets. In Fig. 6, we plot the *Receiver Operating Characteristic* (ROC) curve for the vehicle traffic dataset against the three benchmarks. This figure confirms that CTED performs better than other benchmarks for all the scenarios.

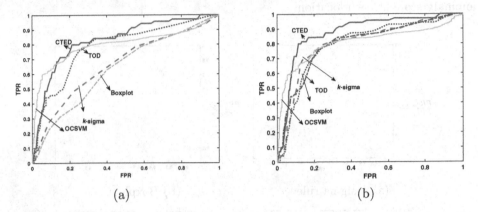

(a)   (b)

Fig. 6. ROC curves for vehicle traffic data for the real anomalous events in Table 2. (a) Legitimate global traffic changes are the normal ground truth, (b) Normal traffic patterns in April are the normal ground truth.

Comparing Fig. 6(a) and (b) reveals that the difference between the performance of CTED and the other benchmarks is mostly larger when legitimate global traffic changes are the normal ground truth. This larger difference stems from the lower false positive rate of CTED because it is more robust to legitimate global traffic changes compared to the benchmark techniques.

Reliable anomaly detection is difficult in practice. Reducing false positives is very important as this makes the anomaly detection system more reliable for city management purposes. A system that is not robust to legitimate global traffic changes generates many false alarms. This makes existing anomaly event detection methods unreliable for use in real applications, such as vehicle accident detection and traffic congestion detection systems.

5.5   Robustness to Legitimate Global Traffic Changes

Figure 7 compares the ratio of the false positive rate of CTED to the other three benchmarks for different values of *True Positive Rate* (TPR). Figure 7 shows that the *False Positive Ratio* (FPR) ratio between CTED and the benchmarks for the legitimate global traffic changes is lower than the local events, which confirms the greater robustness of CTED compared to the benchmarks for legitimate global traffic changes.

Figure 7(a) and (b) highlight that the k-sigma and Boxplot methods produce much higher false positives for legitimate global traffic changes than local events for vehicle traffic data. However, Fig. 7(c) and (d) show that TOD and OCSVM are more robust than the k-sigma and Boxplot methods to legitimate global traffic changes but still less robust than CTED. The greater robustness of TOD and OCSVM to legitimate global traffic changes is mainly due to considering traffic counts at other locations (relative traffic counts) when computing the anomaly score at each location.

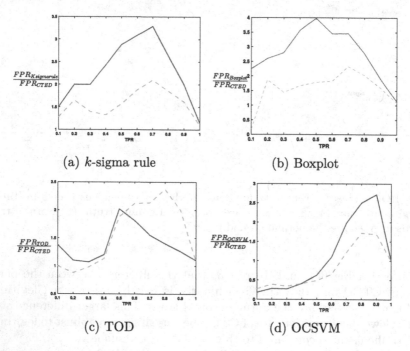

Fig. 7. The ratio of FPR produced by comparison methods compared to CTED for the same values of TPR for the vehicle traffic dataset. The anomaly ground truth is the the White Night event in all the cases. The dashed red lines show the results when the normal traffic patterns in April are considered as the normal ground truth while the blue lines show the results when legitimate global traffic changes are considered as the normal ground truth. (Color figure online)

5.6   Time Complexity

CTED is composed of an offline and an online phase. The complexity of the offline phase in the worst case is $O(m^2 n_{TD}^2)$, where m is the number of locations, and n_{TD} is the number of observation vectors in the training data. The complexity of DBSCAN in the worst case is $O(m^2 n_{TD}^2)$, and the time complexity for building normal linear regression models and weighting them is $O(m^2 n_{TD})$. The offline phase only executed once. Note that $n_{TD} \gg n$, as we discussed in Sect. 4.4.

The online phase is executed whenever a new observation vector arrives. The time complexity for processing each new observation is $O(m^2)$ as we find the estimation error for the current observation in each location based on the other locations.

6   Conclusions

In this paper, we proposed a new unsupervised method for detecting anomalous local traffic events, called CTED. This method that is highly *robust* to legitimate global traffic changes. This method builds normal models for each location by investigating the linear relationships between different locations in the city and uses the models to detect anomalous local events. Our experiments on two real traffic datasets collected in the Melbourne CBD, the pedestrian count and the vehicle traffic count datasets, verify that our simple linear regression-based method accurately detects anomalous real local events while reducing the false positive rate on legitimate global traffic changes compared to four other benchmark methods for anomaly detection in traffic data.

Changes in the city infrastructure can change the normal behaviour of the traffic in several locations of a city. As a future direction of our research, we aim to exploit time series change point detection methods to find the time of these behavioural changes in the traffic, and automatically update CTED when it is necessary.

References

1. Vicroads open traffic volume data. https://vicroadsopendata-vicroadsmaps. opendata.arcgis.com/datasets/147696bb47544a209e0a5e79e165d1b0_0 (2014)
2. City of Melbourne pedestrian counting system. http://www.pedestrian.melbourne. vic.gov.au/ (2015)
3. Baras, K., Moreira, A.: Anomaly detection in university campus WiFi zones. In: 8th IEEE International Conference on Pervasive Computing and Communications Workshops (PERCOM), pp. 202–207 (2010)
4. Dani, M.-C., Jollois, F.-X., Nadif, M., Freixo, C.: Adaptive threshold for anomaly detection using time series segmentation. In: International Conference on Neural Information Processing, pp. 82–89 (2015)
5. Doan, M.T., Rajasegarar, S., Leckie, C.: Profiling pedestrian activity patterns in a dynamic urban environment. In: 4th International Workshop on Urban Computing (UrbComp) (2015)
6. Doan, M.T., Rajasegarar, S., Salehi, M., Moshtaghi, M., Leckie, C.: Profiling pedestrian distribution and anomaly detection in a dynamic environment. In: CIKM, pp. 1827–1830 (2015)
7. Erfani, S.M., Rajasegarar, S., Karunasekera, S., Leckie, C.: High-dimensional and large-scale anomaly detection using a linear one-class SVM with deep learning. Pattern Recogn. **58**, 121–134 (2016)
8. Ester, M., Kriegel, H.-P., Sander, J., Xu, X., et al.: A density-based algorithm for discovering clusters in large spatial databases with noise. In: KDD vol. 34, pp. 226–231 (1996)

9. Frias-Martinez, V., Stolfo, S.J., Keromytis, A.D.: Behavior-profile clustering for false alert reduction in anomaly detection sensors. In: Annual Computer Security Applications Conference (ACSAC), pp. 367–376 (2008)
10. Garcia-Font, V., Garrigues, C., Rifà-Pous, H.: A comparative study of anomaly detection techniques for smart city wireless sensor networks. Sensors 16(6), 868 (2016)
11. Ghafoori, Z., Erfani, S.M., Rajasegarar, S., Bezdek, J.C., Karunasekera, S., Leckie, C.: Efficient unsupervised parameter estimation for one-class support vector machines. IEEE Trans. Neural Netw. Learn. Syst. (2018)
12. Jeong, S.Y., Koh, Y.S., Dobbie, G.: Phishing detection on Twitter streams. In: Cao, H., Li, J., Wang, R. (eds.) PAKDD 2016. LNCS (LNAI), vol. 9794, pp. 141–153. Springer, Cham (2016). https://doi.org/10.1007/978-3-319-42996-0_12
13. Li, X., Li, Z., Han, J., Lee, J.-G.: Temporal outlier detection in vehicle traffic data. In: IEEE 25th International Conference on Data Engineering (ICDE), pp. 1319–1322 (2009)
14. Limthong, K.: Real-time computer network anomaly detection using machine learning techniques. J. Adv. Comput. Netw. 1(1), 1–5 (2013)
15. Nidhal, A. Ngah, U.K., Ismail, W.: Real time traffic congestion detection system. In: 5th International Conference on Intelligent and Advanced Systems (ICIAS), pp. 1–5 (2014)
16. Pukelsheim, F.: The three sigma rule. Am. Stat. 48(2), 88–91 (1994)
17. Rajasegarar, S., Bezdek, J.C., Moshtaghi, M., Leckie, C., Havens, T.C., Palaniswami, M.: Measures for clustering and anomaly detection in sets of higher dimensional ellipsoids. In: International Joint Conference on Neural Networks (IJCNN), pp. 1–8 (2012)
18. Rajasegarar, S., et al.: Ellipsoidal neighbourhood outlier factor for distributed anomaly detection in resource constrained networks. Pattern Recogn. 47(9), 2867–2879 (2014)
19. Reddy, R.R., Ramadevi, Y., Sunitha, K.: Enhanced anomaly detection using ensemble support vector machine. In: ICBDAC, pp. 107–111 (2017)
20. Shi, Y., Deng, M., Yang, X., Gong, J.: Detecting anomalies in spatio-temporal flow data by constructing dynamic neighbourhoods. Comput. Environ. Urban Syst. 67, 80–96 (2018)
21. Tu, J., Duan, Y.: Detecting congestion and detour of taxi trip via GPS data. In: IEEE Second International Conference on Data Science in Cyberspace (DSC), pp. 615–618 (2017)
22. Tukey, J.W.: Exploratory Data Analysis (1977)
23. Witayangkurn, A., Horanont, T., Sekimoto, Y., Shibasaki, R.: Anomalous event detection on large-scale GPS data from mobile phones using Hidden Markov Model and cloud platform. In: Proceedings of the ACM Conference on Pervasive and Ubiquitous Computing Adjunct Publication, pp. 1219–1228 (2013)
24. Zhou, Z., Meerkamp, P., Volinsky, C.: Quantifying urban traffic anomalies. arXiv preprint arXiv:1610.00579 (2016)

Combining Bayesian Inference and Clustering for Transport Mode Detection from Sparse and Noisy Geolocation Data

Danya Bachir[1,2,3]([✉]), Ghazaleh Khodabandelou[2], Vincent Gauthier[2], Mounim El Yacoubi[2], and Eric Vachon[3]

[1] IRT SystemX, Palaiseau, France
danya.bachir@gmail.com
[2] SAMOVAR, Telecom SudParis, CNRS, Université Paris Saclay, Paris, France
[3] Bouygues Telecom Big Data Lab, Meudon, France

Abstract. Large-scale and real-time transport mode detection is an open challenge for smart transport research. Although massive mobility data is collected from smartphones, mining mobile network geolocation is non-trivial as it is a sparse, coarse and noisy data for which real transport labels are unknown. In this study, we process billions of Call Detail Records from the Greater Paris and present the first method for transport mode detection of any traveling device. Cellphones trajectories, which are anonymized and aggregated, are constructed as sequences of visited locations, called sectors. Clustering and Bayesian inference are combined to estimate transport probabilities for each trajectory. First, we apply clustering on sectors. Features are constructed using spatial information from mobile networks and transport networks. Then, we extract a subset of 15% sectors, having road and rail labels (e.g., train stations), while remaining sectors are multi-modal. The proportion of labels per cluster is used to calculate transport probabilities given each visited sector. Thus, with Bayesian inference, each record updates the transport probability of the trajectory, without requiring the exact itinerary. For validation, we use the travel survey to compare daily average trips per user. With Pearson correlations reaching 0.96 for road and rail trips, the model appears performant and robust to noise and sparsity.

Keywords: Mobile phone geolocation · Call Detail Records
Trajectory mining · Transport mode · Clustering · Bayesian inference
Big Data

1 Introduction

The growing use of smartphones generates massive ubiquitous mobility data. With unprecedented penetration rates, mobile networks are supplying the largest geolocation databases. Mobile phone providers collect real-time Call Detail Records (CDR) from calls, text messages or data at no extra-cost for billing

© Springer Nature Switzerland AG 2019
U. Brefeld et al. (Eds.): ECML PKDD 2018, LNAI 11053, pp. 569–584, 2019.
https://doi.org/10.1007/978-3-030-10997-4_35

purposes. Still, traditional transport planning models have so far relied on expensive travel surveys, conducted once a decade. Consequently, surveys are rapidly outdated, while suffering from sampling bias and biased users' responses. Past research used CDR to estimate travel demand [21], optimal locations for new transport infrastructures [7], weekly travel patterns [9], activity-based patterns [12], urban land-use [19], impact of major events or incidents [6] and population dynamics [5,14]. A few studies used triangulation, based on signal strength e.g., in Boston U.S. [4,20]. In Europe, privacy policies restrict triangulation usage to police demands. CDR and GPS data both respect privacy compliance for geolocation. Still GPS data collection requires users to install tracking applications and activate GPS, which has greedy battery consumption. Consequently, GPS samples represent subsets of users' trips while CDR generate locations from larger populations over longer time periods. However CDR geolocation is coarse, noisy and affected by the usage frequency of devices. Raw CDR provide approximate and partial knowledge of true users' paths, hence requiring careful pre-processing. Past methods on transport mode detection mainly involved GPS data and are hardly transposable to CDR. In addition, these studies applied supervised learning [10,18,22] requiring a training dataset of trajectories with transport mode labels. Transport modes were either collected via applications where users consent to enter their travel details, or manually identified using expert knowledge, which is a costly task. In real world scenarios, transport modes of traveling populations are unavailable. Therefore we need new unsupervised approaches to tackle this issue.

This paper presents the first unsupervised learning method for transport mode detection from any CDR trajectory. As this is a first study, we focus on a bi-modal separation between road and rail trips. In collaboration with a mobile phone provider, we process one month trajectories from the Greater Paris, which are anonymized and aggregated for privacy. Trajectories are represented as sequences of visited mobile network areas, called sectors. Our model combines clustering with Bayesian inference to determine the probability that cellphones traveled by road or rail knowing their trajectories on the mobile network. The transport probability of a trajectory is initialized with a prior obtained from the travel survey and updated with each new visited sector. Transport probabilities for sectors are derived after clustering sectors by transport type. Sectors features are constructed using both mobile networks and transport networks spatial properties. Then, for a subset of 15% sectors, we extract transport labels, being road or rail, (e.g., equipments inside train stations, on highways etc.) while the remaining sectors are multimodal. For each cluster, we use the binary labels to calculate continuous transport probabilities as the proportion of labeled sectors among total sectors. Trajectories are thus attributed the most probable mode among road, rail or mixed (i.e., when probabilities are close). For validation, we calculate daily average rail and road trip counts per user and obtain Pearson correlations with the travel survey above 0.96, for the 8 departments of the region. In the next sections, we review the literature in Sect. 2 and describe data engineering in Sect. 3. The methodology steps are presented in Sect. 4. Eventually, we discuss main results in Sect. 5 and provide conclusion.

2   Related Work

Common applications for geolocation data mining are the identification of travel patterns for personal travel recommendation [23,24], anomalous behavior detection [17] and transport planning [12]. Several works used supervised transport mode learning from GPS trajectories. A multilayer perceptron was used to identify car, bus and walkers modes for 114 GPS trajectories in [10]. Features were the average and maximum speed and acceleration, the total and average travel distance, the number of locations divided by travel distance and the number of locations divided by travel time. The best accuracy was 91% using a 10-folds cross validation. In [18], speed and acceleration features were collected from 16 GPS trajectories. Several classification models (Decision Tree, Kmeans, Naïve Bayes, NNeighbor, SVM, Discrete and Continuous HMM) were compared. The Decision tree with Discrete Hidden Markov Model obtained the highest accuracy (74%). Still, supervised approaches with GPS are constrained by the small size of the training data. Moreover, although transport labels can be collected for small GPS datasets, they are unavailable for CDR.

Meanwhile, few studies tackled unsupervised transport mode detection. In [8] fuzzy logic was used as a scoring function calculated between consecutive GPS traces. The transport score was calculated with boolean conditions on speed, distances to transport network and previous mode. Still, this work lacked a performance evaluation. In [15], base stations located inside Paris underground were used to identify underground mode from CDR trips. A record detected by an underground antenna was labeled accordingly. This approach is limited as it relies exclusively on indoor equipment inside the underground. No additional modes were identified. To our knowledge, only one work addressed unsupervised transport mode learning for two modes, road and public transport, using triangulated CDR [20]. The approach applies travel times clustering followed by a comparison with Google travel times. Still, CDR low frequency induces important incertitude and delay on start and end travel times of CDR trips. Consequently a device may not be detected as traveling when the real trip begins and ends. Moreover the presented approach was demonstrated on one unique Origin and Destination (OD) pair which is not sufficient to validate the method. In dense urban areas, travel times can be affected by traffic states (e.g., rush hours) and can be identical for several modes, depending on the OD.

Our work presents a novel method for transport mode detection by combining two unsupervised techniques, namely clustering and Bayesian inference. This model classifies millions of CDR trajectories into road and rail trips. Instead of clustering trajectories with features such as speed or travel time, highly impacted by the imprecision, sparsity and noise of CDR geolocation, we apply clustering on sectors and build spatial features using transport networks. A small subset of road and rail labels is collected for sectors in order to calculate sectors transport probabilities. After the Bayesian inference step, we conduct a large-scale validation for the complete region, using the travel survey. The high Pearson correlations, obtained on daily average trips per user, proves the method is generalizable, performant and robust to noise and sparsity.

3   Data Engineering

For this study, we collect anonymized CDR trajectories from the Greater Paris region, over one month. Sectors features are constructed using the base stations referential jointly with transport networks infrastructures. For data normalization, we introduce a specific procedure accounting for heterogeneous urban density. Label extraction is realized to gather transport labels for a small subset of sectors. For model validation we use the household travel survey from 2010 conducted by Île de France Mobilités-OMNIL-DRIEA [1].

3.1   Mobile Network

Mobile providers do not have access to GPS coordinates of mobile phones. Although we know which base station is connected to a device, it is unlikely to encounter mobile users positioned exactly at the base station. Devices are located inside mobile network areas covered by base stations signal range. For this study, we use the mobile network referential of the Greater Paris region. This region has a 12000 km^2 area with more than 1200 cities and 12 millions inhabitants. It is covered by thousands of mobile network antennas. Each base station emits 2G, 3G or 4G radio signals. Cells are circular areas covered by signals (see Fig. 1). Each cell equipment is oriented toward one direction. The partitions of cells directions are called sectors. The average sector number per antenna is 3 where one sector covers 120° around the base station. A cellular tessellation is composed of a multitude of overlapping areas. We use the sector tessellation to get rid of overlaps and create the voronoï partitions using sectors centroids (see Fig. 2). We associate each mobile phone record to a sector location.

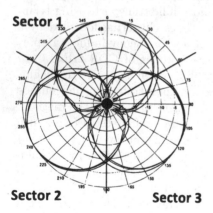

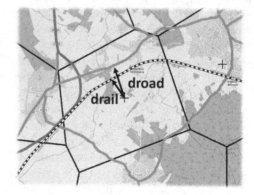

Fig. 1. Schema of a tri-sector antenna. The antenna is represented by the black dot. Circular areas are cells for 2G, 3G and 4G signals.

Fig. 2. Example of a voronoi sector and its associated shortest distance to transports axes. Five roads (colored lines) and one rail line (dashed line) intersect the sector. (Color figure online)

3.2   Transport Networks

Transport networks are used to construct sectors features. We retrieve rails infrastructures for underground, overground, tramway and train stations shared by Île-de-France Mobilité on their platform Open Data STIF [2]. In addition we use OpenStreetMap (OSM) [3] to retrieve highspeed rails and road infrastructures. Roads are categorized by traffic importance. We filter residential roads which have highest road count and lowest traffic.

3.3   Raw Features Construction

We construct our dataset $D = \{d_{rail}, d_{road}, d_{station}, n_{road}, n_{rail}, w_{station}\}$ where features stand for:

- d_{road}: shortest distance between sector centroid and road (see Fig. 2).
- d_{rail}: shortest distance between sector centroid and rail network (see Fig. 2).
- $d_{station}$: shortest distance between sector centroid and train station centroid.
- n_{road}: number of roads intersecting the voronoi.
- n_{rail}: number of rail lines intersecting the voronoi.
- $w_{station}$: weight of train stations calculated as the sum of stations area intersecting the sector voronoi area.

3.4   Data Normalization

We aim to find transport mode usage in sectors. As our raw features are built with spatial information they are impacted by urban density. In the city center the density is higher than in the suburb. Consequently sector areas and distances to transport networks are smaller while there are more transport hubs. We normalize our features to reduce the bias induced by urban density over transport usage. We introduce a normalization specific to our problem:

$$d_{norm,m} = \frac{d_m}{\sum_i d_i} \in [0, 1] \tag{1}$$

$$n_{norm,m} = \frac{n_m}{\sum_i n_i} \in [0, 1] \tag{2}$$

$$w_{norm,station} = \frac{w_{station}}{A_v} \in [0, 1] \tag{3}$$

where $d_m \in \{d_{road}, d_{rail}, d_{station}\}$, $n_m \in \{n_{road}, n_{rail}\}$ and $d_{norm,m}$, resp. $n_{norm,m}$, is the normalized vector for feature d_m, resp. n_m. Feature $w_{norm,station}$ is the normalization of $w_{station}$ by voronoi area A_v.

3.5   Sector Label Extraction

A few base stations are located on transport hubs, such as rail lines, train stations, highways or tunnels. We process this information to construct labels for a small subset of antennas. We assume that each sector inherits from its base station label. We attribute rail labels to indoor equipments located inside the underground and train stations, which represent 4% sectors. We assign road mode to indoor antennas in tunnels, constituting less than 1% sectors. We add outdoor antennas on highways (11% sectors) to increase the size of the road subset. In total we obtain 15% transport labels. In what follows, we use our subset of sectors with categorical transport labels $\{road, rail\}$, as prior knowledge. Still, categorical transport labels are not appropriate for most sectors, including outdoor equipments. In urban areas, such as the Greater Paris, the classic scenario is to encounter several transport modes inside an outdoor sector because of mobile networks' coarse granularity. Thus, we aim to find continuous transport probabilities $P \in [0,1]$ for all sectors, where indoor labeled equipments have maximal probabilities $P \in \{0,1\}$.

3.6 Trajectories Pre-processing

For this study, the mobile provider pre-processed raw anonymized users' positions using noise reduction and segmentation (see Fig. 3). For segmentation, users' locations were separated into stay points i.e., when users remain in the same area, and moving points i.e., when users are assumed traveling. We define a trajectory as a sequence of moving points $T_j^u = \{(X_0, t_0), \ldots, (X_l, t_l)\}$, j being the jth trajectory of the user u. The ith position recorded at timestamp t_i is $X_i = (x_i, y_i)$, where (x_i, y_i) are the centroid coordinates of the visited sector. One trajectory corresponds to one user trip. We construct 95 millions CDR trajectories from 2 millions anonymized users during one month. Similar trajectories are aggregated to respect privacy policies. In order to compare our results with household travel survey, which was conducted for residents of the Greater Paris region, the mobile provider filters users by home department (first two digits of billing address postcode) and exclude visitors.

4   Model

This section presents the unsupervised learning scheme combining clustering and Bayesian inference to estimate transport modes of CDR trajectories. First, the prior transport probability is obtained from the travel survey. Second, the transport likelihood is calculated from the observed records, such as each new visited sector updates the probability. In this perspective, we apply a clustering on sectors. Then, our subset of sectors labels is used to calculate transport probabilities within each cluster. Each sector is assigned a continuous score in $[0, 1]$ reflecting the real transport usage inside i.e., the probability to detect more users on the roads or on the rails. For each trajectory, we assign the mode with highest probability. Eventually, results are validated against the survey.

4.1   Clustering

We aim to find transport clusters for mobile network sectors with an underlying hierarchical structure. Thus we use an agglomerative hierarchical clustering. The

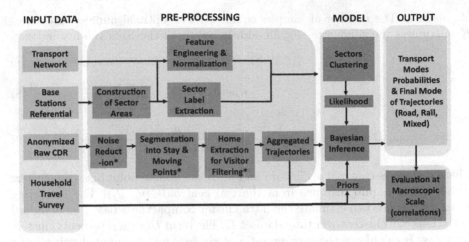

Fig. 3. Transport mode detection workflow applied for this case study. Pre-processing steps annotated with (*) were performed by the mobile operator.

clustering starts with N clusters of size 1, N being the number of sectors. Each sector is recursively merged with its closest neighbor according to a linkage criterion and a distance function. We test three linkage types with three distance functions (euclidean, Manhattan and cosine). Complete linkage minimizes the maximal distance between two points from two clusters. Average linkage minimizes the average distance between clusters points. Ward linkage, with euclidean distance, minimizes the sum of squared error $ESS = \sum_{k,i,j} |X_{ijk} - \bar{x}_{kj}|^2$, where X_{ijk} is the sample value for sector i, feature j and cluster k; $\bar{x}_{kj}$ is the mean value of feature j for cluster k. The agglomerative clustering applies until all data points are merged into a single cluster of size N. A good clustering solution should divide rail transport sectors from road sectors.

4.2   Evaluation Metrics

We use internal evaluation metrics to assess the clustering performance and to identify the optimal cluster number. We used the Silhouette (S) to evaluate clusters separability [13] (see Eq. 4).

$$s_{ik} = \frac{b(i) - a(i)}{max(b(i), a(i))} \tag{4}$$

$$S_k = \frac{1}{N_k} \sum_{i=1}^{N_k} s_{ik} \tag{5}$$

$$S = \frac{1}{N} \sum_k S_k \tag{6}$$

where $a(i)$ is the average intra cluster distances for sector i and $b(i)$ is the lowest value among average inter cluster distances. Here N_k stands for the size of cluster k. The number of samples equals N. The optimal number of clusters K maximizes the silhouette [16]. In addition we used the S_{dbw} validity index.

$$S_{dbw}(k) = Scat(k) + Dens_{db}(k) \tag{7}$$

$$\text{where } Scat(k) = \frac{1}{k} \sum_{i=1}^{k} \frac{\sigma(\nu_i)}{\sigma(D)} \tag{8}$$

$$\text{and } Dens_{db}(k) = \frac{1}{k(k-1)} \sum_{i,j=1}^{k} \frac{dens(u_{ij})}{max(dens(v_i), dens(v_j))} \tag{9}$$

where ν_i denotes centroid of cluster i and u_{ij} is the middle point between clusters i and j i.e., at mid distance from the two centroids (ν_i, ν_j). The scattering index $Scat$ is used to estimate the intra cluster compactness based on standard deviations σ of clusters over total dataset D. The term $Dens_{db}$ represents clusters densities. It calculates the average ratio of clusters middle point densities over clusters centers densities. The underlying assumption is that well defined clusters are denser around their centroids than at their mid distance. This index is a trade-off between clusters densities and variances. It has been depicted as the most performing among internal clustering evaluation metrics in [11,16]. The optimal cluster number is found when the index reaches its minimum.

4.3   Probability Scores of Sectors Transport Mode

For each cluster k we calculate the score $p_{k,m}$ for transport mode $m \in \{rail, road\}$.

$$p_{k,m} = \frac{N_{k,m}}{N_m} \tag{10}$$

where $N_{k,m}$ is the number of labeled sectors belonging to class m in cluster k and N_m is the total number of sectors from class m in the dataset. We normalize $p_{k,m}$ to obtain the probability $P(m|S_i) \in [0, 1]$ of using mode m given a visited sector S_i, belonging to a cluster k.

$$P(m|S_i) = \frac{p_{k,m}}{\sum_j p_{k,j}} \tag{11}$$

Unlabeled sectors obtain transport probabilities according to their cluster. In addition we update the probabilities of outdoor labeled sectors (i.e., highways) using Eqs. 10 and 11. Indoor labeled sectors have binary probabilities in $\{0, 1\}$.

4.4   Bayesian Inference of Trajectories Transport Mode

Bayesian inference is used to determine the main transport mode associated to mobile phone trajectories. In this perspective, we calculate the probability $P(m|T_j^u)$ to take a mode $m \in \{rail, road\}$ knowing the trajectory T_j^u, using Bayes theorem:

$$P(m|T_j^u) = \frac{P(T_j^u|m) * P(m)}{P(T_j^u)} \tag{12}$$

Trajectories are sequences of sectors $\{S_0, \ldots, S_l\}$ visited by mobile phone holders. Thus we have $P(T_j^u|m) = P(S_0, \ldots, S_l|m)$. We assume independence between sectors probabilities such as $P(S_i, S_{i+1}|m) = P(S_i|m)P(S_{i+1}|m)$. This assumption is motivated by the need to reduce the computational cost of the calculation. Thus we can rewrite $P(T_j^u|m) = \prod_{i=0}^{l} P(S_i|m)$. Equation 12 becomes:

$$P(m|T_j^u) = \frac{P(m)}{P(T_j^u)} \prod_{i=0}^{l} P(S_i|m) \tag{13}$$

The term $P(m|S_i)$, previously calculated with Eq. 11, is introduced by applying Bayes theorem a second time, to Eq. 12:

$$P(m|T_j^u) = \frac{\prod_{i=0}^{l} P(S_i)}{P(T_j^u)} P(m)^{1-l} \prod_{i=0}^{l} P(m|S_i) \tag{14}$$

The term $\frac{\prod_{i=0}^{l} P(S_i)}{P(T_j^u)}$ does not influence the mode choice. The prior transport probability $P(m)$ can be seen as the initial guess, before observing records. The prior probability is obtained from the travel survey and is calculated as the average trip counts per user given the home location of cellphone holders, here at the department scale. For rail mode we have $p_{rail,dep} = \frac{AVG_{dep}(c_{rail})}{AVG_{dep}(c_{rail}) + AVG_{dep}(c_{road})} \in [0, 1]$ and $p_{rail,dep} = 1 - p_{road,dep}$, where c_{rail} and c_{road} are the rail and road trip counts, for the day of survey, per user living in the department dep. At last we normalize the posterior transport probability to be in range $[0, 1]$.

$$P(m|T_j^u) \leftarrow \frac{P(m|T_j^u)}{P(rail|T_j^u) + P(road|T_j^u)} \tag{15}$$

Finally we affect the mode obtaining the higher probability to each trajectory. When probabilities are in $[0.4, 0.6]$ the mode is considered mixed.

5 Results

This section summarizes our main results. For the clustering we demonstrate how we determine the number of clusters. We describe clusters according to transport probabilities. From the Bayesian inference of trajectories' transport modes, we visualize transport flows per week day and observe the travel patterns. We provide detailed results comparison with survey, at department scale, using Pearson correlations as evaluation metric.

5.1 Clustering Evaluation

We first compare the three linkage types. Average and complete linkage fail to separate sectors in the city center, with any distance metric. One huge centered cluster is produced with tiny clusters located at the region borders. We retain ward linkage with euclidean distance which produce clusters of comparable size, evenly present across the region. In order to find the optimal number of cluster we draw the dendrogram of the ward agglomerative clustering (see Fig. 4). The latter shows $k = 2$ is a good cluster number as it corresponds to the highest distance gap between merges. A small k leads to a macroscopic partitioning. We look for a higher k to detect finer transport modes tendencies. A clear cut was possible for $k \in \{3, 4, 5, 9\}$, which were therefore also good candidates. We decide to bound the cluster number between 2 and 10. We use additional intra-cluster metrics. We calculate S and S_{dbw} with several k values (see Fig. 5). The silhouette reaches a maximum for $k = 4$, for which separability is the highest.

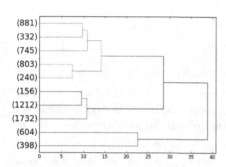

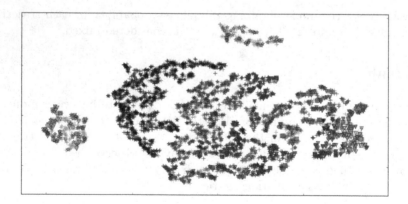

Fig. 4. Dendrogram for $k \in [2, 10]$. The x axis is the height i.e., distances between clusters leaves and nodes. The y axis shows the number of leaves per cluster.

Fig. 5. Silhouette (blue) and S_{dbw} validity index (red) plotted in function of the number of cluster k (Color figure online)

Fig. 6. t-sne projection for dataset D after normalization and z-score transformation. Colors represent clusters for k varying from 1 to 9. The parameters are $n_{component} = 2$, $perplexity = 30$, $learningrate = 200$, $n_{iteration} = 1000$. Stars correspond to road labels, Triangle to rails and crosses to unlabeled sectors. (Color figure online)

According to the S_{dbw} minimization criterion, the optimal number of clusters is $k = 9$, for which clusters are the most compact and dense. For $k \in [5, 10]$ the silhouette reaches a local maximum for $k = 9$. For our problem we favor the larger k hence we select $k = 9$. We visualize the 9 clusters with t-sne (see Fig. 6) and project them on the sectors map (see Fig. 7).

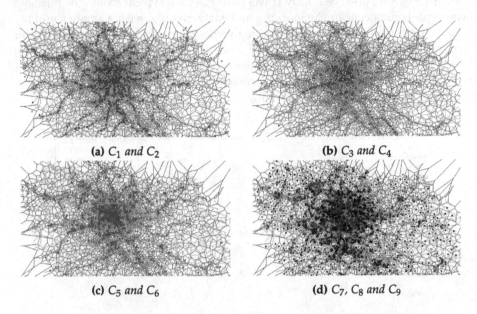

(a) C_1 and C_2

(b) C_3 and C_4

(c) C_5 and C_6

(d) C_7, C_8 and C_9

Fig. 7. QGIS clusters projection (Color figure online)

5.2   Sectors Probabilities and Visualization

We calculate the transport probabilities per cluster (see Table 1). We describe clusters regarding transport usage. Each cluster is displayed in Figs. 6 and 7.

Table 1. Transport mode probabilities and cluster size for $k = 9$

Cluster	C_1	C_2	C_3	C_4	C_5	C_6	C_7	C_8	C_9
Size (%)	14.7	8.50	12.4	4.67	2.20	10.5	24.4	5.60	17.1
P_{RAIL}	0.651	0.567	0.824	0.949	0.421	0.387	0.095	0.071	0.199
P_{ROAD}	0.348	0.432	0.176	0.051	0.579	0.613	0.905	0.929	0.801

– C_1, C_2: mixed-rail clusters with a higher probability for rails, depicted in blue and cyan on Fig. 7a.
– C_3, C_4: rail dominated clusters with many underground sectors located in the city center. It corresponds to the red and yellow cluster on Fig. 7b.
– C_5, C_6: mixed road clusters, shown in magenta and green on Fig. 7c.
– C_7, C_8, C_9: road clusters represented in black, orange and purple on Fig. 7d.

5.3 Trajectories

We infer transport probabilities for one month trajectories, filtering bank holidays. We count the number of rail and road trips (see Fig. 8). Only 3% trips have probabilities in range [0.4, 0.6]. We consider such trips have mixed (or uncertain) mode. In Fig. 8 we observe hourly travel patterns for a typical week. For business days, peak hours occur in the morning and early evening, with a smaller midday peak at lunch time. Morning and evening peaks appear unbalanced. One reason is that mobile phone usage tends to be more important in the evening thus we detect more users and more trips. A second reason could be that users travel more at the end of the day. This phenomenon is more pronounced for road trips, the highest gap being on friday evening.

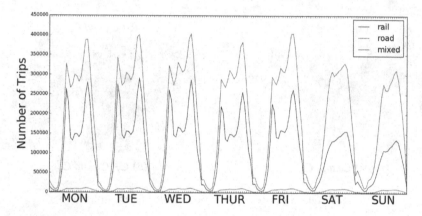

Fig. 8. Estimated trip counts are averaged per week day, per hour and per transport mode. Results are given for 1 month data from the Greater Paris.

5.4 Comparison with Survey

We compare our results with the latest household travel survey, from 2010, for the Greater Paris. About 43000 residents were asked about their travels during their past day, outside holidays. We calculate mobility statistics from survey and MP results (see Table 2). We average survey trip counts per resident: $C^S = \frac{\sum_{i=1}^{k} N_i * w_i}{\sum_{i=1}^{k} w_i}$ where an individual i of weight w_i reported N_i trips for the day he was questioned. The weight w_i was calculated during survey with socio-demographic information to rescale the individual to the entire population. Similarly we average CDR trip counts per day and per device: $C^{MP} = \sum_{i=1}^{U} \sum_{t=1}^{T} \frac{1}{U} \frac{1}{T} n_{u,i}$ where U is the number of phones, T is the number of days and $n_{i,t}$ is the number of trips detected for phone i for day t. In the survey, transport modes are separated in two categories, motorized modes including public transport, cars and motorbikes, and unmotorized modes i.e., walk and bike. Our model outputs the majority mode of a given CDR trajectory, between rail and road. We first examine results for all residents (see Table 2). The survey indicates the average trip number per user during a business day is 4.16 for all modes and 2.45 for motorized trips. We found an average of 2.10 daily trips per person. It seems we were able to detect 86% motorized modes. Because of the coarseness of the mobile network, walkers might be considered as non moving as their movement occurs at a too microscopic scale. In addition, the detection of travels is affected by CDR frequency. When a device is turned-off or unused for a long period of time, users are undetected. Compared to the survey, 14% daily motorized trips are undetected in average. We further analyze results for residents aggregated by home given for the city center, first ring, second ring and department scale (first two digits of postcode). We calculate Pearson correlations between survey and CDR estimates for all trips, motorized, road and rail trips. In addition we calculate the ratio between road and rail trips: $C_{ratio} = \frac{C_{road}}{C_{rail}}$. There is a negative correlation between total survey trips and CDR trips, due to the possible undetection of unmotorized modes. Correlations for rail, road and ratio are all above 0.96 for the three rings scale and the department scale. Still we have smaller ratio than the survey. The department obtaining results most similar with the survey is the city center (Paris). For the latter we detect the same number of motorized trips. This means that all users' trips were detected, suggesting that mobile phone activity of travelers is more important in the city center. From these observations we emit several hypothesis to explain remaining differences. First, because of their cost, surveys are performed on small population samples. Despite the use of weights to scale the sample to the total population, results can still contain sampling bias in addition with users' responses bias. Second, travel surveys are performed every 10 years because of their high cost. The latest complete survey is anterior to our study (seven years difference) which can lead to differences in results. In particular, transport policies over the past years were oriented to favor public transport in the Greater Paris (e.g., introduction of a unique price for transport pass that reduced the price for suburbs). This could have influenced users to take public transports, especially in the suburb. In our opinion trips segmentation might impact results. Indeed our trajectories are segmented based on stay times. Public transport users sometimes experiment waiting times in stations e.g., when users change lines, and signals loss when entering the underground. This could cause higher trip segmentation for CDR rail trips. At last we detect 100% trips in the city center versus 80% in the suburb. In parallel the city center has the highest rail transport usage. This could indicate a bias in mobile phone usage i.e., public transport users are more likely to call, text or navigate on the web than drivers. Therefore some road trips could possibly be undetected (Table 3).

Table 2. Mobility statistics for average trip number per user during a business day (Monday-Friday). Results are given per home location (i.e., 2nd ring, 1st ring and the 8 departments including city center). Left: results for survey (source: EGT 2010-Île de France Mobilités-OMNIL-DRIEA) Right: results with CDR

Home scale	Survey (S)					Mobile Phone (MP)			
	C_{All}^S	C_{Motor}^S	C_{Rail}^S	C_{Road}^S	C_{Ratio}^S	C_{All}^{MP}	C_{Rail}^{MP}	C_{Road}^{MP}	C_{Ratio}^{MP}
All population	4.16	2.45	0.61	1.85	3.03	2.10	0.80	1.30	1.62
City Center (CC)	4.37	1.93	1.11	0.83	0.75	1.94	1.22	0.72	0.59
1st Ring (R1)	4.03	2.25	0.61	1.64	2.69	2.07	0.80	1.27	1.60
2nd Ring (R2)	4.18	2.86	0.38	2.49	6.55	2.24	0.50	1.74	3.45
Dep 77 (D2)	4.12	2.90	0.30	2.60	8.79	2.37	0.49	1.88	3.83
Dep 78 (D3)	4.23	2.88	0.41	2.47	6.03	2.21	0.52	1.69	3.28
Dep 91 (D4)	4.30	3.07	0.34	2.73	7.91	2.15	0.44	1.71	3.92
Dep 92 (D5)	4.18	2.22	0.62	1.60	2.56	1.98	0.83	1.15	1.38
Dep 93 (D6)	3.84	2.20	0.62	1.58	2.57	2.15	0.80	1.35	1.69
Dep 94 (D7)	4.05	2.34	0.60	1.74	2.91	2.11	0.75	1.35	1.79
Dep 95 (D8)	4.06	2.57	0.45	2.13	4.76	2.21	0.57	1.65	2.90

Table 3. Pearson correlation coefficients between survey and results. We calculate correlations across the 3 rings (city center, rings 1 and 2) and across the 8 departments.

Home scale	$(C_{All}^S, C_{All}^{MP})$	$(C_{Motor}^S, C_{All}^{MP})$	$(C_{Road}^S, C_{Road}^{MP})$	$(C_{Rail}^S, C_{Rail}^{MP})$	$(C_{Ratio}^S, C_{Ratio}^{MP})$
Rings (CC, R1-2)	−0.496	0.993	0.995	0.990	0.999
Deps (CC, D2-8)	−0.348	0.751	0.960	0.986	0.978

6   Conclusion

From mobile phone data mining we can capture travel behavior of urban populations on multimodal transport networks. Compared to traditional travel surveys, Call Detail Records are a low-cost and up-to-date knowledge base for smart transport research. In this paper, we have introduced a novel transport mode detection method using CDR trajectories from the Greater Paris. Our model uses three data sources: mobile network data, transport networks and household travel survey. After significant data pre-processing, we combine clustering on mobile network areas, called sectors, with Bayesian inference for trajectories. From the clustering we find 9 clusters best described transport usage in the region. Three clusters exhibit high road probabilities, two had high rail probabilities while four had mixed usage. We compare our final results on trajectories with the household travel survey. Trips are aggregated by users' home location, at the department scale. We calculate the average number of trips per day for each user, averaged over all users. We obtain Pearson correlations above 0.96 for motorized, rail and road modes. It seems we detect exclusively motorized trips, as walkers movements are too microscopic regarding the mobile network scale. To our knowledge this is the first method separating road from rail trips considering all CDR trajectories from all users, with substantial comparison with survey data. Still it is hard to obtain exact same results as the survey. First we might have a different trip segmentation. When users travel, their path on the network are likely to be segmented into subtrips because CDR are affected by waiting times and signals loss. This phenomenon could be more pronounced for public transport travels, as users often change lines and wait in stations. In addition, the detection of travels is impacted by usage frequency of phones. We observe that trips are most likely to be undetected when road usage is predominant. At last, surveys might contain bias, be outdated and miss particular events. This makes validation a difficult task as no available data source is a perfect ground truth. Our work shows encouraging results yet we have several pending issues we want to address in future works. First, although our model proved to be robust to noisy locations, oscillations filtering could be enhanced during CDR pre-processing. Second, as our model outputs one dominant mode, we need to address multi-modal and uncertain behaviors. For future work, we will extend model evaluation with finer scale Origin-Destination trips. We look forward to adding a fourth data source (e.g., travel cards data) for validation. We aim to enrich our model with additional transport modes. Our final model will be implemented by the mobile phone provider for B-2-B with transport operators and urban planners.

Acknowledgments. This research work has been carried out in the framework of IRT SystemX, Paris-Saclay, France, and therefore granted with public funds within the scope of the French Program "Investissements d'Avenir". This work has been conducted in collaboration with Bouygues Telecom Big Data Lab.

References

1. OMNIL. http://www.omnil.fr
2. Open Data STIF. http://opendata.stif.info
3. OpenStreetMap. http://openstreetmap.ord
4. Alexander, L., Jiang, S., Murga, M., González, M.C.: Origin-destination trips by purpose and time of day inferred from mobile phone data. Transp. Res. Part C: Emerg. Technol. **58**, 240–250 (2015)
5. Bachir, D., Gauthier, V., El Yacoubi, M., Khodabandelou, G.: Using mobile phone data analysis for the estimation of daily urban dynamics. In: 2017 IEEE 20th International Conference on Intelligent Transportation Systems (ITSC), pp. 626–632. IEEE (2017)
6. Bagrow, J.P., Wang, D., Barabasi, A.-L.: Collective response of human populations to large-scale emergencies. PloS One **6**(3), e17680 (2011)
7. Berlingerio, M., et al.: Allaboard: a system for exploring urban mobility and optimizing public transport using cellphone data. vol. pt.III. IBM Research, Dublin, Ireland (2013)
8. Biljecki, F., Ledoux, H., Van Oosterom, P.: Transportation mode-based segmentation and classification of movement trajectories. Int. J. Geogr. Inf. Sci. **27**(2), 385–407 (2013)

9. Calabrese, F., Di Lorenzo, G., Liu, L., Ratti, C.: Estimating origin-destination flows using mobile phone location data. IEEE Pervasive Comput. **10**(4), 36–44 (2011)
10. Gonzalez, P., et al.: Automating mode detection using neural networks and assisted GPS data collected using GPS-enabled mobile phones. In: 15th World Congress on Intelligent Transportation Systems (2008)
11. Halkidi, M., Vazirgiannis, M.: Clustering validity assessment: finding the optimal partitioning of a data set. In: Proceedings IEEE International Conference on Data Mining, ICDM 2001, pp. 187–194. IEEE (2001)
12. Jiang, S., Ferreira, J., Gonzalez, M.C.: Activity-based human mobility patterns inferred from mobile phone data: a case study of Singapore. IEEE Trans. Big Data **3**(2), 208–219 (2017)
13. Kaufman, L., Rousseeuw, P.J.: Finding Groups in Data: An Introduction to Cluster Analysis, vol. 344. Wiley, Hoboken (2009)
14. Khodabandelou, G., Gauthier, V., El-Yacoubi, M., Fiore, M.: Population estimation from mobile network traffic metadata. In: 2016 IEEE 17th International Symposium on a World of Wireless, Mobile and Multimedia Networks (WoWMoM), pp. 1–9. IEEE (2016)
15. Larijani, A.N., Olteanu-Raimond, A.-M., Perret, J., Brédif, M., Ziemlicki, C.: Investigating the mobile phone data to estimate the origin destination flow and analysis; case study: Paris region. Transp. Res. Procedia **6**, 64–78 (2015)
16. Liu, Y., Li, Z., Xiong, H., Gao, X., Wu, J.: Understanding of internal clustering validation measures. In: 2010 IEEE 10th International Conference on Data Mining (ICDM), pp. 911–916. IEEE (2010)
17. Pang, L.X., Chawla, S., Liu, W., Zheng, Y.: On detection of emerging anomalous traffic patterns using GPS data. Data Knowl. Eng. **87**, 357–373 (2013)
18. Reddy, S., Mun, M., Burke, J., Estrin, D., Hansen, M., Srivastava, M.: Using mobile phones to determine transportation modes. ACM Trans. Sens. Netw. (TOSN) **6**(2), 13 (2010)
19. Toole, J.L., Ulm, M., González, M.C., Bauer, D.: Inferring land use from mobile phone activity. In: Proceedings of the ACM SIGKDD International Workshop on Urban Computing, pp. 1–8. ACM (2012)
20. Wang, H., Calabrese, F., Di Lorenzo, G., Ratti, C.: Transportation mode inference from anonymized and aggregated mobile phone call detail records. In: 2010 13th International IEEE Conference on Intelligent Transportation Systems (ITSC), pp. 318–323. IEEE (2010)
21. Wang, M.-H., Schrock, S.D., Vander Broek, N., Mulinazzi, T.: Estimating dynamic origin-destination data and travel demand using cell phone network data. Int. J. Intell. Transp. Syst. Res. **11**(2), 76–86 (2013)
22. Zheng, Y., Chen, Y., Li, Q., Xie, X., Ma, W.-Y.: Understanding transportation modes based on GPS data for web applications. ACM Trans. Web (TWEB) **4**(1), 1 (2010)
23. Zheng, Y., Liu, L., Wang, L., Xie, X.: Learning transportation mode from raw GPS data for geographic applications on the web. In: Proceedings of the 17th International Conference on World Wide Web, pp. 247–256. ACM (2008)
24. Zheng, Y., Xie, X.: Learning travel recommendations from user-generated GPS traces. ACM Trans. Intell. Syst. Technol. (TIST) **2**(1), 2 (2011)

CentroidNet: A Deep Neural Network for Joint Object Localization and Counting

K. Dijkstra[1,2]([✉]), J. van de Loosdrecht[1], L. R. B. Schomaker[2],
and M. A. Wiering[2]

[1] Centre of Expertise in Computer Vision and Data Science,
NHL Stenden University of Applied Sciences, Leeuwarden, Netherlands
k.dijkstra@nhl.nl
[2] Department of Artificial Intelligence, Bernoulli Institute, University of Groningen,
Groningen, Netherlands

Abstract. In precision agriculture, counting and precise localization of
crops is important for optimizing crop yield. In this paper CentroidNet is
introduced which is a Fully Convolutional Neural Network (FCNN) architecture specifically designed for object localization and counting. A field
of vectors pointing to the nearest object centroid is trained and combined
with a learned segmentation map to produce accurate object centroids
by majority voting. This is tested on a crop dataset made using a UAV
(drone) and on a cell-nuclei dataset which was provided by a Kaggle
challenge. We define the mean Average F1 score (mAF1) for measuring
the trade-off between precision and recall. CentroidNet is compared to
the state-of-the-art networks YOLOv2 and RetinaNet, which share similar properties. The results show that CentroidNet obtains the best F1
score. We also explicitly show that CentroidNet can seamlessly switch
between patches of images and full-resolution images without the need
for retraining.

1 Introduction

Crop-yield optimization is an important task in precision agriculture. This agri-
cultural output should be maximized while the ecological impact should be minimized. The state of crops needs to be monitored constantly and timely interventions for optimizing crop growth should be applied. The number of plants is
an important indicator for the predicted yield. For a better indication of crop-
yield the plants should be localized and counted during the growth season. This
can potentially be done by using Unmanned Aerial Vehicles (UAVs) to record
images and use Deep Learning to locate the objects.

Deep neural networks are trained using annotated image data to perform a
specific task. Nowadays Convolutional Neural Networks (CNNs) are mainly used
and have shown to achieve state-of-the-art performance. A wide range of applications benefit from deep learning and several general image processing tasks
have emerged. A few examples are segmentation, classification, object detection

© Springer Nature Switzerland AG 2019
U. Brefeld et al. (Eds.): ECML PKDD 2018, LNAI 11053, pp. 585–601, 2019.
https://doi.org/10.1007/978-3-030-10997-4_36

[1] and image tag generation [11]. Recent methods focus on counting and localization [4]. In other cases, object counting is regarded as an object detection task with no explicit focus on counting or as a counting task with no explicit focus on localization [2].

Many object detection architectures exist which vary in several regards. In two-stage detectors like Faster R-CNN [15], a first-stage network produces a sparse set of candidate objects which are then classified in a second stage. One-stage detectors like SSD [8], YOLOv2 [14] and RetinaNet [7] use a single stage to produce bounding boxes in an end-to-end fashion. If an object detection network is fully convolutional it can handle images of varying sizes naturally and is able to adopt a Fully Convolutional Network (FCN) as a backbone [9]. The aforementioned networks are all regarded to be fully convolutional, but rely on special subnetworks to produce bounding boxes.

This paper introduces CentroidNet which has been specifically designed for joint object counting and localization. CentroidNet produces centroids of image objects rather than bounding boxes. The key idea behind CentroidNet is to combine image segmentation and centroid majority voting to regress a vector field with the same resolution as the input image. Each vector in the field points to its relative nearest centroid. This makes the CentroidNet architecture independent of image size and helps to make it a fully-convolutional object counting and localization network. Our idea is inspired by a random-forest based voting algorithm to predict locations of body joints [12] and detect centroids of cells in medical images [5]. By binning the summation of votes and by applying a non-max suppression our method is related to the Hough transform which is known to produce robust results [10].

CentroidNet is a fully-convolutional one-stage detector which can adopt an FCN as a backbone. We choose a U-Net segmentation network as a basis because of its good performance [16]. The output of U-Net is adapted to accommodate CentroidNet. Our approach is compared with state-of-the-art on-stage object-detection networks. YOLOv2 is chosen because of its maturity and popularity. RetinaNet is chosen because it outperforms other similar detectors [7].

A dataset of crops with various sizes and heavy overlap has been created to compare CentroidNet to the other networks. It is produced by a low-cost UAV with limited image quality and limited ground resolution. Our hypothesis is that a small patch of an image of a plant naturally contains information about the location of its centroid which makes a convolutional architecture suitable for the task. Because the leaves of a plant tend to grow outward they implicitly point inward to the location of the centroid of the nearest plant. Our rationale is that this information in the image can be exploited to learn the vectors pointing to the center of the nearest plant. Centroids are calculated from these vectors.

An additional dataset containing microscopic images of cell nuclei for the Kaggle Data Science Bowl 2018[1] is used for evaluating the generality and fully convolutional properties of CentroidNet.

---

[1] https://www.kaggle.com/c/data-science-bowl-2018.

Object detection networks are generally evaluated by the mean Average Precision (mAP) which focuses mainly on minimizing false detections and therefore tends to underestimate object counts. In contrast to this, the mean Average Recall (mAR) can also be used, but tends to overestimate the number of counted objects because of the focus on detecting all instances. We define the mean average F1 (mAF1) score to determine the trade-off between overestimation and underestimation. Similarly to the regular F1 score the mAF1 score is defined as the harmonic mean between the mAP and mAR. For convenience throughout this paper the terms Precision (P), Recall (R) and F1 score are used instead of mAP, mAR and mAF1.

The remainder of this paper is structured as follows. Section 2 introduces the datasets used in the experiments. In Sect. 3, the CentroidNet architecture is explained in detail. Sections 4 and 5 present the experiments and the results. In Sect. 6 the conclusion and directions for future work are discussed.

2   Datasets

The crops dataset is used for comparing CentroidNet to the state-of-the-art networks. The nuclei dataset is used to test the generality of CentroidNet. Both sets will be used to test the fully-convolutional properties of CentroidNet.

2.1   Crops

This dataset contains images of potato plants. It was recorded by us during the growth season of 2017 in the north of the Netherlands. A Yuneec Typhoon 4K Quadrocopter was used to create a video of crops from a height of approximately 10 m. This produced a video with a resolution of 3840×2160 pixels at 24 fps. From this original video, 10 frames were extracted which contain a mix of overlapping plants, distinct plants and empty patches of soil. The borders of each image were removed because the image quality close to borders is quite low because of the wide angle lens mounted on the camera. The cropped images have a resolution of 1800×1500 pixels. Two domain experts annotated bounding boxes of potato plants in each of the images. This set is split into a training and validation set each containing 5 images and over 3000 annotated potato plant locations. These sets are referred to as 'crops-full-training' and 'crops-full-validation'.

The networks are trained using small non-overlapping patches with a resolution of 600×500 pixels. Patches are used because these neural networks use a large amount of memory on the GPU and reducing the image size will also reduce the memory consumption. In our case the original set of 10 images is subdivided into a new set of 90 images. It is well known that by training on more images the neural networks generalize better. Also more randomness can be introduced when drawing mini-batches for training because the pool of images to choose from is larger. An additional advantage of using small patches is that CentroidNet can be trained on small patches and validated on the full-resolution images to measure the impact of using CentroidNet as a fully convolutional network. The sets containing the patches are referred to as 'crops-training' and 'crops-validation'. In Fig. 1 some example images of these datasets are shown. To provide a fair comparison between networks and to reduce the amount of detection errors for partially-observed plants the bounding boxes that are too close to the borders have been removed.

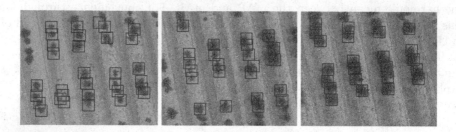

Fig. 1. Three images of the 'crops-training' and 'crops-validation' dataset. The images show an overlay with bounding boxes annotated by one of the experts. The annotations are filtered by removing objects too close to the border.

The bounding-box annotations are converted into a three-channel image that will be used as a target image for training. The first two channels contain the x and y components of vectors pointing to the nearest centroid of a crop (center of its bounding box). The third channel is generated by drawing binary ellipses in the annotated bounding-boxes. More binary maps can be added if more classes are present in the image. This means that the target image contains information about the centroid locations in each pixel and the class of each pixels is known. These three channels help CentroidNet to be a robust centroid detector.

2.2   Kaggle Data Science Bowl 2018

The generality of CentroidNet will be tested on the dataset for the Data Science Bowl 2018 on Kaggle.com. This set is referred to as 'nuclei-full'. The challenge is to create an algorithm that automates the detection of the nucleus in several microscopic images of cells. This dataset contains 673 images with a total of 29,461 annotated nuclei (see Fig. 2). The images vary in resolution, cell type, magnification, and imaging modality. These properties make this dataset particularly interesting for validating CentroidNet. Firstly the variation in color and size of the nuclei makes it ideal for testing a method based on deep learning. Secondly the image sizes vary to a great extent making it suitable for testing our fully-convolutional network by training on smaller images and validation on the full-resolution images.

Because fixed-size tensors are more suitable for training on a GPU the 'nuclei-full' dataset is subdivided into patches of 256×256 pixels. Each patch overlaps with the neighboring patches by a border of 64 pixels in all directions.

This dataset of patches is split into 80% training and 20% validation. These datasets are referred to as 'nuclei-training' and 'nuclei-validation'. The fusion of the results on these patches is not required because the trained network is also applied to the original images in 'nuclei-full' to produce full-resolution results.

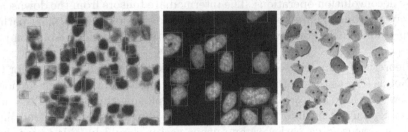

Fig. 2. Three images from 'nuclei-full' set. These images give an indication of the variation encountered in this dataset. The green overlay shows annotated bounding boxes. The right image shows that these bounding boxes can be very small.

While this nuclei-detection application is ideal to validate our method it is difficult to directly compare our approach to the other participants. The original challenge is defined as a pixels-precise segmentation challenge. We redefine this problem as an object localization and counting challenge. This means that the size of the bounding boxes needs to be fixed and that the mean F1 score will be used as an evaluation metric.

A three-channel target image is created for training CentroidNet. The creation method is identical to the one described in the previous subsection.

3   CentroidNet

The input image to CentroidNet can be of any size and can have an arbitrary number of color channels. The output size of CentroidNet is identical to its input size and the number of output channels is fixed. The first two output channels contain the x and the y component of a vector for each pixel. These vectors each point to the nearest centroid of an object and can thus be regarded as votes for where an object centroid is located. All votes are aggregated into bins represented by a smaller image. A pixel in this voting map has a higher value if there is greater chance of centroid presence. Therefore a local maximum represents the presence of a centroid. The remaining channels of the output of CentroidNet contain the logit maps for each class. This is identical to the per-pixel-one-hot output of a semantic segmentation network [9]. For the crops and the nuclei datasets only one such map exists because there is only one class (either object or no object). By combining this map with the voting map an accurate and robust estimation can be made of the centroids of the objects.

In our experiments we use a U-Net implemented in PyTorch as a basis[2] for CentroidNet. In the downscaling pathway the spatial dimensions (width and height) of the input are iteratively reduced in size and the size of the channel dimension is increased. This is achieved by convolution and max-pooling operations. Conversely, in the upscaling pathway the tensor is restored to its original size by deconvolution operations. The intermediate tensors from the downscaling pathway are concatenated to the intermediate tensors of the upscaling pathway to form "horizontal" connections. This helps to retain the high-resolution information.

Theoretically any fully-convolutional network can be used as a basis for CentroidNet as long as the spatial dimensions of the input and the output are identical. However there are certain advantages to employing this specific CNN architecture. By migrating information from spatial dimensions to spectral dimensions (the downscaling pathway) an output vector should be able to vote more accurately over larger distances in the image which helps voting robustness. By concatenating tensors from the downscaling pathway to the upscaling pathway a sharper logit map is created. This also increases the accuracy of the voting vectors and makes the vector field appear sharper which results in more accurate centroid predictions.

The next part explains the details of the CNN architecture. The first part explains the downscaling pathway of CentroidNet and then the upscaling pathway is explained. The final part explains the voting algorithm and the method used to combine the logit map and the voting result to produce final centroid locations.

A CNN consists of multiple layers of convolutional-filter banks that are applied to the input tensor (or input image). A single filter bank is defined as a set of convolutional filters:

$$\mathcal{F}_n^{t \times t} = \{\mathbf{F}_1^{t \times t}, \mathbf{F}_2^{t \times t}, \ldots, \mathbf{F}_n^{t \times t}\} \tag{1}$$

where $\mathcal{F}_n^{t \times t}$ is a set of n filters with a size of $t \times t$. Any convolutional filter in this set is actually a 3-d filter with a depth equal to the number of channels of the input tensor.

The convolutional building block of CentroidNet performs two 3×3 convolution operations on the input tensor and applies the Rectified Linear Unit (ReLU) activation function after each individual operation. The ReLU function clips values below zero and is defined as $\psi(\mathbf{X}_{yx}) = \max(0, \mathbf{X}_{yx})$ where $\mathbf{X}$ is the input tensor. We found that input scaling is not required if ReLU is used as an activation function. Scaling is required if a saturating function like the hyperbolic tangent would be used as an activation function. The convolution operator $\otimes$ takes an input tensor on the left-hand side and a set of convolution filters on the right-hand side. The convolution block is defined by

$$\text{conv}(\mathbf{X}, c) = \psi(\psi(\mathbf{X} \otimes \mathcal{F}_c^{3 \times 3}) \otimes \mathcal{F}_c^{3 \times 3}) \tag{2}$$

---

[2] https://github.com/jaxony/unet-pytorch.

where $\mathbf{X}$ is the input tensor, c is the number of filters in the convolutional layer, and $\otimes$ is the convolution operator.

The downscaling pathway is defined as multiple $\mathrm{conv}(\cdot, \cdot)$ and max-pooling, $\mathrm{pool}(\cdot)$, operations. The initial convolutional operator increases the depth of the input image from 3 (RGB) to 64 channels and reduces the height and width by a max-pooling operation of size 2×2. In subsequent convolutional operations the depth of the input tensor is doubled by increasing the amount of convolutional filters and the height and width of the input tensor is reduced by 1/2 with a max-pooling operation. This is a typical CNN design pattern which results in converting spatial information to semantic information. In CentroidNet this also has the implicit effect of combining voting vectors from distant parts of the image in a hierarchical fashion. The downscaling pathway is mathematically defined as:

$$\mathbf{C}_1 = \mathrm{conv}(\mathbf{X}, 64) \tag{3}$$
$$\mathbf{D}_1 = \mathrm{pool}(\mathbf{C}_1) \tag{4}$$
$$\mathbf{C}_2 = \mathrm{conv}(\mathbf{D}_1, 128) \tag{5}$$
$$\mathbf{D}_2 = \mathrm{pool}(\mathbf{C}_2) \tag{6}$$
$$\mathbf{C}_3 = \mathrm{conv}(\mathbf{D}_2, 256) \tag{7}$$
$$\mathbf{D}_3 = \mathrm{pool}(\mathbf{C}_3) \tag{8}$$
$$\mathbf{C}_4 = \mathrm{conv}(\mathbf{D}_3, 512) \tag{9}$$
$$\mathbf{D}_4 = \mathrm{pool}(\mathbf{C}_4) \tag{10}$$
$$\mathbf{D}_5 = \mathrm{conv}(\mathbf{D}_4, 1024) \tag{11}$$

where $\mathbf{X}$ is the input tensor, $\mathbf{C}_x$ are the convolved tensors, and $\mathbf{D}_x$ are the downscaled tensors. The convolved tensors are needed for the upscaling pathway. The final downscaled tensor $\mathbf{D}_5$ serves as an input to the upscaling pathway.

The upscaling pathway incrementally restores the tensor back to the original image size by deconvolution operations. This is needed because the output tensor should have the same size as the input image to be able to produce one voting vector per pixel. The $\mathrm{up}(\cdot)$ operation first performs the $\mathrm{conv}(\cdot, \cdot)$ operation defined in Eq. 2 and then performs a deconvolution $\oslash$ operation with a filter of 2×2 that doubles the height and width of the input tensor.

$$\mathrm{up}(\mathbf{X}, c) = \psi(\mathbf{X} \oslash \mathcal{F}_c^{2 \times 2}) \tag{12}$$

where $\mathbf{X}$ is the input tensor, c is the number of filters in the deconvolutional layer, and $\oslash$ is the deconvolution operator.

The final part of the CNN is constructed by subsequently upscaling the output tensor of the downscaling pathway $\mathbf{D}_5$. The width and height are doubled and the depth is halved by reducing the amount of convolution filter in the filter bank. The upscaling pathway is given additional high-resolution information in "horizontal" connections between the downscaling and upscaling pathways. This should result in more accurate voting vectors. Before each tensor was subsequently downscaled it was stored as $\mathbf{C}_x$. These intermediate tensors are concatenated to tensors of the same size produced by the upscaling pathway. The operator $\oplus$ concatenates the left-hand-side tensor to the right-hand-side tensor over the depth axis.

$$\mathbf{U}_1 = \text{conv}(\text{up}(\mathbf{D}_5, 512) \oplus \mathbf{C}_4, 512) \tag{13}$$

$$\mathbf{U}_2 = \text{conv}(\text{up}(\mathbf{U}_1, 256) \oplus \mathbf{C}_3, 256) \tag{14}$$

$$\mathbf{U}_3 = \text{conv}(\text{up}(\mathbf{U}_2, 128) \oplus \mathbf{C}_2, 128) \tag{15}$$

$$\mathbf{U}_4 = \text{conv}(\text{up}(\mathbf{U}_3, 64) \oplus \mathbf{C}_1, 64) \tag{16}$$

$$\mathbf{Y} = \mathbf{U}_4 \otimes \mathcal{F}_3^{1 \times 1} \tag{17}$$

where $\mathbf{D}_5$ is the smallest tensor from the downscaling pathway, $\mathbf{U}_x$ are the upscaled tensors, $\oplus$ is the tensor concatenation operator, and the final tensor with its original width and height restored and its number of channels set to 3 is denoted by $\mathbf{Y}$.

The concatenation operator fails to function properly if the size of the input image cannot be divided by 2^5 (e.g. the original input image size cannot be divided by two for five subsequent times). To support arbitrary input-image sizes the tensor concatenation operator $\oplus$ performs an additional bilinear upscaling if the dimensions of the operands are not equal due to rounding errors.

The final 1×1 convolution operation in Eq. 17 is used to set the number of output channels to a fixed size of three (x and y vector components and an additional map of class logits for the object class). It is important that no ReLU activation function is applied after this final convolutional layer because the output contains relative vectors which should be able to have negative values (i.e. a centroid can be located anywhere relative to an image coordinate).

To produce centroid locations from the outputs of the CNN the votes have to be aggregated. The algorithm for this is shown in Algorithm 1. A centroid vote is represented by a relative 2-d vector at each image location. First, all relative vectors in the CNN output $\mathbf{Y}$ are converted to absolute vectors by adding the absolute image location to the vector value. Then for each vote the bin is calculated by performing an integer division of the vector values. In preliminary research we found a bin size of 4 to perform well. By binning the votes the result is more robust and the influence of noise in the vectors is reduced. Finally, the voting matrix $\mathbf{V}$ is incremented at the calculated vector location. When an absolute vector points outside of the image the vote is discarded. The voting map is filtered by a non-max suppression filter which only keeps the voting maxima (the peaks in the voting map) in a specific local neighborhood.

Algorithm 1. Voting algorithm

1: $\mathbf{Y} \leftarrow$ Output of CNN   ▷ The 1st and 2nd channel contain the voting vectors
2: $h, w \leftarrow$ height, width of $\mathbf{Y}$
3: $\mathbf{V} \leftarrow$ zero filled matrix of size (h div 4, w div 4)
4: **for** $y \leftarrow 0$ **to** $h - 1$ **do**
5:     **for** $x \leftarrow 0$ **to** $w - 1$ **do**
6:         $y' \leftarrow (y + \mathbf{Y}[y, x, 0])$ div 4   ▷ Get the absolute-binned y component
7:         $x' \leftarrow (x + \mathbf{Y}[y, x, 1])$ div 4   ▷ Get the absolute-binned x component
8:         $\mathbf{V}[y', x'] \leftarrow \mathbf{V}[y', x'] + 1$ ▷ Aggregate vote
9:     **end for**
10: **end for**

Finally the voting map is thresholded to select high votes. This results is a binary voting map. Similarly the output channel of the CNN which contains the class logits is also thresholded (because we only have one class this is only a single channel). This results in the binary segmentation map. Both binary maps are multiplied to only keep centroids at locations where objects are present. We found that this action reduces the amount of false detections strongly.

$$\mathbf{V}_{y,x} = 1 \text{ if } \mathbf{V}_{y,x} \geq \theta, 0 \text{ otherwise.} \tag{18}$$

$$\mathbf{S}_{y,x} = 1 \text{ if } \mathbf{Y}_{y,x,3} \geq \gamma, 0 \text{ otherwise.} \tag{19}$$

$$\mathbf{V} = \text{enlarge}(\mathbf{V}, 4) \tag{20}$$

$$\mathbf{C}_{y,x} = \mathbf{V}_{y,x} \times \mathbf{S}_{y,x} \tag{21}$$

where $\mathbf{Y}$ is the output of the CNN, $\mathbf{V}$ is the binary voting map, $\mathbf{S}$ is the binary segmentation map, and θ and γ are two threshold parameters. The final output $\mathbf{C}$ is produced by multiplying each element of $\mathbf{V}$ with each element of $\mathbf{S}$ to only accept votes at object locations. The vote image needs to be enlarged because it was reduced by the binning of the votes in Algorithm 1.

4   Experiments

CentroidNet is compared to state-of-the-art object detection networks that share their basic properties of being fully convolutional and the fact that they can be trained in one stage. The YOLOv2 network uses a backbone which is inspired by GoogleNet [13,17] (but without the inception modules). It produces anchor boxes as outputs that are converted into bounding boxes [14]. A cross-platform implementation based on the original DarkNet is used for the YOLOv2 experiments[3]. RetinaNet uses a Feature Pyramid Network [6] and ResNet50 [3] as backbone. Two additional convolutional networks are stacked on top of the backbone to produce bounding boxes and classifications [7]. A Keras implementation of RetinaNet is used in the experiments[4].

The goal of this first experiment is to compare networks. The compared networks are all trained on the 'crops-training' dataset which contains 45 image patches containing a total of 996 plants. The networks are trained to convergence. Early stopping was applied when the loss on the validation set started to increase. The 'crops-validation' set is used as a validation set and contains 45 image patches containing a total of 1090 potato plants. CentroidNet is trained using the Euclidean loss function, the Adam optimizer and a learning rate of 0.001 for 120 epochs. For the other networks mainly the default settings were used. RetinaNet uses the focal loss function proposed in their original paper [7] and an Adam optimizer with a learning rate of 0.00001 for 200 epochs. YOLOv2 was trained with stochastic gradient descent and a learning rate of 0.001 for 1000 epochs. RetinaNet uses pretrained weights from ImageNet and YOLOv2 uses pretrained weights from the Pascal VOC dataset.

---

[3] https://github.com/AlexeyAB/darknet.
[4] https://github.com/fizyr/keras-retinanet.

The goal of the second experiment is to test the full-convolutional properties of CentroidNet. What is the effect when the network is trained on image patches and validated on the set of full-resolution images without retraining? In the third experiment CentroidNet is validated on the nuclei dataset. Training is performed using the 'nuclei-training' dataset which contains 6081 image patches. The 'nuclei-validation' which contains 1520 image patches is used to validate the results. CentroidNet is also validated using all 637 full-resolution images from the 'nuclei-full' dataset without retraining. In this case CentroidNet was trained to convergence with the Adam optimizer with a learning rate of 0.001 during 1000 epochs. The Euclidean loss was used during training. Early stopping was not required because the validation results were stable after 1000 epochs.

The next part explains how the bounding-boxes produced by the networks are compared to the target bounding boxes. The Intersection of Union (IoU) represents the amount of overlap between two bounding boxes, where an IoU of zero means no overlap and a value of one means 100% overlap:

$$\text{IoU}(\mathcal{R}, \mathcal{T}) = \frac{\mathcal{R} \cap \mathcal{T}}{\mathcal{R} \cup \mathcal{T}} \tag{22}$$

where $\mathcal{R}$ is a set of pixels of a detected object and $\mathcal{T}$ is a set of target object pixels. Note that objects are defined by their bounding box, which allows for an efficient calculation of the IoU metric.

When the IoU for a target and a result object have sufficiently high overlap, the object is greedily assigned to this target and counted as a True Positive (TP). False Positives (FP) and False Negatives (FN) are calculated as follows:

$$\text{TP} = \text{count\_if}(\text{IoU}(\mathcal{R}, \mathcal{T}) > \tau) \qquad \forall \mathcal{R}, \mathcal{T} \tag{23}$$
$$\text{FP} = \#\text{results} - \text{TP} \tag{24}$$
$$\text{FN} = \#\text{targets} - \text{TP} \tag{25}$$

where $\#$results and $\#$targets are the number of result and target objects, and τ is a threshold parameters for the minimum IoU value.

All metrics are defined in terms of these basic metrics. Precision (P), Recall (R) and F1 score are defined as

$$P = \frac{\text{TP}}{\text{TP} + \text{FP}} \tag{26}$$

$$R = \frac{\text{TP}}{\text{TP} + \text{FN}} \tag{27}$$

$$F1 = 2 \times \frac{P \times R}{P + R} \tag{28}$$

In Eq. 28 it can be seen that the F1 score gives a trade-off between precision and recall, and thus measures the equilibrium between the overestimation and the underestimation of object counts. The size of the bounding boxes of both the target and the result objects will be set to a fixed size of roughly the size of an object. This fixed size of the bounding box can be interpreted as the size of the neighborhood around a target box in which a result bounding box can still be counted as a true positive. This could possibly be avoided by using other metrics like absolute count and average distance. However, absolute count does not estimate localization accuracy and average distance does not measure counting accuracy. Therefore the F1 score is used to focus the validation on joint object localization and on object counting.

For the crops dataset the bounding-box size is set to 50×50 pixels. For the nuclei dataset a fixed bounding box size of 60×60 pixels is chosen. This size is chosen so that there is not too much overlap between bounding boxes. When boxes are too large the greedy IoU matching algorithm in Eq. 23 could assign result bounding boxes to a target bounding box which is too far away.

The best voting threshold θ (Eq. 18) is determined using the training set. The segmentation threshold γ in Eq. 19 is set to zero for all experiments (result values in the segmentation channel can be negative), and the performance metrics for several IoU thresholds τ (Eq. 23) will be reported for each experiment.

5   Results

This section discusses the results of the experiments. The first subsection shows the comparison between CentroidNet, YOLOv2 and RetinaNet on the 'crops' dataset. The second subsection explores the effect of enlarging the input image size after training CentroidNet on either the 'crops' or the 'nuclei' dataset.

5.1   Comparison with the State-of-the Art on the Crops Dataset

The results for the best F1 score with respect to the voting threshold θ are shown in Table 1. The results show that CentroidNet achieves a higher F1 score on the 'crops-validation' set regardless of the chosen IoU and regardless of the expert (90.4% for expert A, and 93.4% for expert B). For both experts RetinaNet and YOLOv2 obtain lower F1 scores. Interestingly the performance measured for expert B is higher compared to expert A. This is probably because of the higher quality of the annotations produced by expert B. There is an optimal IoU threshold. When the IoU threshold is chosen too low the validation results are adversely affected. This is probably due to the greedy matching scheme involved in calculating the F1 score. Therefore the intuition that a smaller IoU threshold yields higher validation scores seems unfounded.

Table 1. F1 score for both experts on the 'crops-validation' dataset

Using annotations of expert A and using annotations of expert B										
IoU (τ)	0.1	0.2	0.3	0.4	0.5	IoU (τ)	0.1	0.2	0.3	0.4	0.5
CentroidNet	90.0	**90.4**	90.0	89.1	85.7	CentroidNet	93.0	93.2	**93.4**	92.7	90.0
RetinaNet	88.3	89.1	**89.4**	87.7	83.1	RetinaNet	90.5	90.9	**91.1**	90.7	89.2
YOLOv2	87.1	88.4	**88.8**	87.5	82.0	YOLOv2	88.3	88.9	**89.1**	89.0	87.2

The results can be further analyzed by the precision-recall graph shown in Fig. 3. The red curve of CentroidNet is generated by varying the voting threshold θ between 0 and 1024. The curves for YOLOv2 (Blue) and RetinaNet (Green) have been generated by varying the confidence threshold between 0 and 1. The curve of CentroidNet passes the closest to the right-top corner of the graph with a precision of 91.2% and a recall of 95.9% for expert B.

When using the annotated set of expert B (Fig. 3-right), RetinaNet and YOLOv2 show similar recall values when exiting the graph at the most left. CentroidNet and RetinaNet show a similar precision when exiting the graph at the bottom. The precision-recall graph for Expert A (Fig. 3-left) shows that RetinaNet has better precision at the cost of low recall, but the best precision-recall value is observed for CentroidNet.

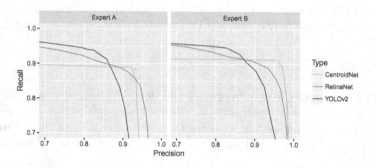

Fig. 3. Precision-recall graphs for expert A (left) and expert B (right). These graphs show that CentroidNet performs best on the trade-off between precision and recall. (Color figure online)

An image from the 'crops-validation' set is used to show detailed results of the inner workings of CentroidNet. The images for the regression output of CentroidNet are shown in the top row of Fig. 4. The left image shows the input image which gives an idea of the challenge this image poses with respect to the image quality and overlap between plants. The middle image of the top row of Fig. 4 shows the magnitude of the target voting vectors (dark is shorter). The right image of the top row shows the magnitude of the learned voting vectors. Important aspects like the length of the vectors and the ridges between objects can be observed in the learned vectors. Interestingly, CentroidNet is also able to learn large vectors for locations which do not contain any green plant pixels.

The bottom row of Fig. 4 shows the final result of CentroidNet. After aggregating the vectors and thresholding the voting map the binary voting map is produced which is shown in the bottom-left image of Fig. 4. The bright dots show where most of the voting vectors point to (the binary voting map). The blue areas show the binary segmentation map which has been used to filter false detections. By converting each centroid to a fixed-size bounding box the bottom-right image is produced. It can be seen that the plants are detected even with heavy overlap (green boxes). In this result a false positive (red box) is caused by an oddly shaped plant group. A false negative is caused by a small undetected plant (blue box).

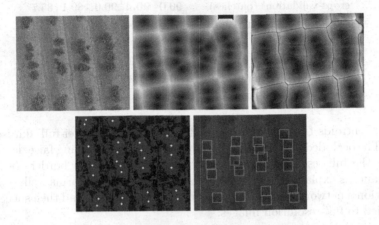

Fig. 4. Result of CentroidNet with the top row showing the input, the target vector magnitudes, and the result vector magnitudes. The bottom-left image shows the binary voting map as bright dots and the binary segmentation map as the blue area (dark). The bottom-right image shows the bounding boxes (Green = true positive, Red = false negative, Blue = false positive). Note that boxes too close to the border are not considered. (Color figure online)

5.2 Testing on Larger Images

A strong feature of CentroidNet is that it is a fully convolutional network. The CentroidNet from the previous subsection which was trained on image patches is used here. Validation is performed on the full resolution images of 1800×1500 pixels. In Table 2 the performance on the image patches in the 'crops-validation' and the full-resolution-image dataset 'crops-full-validation' are shown. The F1 score for expert A goes from 90.4% to 88.4%, which slightly less then the best performance of YOLOv2 in Table 1. The performance on the full-resolution dataset of expert B goes from 93.4% to 91.7%. This is still the best overall performance with respect to the results of the other object detection networks in Table 1. This means that the drop in performance by applying CentroidNet to the full-resolution dataset without retraining is acceptable.

The results of the performance of CentroidNet on the 'nuclei' datasets is shown in Table 3. CentroidNet is trained and validated with the patches from the 'nuclei-training' and 'nuclei-validation' sets. The network is tested with the full-resolution 'nuclei-full' dataset.

CentroidNet shows high precision at the cost of a lower recall. The highest F1 score is obtained on the full-resolution dataset (86.9%). Because the network was not trained on the full-resolution dataset this seems counter intuitive. In Fig. 5

Table 2. CentroidNet F1 score for expert A and B on the crops dataset

IoU (τ)	0.1	0.2	0.3	0.4	0.5
Expert A					
'crops-validation' (patches)	90.0	**90.4**	90.0	89.1	85.7
'crops-full-validation' (full-res)	87.6	**88.4**	87.8	84.8	77.3
Expert B					
'crops-validation' (patches)	93.0	93.2	**93.4**	92.7	90.0
'crops-full-validation' (full-res)	90.2	91.2	**91.7**	89.8	85.0

detected centroids from the 'nuclei-validation' and the 'nuclei-full' datasets are shown. The occluded nuclei at the borders are prone to becoming false detections. Because the full-resolution images have less objects at the border the higher performance is explained. This shows that CentroidNet performs well as a fully-convolutional network that can be trained on image patches and then successfully be applied to full-resolution images.

Table 3. CentroidNet F1 score, Precision and Recall on the nuclei dataset.

IoU (τ)	0.1	0.2	0.3	0.4	0.5	0.6	0.7	0.8	0.9
F1 score									
'nuclei-training' (patches)	81.3	82.0	82.8	83.3	83.5	**83.5**	83.4	74.0	27.8
'nuclei-validation' (patches)	79.2	79.6	79.9	**80.3**	80.1	79.5	77.7	63.4	24.4
'nuclei-full' (full-res)	84.4	85.7	86.5	87.0	86.8	**86.9**	85.7	74.7	33.0
Precision									
'nuclei-training' (patches)	97.3	98.2	99.1	99.7	99.9	**100.0**	99.9	88.6	33.3
'nuclei-validation' (patches)	94.9	95.3	95.8	**96.2**	96.0	95.3	93.1	76.0	29.3
'nuclei-full' (full-res)	94.9	96.3	97.4	**97.9**	97.6	97.8	96.5	84.0	37.1
Recall									
'nuclei-training' (patches)	69.8	70.4	71.1	71.5	71.6	**71.7**	71.6	63.5	23.8
'nuclei-validation' (patches)	68.0	68.3	68.6	**68.9**	68.7	68.2	66.7	54.4	20.9
'nuclei-full' (full-res)	75.9	77.2	77.8	**78.3**	78.2	78.2	77.1	67.3	29.7

6   Discussion and Conclusion

In this paper CentroidNet, a deep neural network for joint localization and counting, was presented. A U-Net [16] architecture is used as a basis. CentroidNet is trained to produce a set of voting vectors which point to the nearest centroid of an object. By aggregating these votes and combining the result with the segmentation mask, also produced by CentroidNet, state-of-the art performance is

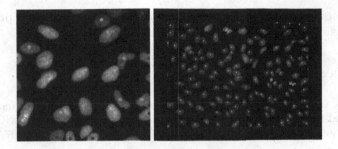

Fig. 5. The left image is a patch from the right image. The left image shows the centroid detection results from and image of the 'nuclei-validation' dataset and the right image shows the centroid detection results on an image from the 'nuclei-full' dataset. Green = true positive, Red = false negative, Blue = false positive (Color figure online)

achieved. Experiments were performed using a dataset containing images of crops made using a UAV and on a dataset containing microscopic images of nuclei. The mean Average F1 score (mAF1) which is the harmonic mean between precision and recall was used as a main evaluation metric because it gives a good indication of the trade-off between underestimation and overestimation in counting and a good estimation of localization performance.

The best performance for the joint localization and counting of objects is obtained using CentroidNet with an F1 score of 93.4% on the crops dataset. In comparison to other object detection networks with similar properties the results were 91.1% for RetinaNet and YOLOv2 obtained and F1 score of 89.1%.

CentroidNet has been tested by training on patches of images and by validating on full-resolution images. On the crops dataset the best F1 score dropped from 93.4% to 91.7%, which still made CentroidNet the best performing network. For the nuclei dataset the F1 score on the full-resolution images was highest, which can be attributed to border effects.

Generally we learned that using a majority voting scheme for detecting object centroids produces robust results with regard to the trade-off between precision and recall. By using a trained segmentation mask to suppress false detection, a higher precision is achieved, especially on the low-quality images produced by drones. A relatively small amount of images can be used for training because votes are largely independent.

Although CentroidNet is the best-performing method with respect to the posed problem. Improvements can still be made in future research. The detection of bounding boxes or instance segmentation maps can be explored, multi-class problems can be investigated or research could focus on reducing the border-effects. Future research could also focus on testing CentroidNet on larger potato-plant datasets or look into localization and counting of other types of vegetation like sugar beets, broccoli or trees using images taken with a UAV. On a more detailed scale the detection of vegetation diseases could be investigated by detecting brown lesions on plant leaves or by looking at detection problems on a microscopic scale that are similar to nuclei detection.

References

1. Cheema, G.S., Anand, S.: Automatic detection and recognition of individuals in patterned species. In: Altun, Y., et al. (eds.) ECML PKDD 2017. LNCS (LNAI), vol. 10536, pp. 27–38. Springer, Cham (2017). https://doi.org/10.1007/978-3-319-71273-4_3
2. Cohen, J.P., Boucher, G., Glastonbury, C.A., Lo, H.Z., Bengio, Y.: Count-ception: counting by fully convolutional redundant counting. In: Proceedings of the IEEE Conference on Computer Vision and Pattern Recognition, pp. 18–26 (2017)
3. He, K., Zhang, X., Ren, S., Sun, J.: Deep residual learning for image recognition. In: Proceedings of the IEEE Conference on Computer Vision and Pattern Recognition, pp. 770–778 (2015)
4. Hsieh, M.R., Lin, Y.L., Hsu, W.H.: Drone-based object counting by spatially regularized regional proposal network. In: The IEEE International Conference on Computer Vision (ICCV), vol. 1 (2017)
5. Kainz, P., Urschler, M., Schulter, S., Wohlhart, P., Lepetit, V.: You should use regression to detect cells. In: Navab, N., Hornegger, J., Wells, W.M., Frangi, A.F. (eds.) MICCAI 2015. LNCS, vol. 9351, pp. 276–283. Springer, Cham (2015). https://doi.org/10.1007/978-3-319-24574-4_33
6. Lin, T.Y., Dollár, P., Girshick, R., He, K., Hariharan, B., Belongie, S.: Feature pyramid networks for object detection. In: Conference on Computer Vision and Pattern Recognition, vol. 1, p. 4 (2017)
7. Lin, T.Y., Goyal, P., Girshick, R., He, K., Dollár, P.: Focal loss for dense object detection. arXiv preprint arXiv:1708.02002 (2017)
8. Liu, W., et al.: SSD: single shot multibox detector. In: Leibe, B., Matas, J., Sebe, N., Welling, M. (eds.) ECCV 2016. LNCS, vol. 9905, pp. 21–37. Springer, Cham (2016). https://doi.org/10.1007/978-3-319-46448-0_2
9. Long, J., Shelhamer, E., Darrell, T.: Fully convolutional networks for semantic segmentation. In: Proceedings of the IEEE Conference on Computer Vision and Pattern Recognition, pp. 3431–3440 (2014)
10. Milletari, F., et al.: Hough-CNN: deep learning for segmentation of deep brain regions in MRI and ultrasound. Comput. Vis. Image Underst. **164**, 92–102 (2017)
11. Nguyen, H.T.H., Wistuba, M., Schmidt-Thieme, L.: Personalized tag recommendation for images using deep transfer learning. In: Ceci, M., Hollmén, J., Todorovski, L., Vens, C., Džeroski, S. (eds.) ECML PKDD 2017. LNCS (LNAI), vol. 10535, pp. 705–720. Springer, Cham (2017). https://doi.org/10.1007/978-3-319-71246-8_43
12. Pietikaäinen, M.: Computer Vision Using Local Binary Patterns. Springer, London (2011). https://doi.org/10.1007/978-0-85729-748-8
13. Redmon, J., Divvala, S., Girshick, R., Farhadi, A.: You only look once: Unified, real-time object detection. In: Proceedings of the IEEE Conference on Computer Vision and Pattern Recognition. pp. 779–788 (2016)
14. Redmon, J., Farhadi, A.: YOLO9000: Better. Stronger. arXiv preprint, Faster (2017)
15. Ren, S., He, K., Girshick, R., Sun, J.: Faster R-CNN: towards real-time object detection with region proposal networks. IEEE Trans. Pattern Anal. Mach. Intell. **39**(6), 1137–1149 (2015)

16. Ronneberger, O., Fischer, P., Brox, T.: U-Net: convolutional networks for biomedical image segmentation. In: Navab, N., Hornegger, J., Wells, W.M., Frangi, A.F. (eds.) MICCAI 2015. LNCS, vol. 9351, pp. 234–241. Springer, Cham (2015). https://doi.org/10.1007/978-3-319-24574-4_28
17. Szegedy, C., et al.: Going deeper with convolutions. In: Computer Vision and Pattern Recognition (CVPR) (2015)

Deep Modular Multimodal Fusion on Multiple Sensors for Volcano Activity Recognition

Hiep V. Le[1][(✉)], Tsuyoshi Murata[1], and Masato Iguchi[2]

[1] Department of Computer Science, Tokyo Institute of Technology, Tokyo, Japan
hiep@net.c.titech.ac.jp, murata@c.titech.ac.jp
[2] Disaster Prevention Research Institute, Kyoto University, Kyoto, Japan
iguchi.masato.8m@kyoto-u.ac.jp

Abstract. Nowadays, with the development of sensor techniques and the growth in a number of volcanic monitoring systems, more and more data about volcanic sensor signals are gathered. This results in a need for mining these data to study the mechanism of the volcanic eruption. This paper focuses on Volcano Activity Recognition (VAR) where the inputs are multiple sensor data obtained from the volcanic monitoring system in the form of time series. And the output of this research is the volcano status which is either *explosive* or *not explosive*. It is hard even for experts to extract handcrafted features from these time series. To solve this problem, we propose a deep neural network architecture called VolNet which adapts Convolutional Neural Network for each time series to extract non-handcrafted feature representation which is considered powerful to discriminate between classes. By taking advantages of VolNet as a building block, we propose a simple but effective fusion model called Deep Modular Multimodal Fusion (DMMF) which adapts data grouping as the guidance to design the architecture of fusion model. Different from conventional multimodal fusion where the features are concatenated all at once at the fusion step, DMMF fuses relevant modalities in different modules separately in a hierarchical fashion. We conducted extensive experiments to demonstrate the effectiveness of VolNet and DMMF on the volcanic sensor datasets obtained from Sakurajima volcano, which are the biggest volcanic sensor datasets in Japan. The experiments showed that DMMF outperformed the current state-of-the-art fusion model with the increase of F-score up to 1.9% on average.

Keywords: Multimodal fusion · Volcano Activity Recognition
Time series · Convolutional Neural Network

1 Introduction

Volcanic eruption causes severe damage to human and society, hence it is one of the main concerns of many people in the world, especially to volcano experts.

© Springer Nature Switzerland AG 2019
U. Brefeld et al. (Eds.): ECML PKDD 2018, LNAI 11053, pp. 602–617, 2019.
https://doi.org/10.1007/978-3-030-10997-4_37

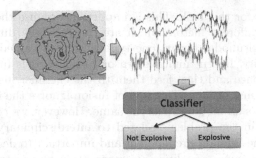

Fig. 1. The overview of Volcano Activity Recognition (VAR). The volcanic monitoring system (left) includes sensors to obtain the signals in the form of time series (right). The goal is to build an explosive eruption classifier. The input is raw time series, and the output in this research is the status of the volcano that is either *explosive* or *not explosive*.

A popular method for volcano research is to analyze sensor signals obtained from the volcanic monitoring system. Multiple sensors are deployed in this system and each sensor is responsible for measuring a specific type of data. Examples of volcanic sensor data are ground deformation, ground surface vibration, and gas emission. These data are represented in the form of time series whose values are numeric and are recorded periodically in real time. As there are correlations between these time series data and volcanic eruption, the data are very valuable for the mining of volcano activities [1]. Because the data is gathered continuously in real time, the amount of data is increasing in size. This opens an opportunity for both volcano and machine learning researchers to mine the data in large scale. Our main focus in this paper is Volcano Activity Recognition (VAR). VAR is the task of classifying time series sensor data into multiple categories of volcano activity. Figure 1 shows the overall structure of VAR. In this paper, we classify the two most important statues of a volcano: *explosive* and *not explosive*. If the classification is successful, this research can give an insight to the mechanism of the eruption. In the context of this paper, the eruption means explosive eruption.

VAR can be solved using raw sensor signal, but the features extracted from this raw data could have more potential in terms of class discrimination than the raw data itself [2]. However, handcrafted feature extraction is time-consuming and is hard to decide even for volcano experts. Recently, deep learning with many layers of nonlinear processing for automatic feature extraction has proven to be effective in many applications such as image classification, speech recognition, and human activity recognition [5]. In this research, we propose a deep neural network architecture called VolNet which adapts Convolutional Neural Network (CNN), a particular type of deep learning model, for VAR on a single sensor. VolNet with its deep architecture is able to learn a hierarchy of abstract features automatically, and then use these features for the classification task. Based on the VolNet, we extend the model to multimodal fusion which takes into account all related sensors for this task. This is important because the eruption is controlled by many different factors which are measured by different sensors. In our context of multimodal fusion, one type of data obtained from a sensor is called as a modality. Recent multiple sensor fusion models fuse the features of all modalities at once and then feed them to the classifier [6–8]. In this paper, we call this "one-time fusion". This way of fusion ignores the properties of each modality and treats all modalities as the same. However, we consider this is not a good approach in the problems related to interdisciplinary study like VAR where the properties of data are different and important to design the solutions. Our assumption is some modalities are more likely to be correlated than others, and hence better to be fused together before they will be fused with other modalities. Based on that idea, we propose a simple but effective fusion model called Deep Modular Multimodal Fusion (DMMF) which uses VolNet as building block. DMMF is able to fuse relevant modalities in each module separately in a hierarchical fashion.

We have conducted extensive experiments for VAR on real world datasets obtained from Sakurajima volcanic monitoring system. Sakurajima is one of the most active volcanoes in Japan. In this paper, we propose two models for VAR: VolNet on a single sensor and DMMF on multiple sensors. First, we compared the performance of VolNet with conventional time series classification on a single sensor. Second, we compared DMMF with the best results obtained from the first experiment on single sensor and the one-time fusion model. The result shows that our proposed VolNet and DMMF outperformed all other state-of-the-art models.

To the best of our knowledge, this work is the first attempt to employ deep neural network for the study of VAR. Our deep model learns various patterns and classifies volcanic eruption accurately. The following are the contributions of this paper:

- Propose an accurate VolNet architecture for VAR on a single sensor.
- Propose a simple but effective fusion model called Deep Modular Multimodal Fusion (DMMF) which fuses the modalities in different modules in a hierarchical fashion.
- Outperform volcano experts on the task of VAR.
- Conduct extensive experiments in real volcanic sensor datasets.

The rest of the paper is organized as follows: We briefly introduce the dataset used in our experiment in Sect. 2. Next, we explain our approaches for VAR in Sect. 3. In the Sect. 4, we show detailed experiments on proposed method and baseline models. Related work will be summarized in Sect. 5. And finally, we conclude the paper in Sect. 6.

2   Datasets

We use volcanic sensor data obtained from Osumi Office of River and National Highway, Kyushu Regional Development Bureau, MLIT[1]. The data is about

---

[1] http://www.mlit.go.jp/en/index.html.

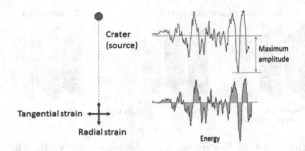

Fig. 2. The explanation of strain data (left) and seismic data (right). Strain data is measured by Strainmeters. These instruments measure linear strain by detecting horizontal contraction or extension in a length. They are installed in the direction of the crater (radial component) and perpendicular to the radial direction (tangential component). Seismic data is measured by Seismometer. This measures ground surface vibration as the velocity of a particle. Square sum of the velocity is proportional to seismic energy to evaluate the intensity of long-term tremor. Maximum amplitude velocity in the seismic records is treated as the instantaneous intensity of the event.

the explosive eruptions of Sakurajima volcano[2] which is one of the most active volcanoes in Japan. There are many explosive eruptions occurring in this volcano every week, so the data is good for VAR. This data includes four types of sensor data arranged into two groups. The first group is the seismic data including "seismic energy" and "maximum amplitude". These data are related to the ground surface vibration. The second group is the strain data including "tangential strain" and "radial strain". These two strains measure the ground deformation horizontally and vertically respectively. The data in each group are correlated with each other as they measure one type of data but in different ways. The details of data measurement and the instruments are shown in Fig. 2.

The data for all sensors are numeric values and recorded in every minute. The total data includes eight years from 2009 to 2016, and this is the biggest dataset about the volcanic monitor in Japan.

3   Proposed Methods

3.1   Problem Definition

VAR takes D sensors as the input and each sensor is a time series of length N. Formally, the input is a matrix of size $D \times N$ with the element x_i^d is the ith element of the time series obtained from sensor d, where $1 \le i \le N$ and $1 \le d \le D$. In case of VAR for single sensor, D is equal to 1 and the input is a time series of length N. In this paper, the output of VAR is the class of the input, which is either *explosive* or *not explosive*. The task is a two-class classification. In VAR, the majority of the input are not explosive, but the explosive cases attract

---

[2] https://en.wikipedia.org/wiki/Sakurajima.

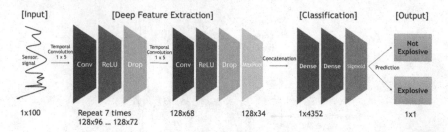

Fig. 3. The architecture of VolNet on a single sensor for VAR. The layers includes Conv (Convolution), ReLU, Drop (Dropout), MaxPool (Max Pooling non-overlapping pooling region with length of 2), Dense (fully-connected), and Sigmoid. The temporal convolution has kernel size of 5. The numbers before and after "x" on the bottom refer to the number of feature maps and the size of a feature map, respectively.

more attention than not explosive cases. Therefore, the goal of this research is not only to optimize the misclassification rate in general, but also maximize the precision and recall of explosive class.

3.2 Challenges

There are two main challenges in this task. The first is extreme class imbalance: Although Sakurajima is an active volcano, it is not explosive for most of the time. This poses a big challenge to the classification task as most classifiers tend to favor the majority class. The second challenge is sensor fusion. Because the eruption is complicated and controlled by many factors, multiple different time series sensor data should be used in order to improve the performance of the classifier. The fusion model should handle multiple data effectively.

3.3 Proposed VolNet for Single Sensor

CNN with its deep architecture is able to learn a hierarchy of abstract features automatically. Given a time series, CNN extracts features from the input using convolution operation with a kernel. Since the convolution operation will spread in different regions of the same sensor, it is feasible for CNN to detect the salient patterns within the input, no matter where they are. Because the nature of time series data is temporal, we use one-dimensional (1D) kernel for each time series independently. The feature map extracted by 1D convolution is obtained as:

$$m_i^{l+1}(x) = \sigma\left(\sum_{f=1}^{F^l} \left[\sum_{p=0}^{P^l-1} K_{if}^l(p) m_f^l(x+p) \right] + b_i^{l+1} \right), \tag{1}$$

where m_i^{l+1} is the feature map i in layer $l+1$, F^l is the number of feature maps in layer l, P^l is the length of the kernel in layer l, K_{if}^l is the convolution kernel of the feature map f in the layer l, and b_i^{l+1} is the bias vector in layer $l+1$. Pooling layer is also used in our model to increase the robustness of features to the small variations. In general, given a feature map, the pooling layer is obtained as:

$$m_i^{l+1}(x) = f_{1 \leq n \leq N^l}\left(m_i^l(x+n) \right), \tag{2}$$

where f is the pooling function, $1 \leq n \leq N^l$ is the range of value function f applies for, and N^l is the length of pooling region.

In this part, we will construct the architecture of VolNet especially for VAR. The overall structure is shown in Fig. 3. There are four main blocks in the architecture. The first block is "Input" and the last one is "Output". The network takes a time series of raw volcanic sensor signal as input and outputs the status of the volcano which is either *explosive* or *not explosive*. The second block called "Deep Feature Extraction" to automatically extract deep features from the time series input. This block includes the following eight small blocks in order: (1) a convolution layer, (2) a rectified linear unit (ReLU) layer that is the activation function mapping the output value using the function $relu(x) = max(0, x)$, and (3) a dropout layer that is a regularization technique [3] where randomly selected neurons are ignored during training, hence reduce over-fitting. We employ a max pooling at the end of this block to decrease the dimension of the feature maps. All the convolution layer has kernel size of 5 and 128 feature maps. Dropout has the probability of 0.5 and we only use dropout for three small blocks (the first, fourth and last small blocks). The reason is more dropout layers can lead to much randomness which is not good in our case. Max pooling has non-overlapping pooling region with length of 2. These hyper parameters are proved to be effective in our task through experiments and chosen using a validation set. The third block called "Classification" is a fully-connected network taking the learned features in previous layer and output the class using sigmoid function $S(x) = \frac{1}{1+e^{-x}}$. We also use dropout layer in this block to reduce over-fitting. One remark in designing the architecture of VolNet especially for VAR is that there are no normalization layers. The experiments showed that adding batch normalization [4] did not improve but worsen the performance.

To train VolNet, we minimize the weighted binary cross entropy loss function, and increase positive weight to deal with class imbalance problem:

$$L = \sum_{i=1}^{batch\_size} y_i \log(y_i') \times weight + (1 - y_i) \log(1 - y_i'),$$

with y is the target and y' is the prediction. The parameter *weight* with value more than 1 is included to the loss function to penalize the cases when the target is 1 (*explosive*) but the prediction is near 0 (*not explosive*). By optimizing the loss function this way, we can force the model to favor explosive class if the ground truth is explosive. The model was trained via Stochastic Gradient Descent with the learning rate of 10^{-3}.

VolNet is designed mainly to deal with one sensor time series. In order to process multiple time series, we proposed a new fusion model built on the top of VolNet called Deep Modular Multimodal Fusion (DMMF).

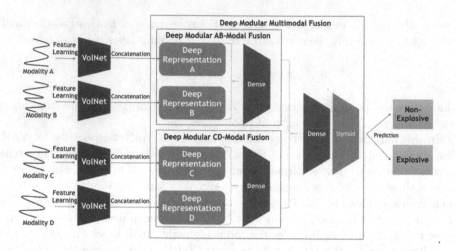

Fig. 4. The proposed Deep Modular Multimodal Fusion (DMMF) for multiple sensors based on VAR. First, we extract the deep representation features using VolNet for each modality independently. Then we group the features from the modalities which are relevant with each other into modules. In this figure, there are two modules A-B and C-D. Each module is followed by a Dense (fully-connected layer) so that they will be fused here. Then, in the next level, two modules will be fused again using one more Dense layer. Finally, we use a Sigmoid layer to compute the classification.

3.4 Proposed Deep Modular Multimodal Fusion for Multiple Sensors

While using one sensor performs well for VAR, fusing multiple sensors can potentially improve the performance. This is especially important in VAR where the eruption is affected by many factors. Also in the monitoring system, each sensor measures a different type of data and data is considered noisy due to earthquakes and typhoons. Therefore, adding more data from different sources can potentially improve the accuracy of VAR.

Recent work on multimodal fusion for sensors adapted CNN for each modality independently and then concatenate all the feature maps in one step [6], [8], which is not appropriate for VAR. In VAR, each modality is the data obtained from one sensor and some related sensors create a group of data. Our assumption is that the modalities in the same group of data are more related than other modalities and related modalities should be fused together before they will be fused with other modalities. For example, the modalities from tangential strain sensor and radial strain sensor make a group called "strain data". They both measure the ground deformation, but in different directions which are horizontally and vertically respectively. Intuitively, the multimodal fusion model that considers fusing these two modalities first is expected to improve the performance. Based on this idea, we propose a simple but effective Deep Modular Multimodal Fusion (DMMF) which is built on the top of our proposed VolNet and is able to fuse relevant modalities in each module separately in a hierarchical fashion.

The overall architecture of DMMF is shown in Fig. 4. First, we use VolNet to extract the feature maps for each modality independently and then concatenate all feature maps to get the deep representation for each modality. Then we group the features from the modalities which are relevant with each other into modules. In each module, we concatenate the features of all modalities into one vector and add a fully-connected network so that these modalities will be fused. We then concatenate all the fused features from all modules and add one more fully-connected layer to fuse all the modules. The final layer is the sigmoid to compute the classification score based on the fused features from all modules. DMMF fuses the modalities in a hierarchical fashion, so all modalities have a chance to fused with each other. Unlike ensemble model where the final classification is made based on the classification of models on different sets of data, DMMF makes the classification based on the feature fusion of different groups of data in a hierarchical fashion.

There are some remarks on the design of DMMF. First, DMMF is built on the top of Volnet, hence takes advantages of deep features extracted from VolNet which is powerful to discriminate between classes. Second, DMMF does not directly concatenate all features at once, but it fuses the modalities in some different modules in a hierarchical fashion. Intuitively, when related modalities are fused together, the fusion will be more robust to noise as relevant modalities tend to complement with each other.

4   Experiments

4.1   Evaluation Metrics

Because of class imbalance problem, we use two type of F-score to do the evaluation. The first metric is F-score of the explosive class (F_E) which is the minority class in VAR. This metric measures the performance of the model in terms of minor but important class. The second metric is the unweighted average F-score (F_{avg}) of the explosive class and the not explosive class. Unweighted F-score is the general metric of the model and the contribution of each class to the score is equal.

4.2   Experiment 1: VAR on a Single Data

In this part, we conduct VAR experiments on each sensor data separately and compare the performance of proposed VolNet with the baseline models. In this experiment, we firstly show the effectiveness of deep feature representation in term of accuracy, and secondly to get insight into the best sensor for VAR.

The data is obtained from Sakurajima volcano as shown in Sect. 2. We use the sliding window technique to segment the time series into sequences. The sliding window length of raw data is 101 min and the sliding step is 10 min in our experiments. The choice of window length is based on the average sum of the inflation and deflation time of all eruptions, which is the domain knowledge from volcano experts. The sliding step of 10 min is chosen to reduce the chance of missing important patterns. In this paper, we use the first order differences of the series $\Delta x_i = x_{i+1} - x_i$ for all experiments. This is based on the characteristics of volcanic sensor signal where the change of two consecutive data points is important. Preliminary experimental result shows that all models using first order difference data outperformed the ones using raw data. Due to using first order differences, the length of each data sample now is 100 data points with the time interval of one minute.

We label each sequence into two classes: *explosive* and *not explosive*. If there is at least one eruption between the starting time and ending time of the sequence, we set it explosive, otherwise not explosive. This strategy considers the importance of the sequence with different numbers of eruption equally, but in fact, it is not common to have a sequence with more than one eruption within its time period. The total number of extracted sequences for eight years from 2009 to 2016 is around 400,000. The ratio of explosive class over not explosive class is approximately 1:10. For each year, we split part of dataset for testing to make sure test set covers all the dataset. In total, we obtain 50,000 sequences for testing and the rest 350,000 for training. There is no overlapping about time between test set and the training set to make sure the accuracy is reliable. We use validation set to pick up the hyper parameters of the models. Validation set is 20% of training data.

We do the experiments on proposed VolNet and the following baseline models:

- VolNet: The network architecture is shown in Fig. 3.
- 1 Nearest Neighbor with Euclidean distance (1NN-ED): Even though 1NN-ED is simple, it is considered one of the best techniques in time series classification [10]. First order difference time series data with normalization is the input of the model.
- 1 Nearest Neighbor with Dynamic Time Warping distance (1NN-DTW): Same as 1NN-ED, but the distance metric is DTW instead. 1NN-DTW is very effective in many time series classification due to its flexibility for computing distance. It also achieves state-of-the-art performance on the task of time series classification together with 1NN-ED [11]. One disadvantage of 1NN-DTW is that testing time is extremely slow, so we only test on 5% of data. We run the test multiple times and take average score. Running in all dataset takes months to finish.
- Means and Variance (MV): Mean and Variance show advantages in some time series classification task [13]. We use the mean and variance of the time series as the features and do the classification using 1NN-ED.
- Symbolic Aggregate Approximation - Vector Space Model (SAX-VSM): Unlike Nearest Neighbor which is distance-based approach, SAX-VSM is a well-known and effective feature-based approach for time series classification [12]. The input is also first order difference of sensor data.
- Support Vector Machine (SVM): The support vector machine with radial basis function (RBF) kernel is used as a classifier [14]. RBF kernel is chosen

Table 1. The results of VolNet and baselines for each type of data. The best model for each data is shown in bold.

	Seismic energy		Maximum amplitude		Radial strain		Tangential strain	
	F_{avg}	F_E	F_{avg}	F_E	F_{avg}	F_E	F_{avg}	F_E
1NN-ED	82.0	67.4	80.4	64.5	70.4	45.3	70.6	45.8
1NN-DTW	82.6	68.3	82.8	69.0	72.8	50.0	57.2	20.6
MV	79.8	63.3	80.5	64.6	58.5	26.9	59.8	28.8
SAX-VSM	74.3	43.4	75.0	44.7	70.6	35.2	71.7	37.0
SVM	85.7	74.2	57.3	18.7	68.6	41.3	72.3	48.4
MLP	84.2	71.6	83.3	70.2	76.0	56.4	76.4	57.2
LSTM	86.0	74.7	87.5	77.4	79.4	62.9	78.9	62.2
VolNet	**92.8**	**87.7**	**93.2**	**88.9**	**89.9**	**88.5**	**92.0**	**86.1**

because it shows the best results among all kernels. The input of the model is the first order difference data and the hyper parameter is carefully tuned using validation set.

- Multilayer Perceptron (MLP): MLP is an effective technique for time series classification [15]. The input is also first order difference sensor data. The architecture of the MLP includes one input layer, two hidden layers, and one output. The architecture and the number of neurons are optimized using validation set. We include this method to show the effectiveness of the feature learning from VolNet.
- Long Short-Term Memory (LSTM): LSTM is well-known to deal with sequential data. The architecture of LSTM includes two layers with the dimension of hidden state is 64. The input is also first order difference of sensor data.

The results of VolNet and the baselines for VAR are shown in Table 1. From the results, VolNet works well on different types of data and consistently outperforms all other baseline models on two evaluation metrics. From the accuracy of all models, we can see that seismic energy and maximum amplitude are the two best for VAR according to the experiments. The accuracy between models are quite different. VolNet is the best model among all models. LSTM is the second best model. The fact that VolNet which is built on CNN works better than LSTM in this case may suggest that the shape of the sequence is more important than the dependency of data through time. MLP also gains good accuracy, but much worse than VolNet. Other baseline models are quite unstable as they only work well on some data. For example, SVM works very well on seismic energy data, but when it comes to maximum amplitude data, it becomes the worst model with very low F_E. Both distance-based and feature-based methods like 1NN-ED, 1NN-DTW, MV and SAX-VSM did not work well on this task. This suggests that the raw signal and handcrafted feature extraction are not as good as deep automatic feature extraction from VolNet.

4.3   Experiment 2: Comparison with Volcano Experts

Volcano experts always want to predict the volcano status by using the sensor data. They try to understand the pattern of explosive eruption extracted from the sensor data. So far, the best way for volcano experts to recognize the eruption at Sakurajima volcano is using tangential strain. The pattern of eruption is gradually increase of tangential strain and then suddenly decrease. The point of eruption is usually the starting point of the decrease [22]. We would like to compare the model from expert and our proposed VolNet in the task of VAR.

We implement the expert model to detect explosive eruption using tangential strain. The dataset for this experiment is different from the previous experiments due to special condition of expert model. The experts need to observe the sensor data prior and after the eruption. The way we create dataset is as follows. For *explosive eruption class* we segment the sequences which has the explosive eruption at the position 80 of a sequence with length 100. This is based on the average inflation length of an eruption is about 80 min and the average deflation length is about 20 min. For *not explosive class* we use all the sequences which do not have any eruption.

Because the common pattern of eruption is an increase of tangential strain until the explosive point and then decrease of tangential strain after the eruption, we can divide the sequence into two parts: inflation (before explosive point) and deflation (after the explosive point). In the case of eruption, if we calculate the first order difference of the values in the inflation part, the amount of change will be positive. And in the case of deflation part, the amount of change will be negative. If there is no eruption, the amount of change in both inflation and deflation will not follow that rule. We call the amount of change in the inflation part is accumulated inflation, and that amount in the case of deflation is accumulated deflation. Expert model classifies the status of volcano based on the accumulated inflation and accumulated deflation.

Let say E is the set of sequences having eruption, and NE is the set of sequences which do not have eruption. For each sequence in E, we calculate the accumulated inflation and accumulated deflation. Then we compute the mean of accumulated inflations and the mean of accumulated deflations over set E. Intuitively, a sequence having the accumulated inflation and deflation near to these means has a high chance to have an eruption. We also do the same calculation for set NE to find the mean of accumulated inflation and deflation for the case of no explosive eruption. In testing phase, given a sequence, we first calculate accumulated inflation and deflation. The sequence will be assigned explosive label if both difference in inflation and deflation are nearer to the explosive case than the not explosive case. Otherwise, we set it not explosive. We use Piecewise Aggregate Approximation (PAA) [9] to transform the time series before calculating the first order difference. This transformation can help to remove the small variations in the sequence. The window size of PAA is decided using validation set. In our experiment, the optimal window size is 4. Table 2 shows the parameters for expert model. We can clearly see that in the case of explosive eruption, accumulated inflation is positive and accumulated deflation is negative.

Table 2. The obtained parameters of expert model

	Explosive eruption	Not explosive eruption
Mean accumulated inflation	0.367	0.019
Median accumulated inflation	0.376	0.023
Standard deviation accumulated inflation	0.569	0.353
Mean accumulated deflation	−2.041	0.038
Median accumulated deflation	−1.622	0.052
Standard deviation accumulated deflation	2.319	0.780

Table 3. The result of expert model and VolNet using tangential strain. The better model is shown in bold.

	F_{avg}	F_E
Expert model	75.1	46.3
VolNet	**86.5**	**73.3**

The results of expert model and our VolNet are shown in Table 3. VolNet outperforms expert model with a wide margin. This once again confirms that deep feature extraction from VolNet is much more powerful than handcrafted feature extraction.

4.4 Experiment 3: Multimodal Fusion

In this part, multiple sensors are used in the experiment. We would like to firstly show the effectiveness of multimodal fusion in term of accuracy compared with the best results obtained using only one sensor, and secondly to show the effectiveness of DMMF over other fusion strategies. We use the same dataset with Experiment 1 and run the experiments on proposed DMMF and the following models:

- Proposed DMMF: The architecture is shown as Sect. 3.4. In the fusion step, modalities are grouped into two modules called "module of strain data" including tangential strain and radial strain, and "module of seismic data" including maximum amplitude and seismic energy. This is based on related sensors from domain knowledge.
- Best model without fusion: We copy the best results with one data from Experiment 1.
- Early fusion: The first convolutional layer accepts data from all sensors using the kernel of $d \times k$, where 4 is the number of sensors and k is the kernel size. In this experiment, $d = 4$ and $k = 5$.
- One-time fusion: The features extracted from different modalities are fused at one step.

The results of fusion models for multiple sensors and the best model with no fusion are shown in Table 4. From the results, proposed DMMF consistently outperforms all other models in both evaluation metrics. Specifically, compared with the best results obtained from one sensor, DMMF improves the performance by 3% and 4.2% on average for F_{avg} and F_E respectively. This proves the effectiveness of fusion model on improving the performance for VAR when more data is used. Early fusion does not improve the accuracy, but worsen the overall accuracy compared with the best model with no fusion. This suggests that fusion strategy is important to improve the accuracy. In contrast, one-time fusion improves the accuracy more than 1% compared to no fusion. However, compared with one-time fusion, DMMF gains an improvement with F_{avg} and F_E increased by 1.9% and 3.7% on average respectively. This result supports our assumption that hierarchical fusion is better than one-time fusion in term of feature learning for VAR.

Table 4. The results of proposed DMMF and compared models. The best model is shown in bold.

	F_{avg}	F_E
Best model without fusion	93.2	88.9
Early fusion	91.0	83.5
One-time fusion	94.3	89.4
DMMF	**96.2**	**93.1**

To further confirm the effectiveness of modular fusion, we run some experiments with some combinations of modalities. Some combinations are:

– Seismic module fusion: Fusion with seismic energy and maximum amplitude.
– Strain module fusion: Fusion with tangential strain and radial strain.
– Maximum Amplitude Tangential Strain fusion: Fusion with maximum amplitude and tangential strain.
– Change group fusion: DMMF fusion with architecture of grouping are {seismic energy, tangential strain} and {maximum amplitude, radial strain}.

The results of experiment are shown in Table 5. We can see that the combination of seismic energy and maximum amplitude (seismic energy data) are better than using seismic energy or maximum amplitude alone. The same thing can apply to strain module fusion. However, when we combine maximum amplitude and tangential strain which belong to two different group of data, the accuracy goes down. This suggests that it is important to consider smart grouping when performing fusion because the relevant modalities in each module can complement each other and improve the accuracy. In change group fusion, we build an architecture exactly the same with DMMF, but try to change the group of data into {seismic energy, tangential strain} and {maximum amplitude, radial strain} which is against data properties. Compared with DMMF, the accuracy of change group fusion goes down. This suggests that the effectiveness of DMMF

Table 5. The results of proposed DMMF and the models with different combinations of modalities. The best model is shown in bold.

	F_{avg}	F_E
Seismic module fusion	95.1	91.1
Strain module fusion	93.8	88.6
Maximum Amplitude Tangential Strain fusion	92.9	87.0
Change group fusion	94.8	90.5
DMMF	**96.2**	**93.1**

is due to smart grouping, not due to deeper architecture because grouping differently worsen the performance. In general, the effectiveness of DMMF comes from hierarchical fusion and smart grouping.

5 Related Work

In this section, we briefly review some related work on volcano activity study and multiple sensor fusion. There is some work using sensor signals from the volcanic monitoring system for volcano-related events. Noticeably, as in [16], the authors applied neural network on seismic signals to classify the volcano events such as landslides, lightning strikes, long-term tremors, earthquakes, and ice quakes. The same research purpose was conducted by [18], but the methodology is based on hidden Markov model instead. The authors in [19] combined seismic data from multiple stations to improve the accuracy of the classifier. One common point of these papers is that the classes of the task are tremors, earthquakes, ice quakes, landslides, and lightning strikes. The concern about the explosive and not explosive status of the volcano is ignored in these work. Our work focuses on the classification of this class using the volcanic sensor data. The closet work to ours is [17]. The author tries to classify the time series of seismic signals to classify the volcano statuses using Support Vector Machine, still the methodology is quite simple and the accuracy is not high. To the best of our knowledge, our work is the first attempt to employ deep neural network for effective VAR on the explosive status of the volcano.

Multimodal deep learning has been successfully applied for many applications like speech recognition, emotion detection [20,21]. In these applications, the modalities are obtained from audio, video, and images. Human activity recognition is one of the most popular applications which uses multiple sensor data for the classification of human activity [6,8]. In these work, CNN is used on multichannel time series. However, in the fusion step, the authors ignored the properties of the modalities and fused all the modalities in one step. Our research considers the properties of the modalities in the fusion step and fuses the modalities in different modules in a hierarchical fashion.

6 Conclusion

In this paper, we demonstrated the advantages of deep architecture based on CNN for VAR. We proposed VolNet and a simple but effective fusion model DMMF which uses VolNet as a building block. DMMF adapts the modality properties to build the deep architecture and form the fusion strategy. The idea is that relevant modalities should be fused together before they will be fused with other less relevant modalities. The key advantages of DMMF are: (1) take advantages of deep non-handcrafted feature extraction and hence powerful to discriminate between classes, (2) relevant modalities in the same module complements with each other and hence is able to deal with noise data. With the extensive experiments, we demonstrated that DMMF consistently outperforms other compared models. This shows the ability of DMMF on combining multiple sensors into one model and the advantages of modular fusion. Moreover, DMMF is not only limited to VAR as it also has the potential to apply for other tasks that require multimodal fusion.

Acknowledgments. We would like to thank Osumi Office of River and National Highway, Kyushu Regional Development Bureau, MLIT for providing volcanic sensor datasets.

References

1. Sparks, R.S.J.: Forecasting volcanic eruptions. Earth Planetary Sci. Lett. **210**(1), 1–15 (2003)
2. Palaz, D., Collobert, R.: Analysis of CNN-based speech recognition system using raw speech as input. EPFL-REPORT-210039. Idiap (2015)
3. Srivastava, N., Hinton, G.E., Krizhevsky, A., Sutskever, I., Salakhutdinov, R.: Dropout: a simple way to prevent neural networks from overfitting. J. Mach. Learn. Res. **15**(1), 1929–1958 (2014)
4. Ioffe, S., Szegedy, C.: Batch normalization: accelerating deep network training by reducing internal covariate shift. In: International Conference on Machine Learning, pp. 448–456 (2015)
5. LeCun, Y., Bengio, Y., Hinton, G.: Deep learning. Nature **521**(7553), 436–444 (2015)
6. Yang, J., Nguyen, M.N., San, P.P., Li, X., Krishnaswamy, S.: Deep convolutional neural networks on multichannel time series for human activity recognition. In: IJCAI, pp. 3995–4001 (2015)
7. Ordóñez, F.J., Roggen, D.: Deep convolutional and LSTM recurrent neural networks for multimodal wearable activity recognition. Sensors **16**(1), 115 (2016)
8. Zheng, Y., Liu, Q., Chen, E., Ge, Y., Zhao, J.L.: Time series classification using multi-channels deep convolutional neural networks. In: Li, F., Li, G., Hwang, S., Yao, B., Zhang, Z. (eds.) WAIM 2014. LNCS, vol. 8485, pp. 298–310. Springer, Cham (2014). https://doi.org/10.1007/978-3-319-08010-9_33
9. Senin, P., Malinchik, S.: SAX-VSM: interpretable time series classification using SAX and vector space model. In: 2013 IEEE 13th International Conference on Data Mining (ICDM), pp. 1175–1180. IEEE (2013)

10. Batista, G.E., Wang, X., Keogh, E.J.: A complexity-invariant distance measure for time series. In: Proceedings of the 2011 SIAM International Conference on Data Mining, pp. 699–710. Society for Industrial and Applied Mathematics (2011)
11. Xi, X., Keogh, E., Shelton, C., Wei, L., Ratanamahatana, C.A.: Fast time series classification using numerosity reduction. In: Proceedings of the 23rd International Conference on Machine Learning, pp. 1033–1040. ACM (2006)
12. Keogh, E., Chakrabarti, K., Pazzani, M., Mehrotra, S.: Dimensionality reduction for fast similarity search in large time series databases. Knowl. Inf. Syst. 3(3), 263–286 (2001)
13. Bulling, A., Blanke, U., Schiele, B.: A tutorial on human activity recognition using body-worn inertial sensors. ACM Comput. Surv. (CSUR) 46(3), 33 (2014)
14. Cao, H., Nguyen, M.N., Phua, C., Krishnaswamy, S., Li, X.: An integrated framework for human activity classification. In: UbiComp, pp. 331–340 (2012)
15. Koskela, T., Lehtokangas, M., Saarinen, J., Kaski, K.: Time series prediction with multilayer perceptron, FIR and Elman neural networks. In: Proceedings of the World Congress on Neural Networks, pp. 491–496. INNS Press, San Diego (1996)
16. Ibs-von Seht, M.: Detection and identification of seismic signals recorded at Krakatau volcano (Indonesia) using artificial neural networks. J. Volcanol. Geothermal Res. 176(4), 448–456 (2008)
17. Malfante, M., Dalla Mura, M., Metaxian, J.-P., Mars, J.I., Macedo, O., Inza, A.: Machine learning for Volcano-seismic signals: challenges and perspectives. IEEE Sig. Process. Mag. 35(2), 20–30 (2018)
18. Benítez, M.C., et al.: Continuous HMM-based seismic-event classification at Deception Island Antarctica. IEEE Trans. Geosci. Remote Sens. 45(1), 138–146 (2007)
19. Duin, R.P.W., Orozco-Alzate, M., Londono-Bonilla, J.M.: Classification of volcano events observed by multiple seismic stations. In: 2010 20th International Conference on Pattern Recognition (ICPR), pp. 1052–1055. IEEE (2010)
20. Ngiam, J., Khosla, A., Kim, M., Nam, J., Lee, H., Ng, A.Y.: Multimodal deep learning. In: Proceedings of the 28th International Conference on Machine Learning (ICML-11), pp. 689–696 (2011)
21. Kahou, S.E., et al.: Emonets: multimodal deep learning approaches for emotion recognition in video. J. Multimodal User Interfaces 10(2), 99–111 (2016)
22. Iguchi, M., Tameguri, T., Ohta, Y., Ueki, S., Nakao, S.: Characteristics of volcanic activity at Sakurajima volcano's Showa crater during the period 2006 to 2011 (special section Sakurajima special issue). Bull. Volcanol. Soc. Japan 58(1), 115–135 (2013)

Nectar Track

Matrix Completion Under Interval Uncertainty: Highlights

Jakub Marecek[1]([⊠]), Peter Richtarik[2,3], and Martin Takac[4]

[1] IBM Research – Ireland, Damastown, Dublin 15, Ireland
jakub.marecek@ie.ibm.com
[2] School of Mathematics, University of Edinburgh, Edinburgh EH9 3FD, UK
[3] KAUST, 2221 Al-Khwarizmi Building, Thuwal 23955-6900,
Kingdom of Saudi Arabia
[4] Department of Industrial and Systems Engineering, Lehigh University,
Bethlehem 18015, USA

Abstract. We present an overview of inequality-constrained matrix completion, with a particular focus on alternating least-squares (ALS) methods. The simple and seemingly obvious addition of inequality constraints to matrix completion seems to improve the statistical performance of matrix completion in a number of applications, such as collaborative filtering under interval uncertainty, robust statistics, event detection, and background modelling in computer vision. An ALS algorithm MACO by Marecek et al. outperforms others, including Sparkler, the implementation of Li et al. Code related to this paper is available at: http://optml.github.io/ac-dc/.

1 Introduction

Matrix completion is a well-known problem: Given dimensions of a matrix X and some of its elements $X_{i,j}, (i,j) \in \mathcal{E}$, the goal is to find the remaining elements. Without imposing any further requirements on X, there are infinitely many solutions. In many applications, however, the matrix completion that minimizes the rank:

$$\min_Y \text{rank}(Y), \text{ subject to } Y_{i,j} = X_{i,j}, (i,j) \in \mathcal{E}, \quad (1)$$

often works as well as the best known solvers for problems in the particular domain. There are literally hundreds of applications of matrix completion, especially in recommender systems [3], where the matrix is composed of ratings, with a row per user and column per product.

Two major challenges remain. The first challenge is related to data quality: when a large proportion of data is missing and one uses matrix completion for data imputation, it may be worth asking whether the remainder data is truly known exactly. The second challenge is related to the rate of convergence and run-time to a fixed precision: many solvers still require hundreds or thousands of CPU-hours to complete a 480189×17770 matrix reasonably well.

U. Brefeld et al. (Eds.): ECML PKDD 2018, LNAI 11053, pp. 621–625, 2019.
https://doi.org/10.1007/978-3-030-10997-4_38

The first challenge has been recently addressed [8] by considering a variant of the problem with explicit uncertainty set around each "supposedly known" value. Formally, let X be an $m \times n$ matrix to be reconstructed. Assume that elements $(i, j) \in \mathcal{E}$ of X we wish to fix, for elements $(i, j) \in \mathcal{L}$ we have lower bounds and for elements $(i, j) \in \mathcal{U}$ we have upper bounds. The variant [8] is:

$$
\begin{aligned}
\min_{X \in \mathbb{R}^{m \times n}} \quad & \mathrm{rank}(X) \\
\text{subject to} \quad & X_{ij} = X_{ij}^{\mathcal{E}}, (i, j) \in \mathcal{E} \\
& X_{ij} \geq X_{ij}^{\mathcal{L}}, (i, j) \in \mathcal{L} \\
& X_{ij} \leq X_{ij}^{\mathcal{U}}, (i, j) \in \mathcal{U}.
\end{aligned}
\tag{2}
$$

We refer to [8] for the discussion of the superior statistical performance.

2   An Algorithm

The second challenge can be addressed using the observation that a rank-r X is a product of two matrices, $X = LR$, where $L \in \mathbb{R}^{m \times r}$ and $R \in \mathbb{R}^{r \times n}$. Let $L_{i:}$ and $R_{:j}$ be the i-th row and j-h column of L and R, respectively. Instead of (2), we shall consider the *smooth, non-convex* problem

$$
\min\{f(L, R) : L \in \mathbb{R}^{m \times r}, R \in \mathbb{R}^{r \times n}\},
\tag{3}
$$

where

$$
\begin{aligned}
f(L, R) := & \tfrac{\mu}{2}\|L\|_F^2 + \tfrac{\mu}{2}\|R\|_F^2 \\
& + f_{\mathcal{E}}(L, R) + f_{\mathcal{L}}(L, R) + f_{\mathcal{U}}(L, R),
\end{aligned}
$$

$$
\begin{aligned}
f_{\mathcal{E}}(L, R) &:= \tfrac{1}{2}\sum\nolimits_{(ij) \in \mathcal{E}} (L_{i:}R_{:j} - X_{ij}^{\mathcal{E}})^2 \\
f_{\mathcal{L}}(L, R) &:= \tfrac{1}{2}\sum\nolimits_{(ij) \in \mathcal{L}} (X_{ij}^{\mathcal{L}} - L_{i:}R_{:j})_+^2 \\
f_{\mathcal{U}}(L, R) &:= \tfrac{1}{2}\sum\nolimits_{(ij) \in \mathcal{U}} (L_{i:}R_{:j} - X_{ij}^{\mathcal{U}})_+^2
\end{aligned}
$$

and $\xi_+ = \max\{0, \xi\}$. The parameter μ helps to prevent scaling issues[1]. We could optionally set μ to zero and then from time to time rescale matrices L and R, so that their product is not changed. The term $f_{\mathcal{E}}$ (resp. $f_{\mathcal{U}}$, $f_{\mathcal{L}}$) encourages the equality (resp. inequality) constraints to hold.

Subsequently, we can apply an alternating parallel coordinate descent method called MACO in [8]. This is based on the observation that although f is not convex jointly in (L, R), it is convex in L for fixed R and in L for fixed R. We

---

[1] Let $X = LR$, then also $X = (cL)(\tfrac{1}{c}R)$ as well, but we see that for $c \to 0$ or $c \to \infty$ we have $\|L\|_F^2 + \|R\|_F^2 \ll \|cL\|_F^2 + \|\tfrac{1}{c}R\|_F^2$.

can hence alternate between fixing R, choosing $\hat{r}$ and $\hat{S}$ of rows of L uniformly at random, updating $L_{i\hat{r}} \leftarrow L_{i\hat{r}} + \delta_{i\hat{r}}$ in parallel for $i \in \hat{S}$, and the respective steps for L. Further, notice that if we fix $i \in \{1, 2, \ldots, m\}$ and $\hat{r} \in \{1, 2, \ldots, r\}$, and view f as a function of $L_{i\hat{r}}$ only, it has a Lipschitz continuous gradient with constant $W_{i\hat{r}}^{\mathcal{L}} = \mu + \sum_{v\,:\,(iv)\in\mathcal{E}} R_{\hat{r}v}^2 + \sum_{v\,:\,(iv)\in\mathcal{L}\cup\mathcal{U}} R_{\hat{r}v}^2$. That is, for all L, R and $\delta \in \mathbb{R}$, we have $f(L + \delta E_{i\hat{r}}, R) \leq f(L, R) + \langle \nabla_L f(L, R), E_{i\hat{r}}\rangle \delta + \frac{W_{i\hat{r}}^{\mathcal{L}}}{2}\delta^2$, where E is the $n \times r$ matrix with 1 in the $(i\hat{r})$ entry and zeros elsewhere. Likewise, one can define V for $R_{\hat{r}j}$. The minimizer of the right hand side of the bound on $f(L + \delta E_{i\hat{r}}, R)$ is hence

$$\delta_{i\hat{r}} := -\tfrac{1}{W_{i\hat{r}}^{\mathcal{L}}}\langle \nabla_L f(L, R), E_{i\hat{r}}\rangle, \tag{4}$$

where $\langle \nabla_L f(L, R), E_{i\hat{r}}\rangle$ equals

$$\mu L_{i\hat{r}} + \sum_{v\,:\,(iv)\in\mathcal{E}}(L_{i:}R_{:v} - X_{iv}^{\mathcal{E}})R_{\hat{r}v}$$
$$+ \sum_{v\,:\,(iv)\in\mathcal{U}\,\&\,L_{i:}R_{:v}>X_{iv}^{\mathcal{U}}}(L_{i:}R_{:v} - X_{iv}^{\mathcal{U}})R_{\hat{r}v}$$
$$+ \sum_{v\,:\,(iv)\in\mathcal{L}\,\&\,L_{i:}R_{:v}<X_{iv}^{\mathcal{L}}}(X_{iv}^{\mathcal{L}} - L_{i:}R_{:v})R_{\hat{r}v}.$$

The minimizer of the right hand side of the bound on $f(L, R + \delta E_{\hat{r}j})$ is derived in an analogous fashion.

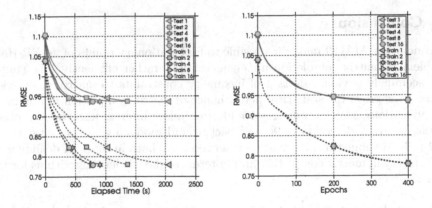

Fig. 1. RMSE as a function of the number of iterations and wall-clock time, respectively, on a well-known 480189×17770 matrix, for $r = 20$ and $\mu = 16$.

3 Numerical Experiments

A particular care has been taken to produce a numerically stable and efficient implementation. Algorithmically, the key insight is that Eq. (4) does not require as much computation as it seemingly does. Let us define matrix $A \in \mathbb{R}^{m \times r}$ and $B \in \mathbb{R}^{r \times n}$ such that $A_{iv} = W_{iv}^{\mathcal{L}}$ and $B_{vj} = V_{vj}^{\mathcal{U}}$. After each update of the solution, we also update those matrices. We also store and update sparse residuals, where

$(\Delta_{\mathcal{E}})_{i,j}$ is $L_{i:}.R_{:j} - X_{ij}^{\mathcal{E}}$ for $(ij) \in \mathcal{E}$ and zero elsewhere, and similarly for $\Delta_{\mathcal{U}}$, $\Delta_{\mathcal{L}}$. Subsequently, the computation of $\delta_{i\hat{r}}$ or $\delta_{\hat{r}j}$ is greatly simplified.

Our C++ implementation stores all data stored in shared memory and uses OpenMP multi-threading. Figure 1 presents the evolution of RMSE over time on the well-known 480189×17770 matrix of rank 20 on a machine with 24 cores of Intel X5650 clocked at 2.67 GHz and 24 GB of RAM. There is an almost linear speed-up visible from 1 to 4 cores and marginally worse speed-up between 4 and 8 cores. The comparison of run-times of algorithms across multiple papers is challenging, especially when some of the implementations are running across clusters of computers in a distributed fashion. Nevertheless, the best distributed implementation, which uses a custom matrix-completion-specific platform for distributed computing [4], requires the wall-clock time of 95.8 s per epoch on a 5-node cluster, for rank 25, and 121.9 s per epoch on a 10-node cluster, again for rank 25, which translates to the use of 47900 to 121900 node-seconds, on the same 480189×17770 matrix (denoted N1). For a recent Spark-based implementation [4], the authors report the execution time of one epoch of 500 s for rank between 25 and 50 on a 10-node cluster, with 8 Intel Xeon cores and 32 GB of RAM per node. A run of 100 epochs, which is required to obtain an acceptable precision, hence takes 50000 to 300000 node-seconds. As can be seen in Fig. 1, our algorithm processes the 100 epochs within 500 node-seconds, while using 8 comparable cores. This illustration suggests an improvement of two orders of magnitude, in terms of run-time.

4   Conclusions

In conclusion, MACO makes it possible to find stationary points of an NP-Hard problem in matrix completion under uncertainty rather efficiently. The simple and seemingly obvious addition of inequality constraints to matrix completion seems to improve the statistical performance of matrix completion in a number of applications, such as collaborative filtering under interval uncertainty, robust statistics, event detection [7,9], and background modelling in computer vision [1,2,5,6]. We hope this may spark further research, both in terms of dealing with uncertainty in matrix completion and in terms of the efficient algorithms for the same.

Acknowledgement. The work of JM received funding from the European Union's Horizon 2020 Programme (Horizon2020/2014-2020) under grant agreement No. 688380. The work of MT was partially supported by the U.S. National Science Foundation, under award numbers NSF:CCF:1618717, NSF:CMMI:1663256, and NSF:CCF:1740796. PR acknowledges support from KAUST Faculty Baseline Research Funding Program.

References

1. Akhriev, A., Marecek, J., Simonetto, A.: Pursuit of low-rank models of time-varying matrices robust to sparse and measurement noise. Preprint arXiv:1809.03550 (2018, submitted)

2. Dutta, A., Li, X., Richtarik, P.: Weighted low-rank approximation of matrices and background modeling. Preprint arXiv:1804.06252 (2018, submitted)
3. Jahrer, M., Töscher, A., Legenstein, R.: Combining predictions for accurate recommender systems. In: KDD, pp. 693–702. ACM (2010)
4. Li, B., Tata, S., Sismanis, Y.: Sparkler: supporting large-scale matrix factorization. In: EDBT, pp. 625–636. ACM (2013)
5. Li, X., Dutta, A.: Weighted low rank approximation for background estimation problems. In: ICCVW, pp. 1853–1861, October 2017
6. Li, X., Dutta, A., Richtarik, P.: A batch-incremental video background estimation model using weighted low-rank approximation of matrices. In: ICCVW, pp. 1835–1843, October 2017
7. Marecek, J., Maroulis, S., Kalogeraki, V., Gunopulos, D.: Low-rank methods in event detection. Preprint arXiv:1802.03649 (2018, submitted)
8. Marecek, J., Richtarik, P., Takac, M.: Matrix completion under interval uncertainty. Eur. J. Oper. Res. **256**(1), 35–43 (2017)
9. Marecek, J., Simonetto, A., Maroulis, S., Kalogeraki, V., Gunopulos, D.: Low-rank subspace pursuit in event detection (2018, submitted)

A Two-Step Approach for the Prediction of Mood Levels Based on Diary Data

Vincent Bremer[1]([📧]) [iD], Dennis Becker[1] [iD], Tobias Genz[1] [iD], Burkhardt Funk[1] [iD], and Dirk Lehr[2] [iD]

[1] Institute of Information Systems, Leuphana University, Lüneburg, Germany
vincent.bremer@leuphana.de
[2] Institute of Psychology, Leuphana University, Lüneburg, Germany

Abstract. The analysis of diary data can increase insights into patients suffering from mental disorders and can help to personalize online interventions. We propose a two-step approach for such an analysis. We first categorize free text diary data into activity categories by applying a bag-of-words approach and explore recurrent neuronal networks to support this task. In a second step, we develop partial ordered logit models with varying levels of heterogeneity among clients to predict their mood. We estimate the parameters of these models by employing MCMC techniques and compare the models regarding their predictive performance. This two-step approach leads to an increased interpretability about the relationships between various activity categories and the individual mood level.

Keywords: Text-mining · Ordinal logit · Diary data

1 Introduction

Mental issues are increasing around the world and access to healthcare programs are limited. Internet-based interventions provide additional access and can close the gap between treatment and demand [8]. In these interventions, participants often provide diary data in which they rank, for example, their mood levels and simultaneously report daily activities. Because various activities from walking a dog, to volunteering, cleaning the house, or having a drink out with friends affect mood in different and complex ways [9], we attempt to analyze the effects that different activities have on the mood level.

In this study, we propose a two-step approach for the analysis of free text diary data that is provided by participants of an online depression treatment [4]. The dataset consists of 440 patients who provided 9,192 diary entries. We utilize text-mining techniques in order to categorize the free text into defined activity categories (exercise, sickness, rumination, work related, recreational, necessary, social, and sleep related activities) and use individualized partial ordered logit models to predict the mood level. This two-step approach allows for interpretability of the effects between the activity categories and the mood level. Thus, besides

U. Brefeld et al. (Eds.): ECML PKDD 2018, LNAI 11053, pp. 626–629, 2019.
https://doi.org/10.1007/978-3-030-10997-4_39

studying these relationships, we contribute to the field of machine learning by proposing a mixed method approach to analyze diary data. This short paper is based on a full paper already published in [3]. Here, more information about the methods, results, and discussion including a full list of references can be found.

2   Method

Figure 1 illustrates the two-step approach. In the first step, we utilize bag-of-words (BoW) categorization and extent the results by applying recurrent neuronal networks (RNN) [5] in order to categorize the free text into activity categories. We split all diary entries into sentences and identify the most frequent ($\geq$10 occurrences) 1- and 2-grams. Next, two of the authors manually associated the frequent 1- and 2-grams with an activity category. Only the 1- and 2-grams that are assigned identically by both authors are utilized for the BoW categorization. The sentences are then assigned to one or multiple activities based on the categorized n-grams. Since 8,032 sentences do not contain any of the n-grams, they cannot be categorized. We then train an Elman network (RNN) on the categorized sentences. The RNN classifies sentences that are not already assigned by the BoW categorization. Some sentences are not associated because these consist of words that do not appear in the training corpus. The results of the BoW categorization and the merged results of both approaches are then utilized as input for the second step.

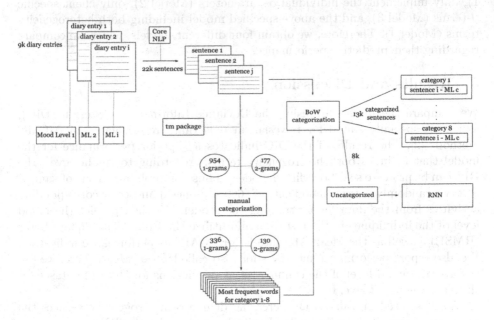

Fig. 1. The process of the two-step approach.

Because the mood level is ranked on a scale from one to ten, we use a partial ordered logit model for the prediction and the analysis of the effects between the assigned activity categories and the mood level. The ordered logit model is based on the proportional odds assumption (POA), which means that independent variables have the same effect on the outcome variable across all ranks of the mood level [7]. The partial ordered logit model, however, allows variables that violate this assumption to vary among the ranks. We test the assumption by a likelihood ratio test. The logit is then calculated as follows:

$$\ln(\theta_{ijt}) = \alpha_{ij} - \left(\underbrace{\sum_{a \in A_1} \beta_{aj}\, x_{ajt}}_{\text{POA holds}} + \underbrace{\sum_{a \in A_2} \beta_{aij}\, x_{ajt}}_{\text{POA violated}} \right),$$

where α_{ij} represents the threshold between the ranks of the mood level for $i = 1, \ldots, I = 9$ and $j = 1, \ldots, J = 440$. The activities of participant j at time t are represented by x_{ajt}, where $A_1 = \{\text{sleep related, recreational activities}\}$ and $A_2 = \{\text{exercise, sickness, rumination, social, work related, necessary activities}\}$. The parameters to be estimated are $\beta_{[\ldots]}$. The index j in α_{ij} addresses the problem of scale usage heterogeneity [6]. Additionally, we hypothesize that the effects of the activities vary among participants. Thus, we also include client specific β-parameters. For a robustness check, we also implement the partial ordered logit model without the consideration of heterogeneity among the participants (Model 1), only implement the individual α-parameters (Model 2), only client specific β-terms (Model 3), and the above specified model including both heterogeneity terms (Model 4). Therefore, we obtain four different models, which we compare regarding their predictive performance.

3   Results and Discussion

We compare the models by using the Deviance Information Criterion (DIC), which is especially suited for Bayesian models that are estimated by MCMC methods [2]. The results of the DIC indicates a superior performance for the model that includes both heterogeneity terms. According to [1], however, the DIC can be prone to select overfitted models. Thus, for applying an out-of-sample test, we randomly extract mood entries (680 sentences) and their corresponding activities from the data before training the model. We then predict the mood level of the individuals in the test data and utilize the Root Mean Square Error (RMSE) as well as the Mean Absolute Error (MAE) as performance indicators. We also report performance measures for a so called *Mean Model*; here, we use the average mood level of the training set as predictions for the test dataset (in this case the mood level 6).

As illustrated in Table 1, an increasing degree of heterogeneity reduces the prediction error. The additionally classified activities by the RNN do not contribute to an increased performance. This can potentially arise because the training data used for the RNN, which is based on the BoW categorization, might

Table 1. Model comparison with levels of heterogeneity for each text-mining approach.

Measure	Model 1	Model 1	Model 2	Model 2	Model 3	Model 3	Model 4	Model 4	Mean
	BoW	RNN	BoW	RNN	BoW	RNN	BoW	RNN	Model
RMSE	2.32	2.33	1.98	1.98	1.87	1.91	1.81	1.86	1.91
MAE	1.78	1.82	1.48	1.49	1.41	1.41	1.37	1.37	1.53

not be accurate enough for the RNN to generate new knowledge. Model 4 for the BoW categorization shows the best predictive performance. Thus, we utilize this model for revealing the relationships between the activities and the mood level.

We find that the category sickness has a strong negative and significant effect on mood. Furthermore, our analysis suggests that the category rumination affects the mood level in a negative way and social activities have a positive effect on the mood level. The other activities are not significant. These results are consistent with literature in the field [9]. During the ECML, we will additionally present the results of a model that directly predicts the mood levels based on the free text data.

References

1. Ando, T.: Bayesian predictive information criterion for the evaluation of hierarchical Bayesian and empirical Bayes models. Biometrika **94**(2), 443–458 (2007)
2. Berg, A., Meyer, R., Yu, J.: Deviance information criterion for comparing stochastic volatility models. J. Bus. Econ. Stat. **22**(1), 107–120 (2004)
3. Bremer, V., Becker, D., Funk, B., Lehr, D.: Predicting the individual mood level based on diary data. In: 25th European Conference on Information Systems, ECIS 2017, Guimarães, Portugal, 5–10 June 2017, p. 75 (2017)
4. Buntrock, C., et al.: Evaluating the efficacy and cost-effectiveness of web-based indicated prevention of major depression: design of a randomised controlled trial. BMC Psychiatry **14**, 25–34 (2014)
5. Elman, J.L.: Finding structure in time. Cogn. Sci. **14**(2), 179–211 (1990)
6. Johnson, T.R.: On the use of heterogeneous thresholds ordinal regression models to account for individual differences in response style. Psychometrika **68**(4), 563–583 (2003)
7. McCullagh, P.: Regression models for ordinal data. J. R. Stat. Soc. **42**(2), 109–142 (1980)
8. Saddichha, S., Al-Desouki, M., Lamia, A., Linden, I.A., Krausz, M.: Online interventions for depression and anxiety - a systematic review. Health Psychol. Behav. Med. **2**(1), 841–881 (2014)
9. Weinstein, S.M., Mermelstein, R.: Relations between daily activities and adolescent mood: the role of autonomy. J. Clin. Child Adolesc. Psychol. **36**(2), 182–194 (2007)

Best Practices to Train Deep Models on Imbalanced Datasets—A Case Study on Animal Detection in Aerial Imagery

Benjamin Kellenberger(✉) [iD], Diego Marcos [iD], and Devis Tuia [iD]

Wageningen University and Research, Wageningen, The Netherlands
{benjamin.kellenberger,diego.marcos,devis.tuia}@wur.nl

Abstract. We introduce recommendations to train a Convolutional Neural Network for grid-based detection on a dataset that has a substantial class imbalance. These include curriculum learning, hard negative mining, a special border class, and more. We evaluate the recommendations on the problem of animal detection in aerial images, where we obtain an increase in precision from 9% to 40% at high recalls, compared to state-of-the-art. Data related to this paper are available at: http://doi.org/10.5281/zenodo.609023.

Keywords: Deep learning · Class imbalance
Unmanned Aerial Vehicles

1 Introduction

Convolutional Neural Networks (CNNs) [5] have led to tremendous accuracy increases in vision tasks like classification [2] and detection [8,9], in part due to the availability of large-scale datasets like ImageNet [11]. Many vision benchmarks feature a controlled situation, with all classes occurring in more or less similar frequencies. However, in practice this isn't always the case. For example, in animal censuses on images from Unmanned Aerial Vehicles (UAVs) [6], the vast majority of images is empty. As a consequence, training a deep model on such datasets like in a classical balanced setting might lead to unusable results.

In this paper, we present a collection of recommendations that allow training deep CNNs on heavily imbalanced datasets (Sect. 2), demonstrated with the application of big mammal detection in UAV imagery. We assess the contribution of each recommendation in a hold-one-out fashion and further compare a CNN trained with all of them to the current state-of-the-art (Sect. 4), where we manage to increase the precision from 9% to 40% for high target recalls. The paper is based on [3].

Supported by the Swiss National Science Foundation (grant PZ00P2-136827).

U. Brefeld et al. (Eds.): ECML PKDD 2018, LNAI 11053, pp. 630–634, 2019.
https://doi.org/10.1007/978-3-030-10997-4_40

2 Proposed Training Practices

The following sections briefly address all the five recommendations that make training on an imbalanced dataset possible:

Curriculum Learning. For the first five training epochs, we sample the training images so that they always contain at least one animal. This is inspired by Curriculum Learning [1] and makes the CNN learn initial representations of *both* animals and background. This provides it with a better starting point for the imbalance problem later on.

Rotational Augmentation. Due to the overhead perspective, we employ 90°-stop image rotations as augmentation. However, we empirically found it to be most effective at a late training stage (from epoch 300 on), where the CNN is starting to converge to a stable solution.

Hard Negative Mining. After epoch 80 we expect the model to have roughly learned the animal and background appearances, and thus focus on reducing the number of false positives. To do so, we amplify the weights of the four most confidently predicted false alarms in every training image for the rest of the training schedule.

Border Class. Due to the CNN's receptive field capturing spatial context, we frequently observed activations in the vicinity of the animals, leading to false alarms. To remedy this effect, we label the 8-neighborhood around true animal locations with a third class (denoted as "border"). This way, the CNN learns to treat the surroundings of the animals separately, providing only high confidence for an animal in its true center. At test time, we simply discard the border class by merging it with the background.

Class Weighting. We balance the gradients during training with constant weights corresponding to the inverse class frequencies observed in the training set.

3 Experiments

3.1 The Kuzikus Dataset

We demonstrate our training recommendations on a dataset of UAV images over the Kuzikus game reserve, Namibia[1]. Kuzikus contains an estimated 3000 large mammals such as the Black Rhino, Zebras, Kudus and more, distributed over $103 \, \text{km}^2$ [10]. The dataset was acquired in May 2014 by the SAVMAP Consortium[2], using a SenseFly eBee[3] with a Canon PowerShot S110 RGB camera as payload. The campaign yielded a total of 654 4000 × 3000 images, covering $13.38 \, \text{km}^2$ with around 4 cm resolution. 1183 animals could be identified in a crowdsourcing campaign [7]. The data were then divided image-wise into 70% training, 10% validation and 20% test sets.

---

[1] http://kuzikus-namibia.de/xe_index.html.

[2] http://lasig.epfl.ch/savmap.

[3] https://www.sensefly.com.

3.2   Model Setup

We employ a CNN that accepts an input image of 512×512 pixels and yields a 32×32 grid of class probability scores. We base it on a pre-trained ResNet-18 [2] and replace the last layer with two new ones that map the 512 activations to 1024, then to the 3 classes, respectively. We add a ReLU and dropout [12] with probability 0.5 in between for further regularization. The model is trained using the Adam optimizer [4] with weight decay and a gradually decreasing learning rate for a total of 400 epochs.

We assess all recommendations in a hold-one-out fashion, and further compare them to a full model and the current state-of-the-art on the dataset, which employs a classifier on proposals and hand-crafted features (see [10] for details).

4   Results and Discussion

Figure 1 shows the precision-recall curves for all the models.

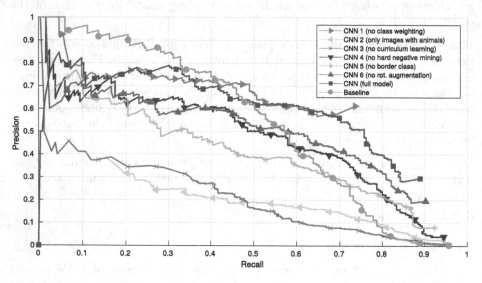

Fig. 1. Precision-recall curves based on the animal confidence scores for the hold-one-out CNNs (first six models), the full model and the baseline

All recommendations boost precision, but with varying strengths. For example, disabling curriculum learning ("CNN 3") yields the worst precision at high recalls—too many background samples from the start seem to severely drown any signal from the few animals. Unsurprisingly, a model trained on only images that contain at least one animal ("CNN 2") is similarly bad: this way, the model only sees a portion of the background samples and yields too many false alarms. The full model provides the highest precision scores of up to 40% at high recalls of 80% and more. At this stage, the baseline reaches less than 10% precision, predicting false alarms virtually everywhere. In numbers, this means that for 80% recall our model predicts 447 false positives, while the baseline produces 2546 false alarms.

5 Conclusion

Many real-world computer vision problems are characterized by significant class imbalances, which in the worst case makes out-of-the-box applications of deep CNNs unfeasible. An example is the detection of large mammals in UAV images, out of which the majority is empty. In this paper, we presented a series of practices that enable training CNNs by limiting the risk of the background class drowning the few positives. We analyzed the contribution of each individual practice (curriculum learning, hard negative mining, etc.) and showed how a CNN, trained with all of them, yields a substantially higher precision if tuned for high recalls.

References

1. Bengio, Y., Louradour, J., Collobert, R., Weston, J.: Curriculum learning. In: Proceedings of the 26th Annual International Conference on Machine Learning, pp. 41–48. ACM, New York (2009)
2. He, K., Zhang, X., Ren, S., Sun, J.: Deep residual learning for image recognition. In: Proceedings of the IEEE conference on Computer Vision and Pattern Recognition, pp. 770–778 (2016)
3. Kellenberger, B., Marcos, D., Tuia, D.: Detecting mammals in UAV images: best practices to address a substantially imbalanced dataset with deep learning. Remote Sensing of Environment (in revision)
4. Kingma, D., Ba, J.: Adam: a method for stochastic optimization. arXiv preprint arXiv:1412.6980 (2014)
5. Krizhevsky, A., Sutskever, I., Hinton, G.E.: ImageNet classification with deep convolutional neural networks. In: Advances in Neural Information Processing Systems, pp. 1097–1105 (2012)
6. Linchant, J., Lisein, J., Semeki, J., Lejeune, P., Vermeulen, C.: Are unmanned aircraft systems (UASs) the future of wildlife monitoring? A review of accomplishments and challenges. Mammal Rev. 45(4), 239–252 (2015)
7. Ofli, F., et al.: Combining human computing and machine learning to make sense of big (aerial) data for disaster response. Big Data 4(1), 47–59 (2016)
8. Redmon, J., Divvala, S., Girshick, R., Farhadi, A.: You only look once: unified, real-time object detection. In: The IEEE Conference on Computer Vision and Pattern Recognition, June 2016
9. Ren, S., He, K., Girshick, R., Sun, J.: Faster R-CNN: towards real-time object detection with region proposal networks. In: Advances in Neural Information Processing Systems, pp. 91–99 (2015)
10. Rey, N., Volpi, M., Joost, S., Tuia, D.: Detecting animals in African Savanna with UAVs and the crowds. Remote Sens. Environ. 200, 341–351 (2017)

11. Russakovsky, O., et al.: Imagenet large scale visual recognition challenge. Int. J. Comput. Vis. **115**(3), 211–252 (2015)
12. Srivastava, N., Hinton, G.E., Krizhevsky, A., Sutskever, I., Salakhutdinov, R.: Dropout: a simple way to prevent neural networks from overfitting. J. Mach. Learn. Res. **15**(1), 1929–1958 (2014)

Deep Query Ranking for Question Answering over Knowledge Bases

Hamid Zafar[1(✉)], Giulio Napolitano[2], and Jens Lehmann[1,2]

[1] Computer Science Institute, University of Bonn, Bonn, Germany
{hzafarta,jens.lehmann}@cs.uni-bonn.de
[2] Fraunhofer IAIS, Sankt Augustin, Germany
{giulio.napolitano,jens.lehmann}@iais.fraunhofer.de

Abstract. We study question answering systems over knowledge graphs which map an input natural language question into candidate formal queries. Often, a ranking mechanism is used to discern the queries with higher similarity to the given question. Considering the intrinsic complexity of the natural language, finding the most accurate formal counter-part is a challenging task. In our recent paper [1], we leveraged Tree-LSTM to exploit the syntactical structure of input question as well as the candidate formal queries to compute the similarities. An empirical study shows that taking the structural information of the input question and candidate query into account enhances the performance, when compared to the baseline system. Code related to this paper is available at: https://github.com/AskNowQA/SQG.

1 Introduction

Question answering (QA) systems provide a convenient interface to enable their users to communicate with the system through natural language questions. QA systems can be seen as advanced information retrieval systems, where (a) users are assumed to have no knowledge of the query language or structure of the underlying information system; (b) the QA system provides a concise answer, as opposed to search engines where users would be presented with a list of related documents. There are three types of source of information being consumed by QA systems, namely unstructured resources (e.g. Wikipedia pages), structured resources and hybrid sources. Given the extensive progress being made in large scale Knowledge Graphs (KGs), we mainly focus on QA systems using KGs as their source of information, since such systems might be able to yield more precise answers than those using a unstructured sources of information.

Given the complexity of the QA over KGs, there is a proclivity to design QA systems by breaking them into various sequential subtasks such as Named Entity Disambiguation (NED), Relation Extraction (RE) and Query Building (QB) among others [2]. Considering the fact that the system might end up with more than one candidate queries due to uncertainty in the linked entities/relations, ambiguity of the input question or complexity of the KGs, a ranking mechanism

© Springer Nature Switzerland AG 2019
U. Brefeld et al. (Eds.): ECML PKDD 2018, LNAI 11053, pp. 635–638, 2019.
https://doi.org/10.1007/978-3-030-10997-4_41

in the final stage of the QA system is required to sort the candidate queries based on their semantic similarity in respect to the given natural language question. Although considerable research has been devoted to QA over KG, rather less attention has been paid to query ranking subtask.

2 Related Work

Bast el al. [3] were inspired by the learning-to-rank approach from the information retrieval community to rank candidate queries using their feature vector, which contains 23 manually crafted features such as *number of entities in the query candidate*. They considered the ranking problem as a preference learning problem where a classifier (e.g. logistic regression) is supposed to pick the better option out of two given options. In a similar line of work, Abujabal et al. [4] hand-picked 16 features and utilized a random forest classifier to learn the preference model. Identifying the feature set requires manual intervention and depends heavily on the dataset at hand. In order to avoid that, Bordes [5] proposed an embedding model, which learns a fixed-size embedding vector representation of the input question and the candidate queries such that a score function produces a high score when the matching question and query are given. Inspired by the success of [5], Yih et al. [6] used deep convolutional neural networks to learn the embeddings and compute semantic similarity of the generated chains of entity/relation with respect to the given question. Despite their advantage to avoid using any manually engineered features, the models introduced by [5,6] failed to exploit the syntactical structure of the input question or the candidate queries. In the next section, we propose to use Tree-LSTM [7] in order to take advantage of the latent information in the structure of question and the candidate queries.

3 Deep Query Ranking

Consider the example question "What are some artists on the show whose opening theme is Send It On?" from [1], the candidate queries of an arbitrary QA pipeline are illustrated in Fig. 3. The candidate queries are similar to each other in the sense that they are made up of a set of entities and relations, which are shared among them. Motivated by the success of embedding models [5,6], we aim to enhance them by considering the structure of input question and candidate queries as well. In this regard, Tai et al. [7] proposed a Tree-LSTM model, which considers the tree representation of the input, as opposed to most RNN based models (e.g. LSTM) which take a sequence of tokens as input. The state of a Tree-LSTM unit depends on the children units (Fig. 2), enabling the model to consume the tree-structure of the input. Consequently, not only the input sequence matters but also how the elements of the input are connected together.

In order to learn the embedding vector we used a similarity function [8] along with two Tree-LSTM models for the input question and the candidate queries. The input to the *Question Tree-LSTM* is the dependency parse tree of the question (Fig. 1), whilst the tree-representation of the candidate queries

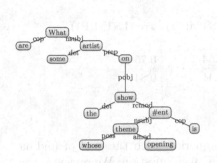

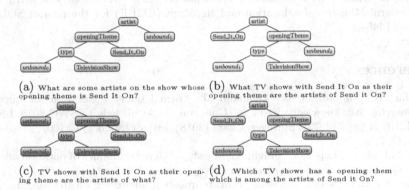

Fig. 1. Dependency parse tree of the running example (from [1])

Fig. 2. The architecture of Tree-LSTM

is fed into the *Query Tree-LSTM* (Fig. 3). The Tree-LSTM models are trained to map their input into a latent vectorized representation such that the pair of question/correct query would have the highest score in respect to the others.

(a) What are some artists on the show whose opening theme is Send It On?

(b) What TV shows with Send It On as their opening theme are the artists of Send it On?

(c) TV shows with Send It On as their opening theme are the artists of what?

(d) Which TV shows has a opening them which is among the artists of Send it On?

Fig. 3. Tree representation of the queries along with their NL meaning (from [1])

4 Empirical Study

We prepared two datasets for the ranking model based on LC-QuAD dataset [9] which consists of 5,000 question-answer pairs. Both datasets consist of questions and candidate queries. The first dataset, *DS-Min* is constructed using only the correct entities/relations, while *DS-Noise* is generated using the correct entities/relations plus four noisy ones per each linked item in the question.

The performance of the Tree-LSTM ranking model is reported in Table 1. The Tree-LSTM outperforms vanilla LSTM in both datasets. While Tree-LSTM performs better in *DS-Noise* in comparison to *DS-Min*, LSTM model degrades in *DS-Noise*. Although there are more training data in *DS-Noise* with balanced distribution of correct/incorrect data items, LSTM is not able to benefit from the information laying in the structure of its input, in contrast to Tree-LSTM.

Table 1. The accuracy of Tree-LSTM vs. LSTM (from [1])

Dataset	Size	Distribution(%) correct/incorrect	LSTM(F1)	Tree-LSTM(F1)
DS-Min	5,930	0.85/0.15	0.54	0.75
DS-Noise	11,257	0.46/0.54	0.41	0.84

5 Conclusions

We presented the problem of ranking formal queries, with the goal of finding the query that truly captures the intention of a given question. We reviewed the recent attempts to the problem and introduced our findings on using Tree-LSTM from our recent paper [1]. The model learns an embedding vector which captures the dependency parsing structure of the question and tree-representation of the queries to compute the similarity of the pairs for improved ranking.

Acknowledgments. This research was supported by EU H2020 grants for the projects HOBBIT (GA no. 688227), WDAqua (GA no. 642795) as well as by German Federal Ministry of Education and Research (BMBF) for the project SOLIDE (no. 13N14456).

References

1. Zafar, H., Napolitano, G., Lehmann, J.: Formal query generation for question answering over knowledge bases. In: Gangemi, A., et al. (eds.) ESWC 2018. LNCS, vol. 10843, pp. 714–728. Springer, Cham (2018). https://doi.org/10.1007/978-3-319-93417-4_46
2. Diefenbach, D., Lopez, V., Singh, K., Maret, P.: Core techniques of question answering systems over knowledge bases: a survey. Knowl. Inf. Syst., pp. 1–41 (2017)
3. Bast, H., Haussmann, E.: More accurate question answering on freebase. In: Proceedings of the 24th ACM International on Conference on Information and Knowledge Management, pp. 1431–1440. ACM (2015)
4. Abujabal, A., Yahya, M., Riedewald, M., Weikum, G.: Automated template generation for question answering over knowledge graphs. In: Proceedings of the 26th International Conference on World Wide Web, pp. 1191–1200 (2017)
5. Bordes, A., Chopra, S., Weston, J.: Question answering with subgraph embeddings. arXiv preprint arXiv:1406.3676 (2014)
6. Yih, S.W.-T., Chang, M.-W., He, X., Gao, J.: Semantic parsing via staged query graph generation: question answering with knowledge base. In: Proceedings of the Joint Conference of ACL and AFNLP (2015)
7. Tai, K.S., Socher, R., Manning, C.D.: Improved semantic representations from tree-structured long short-term memory networks. In: ACL (2015)
8. Yih, W.-T., Richardson, M., Meek, C., Chang, M.-W.: The value of semantic parse labeling for knowledge base question answering. In: 54th Annual Meeting of the Association for Computational Linguistics, pp. 201–206 (2016)
9. Trivedi, P., Maheshwari, G., Dubey, M., Lehmann, J.: LC-QuAD: a corpus for complex question answering over knowledge graphs. In: d'Amato, C., et al. (eds.) ISWC 2017. LNCS, vol. 10588, pp. 210–218. Springer, Cham (2017). https://doi.org/10.1007/978-3-319-68204-4_22

Machine Learning Approaches to Hybrid Music Recommender Systems

Andreu Vall[1(✉)] and Gerhard Widmer[1,2]

[1] Institute of Computational Perception, Johannes Kepler University, Linz, Austria
{andreu.vall,gerhard.widmer}@jku.at
[2] Austrian Research Institute for Artificial Intelligence, Vienna, Austria

Abstract. Music recommender systems have become a key technology supporting the access to increasingly larger music catalogs in on-line music streaming services, on-line music shops, and private collections. The interaction of users with large music catalogs is a complex phenomenon researched from different disciplines. We survey our works investigating the machine learning and data mining aspects of hybrid music recommender systems (i.e., systems that integrate different recommendation techniques). We proposed hybrid music recommender systems robust to the so-called "cold-start problem" for new music items, favoring the discovery of relevant but non-popular music. We thoroughly studied the specific task of music playlist continuation, by analyzing fundamental playlist characteristics, song feature representations, and the relationship between playlists and the songs therein.

Keywords: Music recommender systems
Music playlist continuation · Hybrid recommender systems
Cold-start problem

1 Introduction

Music recommender systems support the interaction of users with large music catalogs. They strongly rely on machine learning and data mining methods to analyze the data describing the users, the music items, the interaction of users with music items, and even the users' context when interacting with the music items [8, Chap. 7]. Even though the techniques at the core of recommender systems are valid for different item domains (e.g., recommending movies, news, or jobs), the inherent characteristics of the music domain must be considered:

- Music can be recommended at different granularity levels (e.g., songs, albums, artists, or even genres or ready-made playlists).
- Music is often consumed in listening sessions, such as albums or playlists. This defines local session contexts that need to be taken into consideration.
- Music recommender systems should adapt their level of interference to the user needs, ranging from simply supporting the exploration of the music catalog, to providing a full *lean-back* experience.

© Springer Nature Switzerland AG 2019
U. Brefeld et al. (Eds.): ECML PKDD 2018, LNAI 11053, pp. 639–642, 2019.
https://doi.org/10.1007/978-3-030-10997-4_42

Music recommender systems have received contributions from different but converging approaches. Research specializing in music information retrieval has often focused on content-based music recommender systems. Music items are represented by features derived from the audio signal, social tags, or web content and recommendations are predicted on the basis of content-wise similarities [5, 7]. On the other hand, research specializing in general recommender systems has usually focused on collaborative filtering techniques, developing statistical models to extract underlying music taste patterns from usage data (e.g., listening logs, or radio stations) [1, 3].

Content-based recommender systems provide fair but limited performance, because the relations derived from content-wise similarities tend to be simple. Collaborative recommender systems capture more abstract music taste patterns, but their performance is heavily affected by the availability of data (users and music items for which few observations are available are poorly represented).

In this paper we survey our recent works on hybrid music recommender systems integrating the strengths of content-based and collaborative recommender systems. We approached two main music recommendation tasks: (1) music artist recommendation, modeling the users' general music preferences over an extended period of time, and (2) music playlist continuation, focusing on next-song recommendations for short listening sessions.

2   Music Artist Recommendation

We proposed in [15] a hybrid extension to a well-established matrix factorization CF approach for implicit feedback datasets [6]. The proposed hybrid CF extension jointly factorizes listening logs and social tags from music streaming services (abundant sources of usage data in the music domain). According to our numerical experiments, the proposed hybrid CF extension yielded more accurate music artist recommendations. Furthermore, we extended the standard evaluation methodology incorporating bootstrap confidence intervals [4] to facilitate the comparison between systems. In the follow-up work [9], we observed that the superior performance of the proposed hybrid CF extension was explained by its robustness to the cold-start problem for new artists (i.e., its ability to better represent music artists for which few observations were available).

3   Music Playlist Continuation

Like previous works on music playlist modeling, we based our research on the exploitation of hand-curated music playlists, which we regard as rich examples from which to learn music compilation patterns.

3.1   Playlist Characteristics and Next-Song Recommendations

Before moving into the design of hybrid music recommender systems for music playlist continuation, we studied which basic playlist characteristics should be considered to effectively predict next-song recommendations. We studied in [13,14] the importance of the song order, the song context, and the song popularity for next-song recommendations. We compared three existing playlist continuation models of increasing complexity on two datasets of hand-curated music playlists. We observed that considering a longer song context has a positive impact on next-song recommendations. We found that the long-tailed nature of the playlist datasets (common in music collections [2]) makes simple models and highly-expressive models appear to perform comparably. However, further analysis revealed the advantage of using highly-expressive models. Our experiments also suggested either that the song order is not crucial for next-song recommendations, or that even highly-expressive models are unable to exploit it. We also proposed an evaluation approach for next-song recommendations that mitigates known issues in the evaluation of music recommender systems.

3.2 Song Representations and Playlist-Song Membership

Given the results of the experiments described in [13,14], we proposed music recommender systems for playlist continuation able to consider the full playlist song context. We also required the proposed systems to hybridize collaborative and content-based recommender systems to ensure robustness to the cold-start problem.

We identified in [11,12] suitable song-level feature representations for music playlist modeling. We investigated features derived from the audio signal, social tags, and independent listening logs. We found that the features derived from independent listening logs are more expressive than those derived from social tags, which in turn outperform those derived from the audio signal. The combination of features from different modalities outperformed the individual features, suggesting that the different modalities indeed carry complementary information.

We further proposed in [12] a hybrid music recommender system for music playlist continuation robust to the cold-start problem for non-popular songs. However, this approach can only extend playlists for which a profile has been pre-computed at training time. In the follow-up work [10], we proposed another hybrid music recommender system for music playlist continuation that regards playlist-song pairs exclusively in terms of feature vectors. This system learns general "playlist-song" membership relationships, which not only make it robust to the cold-start problem for non-popular songs but also enable the extension of playlists not seen at training time.

4 Lessons Learned and Open Challenges

Additional insights and questions arise from the research conducted in the presented works. Importantly, the evaluation of music recommender systems by means of numerical experiments does not reflect the fact that the usefulness of music recommendations is a highly subjective judgment of the end user. Further research on evaluation metrics for music recommender systems is required.

On a related note, the users' subjectivity on the usefulness of music recommendations makes it challenging to anticipate the actual immediate user needs. An interesting research direction to bridge this gap focuses on the development of interactive interfaces that should let users express their current needs.

Acknowledgments. This research has received funding from the European Research Council (ERC) under the European Union's Horizon 2020 research and innovation programme under grant agreement No 670035 (Con Espressione).

References

1. Aizenberg, N., Koren, Y., Somekh, O.: Build your own music recommender by modeling internet radio streams. In: Proceedings of WWW, pp. 1–10 (2012)
2. Celma, Ó.: Music Recommendation and Discovery. Springer, Heidelberg (2010). https://doi.org/10.1007/978-3-642-13287-2
3. Chen, S., Moore, J.L., Turnbull, D., Joachims, T.: Playlist prediction via metric embedding. In: Proceedings of SIGKDD, pp. 714–722 (2012)
4. DiCiccio, T.J., Efron, B.: Bootstrap confidence intervals. Stat. Sci. **11**, 189–212 (1996)
5. Flexer, A., Schnitzer, D., Gasser, M., Widmer, G.: Playlist generation using start and end songs. In: Proceedings of ISMIR, pp. 173–178 (2008)
6. Hu, Y., Koren, Y., Volinsky, C.: Collaborative filtering for implicit feedback datasets. In: Proceedings of ICDM, pp. 263–272 (2008)
7. Knees, P., Pohle, T., Schedl, M., Widmer, G.: Combining audio-based similarity with web-based data to accelerate automatic music playlist generation. In: Proceedings of the International Workshop on Multimedia Information Retrieval, pp. 147–154 (2006)
8. Ricci, F., Rokach, L., Shapira, B.: Recommender Systems Handbook, 2nd edn. Springer, Boston (2015). https://doi.org/10.1007/978-1-4899-7637-6
9. Vall, A.: Listener-inspired automated music playlist generation. In: Proceedings of RecSys, Vienna, Austria (2015)
10. Vall, A., Dorfer, M., Schedl, M., Widmer, G.: A hybrid approach to music playlist continuation based on playlist-song membership. In: Proceedings of SAC, Pau, France (2018)
11. Vall, A., Eghbal-zadeh, H., Dorfer, M., Schedl, M.: Timbral and semantic features for music playlists. In: Machine Learning for Music Discovery Workshop at ICML, New York, NY, USA (2016)
12. Vall, A., Eghbal-zadeh, H., Dorfer, M., Schedl, M., Widmer, G.: Music playlist continuation by learning from hand-curated examples and song features: Alleviating the cold-start problem for rare and out-of-set songs. In: Proceedings of the Workshop on Deep Learning for Recommender Systems at RecSys, Como, Italy (2017)
13. Vall, A., Quadrana, M., Schedl, M., Widmer, G.: The importance of song context and song order in automated music playlist generation. In: Proceedings of ICMPC-ESCOM, Graz, Austria (2018)
14. Vall, A., Schedl, M., Widmer, G., Quadrana, M., Cremonesi, P.: The importance of song context in music playlists. In: RecSys Poster Proceedings, Como, Italy (2017)
15. Vall, A., Skowron, M., Knees, P., Schedl, M.: Improving music recommendations with a weighted factorization of the tagging activity. In: Proceedings of ISMIR, Málaga, Spain (2015)

Demo Track

IDEA: An Interactive Dialogue Translation Demo System Using Furhat Robots

Jinhua Du[✉], Darragh Blake, Longyue Wang, Clare Conran,
Declan Mckibben, and Andy Way

ADAPT Centre, Dublin City University, Dublin, Ireland
{jinhua.du,darragh.blake,longyue.wang,clare.conran,
declan.mckibben,andy.way}@adaptcentre.ie

Abstract. We showcase **IDEA**, an *I*nteractive *D*ialogu*E* tr*A*nslation system using Furhat robots, whose novel contributions are: (i) it is a web service-based application combining translation service, speech recognition service and speech synthesis service; (ii) it is a task-oriented hybrid machine translation system combining statistical and neural machine learning methods for domain-specific named entity (NE) recognition and translation; and (iii) it provides user-friendly interactive interface using Furhat robot with speech input, output, head movement and facial emotions. IDEA is a case-study demo which can efficiently and accurately assist customers and agents in different languages to reach an agreement in a dialogue for the hotel booking.

Keywords: Dialogue translation · Furhat robot · Entity translation

1   Introduction

Applications of machine translation (MT) in many human–human communication scenarios still present many challenges, e.g accurate task-specific entity recognition and translation, ungrammatical word orders in spoken languages etc. Human–human dialogue translation is a more demanding translation task than the general-purpose translation tasks in terms of the recognition and translation of key information in the dialogue, such as the *person (who)*, *location (where)*, *time/date (when)*, and *event (what)* etc. An example of the low accuracy of entity recognition and translation in dialogues between customers and hotel agents from public translation systems is shown below:

Source:	我想定一个{30号}的(大床房)。[五点]可以到酒店。
Reference:	I would like to book a (king room) on {the 30th this month}. I will check in at [five o'clock].
Google:	I want to set a {30th} (bed room). [Five o'clock] can go to the hotel.
Baidu:	I'd like to make a (big bed room) for (No. 30). You can get to the hotel at [five].

© Springer Nature Switzerland AG 2019
U. Brefeld et al. (Eds.): ECML PKDD 2018, LNAI 11053, pp. 645–648, 2019.
https://doi.org/10.1007/978-3-030-10997-4_43

In this example, we use {}, () and [] to highlight the time/date and room type entities and their corresponding translations; use _ to highlight the verb of the event. We can find from this commonly-used dialogue sentence that check-in date, room type and the verb were not translated accurately either by *Google* or *Baidu*. Wrong translations of these key information will impede an effective and efficient communication between the customer and agent.

Accordingly, we carried out a study on task-oriented dialogue machine translation (DMT) with semantics-enhanced NE recognition and translation. As a case study, we developed an interactive DMT demo system for the hotel booking task: **IDEA**, which can assist customers in one language to communicate with hotel agents in another language to reach an agreement in a dialogue. A text input/output-based demo system of IDEAdescribing the working mechanism can be found at https://www.youtube.com/watch?v=5KK6OgMPDpw&t=5s.[1]

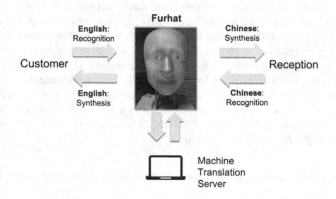

Fig. 1. System architecture of IDEA

2 System Description

2.1 System Architecture and Workflow

The architecture of IDEA is illustrated in Fig. 1. In the hotel booking scenario, Furhat robots provide the speech recognition and speech synthesis services between customers and hotel agents. The DMT Server provides a Web Service for the translation of messages between the customer and the agent through Furhat robots. A key component in our demo system is a Semantic Module which combines statistical and neural machine learning methods for the understanding, extraction and candidate translation of key entities/information in the ongoing dialogue, such as "customer name", "arrival time", "room type" and so on.

---

[1] News regarding the Furhat-based DMT demo system can be found at: https://twitter.com/adaptcentre/status/932957814301044737.

The text input/output-based interface of IDEA is shown in Fig. 2. In the hotel booking scenario, customers and agents speak different languages.[2] Customers can access the hotel website to request a conversation with an agent. Then the agent accepts the customer's request to start the conversion. Messages between the customer and agent will be automatically translated to the customer's or agent's language, and the semantic information (key entities) is automatically recognised, extracted and translated to achieve the intention of the hotel booking.

Agent (Chinese)   Customer (English)

Fig. 2. Interface of IDEA

Figure 3 shows the detailed workflow of IDEA which demonstrates how each module in the system works. In general, customers and agents alternately input texts or speak like a human–human question–answering scenario. We propose a task-oriented semantics-enhanced statistical and neural hybrid method to recognise entities by inferring their specific types based on information such as contexts, speakers etc. Then, the recognised entities will be represented as logical expressions or semantic templates using the grounded semantics module. Thus, the entities can be correctly translated in the contexts. Finally, candidate translations of semantically represented entities will be marked up and fed into a unified bi-directional translation process. Refer to [1] for technical details.

3   Evaluation and Application

We evaluate IDEA on different aspects and results show that: (1) the hybrid NE recognition and translation in the dialogue improves over 20% compared to general-purpose systems regarding the hotel booking scenario; (2) the improved NE translation improves the overall translation performance by over absolute 15% BLEU points on English–Chinese hotel booking scenario; (3) the success rate of booking is improved by over 30% compared to public translation systems.[3]

---

[2] In order to be more understandable, the rest of the paper will assume that customers speak English and agents speak Chinese.

[3] If all pre-defined key information such as *room type, check-in date, room price* etc. can be recognised and translated correctly, then we regard it as a successful dialogue.

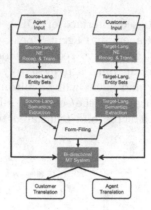

Fig. 3. Workflow of IDEA

The advantages of our task-oriented DMT system include: (1) focusing on the recognition and translation of key information in the dialogue to alleviate misunderstanding; (2) an extra task-oriented semantic module reduces the reliance on large-scale data, and easy to deploy on portable devices; (3) the head movements and facial expressions motivated by speaker's mood and manner of speaking etc. can increase the engagement; (4) the mode of using Furhat robots for interaction (head movements, facial expressions etc.) can be alternatively implemented as an avatar in mobile applications; (5) the techniques proposed and developed in the hotel booking scenario can be quickly adapted and applied to other task-oriented dialogue translations, such as customer services in finance, telecom, retail etc.

4   Conclusion and Future Work

In this paper, we demonstrated IDEA, a task-oriented dialogue translation system using Furhat robots for a hotel booking scenario. Evaluations on different aspects show that our DMT system can significantly improve translation performance and success rate of the task, and can be easily to extended to different task-oriented translation services.

Acknowledgement. We would like to thank the reviewers for their valuable and constructive comments. This research is supported by the ADAPT Centre for Digital Content Technology, funded under the SFI Research Centres Programme (Grant 13/RC/2106), and by SFI Industry Fellowship Programme 2016 (Grant 16/IFB/4490).

Reference

1. Wang, L., Du, J., Li, L., Tu, Z., Way, A., Liu Q.: Semantics-enhanced task-oriented dialogue translation: a case study on hotel booking. In: The Companion Volume of the IJCNLP 2017 Proceedings: System Demonstrations, Taipei, Taiwan, 27 November–1 December , pp. 33–36 (2017)

RAPID: Real-time Analytics Platform for Interactive Data Mining

Kwan Hui Lim[1,3]([✉]), Sachini Jayasekara[1], Shanika Karunasekera[1],
Aaron Harwood[1], Lucia Falzon[2], John Dunn[2], and Glenn Burgess[2]

[1] The University of Melbourne, Parkville, Australia
{kwan.lim,karus,aharwood}@unimelb.edu.au,
w.jayasekara@student.unimelb.edu.au
[2] Defence Science and Technology, Edinburgh, Australia
{lucia.falzon,john.dunn,glenn.burgess}@dst.defence.gov.au
[3] Singapore University of Technology and Design, Singapore, Singapore

Abstract. Twitter is a popular social networking site that generates a large volume and variety of tweets, thus a key challenge is to filter and track relevant tweets and identify the main topics discussed in real-time. For this purpose, we developed the Real-time Analytics Platform for Interactive Data mining (RAPID) system, which provides an effective data collection mechanism through query expansion, numerous analysis and visualization capabilities for understanding user interactions, tweeting behaviours, discussion topics, and other social patterns. Code related to this paper is available at: https://youtu.be/1APLeLT_t8w.

Keywords: Twitter · Social networks · Real-time · Topic tracking

1 Introduction

Social networking sites, such as Twitter, have become a prevalent communication platform in our daily life, with discussions ranging from mainstream topics like TV and music to specialized topics like politics and climate change. Tracking and understanding these discussions provide valuable insights into the general opinions and sentiments towards specific topics and how they change over time, which are useful to researchers, companies, government organizations alike, e.g., advertising, marketing, crisis detection, disaster management. Despite its usefulness, the large volume and

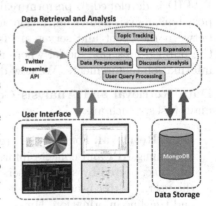

Fig. 1. Overview of RAPID

wide variety of tweets makes it challenging to track and understand the discussions on these topics [2,5]. To address these challenges, we proposed and developed the **R**eal-time **A**nalytics **P**latform for **I**nteractive **D**ata mining (RAPID)

© Springer Nature Switzerland AG 2019
U. Brefeld et al. (Eds.): ECML PKDD 2018, LNAI 11053, pp. 649–653, 2019.
https://doi.org/10.1007/978-3-030-10997-4_44

for topic tracking and analysis on Twitter (Fig. 1). RAPID offers a unique topic-tracking capability using query keyword and user expansion to track topics and related discussions, as well as various analytics capabilities to visualize the collected tweets, users and topics, and understand tweeting and interaction behaviours.

Related Systems and Differences. There has been a number of interesting Twitter-based systems developed for specific application domains such as politics [10], crime and disasters [4], diseases [3], recommendations [11], and they typically utilize a mention/keyword-based retrieval of tweets relating to each domain. Others focus on specific capabilities on Twitter such as a SQL-like query language [6], clustering tweets into broad topics [8], detecting events based on keyword frequency [7]. While these systems provide many interesting capabilities, our RAPID system differs in the following ways: (i) Instead of targetting specific domains, RAPID is designed to be generalizable to any application domain, topic or event; (ii) Many earlier systems retrieve tweets based on user-provided keywords, which may not adequately represent the topic of interest. In contrast, RAPID provides a unique query expansion collection capability that allows for the expansion of seeding keywords and users for a broader collection coverage; (iii) In addition, RAPID allows its users to interact with and control the data stream in real-time, as well as perform a wide and in-depth range of analysis and visualizations techniques, which we further describe in this paper; and (iv) RAPID is highly scalable to the growing volume of tweets generated, by utilizing real-time distributed computing technologies like Apache Storm and Kafka, compared to earlier systems that do not utilize such technologies.

2   System Architecture

RAPID is developed to perform real-time analysis and visualization, as well as post-hoc analysis and visualization on previously collected data. Communication between the client and server are facilitated through Kafka queues, based on the publish-subscribe model where researchers are able to specify their various information requirements. We now describe the main components of RAPID.

Data Retrieval and Analysis Component. This component performs two main tasks, which are:

- **Data Retrieval.** For real-time retrieval, RAPID interfaces with the Twitter Streaming API and collects information such as tweets related to a particular topic, posted by specific users or are within a geo location subscribed by the user, Twitter user details such as the list of followers, profile information and timeline information. For post-hoc processing, RAPID retrieves information stored in the data storage unit based on the researcher's requests. The researcher is able to access all functionalities of the real-time retrieval and in addition, is able to further drill-down on the data by filtering the collected tweets based on specific topics, time periods, locations and set of hashtags.

Unlike many earlier systems, RAPID is designed with an integrated data retrieval and analysis capability such that the data retrieval is continuously expanded for better coverage based on real-time analysis of collected tweets, which we discuss next.

- **Data Analysis.** This includes the sub-tasks of: (i) tweet pre-processing, i.e., tokenizing, topic labelling, extraction of geo-location and other tweet features; (ii) topic tracking via keywords, usernames or bounding boxes, and an enhanced query expansion capability that automatically track topics and related discussions through dynamic expansion of keywords; (iii) user query processing, such as filtering and drilling down the collected data for further analysis based on topics, time periods and/or locations; and (iv) data statistics and analysis, such as updating data storage with latest collection statistics and performing advanced analytics like analyzing hashtags and inferring relationships between hashtags, analyzing word-to-word pairs and word clusters of tweets, tracking discussions through pro-actively fetching tweets replies related to discussions.

Data Storage. The data storage component uses MongoDB for storing meta-data as well as the processed tweets, which can be used later for further post-hoc processing and visualization. RAPID also allows users the freedom to decide the type of processed data that should be persisted in the storage. Meta-data stored in the database includes the details of the users, details of user activities such as commands given by users to the RAPID system and the topics users are subscribed to. In addition to the meta-data, tweets processed by the system, discussions occurred related to tweets can also be stored in the database. One major advantage of having this useful capability is that users can reprocess and visualize the tweets later if such requirement arises, e.g., further drill-down to filter and analyze crisis-related tweets posted on 20 Nov 2017 in Melbourne CBD.

User Interface Component. This component performs three main tasks, namely:

- **User Input.** For topic tracking, researchers can specify a set of keywords, users and/or geo-bounding boxes associated with the topic as the input. The interface also allows users to modify or delete existing tracked topics, with a detailed log of these activities.
- **Real-time Visualization.** Key information and statistics of the tracked topics are visualized using a set of predefined charts, which are updated in real-time as new tweets related to the topic are analyzed by the RAPID system. Screenshots and descriptions of selected charts are shown in Fig. 2.
- **Workbench.** The workbench allows users to visualize tweets that have been stored in the storage component for further analysis. For more flexibility in post-collection analysis, users are able to define a specific time period the tweets have occurred and then the workbench retrieves the related tweets and visualizes them using the same charts used for real-time visualization. Moreover, the workbench summarizes the key statistics of the retrieved tweets

including the number of tweets fetched, number of unique authors, unique hashtags, unique mentions and unique replies.

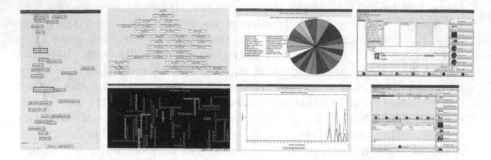

Fig. 2. Screenshots of selected RAPID functionalities, which include (clockwise from left): word-to-word network graph, discussion tree visualization, profile of tweeting locations, overview of RAPID workbench, workbench view on query expansion, real-time tracking of keyword frequency, word cloud of collected tweets.

3   Target Users and Demonstration

We presented the RAPID system for real-time topic tracking and analysis on Twitter, where RAPID offers a unique and effective collection approach via query expansion, numerous analysis capabilities to understand user interactions, tweeting behaviours and discussion cascades, and various visualizations of these types of information. RAPID has been used by researchers from both the Army Research Laboratory in the USA and Defence Science and Technology in Australia [1,9], and will also be of interest to any user interested in tracking, analysing and visualizing topics on Twitter. We will demonstrate the various capabilities of RAPID via use cases of political campaign analysis, monitoring of crises and incidents, in-depth analysis of tweets and users. A demonstration video of RAPID is available at https://youtu.be/1APLeLT_t8w.

Acknowledgments. This research is supported by Defence Science and Technology.

References

1. Falzon, L., McCurrie, C., Dunn, J.: Representation and analysis of Twitter activity: a dynamic network perspective. In: Proceedings of ASONAM 2017 (2017)
2. Kumar, S., Morstatter, F., Liu, H.: Twitter Data Analytics. Springer, New York (2013)
3. Lee, K., Agrawal, A., Choudhary, A.: Real-time disease surveillance using Twitter data: demonstration on flu and cancer. In: Proceedings of KDD 2013 (2013)

4. Li, R., Lei, K.H., Khadiwala, R., Chang, K.C.C.: TEDAS: a Twitter-based event detection and analysis system. In: Proceedings of ICDE 2012 (2012)
5. Liao, Y., et al.: Mining micro-blogs: opportunities and challenges. In: Abraham, A. (ed.) Computational Social Networks. Springer, London (2012). https://doi.org/10.1007/978-1-4471-4054-2_6
6. Marcus, A., Bernstein, M.S., Badar, O., Karger, D.R., Madden, S., Miller, R.C.: Tweets as data: demonstration of TweeQL and TwitInfo. In: SIGMOD 2011 (2011)
7. Mathioudakis, M., Koudas, N.: TwitterMonitor: trend detection over the Twitter stream. In: Proceedings of SIGMOD 2010, pp. 1155–1158 (2010)
8. O'Connor, B., Krieger, M., Ahn, D.: TweetMotif: exploratory search and topic summarization for Twitter. In: Proceedings of ICWSM 2010 (2010)
9. Vanni, M., Kase, S.E., Karunasekara, S., Falzon, L., Harwood, A.: RAPID: real-time analytics platform for interactive data-mining in a decision support scenario. In: Proceedings of SPIE, vol. 10207 (2017)
10. Wang, H., Can, D., Kazemzadeh, A., Bar, F., Narayanan, S.: A system for real-time twitter sentiment analysis of 2012 US presidential election cycle. In: Proceedings of ACL 2012, pp. 115–120 (2012)
11. Wang, J., Feng, Y., Naghizade, E., Rashidi, L., Lim, K.H., Lee, K.E.: Happiness is a choice: sentiment and activity-aware location recommendation. In: Proceedings of WWW 2018 Companion, pp. 1401–1405 (2018)

Interactive Time Series Clustering with COBRAS^TS

Toon Van Craenendonck^(✉), Wannes Meert, Sebastijan Dumančić,
and Hendrik Blockeel

Department of Computer Science, KU Leuven, Leuven, Belgium
{toon.vancraenendonck,wannes.meert,sebastijan.dumancic,
hendrik.blockeel}@kuleuven.be

Abstract. Time series are ubiquitous, resulting in substantial interest in time series data mining. Clustering is one of the most widely used techniques in this setting. Recent work has shown that time series clustering can benefit greatly from small amounts of supervision in the form of pairwise constraints. Such constraints can be obtained by asking the user to answer queries of the following type: *should these two instances be in the same cluster?* Answering "yes" results in a must-link constraint, "no" results in a cannot-link. In this paper we present an *interactive clustering system* that exploits such constraints. It is implemented on top of the recently introduced COBRAS^TS method. The system repeats the following steps until a satisfactory clustering is obtained: it presents several pairwise queries to the user through a visual interface, uses the resulting pairwise constraints to improve the clustering, and shows this new clustering to the user. Our system is readily available and comes with an easy-to-use interface, making it an effective tool for anyone interested in analyzing time series data. Code related to this paper is available at: https://bitbucket.org/toon_vc/cobras_ts/src.

1 Introduction

Clustering is one of the most popular techniques in data analysis, but also inherently subjective [3]: different users might prefer very different clusterings, depending on their goals and background knowledge. Semi-supervised methods deal with this by allowing the user to define constraints that express their subjective interests [4]. Often, these constraints are obtained by querying the user with questions of the following type: *Should these two instances be in the same cluster?* Answering "yes" results in a must-link constraint, "no" in a cannot-link.

In this paper we present an *interactive clustering system* that exploits such constraints. The system is based on COBRAS^TS [2], a recently proposed method for semi-supervised clustering of time series. COBRAS^TS is suitable for interactive clustering as it combines the following three characteristics: (1) it can

Video and code are available at https://dtai.cs.kuleuven.be/software/cobras/.

U. Brefeld et al. (Eds.): ECML PKDD 2018, LNAI 11053, pp. 654–657, 2019.
https://doi.org/10.1007/978-3-030-10997-4_45

present the best clustering obtained so far at *any time*, allowing the user to inspect intermediate results (2) it is *query-efficient*, which means that a good clustering is obtained with only a small number of queries (3) it is *time-efficient*, so the user does not have to wait long between queries. Given small amounts of supervision COBRAS^{TS} has been shown to produce clusterings of much better quality compared to those obtained with unsupervised alternatives [2].

By making our tool readily available and easy to use, we offer any practitioner interested in analyzing time series data the opportunity to exploit the benefits of interactive time series clustering.

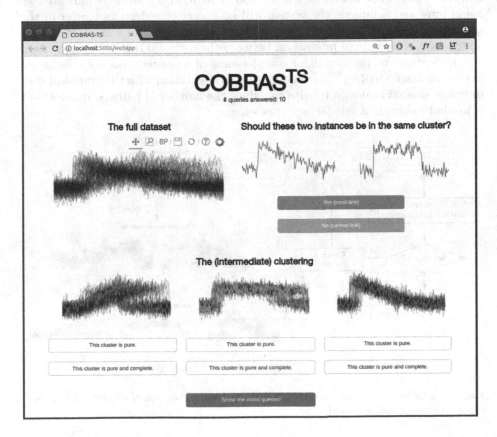

Fig. 1. Screenshot of the web application.

2   System Description

The graphical user interface of the system is shown in Fig. 1. On the top left the full dataset is shown, as a plot with all time series stacked on top of each other. On the top right, the system shows the querying interface. It presents the instances for which the pairwise relation is being queried, and two buttons that the user can click to indicate that these two instances should (not) be in the same cluster. On the bottom, the system shows the intermediate clustering. This clustering is updated after every couple of queries. The main loop that is executed is illustrated in Fig. 2(a): the system repeatedly queries several pairwise relations and uses the resulting constraints to improve the clustering, until the user is satisfied with the produced clustering.

Each time an updated clustering is presented, the user can optionally indicate that a cluster is either pure, or pure and complete. If a cluster is indicated as being pure, the system will no longer try to refine this cluster. It is still possible, however, that other instances will be added to it. If a cluster is indicated as being pure and complete, the system will no longer consider this cluster in the querying process: it will not be refined, and other instances can no longer be added to it. This form of interaction between the user and the clustering system (i.e. indicating the purity and/or completeness of a cluster) was not considered in the original COBRAS^{TS} method, but experimentation with the graphical user interface showed that that it helps to reduce the number of pairwise queries that is needed to obtain a satisfactory clustering.

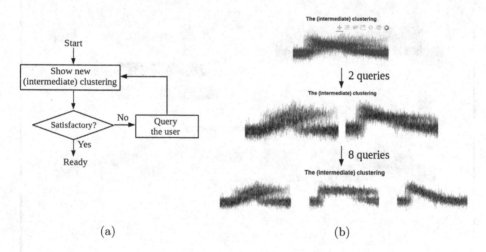

(a)   (b)

Fig. 2. (a) The interactive clustering loop. (b) Demonstration of clustering improvement as queries are answered.

The COBRAS^{TS} system is implemented as a web application that is run locally. It is open source and available online[1]. It is also available on PyPI, allowing installation with a single command[2].

---

[1] https://dtai.cs.kuleuven.be/software/cobras/.

[2] `pip install --find-links` https://dtai.cs.kuleuven.be/software/cobras/datashad er.html `pip cobras_ts[gui]`

3 Example Run

Figure 2(b) shows the sequence of clusterings that is generated by the application for a sample of the CBF dataset [1]. It starts from a single cluster that contains all instances. After the user has answered two pairwise queries, the system presents an updated clustering containing two clusters. The first cluster contains mainly upward clusters, whereas the second cluster contains a mixture of downward and horizontal patterns. As this clustering is not satisfactory yet, more pairwise queries are answered. After 8 more queries, the system again presents an improved clustering. This time, the clustering clearly separates three distinct patterns (upward, horizontal and downward). While distinguishing between these three types of patterns is easy for a user, it is difficult for most existing clustering systems; none of COBRAS^{TS}'s competitors is able to produce a clustering that clearly separates these patterns [2].

4 Conclusion

The proposed demo will present a readily available and easy-to-use web application for interactive time series clustering. Internally, it makes use of the recently developed COBRAS^{TS} approach. The application enables users to exploit minimal supervision to get clusterings that are significantly better than those obtained with traditional approaches.

Acknowledgements. Toon Van Craenendonck is supported by the Agency for Innovation by Science and Technology in Flanders (IWT). This research is supported by Research Fund KU Leuven (GOA/13/010), FWO (G079416N) and FWO-SBO (HYMOP-150033).

References

1. Chen, Y., et al.: The UCR time series classification archive, July 2015. http://www.cs.ucr.edu/eamonn/time_series_data/
2. Van Craenendonck, T., Meert, W., Dumancic, S., Blockeel, H.: COBRAS-TS: a new approach to semi-supervised clustering of time series. https://arxiv.org/abs/1805.00779, under submission, May 2018
3. von Luxburg, U., Williamson, R.C., Guyon, I.: Clustering: science or art? In: Workshop on Unsupervised Learning and Transfer Learning (2014)
4. Wagstaff, K., Cardie, C., Rogers, S., Schroedl, S.: Constrained K-means clustering with background knowledge. In: Proceedings of ICML (2001)

pysubgroup: Easy-to-Use Subgroup Discovery in Python

Florian Lemmerich[1]([⊠]) and Martin Becker[2]

[1] RWTH Aachen University, Aachen, Germany
florian.lemmerich@humtec.rwth-aachen.de
[2] University of Würzburg, Würzburg, Germany
becker@informatik.uni-wuerzburg.de

Abstract. This paper introduces the pysubgroup package for subgroup discovery in Python. Subgroup discovery is a well-established data mining task that aims at identifying describable subsets in the data that show an interesting distribution with respect to a certain target concept. The presented package provides an easy-to-use, compact and extensible implementation of state-of-the-art mining algorithms, interestingness measures, and visualizations. Since it builds directly on the established *pandas* data analysis library—a de-facto standard for data science in Python—it seamlessly integrates into preprocessing and exploratory data analysis steps. Code related to this paper is available at: http://florian.lemmerich.net/pysubgroup.

Subgroup discovery [1,5,7] is a data mining method that assumes a population of individuals and a property of these individuals a researcher is specifically interested in. The goal of subgroup discovery is then to discover the subgroups of the population that are statistically "most interesting" with respect to the distributional characteristics of the property of interest, cf. [12]. A typical subgroup discovery result could for example be stated as *"While only 50% of all students passed the exam, 90% of all female students younger than 21 passed."* Here, "female students younger than 21" describes a subgroup, the exam result is the property of interest specified by the user for this task, and the difference in the passing rate is the interesting distributional characteristic. Subgroup discovery identifies such groups in a large set of candidates. Subgroup discovery has been an active research area in our community for more than two decades in order to find more efficient algorithms, improved measures to identify potentially interesting groups, and interactive mining options. It has also been successfully used in many practical applications, see [5] for an overview.

State-of-the-art implementations of subgroup discovery are available in Java (VIKAMINE [2] and Cortana [10]) and R (rsubgroup[1] and SDEFSR[2]). In Python, however, there is only a basic implementation included in the Orange

---

[1] https://cran.r-project.org/web/packages/rsubgroup/rsubgroup.pdf.
[2] https://cran.r-project.org/web/packages/SDEFSR/vignettes/SDEFSRpackage.pdf.

© Springer Nature Switzerland AG 2019
U. Brefeld et al. (Eds.): ECML PKDD 2018, LNAI 11053, pp. 658–662, 2019.
https://doi.org/10.1007/978-3-030-10997-4_46

workbench.[3] A full featured subgroup discovery implementation that easily integrates with *numpy* and *pandas* libraries, which provide for one of the overall most popular setups for data analysis nowadays, is missing so far. The here presented package *pysubgroup* aims to fill this gap.

1   The pysubgroup Package

The *pysubgroup* package provides a novel implementation of subgroup discovery functions in Python based on the standard *numpy* and *pandas* data analysis libraries. As a design goal, it aims at a concise code base that allows easy access to state-of-the-art subgroup discovery for researchers and practitioners. In terms of algorithms it currently features depth-first-search, an apriori algorithm [6], best-first-search [13], the bsd algorithm [9], and beam search [3]. It includes numerous interestingness measures to score and select subgroups with binary and numeric targets, e.g., weighted relative accuracy, lift, χ^2 measures, (simplified) binomial measures, and extensions to generalization-aware interestingness measures [8]. It also contains specialized methods for post-processing and visualizing results.

Emphasizing usability, subgroup discovery can be performed in just a few lines of intuitive code. Since *pysubgroup* uses the standard pandas DataFrame class as its basic data structure, it is easy to integrate into interactive data exploration and pre-processing with pandas. By defining concise interfaces, *pysubgroup* is also easily extensible and allows for integrating new algorithms and interestingness measures. Based on the Python programming language, *pysubgroup* can be used under Windows, Linux, or macOS. It is 100% open source and available under a permissive Apache license.[4] The source code, documentation and an introductory video is available at http://florian.lemmerich.net/pysubgroup. The package can also be installed via PyPI using `pip install pysubgroup`.

Although *pysubgroup* is currently still in a prototype phase it has already been utilized in practical applications, e.g., for analyzing user motivations in Wikipedia through user surveys and server logs [11].

2   Application Example

Next, we present a basic application example featuring the well-known *titanic* dataset to demonstrate how easy it is to perform subgroup discovery with *pysubgroup*. In this particular example, we will identify subgroups in the data that had a significantly lower chance of survival in the Titanic disaster compared to the average passenger. The complete code required to execute a full subgroup discovery task is the following:

---

[3] http://kt.ijs.si/petra_kralj/SubgroupDiscovery/.
[4] Other licenses can be requested from the authors if necessary.

```
import pysubgroup as ps
import pandas as pd

data = pd.read_csv("../data/titanic.csv")
target = ps.NominalTarget ('survived', True)
searchspace = ps.createSelectors(data, ignore=['survived'])
task = ps.SubgroupDiscoveryTask (data, target, searchspace,
            resultSetSize=5, depth=2, qf=ps.ChiSquaredQF())
result = ps.BeamSearch().execute(task)
```

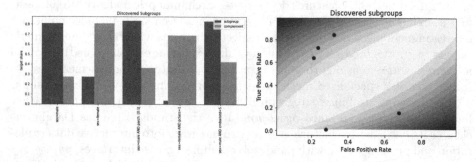

Fig. 1. Visualizations of result subgroups. In the bar visualization on the left, blue bars represent discovered subgroups, green bars their complement in the data. Bar heights indicate the ratio of instances with the property of interest, bar widths show the number of covered instances. On the right, the embedding of the result subgroups in ROC-space is shown. (Color figure online)

The first two lines import the *pandas* data analysis environment and the *pysubgroup* package. The following line loads the data into a standard *pandas* DataFrame object. The next three lines specify a subgroup discovery task. In particular, it defines a target, i.e., the property we are mainly interested in (*'survived'*), the set of basic selectors to build descriptions from (in this case: all), as well as the number of result subgroups returned, the depth of the search (maximum numbers of selectors combined in a subgroup description), and the interestingness measure for candidate scoring (here, the χ^2 measure). The last line executes the defined task by performing a search with an algorithm—in this case beam search. The result is then stored in a list of discovered subgroups associated with their score according to the chosen interestingness measure.

pysubgroup also offers utility functions to inspect and present results. In that direction, the result subgroups and their statistics can be transformed into a separate *pandas* DataFrame that can be resorted, spliced or filtered. Additionally, *pysubgroup* features a visualization component to generate specialized subgroup visualizations with one-line commands, e.g., to create bar visualizations (cf. Fig. 1a) or to show positions of subgroups in ROC-space [4], i.e., the subgroup statistics in a true positive/false positive space (cf. Fig. 1b). Furthermore, *pysubgroup* enables direct export of results into LaTeX via utility functions. For example, a single function call generates the LaTeX sources for Table 1.

Table 1. Example LaTeX table generated by pysubgroup.

Quality	Subgroup	size_sg	target_share_sg
365.887	sex = male	843	19.1%
365.887	sex = female	466	72.7%
304.403	sex = male ∧ parch: [0:1[	709	16.6%
233.201	sex = female ∧ pclass = 1	144	96.5%
225.957	sex = male ∧ embarked = S	623	17.0%

3   Conclusion

This demo paper introduced the *pysubgroup* package that enables subgroup discovery in a *Python/pandas* data analysis environment. It provides a lightweight, easy-to-use, extensible and freely available implementation of state-of-the-art algorithms, interestingness measures and presentation options.

References

1. Atzmueller, M.: Subgroup discovery. Wiley Interdiscipl. Rev. Data Min. Knowl. Discov. **5**(1), 35–49 (2015)
2. Atzmueller, M., Lemmerich, F.: VIKAMINE – open-source subgroup discovery, pattern mining, and analytics. In: Flach, P.A., De Bie, T., Cristianini, N. (eds.) ECML PKDD 2012. LNCS (LNAI), vol. 7524, pp. 842–845. Springer, Heidelberg (2012). https://doi.org/10.1007/978-3-642-33486-3_60
3. Clark, P., Niblett, T.: The CN2 induction algorithm. Mach. Learn. **3**(4), 261–283 (1989)
4. Flach, P.A.: The geometry of ROC space: understanding machine learning metrics through ROC isometrics. In: International Conference on Machine Learning, pp. 194–201 (2003)
5. Herrera, F., Carmona, C.J., González, P., Del Jesus, M.J.: An overview on subgroup discovery: foundations and applications. Knowl. Inf. Syst. **29**(3), 495–525 (2010)
6. Kavšek, B., Lavrač, N.: APRIORI-SD: adapting association rule learning to subgroup discovery. Appl. Artif. Intell. **20**(7), 543–583 (2006)
7. Klösgen, W.: Explora: a multipattern and multistrategy discovery assistant. In: Advances in Knowledge Discovery and Data Mining, pp. 249–271. American Association for Artificial Intelligence (1996)

8. Lemmerich, F., Becker, M., Puppe, F.: Difference-based estimates for generalization-aware subgroup discovery. In: Blockeel, H., Kersting, K., Nijssen, S., Železný, F. (eds.) ECML PKDD 2013. LNCS (LNAI), vol. 8190, pp. 288–303. Springer, Heidelberg (2013). https://doi.org/10.1007/978-3-642-40994-3_19
9. Lemmerich, F., Rohlfs, M., Atzmueller, M.: Fast discovery of relevant subgroup patterns. In: International Florida Artificial Intelligence Research Society Conference (FLAIRS), pp. 428–433 (2010)
10. Meeng, M., Knobbe, A.: Flexible enrichment with Cortana-software demo. In: Proceedings of BeneLearn, pp. 117–119 (2011)
11. Singer, P., et al.: Why we read Wikipedia. In: International Conference on World Wide Web (WWW), pp. 1591–1600 (2017)
12. Wrobel, S.: An algorithm for multi-relational discovery of subgroups. In: Komorowski, J., Zytkow, J. (eds.) PKDD 1997. LNCS, vol. 1263, pp. 78–87. Springer, Heidelberg (1997). https://doi.org/10.1007/3-540-63223-9_108
13. Zimmermann, A., De Raedt, L.: Cluster-grouping: from subgroup discovery to clustering. Mach. Learn. 77(1), 125–159 (2009)

An Advert Creation System for Next-Gen Publicity

Atul Nautiyal[1], Killian McCabe[1], Murhaf Hossari[1], Soumyabrata Dev[1(✉)],
Matthew Nicholson[1], Clare Conran[1], Declan McKibben[1], Jian Tang[3],
Wei Xu[3], and François Pitié[1,2]

[1] The ADAPT SFI Research Centre, Trinity College Dublin,
Dublin, Republic of Ireland
soumyabrata.dev@adaptcentre.ie
[2] Department of Electronic and Electrical Engineering, Trinity College Dublin,
Dublin, Republic of Ireland
[3] Huawei Ireland Research Center, Dublin, Republic of Ireland

Abstract. With the rapid proliferation of multimedia data in the internet, there has been a fast rise in the creation of videos for the viewers. This enables the viewers to skip the advertisement breaks in the videos, using ad blockers and 'skip ad' buttons – bringing online marketing and publicity to a stall. In this paper, we demonstrate a system that can effectively integrate a new advertisement into a video sequence. We use state-of-the-art techniques from deep learning and computational photogrammetry, for effective detection of existing adverts, and seamless integration of new adverts into video sequences. This is helpful for targeted advertisement, paving the path for next-gen publicity. Code related to this paper is available at: https://youtu.be/zaKpJZhBVL4.

Keywords: Advertisement · Online content · Deep learning

1 Introduction

With the ubiquity of multimedia videos, there has been a massive interest from the advertisement and marketing agencies to provide targeted advertisements for the customers. Such targeted advertisements are useful, both from the perspectives of marketing agents and end users. The advertisement agencies can use a powerful media for marketing and publicity; and the users can interact via a personalized consumer experience. In this paper, we attempt to solve this by designing an online advert creation system for next-gen publicity. We develop and implement an end-to-end system for automatically detecting and seamlessly changing an existing billboard in a video by inserting a new advert. This system will be helpful for online marketers and content developers, to develop video contents for targeted audience.

A. Nautiyal, K. McCabe, M. Hossari and S. Dev—Contributed equally and arranged alphabetically.

Fig. 1. New advert integrated into the scene at the place of an existing billboard.

Figure 1 illustrates our system. Our system automatically detects the presence of a billboard in an image frame from the video sequence. Post billboard detection, our system also localizes its position in the image frame. The user is given an opportunity to manually adjust and refine the detected four corners of the billboard. Finally, a new advertisement is integrated into the image, and tracked across all frames of the video sequence. Thereby, we generate a new composite video with the integrated advert.

Currently, there are no such existing framework available in the literature that aid the marketing agents to seamlessly integrate a new advertisement, into an original video sequence. However, a few companies viz. Mirriad [1] uses patented advertisement plantation technique to integrate 3D objects in a video sequence.

2   Technology

The backbone of our advert creation system is based on state-of-the-art techniques from deep learning and image processing. In this section, we briefly describe the underlying techniques used in the various components of the demo system. The different modules of our system are: advert- recognition, localization, and integration.

2.1   Advert Recognition

The first module of our advert creation system is used for the recognition of billboard[1] – does an image frame from the video sequence contain billboard? This helps the system user to automatically detect the presence of billboard in an image frame of the video. We use a deep neural network (DNN) as a binary classifier where classes represent *presence* and *absence* of billboard in video frame respectively. We use a VGG-based network [4] for billboard detection. We use transfer learning with pre-trained ImageNet weights. We freeze the corresponding weights of all layers apart from last 5 layers. We add 3 fully connected layers

---

[1] In this paper, we interchangeably use both the terms, *billboard* and *advert* to indicate a candidate object for new advertisement integration in an image frame.

with a *softmax* layer as the output layer. We train this deep network on our annotated dataset, containing both billboard and non-billboard images, and achieve good accuracy on billboard recognition.

2.2 Advert Localization

The second module of our advert creation system is used for localizing the position of recognized billboard – where is the billboard located in image frame? We use a encoder-decoder based deep neural network that localizes the billboard position in an image. We train this model on our billboard dataset comprising input images (cf. Fig. 2(a)) and corresponding binary ground truth image (cf. Fig. 2(b)). We train the model for several thousands of epochs. The localized billboard is a probabilistic image, that denotes the probability of an image pixel to belong to *billboard* class. We generate the binary threshold image from our computed heatmap using thresholding, and detect the various closed contours on the binary image. Finally, we select the contour with the largest area as our localized billboard position. We thereby compute the initial four corners from the binary image by circumscribing a rectangle on the selected contour with minimum bounding area. The localized advert is shown in Fig. 2.

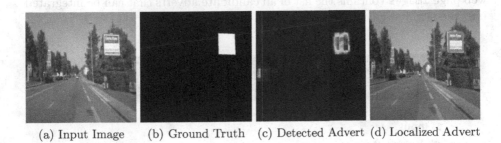

(a) Input Image   (b) Ground Truth   (c) Detected Advert (d) Localized Advert

Fig. 2. Localization of billboard using our advert creation system. We localize the advert from the probabilistic heatmap, by circumscribing a rectangle with minimum bounding area.

2.3 Advert Integration

The third and final module of our system is advert integration – how to integrate a new advert in the video? In this stage, the localized billboard is replaced with a new advert in a seamless and temporally consistent manner. We use Poisson image editing [3] on the new advert, to achieve similar local illumination and local color tone, as the original video sequence. Furthermore, the relative motion of the billboard within the scene is tracked using Kanade-Lucas-Tomasi (KLT) [2] tracking technique.

3   Design and Interface

We have designed an online system to demonstrate the functionalities of the various modules[2]. The web UI interface is designed in `Vue.js` - the progressive JavaScript Framework. The back end is supported via `Express` - Node.js web application framework. The deep neural networks for advert recognition and localization is designed in pure `python`, and the advert integration is implemented in `C++`. The web service to support advert detection is performed in `python flask`. The integration of a new advert into the existing video in the web server is executed via `C++` binary.

Figure 3 illustrates a sample snapshot of our developed web-based tool. The web interface consists of primarily three sections: `Home`, `Demo` and `Images`. The page `Home` provides an overview of the system. The next page `Demo` describes the entire working prototype of our system. The user selects a sample video from the list, runs the billboard detection module to accurately localize the billboard at sample image frames of the video. The detection module estimates the four corners of the billboard. However, the user also gets an option to *refine* the four corners manually, if the detected four corners are not completely accurate. The refined four corners of the billboard are subsequently used for tracking and integration of a new advertisement into the video sequence. The third and final web page `Images` contains the list of all candidate adverts that can be integrated into the selected video sequence.

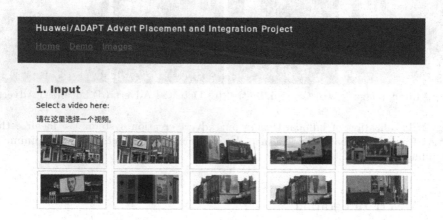

Fig. 3. Interface of the demo for advert detection and integration.

Finally, our system integrates the new advertisement into the detected billboard position, and generates a new composite video with the implanted advertisement.

---

[2] A demonstration video of our advert creation system can be accessed via https://youtu.be/zaKpJZhBVL4.

4   Conclusion and Future Work

In this paper, we have presented an online advert creation system on multimedia videos for a personalized and targeted advertisement. We use techniques from deep neural networks and image processing, for a seamless integration of new adverts into existing videos. Our system is trained on datasets that comprises outdoor scenes and views. Our future work involve further refining the performance of the system, and also generalizing it to other video sequence types.

Acknowledgement. The ADAPT Centre for Digital Content Technology is funded under the SFI Research Centres Programme (Grant 13/RC/2106) and is co-funded under the European Regional Development Fund.

References

1. Mirriad: Scalable, effective campaigns (2018). http://www.mirriad.com/. Accessed 7 May 2018
2. Lucas, B.D., Kanade, T., et al.: An iterative image registration technique with an application to stereo vision (1981)
3. Pérez, P., Gangnet, M., Blake, A.: Poisson image editing. ACM Trans. Graph. (TOG) **22**(3), 313–318 (2003)
4. Simonyan, K., Zisserman, A.: Very deep convolutional networks for large-scale image recognition. CoRR abs/1409.1556 (2014)

VHI: Valve Health Identification for the Maintenance of Subsea Industrial Equipment

M. Atif Qureshi[1]([✉]), Luis Miralles-Pechuán[1], Jing Su[1], Jason Payne[2], and Ronan O'Malley[2]

[1] Centre for Applied Data Analytics Research (CeADAR),
University College Dublin, Dublin, Ireland
{muhammad.qureshi,luis.miralles,jing.su}@ucd.ie
[2] Wood, Galway Technology Park, Parkmore, Galway, Ireland
{jason.payne,ronan.omalley}@woodplc.com

Abstract. Subsea valves are a key piece of equipment in the extraction process of oil and natural gas. Valves control the flow of fluids by opening and closing passageways. A malfunctioning valve can lead to significant operational losses. In this paper, we describe *VHI*, a system designed to assist maintenance engineers with condition-based monitoring services for valves. *VHI* addresses the challenge of maintenance in two ways: a supervised approach that predicts impending valve failure, and an unsupervised approach that identifies and highlights anomalies i.e., an unusual valve behaviour. While the supervised approach is suitable for valves with long operational history, the unsupervised approach is suitable for valves with no operational history.

1 Introduction

Predictive maintenance techniques are designed to identify developing issues in industrial equipment, alerting the need for maintenance before issues become critical [1,2,4]. These techniques aid hardware maintenance engineers in their maintenance tasks and therefore reduce the cost of condition-based monitoring services for large industrial equipment.

In this paper we present *VHI*, a system that aids hardware maintenance engineers manage subsea equipment in the oil and gas industry using supervised and unsupervised machine learning. The supervised approach makes use of the k nearest neighbour algorithm to classify valves as (*healthy* or *unhealthy*). In the unsupervised scenario, we use anomaly detection to capture abrupt changes in valve behaviour by contrasting the sensor readings of consecutive valve opening and closing events.

The innovative aspect of *VHI* is the convenience that it brings to hardware maintenance engineers. Using the supervised approach the condition of valves can be classified and simultaneously explained through simple nearest neighbour

U. Brefeld et al. (Eds.): ECML PKDD 2018, LNAI 11053, pp. 668–671, 2019.
https://doi.org/10.1007/978-3-030-10997-4_48

signature plots. Using the unsupervised approach the condition of new valves can be visualised with anomaly detection plots from different perspectives.

The *VHI* system is beneficial for the industrial sector where maintenance of equipment is often needed. The system requires time-series data generated by the sensors to monitor the condition of the equipment, and it can benefit industries when a history of operational data is present or otherwise.

There are two similar commercial products available in the market for the maintenance of valves: ValveLink Software from Emerson electric (emerson.com) and the VTScada system (trihedral.com). Both of these systems, however, are generic equipment management tools which lead to complex interfaces. *VHI* is tailor-made for the needs of an oil and gas engineers and kept simple and intuitive to use.

2   System Overview

We now present an overview of the *VHI* system. First, we discuss the dataset description, then we discuss supervised and unsupervised approaches.

Data: The dataset is composed of 583 subsea valves which are monitored over multiple years. These valves have a total of $6,648$ open (48.87%) and close (51.12%) events. Each time a valve is opened or closed, the state of the valve is captured by three sensors. Two of the sensors measure pressure, and the third sensor measures cumulative volume. During an event (opening or closing), a sensor records 120 readings at regular intervals and this results in three-time series (one for each sensor).

Supervised Approach: The supervised approach is suitable for valves that have a history of operational data available along with manual assessments. *VHI* uses the k nearest neighbour algorithm [3] for classification and it uses signature plots from the nearest neighbours to help maintenance engineers understand predictions (see Fig. 1).

Unsupervised Approach: The unsupervised approach is suitable for valves for which no operational history is available, making it appropriate for a cold start problem. In the unsupervised approach, we use anomaly detection to capture abrupt changes between consecutive readings from sensors when a valve is either opened or closed. These abrupt changes are calculated by applying distance metrics between consecutive readings. We primarily use dynamic time warping but other distance metrics (e.g. Bray-Curtis and discrete Frechet) are also made available in *VHI*.

3   User Interface

The interface of the supervised approach is composed of four modules (see Fig. 1). The first module is composed of the required inputs. The first input requires a 'csv' file containing the valves' data, where each row contains data captured by

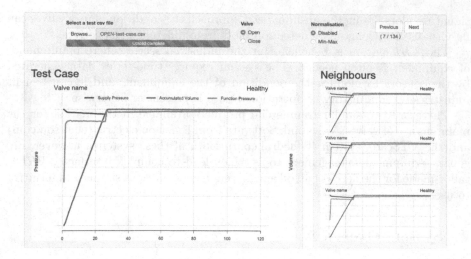

Fig. 1. Screen shot of the supervised approach.

sensors whenever a valve is either opened or closed. The rest of the inputs are to select the event (either open or close), to either apply normalisation to the signals or not, and the last one to navigate between rows of the 'csv' file. The second module (called 'Test Case', see Fig. 1) shows the plots generated by using all three sensors. Each line shows 120 points generated by a sensor. The first plot shows the signature of the predicted valve (on the left side) along with a predicted label (either *healthy* or *unhealthy*). The third module called 'Neighbours' shows the three nearest neighbours (on the right side) along with their respective labels. The prediction is made by majority vote among the three neighbours. The final module is called 'History' which shows a number of previous plots of the valve (being predicted) to demonstrate the evolution of the state of the valve, these plots are not shown due to space limitation.

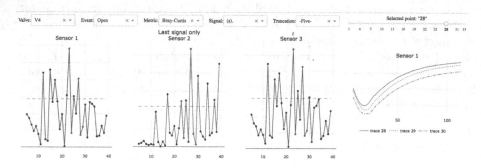

Fig. 2. Screen shot of the unsupervised approach. (Color figure online)

The interface for the unsupervised anomaly detection approach is composed of four modules (see Fig. 2). The first module is composed of the required inputs. The six inputs are to select a valve, an event (opened or closed), a metric (distance), signal transformation (original, first derivative, normalised), truncation of the signal (i.e., eliminating a first few, a last few, or at both sides of the 120 data points), and a slider that controls the instance of a valve progressively over time. The second module shows the anomaly detection plots for each sensor which is calculated by the distance between two consecutive signals (indicated by "Last signal only" in Fig. 2). The colour (which is indicated for demo purposes only) of each point represents the state of the valve at that point (green for *healthy*, blue for *degraded*, and red for *failure*). However, the idea is that a spike in the distance shows an anomaly and needs to be investigated by the engineer. Similar to the second module, the third module shows anomaly detection plots for each sensor but by calculating the distance between the average of three preceding signals and the recent signal (not shown in the Fig. 1 due to space limitation). Finally, the last module shows the original progression of the sensor data over an instance of the valve controlled by the 'Selected point' slider.

The URL https://youtu.be/dueTEovxHqI shows the online video demonstration of the *VHI* system.

4 Conclusions

This paper described *VHI* a novel predictive maintenance system designed to assist oil and gas engineers manage subsea valves. The system uses supervised and unsupervised machine learning techniques to monitor valve health. The system is currently being used by engineers at a number of sites.

Acknowledgements. This publication has emanated from research conducted with the support of Enterprise Ireland (EI), under Grant Number IP20160496 and TC20130013. The data was kindly supplied by BP, supported by Wood.

References

1. Animah, I., Shafiee, M.: Condition assessment, remaining useful life prediction and life extension decision making for offshore oil and gas assets. J. Loss Prev. Process. Ind. **53**, 17–28 (2017)
2. Di Maio, F., Hu, J., Tse, P., Pecht, M., Tsui, K., Zio, E.: Ensemble-approaches for clustering health status of oil sand pumps. Expert. Syst. Appl. **39**(5), 4847–4859 (2012)
3. Kelleher, J.D., Mac Namee, B., D'Arcy, A.: Fundamentals of Machine Learning for Predictive Data Analytics: Algorithms, Worked Examples, and Case Studies. MIT Press, Cambridge (2015)
4. Wu, S., Gebraeel, N., Lawley, M.A., Yih, Y.: A neural network integrated decision support system for condition-based optimal predictive maintenance policy. IEEE Trans. Syst. Man Cybern.-Part A: Syst. Hum. **37**(2), 226–236 (2007)

Tiler: Software for Human-Guided Data Exploration

Andreas Henelius[1,2]([✉]), Emilia Oikarinen[1,2], and Kai Puolamäki[1,2]

[1] Department of Computer Science, Aalto University, Helsinki, Finland
[2] Department of Computer Science, University of Helsinki, Helsinki, Finland
andreas.henelius@helsinki.fi

Abstract. Understanding relations in datasets is important for the successful application of data mining and machine learning methods. This paper describes TILER, a software tool for interactive visual explorative data analysis realising the interactive Human-Guided Data Exploration framework. TILER allows a user to formulate different hypotheses concerning the relations in a dataset. Data samples corresponding to these hypotheses are then compared visually, allowing the user to gain insight into relations in the dataset. The exploration process is iterative and the user gradually builds up his or her understanding of the data. Code related to this paper is available at: https://github.com/aheneliu/tiler.

1 Introduction

An important goal in *Exploratory Data Analysis* (EDA) [10] is to gain insight into different *relations in the data*. Knowledge of relations is essential for successful application of data mining and machine learning methods. Investigating relations can be efficiently performed using interactive visual EDA software, that presents the user different *views* of a dataset, thus leveraging the natural human pattern recognition skills to allow the user to discover interesting relations in the data.

Recently, an *iterative data mining paradigm* [1–3,5] has been presented and also realised in software [7–9] with the emphasis that the user wants to find patterns that are *subjectively interesting* given what she or he currently knows about the data. The system shows the user *maximally informative views*, i.e., views that contrast the most with the user's current knowledge. As the user explores the data and discovers patterns of relations in the data, these patterns are fed back into the system and taken into account during further exploration, so that the user is only shown views displaying currently unknown relations.

Although the knowledge of the user is taken into account one important problem still remains: by design, the user cannot know beforehand which views of the data differ the most from her or his present knowledge. Thus, exploration of the most informative views might seem somewhat random to the user and the views shown might be, even though surprising, not necessarily relevant for the task at hand. The user can have specific ideas (hypotheses) concerning relations in the data at the start of the exploration and such ideas typically also develop

© Springer Nature Switzerland AG 2019
U. Brefeld et al. (Eds.): ECML PKDD 2018, LNAI 11053, pp. 672–676, 2019.
https://doi.org/10.1007/978-3-030-10997-4_49

further during the exploration process. It is hence essential to be able to *focus the exploration process to answer specific questions*. This is realised in our novel EDA paradigm, termed *Human-Guided Data Exploration* (HGDE) [6].

In this paper we present TILER, a software tool for visual EDA that realises the HGDE paradigm for efficient interactive visual EDA. TILER aims to be an easy-to-use tool for exploring relations in datasets by allowing the user to focus the exploration on investigating different hypotheses. TILER is an MIT-licensed R-package available from https://github.com/aheneliu/tiler.

2   Human-Guided Data Exploration

We provide here a high-level description of the key concepts in the HGDE framework, for a complete discussion and theoretical details we refer to [6].

The goal of the user is to *discover relations between the attributes in the data* by a comparison of hypotheses, which can be viewed as a comparison of two distributions with the same known marginal distributions. A permutation-based scheme is used to obtain samples from the distributions, i.e., we permute the given data under a set of constraints defined by the hypotheses. The constraints represent the relations which are assumed to be known about the data: one extreme are unconstrained, column-wise permutations (preserving only the marginals) while the other extreme is the fully constrained case where only the identity permutation satisfies the constraints. In general, the constraints are formulated in terms of *tiles*: tuples of the form $t = (R, C)$, where $R \subseteq [N] = \{1, \ldots, N\}$ and $C \subseteq [M]$ are subsets of the rows (items) and columns (attributes) of an $N \times M$ data matrix. A tile constrains permutations so that all items in a tile are permuted together, i.e., there is a single permutation for a tile operating on each $c \in C$, thus preserving the relations inside t.

Hypotheses are represented in terms of tilings (non-overlapping sets of tiles). For example, the hypotheses can be that either all the attributes in the original dataset are dependent or they are all independent. These hypotheses can be represented with the following two hypothesis tilings: $\mathcal{T}_{\mathcal{H}_1} = \{([N], [M])\}$ and $\mathcal{T}_{\mathcal{H}_2} = \{([N], \{m\}) \mid m \in [M]\}$. A correlation between two variables i and j in a subset of rows R could be studied with the following hypothesis tilings: $\mathcal{T}_{\mathcal{H}_1} = \{(R, \{i, j\})\}$ and $\mathcal{T}_{\mathcal{H}_2} = \{(R, \{i\}), (R, \{j\})\}$. In the general case, the user can focus on specific data items and specific attribute combinations. Focusing allows the user to concentrate on exploring relations in a subset of the data items and attributes, making the interactive exploration more predictable and allowing specific questions to be answered.

In TILER, the user is shown an informative projection of two data samples corresponding to the hypotheses and is tasked with comparing these and drawing conclusions. In an informative projection the two samples differ the most. A *sample* from a distribution corresponding to each hypothesis is obtained by randomly permuting each column in the data, such that the relations between attributes enforced by the tilings are preserved. A tiling hence constrains the permutation of the data. When a user discovers a new pattern, this is added as a constraint (a tile) to both $\mathcal{T}_{\mathcal{H}_1}$ and $\mathcal{T}_{\mathcal{H}_2}$, meaning that the relations expressed by this pattern no longer differ between the two hypotheses. This allows the user to iteratively build up an understanding of the relations in the data.

3 System Design

TILER is developed in R (v. 3.4.4) using SHINY (v. 1.0.5) and runs in a web browser. The tool supports the full HGDE framework and the usage of TILER is described in the video at https://youtu.be/fqKLjMwJHnk.

To explore relations between attributes, the user first specifies the hypotheses being compared. The tool implements different *modes* as shortcuts for typical hypotheses. The *explore*-mode (the default) corresponds to iterative data exploration where the two hypotheses to be compared are that (i) all attributes in the original dataset are dependent or (ii) they are all independent. In the *focus*-mode the exploration is focused on investigating all relations within a particular subset of rows and columns (a focus region). The *compare*-mode implements the general case by allowing the user to specify an arbitrary hypothesis by partitioning the attributes in the focus region into groups.

With TILER, the user visually explores a dataset by comparing two data samples corresponding to the two different hypotheses. The exploration is iterative and the user gradually finds new patterns concerning the relations in the data, which are then added as tiles. Figure 1 shows the main user interface of TILER with the following components:

Tool panel allows the mode (explore, focus, or compare) to be selected and contains tools for selection of points as well as creation of tiles and focus tiles. Points can be selected by brushing in the main view, or by selecting the data from a dropdown menu. Previously added tiles can be selected or deleted. The projection in the main view can be changed and the user can show/hide the original data and the two samples corresponding to the combined effect of the user and hypotheses tilings. The user can also update the distributions after addition of new tiles and then request the next most informative view.

Main view shows the original data (in black) together with samples (in green and blue) corresponding to the two hypotheses being compared. Points on the same row in the sampled data matrices are connected using lines. These lines indicate how points in the data move around due to the randomisation. Since projection of high-dimensional data to lower dimensions can make interpretation complicated, we have here chosen to use 2D axis-aligned projections. The x and y axis are hence directly interpretable on their original scales. We here use correlation as the measure of informativeness, as this is often intuitive and easy to interpret, but other distance measures between the two samples being compared can be used too. This measure is used to show the maximally informative view.

Selection info shows the five largest classes of the selected points (for data with class attributes). This helps the user in understanding what type of points are currently selected and gives insight into the relations in the data.

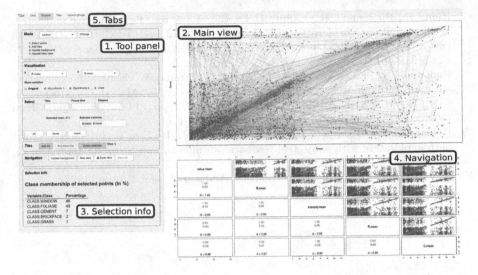

Fig. 1. The main user interface of TILER (showing UCI *image segmentation* dataset [4]). (Color figure online)

Navigation is guided by the scatterplot matrix of the five most interesting attributes in the data, in the bottom right corner. The correlations for both samples and their difference using the correlation-based measure is shown. The scatterplot helps the user to quickly obtain an overview of the data.

Tabs provide functions for loading data, listing tiles, and for defining an attribute grouping in the compare mode.

Acknowledgements. This work has been supported by the Academy of Finland (decisions 319145 and 313513).

References

1. De Bie, T.: Subjective interestingness in exploratory data mining. In: Tucker, A., Höppner, F., Siebes, A., Swift, S. (eds.) IDA 2013. LNCS, vol. 8207, pp. 19–31. Springer, Heidelberg (2013). https://doi.org/10.1007/978-3-642-41398-8_3
2. De Bie, T.: An information theoretic framework for data mining. In: KDD, pp. 564–572 (2011)
3. De Bie, T.: Maximum entropy models and subjective interestingness: an application to tiles in binary databases. Data Min. Knowl. Discov. **23**(3), 407–446 (2011)
4. Dheeru, D., Karra Taniskidou, E.: UCI machine learning repository (2017). http://archive.ics.uci.edu/ml
5. Hanhijärvi, S., Ojala, M., Vuokko, N., Puolamäki, K., Tatti, N., Mannila, H.: Tell me something I don't know: randomization strategies for iterative data mining. In: KDD, pp. 379–388 (2009)
6. Henelius, A., Oikarinen, E., Puolamäki, K.: Human-guided data exploration. arXiv preprint, arXiv:1804.03194 (2018)

7. Kang, B., Puolamäki, K., Lijffijt, J., De Bie, T.: A tool for subjective and interactive visual data exploration. In: Berendt, B., et al. (eds.) ECML PKDD 2016. LNCS (LNAI), vol. 9853, pp. 3–7. Springer, Cham (2016). https://doi.org/10.1007/978-3-319-46131-1_1

8. Puolamäki, K., Kang, B., Lijffijt, J., De Bie, T.: Interactive visual data exploration with subjective feedback. In: Frasconi, P., Landwehr, N., Manco, G., Vreeken, J. (eds.) ECML PKDD 2016. LNCS (LNAI), vol. 9852, pp. 214–229. Springer, Cham (2016). https://doi.org/10.1007/978-3-319-46227-1_14

9. Puolamäki, K., Oikarinen, E., Kang, B., Lijffijt, J., De Bie, T.: Interactive visual data exploration with subjective feedback: an information-theoretic approach. In: ICDE, pp. 1208–1211 (2018)

10. Tukey, J.W.: Exploratory Data Analysis. Addison-Wesley, Boston (1977)

ADAGIO: Interactive Experimentation with Adversarial Attack and Defense for Audio

Nilaksh Das[1](✉), Madhuri Shanbhogue[1], Shang-Tse Chen[1], Li Chen[2], Michael E. Kounavis[2], and Duen Horng Chau[1]

[1] Georgia Institute of Technology, Atlanta, GA, USA
{nilakshdas,madhuri.shanbhogue,schen351,polo}@gatech.edu
[2] Intel Corporation, Hillsboro, OR, USA
{li.chen,michael.e.kounavis}@intel.com

Abstract. Adversarial machine learning research has recently demonstrated the feasibility to confuse automatic speech recognition (ASR) models by introducing acoustically imperceptible perturbations to audio samples. To help researchers and practitioners gain better understanding of the impact of such attacks, and to provide them with tools to help them more easily evaluate and craft strong defenses for their models, we present ADAGIO, the first tool designed to allow interactive experimentation with adversarial attacks and defenses on an ASR model in real time, both visually and aurally. ADAGIO incorporates AMR and MP3 audio compression techniques as defenses, which users can interactively apply to attacked audio samples. We show that these techniques, which are based on psychoacoustic principles, effectively eliminate targeted attacks, reducing the attack success rate from 92.5% to 0%. We will demonstrate ADAGIO and invite the audience to try it on the Mozilla Common Voice dataset. Code related to this paper is available at: https://github.com/nilakshdas/ADAGIO.

Keywords: Adversarial ML · Security · Speech recognition

1 Introduction

Deep neural networks (DNNs) are highly vulnerable to adversarial instances in the image domain [3]. Such instances are crafted by adding small imperceptible perturbations to benign instances to confuse the model into making wrong predictions. Recent work has shown that this vulnerability extends to the audio domain [1], undermining the robustness of state-of-the-art models that leverage DNNs for the task of automatic speech recognition (ASR). The attack manipulates an audio sample by carefully introducing faint "noise" in the background that humans easily dismiss. Such perturbation causes the ASR model to transcribe the manipulated audio sample as a target phrase of the attacker's choosing. Through this research demonstration, we make two major contributions:

© Springer Nature Switzerland AG 2019
U. Brefeld et al. (Eds.): ECML PKDD 2018, LNAI 11053, pp. 677–681, 2019.
https://doi.org/10.1007/978-3-030-10997-4_50

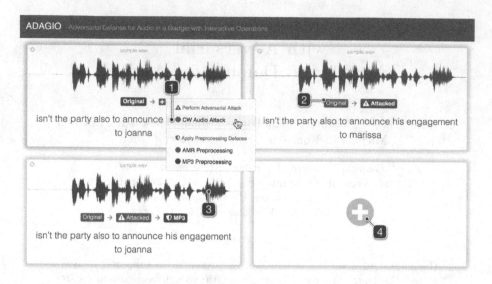

Fig. 1. ADAGIO usage scenario. (1) Jane uploads an audio file that is transcribed by DeepSpeech [5]; then she performs an adversarial attack on the audio in real time by entering a target transcription after selecting the attack option from the dropdown menu, e.g., the state-of-the-art Carlini-Wagner Audio Attack [1]. (2) Jane decides to perturb the audio to change the last word of the sentence from "joanna" to "marissa"; she can listen to the original audio and see the transcription by clicking on the "Original" badge. (3) Jane applies MP3 compression to recover the original, correct transcription from the manipulated audio; clicking on a waveform plays back the audio from the selected position. (4) Jane can experiment with multiple audio samples by adding more cards. For presentation, operations 1, 2 and 3 are shown as separate cards.

1. Interactive exploration of audio attack and defense. We present ADAGIO, the first tool designed to enable researchers and practitioners to interactively experiment with adversarial attack and defenses on an ASR model in real time (see demo: https://youtu.be/0W2BKMwSfVQ). ADAGIO incorporates AMR and MP3 audio compression techniques as defenses for mitigating perturbations introduced by the attack. Figure 1 presents a brief usage scenario showing how users can experiment with their own audio samples. ADAGIO stands for **A**dversarial **D**efense for **A**udio in a **G**adget with **I**nteractive **O**perations.

2. Compression as an effective defense. We demonstrate that non-adaptive adversarial perturbations are extremely fragile, and can be eliminated to a large extent by using audio processing techniques like Adaptive Multi-Rate (AMR) encoding and MP3 compression. We assume a non-adaptive threat model since an adaptive version of the attack is prohibitively slow and often does not converge.

2   ADAGIO: Experimenting with Audio Attack and Defense

We first provide a system overview of ADAGIO, then we describe its primary building blocks and functionality. ADAGIO consists of four major components: (1) an interactive UI (Fig. 1); (2) a speech recognition module; (3) a targeted attack generator module; and (4) an audio preprocessing (defense) module. The three latter components reside on a back-end server that performs the computation. The UI communicates the user intent with the back-end modules through a websocket messaging service, and uses HTTP to upload/download audio files for processing. When the messaging service receives an action to be performed from the front-end, it leverages a custom redis-based job queue to activate the correct back-end module. When the back-end module finishes its job, the server pings back the UI through the websocket messaging service to update the UI with the latest results. Below, we describe the other three components in ADAGIO.

2.1   Speech Recognition

In speech recognition, state-of-the-art systems leverage Recurrent Neural Networks (RNNs) to model audio input. The audio sample is broken up into frames $\{x^{(1)}, \ldots, x^{(T)}\}$ and fed sequentially to the RNN function $f(\cdot)$ which outputs another sequence $\{y^{(1)}, \ldots, y^{(T')}\}$, where each $y^{(t)}$ is a probability distribution over a set of characters. The RNN maintains a hidden state $h^{(t)}$ which is used to characterize the sequence up until the current input $x^{(t)}$, such that, $(y^{(t)}, h^{(t)}) = f(x^{(t-1)}, h^{(t-1)})$. The most likely sequence based on the output probability distributions then becomes the transcription for the audio input. The performance of speech-to-text models is commonly measured in Word Error Rate (WER), which corresponds to the minimum number of word edits required to change the transcription to the ground truth phrase.

ADAGIO uses Mozilla's implementation [5] of DeepSpeech [4], a state-of-the-art speech-to-text DNN model, to transcribe the audio in real time.

2.2   Targeted Audio Adversarial Attacks

Given a model function $m(\cdot)$ that transcribes an audio input x as a sequence of characters y, i.e., $m(x) = y$, the objective of the targeted adversarial attack is to introduce a perturbation δ such that the transcription is now a specific sequence of characters y' of the attacker's choosing, i.e., $m(x + \delta) = y'$. The attack is only considered successful if there is no error in the transcription.

ADAGIO allows users to compute adversarial samples using a state-of-the-art iterative attack [1]. After uploading an audio sample to ADAGIO, the user can click the attack button and enter the target transcription for the audio (see Fig. 1(1)). The system then runs 100 iterations of the attack and updates the transcription displayed on the screen at each step to show progress of the attack.

2.3   Compression as Defense

In the image domain, compression techniques based on psychovisual theory have been shown to mitigate adversarial perturbations of small magnitude [2]. We extend that hypothesis to the audio domain and let users experiment with AMR encoding and MP3 compression on adversarially manipulated audio samples. Since these techniques are based on psychoacoustic principles (AMR was specially developed to encode speech), we posit that these techniques could effectively remove the adversarial components from the audio which are imperceptible to humans, but would confuse the model.

To determine the efficacy of these compression techniques in defending the ASR model, we created targeted adversarial instances from the first 100 test samples of the Mozilla Common Voice dataset using the attack as described in [1]. We constructed five adversarial audio instances for every sample, each transcribing to a phrase randomly picked from the dataset, yielding a total of 500 adversarial samples. We then preprocessed these samples before feeding it to the DeepSpeech model. Table 1 shows the results from this experiment. We see that the preprocessing defenses are able to completely eliminate the targeted success rate of the attack.

Table 1. Word Error Rate (WER) and the targeted attack success rate on the Deep-Speech model (lower is better for both). AMR and MP3 eliminate all targeted attacks, and significantly improves WER.

Defense	WER (no attack)	WER (with attack)	Targeted attack success rate
None	0.369	1.287	92.45%
AMR	0.488	0.666	**0.00%**
MP3	0.400	0.780	**0.00%**

3   Conclusion

We present ADAGIO, an interactive tool that empowers users to experiment with adversarial audio attacks and defenses. We will demonstrate and highlight ADAGIO's features using a few usage scenarios on the Mozilla Common Voice dataset, and invite our audience to try out ADAGIO and freely experiment with their own queries.

References

1. Carlini, N., Wagner, D.: Audio adversarial examples: targeted attacks on speech-to-text. arXiv:1801.01944 (2018)
2. Das, N., et al.: Keeping the bad guys out: protecting and vaccinating deep learning with jpeg compression. arXiv:1705.02900 (2017)

3. Goodfellow, I., Shlens, J., Szegedy, C.: Explaining and harnessing adversarial examples. In: ICLR (2015)
4. Hannun, A., et al.: Deep speech: scaling up end-to-end speech recognition. arXiv:1412.5567 (2014)
5. Mozilla: Deepspeech. https://github.com/mozilla/DeepSpeech

ClaRe: Classification and Regression Tool for Multivariate Time Series

Ricardo Cachucho[1,2]($\boxtimes$), Stylianos Paraschiakos[2], Kaihua Liu[1],
Benjamin van der Burgh[1], and Arno Knobbe[1]

[1] Leiden Institute of Advanced Computer Science, Leiden, The Netherlands
{r.cachucho,b.van.der.burgh,a.j.knobbe}@liacs.leidenuniv.nl,
lkaihua@gmail.com
[2] Leiden University Medical Center, Leiden, The Netherlands
s.paraschiakos@lumc.nl

Abstract. As sensing and monitoring technology becomes more and more common, multiple scientific domains have to deal with big multivariate time series data. Whether one is in the field of finance, life science and health, engineering, sports or child psychology, being able to analyze and model multivariate time series has become of high importance. As a result, there is an increased interest in multivariate time series data methodologies, to which the data mining and machine learning communities respond with a vast literature on new time series methods.

However, there is a major challenge that is commonly overlooked; most of the broad audience of end users lack the knowledge on how to implement and use such methods. To bridge the gap between users and multivariate time series methods, we introduce the ClaRe dashboard. This open source web-based tool, provides to a broad audience a new intuitive data mining methodology for regression and classification tasks over time series. Code related to this paper is available at: https://github.com/parastelios/Accordion-Dashboard.

1 Introduction

Over the past few years, there is an increased interest in the analysis of multivariate time series data. A great deal of this interest is motivated by advances in sensor technology. In many application areas, deploying sensors for continuous monitoring has become a common strategy. Over the last 10 years, sensors are becoming more accurate, with better data communication protocols, smaller and last but not least, cheaper.

From the data science perspective, sensor systems will produce time series data. In the case of sensor networks, multiple variables are collected simultaneously, producing multivariate time series. Adding to that, when collected continuously, these datasets lead to big data challenges. This raized challenges to the data mining community, on how to deal with large multivariate time series. These challenges have attracted the attention of many researcher and lead to a vast literature on time series mining. With the exception of a few good examples

© Springer Nature Switzerland AG 2019
U. Brefeld et al. (Eds.): ECML PKDD 2018, LNAI 11053, pp. 682–686, 2019.
https://doi.org/10.1007/978-3-030-10997-4_51

[1,2], there is still a gap between most of these methods and the potential end users, who may lack a technical background to implement them.

Most of the sciences based on empirical observations have the potential to benefit from technological advances in sensor systems: (1) Children can be monitored continuously to study their social competence; (2) Environmental sciences can benefit from continuous sensing; Civil engineering can develop predictive maintenance of infrastructures using sensor networks; Life sciences and health are already heavily supported by machinery that uses sensors to measure all sort of phenomena. A common link between all the examples mentioned above is that they rely on sensor monitoring systems for their continuous sampling methodologies. The continuous nature of the measurements, lead to large multivariate time series datasets. As a consequence, the traditional data analysis tools based on classical statistics are commonly not applicable to this kind of data. New tools are an opportunity to bridge between data science and empirical sciences.

One could argue that the data mining community is already encouraging the publication of source code and data associated with publications. However, without a deep knowledge on the published method and the language used to implement the code, such released source code targets only a limited audience. Another very significant effort to make machine learning methods more accessible is the release of packages with collections of algorithms, such as Scikit-learn [3] for Phyton or Caret [4] for R. The downside of such packages is the need to be proficient both in the programming language that implements the package of methods and the need to know how to build a data science methodology around the chosen method. At last, there are tools for a broad audience such as Weka [1], MOA [2], Knime [5], JMulTi [6] and SPSS [7], which are intuitive and provide graphical user interfaces. These tools lack on the flexibility to implement new methods and most of them are not designed to analyze multivariate time series.

Our proposal to bridge the gap between new methods and a broad audience, is to build easily accessible web-based tools, with a user interface. we propose *ClaRe*, a *Cla*ssification and *Re*gression tool to model supervized multivariate time series. This Software as a Service (SaaS) tool adopts the *Accordion* algorithm from the previous chapter, to learn informative features and allows users to learn regression and classification models from multivariate time series with mixed sampling rates. Its intuitive web-based interface provides options of importing, pre-processing, modeling and evaluating multivariate time series data. In every step, plotting and saving data or results are allowed. Furthermore, source code, experimental data[1] and video tutorial[2] are available.

2   Tool Overview

ClaRe is a web-based tool that incorporates all the necessary steps for modeling time series with mixed sampling rates. Such time series are often collected from a network of sensors that measures complex phenomena. The output of such

---

[1] https://github.com/parastelios/Accordion-Dashboard.
[2] https://www.youtube.com/watch?v=Vomhr9mBSBU.

sensors are often multiple files that have variables measured at different rates and thus have special needs: (1) Pre-processing needs to include synchronization and merging; (2) Plotting needs to be done using sampling techniques due to the size of such time series; (3) Learning strategies that take into account the temporal nature of the data; (4) Adequate model evaluation strategies that test multiple systems (e.g. people) to reflect the true accuracy of the models.

From a technical perpective, ClaRe also presents benefits in terms of development and deployment. Both front end and server are developed with R, using the *R Shiny* package. This package provides a framework to interact between client and server side through R-scripts. As a result, the tool was easy to implement since only one programming language is used to manage both server and front end. From the deployment perspective, ClaRe's main advantage is its compatibility with all modern web browsers. With ClaRe, one can import and pre-process time series data, build regression or classification models, evaluate them, and export the results. The user can follow the proposed methodology intuitively, using web components that adjust to the user choices and guides the user troughout the data mining methodology. Each panel will be enumerated and explained below, following the CRISP-DM methodology [9].

Import and Pre-processing: When the user accesses the tool online, they are welcomed to the tool by the *Import* panel. To start, the user can upload predictors and target in a single or separate files. In this panel, the user can get a preview of the data and descriptive statistics for all the variables. Having imported the data, the user will be intuitively guided to the *Pre-processing* panel. Here, the user can choose from multiple pre-processing tasks, both generic for all sorts of datasets and specific to sensor-based multivariate time series. The pre-processing tasks include: (1) Selecting the variable the user wants to consider as a target; (2) Normalizing datasets; (3) Removing outliers; (4) Merge multiple files into one dataset, (5) Synchronize time series data with mixed sampling rates; (6) Manage missing values; (7) Plotting inspection as presented in Fig. 1.

Model: After choosing a numeric or nominal target, this panel changes into a regression or classification setup, respectively. The available regression models are a linear regression model and a lag regression model. As for the classification task, the available model is a decision tree. Both classification and regression models construct and select aggregate features using Accordion algorithm [8]. Accordion can be tuned with multiple parameters, which are available in the *Regression* or *Classification* panels. For both classification and regression, one can tune the target's sampling rate, the maximum window size and the number of samples used to perform a greedy search for aggregate features. Additionally, in regression there are multiple options for regression methods (linear and lag).

Evaluation: Having the models learned or loaded, the *Evaluation* panel allows the users to obtain multiple evaluations of the constructed model. For models testing multiple systems, one can use the Leave One Participant Out (LOPO) evaluation. With LOPO, the model is built multiple times, leaving each time one system out of the learning process to validate. This evaluation method is

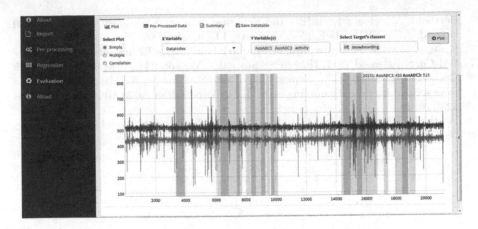

Fig. 1. *ClaRe* dashboard user interface: plotting tab.

especially important to assess the real accuracy of models, once the dataset instances are not independent and identically distributed. Other evaluation functionalities include: (1) statistical summary of the models accuracies/errors; (2) Visualization panel to plot models and predictions; (3) Export models and associated features; (4) Evaluate the model using new datasets; (5) Compare multiple models in new datasets; (6) Flexibility to re-visit these results over multiple user sessions.

3 Conclusion

This paper presents an easily accessible web-tool designated as *ClaRe*. *ClaRe* is a Software as a service (SaaS), which provides any user interested in mining multivariate time series, a methodology for supervized learning. More specifically, it allows users to deal with cases when the multivariate time series data have mixed sampling rates. Making use of intuitive menus, one can easily load one of multiple files, pre-process properly sensor systems data, learn time series models and evaluate the results. At any stage of the mining process, interactive plotting and saving options (for models and data) are available.

References

1. Hall, M., Frank, E., Holmes, G., Pfahringer, B., Reutemann, P., Witten, I.H.: The WEKA data mining software: an update. In: SIGKDD Explorations, vol. 11 (2009)
2. Bifet, A., Holmes, G., Kirkby, R., Pfahringer, B.: MOA: massive online analysis. JMLR **11**, 1601–1604 (2010)
3. Pedregosa, F., et al.: Scikit-learn: machine learning in python. JMLR **12**, 2825–2830 (2011)
4. Kuhn, M.: Caret package. J. Stat. Softw. **28**, 1–26 (2008)

5. Berthold, M., et al.: KNIME - the Konstanz information miner: version 2.0 and beyond. ACM SIGKDD Explor. Newslett. **11**, 26–31 (2009)
6. Krätzig, M.: The Software JMulTi, Applied Time Series Econometrics (2004)
7. BM Corp. Released 2017, IBM SPSS Statistics, Version 24.0, IBM Corp, New York
8. Cachucho, R., Meeng, M., Vespier, U., Nijssen, S., Knobbe, A.: Mining multivariate time series with mixed sampling rates. In: Proceedings of ACM UbiComp, pp. 413–423 (2014)
9. Wirth, R., Hipp, J.: CRISP-DM: Towards a standard process model for data mining. In: Proceedings of PADD, pp. 29–39 (2000)

Industrial Memories: Exploring the Findings of Government Inquiries with Neural Word Embedding and Machine Learning

Susan Leavy(✉), Emilie Pine, and Mark T. Keane

University College Dublin, Dublin, Ireland
{susan.leavy,emilie.pine,mark.keane}@ucd.ie

Abstract. We present a text mining system to support the exploration of large volumes of text detailing the findings of government inquiries. Despite their historical significance and potential societal impact, key findings of inquiries are often hidden within lengthy documents and remain inaccessible to the general public. We transform the findings of the Irish government's inquiry into industrial schools and through the use of word embedding, text classification and visualization, present an interactive web-based platform that enables the exploration of the text to uncover new historical insights. Code related to this paper is available at: https://industrialmemories.ucd.ie.

Keywords: Word embeddings · Text classification · Visualization Government inquiry reports

1 Introduction

The Irish Government published one of the most important documents in the history of the state in 2009. It spanned over 1 million words and detailed the findings of a 9 year investigation into abuse and neglect in Irish industrial schools. The document however, remains largely unread. The structure and style of the report obscures significant information and prevents the kind of system-wide analysis of the industrial school regime which is crucial to identifying recurring patterns and preventing the re-occurrence of such atrocities.

The Industrial Memories project, publicly available online[1], presents a web-based interactive platform where the findings of CICA, commonly known as the Ryan Report, may be explored by researchers and the general public in new ways. The machine learning approach developed as part of this project addressed the challenge of learning within the context of limited volumes of training data. Techniques such as named entity recognition, word embedding and social network analysis are also used to extract information and unearth new insights from the report.

---

[1] https://industrialmemories.ucd.ie.

U. Brefeld et al. (Eds.): ECML PKDD 2018, LNAI 11053, pp. 687–690, 2019.
https://doi.org/10.1007/978-3-030-10997-4_52

2   System Overview

The Industrial Memories project developed a web-based platform where the narrative form of the Ryan Report is deconstructed and key information extracted. In the Ryan Report most of the chapters deal separately with individual industrial schools. This segregation of information prevents a system-wide analysis of common trends and patterns across the industrial school system. To address this, a random forest classifier was employed to annotate the report. An interactive web based platform enables users to extract information and interact with the report using a comprehensive search tool (Fig. 1). Word embeddings were also used to identify actions pertaining to meetings and communication that occurred within the industrial school system facilitating an analysis of the dynamics of power and influence using social network analysis. Each individual named in the report was extracted using the Stanford NE system [1] which for the first time, allows the history of individuals to be traced over the entire industrial school system thus responding to a call for increased open data in government [2].

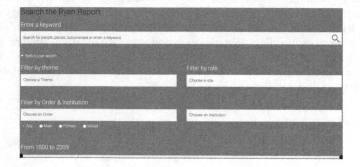

Fig. 1. Text navigation tool

2.1   Semantic Classification of Text

While some semantic categories or themes were conducive to extraction using a rule based approach, categories including the transfer events, descriptions of abuse and witness testimony were extracted using a random forest learning algorithm. The random forest algorithm was chosen because, as an ensemble learner that creates a 'forest' of decision trees by randomly sampling from the training set, it is suited to learning from smaller datasets [4]. The Ryan Report was represented in a relational database with annotated excerpts comprising 6,839 paragraphs (597,651 words). Each paragraph was considered a unit-of-analysis in the Report for the purposes of annotation.

2.2   Feature Extraction with Word Embedding

Feature selection was based on the compilation of domain-specific lexicons in order to address the constraint in this project concerning low volumes of available training data. As with many projects that are based in the humanities, compiling training data is a manual process involving close reading of texts. To address this, domain-specific lexicons were compiled based on the entire dataset yielding high accuracy with low numbers of training examples.

Lexicons were generated using the word2vec algorithm from a number of seed terms [3]. Given that the word2vec algorithm performs best with larger datasets than the Ryan Report, five ensembles were generated and the top thirty synonyms for each seed term were extracted. Terms common across each ensemble were identified to generate domain specific synonyms. Features for the classification process were then derived from these lexicons.

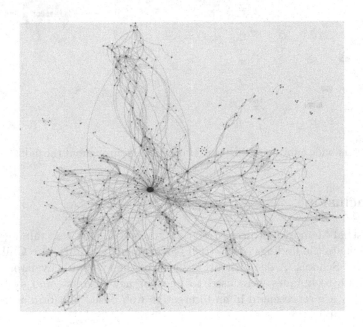

Fig. 2. Collocation network of entities in the Ryan Report

2.3 Exploring Power, Influence and Migration Within the System with Word Embedding and Network Analysis

Neural network analysis was used to generate lexicons describing interactions in the form of meetings, letters and telephone calls to be compiled and excerpts describing these communications extracted from the report. A social network based on these excerpts combined with entities named in the report was constructed along with a network detailing how entities were related in the narrative of the Ryan Report through collocation within paragraphs (Fig. 2). The sub-corpus that was generated by automatically extracting excerpts detailing transfers was examined using association rule analysis revealing the recurring patterns governing the movement of religious staff often in response to allegations of abuse (Fig. 3).

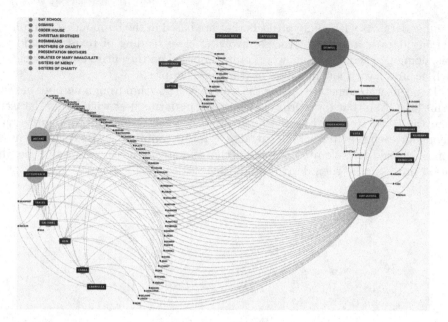

Fig. 3. System-wide analysis of the transfer of religious throughout the industrial school system

3   Conclusion

The Industrial Memories project presents a state-of-the-art text mining application to navigate the findings of the Irish Government's Inquiry into Child Abuse at Industrial Schools. A demonstration video of the system is available online[2]. Text analytic techniques were used to uncover new insights and system-wide patterns that are represented in an interactive web based platform accessible to the general public.

References

1. Finkel, J.R., Grenager, T., Manning, C.: Incorporating non-local information into information extraction systems by Gibbs sampling. In: Proceedings of the 43rd Annual Meeting on Association for Computational Linguistics, pp. 363–370. Association for Computational Linguistics (2005)
2. Kitchin, R.: The Data Revolution: Big Data, Open Data, Data Infrastructures and Their Consequences. Sage, Newcastle upon Tyne (2014)
3. Mikolov, T., Sutskever, I., Chen, K., Corrado, G.S., Dean, J.: Distributed representations of words and phrases and their compositionality. In: Advances in Neural Information Processing Systems, pp. 3111–3119 (2013)
4. Polikar, R.: Ensemble based systems in decision making. IEEE Circ. Syst. Mag. **6**(3), 21–45 (2006)

---

[2] https://youtu.be/cV1xzuJ0dv0.

Monitoring Emergency First Responders' Activities via Gradient Boosting and Inertial Sensor Data

Sebastian Scheurer[1]([email]) [iD], Salvatore Tedesco[2], Òscar Manzano[2],
Kenneth N. Brown[1], and Brendan O'Flynn[2]

[1] Insight Centre for Data Analytics, Department of Computer Science,
University College Cork, Cork, Ireland
sebastian.scheurer@insight-centre.org
[2] Tyndall National Institute, University College Cork, Cork, Ireland

Abstract. Emergency first response teams during operations expend much time to communicate their current location and status with their leader over noisy radio communication systems. We are developing a modular system to provide as much of that information as possible to team leaders. One component of the system is a human activity recognition (HAR) algorithm, which applies an ensemble of gradient boosted decision trees (GBT) to features extracted from inertial data captured by a wireless-enabled device, to infer what activity a first responder is engaged in. An easy-to-use smartphone application can be used to monitor up to four first responders' activities, visualise the current activity, and inspect the GBT output in more detail.

Keywords: Human activity recognition · Machine learning
Boosting · Inertial sensors

1 Introduction

Emergency first responders, such as firefighters, typically enter a building in pairs or small teams, each of which maintains contact to its leader using the portable radio communications device carried by one of its members. As a team makes its way into the building they report their status and progress to, and confirm their next steps with, their leader who tracks their location and status, usually on paper, at every point of the way. Because radio is a notoriously noisy medium, messages have to be repeated frequently to ensure accurate communication, but every minute taken up by these communications adds another minute to the time a person in distress is waiting for help.

The Sensor Technologies for Enhanced Safety and Security of Buildings and its Occupants (SAFESENS) project [8] is developing a novel monitoring system for emergency first responders designed to provide first response team leaders with timely and reliable information about their team's status during emergency response operations, thereby reducing the amount of time taken up by radio

© Springer Nature Switzerland AG 2019
U. Brefeld et al. (Eds.): ECML PKDD 2018, LNAI 11053, pp. 691–694, 2019.
https://doi.org/10.1007/978-3-030-10997-4_53

communications, accelerating response operations, and improving first responders' safety. The system consists of components for monitoring vital signs, indoor
localisation, and human activity recognition (HAR), each with its own set of
sensors—vital signs are captured by an instrumented glove, localisation uses
ranging data from pre-deployed ultra-wideband (UWB) anchors [1,3], and the
HAR component relies on the data captured by the inertial measurement units
(IMU) worn by first responders—and related algorithms. A video which demonstrates the execution of the system can be found at https://bit.ly/iharpkdd.

There is a substantial body of HAR work using wearable IMUs for assisted
living, industrial, or sports applications [7], but these do not necessarily translate to first response operations. The few commercial first responder monitoring
systems that do exist, such as Medtronic's Zephyr^{TM} offerings [4], rely on a
host of wearable sensors built into vests, boots, or gloves, which—in addition to
the fact that first responders are not keen to add to their already cumbersome
equipment—tend to be expensive. Our system demonstrates how the state-of-
the-art in HAR can be used to monitor emergency first responders' activities
using only one wearable device per first responder. This system can be customised to recognise a different set of activities by re-training the classifier with
appropriate sample data.

2 System Architecture and Operation

The system architecture (Fig. 1) shows the different components and how they
can be used during a first response operation. Each first responder wears a

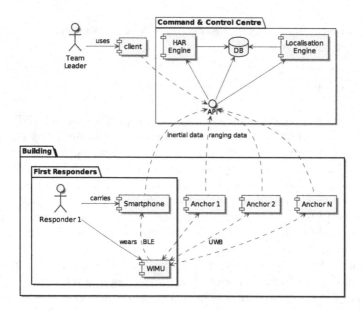

Fig. 1. System architecture. Note only one first responder plus equipment (WIMU and
Smartphone), team leader, and client application are shown.

wireless-enabled IMU (WIMU) that captures inertial data and transmits them via low energy bluetooth (BLE) to the smartphone carried by each first responder. The phone posts the data in batches of configurable duration (default: 10 s) via HTTP to an API exported by the SAFESENS server, which is running on a PC in the Command & Control Centre (CCC). The WIMU also connects to any reachable UWB anchors, which compute ranging data for each connected WIMU, and posts them to the API. The API receives and stores both (inertial and ranging) types of data in a relational database, where they are available to client applications as well as the localisation and HAR algorithm.

3   Recognising and Monitoring First Responder Activities

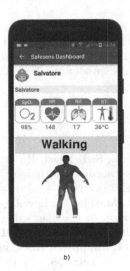

In this demo we shall limit ourselves to the HAR component, for which we have developed an Android application whose initial screen is shown in panel (a) of Fig. 2. This screen illustrates the status of up to four first responders, showing, for each of them, the most recent activity, according to the output of the HAR algorithm, and highlighting it if and when appropriate—e.g., to issue a "firefighter down" alert if a firefighter is thought to have ceased moving and be lying on the ground. Team leaders can tap on a first responder's icon to access a second screen (panel b) that provides more detail about the HAR estimate, and illustrates the most likely activity by means of a 3D model.

Fig. 2. Screenshots of the team leaders' application showing (a) the dashboard, and (b) the screen with details of the current HAR estimate.

The HAR component, once triggered, operates as follows. First, the most recent batch, by default covering 10 s of inertial data, is loaded from the database. Then, the signals are resampled to their mean sampling frequency, using linear interpolation to fill any gaps that might have been introduced by the resampling, before a moving median filter is applied over a window of 3 samples to smooth the signals. Next, the smoothed acceleration signals are separated by means of a low pass filter [2] into their respective gravity and body components, which thenceforth replace the original accelerometer signals. After this, the two components and the smoothed gyroscope signals are segmented into 3 s sliding windows with 1 s overlap, and a set of time- and frequency-domain features is extracted from each window. Finally, the extracted features are passed to the HAR inference algorithm, and the resulting probability estimates for each of the target activities averaged, producing a final estimate for the batch which is returned as response to the client's API request.

The HAR inference algorithm is a gradient boosted ensemble of decision trees (GBT) which has been trained, for the sake of this demo, to recognise six activities, namely standing/sitting, crawling on hands & knees or stomach, walking, falling, and lying down. The GBT hyper-parameters, such as the number of iterations (750), learning rate (0.02), or maximum depth (3) of the trees, have been tuned via leave-one-subject-out cross-validation to minimise the average mean absolute error (MAE) across the target classes. More details on the training data, pre-processing, feature extraction, as well as tuning, training, and evaluation of the GBT can be found in [5] and [6], where this approach achieved a MAE of <4% and Accuracy of >90% when evaluated on data from an unseen individual.

Acknowledgements. This publication emanated from research supported by research grants from Science Foundation Ireland (SFI) and the European Development Fund under grant numbers SFI/12/RC/2289 and 13/RC/2077-CONNECT, and the European funded project SAFESENS under the ENIAC program in association with Enterprise Ireland (IR20140024).

References

1. Brahmi, I.H., Abruzzo, G., Walsh, M., Sedjelmaci, H., O'Flynn, B.: A fuzzy logic approach for improving the tracking accuracy in indoor localisation applications. In: Wireless Days Conference. IEEE, April 2018
2. Karantonis, D.M., Narayanan, M.R., Mathie, M., Lovell, N.H., Celler, B.G.: Implementation of a real-time human movement classifier using a triaxial accelerometer for ambulatory monitoring. Trans. Inf. Technol. Biomed. **10**(1), 156–167 (2006)
3. Khodjaev, J., Tedesco, S., O'Flynn, B.: Improved NLOS error mitigation based on LTS algorithm. Prog. Electromagn. Res. Lett. **58**, 133–139 (2016)
4. Medtronic: Zephyr^{TM} performance systems for first responders and industrial safety (2017). https://www.zephyranywhere.com
5. Scheurer, S., Tedesco, S., Brown, K.N., O'Flynn, B.: Human activity recognition for emergency first responders via body-worn inertial sensors. In: International Conference on Wearable and Implantable Body Sensor Networks. IEEE, May 2017
6. Scheurer, S., Tedesco, S., Brown, K.N., O'Flynn, B.: Sensor and feature selection for an emergency first responders activity recognition system. In: Sensors. IEEE, October 2017
7. Sreenivasan, R.R., Nirmalya, R.: Recent trends in machine learning for human activity recognition–a survey. Wiley Interdiscip. Rev.: Data Min. Knowl. Discov. **8**(4) (2018)
8. Tedesco, S., Khodjaev, J., O'Flynn, B.: A novel first responders location tracking system: architecture and functional requirements. In: Mediterranean Microwave Symposium. IEEE, November 2015

Visualizing Multi-document Semantics via Open Domain Information Extraction

Yongpan Sheng[1], Zenglin Xu[1], Yafang Wang[2], Xiangyu Zhang[1], Jia Jia[2], Zhonghui You[1], and Gerard de Melo[3($\boxtimes$)]

[1] School of Computer Science and Engineering, University of Electronic Science and Technology of China, Chengdu, China
shengyp2011@gmail.com, zenglin@gmail.com, keposmile.z@gmail.com, zhyouns@gmail.com
[2] Shandong University, Jinan, China
wyf181@gmail.com, jiajia911@gmail.com
[3] Rutgers University, New Brunswick, USA
gdm@demelo.org

Abstract. Faced with the overwhelming amounts of data in the 24/7 stream of new articles appearing online, it is often helpful to consider only the key entities and concepts and their relationships. This is challenging, as relevant connections may be spread across a number of disparate articles and sources. In this paper, we present a system that extracts salient entities, concepts, and their relationships from a set of related documents, discovers connections within and across them, and presents the resulting information in a graph-based visualization. We rely on a series of natural language processing methods, including open-domain information extraction, a special filtering method to maintain only meaningful relationships, and a heuristic to form graphs with a high coverage rate of topic entities and concepts. Our graph visualization then allows users to explore these connections. In our experiments, we rely on a large collection of news crawled from the Web and show how connections within this data can be explored. Code related to this paper is available at: https://shengyp.github.io/vmse.

Keywords: Multi-document information extraction
Graph-based visualization

1 Introduction

In today's interconnected world, there is an endless 24/7 stream of new articles appearing online, including news reports, business transactions, digital media, etc. Faced with these overwhelming amounts of information, it is helpful to consider only the key entities and concepts and their relationships. Often, these are spread across a number of disparate articles and sources. Not only do different outlets often cover different aspects of a story. Typically, new information only becomes available over time, so new articles in a developing story need to be connected to previous ones, or to historic documents providing relevant background information.

© Springer Nature Switzerland AG 2019
U. Brefeld et al. (Eds.): ECML PKDD 2018, LNAI 11053, pp. 695–699, 2019.
https://doi.org/10.1007/978-3-030-10997-4_54

In this paper, we present a system[1] that extracts salient entities, concepts, and their relationships from a set of related documents, discovers connections within and across them, and presents the resulting information in a graph-based visualization. Such a system is useful for anyone wishing to drill down into datasets and explore relationships, e.g. analysts and journalists. We rely on a series of natural language processing methods, including open-domain information extraction and coreference resolution, to achieve this while accounting for linguistic phenomena. While previous work on open information extraction has extracted large numbers of subject-predicate-object triples, our method attempts to maintain only those that are most likely to correspond to meaningful relationships. Applying our method within and across multiple documents, we obtain a large conceptual graph. The resulting graph can be filtered such that only the most salient connections are maintained. Our graph visualization then allows users to explore these connections. We show how groups of documents can be selected and showcase interesting new connections that can be explored using our system.

2   Approach and Implementation

2.1   Fact Extraction

The initial phase of extracting facts proceeds as follows:

Document Ranking. The system first select the words appearing in the document collection with sufficiently high frequency as topic words, and computes standard TF-IDF weights for each word. The topic words are used to induce document representations. Documents under the same topic are ranked according to the TF-IDF weights of the topic words in each document. The user can pick such topics, and by default, the top-k documents for every topic are selected for further processing.

Coreference Resolution. Pronouns such as "she" are ubiquitous in language and thus entity names often are not explicitly repeated when new facts are expressed in a text. To nevertheless interpret such textual data appropriately, it is thus necessary to resolve pronouns, for which we rely on the Stanford CoreNLP system [3].

Open-Domain Knowledge Extraction. Different sentences within an article tend to exhibit a high variance with regard to their degree of relevance and contribution towards the core ideas expressed in the article. While some express key notions, others may serve as mere embellishments or anecdotes. Large entity network graphs with countless insignificant edges can be overwhelming for end users. To address this, our system computes document-specific TextRank importance scores for all sentences within a document. It then considers only those sentences with sufficiently high scores. From these, it extracts fact candidates

---

[1] A video presenting the system is available at https://shengyp.github.io/vmse.

as subject-predicate-object triples. Rather than just focusing on named entities (e.g., "Billionaire Donald Trump"), as some previous approaches do, our system supports an unbounded range of noun phrase concepts (e.g., "the snow storm on the East Coast") and relationships with explicit relation labels (e.g., "became mayor of"). The latter are extracted from verb phrases as well as from other constructions. For this, we adopt an open information extraction approach, in which the subject, predicate, and object are natural language phrases extracted from the sentence. These often correspond to syntactic subject, predicate, object, respectively.

2.2 Fact Filtering

The filtering algorithm aims at hiding less representative facts in the visualization, seeking to retain only the most salient, confident, and compatible facts. This is achieved by optimizing for a high degree of coherence between facts with high confidence. The joint optimization problem can be solved via integer linear programming, as follows:

$$\max_{x,y} \quad \boldsymbol{\alpha}^\mathsf{T} \boldsymbol{x} + \boldsymbol{\beta}^\mathsf{T} \boldsymbol{y} \tag{1}$$

$$\text{s.t.} \quad \mathbf{1}^\mathsf{T} \boldsymbol{y} \leq n_{\max} \tag{2}$$

$$x_k \leq \min\{y_i, y_j\} \tag{3}$$

$$\forall \; i < j, i, j \in \{1, \ldots, M\},$$

$$k = (2M - i)(i - 1)/2 + j - i$$

$$x_k, y_i \in \{0, 1\} \, \forall i \in \{1, \ldots, M\}, k \tag{4}$$

Here, $\mathbf{x} \in \mathbb{R}^N$, $\mathbf{y} \in \mathbb{R}^M$ with $N = (M + 1)(M - 2)/2 + 1$. The y_i are indicator variables for facts t_i: If y_i is true, t_i is selected to be retained. x_k represents the compatibility between two facts $t_i, t_j \in T$ $(i, j \leq M, i \neq j)$, where $T = \{t_1, \ldots, t_M\}$ is a set of fact triples containing M elements. β_i denotes the confidence of a fact, and $n_{\max}$ is the number of representative facts desired by the user. α_k is weighted by similarity scores $sim(t_i, t_j)$ between two facts t_i, t_j, defined as $\alpha_k = sim(t_i, t_j) = \gamma \dot{s}_k + (1 - \gamma) \dot{l}_k$. Here, s_k, l_k denote the semantic similarity and literal similarity scores between the facts, respectively. We compute s_k using the *Align, Disambiguate and Walk* algorithm[2], while l_k are computed using the Jaccard index. $\gamma = 0.8$ denotes the relative degree to which the semantic similarity contributes to the overall similarity score, as opposed to the literal similarity. The constraints guarantee that the number of results is not larger than $n_{\max}$. If x_k is true, the two connected facts t_i, t_j should be selected, which entails $y_i = 1$, $y_j = 1$.

2.3 Conceptual Graph Construction

In order to establish a single connected graph that is more consistent, our system provides an interactive user interface, in which expert annotators can merge

---

[2] https://github.com/pilehvar/ADW.

potential entities and concepts stemming from the fact filtering process, whose labels present equivalent meanings. They can discover obvious features in the lexical structure of entities or concepts, e.g., Billionaire Donald Trump, Donald Trump, Donald John Trump, Trump, etc. all refer to the same person. For NER, they can use the powerful entity linking ability from a search engine for deciding on coreference. To support the annotators, once again the *Align, Disambiguate and Walk* tool (see footnote 2) is used for semantically similarity computation between concepts for coreference. After that, on average, there remains not more than 5 subgraphs that can further be connected for different topics. Hence, users were able to add up to three synthetic relations with freely defined labels to connect these subgraphs into a fully connected graph.

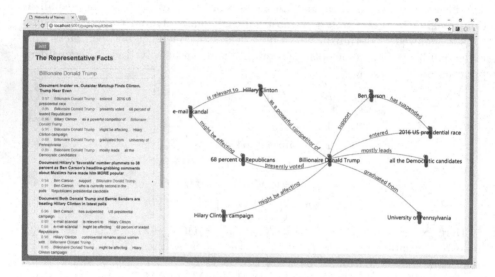

Fig. 1. Example of the user interface: In the left panel, when the user selects the entity "Billionaire Donald Trump" within the set of representative facts extracted from the document topics, the system presents the pertinent entities, concepts, and relations associated with this concept via a graph-based visualization in the right panel, including "Hillary Clinton" as a prominent figure.

The recommended [1] maximum size of a concept graph is 25 concepts, which we use as a constraint. In our evaluation metrics, the coverage rate is the number of topic entities and concepts for which marked as correct divided by the total number of all entities and concepts in the graph. We trained a binary classifier by the topic words with high frequency extracted from different topics to identify the important topic entities and concepts in the set of all potential concepts. We used common features, including frequency, length, language pattern, whether it is named entity, whether it appears in an automatic summarization [2], the ratio of synonyms, with random forests as the model. At inference time for topic concepts, we use the classifier's confidence for a positive classification as the score. We rely on a heuristic to find a full graph that is connected and satisfies the size limit of 25 concepts: We iteratively remove the weakest concepts with relatively lower score until only one connected component of 25 entities and concepts or less remains, which is used as the final conceptual graph. This approach guarantees that the graph is connected with high coverage rate of topic concepts, but might not find the subset of concepts that has the highest total importance score. A concrete example is illustrated in Fig. 1.

Acknowledgments. This paper was partially supported by National Natural Science Foundation of China (Nos. 61572111 and 61876034), and a Fundamental Research Fund for the Central Universities of China (No. ZYGX2016Z003).

References

1. Banko, M., Cafarella, M.J., Soderland, S., Broadhead, M., Etzioni, O.: Open information extraction from the web. In: IJCAI, pp. 2670–2676 (2007)
2. Li, J., Li, L., Li, T.: Multi-document summarization via submodularity. Appl. Intell. **37**(3), 420–430 (2012)
3. Manning, C., Surdeanu, M., Bauer, J., Finkel, J., Bethard, S., McClosky, D.: The Stanford CoreNLP natural language processing toolkit. In: ACL System Demonstrations, pp. 55–60 (2014)

Author Index

Printed in the United States
By Bookmasters